Population Genetics

Population Genetics

Second Edition

Matthew B. Hamilton

WILEY Blackwell

Registered Office
John Wiley & Sons, Inc., 111 River Street, Hoboken, NJ 07030, USA

Editorial Office
9600 Garsington Road, Oxford, OX4 2DQ, UK

For details of our global editorial offices, customer services, and more information about Wiley products, visit us at www.wiley.com.

Wiley also publishes its books in a variety of electronic formats and by print-on-demand. Some content that appears in standard print versions of this book may not be available in other formats.

Library of Congress Cataloging-in-Publication Data

Names: Hamilton, Matthew B., author.
Title: Population genetics / Matthew B. Hamilton.
Description: Second edition. | Hoboken, NJ : Wiley-Blackwell, 2021. |
    Includes bibliographical references and index.
Identifiers: LCCN 2020025434 (print) | LCCN 2020025435 (ebook) | ISBN
    9781118436943 (hardback) | ISBN 9781118436929 (adobe pdf) | ISBN
    9781118436899 (epub)
Subjects: LCSH: Population genetics.
Classification: LCC QH455 .H35 2021 (print) | LCC QH455 (ebook) | DDC
    576.5/8–dc23
LC record available at https://lccn.loc.gov/2020025434
LC ebook record available at https://lccn.loc.gov/2020025435

Cover Design: Wiley
Cover Images: Matthew B. Hamilton

Set in 10/12.5pt Photina by SPi Global, Pondicherry, India

SKY10023806_010821

Dedication

For my wife and best friend, I-Ling

Contents

Preface and acknowledgements

This book was originally born of two desires, one relatively simple and the other more ambitious, both of which were motivated by my experiences learning and teaching population genetics. My first desire was to create an up-to-date survey text of the field of population genetics. At the same time, I set out with the more ambitious goal of offering an alternative body of materials to change the manner in which population genetics is taught and learned. The first edition of the book made progress toward these goals, and the second edition provides updates and refinements in that same vein.

Much of population genetics during the twentieth century was hypothesis-rich but data-poor. The theory developed between about 1920 and 1980 spawned manifold predictions about basic evolutionary processes. However, many of those predictions could be tested with only very limited power for lack of appropriate or sufficient genetic data. With the advancement of high-throughput DNA sequencing and its still widening employment, population genetics has become much less data-limited. Massive amounts of DNA sequence data are being collected for an expanding set of organisms. Polymorphism and divergence data are now available at a scale of many loci to entire genomes per individual. This has led to a new generation in population genetics that is data-rich. Ironically, this abundance of empirical data has reinforced the central role of models and deductive inference in population genetics. Predictive models have grown to support the genetic data that are now available, fostering innovation at the same time.

Coalescent or genealogical branching models are primary among the models employed in population genetics to make predictions and test hypotheses. During the past few decades, coalescent theory has moved from an esoteric problem pursued for purely mathematical reasons to a central conceptual tool of population genetics. Despite this, the teaching of coalescent theory in undergraduate and graduate population genetics courses has not kept pace with its role in prediction and hypothesis testing. A major impediment has been the lack of teaching materials that make coalescent theory truly accessible to students learning population genetics for the first time. One of my goals was to construct a text that will meet this need with a systematic and thorough introduction to the concepts of coalescent theory and its applications in hypothesis testing. The chapter sections on coalescent theory are presented along with the traditional theory of identity by descent on the same topics to help students see the commonality of the two approaches. However, the coalescence chapter sections could easily be assigned as a group. The second edition retains this focus and adds a section on the ancestral recombination graph.

Another of my goals for this text was to offer a range of explanatory styles. Learning the concepts of population genetics in the language of mathematics is often relatively easy for abstract and mathematical learners. However, my aim was to cater to a wide range of learning styles by building a range of features into the text. A key pedagogical feature of the book is boxes set off from the main text that are designed to engage the various learning styles. Problem boxes placed in the text rather than at the end of chapters are designed to provide practice and to reinforce concepts as they are encountered, appealing to experiential learners. These are now augmented in the second edition with additional end-of-chapter problems. Math boxes that explain mathematical derivations will not only appeal to mathematical and logical learners but also provide insight for all readers into the mathematical reasoning employed in population genetics. In addition, the large number of illustrations in the text were designed to appeal and help cultivate visual learning.

A novel feature of the text is Interact boxes that guide students through semi-structured exercises in computer simulations. These Interact boxes utilize web-based simulations developed specifically for this book or public domain software. The simulation problems are an active learning approach and should appeal to experiential or visual learners. Simulations are one of the best ways to demonstrate the outcome of stochastic processes where replication is required before a pattern or generalization can be seen. Because the comprehension of stochastic processes in genetics is a major hurdle for many students, the Interact boxes should aid understanding of central concepts. Additionally, the simulations, spreadsheet models, and scripts provide applications of algorithmic thinking. Algorithmic and computational approaches to problem-solving are now central to prediction and data analysis in population genetics and are useful in most fields of biology and in the sciences more broadly.

The approach to mathematics in the text deserves further explanation. The undergraduate biology curricula employed at most US institutions has students take calculus and applied statistics and usually requires little application of mathematics within biology courses. This leads to students having difficulty in, or avoiding altogether, courses in biological disciplines that require explicit mathematical reasoning. It also leads to courses avoiding explicit mathematical reasoning. Population genetics is built on basic mathematics and probability, and in my experience, students obtain a much deeper understanding of the subject with some comprehension of these mathematical foundations. Therefore, rather than avoid these topics, I have attempted to deconstruct and offer step-by-step explanations of the basic mathematics required for a sound understanding. For those readers with more interest or facility in mathematics, the book presents more detailed derivations in boxes that are separated from the main narrative of the text. There are also some chapter sections containing more mathematically rigorous content. These sections can be assigned or skipped depending on the level and scope of a course supported by this text. This approach will hopefully provide students with the tools to develop their abilities in basic mathematics through application and, at the same time, learn population genetics more fully.

For the second edition, I have tried to incorporate the generous and helpful feedback received from readers of the first edition. John Braverman deserves special mention as a dedicated colleague and friend who has provided sustained suggestions and thoughtful comments. Brent Johnson provided helpful suggestions on statistics topics, and Mak Paranjape helped me understand circuit models. Members of my laboratory and the students who have taken my courses provided feedback on chapter drafts, figures, and effective means to explain the concepts herein. This feedback has been invaluable and has helped me shape the text into a more useful and usable resource for students. The web simulations were developed with the help of Marie Kolawole and Steve Moore, aided by an award from the Georgetown University Initiative on Technology Enhanced Learning.

Many people contributed to the first edition, and their suggestions and input still shapes the book. They include Rachel Adams, Genevieve Croft, John Braverman, Paulo Nuin, James Crow, A.W.F. Edwards, Sivan Rottenstreich Leviyang, Judy Miller, John Dudley, Stephen Moose, Michel Veuille, Eric Delwart, John Epifanio, Robert J. Robbins, Peter Armbruster, Ronda Rolfes, and Martha Weiss. I also thank the anonymous reviewers of the first edition from Aberdeen University, Arkansas State University, Cambridge University, Michigan State University, University of North Carolina, and University of Nottingham. Nancy Wilton, Elizabeth Frank, Haze Humbert, Karen Chambers, and Nik Prowse of Wiley-Blackwell helped bring the first edition to fruition.

Matthew B. Hamilton
October 2020

About the companion websites

This book is accompanied by companion websites for Instructors and Students:

www.wiley.com/go/hamilton/populationgenetics

The Instructor website includes:

- Solutions to the end-of-chapter exercises
- Powerpoints of all figures from the book for downloading, to aid teaching

The Student website includes:

- Chapter resources for Interact Boxes, Problem Boxes, and end-of-chapter exercises

CHAPTER 1

Thinking like a population geneticist

All scientific fields possess a body of concepts as well as a specialized vocabulary used to express these concepts precisely. Population genetics is no different, and the entirety of this book is designed to introduce, explain, and demonstrate these concepts and vocabulary. What may be unique about population genetics among the natural sciences is the way that its practitioners approach questions about the biological world. Population genetics is a dialog between predictions based on the principles of Mendelian inheritance and observations from the empirical measurement of genotype and allele frequencies. Idealized predictions stemming from general principles form the basis of hypotheses that can be tested through observation, experiment, and comparison. At the same time, empirical patterns observed within and among populations are evaluated for evidence of their causes via predictive models. This first chapter will explore some of the ways that population genetics approaches and defines problems that are relevant to the topics in all chapters. The chapter is also intended to give some insight into how to approach the study of population genetics.

1.1 Expectations

· What Do We Expect to Happen?
· Expectations Are the Basis of Understanding Cause and Effect

In our everyday lives, there are many things that we expect to occur or not to occur based on the knowledge of our surroundings and past experience. For example, you probably do not expect to get hit by a meteorite while walking to your next population genetics class. Why not? Meteorites *do* impact the surface of the Earth and, on occasion, strike something noticeable to people nearby. A few times in the distant past, in fact, large meteors have hit the Earth and left evidence like the Chicxulub impact crater on the Yucatán Peninsula in Mexico. What influences your lack of concern? It is probably a combination of basic knowledge of the principles of physics that apply to meteors as well as your empirical observations of the frequency and location of meteor strikes. Basic physics tells us that a small meteor on a collision course with the Earth is unlikely to hit the surface since most objects burn up from the friction they experience traveling through the Earth's atmosphere. You might also reason that even if the object is big enough to pass through the atmosphere intact, and there are far fewer of these, then the Earth is a large place and, just by chance, the impact is unlikely to be even remotely near you. Finally, you have most probably never witnessed a large meteorite impact or even heard of one occurring during your lifetime. You have combined your knowledge of the physical world and your experience to arrive (perhaps unconsciously) at a prediction or an expectation: meteorite strikes are possible but are so infrequent that the risk of being struck while on the way to class is miniscule. In this very same way, you have constructed models of many events and processes in your physical and social world and used the resulting predictions to make comparisons and decisions.

> **Expectation:** The expected value of a random variable, especially the average; a prediction or forecast.

The study of population genetics similarly revolves around constructing and testing expectations for genetic variation in populations of individual

Population Genetics, Second Edition. Matthew B. Hamilton.
© 2021 John Wiley & Sons, Inc. Published 2021 by John Wiley & Sons, Inc.
Companion website: www.wiley.com/go/hamilton/populationgenetics

organisms. Expectations attempt to predict things like how much genetic variation is present in a population, how genetic variation in a population changes over time, and the pattern of genetic variation that might be left behind by a given biological process that acts over time or through space. Building these expectations involves the use of first principles or the set of very basic rules and assumptions that define how natural systems work at their lowest, most basic levels. A first principle in physics is the force of gravity. In population genetics, first principles are the very basic mechanisms of Mendelian particulate inheritance and processes such as mutation, mating patterns, gene flow, and natural selection that increase, decrease, and shape genetic variation. These foundational rules and processes are used and combined in population genetics with the ultimate goal of building a comprehensive set of predictions that can be applied to any species and any genetic system.

Empirical study in population genetics also plays a central role in constructing and evaluating predictions. In population genetics as in all sciences, empirical evidence is drawn from intentional observations, cleverly constructed comparisons, and experiments. Genetic patterns observed in actual populations are compared with expected patterns to test models constructed using general principles and assumptions. For example, we could construct a mathematical or computer simulation model of random genetic drift (change in allele frequency due to sampling from finite populations) based on abstract principles of sampling from a finite population and biological reproduction. We could then compare the predictions of such a model to the observed change in allele frequency through time in a laboratory population of *Drosophila melanogaster* (fruit flies). If the change in allele frequency in the fruit fly population matched the change in allele frequency predicted using the model of genetic drift, then we could conclude that the model effectively summarizes the biological sampling processes that take place in fruit fly populations.

It is also possible to use well-tested and accepted model expectations as a basis to hypothesize what processes caused an observed pattern in a biological population. Again, to use a *D. melanogaster* population as an example, we might ask whether an observed change in allele frequency over some generations in a wild population could be explained by genetic drift. If the observed allele frequency change is within the range of the predicted change in allele frequencies based on a model of genetic drift, then we have identified a possible *cause* of the observed pattern. Comparing observed genetic patterns in

populations often requires modifications to existing models or the construction of novel models in order to develop appropriate expectations. For example, a model of genetic drift constructed for *D. melanogaster* might naturally assume that all individuals in the population are diploid (individuals that possess paired sets of homologous chromosomes). If we wanted to use that same model to predict genetic drift in a population of honeybees, we would have to account for the fact that their males are haploid (individuals that possess single copies of each chromosome) while females are diploid. This change in reproductive biology could be taken into account by altering the assumptions of the model of genetic drift to make predictions appropriate for honeybee populations. Note that without some modifications, a single model of genetic drift would not accurately predict allele frequencies over time in both fruit flies and honeybees since their patterns of reproduction and chromosomal inheritance are different.

Parameters and parameter estimates

While developing the expectations of population genetics in this book, we will most often be working with idealized quantities. For example, allele frequency in a population is a fundamental quantity. For a genetic locus with two alleles, A and a, it is common to say that p equals the frequency of the A allele and q equals the frequency of the a allele. In mathematics, **parameter** is another term for an idealized quantity like an allele frequency. It is assumed that parameters have an exact value. Put another way, parameters are idealized quantities where the messy, real-life details of how to measure the quantities they represent are completely ignored.

Empirical population genetics measures quantities such as allele frequencies to give **parameter estimates** by sampling and then measuring the alleles and genotypes present in actual populations. All experiments, observations, and even simulations in population genetics produce parameter estimates of some sort. There is a subtle notational convention used to indicate an estimate, that is, the hat or ^ character above a variable. Estimates wear hats whereas parameters do not. Using allele frequency as an example, we would say $\hat{p}$ (pronounced "p hat") equals the number of A alleles sampled divided by the total number of alleles sampled. Intuitively, we can see from the denominator in the expression for $\hat{p}$ that the allele frequency estimate will depend on the sample we gather to make the estimate.

In actual populations, a parameter has a true value. For the allele frequency p, knowing this true value would require examining the genotype of every individual and counting *all* A and a alleles to determine their frequency in the population. This task is impractical or impossible in most cases. Instead, we rely on an estimate of allele frequency, $\hat{p}$, obtained from a sample of individuals from the population. Sampling leads to some uncertainty in parameter estimates because repeating the sampling and parameter estimate process would likely lead to a somewhat different parameter estimate each time. Quantifying this uncertainty is important to determine whether repeated sampling might change a parameter estimate by just a little or change it by a lot. When dealing with parameters, we might expect that $p + q = 1$ exactly if there are only two alleles with allele frequencies p and q. However, if we are dealing with estimates, we might say the two allele frequency estimates should sum to approximately one ($\hat{p} + \hat{q} \approx 1$) since each allele frequency is estimated with some errors. The more uncertain the estimates of $\hat{p}$ and $\hat{q}$, the less we should be surprised to find that their sum does not equal the expected value of one.

> **Parameter:** A variable or constant appearing in a mathematical expression; a value (usually unknown) used to represent a certain population characteristic; any factor that defines a system and determines or limits its performance.
> **Estimate:** An indication of the value of an unknown quantity based on observed data; an approximation of a true score, parameter, or value; a statistical estimate of the value of a parameter.

It could be said that statistics sits at the intersection of theoretical and empirical population genetics. Parameters and parameter estimates are fundamentally different things. Estimation requires effort to understand sampling variation and quantify sources of error and bias in samples and estimates. The distinction between parameters and estimates is critical when comparing actual populations with expectations to test hypotheses. When large, random samples can be taken, estimates are likely to have minimal errors. However, there are many cases

where estimates have a great deal of uncertainty, which limits the ability to evaluate expectations. There are also instances where very different processes may produce very similar expected results. In such cases, it may be difficult or impossible to distinguish the different potential causes of a pattern due to the approximate nature of estimates. While this book focuses mostly on parameters, it is useful to bear in mind that testing or comparing expectations requires the use of parameter estimates and statistics that quantify sampling error. The Appendix provides a review of some basic statistics that are used in the text.

Inductive and deductive reasoning

Population genetics employs both **inductive** and **deductive reasoning** in an effort to understand the biological processes operating in actual populations as well as to elucidate the general processes that cause population genetic phenomena. The inductive approach to population genetics involves assembling measures of genetic variation (parameter estimates) from various populations to build up evidence that can be used to identify the underlying processes that produced the observed patterns. This approach is logically identical to that used by Isaac Newton, who used knowledge of how objects fall to the surface of the Earth as well as knowledge of the movement of planets to arrive at the general principles of gravity. Application of inductive reasoning requires detailed familiarity with the various empirical data types in population genetics, such as DNA sequences, along with the results of studies that report observed patterns of genetic variation. From this accumulated empirical information, it is then possible to draw more general conclusions about the qualities and quantities of genetic variation in populations. Model organisms like *D. melanogaster* and *Arabidopsis thaliana* play a large role in population genetic conclusions reached by inductive reasoning. Because model organisms receive a large amount of scientific effort, for example, to completely sequence and annotate their genomes, a great deal of available genetic data are accumulated for these species. Based on this evidence, many inferences have been made about population genetic processes. Although model organisms are very rich sources of empirical information, the number of species is limited by definition so that any generalizations may not apply universally to all species.

> **Deductive reasoning:** Using general principles to reach conclusions about specific instances.
> **Inductive reasoning:** Utilizing the knowledge of specific instances or cases to arrive at general principles.

The study of population genetics can also be approached using deductive reasoning. The actions of general processes such as genetic drift, mutation, and natural selection are represented by parameters in the mathematical equations that make up population genetic models. These models can then be used to make predictions about the quantity of genetic variation and patterns of genetic variation in space and time. Such population genetic models make general predictions about things like rates of change in allele frequency, the eventual equilibrium of allele or genotype frequencies, and the net outcome of several processes operating at the same time. These predictions are very general in that they apply to any population of any species since the predictions arose from general principles in the first place. At the same time, such general predictions may not be directly applicable to a specific population because the general principles and assumptions used to make the prediction are not specific enough to match an actual population.

Historically, the field of population genetics has developed from an interplay between arguments and evidence developed using both inductive and deductive reasoning approaches. Nonetheless, most of the major ideas in population genetics can be first approached with deductive reasoning by learning and understanding the expectations that arise from the principles of Mendelian heredity. This book stresses on the process of deductive reasoning to arrive at these fundamental predictions. Empirical evidence related to expectations is included to illustrate predictions and to demonstrate hypothesis tests that result from expectations. Because the body of empirical results in population genetics is very large, readers should resist the temptation to generalize too much from the limited number of empirical studies that are presented. Detailed reviews of particular areas of population genetics, many of which are cited, are a better source for comprehensive summaries of empirical studies.

In the next chapter, we will start by building expectations for the frequencies of diploid genotypes based on the foundation of particulate inheritance: that alleles are passed unaltered from parents to offspring. There is ample support for particulate inheritance from both molecular biology, which identifies DNA as the hereditary molecule, and from allele and genotype frequencies that can be observed in actual populations. The general principle of particulate inheritance has been used to formulate a wide array of expectations about allele and genotype frequencies in populations.

1.2 Theory and assumptions

- What Is a Theory and What Are Assumptions?
- How Can Theories Be Useful with So Many Assumptions?

In colloquial usage, the word *theory* refers to something that is known with uncertainty, or a quantity that is approximate. On a day you are running late leaving work, you might say, "In theory, I am supposed to depart at 6:00 pm." In science, theory has a very different meaning. Theory is the accumulation of expectations and observations that have withstood tests and critical scrutiny and are accepted by at least some practitioners of a scientific field. Theory is the collection of all of the expectations developed for specific cases or individual biological processes that together form a more comprehensive set of general principles. The combination of Darwin's hypothesis of natural selection with the laws of Mendelian particulate inheritance is often called the *modern synthesis* of evolutionary biology since it is a comprehensive theory to explain the causes of evolutionary change. The modern synthesis can offer causal explanations for biological phenomena ranging from antibiotic resistance in bacteria to the behavior of elephants to the rate of DNA sequence change, as well as make predictions to guide animal and plant breeders. In all of the modern synthesis, population genetics plays a central role.

It is common for the uninitiated to ask the question "what good is theory if it is based on so many assumptions?" A body of theory is a useful tool to articulate assumptions and generate testable predictions. Theory that generates many testable predictions about the world also offers many opportunities to falsify its predictions and assumptions. Since hypotheses cannot be proven directly, but alternative hypotheses can be disproven, the generation of plausible, testable alternative hypotheses is a requirement for scientific inquiry. Strong theories are able to make accurate predictions, offer causal explanations for diverse observations, and generate alternative hypotheses based on revised assumptions.

The words *theory* and *assumption* can seem abstract, but you should not be intimidated by them. Theories are just collections of expectations, each with a set of assumptions that place bounds on the prediction being made. If you understand what motivates an expectation, its predictions, and its assumptions, then you understand theory. Most expectations in population genetics will have at least a few, and often many, assumptions used to define and bound the situation. For example, we might assume something about the size of a population or the absence of mutation, or that all genotypes are diploid with two alleles. This is a way of limiting the prediction to appropriate circumstances and a way of defining which quantities and conditions can vary and which are fixed. Each of these assumptions can influence the generality of an expectation. Each assumption can also be relaxed or altered to see how strongly it influences the expectation. To return to the example in the preceding section, if, one day, meteorites start falling around us with regularity, we would be forced to call into question some of the basic assumptions originally used to formulate our expectation that meteorite strikes should be rare events. In this way, assumptions are useful tools to ask "what if...?" as part of the process of developing a prediction. If our initial "what if...?" conditions do not match a situation, then the resulting prediction will probably be inaccurate.

In population genetics, as in much of science where theory and expectations are involved, empirical data and model expectations are routinely compared. Imagine observing a set of genotype frequencies in a biological population. It would then be natural to construct an idealized population by using theory that approximates the biological population. This is an attempt to construct an idealized population that is *equivalent* to the actual population from the perspective of the processes influencing genotype frequencies. For example, a large population may behave exactly like a small, randomly mating ideal population in terms of genotype frequencies. This equivalence allows us to use expectations for ideal populations with one or a few variables specified in order to describe an actual population where there are many more, usually unknown, parameters. What we strive to do is to focus on those variables that strongly influence genotype frequencies in the actual population. In this way, it is often possible to reduce the complexity of a real population and determine the key

variables that strongly influence a property like genotype frequencies. The ideal population is not meant to match the actual population in every detail.

Theory: A scheme or system of ideas or statements held as an explanation or account of a group of facts or phenomena; the general laws, principles, or causes of something known or observed.
Infer: To draw a conclusion or make a deduction based on facts or indications; to have as a logical consequence.

From the comparison of expectation and observation, we infer that the first principles used to construct the expectation are sound if they can be used to explain patterns observed in the biological world. However, there is a major distinction between considering an actual and idealized population *equivalent* and considering them *identical*. This is seen in cases where the observed pattern in an actual population is consistent with the expectations from several model populations built around distinct and incompatible assumptions. In such cases, it is not possible to infer the processes that cause a given pattern without additional information. A common example in population genetics are cases of genetic patterns that are potentially consistent with the random process of genetic drift and, at the same time, consistent with some form of the deterministic process of natural selection. In such cases, unambiguous inference of the underlying cause of a pattern is not possible without additional empirical information or more precise expectations.

1.3 Simulation

    A Method of Practice, Trial and Error Learning, and Exploration

Imagine learning to play the piano without ever touching a piano or practicing the hand movements required to play. What if you were expected to play a difficult concerto after extensive exposure (perhaps a semester) to only verbal and written descriptions of how other people play? Such a teaching style would make learning to play the piano very difficult because there would be no opportunity for practice, trial and error, or exploration. You would not have

Interact box 1.1 The textbook website

Throughout this book, you will encounter Interact boxes. These boxes contain opportunities for you to interact directly with the material in the text by using computer simulations designed to demonstrate fundamental concepts of population genetics. Each box will contain step-by-step instructions for you to follow in order to carry out a simulation. By following the instructions, you will get started with the simulation. However, always feel free to use your own imagination and intuition. After following the instructions in the Interact box and understanding the point at hand, enter different values, push more buttons, and even read the documentation. You can also return to Interact boxes at a later time, perhaps after you have read and understood more of the text, to reconsider a simulation or view it in a different light. You can also use the simulations to answer questions that may occur to you or to test hypotheses that you may have. Questions in population genetics that start off "What would happen if…?" can often be answered with simulation.

The book's website gives you the worldwide web address (URL) for each interact box. This prevents problems in case web addresses change because the website can be updated while your copy of the text cannot be updated.

Step 1 Open a web browser and enter http://www.wiley.com/go/hamiltongenetics
Step 2 Click on the Chapter resources link that is associated with Interact boxes.
Step 3 Verify that the page gives links for each of the Interact boxes listed by their number. You could also bookmark this page so you can access it directly in the future.

Congratulations! You have completed the first Interact box.

the opportunity for direct experience nor incremental improvement of your understanding. Unfortunately, this is exactly how science courses are taught to some degree. You are expected to learn and remember concepts with only limited opportunities for directly observing principles in action. In fairness, this is partly due to the difficulty of carrying out some of the experiments or observations that originally lead someone to discover and understand an important principle.

In the field of population genetics, computer simulations can be used to effectively demonstrate many fundamental genetic processes. In fact, computer simulations are an important research tool in population genetics. Therefore, when you conduct simulations, you are both learning by direct experience and learning using the same methods that are used by researchers. Simulations allow us to view how quantities like allele frequencies change over time, observe their dynamics, and determine whether a stable end point is reached: an equilibrium. With simulations, we can view dynamics (change over time) and equilibria over very long periods of time and under a vast array of conditions in an effort to reach general conclusions. Without simulations, it would be impossible for us to directly observe allele

frequencies over such long periods of time and in such diverse biological situations.

Simulations are an effective means to understand some of the fundamental predictions of populations genetics. Mathematical expressions are frequently used to express dynamics and equilibria in population genetics, but the equations alone can be opaque at first. Simulations provide a means to explore the relationships among variables that are summarized in the compact language of mathematics. Many people feel that a set of mathematical equations is much more meaningful after having the chance to explore what they describe with some actual numerical values. Simulation provides the means to explore what equations predict and can make learning population genetics an easier, more rewarding experience.

Carrying out simulations has the potential to make the expectations of population genetics much more accessible and understandable. Conducting simulations is not much extra work, especially once you get into the practice of using the text and simulation software in concert. You can approach simulations as if they are games, where each one shows a visual scene that helps to solve a puzzle. In addition, simulations can help you develop a more intuitive understanding of population genetic predictions so you do not have to approach the expectations of population genetics as disembodied or unanimated "facts."

It is important to approach simulations in a systematic and organized fashion, not as just a collection of buttons to press and text entry boxes to be filled in on a whim. It is absolutely imperative that you understand the meaning behind each variable that you can control as well as the meaning of the results you obtain. To do so successfully, you will need to be aware of both specific details and larger patterns, or both the individual trees and the forest that they compose. For example, in a simulation that presents results as a graph, it is important that you understand the details of what variables are represented on each axis and the range of axis values. Sometimes these details are not always completely obvious in simulation software, requiring you to use both your intuition and knowledge of the population genetic processes being simulated.

Once you are comfortable with the details of a simulation, you will also want to keep track of the "big picture" patterns that emerge as you view simulation results. Seeing these patterns will often require that you examine the results over a range of conditions. Try approaching simulations as experiments by changing only one variable at a time until you understand its effects on the outcome. Changing several things all at once can lead to confusion and an inability to see cause-and-effect relationships, unless you have fully understood the effects of individual variables. Finally, try writing down parameter values you have tried in a simulation and sketching or tabulating results on paper as you work with a simulation. Use all of your skills as a scientist and student when conducting simulations, and they will become a powerful learning tool. Eventually, you may even use scripting and programming to carry out your own simulations specifically designed to explore your own genetic hypotheses.

Chapter 1 review

- Both general principles and direct measurements taken in actual populations combine to form comprehensive expectations about amounts, patterns, and cause-and-effect relationships in population genetics.

- The theory of population genetics is the collection of well-accepted expectations used to articulate a wide array of predictions about the biological processes that shape genetic variation.
- Parameters are idealized quantities that are exact, while parameter estimates wear notational "hats" to remind us that they have statistical uncertainty.
- Population genetics uses both inductive reasoning to generalize from the knowledge of specifics and deductive reasoning to build up predictions from general principles that can be applied to specific situations.
- Population genetics is not a spectator sport! Direct participation through computer simulation provides the opportunity to see population genetic processes in action. You can learn by trial and error and test your own understanding by making predictions and then comparing them with simulation results.

Further reading

For a history of population genetics from Darwin to the 1930s, see:

Provine, W.B. (1971). *The Origins of Theoretical Population Genetics*. Chicago, IL: University of Chicago Press.

For a concise history of population genetics since the mid-1960s that highlights major conceptual advances as well as technical innovations to measure genetic variation, see:

Charlesworth, B. and Charlesworth, D. (2017). Population genetics from 1966 to 2016. *Heredity* 118: 2–9.

For two personal and historical essays on the past, present, and assumptions of theoretical population genetics, see:

Lewontin, R.C. (1985). Population genetics. In: *Evolution: Essays in Honour of John Maynard Smith* (eds. P.J. Greenwood, P.H. Harvey and M. Slatkin), 3–18. Cambridge: Cambridge University Press.
Wakeley, J. (2005). The limits of theoretical population genetics. *Genetics* 169: 1–7.

CHAPTER 2

Genotype frequencies

2.1 Mendel's model of particulate genetics

- Mendel's breeding experiments.
- Independent assortment of alleles.
- Independent segregation of loci.
- Some common genetic terminology.

In the nineteenth century, there were several theories of heredity, including inheritance of acquired characteristics and blending inheritance. Jean-Baptiste Lamarck is most commonly associated with the discredited hypothesis of inheritance of acquired characteristics (although it is important to recognize his efforts in seeking general causal explanations of evolutionary change). He argued that individuals contain "nervous fluid" and that organs or features (phenotypes) employed or exercised more frequently attract more nervous fluid, causing the trait to become more developed in their offspring. His widely known example is the long neck of the giraffe, which he said developed because individuals continually stretched to reach leaves at the tops of trees. Later, Charles Darwin and many of his contemporaries subscribed to the idea of blending inheritance. Under blending inheritance, offspring display phenotypes that are an intermediate combination of parental phenotypes (Figure 2.1).

From 1856 to 1863, the Augustinian monk Gregor Mendel carried out experiments with pea plants that demonstrated the concept of particulate inheritance. Mendel showed that phenotypes are determined by discrete units that are inherited intact and unchanged through generations. His hypothesis was sufficient to explain three common observations: (i) phenotype is sometimes identical between parents and offspring; (ii) offspring phenotype can differ from that of the parents; and (iii) "pure" phenotypes of earlier generations could skip generations and reappear in later generations. Neither blending inheritance nor inheritance of acquired characteristics are satisfactory explanations for all of these observations. It is hard for us to fully appreciate now, but Mendel's results were truly revolutionary and served as the very foundation of population genetics. The lack of an accurate mechanistic model of heredity severely constrained biological explanations of cause and effect up to the point that Mendel's results were "rediscovered" in the year 1900.

It is worthwhile to briefly review the experiments with pea plants that Mendel used to demonstrate independent assortment of both alleles within a locus and of multiple loci, sometimes dubbed Mendel's first and second laws. We need to remember that this was well before the Punnett square, which originated in about 1905. Therefore, the conceptual tool we would use now to predict progeny genotypes from parental genotypes was a thing of the future. So, in revisiting Mendel's experiments, we will not use the Punnett square in an attempt to follow his logic. Mendel only observed the phenotypes of generations of pea plants that he had hand-pollinated. From these phenotypes and their patterns of inheritance, he inferred the existence of heritable factors. His experiments were actually both logical and clever, but are now taken for granted since the basic mechanism of particulate inheritance has long since ceased to be an open question. It was Mendel who established the first and most fundamental prediction of population genetics: expected genotype frequencies.

Mendel used pea seed coat color as a phenotype he could track across generations. His goal was to determine, if possible, the general rules governing the

Population Genetics, Second Edition. Matthew B. Hamilton.
© 2021 John Wiley & Sons, Inc. Published 2021 by John Wiley & Sons, Inc.
Companion website: www.wiley.com/go/hamilton/populationgenetics

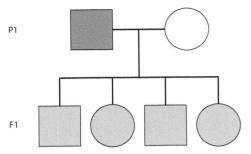

Figure 2.1 The model of blending inheritance predicts that progeny have phenotypes that are the intermediate of their parents. Here, "pure" blue and white parents yield light blue progeny, but these intermediate progeny could never themselves be parents of progeny with pure blue or white phenotypes identical to those in the P1 generation. Crossing any shade of blue with a pure white or blue phenotype would always lead to some intermediate shade of blue. By convention, in pedigrees, females are indicated by circles and males by squares while "P" refers to parental and "F" to filial.

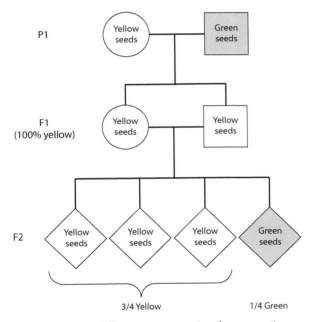

Figure 2.2 Mendel's crosses to examine the segregation ratio in the seed coat color of pea plants. The parental plants (P1 generation) were pure breeding, meaning that if self-fertilized all resulting progeny had a phenotype identical to the parent. Some individuals are represented by diamonds since pea plants are hermaphrodites and can act as a mother, a father, or can self-fertilize.

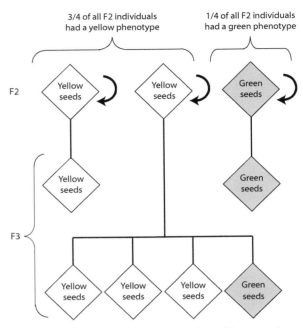

Figure 2.3 Mendel self-pollinated (indicated by curved arrows) the F2 progeny produced by the cross shown in Figure 2.2. Of the F2 progeny that had a yellow phenotype (3/4 of the total), 1/3 produced all progeny with a yellow phenotype and 2/3 produced progeny with a 3 : 1 ratio of yellow and green progeny (or 3/4 yellow progeny). Individuals are represented by diamonds since pea plants are hermaphrodites.

inheritance of pea phenotypes. He established "pure"-breeding lines (meaning plants that always produced progeny with phenotypes like themselves) of peas with both yellow and green seeds. Using these pure-breeding lines as parents, he crossed a yellow-

and a green-seeded plant. The parental cross and the next two generations of the progeny are shown in Figure 2.2. Mendel recognized that the F1 plants had an "impure" phenotype because of the F2 generation plants, of which three-quarters had yellow and one-quarter had green seed coats.

His insightful next step was to self-pollinate a sample of the plants from the F2 generation (Figure 2.3). He considered the F2 individuals with yellow and green seed coats separately. All green-seeded F2 plants produced green progeny and thus were "pure" green. However, the yellow-seeded F2 plants were of two kinds. Considering just the yellow F2 seeds, one-third were pure and produced only yellow-seeded progeny, whereas two-thirds were "impure" yellow since they produced both yellow- and green-seeded progeny. Mendel combined the frequencies of the F2 yellow and green phenotypes along with the frequencies of the F3 progeny. He reasoned that three-quarters of all F2 plants had yellow seeds, but these could be divided into plants that produced pure yellow F3 progeny (one-third) and plants that produced both yellow and green F3 progeny (two-thirds). So, the ratio of pure yellow to impure yellow in the F2 was (1/3 × 3/4 =) 1/4 pure yellow to (2/3 × 3/4 =) 1/2 "impure" yellow. The green-seeded progeny comprised one-quarter of the F2 generation and all produced green-seeded progeny when self-fertilized, so that (1 × 1/4 green =) 1/4 pure green. In total, the ratios of phenotypes in the F2 generation were 1 pure yellow : 2 impure yellow : 1 pure green or 1 : 2 : 1. Mendel reasoned that "the ratio of 3 : 1 in which the distribution of the dominating and recessive traits take place in the first generation therefore resolves itself into the ratio of 1 : 2 : 1 if one differentiates the meaning of the dominating trait as a hybrid and as a parental trait" (quoted in Orel 1996). During his work, Mendel employed the terms "dominating" (which became dominant) and "recessive" to describe the manifestation of traits in impure or heterozygous individuals.

With the benefit of modern symbols of particulate heredity, we could diagram Mendel's monohybrid cross with pea color in the following way.

P1	Phenotype	Yellow × green
	Genotype	GG Gg
	Gametes produced	G G
F1	Phenotype	All "impure"yellow
	Genotype	Gg
	Gametes produced	G, g

A Punnet square could be used to predict the phenotypic ratios of the F2 plants

	G	G
G	GG	Gg
G	Gg	Gg

F2	Phenotype	3 Yellow : 1 green
	Genotype	GG Gg Gg
	Gametes produced	G G, g G

and another Punnet square could be used to predict the genotypic ratios of the two-thirds of the yellow F2 plants

	G	G
G	GG	Gg
G	Gg	Gg

Mendel's first "law": Predicts independent segregation of alleles at a single locus: two copies of a diploid locus (a pair of alleles that make a diploid genotype) segregate independently into gametes so that in a large number of gametes half carry one allele and the other half carry the other allele.

Individual pea plants obviously have more than a single phenotype, and Mendel followed the inheritance of other characters in addition to seed coat color. In one example of his crossing experiments, Mendel tracked the simultaneous inheritance of both seed coat color and seed surface condition (either wrinkled ["angular"] or smooth). He constructed an initial cross among pure-breeding lines identical to what he had done when tracking seed color inheritance, except now there were two phenotypes (Figure 2.4). The F2 progeny appeared in the phenotypic ratio of 9 round/yellow : 3 round/green : 3 wrinkled/yellow : 1 wrinkled/green.

How did Mendel go from this F2 phenotypic ratio to the second law? He ignored the wrinkled/smooth phenotype and just considered the yellow/green seed color phenotype in self-pollination crosses of F2 plants just like those for the first law. In the F2 progeny, 12/16 or three-quarters had a yellow seed coat and 4/16 or one-quarter had a green seed coat, or a 3 yellow : 1 green phenotypic ratio. Again using self-pollination of F2 plants like those in Figure 2.3, he showed that the yellow phenotypes were (1/3 × ¾) one-quarter pure and (2/3 × ¾) one-half impure yellow. Thus, the segregation ratio for seed color was 1 : 2 : 1 and the wrinkled/smooth phenotype did not alter this result. Mendel obtained an identical result when considering instead only the wrinkled/smooth phenotype and ignoring the seed color phenotype.

Mendel concluded that a phenotypic segregation ratio of 9 : 3 : 3 : 1 is the same as combining two independent 3 : 1 segregation ratios of two phenotypes since (3 : 1) × (3 : 1) = 9 : 3 : 3 : 1. Similarly, the multiplication of two (1 : 2 : 1) phenotypic ratios will predict the two phenotype ratios (1 : 2 : 1) × (1 : 2 : 1) = 1 : 2 : 1 : 2 : 4 : 2 : 1 : 2 : 1. We now recognize that dominance in the first two phenotype ratios masks the ability to distinguish some of the homozygous and heterozygous genotypes, whereas the ratio in the second case would result if there was no

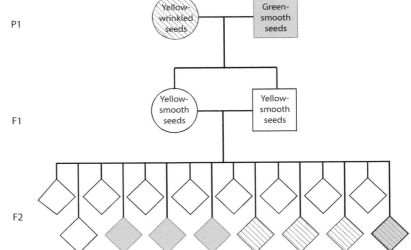

Figure 2.4 Mendel's crosses to examine the segregation ratios of two phenotypes, seed coat color (yellow or green) and seed coat surface (smooth or wrinkled), in pea plants. The stippled pattern indicates wrinkled seeds, while the solid color indicates smooth seeds. The F2 individuals exhibited a phenotypic ratio of 9 round-yellow: 3 round-green: 3 wrinkled-yellow: 1 wrinkled-green.

dominance. You can confirm these conclusions by working out a Punnett square for the F2 progeny in the two-locus case.

Mendel's second "law": Predicts independent assortment of multiple loci: during gamete formation, the segregation of alleles of one locus is independent of the segregation of alleles of another locus.

Mendel performed similar breeding experiments with numerous other pea phenotypes and obtained similar results. Mendel described his work with peas and other plants in lectures and published it in 1866 in the *Proceedings of the Natural Science Society of Brünn* in German where it went unnoticed for nearly 35 years. However, Mendel's results were eventually recognized, and his paper was translated into several languages. Mendel's rediscovered the hypothesis of particulate inheritance was also bolstered by evidence from microscopic observations of chromosomes during cell division that led Walter Sutton to propose in 1902 that chromosomes are the physical basis of heredity, supported by results obtained independently by Theodor Boveri at around the same time (see Crow and Crow 2002).

Much of the currently used terminology was coined as the field of particulate genetics initially developed. Therefore, many of the critical terms in genetics have remained in use for long periods of time. However, the meanings and connotations of these terms have often changed as our understanding of genetics has also changed.

Unfortunately, this has led to a situation where words can sometimes mislead. A common example is equating *gene* and *allele*. For example, it is commonplace for news media to report scientific breakthroughs where a "gene" has been identified as causing a particular phenotype, often a debilitating disease. Very often what is meant in these cases is that a genotype or an *allele* with the phenotypic effect has been identified. Both unaffected and affected individuals all possess the gene, but they differ in their alleles and therefore in their genotype. If individuals of the same species really differed in their gene content (or loci they possessed), that would provide evidence of additions or deletions to genomes. For an interesting discussion of how terminology in genetics has changed – and some of the misunderstandings this can cause, see Judson (2001).

Gene: A unit of particulate inheritance; in contemporary usage, it usually means an exon or series of exons, or a DNA sequence that codes for an RNA or protein.

Locus (plural **loci**, pronounced "low-sigh"): Literally "place" or location in the genome; in contemporary usage, it is the most general reference to *any* sequence or genomic region, including non-coding regions.

Allele: A variant or alternative form of the DNA sequence at a given locus.

Genotype: The set of alleles possessed by an individual at one locus; the genetic composition of an individual at one locus or many loci.

Phenotype: The morphological, biochemical, physiological, and behavioral attributes of an individual; synonymous with character and trait.

Dominant: Where the expressed phenotype of one allele takes precedence over the expressed phenotype of another allele. The allele associated with the expressed phenotype is said to be dominant. Dominance is seen on a continuous scale that includes "complete" dominance (one allele completely masks the phenotype of another allele so that the phenotype of a heterozygote is identical to a homozygote for the dominant allele) and "partial" or "incomplete" dominance (masking effect is incomplete so that the phenotype of a heterozygote is intermediate to both homozygotes) and includes over- and under-dominance (phenotype is outside the range of phenotypes seen in the homozygous genotypes). The lack of dominance (heterozygote is exactly intermediate to the phenotypes of both homozygotes) is when the effects of alleles are additive, a situation sometimes termed "codominance" or "semi-dominance."

Recessive: The expressed phenotype of one allele is masked by the expressed phenotype of another allele. The allele associated with the concealed phenotype is said to be recessive.

2.2 Hardy–Weinberg expected genotype frequencies

- Hardy–Weinberg and its assumptions.
- Each assumption is a population genetic process.
- Hardy–Weinberg is a null model.
- Hardy–Weinberg in haplo-diploid systems.

Mendel's "laws" could be called the original expectations in population genetics. With the concept of particulate genetics established, it was possible to make a wide array of predictions about genotype and allele frequencies as well as the frequency of phenotypes with a one-locus basis. Still, progress and insight into particulate genetics were gradual. Until 1914, it was generally believed that rare (infrequent) alleles would disappear from populations over time. Godfrey H. Hardy (1908) and Wilhelm Weinberg (1908) worked independently to show that the laws of Mendelian heredity did not predict such a phenomenon (see Crow 1988). In 1908, they both formulated the relationship that can be used to predict allele frequencies given genotype frequencies or predict genotype frequencies given allele frequencies. This relationship is the well-known Hardy–Weinberg equation.

$$p^2 + 2pq + q^2 = 1 \qquad (2.1)$$

where p and q are allele frequencies for a genetic locus with two alleles.

Genotype frequencies predicted by the Hardy–Weinberg equation can be summarized graphically. Figure 2.5 shows Hardy–Weinberg expected genotype frequencies on the y axis for each genotype for any given value of the allele frequency on the x axis. Another graphical tool to depict genotype and allele frequencies simultaneously for a single locus with two alleles is the de Finetti diagram (Figure 2.6). As we will see, de Finetti diagrams are helpful when examining how population genetic processes dictate allele and genotype frequencies. In both graphs, it is apparent that heterozygotes are most frequent when the frequency of the two alleles is equal to 0.5. You can also see that when an allele is rare, the corresponding homozygote genotype is even rarer since the genotype frequency is the square of the allele frequency.

A single generation of reproduction where a set of conditions, or assumptions, is met will result in a

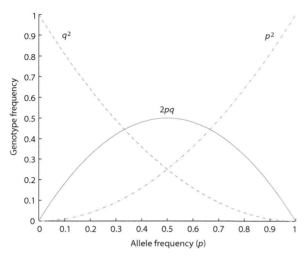

Figure 2.5 Hardy–Weinberg expected genotype frequencies for AA, Aa, and aa genotypes (I-axis) for any given value of the allele frequency (*x*-axis). Note that the value of the allele frequency not graphed can be determined by $q = 1 - p$.

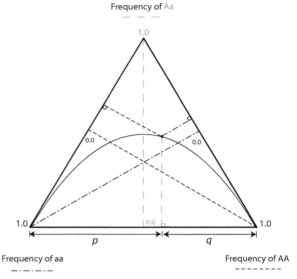

Figure 2.6 A de Finetti diagram for one locus with two alleles. The triangular coordinate system results from the requirement that the frequencies of all three genotypes must sum to one. Any point inside or on the edge of the triangle represents all three genotype frequencies of a population. The parabola describes Hardy–Weinberg expected genotype frequencies. The dashed lines represent the frequencies of each of the three genotypes between zero and one. Genotype frequencies at any point can be determined by the length of lines that are perpendicular to each of the sides of the triangle. A practical way to estimate genotype frequencies on the diagram is to hold a ruler parallel to one of the sides of the triangle and mark off the distance on one of the frequency axes. The point on the parabola is a population in Hardy–Weinberg equilibrium where the frequency of AA is 0.36, the frequency of aa is 0.16, and the frequency of Aa is 0.48. The perpendicular line to the base of the triangle also divides the bases into regions corresponding in length to the allele frequencies. Any population with genotype frequencies not on the parabola has an excess (above the parabola) or deficit (below the parabola) of heterozygotes compared to Hardy–Weinberg expected genotype frequencies.

population that meets Hardy–Weinberg expected genotype frequencies, often called Hardy–Weinberg equilibrium. The list of assumptions associated with this prediction for genotype frequencies is long. The set of assumptions includes:

- the organism is diploid,
- reproduction is sexual (as opposed to clonal),
- generations are discrete and non-overlapping,
- the locus under consideration has two alleles,
- allele frequencies are identical among all mating types (i.e. sexes),
- mating is random (as opposed to assortative),
- there is random union of gametes,
- population size is very large, effectively infinite,
- migration is negligible (no population structure, no gene flow),
- mutation does not occur or its rate is very low,
- natural selection does not act (all individuals and gametes have equal fitness).

These assumptions make intuitive sense when each is examined in detail (although this will probably be more apparent after more reading and simulation). As we will see later, Hardy–Weinberg holds for any number of alleles, although Eq. 2.1 is valid for only two alleles. Many of the assumptions can be thought of as assuring random mating and production of all possible progeny genotypes. Hardy–Weinberg genotype frequencies in progeny would not be realized if the two sexes have different allele frequencies even if matings take place between random pairs of parents. It is also possible that just by

chance not all genotypes would be produced if only a small number of parents mated, just like flipping a fair coin only a few times may not produce an equal number of heads and tails. Natural selection is a process that causes some genotypes in either the parental or progeny generations to be more frequent than others. So, it is logical that Hardy–Weinberg expectations would not be met if natural selection were acting. In a sense, these assumptions define the biological processes that make up the field of population genetics. Each assumption represents one of the conceptual areas where population genetics can make testable predictions via expectation in order to distinguish the biological processes operating in populations. This is quite a set of accomplishments for an equation with just three terms!

Interact box 2.1
Genotype frequencies for one locus with two alleles

You can use the simulation website associated with this book to explore an interactive version of Figure 2.6. Find Interact Box 2.1 on the text web page and click on the link for the simulation (web URLs are not provided in the text since they may change over time).

Once you are at the Simulations website home page, use the **Simulations** menu to select the **de Finetti** simulation. The simulation is based on a triangular graph like that in Figure 2.6 with a control pane at the left where you can set parameters. With **Mating Model** set on **Random Mating**, use the sliders to set genotype frequencies. The parabola defines Hardy–Weinberg expected genotype frequencies, so try to adjust the genotype frequencies to fall at different locations along the parabola. Also, try genotype frequencies that are located above and below the parabola.

Null model: A testable model of no effect or a background effect. A prediction or expectation based on the simplest assumptions to predict outcomes. Often, population genetic null models make predictions based on purely random processes such as random mating or genetic drift, random samples or combinations, or variables having background effects on allele or genotype frequencies.

Despite all of this praise, you might ask: what good is a model with so many restrictive assumptions? Are all these assumptions likely to be met in actual populations? The Hardy–Weinberg model is not necessarily meant to be an exact description of any actual population, although actual populations often exhibit genotype frequencies predicted by Hardy–Weinberg. Hardy–Weinberg provides a **null model**, a prediction based on a simplified or idealized situation where no biological processes are acting and genotype frequencies are the result of random combination. Actual populations can be compared with this null model to test hypotheses about the evolutionary forces acting on allele and genotype frequencies. The important point and the original motivation for Hardy and Weinberg was to show that the process of particulate inheritance itself does not cause any changes in allele frequencies across generations. Thus, changes in allele frequency or departures from Hardy–Weinberg expected genotype frequencies must be caused by processes that alter the outcome of basic inheritance.

In the final part of this section, we will explore genotype frequency expectations adjusted to account for ploidy (the number of homologous chromosomes) differences between males and females as seen in chromosomal sex determination and haplo-diploid organisms. In chromosomal sex determination as seen in mammals, birds, and Lepidoptera (butterflies), one sex is determined by possession of two identical chromosomes (the homogametic sex) and the other sex determined by possession of two different chromosomes (the heterogametic sex). In mammals, females are homogametic (XX) and males heterogametic (XY), whereas, in birds, the opposite is true, with heterogametic females (ZW) and homogametic males (ZZ). In haplo-diploid species such as bees and wasps (Hymenoptera), males are haploid (hemizygous) for all chromosomes, whereas females are diploid for all chromosomes.

Predicting genotype frequencies at one locus in these cases under random mating and the other assumptions of Hardy–Weinberg requires keeping track of allele or genotype frequencies in both sexes and loci on specific chromosomes. An effective method is to draw a Punnett square that distinguishes the sex of an individual as well as the gamete types that can be generated at mating (Table 2.1). The Punnett square shows that genotype frequencies in the diploid sex are identical to Hardy–Weinberg expectations for autosomes, whereas genotype frequencies are equivalent to allele frequencies in the haploid sex. One consequence of different chromosome types between the sexes is that fully recessive phenotypes are more common in the heterogametic sex, where a single chromosome determines the phenotype and recessive phenotypes appear at the allele frequency. However, in the homogametic sex, fully recessive phenotypes appear at the frequency of the recessive genotype (e.g. q^2) since they are masked in heterozygotes. Some types of color blindness in humans are examples of traits

Table 2.1 Punnett square to predict genotype frequencies for loci on sex chromosomes and for all loci in males and females of haploid-diploid species. Notation in this table is based on birds where the sex chromosomes are Z and W (*ZZ* males and *ZW* females) with a diallelic locus on the Z chromosome possessing alleles A and a at frequencies *p* and *q*, respectively. In general, genotype frequencies in the homogametic or diploid sex are identical to Hardy–Weinberg expectations for autosomes, while genotype frequencies are equal to allele frequencies in the homogametic or haploid sex.

Homozygotic or diploid sex

Genotype		*ZZ*	
Gamete		Z-A	Z-a
Frequency		*p*	*q*
		Z-A Z-A	Z-A Z-a
		p^2	*pq*
		Z-A Z-a	Z-a Z-a
		Pq	q^2
		Z-A W	Z-a W
		p	*q*

Heterozygotic or haploid sex

Genotype	Gamete	Frequency
	Z-A	*p*
ZW	Z-a	*q*
	W	

Expected genotype frequencies under random mating

Homogametic sex		**Homogametic sex**	
Z-A Z-A	p^2	Z-A W	*p*
Z-A Z-a	*2pq*		
Z-a Z-a	q^2	Z-a W	*q*

due to genes on the X chromosome (called "X-linked" traits) that are more common in men than in women due to haplo-diploid inheritance.

Later, in Section 2.4, we will examine two categories of applications of Hardy–Weinberg expected genotype frequencies. The first set of applications arises when we assume (often with supporting evidence) that the assumptions of Hardy–Weinberg are true. We can then compare several expectations for genotype frequencies with actual genotype frequencies to distinguish between several alternative hypotheses. The second type of application is where we examine what results when assumptions of Hardy–Weinberg are not met. There are many cases where population genotype frequencies can be used to reveal the action of various population genetic processes. Before that, the next section builds a proof of the Hardy–Weinberg prediction that inheritance per se will not alter allele frequencies.

2.3 Why does Hardy–Weinberg work?

- A proof of Hardy–Weinberg.
- Hardy–Weinberg with more than two alleles.

The Hardy–Weinberg equation is one of the most basic expectations we have in population genetics. It is very likely that you were already familiar with the Hardy–Weinberg equation before you picked up this book. But where does Hardy–Weinberg actually come from? What is the logic behind it? Let's develop a simple proof that Hardy–Weinberg is actually true. This will also be our first real foray into the type of the algebraic argument that much of population genetics in built on. Given that you start out knowing the conclusion of the Hardy–Weinberg tale, this gives you the opportunity to focus on the style in which it is told. Algebraic or quantitative arguments are a central part of the language and vocabulary of population genetics, so part of the task of learning population genetics is becoming accustomed to this mode of discourse.

We would like to prove that $p^2 + 2pq + q^2 = 1$ accurately predicts genotype frequencies given the values of allele frequencies. Let's start off by making some explicit assumptions to bound the problem.

The assumptions, in no particular order, are:

1 mating is random (parents meet and mate according to their frequencies);
2 all parents have the same number of offspring (equivalent to no natural selection on fecundity);
3 all progeny are equally fit (equivalent to no natural selection on viability);
4 there is no mutation that could act to change an A to a or an a to A;
5 it is a single population that is very large;
6 there are two and only two mating types.

Now, let's define the variables we will need for a case with one locus that has two alleles (A and a).

N = Population size of individuals (N diploid individuals have $2N$ alleles)

Allele frequencies
$$p = \text{frequency(A allele)}$$
$$= \text{(total number of A alleles)}/2N$$

$$q = \text{frequency(a allele)}$$
$$= \text{(total number of a alleles)}/2N$$

$$p + q = 1$$

Genotype frequencies
$$X = \text{frequency(AA genotype)}$$
$$= \text{(total number of AA genotypes)}/N$$

$$Y = \text{frequency(Aa genotype)}$$
$$= \text{(total number of Aa genotypes)}/N$$

$$Z = \text{frequency(aa genotype)}$$
$$= \text{(total number of aa genotypes)}/N$$

$$X + Y + Z = 1$$

We do not distinguish between the heterozygotes Aa and aA and treat them as being equivalent genotypes. Therefore, we can express allele frequencies in terms of genotype frequencies by adding together the frequencies of A-containing and a-containing genotypes:

$$p = X + \tfrac{1}{2}Y \qquad (2.2)$$
$$q = Z + \tfrac{1}{2}Y \qquad (2.3)$$

Each homozygote contains two alleles of the same type, while each heterozygote contains one allele of each type so the heterozygote genotypes are each weighted by half.

With the variables defined, we can then follow allele frequencies across one generation of

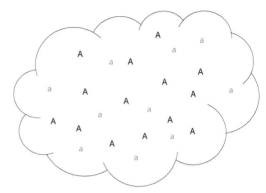

Figure 2.7 A schematic representation of random mating as a cloud of gas where the frequency of A's is 14/24 and the frequency of a's 10/24. Any given A has a frequency of 12/20 and will encounter another A with probability of 14/24 or an a with the probability of 10/24. This makes the frequency of an A-A collision $(14/24)^2$ and an A-a or a-A collision $2(14/24)(10/24)$, just as the probability of two independent events is the product of their individual probabilities. The population of A's and a's is assumed to be large enough so that taking one out of the cloud will make almost no change in the overall frequency of its type.

reproduction. The first step is to calculate the probability that parents of any two particular genotypes will mate. Since mating is assumed to be random, the chance that two genotypes will mate is just the product of their individual frequencies. As shown in Figure 2.7, random mating can be thought of as being like gas atoms in a balloon. As with gas atoms, each genotype or gamete bumps into others at random, with the probability of a collision (or mating or union) being the product of the frequencies of the two objects colliding. To calculate the probabilities of mating among the three different genotypes, we can make a table to organize the resulting mating frequencies. This table will predict the mating frequencies among genotypes in the initial generation, which we will call generation t.

A parental mating frequency table (generation t) is shown below.

		Dads		
		AA	Aa	aa
Moms	Frequency	X	Y	Z
AA	X	X^2	XY	XZ
Aa	Y	XY	Y^2	YZ
Aa	Z	ZX	ZY	Z^2

The table expresses parental mating frequencies in the currency of genotype frequencies. For example, we expect matings between AA moms and Aa dads to occur with a frequency of XY.

Next, we need to determine the frequency of each genotype in the offspring of any given parental mating pair. This will require that we predict the offspring genotypes resulting from each possible parental mating. We can do this easily with a Punnett square. We will use the frequencies of each parental mating (above) together with the frequencies of the offspring genotypes. Summed for all possible parental matings, this gives the frequency of offspring genotypes one generation later, or in generation $t + 1$. A table will help organize all the frequencies, like the offspring frequency table (generation $t + 1$) shown below.

Parental mating	Total frequency	Offspring genotype frequencies		
		AA	**Aa**	**aa**
AA × AA	X^2	X^2	0	0
AA × Aa	$2XY$	XY	XY	0
AA × aa	$2XZ$	0	$2XZ$	0
Aa × Aa	Y^2	$Y^2/4$	$(2Y^2)/4$	$Y^2/4$
Aa × aa	$2YZ$	0	YZ	YZ
aa × aa	Z^2	0	0	Z^2

In this table, the total frequency is just the frequency of each parental mating pair taken from the parental mating frequency table. We now need to partition this total frequency of each parental mating into the frequencies of the three progeny genotypes produced. Let's look at an example. Parents with AA and Aa genotypes will produce progeny with two genotypes: half AA and half Aa (you can use a Punnett square to show this is true). Therefore, the AA × Aa parental matings, which have a total frequency of $2XY$ under random mating, are expected to produce ($\frac{1}{2}$) $2XY = XY$ of each of AA and Aa progeny. The same logic applies to all of the other parental matings. Notice that each row in the offspring genotype frequency table sums to the total frequency of each parental mating.

The columns in the offspring genotype frequency table are the basis of the final step. The sum of each

column gives the total frequencies of each progeny genotype expected in generation $t + 1$. Let's take the sum of each column, again expressed in the currency of genotype frequencies, and then simplify the algebra to see whether Hardy and Weinberg were correct.

$$
\begin{aligned}
AA &= X^2 + XY + Y^2/4 \\
&= (X + \tfrac{1}{2}Y)^2 \text{ (recall that } p = X + \tfrac{1}{2}Y) \\
&= p^2 \\
aa &= Y^2/4 + YZ + Z^2 \\
&= (Z + \tfrac{1}{2}Y)^2 \text{ (recall that } q = Z + \tfrac{1}{2}Y) \\
&= q^2 \\
Aa &= XY + 2XZ + 2Y^2/4 + YZ \\
&= 2(XY/2 + XZ + Y^2/4 + YZ/2) \\
&= 2(X + Y/2)(Z + Y/2) \\
&= 2pq
\end{aligned}
\qquad (2.4)
$$

So, we have proved that progeny genotype and allele frequencies are identical to parental genotype and allele frequencies over one generation or that $f(A)_t = f(A)_{t+1}$. The major conclusion here is that *genotype frequencies remain constant over generations as long as the assumptions of Hardy–Weinberg are met*. In fact, we have just proved that under Mendelian heredity, genotype and allele frequencies should not change over time unless one or more of our assumptions is not met. This simple model of expected genotype frequencies has profound conclusions. In fact, Hardy–Weinberg expected genotype frequencies serve as one of the most basic tools to test for the action of biological processes that alter genotype and allele frequencies.

You might wonder whether Hardy–Weinberg applies to loci with more than two alleles. For the last point in this section, let's explore that question. With three alleles at one locus (allele frequencies symbolized by p, q, and r), Hardy–Weinberg expected genotype frequencies are $p^2 + q^2 + r^2 + 2pq + 2pr + 2qr = 1$. These genotype frequencies are obtained by expanding $(p + q + r)^2$, a method that can be applied to any number of alleles at one locus. In general, expanding the squared sum of the allele frequencies will show:

- the frequency of any homozygous genotype is the squared frequency of the single allele that composes the genotype ([allele frequency]^2);
- the frequency of any heterozygous genotype is twice the product of the two allele frequencies

that comprise the genotype (2[allele 1 frequency] [allele 2 frequency]), and

- there are as many homozygous genotypes as there are alleles and $\frac{N(N-1)}{2}$ heterozygous genotypes where N is the number of alleles.

Do you think it would be possible to prove Hardy–Weinberg for more than two alleles at one locus? The answer is absolutely, yes. This would just require constructing larger versions of the parental genotype mating table and expected offspring frequency table as we did for two alleles at one locus.

2.4 Applications of Hardy–Weinberg

- Estimate the frequency of an observed genotype in a forensic DNA typing case.
- Test the null hypothesis that observed and expected genotype frequencies are identical.
- Use Hardy–Weinberg to compare two genetic models for observed phenotypes.

In the previous two sections, we established the Hardy–Weinberg expectations for genotype frequencies. In this section, we will examine three ways that expected genotype frequencies are employed in practice. The goal of this section is to become familiar with realistic applications as well as hypothesis tests that compare observed and Hardy–Weinberg expected genotype frequencies. In this process, we will also look at a specific method to account for sampling error (see Appendix).

Forensic DNA profiling

Our first application of Hardy–Weinberg can be found in newspapers on a regular basis and commonly dramatized on television. A terrible crime has been committed. Left at the crime scene was a biological sample that law enforcement authorities use to obtain a multilocus genotype or DNA profile. A suspect in the crime has been identified and subpoenaed to provide a tissue sample for DNA profiling. The DNA profile from the suspect and from the crime scene are identical. The DNA profile is shown in Table 2.2. Should we conclude that the suspect left the biological sample found at the crime scene?

To answer this critical question, we will employ Hardy–Weinberg to predict the expected frequency of the DNA profile or genotype. Just because two DNA profiles match, there is not necessarily strong evidence that the individual who left the evidence DNA and the suspect are the same person. It is possible that there are actually two or more people with identical

Table 2.2 An example DNA profile for three STR ("simple tandem repeat") loci commonly used in human forensic cases. Locus names refer to the human chromosome (e.g. D3 = third chromosome) and chromosome region where the SRT locus is found. The allele states are the numbers of repeats at that locus (see Box 2.1).

Locus	D3S1358	D21S11	D18S51
Genotype	17, 18	29, 30	18, 18

DNA profiles. Hardy–Weinberg and Mendel's second law will serve as the bases for us to estimate just how frequently a given DNA profile should be observed. Then, we can determine whether two unrelated individuals sharing an identical DNA profile is a likely occurrence.

To determine the expected frequency of a one-locus genotype, we employ the Hardy–Weinberg Eq. (2.1). In doing so, *we are implicitly accepting that all of the assumptions of Hardy–Weinberg are approximately met*. If these assumptions were not met, then the Hardy–Weinberg equation would not provide an accurate expectation for the genotype frequencies! To determine the frequency of the three-locus genotype in Table 2.2, we need allele frequencies for those loci, which are found in Table 2.3. Starting with the locus D3S1358, we see in Table 2.3 that the 17-repeat allele has a frequency of 0.2118 and the 18-repeat allele a frequency of 0.1626. Then, using Hardy–Weinberg, the 17, 18 genotype has an expected frequency of 2(0.2118)(0.1626) = 0.0689 or 6.89%. For the two other loci in the DNA profile of Table 2.2, we carry out the same steps.

D21S11   29-Repeat allele frequency = 0.1811
          30-Repeat allele frequency = 0.2321
          Genotype frequency
          = 2(0.1811)(0.2321) = 0.0841 or 8.41%

D18S51   18-Repeat allele frequency = 0.0918
          Genotype frequency = $(0.0918)^2$ = 0.0084 or 0.84%

The genotype for each locus has a relatively large chance of being observed in a population. For example, a little less than 1% of Caucasian U.S. citizens (or about 1 in 119) are expected to be homozygous for the 18-repeat allele at locus D18S51. Therefore, a match between evidence and suspect DNA profiles homozygous for the 18 repeat at that locus would not be strong evidence that the samples came from the same individual.

Table 2.3 Allele frequencies for nine STR loci commonly used in forensic cases estimated from 196 US Caucasians sampled randomly with respect to geographic location. The allele states are the numbers of repeats at that locus (see Box 2.1). Allele frequencies (Freq) are as reported in Budowle et al. (2001). Table 1 from FBI sample population.

D3S1358		vWA		D21S11		D18S51		D13S317	
Allele	Freq	Allele	Freq	Allele	Freq	Allele	Freq	Allele	Freq
12	0.0000	13	0.0051	27	0.0459	<11	0.0128	8	0.0995
13	0.0025	14	0.1020	28	0.1658	11	0.0128	9	0.0765
14	0.1404	15	0.1122	29	0.1811	12	0.1276	10	0.0510
15	0.2463	16	0.2015	30	0.2321	13	0.1224	11	0.3189
16	0.2315	17	0.2628	30.2	0.0383	14	0.1735	12	0.3087
17	0.2118	18	0.2219	31	0.0714	15	0.1276	13	0.1097
18	0.1626	19	0.0842	31.2	0.0995	16	0.1071	14	0.0357
19	0.0049	20	0.0102	32	0.0153	17	0.1556		
				32.2	0.1122	18	0.0918		
				33.2	0.0306	19	0.0357		
				35.2	0.0026	20	0.0255		
						21	0.0051		
						22	0.0026		

FGA		D8S1179		D5S818		D7S820	
Allele	freq	Allele	freq	Allele	freq	Allele	Freq
18	0.0306	<9	0.0179	9	0.0308	6	0.0025
19	0.0561	9	0.1020	10	0.0487	7	0.0172
20	0.1454	10	0.1020	11	0.4103	8	0.1626
20.2	0.0026	11	0.0587	12	0.3538	9	0.1478
21	0.1735	12	0.1454	13	0.1462	10	0.2906
22	0.1888	13	0.3393	14	0.0077	11	0.2020
22.2	0.0102	14	0.2015	15	0.0026	12	0.1404
23	0.1582	15	0.1097			13	0.0296
24	0.1378	16	0.0128			14	0.0074
25	0.0689	17	0.0026				
26	0.0179						
27	0.0102						

Fortunately, we can combine the information from all three loci. To do this, we use the **product rule**, which states that the probability of observing multiple independent events is just the product of each individual event. We already used the product rule in the last section to calculate the expected frequency of each genotype under Hardy–Weinberg by treating each allele as an independent probability. Now, we just extend the product rule to cover multiple genotypes, *under the assumption that each of the loci is independent by Mendel's second law* (the assumption is justified here since each of the loci is on a separate chromosome). The expected frequency of the three-locus genotype (sometimes called the *probability of identity*) is then $0.0689 \times 0.0841 \times 0.0084 = 0.000049$ or 0.0049%. Another way to express this probability is as an **odds ratio**, or the reciprocal of the probability (an approximation that holds when

the probability is very small). Here, the odds ratio is $1/0.000049 = 20\,408$, meaning that we would expect to observe the three locus DNA profile once in 20 408 Caucasian Americans.

> **Product rule:** The probability of two (or more) independent events occurring simultaneously is the product of their individual probabilities.

Now, we can return to the question of whether two unrelated individuals are likely to share an identical three-locus DNA profile by chance. One out of every 20 408 Caucasian Americans is expected to have the genotype in Table 2.2. Although the three-locus DNA profile is considerably less frequent than a genotype for a single locus, it still does not approach a unique, individual identifier. Therefore, there is a finite chance that a suspect will match an evidence DNA profile by chance alone. Such DNA profile matches, or "inclusions," require additional evidence to ascertain guilt or innocence. In fact, the term prosecutor's fallacy was coined to describe failure to recognize the difference between a DNA match and guilt (for example, a person can be present at a location and not involved in a crime). Only when DNA profiles do not match, called an "exclusion," can a suspect be unambiguously and absolutely ruled out as the source of a biological sample at a crime scene.

Current forensic DNA profiles use 10–13 loci to estimate expected genotype frequencies. Problem 2.1 gives a 10-locus genotype for the same individual in Table 2.2, allowing you to calculate the odds ratio for a realistic example. In Chapter 4, we will reconsider the expected frequency of a DNA profile with the added complication of allele frequency differentiation among human racial groups.

Problem box 2.1
The expected genotype frequency for a DNA profile

Calculate the expected genotype frequency and odds ratio for the 10-locus DNA profile below. Allele frequencies are given in Table 2.3.

D3S1358	17, 18
vWA	17, 17
FGA	24, 25
Amelogenin	X, Y
D8S1179	13, 14
D21S11	29, 30
D18S51	18, 18
D5S818	12, 13
D13S317	9, 12
D7S820	11, 12

What does the amelogenin locus tell us and how did you assign an expected frequency to the observed genotype? Is it likely that two unrelated individuals would share this 10-locus genotype by chance? For this genotype, would a match between a crime scene sample and a suspect be convincing evidence that the person was present at the crime scene?

Testing Hardy–Weinberg expected genotype frequencies

A common use of Hardy–Weinberg expectations is to test for deviations from its null model. Populations with genotype frequencies that do not fit Hardy–Weinberg expectations are evidence that one or more of the evolutionary processes embodied in the assumptions of Hardy–Weinberg are acting to determine genotype frequencies. Our null hypothesis is that genotype frequencies meet Hardy–Weinberg expectations within some degree of estimation error. Genotype frequencies that are not close to Hardy–Weinberg expectations allow us to reject this null hypothesis. The processes in the list of assumptions then become possible alternative hypotheses to explain observed genotype frequencies. In this section, we will work through a hypothesis test for Hardy–Weinberg equilibrium.

The first example uses observed genotypes for the MN blood group, a single locus in humans that has two alleles (Table 2.4). First, we need to estimate the frequency of the M allele, using the notation that the estimated frequency of M is $\hat{p}$ and the frequency of N is $\hat{q}$. Note that the "hat" superscripts indicate that these are allele frequency *estimates* (see Chapter 1). The total number of alleles is $2N$ given a sample of N diploid individuals. We can then count up all of the alleles of one type to estimate the frequency of that allele.

$$\hat{p} = \frac{2 \times \text{Frequency(MM)} + \text{frequency(MN)}}{2N} \quad (2.5)$$

$$\hat{p} = \frac{2 \times 165 + 562}{2 \times 1066} = \frac{892}{2132} = 0.4184 \quad (2.6)$$

Since $\hat{p} + \hat{q} \approx 1$, we can estimate the frequency of the N allele by subtraction as $\hat{q} = 1 - \hat{p} = 1 - 0.4184 = 0.5816$.

Using these allele frequencies allows calculation of the Hardy–Weinberg expected genotype frequency and number of individuals with each genotype, as shown in Table 2.4. In Table 2.4, we can see that the match between the observed and expected is not perfect, but we need some method to ask whether the difference is actually large enough to conclude that Hardy–Weinberg equilibrium does not hold in the sample of 1066 genotypes. Remember that any allele frequency estimate ($\hat{p}$) could differ slightly from the true parameter (p) due to chance events as well as due to random sampling in the group of genotypes used to estimate the allele

Box 2.1 DNA profiling

The loci used for human DNA profiling are a general class of DNA sequence marker known as simple tandem repeat (STR), simple sequence repeat (SSR), or microsatellite loci. These loci feature tandemly repeated DNA sequences of one to six base pairs (bp) and often exhibit many alleles per locus and high levels of heterozygosity. Allelic states are simply the number of repeats present at the locus, which can be determined by electrophoresis of polymerase chain reaction (PCR) amplified DNA fragments. STR loci used in human DNA profiling generally exhibit Hardy–Weinberg expected genotype frequencies; there is

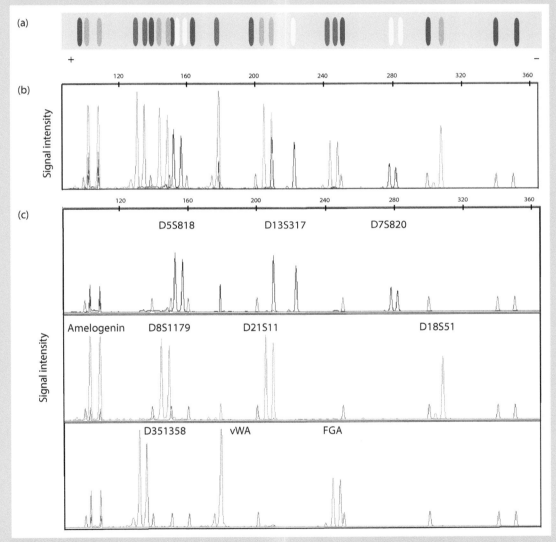

Figure 2.8 The original data for the DNA profile given in Table 2.2 and **Problem Box 2.1** obtained by capillary electrophoresis. The PCR oligonucleotide primers used to amplify each locus are labeled with a molecule that emits blue, green, or yellow light when exposed to laser light. Thus, the DNA fragments for each locus are identified by their label color as well as their size range in base pairs. Panel A shows a simulation of the DNA profile as it would appear on an electrophoretic gel (+ indicates the anode side). Blue, green, and yellow label the 10 DNA profiling loci, shown here in grayscale. The red DNA fragments are size standards with a known molecular weight used to estimate the size in base pairs of the other DNA fragments in the profile. Panel B shows the DNA profile for all loci and the size standard DNA fragments as a graph of color signal intensity by size of DNA fragment in base pairs. Panel C shows a simpler view of trace data for each label color independently with the individual loci labeled above the trace peaks. A few shorter peaks are visible in the yellow, green, and blue traces of Panel C that are not labeled as loci. These artifacts, called "pull up" peaks, are caused by intense signal from a locus labeled with another color (e.g. the yellow and blue peaks in the location of the green labeled *amelogenin locus*). A full color version of this figure is available on the textbook website. (*continued*)

Box 2.1 (continued)

evidence that the genotypes are selectively "neutral" (e.g. not affected by natural selection), and the loci meet the other assumptions of Hardy–Weinberg. STR loci are employed widely in population genetic studies and in genetic mapping (see reviews by Goldstein and Pollock 1997; McDonald and Potts 1997).

This is an example of the DNA sequence found at a microsatellite locus. This sequence is the 24.1 allele from the fibrinogen alpha chain gene, or FGA locus (Genbank accession no. AY749636; see Figure 2.8). The integral repeat is the 4 bp sequence CTTT, and most alleles have sequences that differ by some number of

full CTTT repeats. However, there are exceptions where alleles have sequences with partial repeats or stutters in the repeat pattern, for example, the TTTCT and CTC sequences imbedded in the perfect CTTT repeats. In this case, the 24.1 allele is 1 bp longer than the 24-allele sequence.

GCCCCATAGGTTTTGAACTCACAGATTAAA
CTGTAACCAAAATAAAATTAGGCATTAT
TTACAAGCTAGTTT CTTT CTTT CTTT TTTCT CTTT
CTTT CTTT CTTT CTTT CTTT CTTT CTTT CTTT CTTT
CTTT CTTT CTTT CTTT CTTT CTTT CTC CTTC CTTC
CTTT CTTC CTTT CTTT TTTGCTGGCA
ATTACAGACAAATCAA

Table 2.4 Expected numbers of each of the three MN blood group genotypes under the null hypotheses of Hardy–Weinberg. Genotype frequencies are based on a sample of 1066 Chukchi individuals, a native people of eastern Siberia (Roychoudhury and Nei 1988).

Frequency of M = $\hat{p}$ = 0.4184
Frequency of N = $\hat{q}$ = 0.5816

Genotype	Observed	Expected number of genotypes	Observed − Expected
MM	165	$N \times \hat{p}^2 = 1066 \times (0.4184)^2 = 186.61$	−21.6
MN	562	$N \times 2\hat{p}\hat{q} = 1066 \times 2(0.4184)(0.5816) = 518.80$	43.2
NN	339	$N \times \hat{q}^2 = 1066 \times (0.5816)^2 = 360.58$	−21.6

frequencies. Asking whether genotypes are in Hardy–Weinberg proportions is actually the same as asking whether a coin is "fair." With a fair coin, we expect one-half heads and one-half tails if we flip it a large number of times. But even with a fair coin, we can get something other than exactly 50 : 50 even if the sample size is large. We would consider a coin fair if in 1000 flips it produced 510 heads and 490 tails. However, the hypothesis that a coin is fair would be in doubt if we observed 250 heads and 750 tails given that we expect 500 of each.

In more general terms, the expected frequency of an event, p, times the number of trials or samples, n, gives the expected number of events or np. To test the hypothesis that p is the frequency of an event in an actual population, we compare np with $n\hat{p}$. Close agreement suggests that the parameter and the

estimate are the same quantity. But a large disagreement instead suggests that p and $\hat{p}$ are likely to be different probabilities. The chi-squared (χ^2) distribution is a statistical test commonly used to compare np and $n\hat{p}$. The χ^2 test *provides the probability of obtaining the difference (or more) between the observed $n\hat{p}$ and expected (np) number of outcomes by chance alone if the null hypothesis is true.* As the difference between the observed and expected grows larger, it becomes less probable that the parameter and the parameter estimate are actually the same but differ in a given sample due to chance. The χ^2 statistic is:

$$\chi^2 = \sum \frac{(\text{observed} - \text{expected})^2}{\text{expected}} \quad (2.7)$$

where Σ (pronounced "sigma") indicates taking the sum of multiple terms.

The χ^2 formula makes intuitive sense. In the numerator, there is a difference between the observed and Hardy–Weinberg expected number of individuals. This difference is squared, like a variance, since we do not care about the direction of the difference but only the magnitude of the difference. Then, in the denominator, we divide by the expected number of individuals to make the squared difference relative. For example, a squared difference of 4 is small if the expected number is 100 (it is 4%) but relatively larger if the expected number is 8 (it is 50%). Adding all of these relative squared differences gives the total relative squared deviation observed over all genotypes.

$$\chi^2 = \frac{(-21.6)^2}{186.61} + \frac{(43.2)^2}{518.80} + \frac{(-21.6)^2}{360.58} = 7.39$$
$$(2.8)$$

We need to compare our statistic to values from the χ^2 distribution. But, first, we need to know how much information, or the degrees of freedom (commonly abbreviated as df), was used to estimate the χ^2 statistic. In general, degrees of freedom are based on the number of categories of data: df = no. of classes compared − no. of parameters estimated −1 for the χ^2 test itself. In this case, df = 3−1 − 1 = 1 for three genotypes and one estimated allele frequency (with two alleles: the other allele frequency is fixed once the first has been estimated).

Figure 2.9 shows a χ^2 distribution for one degree of freedom. Small deviations of the observed from the expected are more probable since they leave more area of the distribution to the right of the χ^2 value. As the χ^2 value gets larger, the probability that the difference between the observed and expected is just due to chance sampling decreases (the area under the curve to the right gets smaller). Another way

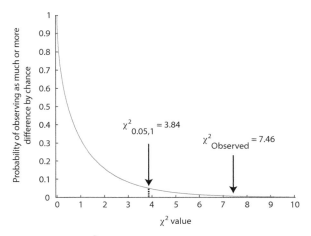

Figure 2.9 A χ^2 distribution with one degree of freedom. The χ^2 value for the Hardy–Weinberg test with MN blood group genotypes as well as the critical value to reject the null hypothesis are shown. The area under the curve to the right of the arrow indicates the probability of observing that much or more difference between the observed and expected outcomes.

of saying this is that as the observed and expected get increasingly different, it becomes more improbable that our null hypothesis of Hardy–Weinberg is actually the process that is determining genotype frequencies. Using Table 2.5, we see that a χ^2 value of 7.46 with 1 df has a probability between 0.01 and 0.001. The conclusion is that the observed genotype frequencies would be observed less than 1% of the time in a population that actually had Hardy–Weinberg expected genotype frequencies. Under the null hypothesis, we do not expect this much difference or more from Hardy–Weinberg expectations to occur often. By convention, we would reject chance as the explanation for the differences if the χ^2 value had a probability of 0.05 or less. In other words, if chance explains the difference in five trials out of 100 or less, then we reject the hypothesis that

Table 2.5 χ^2 values and associated cumulative probabilities in the right-hand tail of the distribution for one through five degrees of freedom.

df	Probability					
	0.5	0.25	0.10	0.05	0.01	0.001
1	0.4549	1.3233	2.7055	3.8415	6.6349	10.8276
2	1.3863	2.7726	4.6052	5.9915	9.2103	13.8155
3	2.3660	4.1083	6.2514	7.8147	11.3449	16.2662
4	3.3567	5.3853	7.7794	9.4877	13.2767	18.4668
5	4.3515	6.6257	9.2364	11.0705	15.0863	20.5150

the observed and expected patterns are the same. The critical value above which we reject the null hypothesis for a χ^2 test is 3.84 with 1 df, or in notation $\chi^2_{0.05,\ 1} = 3.84$. In this case, we can clearly see an excess of heterozygotes and deficits of homozygotes, and employing the χ^2 test allows us to conclude that Hardy–Weinberg expected genotype frequencies are not present in the population.

Assuming Hardy–Weinberg to test alternative models of inheritance

Biologists are all probably familiar with the ABO blood group and are aware that mixing blood of different types can cause blood cell lysis and possibly result in death. Although we take this for granted now, there was a time when blood types and their patterns of inheritance defined an active area of clinical research. It was in 1900 that Karl Landsteiner of the University of Vienna mixed the blood of the people in his laboratory to study the patterns of blood cell agglutination (clumping). Landsteiner was awarded the Nobel Prize for Medicine in 1930 for his discovery of human ABO blood groups. Not until 1925, due to the research of Felix Bernstein, was the genetic basis of the ABO blood groups resolved (see Crow 1993a).

Landsteiner observed the presence of four blood phenotypes A, B, AB, and O. A logical question was then, "what is the genetic basis of these four blood group phenotypes?" We will test two hypotheses (or models) to explain the inheritance of ABO blood groups that coexisted for 25 years. The approach will use the frequency of genotypes in a sample population to test the two hypotheses rather than an approach such as examining pedigrees. The hypotheses are that the four blood group phenotypes are explained by either two independent loci with two alleles each with one allele completely dominant at each locus (hypothesis 1) or a single locus with three alleles where two of the alleles show no dominance with each other but both are completely dominant over a third allele (hypothesis 2). Throughout, we will assume that Hardy–Weinberg expected genotype frequencies are met in order to determine which hypothesis best fits the available data.

Our first task is a straightforward application of Hardy–Weinberg in order to determine the expected frequencies of the blood group genotypes. The genotypes and the expected genotype frequencies are shown in Table 2.6. Look at the table but cover up the expected frequencies with a sheet of paper. The genotypes given for the two hypotheses would both explain the observed pattern of four blood groups. Hypothesis 1 requires complete dominance of the A and B alleles at their respective loci. Hypothesis 2 requires A and B to have no dominance with each other but complete dominance when paired with the O allele.

Now, let's construct several of the expected genotype frequencies (before you lift that sheet of paper). The O blood group under hypothesis 1 is the frequency of a homozygous genotype at two loci (aa bb). The frequency of one homozygote is the square of the allele frequency: fa^2 and fb^2 if we use fx to indicate the frequency of allele x. Using the product rule or Mendel's second law, the expected frequency of the two-locus genotype is the product of frequencies of the one-locus genotypes, fa^2 and fb^2. For the next

Table 2.6 Hardy–Weinberg expected genotype frequencies for the ABO blood groups under the hypotheses of 1) two loci with two alleles each, and 2) one locus with three alleles. Both hypotheses have the potential to explain the observation of four blood group phenotypes. The notation "fx" is used to refer to the frequency of allele x. The underscore ("_") indicates any allele, for example, A_ means both AA and Aa genotypes. The observed blood type frequencies were determined for Japanese people living in Korea (from Berstein (1925) as reported in Crow (1993b)).

Blood	Genotype		Expected genotype frequency		Observed
Type	Hypothesis 1	Hypothesis 2	Hypothesis 1	Hypothesis 2	(total = 502)
O	aa bb	OO	fa^2fb^2	$(fO)^2$	148
A	A_ bb	AA, AO	$(1-fa^2)(fb)^2$	$fA^2 + 2fAfO$	212
B	aa B_	BB, BO	$fa^2(1-fb^2)$	$fB^2 + 2fBfO$	103
AB	A_ B_	AB	$(1-fa^2)(1-fb^2)$	$2fAfB$	39

genotype under hypothesis 1 (A_ bb), we use a little trick to simplify the amount of notation. The genotype A_ means AA or Aa: in other words, any genotype but aa. Since the frequencies of the three genotypes at one locus must sum to 1, we can write fA_ as $1 - $ faa or $1 - $ fa^2. Then, the frequency of the A_ bb genotype is $(1 - $ fa$^2)($ fb$)^2$. You should now work out and write down the other six expected genotype frequency expressions: then lift the paper and compare your work with Table 2.6.

The next step is to compare the expected genotype frequencies for the two hypotheses with observed genotype frequencies. To do this, we will need to estimate allele frequencies under each hypothesis and use these to compute the expected genotype frequencies. (Although these allele frequencies are parameter estimates, the "hat" notation is not used for readability.) For the hypothesis of two loci (hypothesis 1), fb^2 = $(148 + 212)/502 = 0.717$, so we can estimate the allele frequency as fb = $\sqrt{\text{fb}^2}$ = $\sqrt{0.717}$ = 0.847. The other allele frequency at that locus is then determined by subtraction fB = $1 - 0.847 = 0.153$. Similarly, for the second locus fa^2 = $(148 + 103)/502$ = 0.50 and fa = $\sqrt{\text{fa}^2}$ = $\sqrt{0.50}$ = 0.707, giving fA = $1 - 0.707$ = 0.293 by subtraction.

For the hypothesis of one locus with three alleles (hypothesis 2), we estimate the frequency of any of the alleles by using the relationship that the three allele frequencies sum to 1. This basic relationship can be reworked to obtain the expected genotype

frequency expressions into expressions that allow us to estimate the allele frequencies (see Problem Box 2.2). It turns out that adding together all expected genotype frequency terms for two of the alleles estimates the square of one minus the other allele. For example, $(1 - $ fB$)^2$ = fO^2 + fA^2 + 2fAfO; and, checking in Table 2.7, this corresponds to $(148 + 212)/502 = 0.717$. Therefore, $1 - $ fB = 0.847 and fB = 0.153. Using similar steps, $(1 - $ fA$)^2$ = fO^2 + fB^2 + 2fBfO = $(148 + 103)/502 = 0.50$. Therefore, $1 - $ fA = 0.707 and fA = 0.293. Finally, by subtraction, fO = $1 - $ fB $- $ fA = $1 - 0.153 - 0.293 = 0.554$.

Problem box 2.2
Proving allele frequencies are obtained from expected genotype frequencies

Can you use algebra to prove that adding together expected genotype frequencies under hypotheses 1 and 2 in Table 2.7 gives the allele frequencies shown in the text? For the genotypes of hypothesis 1, show that f(aa bb) + f(A_ bb) = fbb. For hypothesis 2, show the observed genotype frequencies that can be used to estimate the frequency of the B allele starting off with the relationship fA + fB + fO = 1 and then solving for fB in terms of fA and fO.

Table 2.7 Expected numbers of each of the four blood group genotypes under the hypotheses of 1) two loci with two alleles each, and 2) one locus with three alleles. Estimated allele frequencies are based on a sample of 502 individuals.

Blood	Observed	Expected number of genotypes	Observed – Expected	(Observed – Expected)^2/ Expected
Hypothesis 1: fA = 0.293, fa = 0.707, fB = 0.153, fb = 0.847				
O	148	$502(0.707)^2(0.847)^2 = 180.02$	−32.02	5.69
A	212	$502(0.500)(0.847)^2 = 180.07$	31.93	5.66
B	103	$502(0.707)^2(0.282) = 70.76$	32.24	14.69
AB	39	$502(0.500)(0.282) = 70.78$	−31.78	14.27
Hypothesis 2: fA = 0.293, fB = 0.153, fO = 0.554				
O	148	$502(0.554)^2 = 154.07$	−6.07	0.24
A	212	$502[(0.293)^2 + 2(0.293)(0.554)] = 206.07$	5.93	0.17
B	103	$502[(0.153)^2 + 2(0.153)(0.554)] = 96.85$	6.15	0.39
AB	39	$502[2(0.293)(0.153)] = 45.01$	−6.01	0.80

The number of genotypes under each hypothesis can then be found using the expected genotype frequencies in Table 2.6 and the estimated allele frequencies. Table 2.7 gives the calculation for the expected numbers of each genotype under both hypotheses. We can also calculate a chi-squared value associated with each hypothesis based on the difference between the observed and expected genotype frequencies. For hypothesis 1, $\chi^2 = 40.32$, whereas, for hypothesis 2, $\chi^2 = 1.60$. Both of these tests have one degree of freedom (4 genotypes -2 for estimated allele frequencies -1 for the test), giving a critical value of $\chi^2_{0.05,1} = 3.84$. Clearly, the hypothesis of three alleles at one locus is the better fit to the observed data. Thus, we have just used genotype frequency data sampled from a population with the assumptions of Hardy–Weinberg equilibrium as a means to distinguish between two hypotheses for the genetic basis of blood groups.

Problem box 2.3
Inheritance for corn kernel phenotypes

Corn kernels are individual seeds that display a wide diversity of phenotypes (see Figure 2.10 and Plate 2.10). In a total of 3816 corn seeds, the following phenotypes were observed:

Purple, smooth 2058
Purple, wrinkled 728
Yellow, smooth 769
Yellow, wrinkled 261

Are these genotype frequencies consistent with inheritance due to one locus with three alleles or two loci each with two alleles?

Figure 2.10 Corn cobs demonstrating yellow and purple seeds that are either wrinkled or smooth.

2.5 The fixation index and heterozygosity

- The fixation index (F) measures deviation from Hardy–Weinberg expected heterozygote frequencies.
- Examples of mating systems and F in wild populations.
- Observed and expected heterozygosity.

The mating patterns of actual organisms frequently do not exhibit the random mating assumed by Hardy–Weinberg. In fact, many species exhibit mating systems that create predictable deviations from Hardy–Weinberg expected genotype frequencies. The term **assortative mating** is used to describe patterns of non-random mating. **Positive assortative mating** describes the case when individuals with like genotypes or phenotypes tend to mate. **Negative assortative mating** (also called disassortative mating) occurs when individuals with unlike genotypes or phenotypes tend to mate. Both of these general types of non-random mating will impact expected genotype frequencies in a population. This section describes the impacts of non-random mating on genotype frequencies and introduces a commonly used measure of non-random mating that can be utilized to estimate mating patterns in natural populations.

Mating among related individuals, termed **consanguineous mating** or **biparental inbreeding**, increases the probability that the resulting progeny are homozygous compared to random mating. This occurs since relatives, by definition, are more likely than two random individuals to share one or two alleles that were inherited from ancestors they share in common (this makes mating among relatives a form of assortative mating). Therefore, when related individuals mate, their progeny have a higher chance of receiving the same allele from both parents, giving them a greater chance of having a homozygous genotype. **Sexual autogamy** or **self-fertilization** is an extreme example of consanguineous mating where an individual can mate with itself by virtue of possessing reproductive organs of both sexes. Many plants and some animals, such as the nematode *Caenorhabditis elegans*, are hermaphrodites that can mate with themselves.

There are also cases of disassortative mating, where individuals with unlike genotypes have a higher probability of mating. A classic example in mammals is mating based on genotypes at major histocompatibility complex (MHC) loci, which produce proteins involved in self/non-self recognition in immune response. Mice are able to recognize individuals with similar MHC genotypes via odor, and based on these odors, avoid mating with individuals possessing a similar MHC genotype. Experiments where young mice were raised in nests of either their true parents or foster parents (called cross-fostering) showed that mice learn to avoid mating with individuals possessing odor cues similar to their nest-mates' rather than avoiding MHC-similar individuals per se (Penn and Potts 1998). This suggests that mice learn the odor of family members in the nest and avoid mating with individuals with similar odors, indirectly leading to disassortative mating at MHC loci as well as the avoidance of consanguineous mating. One hypothesis to explain the evolution of disassortative mating at MHC loci is that the behavior is adaptive since progeny with higher heterozygosity at MHC loci may have more effective immune response. There is evidence that some animals prefer mates with dissimilar MHC genotypes (e.g. Miller et al. 2009), while, in humans, the possibility remains controversial (Qiao et al. 2018).

The effects of non-random mating on genotype frequencies can be measured by comparing Hardy–Weinberg expected frequency of heterozygotes, which assumes random mating, with observed heterozygote frequencies in a population. A quantity called the **fixation index**, symbolized by F (f is reserved for the coancestry coefficient introduced later in Section 2.6.), is commonly used to compare how much heterozygosity is present in an actual population relative to the expected levels of heterozygosity under random mating

$$F = \frac{H_e - H_o}{H_e} \qquad (2.9)$$

where H_e is the Hardy–Weinberg expected frequency of heterozygotes based on population allele frequencies and H_o is the observed frequency of heterozygotes. Dividing the difference between the expected and observed heterozygosity by the expected heterozygosity expresses the difference in the numerator as a percentage of the expected heterozygosity. Even if the difference in the numerator may seem small, it may be large relative to the expected heterozygosity. Dividing by the expected heterozygosity also puts F on a convenient scale of -1 and $+1$. Negative values indicate heterozygote excess and positive values indicate homozygote excess relative to Hardy–Weinberg

---

Interact box 2.2 Assortative mating and genotype frequencies

The impact of assortative mating on genotype and allele frequencies can be simulated on the text simulation website. Use the **Simulation** menu and select **de Finetti**. The program models several non-random mating scenarios based on the settings in the **Mating Model** box. Start with Random Mating, set the initial genotype frequencies using the sliders for the frequencies of AA and Aa, and set **Generations to simulate** to 20. The genotype frequencies over time will be plotted on the triangle. Recall that if the points for each generation change position only vertically, then only genotype frequency is changing, while a movement to the left or right means that allele frequencies have changed. Try a set of three or four initial genotype frequencies that vary both allele and genotype frequencies. Under random mating, why does it appear that there are only two points even though 20 generations are simulated? How long does it take for a population to reach equilibrium with random mating?

Select the **Positive Assortative** radio button and repeat the simulations using the same initial genotype frequencies you used for random mating. Then, select the **Negative Assortative** radio button and again run the simulation using the same initial genotype frequencies that you employed for the other two mating models. How do the two types of non-random mating affect genotype frequencies? Allele frequencies?

expectations. In fact, the fixation index can be interpreted as the correlation between the two alleles sampled to make a diploid genotype (see the Appendix for an introduction to correlation if necessary). Given that one allele has been sampled from the population, if the second allele tends to be identical, there is a positive correlation (e.g. A and then A or a and then a); if the second allele tends to be different, there is a negative correlation (e.g. A and then a or a and then A); and if the second allele is independent, there is no correlation (e.g. equally likely to be A or a). With random mating, no correlation is expected between the first and second allele sampled to make a diploid genotype.

Assortative mating: Mating patterns where individuals do not mate in proportion to their genotype frequencies in the population; mating that is more (positive assortative mating) or less (negative assortative mating) frequent with respect to genotype or genetically based phenotype than expected by random combination.
Consanguineous mating: Mating between related individuals that can take the form of biparental inbreeding (mating between two related individuals) or sexual autogamy (self-fertilization).
Fixation index (*F*): The proportion by which heterozygosity is reduced or increased relative to the heterozygosity in a randomly mating population with the same allele frequencies.

Let's work through an example of genotype data for one locus with two alleles that can be used to estimate the fixation index. Table 2.8 gives observed counts and frequencies of the three genotypes in a sample of 200 individuals. To estimate the fixation index from these data requires an estimate of allele frequencies first. The allele frequencies can then be used to determine expected heterozygosity under the assumptions of Hardy–Weinberg. If p represents the frequency of the B allele,

$$\hat{p} = \frac{142 + \frac{1}{2}(28)}{200} = 0.78 \qquad (2.10)$$

using the genotype counting method to estimate allele frequency (Table 2.8 uses the allele counting method). The frequency of the b allele, q, can be estimated directly in a similar fashion or by subtraction ($\hat{q} = 1 - \hat{p} = 1 - 0.78 = 0.22$) since there are only two alleles in this case. The Hardy–Weinberg expected frequency of heterozygotes is $H_e = 2\hat{p}\hat{q} = 2(0.78)(0.22) = 0.343$. It is then simple to estimate the fixation index using the observed and expected heterozygosities.

$$\hat{F} = \frac{0.343 - 0.14}{0.343} = 0.59 \qquad (2.11)$$

In this example, there is a clear deficit of heterozygotes relative to Hardy–Weinberg expectations. The population contains 59% fewer heterozygotes than would be expected in a population with the same allele frequencies that was experiencing random mating and the other conditions set out in the assumptions of Hardy–Weinberg. Interpreted as a

---

Table 2.8 Observed genotype counts and frequencies in a sample of $N = 200$ individuals for a single locus with two alleles. Allele frequencies in the population can be estimated from the genotype frequencies by summing the total count of each allele and dividing it by the total number of alleles in the sample ($2N$).

Genotype	Observed	Observed frequency	Allele count	Allele frequency
BB	142	$\frac{142}{200} = 0.71$	284 B	$\hat{p} = \frac{284 + 28}{400} = 0.78$
Bb	28	$\frac{28}{200} = 0.14$	28 B, 28 b	
bb	30	$\frac{30}{200} = 0.15$	60 b	$\hat{q} = \frac{60 + 28}{400} = 0.22$

correlation between the allelic states of the two alleles in a genotype, this value of the fixation index tells us that the two alleles in a genotype are much more frequently of the same state than expected by chance.

In biological populations, a wide range of values have been observed for the fixation index (Table 2.9). Fixation indices have frequently been estimated with allozyme data (see Box 2.2). Estimates of $\hat{F}$ are generally correlated with mating system. Even in species where individuals possess reproductive organs of one sex only (termed **dioecious** individuals), mating among relatives can be common and ranges from infrequent to almost invariant. In other cases, mating is essentially random or complex mating and social systems have evolved to prevent consanguineous mating. Pure-breed dogs are an example where mating among relatives has been enforced by humans to develop lineages with specific phenotypes and behaviors, resulting in high fixation indices in some breeds. Many plant species possess both male and female sexual functions (hermaphrodites) and exhibit an extreme form of consanguineous mating, self-fertilization, that causes rapid loss of heterozygosity. In the case of Ponderosa pines in Table 2.9, the excess of heterozygotes may be due to natural selection against homozygotes at some loci (inbreeding depression). This makes the important point that departures from Hardy–Weinberg expected genotype frequencies estimated by the fixation index are potentially influenced by processes in addition to the mating system. Genetic loci free of the influence of other processes such as natural selection are often sought to estimate $\hat{F}$. In addition, $\hat{F}$ can be estimated using the average of multiple loci, which will tend to reduce bias since loci will differ in the degree they are influenced by other processes and outliers will be apparent.

The fixation index can be understood as a measure of the correlation between the states of the two alleles in a diploid genotype. When $F = 0$ there is no correlation between the two alleles in a genotype, the states of the two alleles are independent as we expect under Mendel's first law. If $F > 0$ there is a positive correlation such that if one of the alleles in a genotype is an A, for example, then the other allele will have a correlated state and also be an A. When $F < 0$ there is a negative correlation between the states of the two alleles in a genotype and heterozygotes are more common since the two alleles tend to have different states.

Extending the fixation index to loci with more than two alleles requires a means to calculate the expected frequency of genotypes with identical alleles (or with non-identical alleles) for an arbitrary number of alleles at one diploid locus. This can be accomplished by adding up all of the expected frequencies of each possible homozygous genotype

Table 2.9 Estimates of the fixation index ($\hat{F}$) for various species based on pedigree or molecular genetic marker data.

Species	Mating system	$\hat{F}$	Method	References
Humans				
Homo sapiens	outcrossed	0.0001–0.046	pedigree	Jorde (1997)
Snail				
Bulinus truncates	selfed & outcrossed	0.6–1.0	microsatellites	Viard et al. (1997)
Domestic dogs				
Breeds combined	outcrossed	0.33	allozyme	Christensen et al. (1985)
German Shepard	outcrossed	0.10		
Mongrels	outcrossed	0.06		
Plants				
Arabidopsis thaliana	Selfed	0.99	allozyme	Abbott et al. (1989)
Pinus ponderosa	outcrossed	−0.37	allozyme	Brown (1979)

Box 2.2 Protein locus or allozyme genotyping

Determining the genotypes of individuals at enzymatic protein loci is a rapid technique to estimate genotype frequencies in populations. Protein analysis was the primary molecular genotyping technique for several decades before DNA-based techniques became widely available. Alleles at loci that code for proteins with enzymatic function can be ascertained in a multi-step process. First, fresh tissue samples are ground up under conditions that preserve the function of proteins. Next, these protein extracts are loaded onto starch gels and exposed to an electric field. The electrical current results in electrophoresis where proteins are separated based on their ratio of molecular charge to molecular weight. Once electrophoresis is complete, the gel is then "stained" to visualize specific enzymes. The primary biochemical products of protein enzymes are not themselves visible. However,

a series of biochemical reactions in a process called enzymatic coupling can be used to eventually produce a visible product (often nitro blue tetrazolium or NBT) at the site where the enzyme is active (see Figure 2.11). If different DNA sequences at a protein enzyme locus result in different amino acid sequences that differ in net charge, then multiple alleles will appear in the gel after staining. The term allozyme (also known as isozyme) is used to describe the multiple allelic staining variants at a single protein locus. Allozyme electrophoresis and staining detects only a subset of genetic variation at protein coding loci. Amino acid changes that are charge neutral and nucleotide changes that are synonymous (do not alter the amino acid sequence) cannot be detected by allozyme electrophoresis methods. Refer to Manchenko (2003) for a technical introduction and detailed methods of allozyme detection.

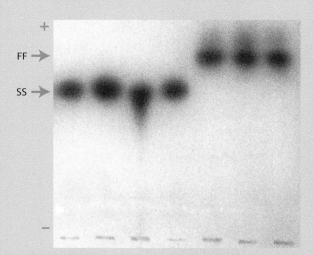

Figure 2.11 An allozyme gel stained to show alleles at the phosphoglucomutase or PGM locus in striped bass and white bass. The right-most three individuals are homozygous for the faster migrating allele (FF genotype), while the left-most four individuals are homozygous for the slower migrating allele (SS genotype). No double-banded heterozygotes (FS genotype) are visible on this gel. The + and − indicate the anode and cathode, respectively, ends of the gel. Wells where the individual samples were loaded into the gel can be seen at the bottom of the picture. Gel picture kindly provided by J. Epifanio.

and subtracting this total from 1 or summing the expected frequencies of all heterozygous genotypes:

$$H_e = 1 - \sum_{i=1}^{k} p_i^2 \qquad (2.12)$$

where k is the number of alleles at the locus, the p_i^2 and $2p_ip_j$ terms represent the expected homozygote genotype frequencies with random mating based on allele frequencies, and $\sum_{i=1}^{k}$ indicates summation of the frequencies of the k homozygous genotypes. Under random mating, $H_e = \sum_{i=1}^{k-1} \sum_{j=i+1}^{k} 2p_ip_j$. This quantity was called the **gene diversity** by Nei (1973) to distinguish it from the heterozygosity when there is non-random mating within populations and to recognize that it is a quantity that can be applied to polyploids (see Meirmans et al. 2018). The expected heterozygosity can be adjusted for small samples by multiplying H_e by $2N/(2N-1)$ where N is the total number of genotypes (Nei and Roychoudhury 1974), a correction that makes little difference unless N is about 50 or fewer individuals. In a similar manner, the observed heterozygosity (H_o) is the sum of the frequencies of all heterozygotes observed in a sample of genotypes:

$$\hat{H}_o = \sum_{i=1}^{h} H_i \qquad (2.13)$$

where the observed frequency of each heterozygous genotype H_i is summed over the $h = k(k-1)/2$ heterozygous genotypes possible with k alleles. Both H_e and H_o can be averaged over multiple loci to obtain mean heterozygosity estimates for two or more loci. Heterozygosity provides one of the basic measures of genetic variation, or more formally **genetic polymorphism**, in population genetics.

The fixation index as a measure of deviation from expected levels of heterozygosity is a critical concept that will appear in several places later in this text. The fixation index plays a conceptual role in understanding the effects of population size on heterozygosity (Chapter 3) and also serves as an estimator of the impact of population structure on the distribution of genetic variation (Chapter 4).

2.6 Mating among relatives

- Mating among relatives alters genotype frequencies but not allele frequencies.
- Mating among relatives and the probability that two alleles are identical by descent.
- The coancestry coefficient and autozygosity.
- Phenotypic consequences of mating among relatives.
- Inbreeding depression and its possible causes.
- The many meanings of inbreeding.

The previous section of this chapter showed how non-random mating can increase or decrease the frequency of heterozygote genotypes compared to the frequency that is expected with random mating. The last section also introduced the fixation index as well as ways to quantify heterozygosity in a population. This section will build on that foundation to show two concepts: (i) the consequences of non-random mating on allele and genotype frequencies in a population and (ii) the probability that two alleles are identical by descent. The focus will be on positive genotypic assortative mating (like genotypes mate) or inbreeding since this will eventually be helpful to understand genotype frequencies in small populations. The end of this section will consider some of the consequences of inbreeding and the evolution of autogamy.

Impacts of non-random mating on genotype and allele frequencies

Let's develop an example to understand the impact of mating among relatives on genotype and allele frequencies in a population. Under complete positive assortative mating or selfing, an individual mates with another individual possessing an identical genotype. Figure 2.12 diagrams the process of positive genotypic assortative mating for a diallelic locus, following the frequencies of each genotype through time. Initially, the frequency of the heterozygote is H but this frequency will be halved each generation. A Punnett square for two heterozygotes shows that half of the progeny are heterozygotes ($H/2$). The other half of the progeny are homozygotes ($H/2$), composed of one-quarter of the original heterozygote frequency of each homozygote genotype ($H/2[1-1/2]$). It is obvious that matings among like homozygotes will produce only identical homozygotes, so the homozygote genotypes each yield a constant frequency of homozygous progeny each generation. In total, however, the frequency of the homozygous genotypes increases by a factor of $\frac{H}{2}\left(1 - \frac{1}{2}\right)$ each generation due to homozygous progeny of the heterozygous genotypes. If the process of complete assortative mating continues, the population rapidly loses heterozygosity and approaches a state where the frequency of heterozygotes is zero.

As an example, imagine a population where $p = q = 0.5$ that has Hardy–Weinberg genotype frequencies $D = 0.25$, $H = 0.5$, and $R = 0.25$. Under complete positive assortative mating, what would be the frequency of heterozygotes after five generations? Using Figure 2.12, at time $t = 5$, heterozygosity would be $H(1/2)^5 = H(1/32) = 1/64$ or 0.016. This is a drastic reduction in only five generations.

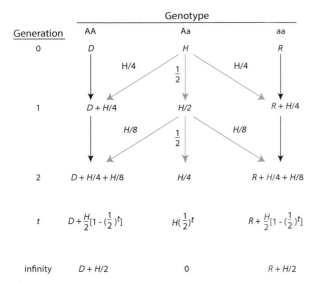

Figure 2.12 The impact of complete positive genotypic assortative mating (like genotypes mate) or self-fertilization on genotype frequencies. The initial genotype frequencies are represented by D, H, and R. When either of the homozygotes mates with an individual with the same genotype, all progeny bear their parent's homozygous genotype. When two heterozygote individuals mate, the expected genotype frequencies among the progeny are one half heterozygous genotypes and one quarter of each homozygous genotype. Every generation, the frequency of the heterozygotes declines by one half while one quarter of the heterozygote frequency is added to the frequencies of each homozygote (diagonal arrows). Eventually, the population will lose all heterozygosity, although allele frequencies will remain constant. Therefore, assortative mating or self-fertilization changes the pairing of alleles in genotypes but not the allele frequencies themselves.

Genotype frequencies change quite rapidly under complete assortative mating, but what about allele frequencies? Let's employ the same example population with $p = q = 0.5$ and Hardy–Weinberg genotype frequencies $D = 0.25$, $H = 0.5$, and $R = 0.25$ to answer the question. For both of the homozygous genotypes, the initial frequencies would be $D = R = (0.5)^2 = 0.25$. In Figure 2.12, the contribution of each homozygote genotype frequency from mating among heterozygotes after five generations is $H/2(1-(1/2)^5) = H/2(1–1/32) = H/2(31/32)$. With the initial frequency of $H = 0.5$, $H/2(31/32) = 0.242$. Therefore, the frequencies of both homozygous genotypes are $0.25 + 0.242 = 0.492$ after five generations. It is also apparent that the total increase in homozygotes $(31/32)$ is exactly the same as the total decrease in heterozygotes $(31/32)$, so the allele frequencies in the population have remained constant. After five generations of assortative mating in this example, genotypes are much more likely to contain two identical alleles than they are to contain two unlike alleles. This conclusion is also reflected in the value of the fixation index for this example, $\hat{F} = (0.5 - 0.016)/0.5 = 0.968$. In general, positive assortative mating or *inbreeding changes the way in which alleles are "packaged" into genotypes*, increasing the frequencies of all homozygous genotypes by the same total amount that heterozygosity is decreased, but allele frequencies in a population do not change.

The fact that allele frequencies do not change over time can also be shown elegantly with some simple algebra. Using the notation in Figure 2.12 and defining the frequency of the A allele as p and the a allele as q with subscripts to indicate generation, allele frequencies can be determined by the genotype counting method as $p_0 = D_0 + \frac{1}{2}H_0$ and $q_0 = R_0 + \frac{1}{2}H_0$. Figure 2.12 also provides the expressions for genotype frequencies from one generation to the next $D_1 = D_0 + \frac{1}{4}H_0$, $H_1 = \frac{1}{2}H_0$, and $R_1 = R_0 + \frac{1}{4}H_0$. We can then use these expressions to predict allele frequency in one generation:

$$p_1 = D_1 + \frac{1}{2}H_1 \tag{2.14}$$

as a function of genotype frequencies in the previous generation using substitution for D_1 and H_1:

$$p_1 = D_0 + \frac{1}{4}H_0 + \frac{1}{2}\left(\frac{1}{2}H_0\right) \tag{2.15}$$

which simplifies to:

$$p_1 = D_0 + \frac{1}{2}H_0 \qquad (2.16)$$

and then recognizing that the right-hand side is equal to the frequency of A in generation 0:

$$p_1 = p_0 \qquad (2.17)$$

Thus, allele frequencies remain constant under complete assortative mating. As practice, you should carry out the algebra for the frequency of the a allele.

Under complete self-fertilization heterozygosity declines very rapidly. There can also be partial self-fertilization in a population (termed **mixed mating**), where some matings are self-fertilization and others are between two individuals (called **outcrossing**). In addition, many organisms are not capable of self-fertilization but instead engage in biparental inbreeding (mating between two different but related individuals) to some degree. In general, these forms of mating among relatives will reduce heterozygosity

compared to random mating, although they will not drive heterozygosity toward zero as in the case of complete selfing. The rate of decline in heterozygosity can be determined for many possible types of mating systems, and a few examples are shown in Figure 2.13. Regardless of the specifics of the form of consanguineous mating that occurs, it remains true that mating among relatives causes alleles to be packaged more frequently as homozygotes (heterozygosity declines) and most forms of mating among relatives do not alter allele frequencies in a population. Negative assortative mating is an exception where allele frequencies can change depending on the initial allele frequencies (Workman 1964).

Coancestry coefficient and autozygosity

The effects of consanguineous mating can also be thought of as increasing the probability that two alleles at one locus in an individual are inherited from the same ancestor. Such a genotype would be homozygous and considered **autozygous** since the

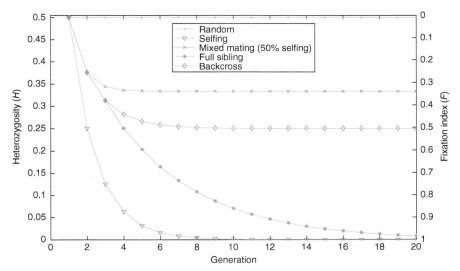

Figure 2.13 The impact of various systems of mating on heterozygosity (H) and the fixation index (F) over time. All populations have allele frequencies of $p = q = 0.5$ and initially are mating at random so heterozygosity equals 0.5 and remains at that level with random mating. As different patterns of mating among relatives occur in the four independent populations, observed heterozygosity declines and the fixation index increases at different rates depending on the coancestry coefficient (f) of each mating type. Selfing was 100% self-fertilization, while mixed mating was 50% of the population self-fertilizing and 50% mating at random. Full sibling is brother–sister or parent–offspring mating. Backcross is one individual mated to its progeny, then to its grand progeny, then to its great-grand progeny and so on, a mating scheme that is difficult to carry on for many generations. The fixation index at each generation for each mating scheme is based on the following recursion equations: selfing $F_{t+1} = \frac{1}{2}(1 + F_t)$; mixed $F_{t+1} = \frac{1}{2}(1 + F_t)$ (s) where s is the selfing rate; full sibling $F_{t+2} = \frac{1}{4}(1 + 2F_{t+1} + F_t)$; backcross $F_{t+1} = \frac{1}{4}(1 + 2F_t)$.

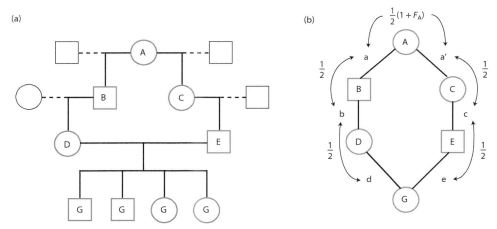

Figure 2.14   Average relatedness and autozygosity as the probability that two alleles at one locus are identical by descent. Panel A shows a pedigree where individual A has progeny that are half-siblings (B and C). B and C then produce progeny D and E, which in turn produce offspring G. Panel B shows only the paths of relatedness where alleles could be inherited from A, with curved arrows to indicate the probability that gametes carry alleles identical by descent. Upper case letters for individuals represent diploid genotypes and lower case letters indicate allele copies within the gametes produced by the genotypes. The probability that A transmits a copy of the same allele to B and C depends on the degree of inbreeding for individual A or F_A.

alleles were inherited from a common ancestor. If the two alleles are not inherited from the same ancestor in the recent past, we would call the genotype **allozygous** (*allo-* means other). You are probably already familiar with autozygosity, although you may not recognize it as such. Two times the probability of autozygosity (since diploid individuals have two alleles) is commonly expressed as the **degree of relatedness** among relatives. For example, full siblings (full brothers and sisters) are one-half related and first cousins are one-eighth related. Using a pedigree and tracing the probabilities of inheritance of an allele, the autozygosity and the basis of average relatedness can be seen.

Autozygosity is measured by the **coefficient of coancestry** (sometimes called the coefficient of kinship) and symbolized as *f*, can be seen in a pedigree such as that shown in Figure 2.14. Figure 2.14a gives a hypothetical pedigree for four generations. The pedigree can be used to determine the probability that the fourth-generation progeny, labeled G, have autozygous genotypes due to individual A being a common ancestor of both their maternal and paternal parents. To make the process simpler, Figure 2.14b strips away all of the external ancestors and shows only the paths where alleles could be inherited in the progeny from individual A.

Allozygous genotype: A homozygous or heterozygous genotype composed of two alleles not inherited from a recent common ancestor.
Autozygosity (*f*): The probability that two alleles in a homozygous genotype are identical by descent.
Autozygous genotype: A homozygous genotype composed of two identical alleles that are inherited from a common ancestor.
Coancestry coefficient (Θ): The probability that two randomly sampled gametes, one from each of two individuals, both carry a given allele that is identical by descent.
Identity by descent (IBD): Sharing the same state because of transmission from a common ancestor.
Relatedness: The expected proportion of alleles between two individuals that are identical by descent; twice the autozygosity.

To begin the process of determining the autozygosity for G, it is necessary to determine the probability that A transmitted the same allele to individuals B and C,

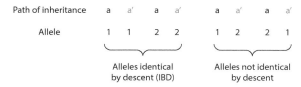

Figure 2.15 The possible patterns of transmission from one parent to two progeny for a locus with two alleles. Half of the outcomes result in the two progeny inheriting an allele that is identical by descent. The a and a' refer to paths of inheritance in the pedigree in panel B of Figure 2.14.

or in notation $P(a = a')$. With two alleles designated 1 and 2, there are only four possible patterns of allelic transmission from A to B and C, as shown in Figure 2.15. In only half of these cases do B and C inherit an identical allele from A, so $P(a = a') = 1/2$. This probability would still be ½ no matter how many alleles were present in the population, since the probability arises from the fact that diploid genotypes have only two alleles.

To have a complete account of the probability that B and C inherit an identical allele from A, we also need to take into account the past history of A's genotype since it is possible that A was itself the product of mating among relatives. If A was the product of some level of biparental inbreeding, then the chance that it transmits alleles identical by descent to B and C is greater than if A was from a randomly mating population. Another way to think of it is, with A being the product of some level of inbreeding instead of random mating, the chances that the alleles transmitted to B and C are not identical (see Figure. 2.14b) will be less than ½ by the amount that A is inbred. If the degree to which A is inbred (or the probability that A is autozygous) is F_A, then the total probability that B and C inherit the same allele is:

$$P(a = a') = ½ + ½F_A = ½(1 + F_A) \qquad (2.18)$$

If the parents of individual A are unrelated, then F_A is 0 in Eq. 2.18, and then the chance of transmitting the same allele to B and C reduces to the ½ expected in a randomly mating population.

For the other paths of inheritance in Figure 2.14, the logic is similar to determine the probability that an allele is identical by descent. For example, what is the probability that the allele in gamete d is identical by descent to the allele in gamete b, or $P(b = d)$? When D mated, it passed on one of two alleles, with a probability of ½ for each allele. One allele was inherited from each parent, so there is a ½ chance of transmitting a maternal or paternal allele. This makes $P(b = d) = ½$. (Just like with individual A, P

$(b = d)$ could also be increased to the extent that B was inbred, although random mating for all genotypes but A is assumed here for simplicity.) This same logic applies to all other paths in the pedigree that connect A and the progeny G. The probability of a given allele being transmitted along a path is independent of the probability along any other path, so the probability of autozygosity (symbolized as f to distinguish it from the preexisting homozygote excess or deficit of the population individual A belongs to, or F_A) over the entire pedigree for any of the G progeny is:

$$f_G = f_{DE} \quad ½ \times \quad ½ \times \quad ½(1 + F_A) \quad ½ \times \quad ½$$
$$ P(b = d) \; P(a = b) \; P(a = a') \; P(a' = c) \; P(c = e)$$
$$= (½)^5 (1 + F_A) = 1/32(1 + F_A)$$

$$(2.19)$$

since independent probabilities can be multiplied to find the total probability of an event. This is equivalent to the average relatedness among half-cousins. In general, for pedigrees, $f = (½)^i(1 + F_A)$ where A is the common ancestor and i is the number of paths or individuals over which alleles are transmitted. By writing down the chain of individuals and counting the individuals along paths of inheritance, we can determine the probability that a sample of two alleles, one from each individual, would exhibit both alleles identical by descent. That method gives G̶DBA̲CEG̶ or five ancestors for $\left(\frac{1}{2}\right)^5$, yielding a result identical to Eq. 2.19.

We can use the method of tracing paths between ancestors to determine the coancestry coefficient, often symbolized by **Θ**, for any type of relationship. The pedigree in Figure 2.16 provides a set of examples of close relatives where we can determine the coancestry coefficient using paths of inheritance.

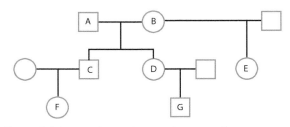

Figure 2.16 A pedigree showing first (A and C are parent and offspring, C and D are full siblings), second (A and B are the grandparents of F and G, D, and E are half siblings), and third (F and G are cousins) degree relatives. The coancestry coefficient gives the probability that an allele sampled from each of two related individuals is identical by descent, defining the degree of relatedness.

In general, coefficients of coancestry for two individuals A and B can be determined using

$$\Theta_{AB} = \sum_{1}^{\#ancestors} \left(\frac{1}{2}\right)^{P+1} (1 + F) \qquad (2.20)$$

where P is the number of ancestor–descendant paths connecting A and B, and F is the homozygote excess or deficit of the ancestor (Wright 1922; Thompson 1988). The **inbreeding coefficient** of an individual, or the probability of an individual inheriting two copies of the same allele, is a function of the coancestry of its parents. One example is the coancestry between one parent and an offspring, such as individuals A and C in Figure 2.16. There is one ancestor–descendant link between A and C, so the coancestry coefficient is

$$\Theta_{AC} = \left(\frac{1}{2}\right)^{2} = \frac{1}{4} \qquad (2.21)$$

assuming that the parent A has $F = 0$. For the parent, the chance of sampling the allele that was transmitted to the offspring is ½. For the offspring too, the chance of sampling the allele inherited from that parent is ½. When combined, the chance of sampling the one allele that is identical by descent (IBD) between parents and offspring is $(1/2)^2 = 1/4$. When considering *both* alleles in the offspring, one of them is IBD to one parent, so a parent and an offspring are ¼ + ¼ = ½ related.

The coancestry coefficient for full siblings, such as individuals C and D in Figure 2.16, is a case where individuals share two parents in common and therefore have two common ancestors to account for. Counting the paths C–A–D and C–B–D, we obtain

$$\Theta_{CD} = \left(\frac{1}{2}\right)^{3} + \left(\frac{1}{2}\right)^{3} = \frac{1}{4} \qquad (2.22)$$

The chance that a given allele was transmitted from one parent to one offspring is ½, with a probability of $(1/2)^2 = ¼$ of *both* full siblings inheriting the same allele from one parent. Because C and D share both parents and can inherit alleles identical by descent from both, we add the coancestries of each allele to give a relatedness of $1/4 + 1/4 = ½$.

The coancestry coefficients for self-fertilization and for full siblings explain the pattern of decreasing heterozygosity and increasing fixation indices over generations seen in Figure 2.13. For self-fertilization,

each generation has a coancestry coefficient of ½ when a self-fertilized parent descended from unrelated individuals, making the fixation index equal to ½ after one generation of selfing. For a second generation of self-fertilization, the coancestry coefficient remains ½ but now the parent has a higher probability of being homozygous because of the first generation of selfing. This makes the second-generation fixation index $F_3 = ½(1 + ½) = 3/4$.

In the case of full siblings, the recursion equation is $F_{t+2} = ¼(1 + 2F_{t+1} + F_t)$. For full siblings, the coancestry coefficient is ¼ when their parents have no history of consanguineous mating among their ancestors, resulting in $F_2 = ¼$. Taking two individuals from the second generation of full sibling mating as parents means that they each have a higher probability of being homozygous that is added to the constant ¼ probability of coancestry for full siblings. This gives a fixation index after three generations of full sibling mating of $F_3 = ¼(1 + 2[¼] + 0) = 3/8$ (it is not until the fourth generation that the grandparents in generation t have a fixation index greater than zero).

It is useful to determine the coancestry coefficient for a specific set of relatives or pedigree as well as the change in the fixation index over generations for mating systems, especially in quantitative genetics. With the recent ability to obtain genome-scale DNA sequences of numerous individuals, it is now possible to estimate coancestry directly from observed DNA polymorphism. Direct estimation of shared DNA using sequences highlights that traditional coancestry coefficients obtained from pedigrees are an average for a large sample of individuals and there can be substantial variation around these averages in the DNA-level relatedness of an observed set of individuals (Ackerman et al. 2017). The general point is to understand that mating among relatives as a process that increases autozygosity in a population. When individuals have common relatives, the chance that they share alleles identical by descent is increased as is the chance that the genotype of one individual is homozygous.

The departure from Hardy–Weinberg expected genotype frequencies, the coancestry coefficient, the autozygosity, and the fixation index are interrelated. Another way of stating the results that were developed in Figure 2.12 is that F measures the degree to which Hardy–Weinberg genotype frequencies are not met due to departure from purely random union of alleles in diploid genotypes as expressed by f. To see this, imagine starting with a one locus with two alleles in a population at

Box 2.3 Locating relatives using genetic genealogy methods

Genetic genealogy is the use of DNA testing to augment historical records with the goal of identifying ancestors and descendants. The approach is to use autosomal, Y-chromosome, and mitochondrial DNA genetic markers to locate relatives based on observed sharing of alleles and haplotypes. With enough genetic loci, even distant relatives who have small coancestry coefficients can be identified if their genotypes are available for comparison. Genetic profile data has been collected for millions of individuals who have participated in direct-to-consumer DNA testing. Applications include reconstruction of family histories, identification of ancestral population origins, and identifying the biological parents of adoptees.

Genetic genealogy or familial searching methods are used to identify suspects in unsolved serial violent crimes. While DNA profiles can be used as evidence to link a known suspect to a crime, direct identification of an unknown suspect is impossible if that person's DNA profile is not already present in a forensic or offender genetic database. A recent innovation is to use DNA profiles obtained from crime scenes to identify the relatives of suspects through searches in genetic genealogy databases. This approach initially utilized public databases to search for relatives and led to numerous cold cases being solved. At the same time, law enforcement and other uses of genealogical databases have generated concern over the privacy of genetic information.

Hardy–Weinberg expected genotype frequencies with F equal zero. The expected frequencies of the three genotypes in progeny could be expressed as

$$freq(AA) = p^2 + fpq$$
$$freq(Aa) = 2pq - f2pq \qquad (2.23)$$
$$freq(aa) = q^2 + fpq$$

With consanguineous mating, the decline in realized heterozygosity (substituting H for $freq[Aa]$) in the progeny is proportional to the coancestry coefficient which gives the probability that alleles identical in state are inherited because of relatedness among parents, shown by substituting H_e for $2pq$ in Eq. 2.20 and rearranging to give

$$H = H_e(1 - f) \qquad (2.24)$$

where H_e is the Hardy–Weinberg expected heterozygosity based on population allele frequencies. Rearranging Eq. 2.21 in terms of the autozygosity gives

$$f = 1 - \frac{H}{H_e} \qquad (2.25)$$

This is really exactly the same quantity as the fixation index (Eq. 2.9)

$$f = 1 - \frac{H}{H_e} = \frac{H_e}{H_e} - \frac{H}{H_e} = \frac{H_e - H}{H_e} \qquad (2.26)$$

With the understanding that we are considering only a single generation of mating and that the population initially had Hardy–Weinberg expected genotype frequencies, we see that the fixation index measures an excess of homozygosity exactly equal to the autozygosity. Sustained non-random mating over multiple generations will continue to increase the homozygosity, and therefore, F will reflect the *cumulative* deficit of heterozygosity from all past mating among relatives. In contrast, f is defined on a per generation basis and reflects only the probability of identity by descent at mating in the most recent generation. This distinction can be observed in Figure 2.13 where F increases over time toward an equilibrium as a function of the coancestry coefficient due to each mating system.

We can also use the fixation index in Eq. 2.23 rather than the autozygosity to express deviations from Hardy–Weinberg expected genotype frequencies due to accumulated autozygosity caused by mating patterns such as $freq(AA) = p^2 + Fpq$ and $freq(Aa) = 2pq - F2pq$ where $F2pq$ is the amount of heterozygosity missing from (or added to) the Hardy–Weinberg expected heterozygote frequency with half of that total or *Fpq* being added to (or subtracted from) each of the homozygotes. It is important to note that this time *F* represents the correlation of allelic state within genotypes from all causes accumulated over time that alters genotype frequencies from their Hardy–Weinberg expected values. While the difference in notation between *F* and *f* seems minor, their biological interpretations differ substantially in this example.

Returning to Figure 2.13 reinforces the relationship of the coancestry coefficient, the fixation index, and the decline in heterozygosity in several specific cases of regular consanguineous mating. Remember that in all cases in Figure 2.13, the Hardy–Weinberg expected heterozygosity is 0.5 when mating is random and *f* = 0.

Phenotypic consequences of mating among relatives

The process of consanguineous mating is associated with changes in the mean phenotype within a population. These changes arise from two general causes: changes in genotype frequencies in a population per se and fitness effects associated with changes in genotype frequencies.

The mean phenotype of a population will be impacted by any changes in genotype frequency. To show this, it is necessary to introduce terminology to express the phenotype associated with a given genotype, a topic covered in much greater detail and explained more fully in Chapters 9 and 10. We will assign AA genotypes the phenotype +*a*, heterozygotes the phenotype *d*, and aa homozygotes the phenotype −*a*. Each genotype contributes to the overall phenotype based on how frequent it is in the population. The mean phenotype in a population is then the sum of each genotype-frequency-weighted phenotype (Table 2.10). When there is no dominance, the phenotype of the heterozygotes is exactly intermediate between the phenotypes of the two homozygotes and *d* = 0. In that case, it is easy to see that mating among relatives will not change the mean phenotype in the population since both homozygous genotypes increase by the same amount and their effects on the mean phenotype cancel out (mean = $ap^2 + aFpq$ + ~~d2pq~~ − ~~dF2pq~~ − aq^2 − aFpq, where the heterozygote terms are crossed out since *d* = 0). When there is some degree of dominance (positive *d* indicates the phenotype of Aa is like that of AA while negative *d* indicates the phenotype of Aa is like that of aa), then the mean phenotype of the population will change with consanguineous mating since heterozygotes will become less frequent. If dominance is in the direction of the +*a* phenotype (*d* > 0), then mating among relatives will reduce the population mean because the heterozygote frequency will drop. Similarly, if dominance is in the direction of −*a* (*d* < 0), then mating among relatives will increase the population mean again because the heterozygote frequency decreases. It is also true in the case of dominance that a return to random mating will restore the frequencies of heterozygotes and return the population mean to its original value mating among relatives. These changes in the population mean phenotype are simply a consequence of changing the genotype frequencies when there is no change in the allele frequencies.

There is a wealth of evidence that the increase of homozygosity caused by mating among relatives has **deleterious** (harmful or damaging) consequences and is associated with a decline in the average phenotype in a population, a phenomenon referred to as **inbreeding depression**. Since the early twentieth century, studies in animals and plants that have been intentionally inbred provide ample evidence that decreased performance, growth,

Table 2.10 The mean phenotype in a population that is experiencing consanguineous mating. The fixation index quantifying deviation from Hardy–Weinberg expected genotype frequencies is *F*, and *d* = 0 when there is no dominance.

Genotype	Phenotype	Frequency	Contribution to population mean
AA	+*a*	$p^2 + Fpq$	$ap^2 + aFpq$
Aa	*D*	$2pq - F2pq$	$d2pq - dF2pq$
aa	−*a*	$q^2 + Fpq$	$-aq^2 - aFpq$

population mean: $ap^2 + d2pq - dF2pq - aq^2 = a(p-q) + d2pq(1-F)$

reproduction, viability (all measures of fitness), and abnormal phenotypes are associated with consanguineous mating. A related phenomenon is **heterosis** or hybrid vigor, characterized by beneficial consequences of increased heterozygosity such as increased viability and reproduction, or the reverse of inbreeding depression. One example is the heterosis exhibited in corn, which has led to the widespread use of F1 hybrid seed in industrial agriculture.

> **Heterosis**: The increase in performance, survival, and ability to reproduce of individuals possessing heterozygous loci (hybrid vigor); increase in the population average phenotype associated with increased heterozygosity.
>
> **Inbreeding depression**: The reduction in performance, survival, and ability to reproduce of individuals possessing homozygous genotypes; decrease in the population average phenotype associated with mating among relatives that increases homozygosity.

There is evidence that humans possess homozygosity because of mating among relatives and also experience inbreeding depression. Genome-scale genetic marker data in humans has revealed stretches of the genome where both chromosomes possess identical alleles called **runs of homozygosity** that are explained by both recent family members being related as well as by the history of population size and mixing (reviewed by Ceballos et al. 2018). There is also evidence for inbreeding depression in humans

based on observed phenotypes in the offspring of couples with known consanguinity. For example, mortality among children of first-cousin marriages was about 3.5% greater than for marriages between unrelated individuals measured in a range of human populations (Bittles and Black 2010). Human studies have utilized existing parental pairs with relatively low levels of inbreeding, such as uncle/niece, first cousins, or second cousins, in contrast to animal and plant studies where both very high levels and a broad range of coancestry coefficients are achieved intentionally. Drawing conclusions about the causes of variation in phenotypes from such observational studies requires caution, since the prevalence of consanguineous mating in humans is also correlated with social and economic variables such as illiteracy, age at marriage, duration of marriage, and income. These latter variables are therefore not independent of consanguinity and can themselves contribute to variation in phenotypes such as fertility and infant mortality (see Bittles and Black 2010).

The Mendelian genetic causes of inbreeding depression have been a topic of population genetics research for more than a century. None other than Charles Darwin carried out experimental pollinations in numerous plant species and observed that progeny of self-fertilization were shorter and produced fewer seeds than outcrossed progeny (Darwin 1876). There are two classical hypotheses to explain inbreeding depression and changes in fitness as the fixation index increases (Charlesworth and Charlesworth 1999; Carr et al. 2003). Both hypotheses predict that levels of inbreeding depression will increase along with consanguineous mating that increases homozygosity, although for

Table 2.11 A summary of the Mendelian basis of inbreeding depression under the dominance and overdominance hypotheses along with predicted patterns of inbreeding depression with continued consanguineous mating.

Hypothesis	Mendelian basis	Low fitness genotypes	Changes in inbreeding depression w/ continued consanguineous mating
Dominance	recessive and partly recessive deleterious alleles	only homozygotes for deleterious recessive alleles	purging of deleterious alleles that is increasingly effective as degree of recessiveness increases
Overdominance	heterozygote advantage or heterosis	all homozygotes	no changes as long as consanguineous mating keeps heterozygosity low

different reasons (Table 2.11). The first hypothesis, often called the **dominance hypothesis**, is that increasing homozygosity increases the phenotypic expression of fully and partly recessive alleles with deleterious effects. The second hypothesis is that inbreeding depression is the result of the decrease in the frequency of heterozygotes that occurs with consanguineous mating. This explanation supposes that heterozygotes have higher fitness than homozygotes (heterosis) and is called the **overdominance hypothesis**. In addition, the fitness interactions of alleles at different loci (epistasis; see Chapter 9) may also cause inbreeding depression, a hypothesis that is particularly difficult to test (see Carr and Dudash 2003). These causes of inbreeding depression may all operate simultaneously.

These dominance and overdominance hypotheses make different testable predictions about how inbreeding depression (measured as the average phenotype of a population) will change over time with continued consanguineous mating. Under the dominance hypothesis, recessive alleles that cause lowered fitness are more frequently found in homozygous genotypes under consanguineous mating. This exposes the deleterious phenotype and the genotype will decrease in frequency in a population by natural selection (individuals homozygous for such alleles have lower survivorship and reproduction). This reduction in the frequency of deleterious alleles by natural selection is referred to as **purging of genetic load**. Purging increases the frequency of alleles that do not have deleterious effects when homozygous, so that the average phenotype in a population then returns to the initial average it had before the onset of consanguineous mating. In contrast, the overdominance hypothesis does not predict a purging effect with consanguineous mating. With consanguineous mating, the frequency of heterozygotes will decrease and not recover until mating patterns change (see Figure 2.12). Even if heterozygotes are frequent and have a fitness advantage, each generation of mating and Mendelian segregation will reconstitute the two homozygous genotypes so purging cannot occur. These predictions highlight the major difference between the hypotheses. Inbreeding depression with overdominance arises from genotype frequencies in a population, while inbreeding depression with dominance is caused by the frequency of deleterious recessive alleles in a population. Models of natural selection that are relevant to inbreeding depression on population genotype and allele frequencies receive detailed coverage in Chapter 6.

Inbreeding depression in many animals and plants appears to be caused, at least in part, by deleterious recessive alleles consistent with the dominance hypothesis (Byers and Waller 1999; Charlesworth and Charlesworth 1999; Crnokrak and Barrett 2002). A classic example of inbreeding depression and recovery of the population mean for litter size in mice is shown in Figure 2.17. Model research organisms such as mice, rats, and *Drosophila*, intentionally bred by schemes such as full sibling mating for 10s or 100s of generations to create highly homozygous, so-called pure-breeding lines, are also not immune to inbreeding depression. Such inbred lines are often founded from multiple families, and many of these family lines go extinct from low viability or reproductive failure with habitual inbreeding. This is another type of purging effect due to natural selection that leaves only those lines that exhibit less inbreeding depression, which could be due to dominance, overdominance, or epistasis.

There is evidence that inbreeding depression exhibits environmental dependence due to variation among environments in phenotypic expression, dominance, and natural selection (Armbruster and

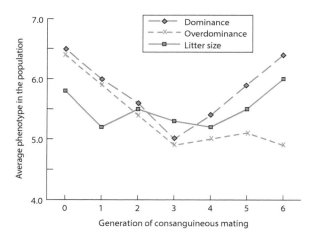

Figure 2.17 A graphical depiction of the predictions of the dominance and overdominance hypotheses for the genetic basis of inbreeding depression. The line for dominance shows purging and recovery of the population mean under continued consanguineous mating expected if deleterious recessive alleles cause inbreeding depression. However, the line for overdominance as the basis of inbreeding depression shows no purging effect since heterozygotes continue to decrease in frequency. The results of an inbreeding depression experiment with mice show that litter size recovers under continued brother–sister mating as expected under the dominance hypothesis (Lynch 1977). Only two of the original 14 pairs of wild-caught mice were left at the sixth generation. Not all of the mouse phenotypes showed patterns consistent with the dominance hypothesis.

Reed 2005; Cheptou and Donohue 2011). The social and economic correlates of inbreeding depression in humans mentioned above are a specific example of environmental effects on phenotypes. Inbreeding depression can be more pronounced when environmental conditions are more severe or limiting. For example, in the plant rose pink (*Sabatia angularis*) progeny from self-fertilizations showed decreasing performance when grown in the greenhouse, a garden, and their native habitat, consistent with environmental contributions to the expression of inbreeding depression (Dudash 1990). In another study, the number of surviving progeny for inbred and random-bred male wild mice (*Mus domesticus*) was similar under laboratory conditions, but male progeny of matings between relatives sired only 20% of the surviving progeny that males from matings between unrelated individuals did when under semi-natural conditions due to male–male competition (Meagher et al. 2000). However, not all studies show environmental differences in the expression of inbreeding depression. As an example, uniform levels of inbreeding depression were shown by mosquitoes grown in the laboratory and in natural tree holes where they develop as larvae and pupae in the wild (Armbruster et al. 2000).

The degree of inbreeding depression also depends on the phenotype being considered. In plants, traits early in the life cycle such as germination less often show inbreeding depression than traits later in the life cycle such as growth and reproduction (Husband and Schemske 1996). A similar pattern is apparent in animals, with inbreeding depression most often observed for traits related to survival and reproduction.

Inbreeding depression is a critical concept when thinking about the evolution of mating patterns in plants and animals. Suppose that a single locus determines whether an individual will self or outcross and the only allele present in a population is the outcrossing allele. Then imagine that mutation produces an allele at that locus, which, when homozygous, causes an individual to self-fertilize. Such a selfing allele would have a transmission advantage over outcrossing alleles in the population. To see this, consider the number of allele copies at the mating locus transmitted from parents to progeny. Parents with outcrossing alleles mate with another individual and transmit one allele to their progeny. Self-fertilizing parents, however, are both mom and dad to their offspring and transmit two alleles to their progeny. In a population of constant size where each

individual contributes an average of one progeny to the next generation, the selfing allele is reproduced twice as fast as an outcrossing allele and would rapidly become fixed in the population (see Lande and Schemske 1985; Fisher 1999). Based on this twofold higher rate of increase of the selfing allele, complete self-fertilization would eventually evolve unless some disadvantage counteracted the increase of selfed progeny in the population. Inbreeding depression where the average fitness of outcrossed progeny exceeds the average fitness of selfed progeny by a factor of two could play this role. If outcrossed progeny are at a twofold advantage due to inbreeding depression, then complete outcrossing would evolve. Explaining the existence of populations that engage in intermediate levels of selfing and outcrossing, a mating system common in plants, remains a challenge under these predictions (Byers and Waller 1999).

The many meanings of inbreeding

Unfortunately, the word inbreeding is used as a generic term to describe multiple distinct, although interrelated, concepts in population genetics (Jacquard 1975; Templeton and Read 1994). Inbreeding can apply to:

- consanguinity or kinship of two different individuals based on two alleles sampled at random;
- autozygosity of two alleles within an individual;
- the fixation index and Hardy–Weinberg expected and observed genotype frequencies, especially when there is an excess of homozygotes;
- inbreeding depression caused by deleterious recessive alleles or by overdominance;
- the description of the mating system of a population or species (as in inbred);
- genetic subdivision of a species into populations that exchange limited levels of gene flow such that individual populations increase in autozygosity;
- the increase in homozygosity in a population due to its finite size.

These different concepts all relate in some way to either the probability of allele being identical by descent or to expected genotype frequencies in a population, so the connection to inbreeding is clear. Awareness of the different ways the word inbreeding is used as well as an understanding of these different uses will prevent confusion, which can often be avoided simply by using more specific terminology. Remembering that the concepts are interrelated under the general umbrella of inbreeding can also help in realizing the equivalence of the population genetic processes in operation. The next chapter will show how finite population size is equivalent in its effects to inbreeding. Chapter 4 will take up the topic of population subdivision.

2.7 Hardy–Weinberg for two loci

- Expected genotype frequencies with two loci.
- Quantifying gametic disequilibrium with *D*.
- Approach to gametic equilibrium over time.
- Causes of gametic disequilibrium.

Gametic disequilibrium

We saw earlier in the chapter that Hardy–Weinberg could be extended to give expected genotype frequencies for two loci using via the product rule. While this is accepted without question now, in the early days of population genetics, it was a challenge to explain. In 1902, Walter Sutton and Theodor Boveri advanced the chromosome theory of heredity. They observed cell division and hypothesized that the discrete bodies seen separating into sets at meiosis and mitosis contained hereditary material that was transmitted from parents to offspring. At the time, the concept of chromosomal inheritance presented a paradox. Mendel's second law says that gamete **haplotypes** (**haploid genotype**) should appear in frequencies proportional to the product of allele frequencies. This prediction conflicted with the chromosome theory of heredity since there are not enough chromosomes to represent each hereditary trait.

To see the problem, take the example of *Homo sapiens* with a current estimate of about 20 000 protein coding genes in the nuclear genome. However, humans have only 23 pairs of chromosomes, or a large number of loci but only a small number of chromosomes. So, if chromosomes are indeed hereditary molecules, many genes must be on the same chromosome (on average about 870 genes per chromosome for humans if there are 20 000 genes). This means that some genes are physically **linked** by being located on the same chromosome. The solution to the paradox is the process of **recombination**. Sister chromatids touch at random points during

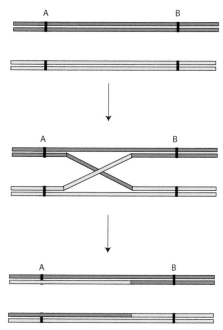

Figure 2.18 A schematic diagram of the process of recombination between two loci, A and B. Two double-stranded chromosomes (drawn in color and gray) exchange strands and form a Holliday structure. The cross over event can resolve into either of two recombinant chromosomes that generate new combinations of alleles at the two loci. The chance of a cross over event occurring generally increases as the distance between loci increases. Two loci are independent when the probability of recombination and non-recombination are both equal to ½. Gene conversion, a double cross over event without exchange of flanking strands, is not shown.

meiosis and exchange short segments, a process known as **crossing-over** (Figure 2.18).

Linkage of loci has the potential to impact multi-locus genotype frequencies and violate Mendel's law of independent segregation, which assumes the absence of linkage. To generalize expectations for genotype frequencies for two (or more) loci requires a model that accounts explicitly for linkage by including the rate of recombination between loci. The effects of linkage and recombination are important determinants of whether or not expected genotype frequencies under independent segregation of two loci (Mendel's second law) are met. Autosomal linkage is the general case that will be used to develop expectations for genotype frequencies under linkage.

The frequency of a two-locus gamete haplotype will depend on two factors: (i) allele frequencies and (ii) the amount of recombination between the two loci. We can begin to construct a model based on the recombination rate by asking what gametes are generated by the genotype $A_1A_2B_1B_2$. Throughout this section, loci are indicated by the letters, alleles at the loci by the numerical subscripts, and allele frequencies by p_1 and p_2 for locus A and q_1 and q_2 for locus B. The problem is easier to conceptualize if we draw the two-locus genotype as being on two lines akin to chromosomal strands

$$\begin{array}{cc} A_1 & B_1 \\ \hline A_2 & B_2 \end{array}$$

This shows a genotype as two haplotypes and reveals **phase** or the sets of alleles packaged together on the same chromosomal strand (in contrast to writing the genotype as $A_1A_2B_1B_2$ where phase would be unknown). Given this physical arrangement of the two loci, what are the gametes produced during meiosis with and without recombination events?

A_1B_1 and A_2B_2	"Coupling" gametes: alleles on the same chromosome remain together (a term coined by Bateson and Punnett).
A_1B_2 and A_2B_1	"Repulsion" gametes: alleles on the same chromosome seem repulsed by each other and pair with alleles on the opposite strand (a term coined by Thomas Morgan Hunt).

The **recombination fraction**, symbolized as c (or sometimes r), refers to the total frequency of gametes resulting from recombination events between two loci. Using c to express an arbitrary recombination fraction, let's build an expectation for the frequency of coupling and repulsion gametes. If c is the rate of recombination, then $1 - c$ is the rate of non-recombination since the frequency of all gametes is 1, or 100%. Within each of these two categories of gametes (coupling and repulsion), two types of gametes are produced so the frequency of each gamete type is half that of the total frequency for the gamete category. We can also determine the expected frequencies of each gamete under random association of the alleles at the two loci based on Mendel's law of independent segregation.

Gamete	Frequency		
	Expected	**Observed**	
A_1B_1	p_1q_1	$g_{11} = (1-c)/2$	$1 - c$ is the frequency of *all* coupling gametes.
A_2B_2	p_2q_2	$g_{22} = (1-c)/2$	
A_1B_2	p_1q_2	$g_{12} = c/2$	c is the frequency of *all* recombinant gametes.
A_2B_1	p_2q_1	$g_{21} = c/2$	

The recombination fraction, c, can be thought of as the probability that a recombination event will occur between two loci. With independent assortment, the coupling and repulsion gametes are in equal frequencies and c equals ½ (like the chances of getting heads when flipping a coin). Values of c less than ½ indicate that recombination is less likely than non-recombination, so coupling gametes are more frequent. Values of c greater than ½ are possible and would indicate that recombinant gametes are more frequent than non-recombinant gametes (although such a pattern would likely be due to a process such as natural selection eliminating coupling gametes from the population rather than recombination exclusively).

We can utilize observed gamete frequencies to develop a measure of the degree to which alleles are associated within gamete haplotypes. This quantity is called the **gametic disequilibrium** (or sometimes linkage disequilibrium) **parameter** and can be expressed by:

$$D = \underset{\text{(coupling term)}}{g_{11}g_{22}} - \underset{\text{(repulsion term)}}{g_{12}g_{21}} \quad (2.27)$$

where g_{xy} stands for a gamete frequency. D is the difference between the product of the coupling gamete frequencies and the product of the repulsion gamete frequencies. This makes intuitive sense: with independent assortment, the frequencies of the coupling and repulsion gamete types are identical and cancel out to give $D = 0$, or **gametic equilibrium**. Another way to think of the gametic disequilibrium parameter is as a measure of the difference between observed and expected gamete frequencies: $g_{11} = p_1q_1 + D$, $g_{22} = p_2q_2 + D$, $g_{12} = p_1q_2 - D$, and $g_{21} = p_2q_1 - D$ (note that observed and expected gamete frequencies cannot be negative). In this sense, D measures the deviation of gamete frequencies from what is expected under independent assortment. Since D can be either positive or negative, both coupling and repulsion gametes can be in excess or deficit relative to the expectations of independent assortment.

Different estimators of gametic disequilibrium have different strengths and weaknesses (see Hedrick 1987; Flint-Garcia et al. 2003). The discussion here will focus on the classical parameter and estimator D to develop the conceptual basis of measuring gametic disequilibrium and to understand the genetic processes that cause it.

> **Gametic disequilibrium**: An excess or deficit or absence of all possible combinations of alleles at a pair of loci in a sample of gametes or haplotypes.
> **Linkage**: Co-inheritance of loci caused by physical location on the same chromosome.
> **Recombination fraction**: The proportion of "repulsion" or recombinant gametes produced by a double heterozygote genotype each generation.

Now that we have developed an estimator of gametic disequilibrium, it can be used to understand how allelic association at two loci changes over time or its dynamic behavior. If a very large population without natural selection or mutation starts out with some level of gametic disequilibrium, what happens to D over time with recombination? Imagine a population with a given level of gametic disequilibrium at the present time ($D_{t\,=\,n}$). How much gametic disequilibrium was there a single generation before the present at generation $n - 1$? Recombination will produce c recombinant gametes each generation so that:

$$D_{tn} = (1 - c)D_{tn-1} \qquad (2.28)$$

Since gametic disequilibrium decays by a factor of $1 - c$ each generation,

$$D_{tn} = (1 - c)D_{tn-1} = (1 - c)^2 D_{tn-2} = (1 - c)^3 D_{tn-3}... \qquad (2.29)$$

We can predict the amount of gametic disequilibrium over time by using the amount of disequilibrium initially present (D_{t0}) and multiplying it by $(1 - c)$ raised to the power of the number of generations that have elapsed:

$$D_{tn} = D_{t0}(1 - c)^n \qquad (2.30)$$

Figure 2.19 shows the decay of gametic disequilibrium over time using Eq. 2.30. Initially, there are only coupling gametes in the population and no repulsion gametes, giving a maximum amount of gametic disequilibrium. As c increases, the approach to gametic equilibrium ($D = 0$) is more rapid. Eq. 2.30 and Figure 2.20 both assume that there are no other processes acting to counter the mixing effect of recombination. Therefore, the steady-state will always be equal frequencies of all gametes ($D = 0$), with the recombination rate determining how rapidly gametic equilibrium is attained.

A hypothesis test that the observed level of gametic disequilibrium is significantly different than expected under random segregation can be carried out with:

$$\chi^2 = \frac{\hat{D}^2 N}{p_1 p_2 q_1 q_2} \qquad (2.31)$$

where N is the total sample size of gametes, $\hat{D}$ is a gametic disequilibrium estimate, and p and q are

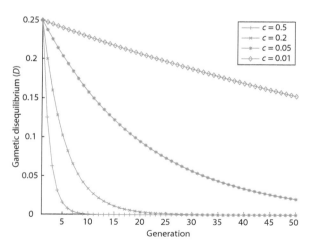

Figure 2.19 The decay of gametic disequilibrium (D) over time for four recombination rates. Initially, there are only coupling ($P_{11} = P_{22} = ½$) and no repulsion gametes ($P_{12} = P_{21} = 0$). Gametic disequilibrium decays as a function of time and the recombination rate ($D_{t\,=\,n} = D_{t\,=\,0}[1-c]^n$) assuming a single large population, random mating and no counteracting genetic processes. If all gametes were initially repulsion, gametic disequilibrium would initially equal -0.25 and decay to zero in an identical fashion.

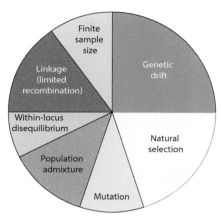

Figure 2.20 A hypothetical partitioning of the contributions to the total population gametic disequilibrium (D) in a population caused by numerous population genetic processes. The finite sample of gametes or genotypes used to measure D can itself contribute to the disequilibrium observed, as can departure from Hardy–Weinberg expected genotype frequencies at single loci or within-locus disequilibrium. The fractions of the total gametic disequilibrium attributable to each cause will vary depending on history of a population and the relative strengths of the multiple processes acting in a population.

the allele frequencies at two diallelic loci. The χ^2 value has 1 degree of freedom and can be compared with the critical value found in Table 2.5.

One potential drawback of D in Eq. 2.27 is that its maximum value depends on the allele frequencies in the population. This can make interpreting an estimate of D or comparing estimates of D from different populations problematic. For example, it is possible that two populations have very strong association among alleles within gametes (e.g. no repulsion gametes), but the two populations differ in allele frequency so that the maximum value of D in each population is also different. If all alleles are not at equal frequencies in a population, then the frequencies of the two coupling or the two repulsion gametes are also not equal. When $D < 0$, D_{max} is the value of $-p_1q_1$ or $-p_2q_2$ that is closer to zero, whereas when $D > 0$, D_{max} is the value of p_1q_2 or p_2q_1 that is closer to zero.

A way to avoid these problems is to express D as the percentage of its largest value:

$$D' = D/D_{max} \qquad (2.32)$$

This gives a measure of gametic disequilibrium that is normalized by the maximum or minimum value D

can assume given population allele frequencies. Even though a given value of D may seem small in the absolute, it may be large relative to D_{max} given the population allele frequencies. A related and more commonly employed measure expresses disequilibrium between two loci as a correlation:

$$\rho = \frac{D}{\sqrt{p_1p_2q_1q_2}} \qquad (2.33)$$

where ρ (pronounced "roe") takes the familiar and more easily interpreted range of -1 to $+1$ (the disequilibrium correlation is sometimes given as ρ^2 with $0 \le \rho^2 \le 1$) (Lewontin 1988). Analogous to the fixation index, the two locus disequilibrium correlation can be understood as a measure of the correlation between the states of the two alleles found together in a two locus haplotypes. When $\rho = 0$ there is no correlation between the alleles at two loci that are found paired in gametes or on the same chromosome – the allelic states are independent as expected under Mendel's second law. If $\rho > 0$ there is a positive correlation such that if one of the alleles at one locus is an A, for example, then the allele at the second locus will have a correlated state and might often be a B allele. When $\rho < 0$ there is a negative correlation between the states of two alleles in a haplotype, such as if A is infrequently paired with B.

Thus far, we have approached gametic disequilibrium by focusing on the frequency of four gamete haplotypes. A helpful complement is to consider the gametes made by all possible two locus genotypes as shown in Table 2.12. This table is somewhat like the table of parental matings and their offspring genotype frequencies we made to prove Hardy–Weinberg for one locus, except Table 2.12 predicts the frequencies of gametes that will make up the next generation rather than genotype frequencies in the next generation. Most genotypes produce recombinant gametes that are identical to non-recombinant gametes (e.g. the A_1B_1/A_1B_2 genotype produces A_1B_1 and A_1B_2 coupling gametes and A_1B_1 and A_1B_2 repulsion gametes). Only two genotypes – both types of double heterozygotes – will produce recombinant gametes that are different than parental haplotypes. These are the only two places where c enters into the expressions for expected gamete frequencies because recombination does not change the gametes produced by the other eight two locus genotypes.

Table 2.12 Expected frequencies of gametes for two diallelic loci in a randomly mating population with a recombination rate between the two loci of c. The first eight genotypes have non-recombinant and recombinant gametes that are identical. The last two genotypes produce novel recombinant gametes, requiring inclusion of the recombination rate to predict gamete frequencies. Summing down each column of the table gives the total frequency of each gamete in the next generation.

Parental mating	Expected frequency of mating	Frequency of gametes in next generation			
		A_1B_1	A_2B_2	A_1B_2	A_2B_1
A_1B_1/A_1B_1	$(p_1q_1)^2$	$(p_1q_1)^2$			
A_2B_2/A_2B_2	$(p_2q_2)^2$		$(p_2q_2)^2$		
A_1B_1/A_1B_2	$2(p_1q_1)(p_1q_2)$	$(p_1q_1)(p_1q_2)$		$(p_1q_1)(p_1q_2)$	
A_1B_1/A_2B_1	$2(p_1q_1)(p_2q_1)$	$(p_1q_1)(p_2q_1)$			$(p_1q_1)(p_2q_1)$
A_2B_2/A_1B_2	$2(p_2q_2)(p_1q_2)$		$(p_2q_2)(p_1q_2)$	$(p_2q_2)(p_1q_2)$	
A_2B_2/A_2B_1	$2(p_2q_2)(p_2q_1)$		$(p_2q_2)(p_2q_1)$		$(p_2q_2)(p_2q_1)$
A_1B_2/A_1B_2	$(p_1q_2)^2$			$(p_1q_2)^2$	
A_2B_1/A_2B_1	$(p_2q_1)^2$				$(p_2q_1)^2$
A_2B_2/A_1B_1	$2(p_2q_2)(p_1q_1)$	$(1-c)(p_2q_2)(p_1q_1)$	$(1-c)(p_2q_2)(p_1q_1)$	$c(p_2q_2)(p_1q_1)$	$c(p_2q_2)(p_1q_1)$
A_1B_2/A_2B_1	$2(p_1q_2)(p_2q_1)$	$c(p_1q_2)(p_2q_1)$	$c(p_1q_2)(p_2q_1)$	$(1-c)(p_1q_2)(p_2q_1)$	$(1-c)(p_1q_2)(p_2q_1)$

We can relate two locus Hardy–Weinberg expected genotype frequencies to the recombination rate and two locus disequilibrium if we sum the columns to determine the expected gamete frequencies with the possibility of recombination. Focus on the column for the gamete A_1B_1. Summing the five terms in that column, we get

$$g_{11}{}^2 + g_{11}g_{22} + g_{11}g_{21} + (1-c)g_{22}g_{11} + (c)g_{12}g_{21} \tag{2.34}$$

And expanding the two terms on the right gives

$$g_{11}{}^2 + g_{11}g_{22} + g_{11}g_{21} + g_{22}g_{11} - (c)g_{22}g_{11} + (c)g_{12}g_{21} \tag{2.35}$$

which can be rearranged by noticing the first four terms all contain g_{11} which can be factored out to give

$$g_{11}(g_{11} + g_{22} + g_{21} + g_{22}) - cg_{22}g_{11} + cg_{12}g_{21} \tag{2.36}$$

Recall that $D = g_{11}g_{22} - g_{12}g_{21}$ and make the substitution to obtain

$$g_{11}(g_{11} + g_{22} + g_{21} + g_{22}) - cD \tag{2.37}$$

Next, notice that $(g_{11} + g_{22} + g_{21} + g_{22})$ is the sum of all gamete frequencies and equals one. Making that substitution, we obtain

$$g_{11} - cD \tag{2.38}$$

This final result shows that gamete frequencies in the second generation are a function of the gamete frequency we expect from multiplying the respective allele frequencies, increased or decreased by the product of the recombination rate and the amount of two locus disequilibrium. The expected frequency of the $A_1A_1B_1B_1$ genotype, for example, in the next generation is then $(g_{11} - cD)^2$, and it is not just a function of the product of the allele frequencies but also depends on the recombination rate and the amount of two locus disequilibrium. This is analogous to adjusting single locus H-W expected genotype frequencies using F to account for one locus disequilibrium.

It is helpful to keep in mind that the term **linkage disequilibrium** is widely employed in the literature and has deep historic roots (e.g. Lewontin 1964), even though it is an imprecise label that confounds a pattern (two locus haplotypes or genotypes departing from the frequencies expected by the product of frequencies of alleles) and a process. Linkage disequilibrium is a misnomer since physical linkage only dictates the rate at which allelic combinations approach independent assortment or equilibrium. Processes other than linkage are responsible for the production of deviations from independent assortment of alleles at multiple loci in gametes. Using terms like gametic disequilibrium or two-locus disequilibrium reminds us that the deviation from random association of alleles at two loci is a pattern seen in gametes or haplotypes. Although linkage can certainly contribute to this pattern, so can many other population genetic processes. It is likely that several processes operating simultaneously produce the two-locus disequilibrium observed in any population, as illustrated by the pie chart in Figure 2.20.

Gametic disequilibrium is a central concept in formulating predictions for multiple locus genotype and haplotype frequencies in populations. Observations of the amount of gametic disequilibrium present in populations can then be used to identify the fundamental population genetic processes operating in populations. Thus, gametic disequilibrium forms the basis for a wide range of hypotheses to explain multiple locus genotype and haplotype frequencies, with gametic equilibrium or Mendel's second law serving as the null hypothesis. The numerous processes that maintain or increase gametic disequilibrium include those discussed in more detail the following sections.

Physical linkage

Linkage is the physical association of loci on a chromosome that causes alleles at the loci to be inherited in their original combinations. This association of alleles at loci on the same chromosome is broken down by crossing over and recombination. The probability that a recombination event occurs between two loci is a function of the distance along the chromosome between two loci. Loci that are very far apart (or on separate chromosomes) have recombination rates approaching 50% and are said to be unlinked. Loci located very near each other on the same chromosome might have recombination rates of 5 or 1% and would be described as tightly linked. Therefore, the degree of physical linkage of loci dictates the recombination rate and thereby the decay of gametic disequilibrium. Genome locations are often mapped in terms of their recombination frequencies with the measure **centimorgan** (abbreviated **cM**) or **map unit** (m.u.) where 1 cM is equivalent to a 1% recombination rate under a model that corrects for multiple crossovers called Haldane's map function (see Casares 2007).

Linkage-like effects can be seen in some chromosomes and genomes where gametic disequilibrium is expected to persist over longer time scales due to exceptional inheritance or recombination patterns. Organisms such as birds and primates have chromosomal sex determination, with the well-known X and Y sex chromosome system in humans. Loci located on X chromosomes experience recombination, whereas those on Y chromosomes experience no recombination. This is caused by the Y chromosome lacking a homologous chromosome to pair with at meiosis since YY genotypes do not exist. In addition, we would expect that the rate of decay of gametic disequilibrium for X chromosomes is about half that of autosomes with comparable recombination rates, since X recombination takes place only in females (XX) at meiosis, and not at all in males (XY). Organelle genomes found in mitochondria and chloroplasts are a case where gametic disequilibrium persists indefinitely since these genomes are uniparentally inherited and do not experience observable levels of recombination. There is variation in the rate of recombination among species, within and among populations and between sexes (Stapley et al. 2017), with "hotspots" that show elevated rates as well as areas of restricted recombination such that genomes may have marked heterogeneity in recombination rates. In humans, for example, patterns of haplotype polymorphism suggest that about 80% of all recombination events take place in a subset of only about 15% of the genome (Myers et al. 2006).

Natural selection

Natural selection is a process that can continuously counteract the randomizing effects of recombination. Imagine a case where two locus genotypes confer different rates of survival or different levels of reproduction. In such a case, natural selection will reduce the frequency of lower fitness genotypes, which will also reduce the number of gametes these genotypes contribute to forming the next generation. At the same time that natural selection is acting, recombination is also working to randomize the associations of alleles at the two loci. Figure 2.21 shows an example of epistatic natural selection acting to maintain gametic disequilibrium in opposition to recombination acting to establish gametic equilibrium. Natural selection at one locus can also impact the frequency

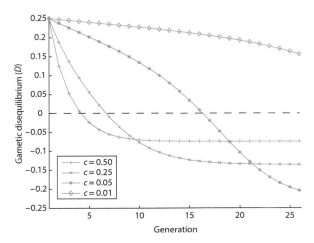

Figure 2.21 The decay of gametic disequilibrium (D) over time when both strong natural selection and recombination are acting. Initially, there are only coupling ($P_{11} = P_{22} = \frac{1}{2}$) and no repulsion gametes ($P_{12} = P_{21} = 0$). The relative fitness values of the AAbb and aaBB genotypes are one, while all other genotypes have a fitness of 0.5, a form of epistasis for relative fitness. Unlike in Figure 2.19, gametic disequilibrium does not decay to zero over time due to natural selection that is stronger than recombination.

of nearby loci that experience limited recombination. Both **hitchhiking,** where natural selection rapidly increasing the frequency of a beneficial genotype alters the frequency of linked loci, and **background selection,** where natural selection eliminates low fitness genotypes, can lead to linkage disequilibrium at loci adjacent to the loci experiencing selection. This is because natural selection has the potential to change the frequency of haplotypes more quickly than recombination can act to randomize the arrangement of alleles found together in the same haplotype. More on these natural selection topics can be found in later chapters.

The action of natural selection acting on differences in gamete fitness can produce steady-states other than $D = 0$ even with free recombination. In such cases, the population reaches a balance where the action of natural selection to increase the absolute value of D and the action of recombination to bring D back to zero cancel each other out. The point where the two processes are exactly equal in magnitude but opposite in their effects is where gametic disequilibrium will be maintained in a population. It is important to recognize that the amount of steady-state gametic disequilibrium depends on which genotypes have high fitness values, so natural selection and recombination could also act in concert to accelerate the decay of gametic disequilibrium more rapidly than just recombination alone.

Mutation

Alleles change from one form to another by the random process of mutation, which can either increase or decrease gametic disequilibrium. First consider the case of mutation producing a novel allele not found previously in the population. Since a new allele is present in the population as only a single copy, it is found only in association with the other alleles on the chromosome strand where it originated. Thus, a novel allele produced by mutation would initially increase gametic disequilibrium. Should the novel allele persist in the population and increase in frequency, then recombination will work to randomize the other alleles found with the novel allele and eventually dissipate the gametic disequilibrium. Mutation can also produce alleles identical to those currently present in a population. In that case, mutation can contribute to randomizing the combinations of alleles at different loci and thereby decrease levels of gametic disequilibrium. On the other hand, if the population is at gametic equilibrium mutation can create gametic disequilibrium by changing the frequencies of gamete haplotypes. However, it is important to recognize that mutation rates are often very low and the gamete frequency changes caused by mutation are inversely proportional to population size, so that mutation usually makes a modest contribution to overall levels of gametic disequilibrium. A simulation study showed that excluding any alleles at a frequency of less than 5–10% from estimates of D can eliminate most of the gametic disequilibrium attributable to recent mutations (Hudson 1985).

Mixing of diverged populations

The mixing of two genetically diverged populations, often termed **admixture**, can produce substantial levels of gametic disequilibrium. This is caused by different allele frequencies in the two source populations that result in different gamete frequencies at gametic equilibrium. Recombination acts to produce independent segregation but it does so only based on the allele frequencies within a group of mating individuals. Table 2.13 gives an example of gametic disequilibrium produced when two populations with diverged allele frequencies are mixed equally to form

a third population. In the example, the allele frequency divergence is large, and admixture produces a new population where gametic disequilibrium is 64% of its maximum value. In general, gametic disequilibrium due to the admixture of two diverged populations increases as allele frequencies become more diverged between the source populations, and the initial composition of the mixture population approaches equal proportions of the source populations.

Mating system

As covered earlier in this chapter, self-fertilization and mating between relatives increase homozygosity at the expense of heterozygosity. An increase in homozygosity causes a reduction in the effective rate of recombination because crossing over between two homozygous loci does not alter the gamete haplotypes produced by that genotype. The effective recombination fraction under self-fertilization is

$$c_{effective} = c\left(1 - \frac{s}{2-s}\right) \tag{2.39}$$

where s is the proportion of progeny produced by self-fertilization each generation. This is based on the expected fixation index at equilibrium $F_{eq} = \frac{s}{2-s}$ (Haldane 1924; see Figure 2.13). Figure 2.22 shows the decay in gametic disequilibrium predicted by Eq. 2.39 for four self-fertilization rates in the cases of free recombination ($c = 0.5$) and well as tight linkage ($c = 0.05$). Self-fertilization clearly increases the persistence of gametic disequilibrium, with marked effects at high selfing rates.

Table 2.13 Example of the effect of population admixture on gametic disequilibrium. In this case, the two populations are each at gametic equilibrium given their respective allele frequencies. When an equal number of gametes from each of these two genetically diverged populations are combined to form a new population, gametic disequilibrium results from the diverged gamete frequencies in the founding populations. The allele frequencies are: population 1 $p_1 = 0.1$, $p_2 = 0.9$, $q_1 = 0.1$, $q_2 = 0.9$; population 2 $p_1 = 0.9$, $p_2 = 0.1$, $q_1 = 0.9$, $q_2 = 0.1$. In population 1 and population 2, gamete frequencies are the product of their respective allele frequencies as expected under independent segregation. In the mixture population, all allele frequencies become the average of the two source populations (0.5) with $D_{max} = 0.25$.

Gamete	Gamete frequency	Population 1	Population 2	Mixture population
A_1B_1	g_{11}	0.01	0.81	0.41
A_2B_2	g_{22}	0.81	0.01	0.41
A_1B_2	g_{12}	0.09	0.09	0.09
A_2B_1	g_{21}	0.09	0.09	0.09
D		0.0	0.0	0.16
D'		0.0	0.0	0.16/0.25 = 0.64

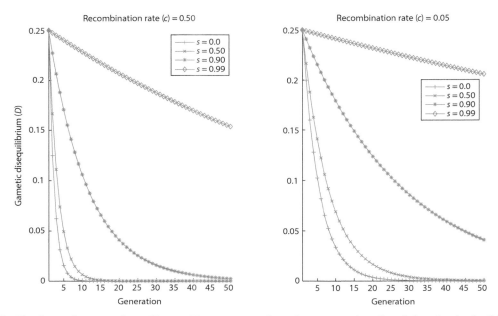

Figure 2.22 The decay of gametic disequilibrium (*D*) over time with random mating (*s* = 0) and three levels of self-fertilization. Initially, there are only coupling ($P_{11} = P_{22} = ½$) and no repulsion gametes ($P_{12} = P_{21} = 0$). Self-fertilization slows the decay of gametic disequilibrium appreciably even when there is free recombination because double heterozygote genotypes are infrequent.

In fact, the predominantly self-fertilizing plant *Arabidopsis thaliana* exhibits gametic disequilibrium over much longer regions of chromosome compared to outcrossing plants and animals (see review by Flint-Garcia et al. 2003, Kim et al. 2007).

Population size

It is possible to observe gametic disequilibrium just by chance in small populations or small samples of gametes. Recombination itself is a random process in terms of where crossing over events occur in the genome. As shown in the Appendix, estimates are more likely to approach their true values as larger samples are taken. This applies to mating patterns and the number of gametes that contribute to surviving progeny in biological populations. If only a few individuals mate (even at random) or only a few gametes found the next generation, then this is a small "sample" of possible gametes that could deviate from independent segregation just by chance. When the chance effects due to population size and recombination are in equilibrium, the effects of population size can be summarized approximately by

$$\rho^2 = \frac{1}{1 + 4N_e c} \qquad (2.40)$$

where N_e is the genetic effective population size and *c* is the recombination fraction per generation (Hill

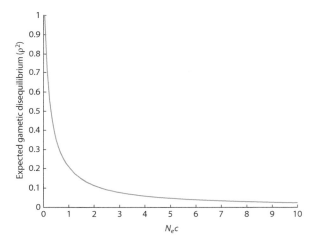

Figure 2.23 Expected levels of the squared gametic disequilibrium correlation (ρ^2) due to the combination of finite effective population size (N_e) and recombination at rate (*c*). Gametic disequilibrium is greater when fewer recombinant haplotypes are produced each generation (small *c*), the population is small causing haplotype frequencies to fluctuate by chance (small N_e), or if both factors are acting in combination (small $N_e c$).

and Robertson 1968; Ohta and Kimura 1969a, b; the basis of this type of equation is derived in Chapter 4). As shown in Figure 2.23, when the product of N_e and *c* is small, chance sampling contributes to maintaining some gametic disequilibrium since only a few gametes contribute to the next generation when N_e is small and genetic drift is strong, or only a

Interact box 2.4 Estimating genotypic disequilibrium

In practice, the recombination fraction for two loci can be measured by crossing a double heterozygote with a double homozygote and then counting the recombinant gametes. However, this basic experiment cannot be carried out unless individuals can be mated in controlled crosses, excluding many, if not most, species. For two diploid loci, disequilibrium can occur between loci at two alleles positioned on the same chromosome, as well as between loci at two alleles positioned on different chromosomes (Weir 1996). With observed genotype data from pairs of loci where the phase of alleles or gamete organization is unknown, these latter two types of between-locus disequilibrium cannot be distinguished but they can be considered together as genotypic disequilibrium (Rogers and Huff 2009).

An approximate means to test for gametic equilibrium is to examine the joint frequencies of genotypes at pairs of loci. If there is independent segregation at the two loci, then the genotypes observed at one locus should be independent of the genotypes at the other locus. Such contingency table tests are commonly employed to determine whether genotypes at one locus are independent of genotypes at another locus.

Contingency table tests involve tabulating counts of all genotypes for pairs of loci. In Table 2.14, genotypes observed at two microsatellite loci (AC25-6#10 and AT150-2#4) within a single population (the Choptank River) of the fish *Morone saxatilis* are given. The genotypes of 50 individuals are tabulated with alleles at each locus are represented with numbers. For example, there were 15 fishes that had a 22 homozygous genotype for locus AC25-6#10 and also had a 44 homozygous genotype for locus AT150-2#4. This joint frequency of homozygous genotypes is unlikely if genotypes at the two loci are independent, in which case the counts should be distributed randomly with respect to genotypes.

In the striped bass case shown here, null alleles (microsatellite alleles that are present in the genome but not reliably amplified by PCR) are probably the cause of fewer than expected heterozygotes that lead to a non-random joint distribution of genotypes (Brown et al. 2005). Thus, the perception of gametic disequilibrium can be due to technical limitations of genotyping techniques in addition to population genetic processes such as reduced recombination (or linkage), self-fertilization, consanguineous mating, and mixing of diverged populations that cause actual gametic disequilibrium.

Genepop on the Web can be used to construct genotype count tables for pairs of loci and carry out statistical tests that compare observed to those expected by chance. Instructions on how to use Genepop and an example of striped bass microsatellite genotype data set in the Genepop format are available on the text website along with a link to the Genepop site.

Table 2.14 Joint counts of genotype frequencies observed at two microsatellite loci in the fish *Morone saxatilis*. Alleles at each locus are indicated by numbers (e.g. 12 is a heterozygote and 22 is a homozygote).

	Genotype at locus AC25-6#10					
Genotype at locus AT150-2#4	**12**	**22**	**33**	**24**	**44**	**Row totals**
22	0	0	1	0	0	1
24	1	4	0	4	1	10
44	2	15	0	0	0	17
25	0	3	0	0	0	3
45	0	8	0	1	0	9
55	1	1	0	0	0	2
26	0	1	0	2	0	3
46	1	3	0	0	0	4
56	0	0	0	1	0	1
Column totals	5	35	1	8	1	50

few recombinant gametes exist when c is small. The lesson is that D as we have used it in this section assumes a large population size (similar to Hardy–Weinberg) so that actual gamete frequencies approach those expected based on allele frequencies, an assumption that is not met in actual populations to some degree because they are finite. Strong growth in population size over time can also alter the rate of decay of gametic disequilibrium compared to that seen in a population of constant size through time (Pritchard and Przeworski 2001; Rogers 2014).

Chapter 2 review

- Mendel's experiments with peas lead him to hypothesize particulate inheritance with independent segregation of alleles within loci and independent assortment of multiple loci.
- Expected genotype frequencies predicted by the Hardy–Weinberg equation (for any number of alleles) show that Mendelian inheritance should lead constant allele frequencies across generations. This prediction has a large set of assumptions about the absence of many population genetic processes. Hardy–Weinberg expected genotype frequencies therefore serve as a null model used as a standard of reference.
- The null model of Hardy–Weinberg expected genotype frequencies can be tested directly or assumed to be approximately true in order to test other hypotheses about Mendelian inheritance.
- The fixation index (F) measures departures from Hardy–Weinberg expected genotype frequencies (excess or deficit of heterozygotes) that can be caused by patterns of mating.
- Mating among relatives or consanguineous mating causes changes in genotype frequencies (specifically a decrease in heterozygosity) but usually no changes in allele frequencies.
- Mating among relatives is a process that increases the autozygosity or chance that alleles descended from a common ancestor are found together in a diploid genotype.
- The coancestry coefficient gives the probability that an allele sampled from each of two individuals is identical by descent, defining relatedness between individuals.
- The fixation index, the coancestry coefficient, and autozygosity are all interrelated measures of changes in genotype frequencies due to consanguineous mating.

- Mating among relatives alters mean phenotypes because homozygosity increases.
- An increase in homozygosity leads to inbreeding depression, which ultimately is caused by overdominance (heterozygote advantage) or dominance (deleterious recessive alleles).
- The gametic disequilibrium parameter (D) and its correlation version (ρ) measure the degree of association of alleles paired at two loci compared with random pairing. Gametic disequilibrium is broken down by recombination, decaying by the maximum of 50% per generation when loci experience free recombination.
- A wide variety of population genetic processes – natural selection, chance, admixture of populations, mating system, and mutation – can maintain and increase gametic disequilibrium even between loci without physical linkage to reduce recombination.

Further reading

For a detailed history of Gregor Mendel's research in the context of early theories of heredity as well as the analysis of Mendel's results by subsequent generations of scientists, see:

Orel, V. (1996). *Gregor Mendel: The First Geneticist*. Oxford: Oxford University Press.

For perspectives on whether or not Gregor Mendel may have fudged his data, see a set of articles published together:

Myers, J.R. (2004). An alternative possibility for seed coat color determination in Mendel's experiment. *Genetics* 166: 1137.
Novitiski, C.E. (2004). Revision of Fisher's analysis of Mendel's garden pea experiments. *Genetics* 166: 1139–1140.
Novitiski, E. (2004). On Fisher's criticism of Mendel's results with the garden pea. *Genetics* 166: 1133–1136.

More on the history of GH Hardy's contributions to expected genotype frequencies is explained in:

Edwards, A.W.F. (2008). G. H. Hardy (1908) and Hardy–Weinberg equilibrium. *Genetics* 179: 1143–1150.

For a brief biography of Reginald Punnett and his work on expected genotype frequencies, see:

Edwards, A.W.F. (2012). Reginald Crundall Punnett: first Arthur Balfour professor of genetics, Cambridge, 1912. *Genetics* 192: 3–13.

To learn more about the short tandem repeat (STR) genetic marker loci used in forensic investigation, consult:

Butler, J.M. (2006). Genetics and genomics of core short tandem repeat loci used in human identity testing. *Journal of Forensic Sciences* 51: 253–265.

To learn more about the Mendelian genetics of the ABO blood group, see the brief history:

Crow, J.F. (1993). Felix Bernstein and the first human marker locus. *Genetics* 133: 4–7.

Genomic sequence data has provided new insights on the classical concept of coancestry and relatedness through explicit tracing of DNA segments that are identical by descent. Consult this review to learn more:

Speed, D. and Balding, D.J. (2015). Relatedness in the post-genomic era: is it still useful? *Nature Reviews Genetics* 16 (1): 33–44.

For more detail on ways to estimate gametic disequilibrium and its applications, consult:

Slatkin, M. (2008). Linkage disequilibrium – understanding the evolutionary past and mapping the medical future. *Nature Reviews Genetics* 9 (6): 477–485.

End-of-chapter exercises

1   Estimate allele frequencies, compute genotype frequencies expected under the null hypothesis of Hardy–Weinberg genotype frequencies, and then use a chi-square to test the hypothesis of Hardy–Weinberg for these data from a sample of 459 Yugoslavians: MM, 144; MN, 201; NN, 114.

2   Allozyme genotype data has been collected from 1000 diploid individuals in a single population. The observed genotype frequencies for one polymorphic locus exhibited three alleles – F = fast, M = medium, and S = slow based on band migration rates in starch gels.

Genotypes	Observed #
FF	320
MM	120
SS	235
FM	80
FS	125
MS	120

Estimate the allele frequencies, the Hardy–Weinberg expected genotype frequencies, and the fixation index, F. Is there random mating in this population? What are the *most likely* possible causes of the observed fixation index? Interpret the estimated fixation index (F) from the perspectives of an excess or deficit of heterozygosity, the probability of autozygosity, and the correlation between the allelic states of uniting gametes.

3   Using Hardy–Weinberg expected frequencies for each of the six genotypes in problem 3, show that the fixation index serves to "adjust" the expected frequencies up or down in frequency to give the observed genotype frequencies. In other words, for each genotype, use the H-W expected genotype frequency and the fixation index to obtain the observed genotype frequency to within rounding error.

4   Draw a pedigree and use it explain why you are 25% related to an aunt or an uncle. Show the complete computations for identity by descent.

5   What is the coancestry coefficient for a parent and its progeny produced by self-fertilization?

6   What is the coancestry coefficient of half-siblings, such as individuals D and E in Figure 2.16? How does that compare to the coancestry of full siblings and why are the two coancestries different?

7   What is the coancestry coefficient between a grandparent and a grand-offspring, such as individuals A and F in Figure 2.16? What is the coancestry assuming the grandparent has $F = 0$ or $F = 0.5$. Explain the probabilities of allele sampling that explain the coancestry coefficient and why a non-zero value of F changes the coancestry coefficient.

8   Suppose you observed these haplotype frequencies: freq(A_1B_1) = 0.18, freq(A_1B_2) = 0.26,

freq(A_2B_1) = 0.20, and freq(A_2B_2) = 0.36. Estimate the allele frequencies in this population using the observed haplotype frequencies. Then, estimate haplotype frequencies expected under Mendel's second law. Calculate the gametic disequilibrium parameter D, D', and the gametic correlation ρ for this population. If the frequencies are based on a sample of 100 gametes, is the value of D statistically different than zero?

9 Using the information in problem 8, show that expected haplotype frequencies under Mendel's second law are equal to a function of the observed haplotype frequency and D.

10 In Table 2.12, sum up the terms for the gametes expected from each type of parental mating to obtain the total expected frequency of the A_1B_2 gamete in the progeny.

11 Assuming that haplotype phase is identical to that in Table 2.12, the loci experience recombination at the rate $c = 0.40$, the allele frequencies are $A_1 = B_1 = 0.7$, and the population currently exhibits $D = -0.09$ What is the expected frequency of the A_1B_1/A_1B_2 genotype after one generation of random mating?

12 Search the literature for a recent research paper that utilizes one or more of the population genetic predictions covered in this chapter. The topic can be any organism, application, or process, but the paper must include a hypothesis test involving a topic such as Hardy–Weinberg expected genotype frequencies, the fixation index (F), inbreeding depression, or gametic disequilibrium. Summarize the main hypothesis, goal, or rationale of the paper. Then explain how the paper utilized a population genetic prediction from this chapter and summarize the results and the conclusions based on the prediction.

13 Construct a simulation model of two locus genotype frequencies and gametic disequilibrium. Instructions to build a spreadsheet model can be found on the text web site. These instructions can also be implemented in a programming language such as Python or R.

Problem box answers

Problem box 2.1 answer

Using the allele frequencies in Table 2.3 we can calculate the expected genotype frequencies for each locus:

D3S1358: 2(0.2118)(0.1626) = 0.0689;
D21S11: 2(0.1811)(0.2321) = 0.0841;
D18S51: $(0.0918)^2$ = 0.0084;
vWA: $(0.2628)^2$ = 0.0691;
FGA: 2(0.1378)(0.0689) = 0.0190;
D8S1179: 2(0.3393)(0.2015) = 0.1367;
D5S818: 2(0.3538)(0.1462) = 0.1035;
D13S317: 2(0.0765)(0.3087) = 0.0472;
D7S820: 2(0.2020)(0.1404) = 0.0567.

As is evident from the allele designations, the amelogenin locus resides on the sex chromosomes and can be used to distinguish chromosomal males and females. It is a reasonable approximation to say that half of the population is male and assign a frequency of 0.5 to the amelogenin genotype. The expected frequency of the 10-locus genotype is therefore $0.0689 \times 0.0841 \times 0.0084 \times 0.0691 \times 0.0190 \times 0.1367 \times 0.1035 \times 0.0472 \times 0.0567 \times 0.5 = 1.210 \times 10^{-12}$. The odds ratio is one in 862 551 506 311. This 10-locus DNA profile is effectively a unique identifier since the human population in 2018 was approximately 7.7 billion, and we would expect to observe this exact 10-locus genotype only once in a population 112 times larger than the human population in 2018. In fact, it is probable that this 10-locus genotype has only occurred once in all of the humans who have ever lived.

Problem box 2.2 answer

Here, p is used to indicate allele frequency. For hypothesis 1, the observed frequency of the bb genotype is given by:

$$p(A\_bb) + p = (pa)^2(pb)^2 + \left(1 - (pa)^2\right)(pb)^2$$

Expanding the second term gives:

$$= (pa)^2(pb)^2 + (pb)^2 - (pa)^2(pb)^2 = (pb)^2$$

The other allele frequencies can be obtained by similar steps. This must be true under Mendel's second law if the two loci are truly independent.

For hypothesis 2, the frequency of the B allele can be obtained from the fact that all the allele frequencies must sum to 1:

$$pA + pB + pO = 1$$

Then subtracting pB from each side gives:

$$pA + pO = 1 - pB$$

Squaring both sides gives:

$$(pA + pO)^2 = (1 - pB)^2$$

The left side of which can be expanded to:

$$(pA + pO)^2 = (pA + pO)(pA + pO) = pO^2$$
$$+ 2pApO + pA^2$$

And then:

$$(1 - pB)^2 = pO^2 + 2pApO + pA^2$$

The last expression on the right is identical to the sum of the first and second expected genotype frequencies for hypothesis 2 in Table 2.3. Expressions for the frequency of the A and O alleles can also be obtained in this fashion.

Problem box 2.3 answer

The first step is to hypothesize genotypes under the two models of inheritance, as shown in Tables 2.6 and 2.7 for blood groups. Then, these genotypes can be used to estimate allele frequencies (here, p indicates probability or frequency). For the hypothesis of two loci with two alleles each:

$$pa^2 = \frac{769 + 261}{3816} = 0.270, pa = 0.52$$
$$pb^2 = \frac{728 + 261}{3816} = 0.260, pb = 0.51$$
$$pA = 1 - pa = 0.48, pB = 1 - pb = 0.49$$

For the hypothesis of one locus with three alleles:

$$(1 - pB)^2 = (pA + pC)2 = pA^2 + 2pApC + pC^2$$

$$= \frac{728 + 261}{3816} = 0.260$$

$$1 - pB = \sqrt{0.260} = 0.51$$

$$pB = 0.49(1 - pA)^2 = (pB + pC)^2 = pB^2 + 2pBpC$$

$$+ pC^2 = \frac{769 + 261}{3816} = 0.270$$

$$1 - pA = \sqrt{0.270} = 0.52$$

$$pA = 0.48$$

$$pC = 1 - pA - pB = 0.03$$

The expected numbers of each genotype as well as the differences between the observed and expected genotype frequencies are worked out in the tables. For the hypothesis of two loci with two alleles each, $\chi^2 = 0.266$, whereas $\chi^2 = 19\,688$ for the hypothesis of one locus with three alleles. Both of these tests have one degree of freedom (4 genotypes −2 for estimated allele frequencies −1 for the test), giving a critical value of $\chi^2_{0.05,1} = 3.84$ from Table 2.5. The deviations between observed and expected genotype frequencies could easily be due to chance under the hypothesis of two loci with two alleles each. However, the observed genotype frequencies are extremely unlikely under the hypothesis of three alleles at one locus since the deviations between observed and expected genotype frequencies are very large.

Phenotype	Genotype	Observed	Expected number of genotypes	Observed − expected	(Observed − expected)^2/expected
Hypothesis 1: two loci with two alleles each					
Purple/smooth	A_B_	2058	$3816\,(1-0.52^2)$ $(1-0.51^2) = 2060.0$	2.0	0.002
Purple/wrinkled	A_bb	728	$3816\,(1-0.52^2)$ $(0.51)^2 = 724.2$	3.8	0.020
Yellow/smooth	aaB_	769	3816 $(0.52)^2(1-0.51^2) = 763.5$	5.5	0.040
Yellow/wrinkled	Aabb	261	3816 $(0.52)^2(0.51)^2 = 268.4$	−7.4	0.204
Hypothesis 2: one locus with three alleles					
Purple/smooth	AB	2058	$3816\,(2(0.48)$ $(0.49)) = 1795$	63.0	38.5
Purple/wrinkled	AA, AC	728	$3816\,((0.48)^2 + 2(0.48)$ $(0.03)) = 989.1$	−261.1	68.9
Yellow/smooth	BB, BC	769	$3816\,((0.49)^2 + 2(0.49)$ $(0.03)) = 1028.4$	− 259.4	63.7
Yellow/wrinkled	CC	261	$3816\,(0.03)^2 = 3.4$	257.6	19 517.0

CHAPTER 3

Genetic drift and effective population size

3.1 The effects of sampling lead to genetic drift

- Biological populations are finite.
- A simple sampling experiment with micro-centrifuge tube populations.
- The Wright–Fisher model of sampling.
- Sampling error and genetic drift in biological populations.

In Chapter 2, that population size is very large, effectively infinite, was among the assumptions listed for Hardy–Weinberg expected genotype frequencies to be realized. This entire chapter will be devoted to the changes in allele and genotype frequency that occur when this assumption is not met and populations are small or at least finite. Population size has profound effects on allele frequencies in biological populations and has a specific definition in the context of population genetics. A variable for population size in one form or another appears in many of the fundamental equations used to predict genotype or allele frequencies in populations. In those expectations where no explicit variable for population size appears, there is an assumption instead, just as in the Hardy–Weinberg expectation for genotype frequencies. There is a strong biological motivation behind this attention to population size. All biological populations, without exception, are finite. Therefore, no actual population ever exactly meets the population size assumption of Hardy–Weinberg, although some may be large enough to show few genetic effects of finite size over relatively short periods of time. There is also a tremendous range of population sizes in the biological world. An understanding of the population genetic effects of population size will help to explain why some populations and species violate the assumptions to a greater degree than others, making sense of both the factors that cause differences in population size and the consequences of such differences. The causes and allele frequency consequences of finite population size can be understood and modeled in a variety of ways. Those models and concepts critical to understanding the population genetic impacts of finite population size will be the topics of this chapter.

A simple, hands-on demonstration can be used to show the role that population size plays in allele frequency in a population from one generation to the next. A plastic beaker filled with micro-centrifuge tubes can be used to represent gametes (Figure 3.1). The micro-centrifuge tubes are of two different colors, say blue and clear, and there are 50 of each per beaker. Each beaker approximates a population with one diallelic locus where the allele frequencies are $p = q = 0.5$. Imagine sampling four tubes from a beaker and recording the resulting frequencies of the blue and clear tubes. Then, imagine (after returning the four tubes and mixing the contents of the beaker) drawing out a sample of 20 tubes and recording the frequency of the blue and clear tubes. These handfuls of micro-centrifuge tubes represent the sampling process that occurs during reproduction and can be used to understand what happens to allele frequencies over time in a finite population.

Some results typical for sampling from these micro-centrifuge tube populations are given in Table 3.1. The results have several striking patterns. First, micro-centrifuge tube "allele" frequencies fluctuate quite a bit in the samples. In some cases, the results are 0.50/0.50 like in the ancestral population, but the results range from 0.35/0.65 in the sample of 20 to 0.0/1.0 in the sample of four. This latter result is called **fixation** or **loss** since one allele

Population Genetics, Second Edition. Matthew B. Hamilton.
© 2021 John Wiley & Sons, Inc. Published 2021 by John Wiley & Sons, Inc.
Companion website: www.wiley.com/go/hamilton/populationgenetics

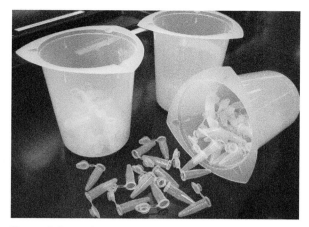

Figure 3.1 Beakers filled with microfuge tubes can be used to simulate the process of sampling and genetic drift.

composed the entire sample (its frequency went to 1) and the other allele was not sampled at all (its frequency went to 0). Second, the amount of fluctuation in the allele frequencies appears to be related to the size of the sample that was taken from the original population. The samples of four had greater fluctuations in allele frequency, including a case of fixation and loss. The samples of 20 deviated somewhat less from the original allele frequencies of 0.50/0.50, and in those 10 trials, no fixation/loss events were observed.

Compare these micro-centrifuge tube sampling results with what would be expected in an infinite population with $p = q = 0.5$. In the infinite population, there would not be a sample of four or 20 drawn to found the next generation, the entire population would

be used to found the next generation. Within the bounds of the micro-centrifuge tube population analogy, that would be like taking the entire beaker and just pouring it into another beaker to found the next generation: the allele frequencies would remain identical to the original frequency. This would also mean that if all other assumptions of Hardy–Weinberg were met, genotype frequencies would also remain constant (1/4 AA, 1/2 Aa, and 1/4 aa with $p = q = 0.5$).

Now, return to the finite micro-centrifuge tube populations with a sample of two individuals, or four gametes, drawn to found the population in the next generation. What are the chances that this next generation will consist of only AA genotypes? This is the same as asking what is the probability of sampling four blues or clear tubes in a handful of four. Since drawing one clear or blue tube has an independent probability of ½, the probability of getting four is $(\frac{1}{2})^4 = 1/16$. The same result can be seen from the perspective of genotypes by asking what are the chances of founding a population with two homozygous genotypes. If the source population is in Hardy–Weinberg equilibrium, then 1/4 of all genotypes are one of the two homozygotes. The chance of drawing two identical homozygous genotypes is the product of their independent probabilities, or $(1/4)^2 = 1/16$.

The micro-centrifuge tube populations are a low-tech demonstration that genotype and allele frequencies fluctuate from one generation to the next due to small samples, or **sampling error**, in a process called **genetic drift**. The amount of genetic drift increases as the size of the sample used to found the

Table 3.1 Typical results of sampling from beaker populations of micro-centrifuge tubes where the frequency of both blue (*p* below) and clear tubes is ½. After each draw, all tubes are replaced, and the beaker is mixed to randomize the tubes for the next draw.

Trial	N = 4			N = 20		
	Blue	clear	p	blue	clear	P
1	1	3	0.25	12	8	0.60
2	2	2	0.50	10	10	0.50
3	3	1	0.75	9	11	0.45
4	0	4	0.0	7	13	0.35
5	2	2	0.50	8	12	0.40
6	1	3	0.25	11	9	0.55
7	2	2	0.50	11	9	0.55
8	3	1	0.75	12	8	0.60
9	2	2	0.50	10	10	0.50
10	1	3	0.25	9	11	0.45

next generation decreases. Another way to restate the population size assumption of Hardy–Weinberg is to say instead that Hardy–Weinberg assumes that there is very little or no genetic drift occurring.

Sampling error: The difference between the value found in a finite sample from a population and the true value in the population.
Genetic drift: Random changes in allele frequency from one generation to the next in biological populations due to the finite samples of individuals, gametes, and ultimately alleles that contribute to the next generation. The amount of genetic drift increases as the size of the sample used to found the next generation decreases.
Stochastic process: A process where individual outcomes are dictated by chance but the average of a large number of outcomes can be described as a probability distribution based on initial conditions.
Wright–Fisher model: A simplified version of the biological life cycle where all sampling to found the next generation occurs from an infinite pool of gametes built from equal contributions of all individuals. This approximation is commonly employed to model genetic drift.

To extend and generalize the model of genetic drift started with the micro-centrifuge tube populations, a model of the biological process of reproduction is helpful. To do this, let us consider the process of reproduction in populations. During reproduction, individual adult organisms produce gametes. These gametes are exchanged with mates and fuse to form zygotes, and these zygotes develop into a new generation of adult organisms (Figure 3.2). This schematic of the biological life cycle is called the Wright–Fisher model of sampling (introduced by Sewall Wright (1931) and Ronald A. Fisher (1999; originally published in 1930)). It is not completely biologically realistic. There is obviously not an infinite number of gametes in any real population, and sampling events can take place at many points during the life history of a population of organisms. But it allows the process of genetic drift to be reduced to a point that it can be modeled in a simple fashion. The Wright–Fisher model makes assumptions identical to those of Hardy–Weinberg (see Section 2.2), with the exception that the population is finite rather than approaching infinite. Particularly critical assumptions include:

- generations are discrete and do not overlap, equivalent to adults that reproduce synchronously but only once during their lifetime;
- the numbers of females and males are equal;
- the size of the population (*N* individuals) remains constant through time; and

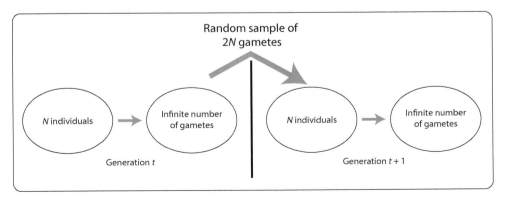

Figure 3.2 The Wright–Fisher model of genetic drift uses a simplified view of biological reproduction where all sampling occurs at one point in the life cycle of each generation – sampling 2*N* gametes from an infinite gamete pool. In this case, *N* diploid individuals (*N*/2 of each sex) generate an infinite pool of gametes where allele frequencies are perfectly represented, a finite sample of 2*N* alleles is drawn from this gamete pool to form *N* new diploid individuals in the next generation. Genetic drift takes place only in the random sample of 2*N* gametes to form the next generation. Major assumptions include non-overlapping generations, equal fitness of all individuals and constant population size through time. The model can easily be adjusted for haploid individuals or loci by assuming 2*N* individuals or sampling *N* gametes to form the next generation.

•   all individuals are equal in their production of gametes and all gametes are equally viable, equivalent to no natural selection.

These assumptions reduce the complexity of sampling error in biological populations, concentrating all samplings into a single step as an approximation. This simplification approximates the genetic drift that occurs in biological populations. For example, sampling at several points in the life cycle can be equivalent in its effect on allele frequencies as the same total amount of sampling at a single point in the life cycle. As will be shown later in this chapter, sampling events may occur at many stages of the life cycle in actual biological populations. The implications of relaxing other assumptions such as constant population size are topics of later chapters.

A major limitation of the micro-centrifuge tube demonstration is that it shows the effects of genetic drift over only one generation. A more general model of the effect of sampling error is needed to predict what may happen to allele frequencies over many generations. A general model would sample from one generation to found the next generation, then build a large pool of gametes (like the beakers of micro-centrifuge tubes) with those new allele frequencies. The sampling process would then be continued for many generations. Figure 3.3 shows computer-simulated allele frequencies based on this more general model for many generations under the assumptions of the Wright–Fisher model (note that the populations in Figure 3.3 are twice as large as the samples of micro-centrifuge tubes). The effects of genetic drift are more obvious over longer periods of time. Allele frequencies over a few generations change at random, both increasing and decreasing, sometimes changing very little in one generation and other times changing more substantially. There is a clear trend that over time in these genetic drift simulations that the frequency of one allele reaches either fixation ($p = 1.0$) or loss ($p = 0.0$), identical to loss ($q = 0.0$) and fixation ($q = 1.0$) for the alternate allele. There is also a trend that fixation or loss occurs in fewer generations with the smaller between generation sample size and more slowly with the larger between generation sample size. Since these

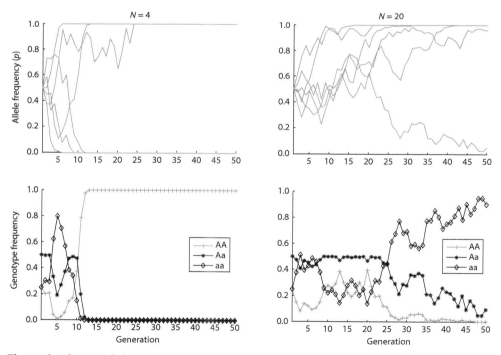

Figure 3.3   The results of genetic drift continued every generation in populations of $2N = 4$ and $2N = 20$. In the top panels, the six lines represent independent replicates or independent populations experiencing genetic drift starting at the same initial allele frequency ($p = 0.5$). The random nature of genetic drift can be seen by the zig-zag changes in allele frequency that have no apparent direction. Allele frequencies that reach the upper or lower axes represent cases of fixation or loss. In the bottom panels, the genotype frequencies are shown for the allele frequencies represented by the black line under the assumption of random mating within each generation. The changes in genotype frequencies are a consequence of changes in allele frequencies due to genetic drift.

simulations do not include any processes that could reintroduce genetic variation, once a population has reached fixation or loss, there can be no further change in allele frequency.

The bottom panels of Figure 3.3 show genotype frequencies based on random mating for one of the populations represented in the top panels. Genetic drift clearly causes genotype frequencies to change over time along with the changes in allele frequency. This is in contrast to the processes considered in Chapter 2, such as consanguineous mating, that result in changes in genotype frequency only but do not alter allele frequency. Genetic drift is most commonly modeled and demonstrated from the perspective of allele frequencies since it is easier to summarize a diallelic locus in a population as two allele frequencies rather than three genotype frequencies. But, remember that genotype frequencies are affected by genetic drift too, and genotype frequencies can always be obtained by multiplication given allele frequencies under the assumption of random mating.

Up to this point, the beakers of micro-centrifuge tubes and computer simulations only considered cases of alleles at equal initial frequencies ($p = q = 0.5$). Figure 3.4 shows results of computer simulations for genetic drift where initial allele frequencies are $p = 0.2$ and $p = 0.8$ with identical population sizes of $N = 25$. Identical simulations under the assumptions of the Wright–Fisher model were used to produce both Figures 3.3 and 3.4, so the results in the two cases can be compared directly. Initial allele frequencies do influence the outcome of genetic drift in the simulations shown in Figure 3.4. The lower initial allele frequency is associated with more frequent loss of the allele (five of the six replicates), while the higher initial allele frequency is fixed in five of the six replicates. These results are consistent with what would occur in a much larger number of replicate simulations with the same initial allele frequencies and population size. A larger sample would show that the probability that an allele reaches fixation under genetic drift is the same as the initial allele frequency. This makes intuitive sense, since with a lower initial frequency a population is closer in frequency to loss than to fixation. If the direction and magnitude of genetic drift in allele frequencies is random, there is a better chance of reaching 0 on average than reaching 1. This same pattern would be true for any initial values of the allele frequency closer to 0 or to 1, except in the special case of equal allele frequencies where the chances of fixation or loss for an allele would be equal.

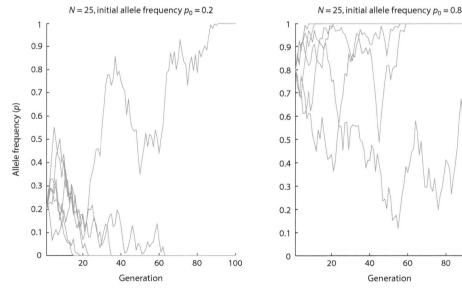

Figure 3.4 The results of genetic drift with different initial allele frequencies. The two panels have identical population sizes ($N = 25$) but initial allele frequency is $p = 0.2$ on the right and $p = 0.8$ on the left. The chances of fixation are equal to the initial allele frequency, a generalization that can be seen by examining a large number of replicates of genetic drift. Consistent with this expectation, more replicates go to loss on the left and more reach fixation on the right. Even with the difference in initial allele frequencies, the random trajectory of allele frequencies is apparent.

Interact box 3.1 Genetic drift

Genetic drift can be simulated readily using a computer program. Use the text simulation website (or a similar tool) to simulate genetic drift.

- Set the parameters for 10 loci, an initial allele frequency of 0.5 and view the results for 500 generations so that all fixation/loss events are visible. Examine drift for population sizes of 4, 20, 50, and 100. Record the generation of fixation or loss for about 20 replicates of each population size. What is the average number of generations to fixation or loss? Are there equal numbers of fixation and loss events?
- Progressively reduce (or increase) the initial allele frequencies by intervals of 0.1 for a single population size. Record the generation of fixation or loss for about 50 replicates of initial allele frequency. What is the observed relationship between initial allele frequency and probability of fixation or loss? Do these averages change if the population size changes?

Using a spreadsheet program like Microsoft Excel can speed the calculation of averages for a list of values. Enter the values in columns and then use the average function ("=AVERAGE( )" with the range specified in the parentheses, such as "C1:C10," to indicate the values of cells 1 through 10 in column C). A wide range of other useful functions are provided under **Functions ...** in the **Insert** menu.

Under the Wright–Fisher model, the following are general conclusions about the action of genetic drift in finite populations:

- the direction of changes in allele frequency is random;
- the magnitude of random fluctuations in allele frequencies from generation to generation increases as the population size decreases;
- fixation or loss is the equilibrium state if there are no other processes acting to counteract genetic drift or reintroduce genetic variation;
- genetic drift changes allele frequencies and thereby genotype frequencies; and
- the probability of eventual fixation of an allele is equal to its initial frequency (or the probability of ultimate loss of an allele is equal to 1 minus its initial frequency).

The next section of the chapter will develop two probability models of sampling error to provide more rigorous evidence for the conclusions reached in this section and to reach additional generalizations about the action of genetic drift.

3.2 Models of genetic drift

- The binomial probability distribution.
- Markov chains.
- The diffusion approximation of genetic drift.

The last section demonstrated the phenomenon of genetic drift caused by sampling error and drew some general conclusions based on the results of computer simulations. Building on this foundation, this section will introduce three probability models that can be used to confirm these and other general properties of the process of genetic drift. The first model, the binomial distribution, will be used to show that the magnitude of genetic drift from one generation to the next depends on allele frequencies in the population. The second model, the Markov chain, will be used to show the rate of change of allele frequencies under genetic drift. The third model, a continuous time approximation to the Markov chain, will be introduced to show how genetic drift can be modeled as the diffusion of particles.

The binomial probability distribution

To develop the first model, let us return to the microcentrifuge tube populations from the last section. When sampling a tube from the beaker, there are only two outcomes, a blue tube or a clear tube, which are used to represent the two alleles at one locus. The tubes are a specific case of a **Bernoulli random variable** (sometimes called a binomial random variable), or a variable representing a trial or sample that can have only two outcomes. Coin flips with either heads or tails outcomes are another example of a Bernoulli random variable. What we often want to know is, what are the chances of obtaining a given set of Bernoulli outcomes? For example, what are the chances of obtaining four heads when flipping a coin four times? In our micro-centrifuge tube samples, what are the chances of one of the possible outcomes (20 blue, 19, blue, 18 blue, ..., 0 blue) when sampling 20 tubes from the beaker? Answers to these types of questions require a means to estimate a probability distribution.

The binomial (literally, "two names") formula defines the probability distribution for the sum of N independent samples of a Bernoulli variable:

$$P_{(i = A)} = \binom{2N}{i} p^i q^{2N-i} \text{ where } \binom{2N}{i} = \frac{(2N)!}{i!(2N-i)!}$$

$$(3.1)$$

The binomial formula gives the probability of sampling i A alleles in a sample of $2N$ from a population where the A allele has a frequency of p and the alternate a allele is at a frequency of q. The p^i and q^{2N-i} terms estimate the probability of observing i and $2N - i$ independent events, each with probability p and q, respectively. The term $\binom{2N}{i}$ (pronounced "two N draw i") serves as a way to enumerate the different ways (or permutations) of obtaining i A's in a sample of $2N$.

Applying the binomial to the micro-centrifuge tube sampling results from the last section will illustrate how the binomial provides the probability for a specific sampling outcome. In the beaker, the blue and clear tubes were both at a frequency of $p = q = \frac{1}{2}$. In the draws of $N = 4$, a result of two blue and two clear tubes occurred in four out of 10 draws or 40% of the time (Table 3.1). When drawing samples of tubes, there are 2^N possible combinations. So, for samples of $N = 4$, there are $2^4 = 16$ combinations (like the number of genotypes in a Punnet square for four alleles at a locus). Of these 16 combinations, there are exactly six (bbcc, bcbc, bccb, cbcb, cbbc, ccbb) that yield two blue and two clear tubes. This same result can be obtained by using

$$\binom{4}{2} = \frac{4!}{2!2!} = 6 \qquad (3.2)$$

to enumerate the number of possible permutations of outcomes of one type in a sample of $2N$ objects.

The ! notation stands for factorial, and $n!$ equals $1 \times 2 \times 3 \times ... \times n - 1 \times n$ and $0! = 1$. The other part of the binomial formula calculates the probability of obtaining a sample of i blue tubes and $2N - i$ clear tubes. The blue tubes are at a frequency of p in the population, so the probability of sampling i of them is p^i since each is an independent event. The same logic applies to the clear tubes, whose frequency in the population is q and the number sampled is the remaining sample size not made up of blue tubes or $2N - i$, to give a probability of q^{2N-i}. Bringing both of these components of the binomial formula together,

$$P_{b=2} = \frac{4!}{2!(4-2)!} \left(\frac{1}{2}\right)^2 \left(\frac{1}{2}\right)^2 = \frac{24}{4} \left(\frac{1}{4}\right) \left(\frac{1}{4}\right)$$

$$= \frac{6}{16} \text{ or } 0.375$$

$$(3.3)$$

gives the expected frequency of draws of two blue and two clear tubes when sampling four tubes. This expected value is very close to what was observed in 10 draws of four tubes in Table 3.1.

The binomial formula can be used to calculate the expected probability of observing each of the possible outcomes when drawing samples of $2N = 4$ and $2N = 20$ micro-centrifuge tubes from beaker populations. These probability distributions (Figure 3.5) summarize what we would expect to find if we drew many independent samples and then tabulated the results. The probability for each bar in the histograms of Figure 3.5 was determined using the binomial formula. For example, the expected frequency of sampling 12 blue tubes in a total sample of 20 tubes is

$$P_{b=12} = \frac{20!}{(12!)(8!)} (0.5)^{12} (0.5)^8 = 0.1201 \quad (3.4)$$

These probability distributions explain why a fixation/loss event was observed when $2N = 4$ but not for $2N = 20$, since the former outcome is expected in one out of 16 draws but the latter only once in 1 048 576 draws. With knowledge of the binomial probability distribution, the Wright–Fisher model of genetic drift (Figure 3.2) makes a lot of sense. It was constructed, in fact, to articulate the assumptions that underlie the use of the binomial formula and binomial probability distributions to model genetic drift.

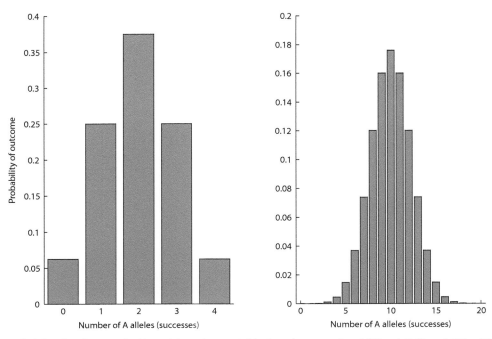

Figure 3.5 Probability distributions for binomial random variables based on samples of $2N = 4$ (left) and $2N = 20$ (right) from populations where $p = q = 0.5$. These distributions describe the expected probability of each of the possible outcomes of the microcentrifuge sampling experiment described in the text.

Problem box 3.1
Applying the binomial formula

Two independent laboratory populations of the fruit fly *Drosophila melanogaster* were observed for two generations. The populations each had a size of $N = 24$ individuals with an equal number of males and females. In the first generation, both populations were founded with $fA = p = 0.5$. In the second generation, one population showed $fA = p = 0.458$ and the other $fA = p = 0.521$. What are the chances of observing these allele frequencies after one generation of genetic drift?

Bernoulli or binomial random variable: A variable representing a trial or sample that can have only two possible outcomes, such as 0 or 1.

Applying the binomial formula to determine expected probabilities associated with particular sampling outcomes is useful, and there is an even

broader lesson that can be learned from the binomial. The examples up to this point have focused on the expected value. Shifting our perspective, we can use the binomial to explore how variable allele frequency changes under genetic drift should be. The variance of a Bernoulli or binomial random variable is:

$$\sigma^2 = pq \qquad (3.5)$$

This result is derived in Math Box 3.1 for those readers who would like to work through the details. The maximum variability will occur when $p = q = \frac{1}{2}$. The standard deviation ($\sigma = \sqrt{pq}$) and standard error of the allele frequency,

$$SE = \sqrt{\frac{pq}{2N}} \qquad (3.6)$$

are also easy to obtain (see Appendix). The standard error is the standard deviation of a mean, and the mean in this case is the expected value or allele frequency p or q. For genetic drift under the Wright–Fisher model, equally frequent alleles will give the widest range of outcomes for a given sample size (Figure 3.6). The variability in allele frequency caused by genetic drift decreases as a population approaches fixation or loss, causing pq to approach 0 (Figure 3.7). This result makes intuitive sense.

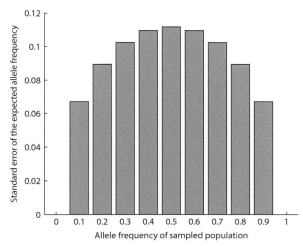

Figure 3.6   The standard error of the allele frequency ($\sigma = \sqrt{\frac{pq}{2N}}$) for a binomial random variable for a sample size of $N = 10$ for a range of allele frequencies. The standard error of the allele frequency decreases as the allele frequency approaches fixation or loss. In the same way, genetic drift is less effective at spreading out the distribution of allele frequencies as alleles approach fixation or loss. The standard deviation is zero when the allele frequencies are zero or one since there is no genetic variation and any size sample will faithfully reproduce the allele frequencies in the source population.

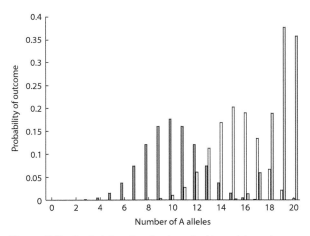

Figure 3.7   Probability distributions for binomial random variables based on samples of $2N = 20$ from populations where the allele frequency is 0.50, 0.75, or 0.95. The range of probable outcomes with sampling depends on the allele frequency. As allele frequencies approach the boundaries of fixation or loss, there is a decreasing number of outcomes other than fixation or loss that are probable due to sampling error.

When alleles are equally frequent, sampling error is equally likely to increase or decrease allele frequency and could produce an outcome anywhere along the spectrum of possible allele frequencies. At the other extreme, when one allele is very nearly fixed except for one copy of the alternate allele ($p = 1 - \frac{1}{2N}$ and $q = \frac{1}{2N}$), drift has a reasonable chance of only several sampling errors, such as no, one, two, or three copies of the low-frequency allele. Sampling error that causes fixation of high frequency allele is quite likely. However, sampling error that results in greatly increased frequencies of the low-frequency allele in one generation would be very, very unlikely.

There is a graphical metaphor to summarize the consequences of initial allele frequency for the range of outcomes in allele frequency under genetic drift. Figure 3.8 shows the range of possible allele frequencies (0–1) in a population and indicates the effects of genetic drift by the width of the arrows and the vertical scale. The range of outcomes probable under drift depends on the allele frequency in a population. When both alleles are equally frequent ($pq = 0.25$, its maximum), the sampling error is the largest so that the rate of genetic drift in changing allele frequencies is greatest. When the allele frequency is closer to fixation or loss ($pq < 0.25$), the sampling error is smaller and the rate of genetic drift in changing allele frequencies is also smaller. This explains the tendency of populations to go to fixation or loss under genetic drift. The sampling effect is greatest when the genetic variation is greatest, but also weakest when genetic variation is least (Figure 3.8). A population is most likely to experience larger changes in allele frequencies, toward fixation or loss, due to drift when both alleles are near equal frequencies. However, a population with strongly unequal allele frequencies is less likely to experience genetic drift of a magnitude that would equalize allele frequencies.

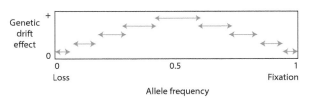

Figure 3.8   A schematic illustration of how the effects of genetic drift due to sampling error depend on allele frequencies in a population. The horizontal axis represents allele frequency and the width of the arrows represents the standard error of allele frequency ($\sigma = \sqrt{\frac{pq}{2N}}$) at a given allele frequency. Larger standard errors for allele frequency are another way of saying that sampling error will cause a greater range of outcomes, equivalent to a larger effect of genetic drift.

Math box 3.1 Variance of a binomial variable

The variance in the outcome of binomial sampling over many trials is surprisingly easy to derive. Assuming a diallelic locus, let p be the fraction of successes (such as sampling the A allele) and q be the fraction of failures (such as failing to sample the A allele and getting an a allele instead) so that $p + q = 1$. In the Appendix, a variance was defined as the average of the squared differences between each estimate and the average. If 1 is used to represent a success and 0 a failure and p and q are used to represent the frequency of each outcome, the variance in successes is:

$$\sigma^2 = p(1 - \overline{x})^2 + q(0 - \overline{x})^2 \qquad (3.7)$$

The average of a binomial variable ($\overline{x}$) is simply the probability of a success or p, just as when flipping a fair coin a large number of times the number of successes would approach the expected value of one-half heads or tails. Substituting p for $\overline{x}$ gives:

$$\sigma^2 = p(1 - p)^2 + q(0 - p)^2 \qquad (3.8)$$

which can then be simplified by substituting $1 - p = q$ into the left-hand term:

$$\sigma^2 = p(q)^2 + q(0 - p)^2 \qquad (3.9)$$

and multiplying out the right-hand term to give

$$\sigma^2 = pq^2 + qp^2 \qquad (3.10)$$

This result can then be rearranged by finding a factor common to both terms

$$\sigma^2 = pq(q + p) \qquad (3.11)$$

which simplifies after noticing that $q + p = 1$

$$\sigma^2 = pq \qquad (3.12)$$

Markov chains

The next step in understanding genetic drift is to consider its effects in a large number of replicate populations. Instead of focusing on allele frequency in just a single population like in the last section, let us now explore the case where there is a collection of numerous independent but identical populations (an infinite number of replicate populations is sometimes called an *ensemble* in physics and mathematics). Using the approach of genetic drift in multiple finite populations, this section will cultivate an understanding of how drift works on average among many populations and will develop a prediction of how rapidly genetic drift causes populations to reach fixation and loss.

To get started, consider populations composed of a diallelic locus in a single diploid individual. Since there are only two alleles in a population, there are three possibilities for the numbers of one of the alleles: zero, one, and two copies. Each of these possible states in a population could be referred to by the number of A alleles, 0 through 2, which can be summarized in notation as $P(0)$, $P(1)$, and $P(2)$. With this very basic type of population, we can ask: what are the chances that a population starting out in one of these three states ends up in one of these three states due to sampling error? For example, what is the chance of starting out with two copies of A and ending up with one copy of A with a sample size of one individual (two gametes) between generations? This chance is known as the **transition probability** for allelic states. The transition probability is determined with the binomial formula:

$$P_{i \to j} = \binom{2N}{j} p^j q^{2N - j} \qquad (3.13)$$

where i is the initial number of alleles, j is the number of alleles after sampling, and N is the sample size of diploid individuals. As before, $\binom{2N}{j} = \dfrac{2N!}{j!(2N - j)!}$ enumerates the possible draws that yield j copies of the allele and $p^j q^{2N - j}$ is the probability of sampling j copies of the allele given the allele frequencies in the initial population.

Equation 3.13 can be used to determine the expected frequencies of populations with a given allelic state in one generation based on the frequencies of populations in each allelic state in the previous generation. To predict the frequency of populations with one allelic state, we need to add up the chances that populations in *all* states in the previous generation transition to this state. Let us work through what is essentially bookkeeping to see this. The expected frequency of populations with two A alleles in generation one is the sum of the probabilities that populations a generation before with zero, one, and two A alleles become populations with two A alleles through sampling error. This can be stated in an equation as:

$$P_{t=1}(2) = (P_{2\rightarrow2})P_{t=0}(2) + (P_{1\rightarrow2})P_{t=0}(1) + (P_{0\rightarrow2})P_{t=0}(0) \quad (3.14)$$

for the case of populations of a single diploid individual. In Eq. 3.14, the probability or frequency of a given allelic state is indicated by $P(x)$ with subscripts to indicate the generation. In a population with one A and one a allele, the chances of sampling two A alleles, $P_{1\rightarrow2}$, is

$$P_{1\rightarrow2} = \frac{2!}{2!0!}\left(\frac{1}{2}\right)^2\left(\frac{1}{2}\right)^0 = \frac{2}{2}\left(\frac{1}{4}\right)(1) = \frac{1}{4} \quad (3.15)$$

using Eq. 3.13. For populations that are at fixation or loss, sampling cannot change the allele frequency. Therefore, populations initially fixed for A ($P_{t=0}(2)$) all transition to populations fixed for A ($P_{2\rightarrow2} = 1$) but none of the populations initially lost for A ($P_{t=0}(0)$) can transition to any other state, so $P_{0\rightarrow2}$ is zero. Lastly, the $P_{t=0}(2)$, $P_{t=0}(1)$, and $P_{t=0}(0)$ terms each represent the frequencies of populations that possess a given allelic state. This means that the transition probabilities are multiplied by the frequency of populations in a given allelic

state to determine the expected frequency of populations with a given allelic state in the next generation.

Using this same logic, the expected frequencies of populations with zero, one, and two A alleles after one generation of sampling error are shown in Table 3.2. Using these transition probabilities, the expected population frequencies over four generations of sampling are shown in Figure 3.9. The result of the equations in Table 3.2 is a generation-by-generation prediction for the *average* behavior of populations under genetic drift when there is an infinite number of replicate populations. This method of modeling the action of genetic drift is known as a **Markov chain** model. It is important to recognize that the outcome of genetic drift for a single population cannot be predicted. Rather, we can only know the probability that a single population experiences a given change in allele frequency such as going from 1 to 0 copies of the A allele. If many replicate populations are experiencing genetic drift, then the Markov chain predicts the proportion of populations that have a given allelic state in each generation.

Figure 3.10 shows the first two steps in the Markov chain for a population of two diploid individuals or four gametes, similar to the micro-centrifuge tube sampling experiment from the first section of this chapter. With a slightly larger population size than in Figure 3.9, there are a larger number of allelic state transitions to account for between generations. However, the transition probabilities from each possible allelic state to each possible allelic state are still determined with the binomial formula in Eq. 3.13. To obtain the proportion of populations that transition from one state to any state a generation later, the binomial transition probability is multiplied by the proportion of populations in a given allelic state.

Table 3.2 The equations used to calculate the expected frequency of populations with zero, one or two A alleles in generation one ($t = 1$) based on the previous generation ($t = 0$). Frequencies at $t = 1$ depend on both transition probabilities due to sampling error (constant terms like 0, 1, or ½) and population frequencies in the previous generation ($P_{t=0}(x)$ terms). The transition probabilities are calculated with the binomial formula $(P_{i\rightarrow j} = \binom{2N}{j}p^j q^{2N-j})$. Since sampling error cannot change the allele frequency of a population at fixation or loss, $P_{2\rightarrow2}=1$ and $P_{0\rightarrow0}=1$ while the other possibilities have a probability of zero.

One generation later ($t = 1$)			Initial state: number of A alleles ($t = 0$)				
A alleles	Expected frequency		2		1		0
2	$P_{t=1}(2)$	=	$(P_{2\rightarrow2})P_{t=0}(2)$	+	$(P_{1\rightarrow2})P_{t=0}(1)$	+	$(0)P_{t=0}(0)$
1	$P_{t=1}(1)$	=	$(0)P_{t=0}(2)$	+	$(P_{1\rightarrow1})P_{t=0}(1)$	+	$(0)P_{t=0}(0)$
0	$P_{t=1}(0)$	=	$(0)P_{t=0}(2)$	+	$(P_{1\rightarrow0})P_{t=0}(1)$	+	$(P_{0\rightarrow0})P_{t=0}(0)$

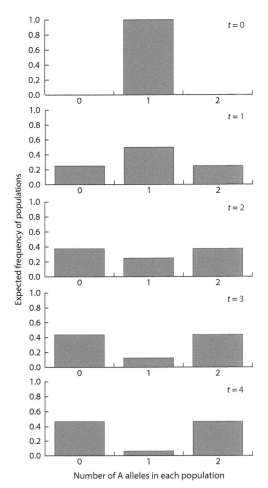

Figure 3.9 The expected frequencies of populations with zero, one or two A alleles over five generations genetic drift. Initially, all populations have one A and one a allele ($p = q = 0.5$). Each generation two gametes are sampled from each population under the Wright–Fisher model to found a new population. This distribution assumes a very large number of independent replicate populations.

Using one of the transitions in Figure 3.10 as an example, the chance that a single population with one A allele at $t = 1$ transitions to the same state of one A allele is

$$P_{1\rightarrow1} = \frac{4!}{1!3!}\left(\frac{1}{4}\right)^1\left(\frac{3}{4}\right)^3 = 0.422 \qquad (3.16)$$

The proportion of all populations with one A allele at $t = 1$ is 4/16 or 0.25. Therefore, (0.422)(0.25) = 0.1055 is the proportion of many replicate populations that should transition from one A allele at $t = 1$ to one A allele at $t = 2$. Figure 3.10 shows how all such transitions over three generations add together to determine the overall proportions of populations with each allelic state.

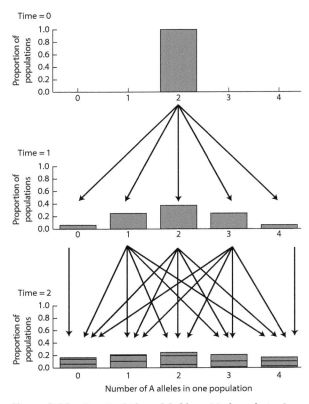

Figure 3.10 Genetic drift modeled by a Markov chain. In this case, the sample size is two diploid genotypes ($2N = 4$) or four gametes per generation. Initial allele frequencies in all populations are $p = q = 0.5$. In one generation, sampling error shifts some proportions of the initial populations that contain two copies of each allele to states of 0, 1, 2, 3, or 4 copies of one allele. Between generations 1 and 2, sampling error again shifts some proportion of the initial populations to states of 0, 1, 2, 3, or 4 copies of one allele. However, in generation 1, there are populations present with all allelic states. The arrows represent the possible allelic states produced by sampling error in the third generation for each of the states in the second generation. The bars in the histogram for the third generation are segmented with horizontal lines to show the contributions of each second-generation allelic state to the total frequency of populations with a given allelic state (some contributions are very small and are difficult to see). As the Markov process continues, the frequency distribution accumulates more and more of the populations at states of zero and four alleles, eventually reaching fixation or loss for all populations.

Markov chains are convenient to model genetic drift because the frequency of populations in a given allelic state depends only on the frequencies in the previous generation (a quality called the **Markov property**). Table 3.2 can be used as a matrix of transition probabilities for any one generation of genetic drift, giving the frequency populations in each allelic state based on the transition probabilities for the number of alleles sampled and the frequencies of populations in each allelic state in the previous generation. Although a population of one diploid

Interact box 3.2 Genetic drift simulated with a markov chain model

A computer program is an easy way to simulate genetic drift with a Markov chain model. The text website has a link to an R script that simulates genetic drift using a Markov chain for a population of 2N = 4 and graphs the results for three different time points.

Get the code to run and compare the graphs to those in Figure 3.10.

The code is written so that all populations initially have a frequency of 0.5 for the A allele. Read the comments (lines that begin with #) to identify the line where the vector of initial frequencies is defined. Modify the code so that the initial proportion of populations at each allele frequency is different, such as a uniform distribution of 20% of populations at each of the allele frequencies. How does this alter time to fixation and loss? Why?

Try extending the number of generations or plotting the proportion of populations at each allele frequency for different generations.

There is also a spreadsheet model of a simple Markov chain that you can view and modify.

individual is not very interesting in biological terms, it is a convenient case to study mathematically. Using techniques of matrix algebra to determine eigenvalues for the transition probabilities matrix represented by Table 3.2 (see Otto and Day 2007 for matrix algebra background), it is possible to show that the rate at which genetic variation is lost from the collection of many populations is:

$$1 - \frac{1}{2N} \qquad (3.17)$$

This says that genetic drift reduces genetic variation by an amount equal to the inverse of twice the effective population size every generation due to sampling error. (This same conclusion can be reached using the approach of consanguineous mating, as shown in Section 3.5.) This rate of loss of genetic variation can clearly be seen in Figure 3.9, where the frequency of genetically variable populations (those with one A allele) halves each generation since the population size is one. This result applies to any population size and shows us that the effects of genetic drift relate directly to the size of a population.

Markov chain: A sequence of discrete random variables in which the probability distribution of states at time $t+1$ depends only on the states at time t.

Markov property: Probability of a given outcome in the next step or time interval depends only on the present state and has no "memory" of states or events before the present time.

Problem box 3.2 Constructing a transition probability matrix

Understanding Markov chains is easier with some practice constructing them. Try constructing the transition matrix for a diploid population size of two (identical to the micro-centrifuge tube sampling experiment with a sample size of four tubes). Similar to Table 3.2, set up a matrix where columns represent the initial allelic state and the terms in rows add together to determine the proportion of populations with a given state one generation later. Then use the binomial formula to calculate the chance that a single population makes each of the allelic state transitions. Indicate the frequency of populations in the initial generation ($t = 0$) with a given allelic state by the variable $Pt = 0(x)$ where x is the number of alleles.

Think about the problem before carrying out any calculations. It is less work than it may appear at first. Two columns have probabilities of either zero or one. Two other columns have the same probabilities but in reversed order.

Biological populations that closely mimic the ensemble population of Markov chain models are relatively easy to construct and maintain given the right choice of organism and some persistent effort. In fact, the first studies of allele frequencies in many identical replicate biological populations were carried out in the 1950s (for example Kerr and Wright 1954; Wright and Kerr 1954). The organisms of choice were fruit flies (*Drosophila* species) since many individuals can be raised in a small space, generation times are short, and a population can be unambiguously defined as one bottle (containing food) of flies. To rear flies, males and females are put together in a bottle and allowed to mate. The adults are removed from the bottle after the females have time to lay eggs on the food. The larvae that emerge from these eggs and become mature flies can then be sampled to found a new generation in a fresh bottle. Figure 3.11A shows the results of one such classic experiment that followed allele frequencies in 107 replicate populations for 19 generations (Buri 1956). All of the populations were constructed to have initial allele frequencies of $p = q = 0.5$ at a diallelic locus (alleles were wild type and bw^{75}). The distribution of allele frequencies in the 107 populations quickly spread out from the initial frequency. Around the fifth generation, a few populations have reached either fixation or loss for the bw^{75} allele. As more generations elapse, the distribution becomes flatter with more and more populations reaching fixation and loss.

The overall shape of the distribution of population allele frequencies for the fly populations closely matches the expected population frequencies according to a Markov chain model of genetic drift for a population of 16 individuals, as shown in Figure 3.11B. In particular, the fly populations and the Markov chain model both show a rapid spread from the initial frequency and an equal number of populations that reach fixation or loss. However, notice that the fly populations have a less even distribution of allele frequencies due to the relatively small number of populations compared to the smoothly continuous distribution of the model which assumes an infinite number of populations. Another difference is that the fly populations showed more rapid accumulation of fixation and loss compared to the model predictions. Even though the fly populations were founded with eight males and eight females every generation, perhaps not all individuals contributed to reproduction, making the effective population size smaller than it seemed. We will explore why the populations might have behaved as though they were smaller than 16 individuals in the next section of the chapter.

The diffusion approximation of genetic drift

The Markov chain model has discrete allelic states and time advances from the initial conditions in individual, discrete generations, as is the case in actual biological populations. This discrete step process can be approximated using mathematical expressions where time and allele frequency are continuous variables. This class of model is based on the processes of molecular diffusion and so is termed the **diffusion approximation of genetic drift** (often called the diffusion equation), first solved by Motoo Kimura (1955). The diffusion approximation is based on partial differential equations and advanced mathematical techniques, so a complete explanation is beyond the scope of this text. However, the general principles behind diffusion equations can be understood readily, especially with the aid of a physical metaphor. The goal of this section is to introduce the situation that diffusion equations model using a particle metaphor and then to cover some of the conclusions about the process of genetic drift that have been reached using the diffusion equation. Readers can see Otto and Day (2007) as well as Denny and Gaines (2000) for more on diffusion in time and space.

Diffusion is the process where particles, moving in random directions, spread out and eventually reach a uniform concentration within the physical boundaries that limit their movement. For example, imagine putting a drop of ink in the center of a Petri dish filled with water. Initially, the concentration of ink is very uneven, but, as time passes, the ink will diffuse and eventually reach a uniform concentration everywhere in the Petri dish. The rate at which the ink spreads out depends on what is called a **diffusion coefficient**. To understand the diffusion coefficient, we have to examine random movement of particles in some detail. This will lead to an improved understanding of genetic drift, so please be patient.

Let us modify the ink-diffusion example by substituting a special Petri dish. In this imaginary dish, the ink particles can move only to the left or to the right from their current position, a situation diagrammed in Figure 3.12. The particles have a constant velocity, so each will move the same distance in a fixed amount of time. We can call this distance moved per unit time δ (pronounced "delta") since it is the change in position to the left or right. The direction of particle movement is random, with equal

(a)

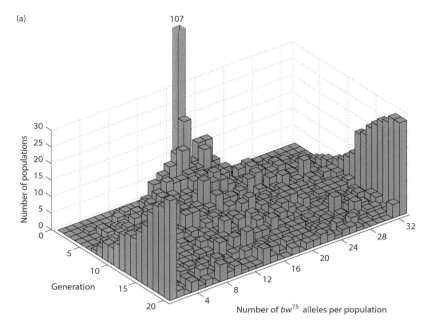

Figure 3.11 Allelic states (or allele frequencies) for 107 *Drosophila melanogaster* populations where 16 individuals (8 of each sex) were randomly chosen to start each new generation (panel A). Initially, all 107 populations had equal numbers of the wild type and bw^{75} alleles (the latter causes homozygotes to have a red-orange and heterozygotes an orange body color so genotypes can be determined visually). The allelic states of the population rapidly spread out, and many populations reached fixation or loss by the 19th generation. The expected frequency of populations in each allelic state determined with a Markov chain model for a population size of 16 with 107 populations that initially have equal frequencies of two alleles (panel B). The *D. melanogaster* populations show a higher rate of fixation and loss than the model populations, suggesting that the population size was actually less than 16 individuals each generation. The *D. melanogaster* data come from Table 13 in Buri (1956).

(b)

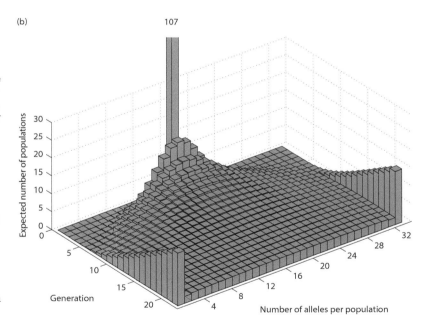

probability of moving to the left ($p = 1/2$) or right ($q = 1/2$) at any moment in time. Let us pick a point of reference somewhere along this axis of movement, call it x, and then track the movement of the ink particles relative to that point. First, what is the average movement of N particles between two time points? There are p of the particles traveling toward x that each move the distance $+\delta$. There are also q of the particles traveling away from x that each move the distance $-\delta$. The average movement of the particles is then

$$\bar{x} = p(\delta) + q(-\delta) \qquad (3.18)$$

Since the particles have equal chances of moving left or right ($p = q = 1/2$)

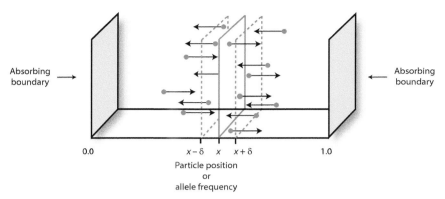

Figure 3.12 An imaginary Petri dish that confines ink particles such that they can move only to the left or to the right from their current position. The particles have a constant velocity, so each will move the distance δ in a fixed amount of time. If the direction of particle movement is random (equal probability of moving left or right at any moment in time), the mean position of particles does not change but the variance in particle position increases with time. The frequency of particles passing through an area such as the plane at x depends on the net balance of particles arriving minus those that are leaving called the flux of particles. The flux is determined by both the rate of diffusion of particles and gradients in the concentration of particles (the net movement of particles is from areas of higher concentration to areas of lower concentration). If the left and right boundaries capture particles, then the diffusion coefficient drops to zero at those points and particles will accumulate. The process of diffusion for particles is analogous to the process of genetic drift for allele frequencies in an ensemble population where allele frequencies "diffuse" because of sampling error.

$$\bar{x} = \frac{1}{2}(\delta) - \frac{1}{2}(\delta) = 0 \qquad (3.19)$$

which means that the average or net movement of particles is 0. To relate this to genetic drift, if populations are like the ink particles but moving randomly in the one dimension of allele frequency, then we expect equal numbers of populations to move toward fixation and toward loss. The *average* change in allele frequency among all populations is expected to be 0.

The next thing we could do is to describe the variance in the position of ink particles over time, a measure of how spread out the particles become. Intuition suggests that even though the average is 0 the variance should not be 0: spreading out of particles is what occurs during diffusion after all. Equation A.2 shows that the variance is the average squared deviation from the mean. We just showed the mean particle location is zero. So, the variance in the location of particles is then just the average square of their positions after one time step. The location of one particle, call it particle i, at time $t = 1$ can be expressed as its location at time $t = 0$ plus the amount a particle moves during one time step

$$x_{i(t=1)} = x_{i(t=0)} + \delta \qquad (3.20)$$

To get an expression for the variance in particle location, we need to start out by squaring this expression for particle location

$$x^2_{i(t=1)} = \left(x_{i(t=0)} + \delta\right)^2 \qquad (3.21)$$

and expanding the right side to get

$$x^2_{i(t=1)} = x^2_{i(t=0)} + 2x_{i(t=0)}\delta + \delta^2 \qquad (3.22)$$

Using Eq. 3.22, which is the squared position of one particle, we can average over all N particles to get the variance in particle position

$$\sigma^2\left(x_{i(t=1)}\right) = \frac{1}{N}\sum_{i=1}^{N}\left(x^2_{i(t=0)} + 2x_{i(t=0)}\delta + \delta^2\right)$$

$$(3.23)$$

This expression simplifies considerably. The value of δ for a large number of particles should be 0, using the same reasoning as when determining average particle position, since an equal number are moving left and right (δ² is *not* zero because the *squared* change in position will always be positive). The middle term in Eq. 3.23 then drops out since it is multiplied by 0. This leaves the variance in particle position as

$$\sigma^2 \left(x_{i(t=1)} \right) = \overline{x^2}_{i(t=0)} + \delta^2 \qquad (3.24)$$

The first term is the average of the squared particle position at time $t = 0$. The second term is the square of step length that particles take between time points. If a group of particles all started out at position 0 (meaning $\overline{x^2}_{i(t=1)} = 0$), then the variance in particle position increases by δ^2 every time interval. If t is the number of time steps that have elapsed for particles that started out at position 0, the variance in particle position is $t\delta^2$. As intuition suggests after watching things like ink diffuse in water, the variance in particle position is not 0 and increases with time.

Now, we are ready to return to the diffusion coefficient. The diffusion coefficient (D) is defined as half the rate at which the variance in particle position changes as time advances. In symbols, this is

$$D = \frac{1}{2} \frac{d\sigma^2}{dt} \qquad (3.25)$$

The source of the factor of ½ can be seen in Figure 3.12. Only half of the particles near the point x (within δ or one step of the plane) will be headed away and increasing their dispersion while the other half will be headed toward x and not dispersing. Half the variance in particle position, $\frac{\delta^2}{2}$, is, therefore, the diffusion coefficient for physical molecules. The diffusion coefficient tells us how fast particles spread out around some point due to random movement.

Allele frequency in an ensemble population has an analog of the diffusion coefficient. Allele frequency "diffusion" is the spreading out and flattening of the allelic state distributions over time as seen in Markov chain models (see Figure 3.11). Recall from Eq. 3.6 that the standard error of the allele frequency is $\sqrt{\frac{pq}{2N}}$, which can also be thought of as the standard deviation of the mean allele frequency. The variance of the mean allele frequency is then the square root of the standard deviation or $\frac{pq}{2N}$. This later quantity is the variance per time period or per generation, which could be expressed as

$$\frac{pq}{2N} = \frac{d\sigma^2}{dt} \qquad (3.26)$$

By substituting Eq. 3.26 into Eq. 3.25, we obtain an expression for the diffusion coefficient of allele frequency

$$D = \frac{1}{2} \frac{pq}{2N} \qquad (3.27)$$

along the one-dimensional axis of allele frequency. By substituting $q = 1 - p$, we obtain

$$D = \frac{p - p^2}{4N} \qquad (3.28)$$

The genetic diffusion coefficient depends on both the allele frequency and the size of the population. Diffusion of allele frequency is the greatest when $p = q = 1/2$ and declines to 0 as p approaches 0 or 1. Populations (as with particles) tend to diffuse to areas where the diffusion coefficient is the lowest and then get stuck there since the rate of spread of particles (the variance in position per time step) is reduced. The rate of diffusion also depends on the size of the population, decreasing as N increases. For particles, this is due to more frequent collisions that reduce the ability to move as the concentration of particles increases (think of trying to walk in a straight line while in a large crowd of people). In biological populations, the diffusion coefficient depends on N since the population size determines the amount of sampling error from generation to generation. It is satisfying that both of these features of allele frequency diffusion agree with our previous generalizations about genetic drift obtained with distinct approaches to the problem.

Next, we would like to keep track of the chance that a particle is at a given location along the axis of diffusion. This probability distribution shows how many particles out of a large number should be at each point along the axis, just as Markov chain models show the expected number of populations at each allelic state. This requires that we know the **flux** or the net number of particles moving through a defined area per time interval. Let us define the area, call it A, where we will determine the flux through the plane at the point x (Figure 3.12). The particles that will move through plane A in one time step must be within plus or minus δ of x because a particle travels the distance δ in one time step. The net number of particles moving to the right through A is the same thing as the difference in the number of particles moving from the left and from the right

$$\text{Net } N(\text{R}) = \frac{1}{2}N(\text{L}) - \frac{1}{2}N(\text{R}) \qquad (3.29)$$

where N represents the number of particles moving left (L) or right (R) through A and the $\frac{1}{2}$ is because only half of the particles on each side of x will move toward x each time step. Factoring and rearranging gives

$$\text{Net } N(\text{R}) = -\frac{1}{2}[N(\text{R}) - N(\text{L})] \qquad (3.30)$$

The flux is defined per area per time, so we need to divide it by area (A) and time (t):

$$J_x = -\frac{1}{2}\frac{[N(\text{R}) - N(\text{L})]}{At} \qquad (3.31)$$

where J_x represents the flux at point x. We can multiply Eq. 3.31 by δ^2/δ^2 (or 1) and then rearrange it to get

$$J_x = -\frac{\delta^2}{2t}\frac{1}{\delta}\left[\frac{N(\text{R})}{A\delta} - \frac{N(\text{L})}{A\delta}\right] \qquad (3.32)$$

Now, notice that the number of particles moving left and moving right are both divided by an area times a distance along the axis of diffusion ($A\delta$). As you can see in Figure 3.12, this defines a three-dimensional volume. The number of particles per volume is equivalent to a concentration, so we can use C to represent the concentration of particles. Also, notice that $\frac{\delta^2}{2}$ is the diffusion coefficient for one time step (Eq. 3.25). Making these substitutions gives

$$J_x = -D\frac{C(\text{R}) - C(\text{L})}{\delta} \qquad (3.33)$$

If we say that the concentration of particles to the right of area A is taken at location $x + \delta$ [$C(\text{R}) = C(x + \delta)$] and the concentration of particles to the left is actually taken at area A [$C(\text{L}) = C(x)$], the fraction in Eq. 3.33 takes the form of a first derivative

$$J_x = -D\frac{C(x + \delta) - C(x)}{\delta} \qquad (3.34)$$

(A straightforward refresher on methods and interpretations for derivatives can be found in Newby (1980).) As the distance between x and δ shrinks toward 0 ($\lim_{\delta \to 0}$), the flux at any point x becomes

$$J_x = -D\frac{dC}{dx} \qquad (3.35)$$

In total, the flux is determined by the product of the rate at which particles spread out from a given point and the rate of change in concentration along the axis of diffusion. The sign of the flux tells the direction of net movement of particles. The flux is positive, meaning that the number of particles in the area to the right of A will increase after one time step, if more particles move in from the left than move out going right. As we would expect, there is a net movement of particles from areas of higher concentration into areas of lower concentration. For example, a higher concentration of particles to the left of point x in Figure 3.12 will result in a positive flux, meaning that after one time step of diffusion there will be an increase in the number of particles at point x. When the concentration of particles is the same everywhere along the axis of diffusion, the flux must be zero since the numbers that move into and out of any area are equal.

Diffusion coefficient (D): Half the rate at which the variance in particle position (or allele frequency in a single population) changes as time advances.
Flux (J_x): The net number of particles (or populations) moving through a defined area (or allele frequency) per time interval.

Let us now take all the concepts of particle diffusion and apply them to genetic drift in an ensemble population. We want to predict the change in the chance that a population has a given allele frequency, just as we did with the Markov chain model. In this case, allele frequency is a continuous variable and the chance that a population has an allele frequency between x and $x + \delta$ at a given time t is called the probability density, symbolized by $\varphi(x,t)$. Probability density for populations is just like concentration for particles, so $\varphi(x,t)$ is the analog of $C(x,t)$. The probability density $\varphi(x,t)$ at any point along the axis of allele frequency will depend on the net difference between populations which drift into allele frequency x and those which drift out of allele frequency x. This is the flux in allele frequency at point x and time t. To know the probability at all points along the axis of diffusion, we need the rate of change in the probability density with change in allele frequency. This is the rate of change in the flux of populations with change in allele frequency

$$\frac{\partial \varphi(x,t)}{\partial t} = -\frac{\partial}{\partial x} J(x,t) \qquad (3.36)$$

where ∂ is the symbol for a partial differential. The derivation for particle flux hid one detail that we now need to reveal in order to continue. The flux depends on both the mean movement and the net movement of particles. Imagine, for example, that the ink particles in Figure 3.12 were positively charged and one of the boundaries was negatively charged. The ink particles would diffuse, but, at the same time, the whole cloud of particles would be moving on average toward the negatively charged boundary. In such a case, the flux at any point x would also need to account for changes to the mean position of particles $M(x)$ so that $J_x = M(x)\mathrm{d}t - D\frac{\mathrm{d}C}{\mathrm{d}x}$. This mean change was neglected earlier since the mean position is 0 if particles are moving left or right at random and are not influenced by some force changing the mean location of all particles. Substituting this full version of the flux (remember that $C(x,t)$ is now $\varphi(x,t)$) gives

$$\frac{\partial}{\partial t} \varphi(x,t) = -\frac{\partial}{\partial x} \left[M(x)\varphi(x,t) - D\frac{\partial \varphi(x,t)}{\partial x} \right] \quad (3.37)$$

We can also substitute the flux in allele frequency from Eq. 3.27 (using x to represent p and $1 - x$ to represent q)

$$\frac{\partial}{\partial t} \varphi(x,t) = -\frac{\partial}{\partial x} \left[M(x)\varphi(x,t) - \left(\frac{1}{2}\frac{x(1-x)}{2N} \right) \frac{\partial \varphi(x,t)}{\partial x} \right]$$
$$(3.38)$$

This is called the forward Kolmogorov equation after Andrei Kolmogorov who developed it to describe a continuous time Markov processes.

With only random sampling error acting to change allele frequency, ($M(x)\partial t = 0$), this rearranges to the diffusion equation for genetic drift

$$\frac{\partial \varphi(x,t)}{\partial t} = \frac{1}{4N}\frac{\partial^2}{\partial x^2} [x(1-x)\varphi(x,t)] \qquad (3.39)$$

The diffusion equation predicts the probability distribution of allele frequencies in many populations over time, and some examples are given in Figure 3.13. Compare Figure 3.13 with Figure 3.10, and it is apparent that the diffusion equation and the Markov chain model both make similar predictions for the outcome of genetic drift. A final point is that the term diffusion *approximation* implies that the diffusion equation makes some assumptions. Noteworthy assumptions are that the number of populations is very large, approaching infinity, and the allele frequency distribution is continuous so that the distribution of allele frequencies is a smooth curve (compare these assumptions with the allelic state distribution in Figure 3.11A with its discrete allelic states and finite number of populations).

The diffusion equation has been used to arrive at a number of generalizations about genetic drift. A widely used set of generalizations is the average time to fixation for alleles that eventually fix in a population and the average time to loss for alleles that eventually are lost from a population

$$\bar{T}_{fix} = -4N\frac{(1-p)\ln(1-p)}{p} \text{ and } \bar{T}_{loss} = -4N\frac{p\ln(p)}{1-p}$$
$$(3.40)$$

where p is the initial allele frequency (Kimura and Ohta 1969a, b). (Note that the natural log of a number less than 1 is always negative: $\ln(1) = 0$ and $\ln(x) \to -\infty$ as x approaches 0, so that the average time will always be a positive number.) These two expressions can be combined to obtain the weighted average time that an allele segregates in a population (the allele is neither fixed nor lost)

$$\bar{T}_{segregate} = -4N[p\ln(p) + (1-p)\ln(1-p)]$$
$$(3.41)$$

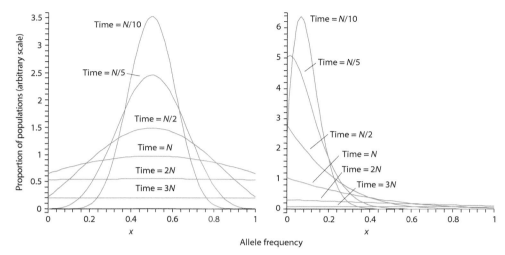

Figure 3.13 Probability densities of allele frequency for many replicate populations predicted using the diffusion equation. The initial allele frequency is 0.5 on the left and 0.1 on the right. Each curve represents the probability that a single population would have a given allele frequency after some intervals of time has passed. The area under each curve is the proportion of alleles that are not fixed. Time is scaled in multiples of the effective population size, N. Both small and large populations have identically shaped distributions, although small populations reach fixation and loss in less time than large populations. The populations that have reached fixation or loss are not shown for each curve.

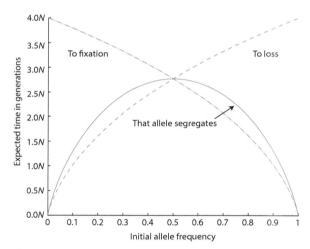

Figure 3.14 Average time that an allele segregates, takes to reach fixation, or takes to reach loss, depending on its initial frequency when under the influence of genetic drift alone. Alleles remain segregating (persist) for an average of $2.8N$ generations when their initial frequency is ½. Fixation or loss takes up to an average of $4N$ generations when alleles are initially very rare or nearly fixed, respectively. Since these are average times, alleles in individual populations experience longer and shorter fixation, loss, and segregation times. Time is scaled in multiples of the population size.

The predictions from these equations quantify our intuition about the action of genetic drift (Figure 3.14). Alleles close to fixation or loss do not take long to reach fixation or loss. Alternatively, an allele initially very close to fixation (or loss) would take a long time, about $4N$ generations if N is large, if

it were to reach the opposite condition of being very close to loss (or fixation). Also, the closer an allele is to the initial frequency of ½, the longer it will segregate before reaching fixation or loss up to a maximum of about $2.8N$ generations. The curves for times to fixation or loss and segregation times have an identical shape no matter what the population size is. The population size plays a role only in the absolute average number of generations that will elapse.

If you have worked your way through this section, you deserve congratulations for your persistence. The basis of the diffusion equation is definitely more abstract than the basis of Markov chains, but the overall results provided by the two models are very similar. Those who would like to learn more about the diffusion equation, its assumptions, and how it can be extended to include processes such as mutation, migration, and natural selection along with genetic drift, can consult Roughgarden (1996) and Otto and Day (2007).

3.3 Effective population size

- Defining genetic populations.
- Census and effective population size.
- Example of bottleneck and harmonic mean to demonstrate effective population size and census size.
- Effective population size due to unequal sex ratio and variation in family size.

Up to this point, we have used the term population size without much fanfare to indicate how many individuals a population contains. We now need to focus additional attention on the idea of population size. The number of individuals in a population seems like a straightforward quantity that can be determined easily. In the context of the Wright–Fisher model, the population size is an unambiguous quantity. Unfortunately, in most biological populations, it is difficult or impossible to determine the number of gametes that contribute to the next generation. We need another way to define the size of populations.

The definition of the population size in population genetics relies on the dynamics of genetic variation in the population. This definition means that the size of a population is defined by the way genetic variation in the population *behaves*. The notion that "if it walks like a duck and quacks like a duck, it probably is a duck" is also applied to the size of populations. The size of a population depends on how genetic variation changes over time. If a population shows allele frequencies changing slowly over time under the exclusive influence of genetic drift, then the population has the dynamics associated with relatively large size. It "quacks" like a big population. In the same way, a population with a large number of individuals might show rapid genetic drift, indicating that it is really a small population from the perspective of genetic variation. It looks big, but its "quack" gives it away as a small population.

Making a distinction between the dynamics of genetic variation in a population and the number of individuals in a population suggests that there are really two types of population size. One is the head count of individuals in a population, called the **census population size**, symbolized by *N*. The other is the genetic size of a population. This genetic size is determined by comparing the rate of genetic drift in an actual population with the rate of genetic drift in an ideal population meeting the assumptions of the Wright–Fisher model. The population size in the model that produces that same rate of genetic drift as seen in an actual population is the genetic size of the actual population. In comparing an actual population with an ideal model population, we are asking about the overall genetic effects of the census size. Thus, we also recognize the **effective population size**, N_e, as the size of an ideal

Figure 3.15 Sewall Wright (1889–1988) with a guinea pig in an undated photograph taken during his years as a professor at the University of Chicago. Starting in 1912 and throughout his career, Wright studied the genetic basis of coat colors and physiological traits in guinea pigs. Wright, along with J. B. S. Haldane and R. A. Fisher, established many of the early expectations of population genetics using mathematical models to make predictions. Many of the conceptual frameworks in population genetics today were originated by Wright, especially those related to mating among relatives, genetic drift, and structured populations. An often retold (although mythical) story was that Wright, who would sometimes carry a guinea pig with him, would, on occasion, absent-mindedly employ the animal to erase the chalk board while lecturing. Provine's (1986) biography details Wright's manifold contributions to population genetics and his interactions with other major figures such as Fisher. Source: Special Collections Research Center/University of Chicago Library.

population that experiences as much genetic drift as an actual population regardless of its census size. This concept was originally introduced by Sewall Wright (1931), who is shown in Figure 3.15. An approximate way to think of the difference between the two population sizes is that the census size is the total number of individuals and the effective size is the number of individuals that actually contribute gametes to the next generation. We will refine this definition throughout this rest of the chapter.

Census population size (*N*): The number of individuals in a population; the head count size of a population.
Effective population size (*N*ₑ): The size of an ideal Wright–Fisher population that maintains as much genetic variation or experiences as much genetic drift as an actual population regardless of census size.

Let us examine several biological phenomena that cause effective population size and census population size to be different. This will help to illustrate the effective population size and make its definition more intuitive.

Actual populations often fluctuate in the number of individuals present over time. A classic example is rabbit/lynx population cycles due to predator/prey dynamics, where census population sizes of both species fluctuate over a fairly wide range on about a 10 year cycle. Another category of example is the establishment of a new population by a small number of individuals, called a **founder event**. One well-documented founder event was the introduction of European starlings in the New World. These birds, now very common throughout North America, can all be traced to approximately 15 pairs that survived from a larger group released in New York's Central Park in 1890. What is now a very large population descended from a sample of 60 alleles in the small number of founding individuals, which were sampled from a very large population of birds in Europe.

To model the genetic effects of this type of fluctuation in population size over time, suppose a population starts out with 100 individuals, experiences a reduction in size to 10 individuals for one generation, and then recovers to 100 individuals in the third generation (Figure 3.16). (Recall from earlier in the chapter that this situation violates the constant population size assumption of the Wright–Fisher model of genetic drift.) This will cause an increased chance of fixation or loss of alleles (variance in allele frequency will increase) and thereby increase the rate of genetic drift in that one generation. But what is the effective size of this population after it recovers to 100 individuals? We can estimate the effect of fluctuations in populations on the overall effective size using the **harmonic mean**:

$$\frac{1}{N_e} = \frac{1}{t}\left[\frac{1}{N_{e(t=1)}} + \frac{1}{N_{e(t=2)}} + \cdots + \frac{1}{N_{e(t)}}\right] \quad (3.42)$$

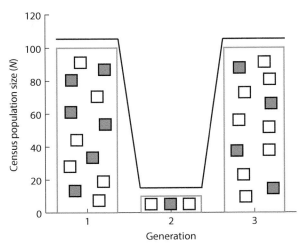

Figure 3.16 A schematic representation of a genetic bottleneck where census population size fluctuates across generations. The harmonic mean of census population size (N) is 25 and provides an estimate of the effects of genetic drift over three generations or the effective population size (N_e). In other words, this population of fluctuating size would experience as much genetic drift as an ideal Wright–Fisher population with a constant population size of 25. The colored squares represent alleles present in each generation. In the first generation, the alleles are equally frequent and end up at frequencies of 25 and 75%. Such an allelic state transition would be extremely unlikely if 200 gametes were sampled to found generations 2 and 3. However, the observed allelic transition is expected 25% of the time in a sample of 4 gametes.

where *t* indicates the total number of generations (all other assumptions of Wright–Fisher populations are met). The harmonic mean gives more weight to small values by virtue of summing the inverses of population size. It also serves as an approximation of $\left(1 - \frac{1}{2N_e}\right)^t$, which we saw was the rate of decrease in genetic variation in Markov chain models of genetic drift earlier in the chapter. In this example:

$$\frac{1}{N_e} = \frac{1}{3}\left[\frac{1}{100} + \frac{1}{10} + \frac{1}{100}\right] \quad (3.43)$$

So, $N_e = 1/0.04 = 25$. Contrast this with the arithmetic mean of the census population size, which is 70. Only those alleles that actually pass through the **genetic bottleneck** of 10 individuals are represented in later generations, regardless of how large the census size is, so the mean census size is much too high to use to predict the behavior of allele frequencies since it will underestimate genetic drift. In this case, we expect allele frequencies in the population to behave similarly to allele frequencies in an ideal Wright–Fisher population with a constant size of 25 over three generations. Like finite population size, fluctuations in population size through time are a universal feature of biological populations. Populations obviously vary greatly in the degree of size fluctuations and the time scale of these fluctuations, but $N_e < N$ caused by temporal fluctuations in N is a widespread phenomenon.

Founder effect: The establishment of a population by one or a few individuals, resulting in small effective population size in a newly founded population.

Genetic bottleneck: A sharp but often transient reduction in the size of a population that increases allele frequency sampling error and has a disproportionate impact on the effective population size in later generations even if census sizes increase.

Although the term bottleneck is usually associated with sharp reductions in the overall population size, there are other aspects of biological populations that have the same impact by increasing the sampling error in allele frequency across generations. Mating patterns can have a major impact on effective population size when individuals of different sexes make unequal contributions to reproduction. This occurs in populations where individuals of one sex compete for mating access to individuals of the other sex. In such a situation, the numbers of females and males that breed, or the breeding sex ratio, may not be equal (even if the population sex ratio is equal). The leads to increased genetic drift compared with the case of a breeding sex ratio of 1 : 1, since the pool of alleles passed to the next generation will be sampled from fewer individuals in one sex. Thus, the less frequent sex becomes an allelic bottleneck of sorts. The effective size in such cases is:

$$N_e = 4\frac{N_m N_f}{N_m + N_f} \qquad (3.44)$$

where N_f is the number of females and N_m is the number of males *breeding* in the population and all other assumptions of Wright–Fisher populations are met. Equation 3.44 shows that the effective population size approaches four when the rarer sex approaches a single individual and that the effective size is maximized when there is a breeding sex ratio of 1 : 1.

Let us look at a case where N_m does not equal N_f to see the impact on the effective population size. Elephant seals (*Mirounga leonina*) are a classic example of highly unequal breeding sex ratios since the mating system is harem polygyny. In one study of breeding patterns on Sea Lion Island in the Falkland Islands, about 550 females and 75 males were observed on land where mating takes place (Fabiani et al. 2004). Using genetic markers to ascertain the parentage of pups, it was determined that only 28% of the males fathered offspring during the course of two breeding seasons. Therefore, the breeding sex ratio was about $N_m = 21$ and $N_f = 550$. The effective population size during each breeding season was:

$$N_e = 4\frac{(21)(550)}{21 + 550} = 80.91 \qquad (3.45)$$

or was equivalent to an ideal population of 40 females and 40 males where breeding sex ratio is 1 : 1. The strongly unequal breeding sex ratio for elephant seals results in an effective population size an order of magnitude less than the census size of 625 individuals.

A third factor that distinguishes the census and effective population sizes is the degree to which adult individuals in the population contribute to the next generation. One of the assumptions of Wright–Fisher populations is that all individuals contribute an equal number of gametes to the infinite gamete pool. To maintain a population that is not changing in size across generations, each individual must produce one surviving progeny, on average, to replace itself. In outcrossing species, when each pair of individuals produces an average of two progeny, then the population will be stable in size over time. In terms of the Wright–Fisher population, that means that each individual contributes an *average* of two gametes to the next generation from the infinite gamete pool.

There are many patterns of individual reproduction within an outcrossing population that can achieve a *mean* rate of reproduction that results in a stable population through time. It might be the case that all individuals produce exactly two progeny. Another possibility is that a few parents produce no offspring, whereas most parents produce two offspring, and a few parents produce four offspring that offset the reduction in the average caused by the parents with no progeny. In the extreme, one pair of parents could produce all N offspring and all other pairs fail to reproduce successfully. The variance in family size can be used to describe these different patterns of individual reproduction. As variance in family size increases, the alleles passed to the next generation come increasingly from those parents producing more offspring.

The effective population size due to variation in family size is:

$$N_e = \frac{4N_{t-1}}{\mathrm{var}(k) + \overline{k}^2 - \overline{k}} \qquad (3.46)$$

where N_{t-1} is the size of the parental population and k is the number of gametes that result in progeny, or family size for outcrossing organisms (Crow and Denniston 1988). The equation shows that for a stable population ($\overline{k} = 2$) when the variance in family size is equal to the average family size, then there is no "bottleneck" due to family size variation: the population size of parents is the effective population size. The Wright–Fisher model assumes that the production of progeny has exactly this quality of the mean family size being equal to the variance in family size. This is because the binomial sampling model of the Wright–Fisher model converges to a Poisson distribution as the number of trials grows very large

and the chance of any given outcome becomes very small. (The Poisson distribution gives the probability of observing a given number of events in a fixed time period when there are discrete outcomes, and events are independent and occur at random; see Otto and Day 2007).

We can explore the consequences of variance in family size with some examples. Figure 3.17 shows three hypothetical distributions of family size. The first is an ideal Poisson distribution where the mean family size is equal to the variance in family size. This is the standard assumption used in the Wright–Fisher model of genetic drift. The next distribution is an example of highly skewed family sizes where a relatively small proportion of the population contributes most of the progeny. The final distribution shows family size variation that is less than expected for a Poisson distribution. If the Poisson distribution is used as the standard, the other populations with differing distributions of family size show more or less genetic drift, respectively, due to modification of the bottleneck-like effect of unequal family sizes. These distributions also illustrate that the effective population size based on variance in family size has the

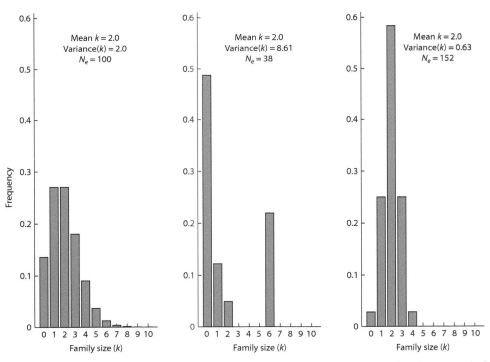

Figure 3.17 Distributions of family size. The variance equals the mean as expected for a Poisson distribution on the left. However, the center distribution has a few families that are very prolific while 75% of the families produce 2 or fewer progeny with most individuals failing to reproduce. The distribution on the right has less variance in family size than expected for a Poisson distribution with most families of size two. The Poisson distribution is taken as the standard with an effective size of 100. By comparison, the center distribution has a smaller effective population size and the distribution on the right a larger effective population size.

unique quality that N_e can actually be larger than N if the variance in family size is less than the mean of family size. This stands in contrast to population size fluctuations through time and unequal breeding sex ratios, where N_e can only be less than or equal to N but not greater than N.

These family size distributions are not just theoretical entities. Many annual plants show variation in reproductive success that exceeds the mean, demonstrating that variation in family size contributes to overall rates of genetic drift (Heywood 1986). In one study of salmon, the large variance in reproductive success among anadromous males had a greater impact on the effective population size than breeding sex ratio (Jones and Hutchings 2002). In contrast, Poisson-distributed male reproductive success has been observed in laboratory populations of *D. melanogaster*, partly supporting the effective population size assumption behind many genetic experiments that have used fruit flies reared in the laboratory (Joshi et al. 1999).

Problem box 3.3
Estimating N_e from information about N

Imagine that a conservation biologist approaches you asking for assistance in estimating the genetic impacts of a recent event in a captive population of animals housed in a zoo. The zoo building where the animals were kept experienced a fire, killing some animals outright and requiring the survivors to be relocated to a new enclosure that is not ideal for breeding. Before the fire, the population was stable at 30 males and 30 females for many generations with an effective population size of 60. After the fire, there were 15 females and 10 males. Due to the disruption and relocation of the animals, breeding behavior changed. Before the fire, variation in family size was Poisson distributed with a mean of 2.0. In the one generation after the fire, family size has a mean of 4.0 and variance of 6.5. What are the genetic impacts of the fire on the effective population size? What are some of the assumptions specific to this case used in your estimate of N_e?

In the last section of the chapter, we compared allele frequencies over time in 107 *Drosophila* populations to allele frequencies expected from the Markov chain model (Figure 3.11). The fly populations, founded each generation with eight female and eight male flies, experienced a faster rate of fixation or loss than expected for an effective population size of 16. In fact, the fly populations reached fixation and loss at a rate comparable to a population with an effective size of about 10 or 11 (try modifying the R script in Interact box 3.2 to show this). The concepts in this section that distinguish between census and effective population sizes can be used to explain Buri's results. Although there was a census size of 16 flies in each bottle, an unequal breeding sex ratio in each bottle could explain the higher rate of fixation and loss. For example, a breeding sex ratio of eight females and six males due to the failure of some males to mate successfully each generation would give an effective population size of $N_e \approx 14$ using Eq. 3.44. It is also possible that there was a relatively high degree of variation in reproduction among the females. For example, if variance in family size was 3.5 and was combined with the effects of the unequal breeding sex ratio (using 14 instead of 16 for N_{t-1}), Eq. 3.46 estimates that $N_e \approx 10$. It might also be that, in a few of the generations, the population size was smaller than intended due to mistakes when handling and transferring flies to new bottles. However, Eq. 3.42 shows that the effect of a population size of 14 for one generation out of 19 is slight ($N_e = 15.88$). Therefore, infrequent fluctuating population sizes would probably have had only a minor impact on the results. Thus, the difference between the rate of fixation and loss in the Markov chain model and in the actual fly populations can be explained by several plausible factors that distinguish the census and effective population sizes.

3.4 Parallelism between Drift and mating among relatives

- Autozygosity due to sampling in a finite gamete population.
- The relationship between the fixation index (F) and heterozygosity (H).
- Decline in heterozygosity over time due to genetic drift.
- Heterozygosity in island and mainland populations.

The chapter up to this point has focused on population size and genetic drift. This section will demonstrate that finite population size can also be thought of as a form of mating among relatives. In large populations with random mating, chance biparental inbreeding is unlikely to occur often. However, in small populations, the chance of mating with a relative is larger since the number of possible mates is limited. As populations get smaller, the probability of chance matings between related individuals should increase. Genetic drift also occurs due to finite population size. Therefore, genetic drift and the tendency for inbreeding are interrelated phenomena connected to the size of a population. Both have the result of increasing the homozygosity in a population over time.

Before we can reach the goal of showing that genetic drift and inbreeding are really equivalent genetic processes, it is necessary to develop a bit of conceptual machinery. In Chapter 2, autozygosity was defined as the probability that two alleles are identical by descent (IBD) and demonstrated using a pedigree. We now need to revisit the autozygosity from the perspective of a finite population. Figure 3.18 shows three possible ways that alleles could be sampled from a finite population of gametes when constructing diploid genotypes. This gamete population meets Wright–Fisher assumptions with the exception that it is finite and contains alleles that are identical in state but are not IBD. To make offspring in the next generation, alleles are sampled with replacement from the population of $2N$ gametes. What is the probability that two alleles in a genotype in the next generation are IBD? Given that one allele has been sampled, say an A_1 allele, what is the probability of sampling the *same* allele on the next draw? Since there is only one copy of this allele in gamete population, there is only one of the $2N$ alleles that are the same. Therefore, the chances of sampling the same allele to make a genotype are $\frac{1}{2N_e}$, which is also the probability that the alleles in a genotype are IBD or are autozygous. The probability that two alleles in a genotype are not IBD or are allozygous is then $1 - \frac{1}{2N_e}$. Through the process of random sampling, populations can accumulate autozygous genotypes, as do populations where mating takes place among relatives.

We can use the probability of autozygosity in a finite population to define the fixation index (recall that it was defined as expected heterozygosity minus observed heterozygosity all over expected heterozygosity in Chapter 2) as

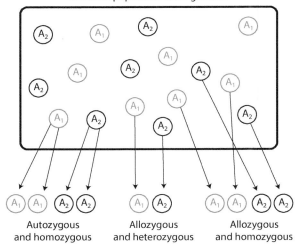

Ancestral population of 2N gametes

Possible genotypes in next generation
when sampling with replacement

Figure 3.18 Autozygosity and allozygosity in a finite population where identity by descent is related to the size of the population. Finite populations accumulate genotypes containing alleles IBD through random sampling in a manner akin to mating among relatives. In this example, alleles in the ancestral gamete pool identical in state are not IBD. Sampling of alleles takes place to form the diploid genotypes of the next generation. By chance, the same allele can be sampled twice to form an autozygous genotype with probability $\frac{1}{2N_e}$. The chance of not sampling the same allele twice is the probability of all other outcomes or $1 - \frac{1}{2N_e}$. Autozygous genotypes must be homozygous, but allozygous genotypes can be either homozygous or heterozygous.

$$F_t = \frac{1}{2N_e} \qquad (3.47)$$

for generation t under the assumption that none of the alleles in the gamete pool in generation $t - 1$ are IBD. To make this more general, we could also account for the possibility that some of the gametes in the ancestral population of generation $t - 1$ have alleles that are IBD from past inbreeding or random sampling. To do this, we need to reexamine the probability that alleles are allozygous. Although two distinct alleles in the gamete pool of Figure 3.18 may be sampled to form what appears to be an allozygous homozygote, it is possible that these two alleles sampled are actually IBD. That would mean that the gamete pool would be inbred to some degree F_{t-1} instead of containing only allozygous alleles. The fixation index then becomes

$$F_t = \frac{1}{2N_e} + \left(1 - \frac{1}{2N_e}\right)F_{t-1} \qquad (3.48)$$

where the first term is due to sampling between generations and the second term is the proportion of apparently allozygous alleles in the gamete population that are actually autozygous due to past sampling or mating among relatives.

By definition, F is the reduction in heterozygosity as well as the increase in homozygosity compared to Hardy–Weinberg expected genotype frequencies. If F is proportional to the homozygosity and amount of inbreeding, then $1 - F$ is proportional to the amount of heterozygosity and random mating. If we are interested in predicting levels of heterozygosity, multiplying both sides of Eq. 3.48 by -1 and then adding 1 to both sides give

$$1 - F_t = \left(1 - \frac{1}{2N_e}\right)(1 - F_{t-1}) \qquad (3.49)$$

or an expression for one minus the expected homozygosity due to random sampling from a finite gamete population. To express this in terms of the heterozygosity, we can use $H_t = 2pq(1 - F_{t-1})$ from Eq. 2.20 to obtain $1 - F_{t-1} = \frac{H_t}{2pq}$. Substituting this into Eq. 3.49 gives:

$$\frac{H_t}{2pq} = \left(1 - \frac{1}{2N_e}\right)\left(\frac{H_{t-1}}{2pq}\right) \qquad (3.50)$$

which after multiplying both sides by $2pq$ gives:

$$H_t = \left(1 - \frac{1}{2N_e}\right)H_{t-1} \qquad (3.51)$$

Taking this relationship over an arbitrary number of generations gives

$$H_t = \left(1 - \frac{1}{2N_e}\right)^t H_0 \qquad (3.52)$$

where H_0 is the initial heterozygosity and H_t is the heterozygosity after t generations have elapsed.

There is a very general and biologically meaningful relationship contained in Eq. 3.52. It shows that heterozygosity declines by a factor of $1 - \frac{1}{2N_e}$ every generation caused simply by sampling from a finite population that results in some autozygous genotypes every generation. Recall that this is exactly

the same result that was obtained for the rate of loss of genetic variation with the Markov chain model. The degree of sampling varies directly with the effective population size, so that the rate of increase in autozygous genotypes also depends directly on the effective population size. The expected heterozygosity from Eq. 3.52 is shown in Figure 3.19 for four different effective population sizes over 50 generations. For comparison, heterozygosity in six independent replicate populations experiencing genetic drift are also plotted. The random trajectories of heterozygosity in these individual populations make clear that Eq. 3.52 provides an expectation for average heterozygosity taken across a large number of replicate populations or numerous independent neutral loci if applied to a single population.

There are two conclusions that can be drawn from the interrelationship between autozygosity and the effective population size. First, genetic drift causes populations to become more like those that have experienced mating among relatives in the sense that autozygosity and homozygosity increase *even though mating is random*. An important distinction is that genetic drift causes heterozygosity to decrease due to the fixation and loss of alleles. In contrast, mating among relatives decreases heterozygosity by changing genotype frequencies but does not impact allele frequencies. Genetic drift produces homozygosity since ultimately one allele reaches fixation while consanguineous mating produces homozygous genotypes for all alleles in the population. Second, mating systems where there is regular mating among relatives cause genetic variation in populations to behave, from the perspective of heterozygosity, as if the effective population size were smaller than it would be under complete random mating. For example, when a parent self-fertilizes, the alleles transmitted to its progeny are sampled from a gamete pool containing only two possible alleles. Compare that with the case of outcrossing between unrelated individuals, where the alleles transmitted to progeny are sampled from a pool of four possible alleles. The probability of autozygosity is clearly higher with the smaller parental gamete pool associated with consanguineous mating just as it is also higher in a smaller population where mating is random.

The faster decrease in heterozygosity in small compared to large populations can be seen in the wild. One study used the reasoning that

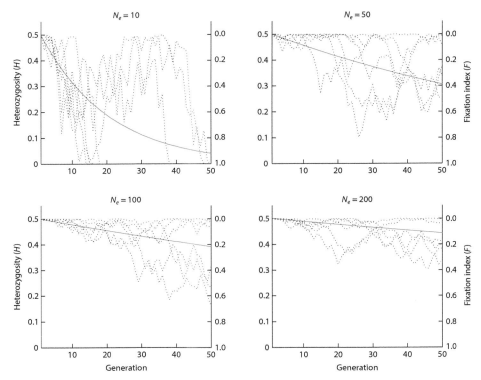

Figure 3.19 The decline in heterozygosity as a consequence of genetic drift in finite populations. The solid lines show expected heterozygosity over time according to $H_t = \left(1 - \frac{1}{2N_e}\right)H_{t-1}$. The decrease in heterozygosity can also be thought of as an increase in autozygosity or the fixation index (F) through time under genetic drift. The dotted lines in each panel are levels of heterozygosity ($2pq$) in six replicate finite populations experiencing genetic drift. There is substantial random fluctuation around the expected value for any individual population.

Interact box 3.3 Heterozygosity over time in a finite population

Heterozygosity for multiple loci under genetic drift can be simulated using the text simulation web site.

Populus simulates genetic drift in a finite population and then tracks values of the inbreeding coefficient over time. In Populus, click on the **Model** menu and select **Mendelian Genetics** and then **Inbreeding**. A dialog box will open that has entry fields for the effective population size and the initial level of the inbreeding coefficient for a diallelic locus. You can also set the number of generations to run the simulation. To get started, set **Population** = 30, **Initial Frequency** = 0.0, and **Generations** = 120. A graph of the results will appear after entering the simulation parameter values. Despite the name, **Initial Frequency** is actually the initial level of departure from Hardy–Weinberg genotype frequencies (F) in the population. A value of zero means that the population is in Hardy–Weinberg equilibrium. Pressing the **View** button in the model dialog will generate a new data set and redraw the graph.

The graph will show three types of fixation coefficients. F_t is the "theoretical" inbreeding coefficient based on the decline in heterozygosity over time (Eq. 3.52) or $F_t = 1 - \left(1 - \frac{1}{2N_e}\right)^t$. The other two lines are both based on the observed frequency of homozygous genotypes in the simulated population. F_a (the blue line) is the actual frequency of homozygous individuals in the population (often called F_i outside of Populus). F_f (the green line) is the population homozygosity. F_a and F_f can be different because the individual homozygosity tracks the combined frequency of homozygotes for either allele while the population homozygosity tracks how close the population is to fixation or loss (global homozygosity). When the population homozygosity is 1, there can be only one homozygous genotype in the population.

Why do the individual and population homozygosity values fluctuate? Is the amount of fluctuation related to the population size? Although the graph does not show the heterozygosity over time, what would lines for the theoretical and individual and population heterozygosity look like? Try graphing each of these quantities on paper for a given run based on the three inbreeding coefficients.

organisms restricted to islands should have smaller census population sizes than the same species found on adjacent mainland areas (Frankham 1998). Based on the relationship between autozygosity and effective population size just derived, the expectation is that island populations should show lower levels of heterozygosity than mainland populations. Although there were some exceptions, the general pattern was that island populations had lower levels of heterozygosity than did mainland populations (Table 3.3). This is exactly the pattern expected due to the faster rate of decrease in heterozygosity in smaller populations even when mating is random. The comparison

assumed that the island and mainland populations remained very similar in terms of the degree of consanguineous mating as well as genetic parameters that influence the input of genetic variation such as the migration and mutation rates. Since the comparisons were between populations of the same species that are, therefore, very closely related, these assumptions seem likely to be met at least approximately.

3.5 Estimating effective population size

- Variance and inbreeding effective population size.
- Estimating effective population size from temporal changes in allele frequencies, decline in heterozygosity, or gametic disequilibrium.
- Breeding effective population size in continuous populations and genetic neighborhoods.
- Effective population sizes for different genomes.

The concept of effective population size should now seem comfortable. This section will focus on putting models into practice in order to estimate the effective population size. This requires a shift in perspective from using models to develop expectations about allele frequencies to using models to explain past patterns in allele frequencies. Working through a detailed example of estimating the effective population size will highlight the distinctions between two definitions of the effective population size. This section will also introduce a new definition of the effective population size used widely for species with continuous distributions that lack discrete geographic population boundaries. In addition, the effective population sizes of nuclear and organelle genomes will be compared.

Table 3.3 The level of heterozygosity found in island and mainland populations of the same species demonstrates that small population size has effects akin to inbreeding. Heterozygosity in island and mainland populations is compared using the effective inbreeding coefficient $F_e = 1 - \dfrac{H_{island}}{H_{mainland}}$. $F_e > 0$ when the mainland populations exhibit more heterozygosity, $F_e < 0$ when the island populations exhibit more heterozygosity, and F_e is zero when levels of heterozygosity are equal. Values given are ranges when more than one set of comparisons was reported from a single source.

Species	F_e
Mammals	
Wolf (*Canis lupis*)	0.552
Lemur (*Lemur macaco*)	0.518
Mouse (*Mus musculus*)	−0.048 – 1.000
Norway rat (*Rattus rattus*)	−0.355 – 0.710
Leopard (*Panthera pardus*)	0.548
Cactus mouse (*Peromyscus eremicus*)	0.445 – 0.899
Shrew (*Sorex cinereus*)	−0.241 – 0.468
Black bear (*Ursus americanus*)	0.545
Birds	
Singing starling (*Aplonis cantoroides*)	0.231 – 0.833
Chaffinch (*Fringilla coelebs*)	−0.164 – 0.504
Reptiles	
Shingleback lizard (*Trachydosaurus rugosus*)	0.069 – 0.311

Source: Data from Frankham (1998).

Different types of effective population size

Just as there are several models constructed on different foundations used to describe how genetic variation changes due to finite population size, there are several ways to define the effective population size such that it can be estimated. This is because the effective population size is really defined as a consequence of the models constructed to predict the behavior of genetic variation in populations. To see this, let us consider the **inbreeding** and **variance** effective population sizes.

Inbreeding effective population size ($N_e{}^i$): The size of an ideal population that would show the same probability of allele copies being IBD as an actual population.
Variance effective population size ($N_e{}^v$): The size of an ideal population that would show the same sampling variance in allele frequency as an actual population.

In an ideal finite population (under the infinite alleles model), the chance of sampling two copies of the same allele depend on the number of gene copies that pass from one generation to the next and is $\frac{1}{2N_e}$ (see Section 3.3). This is the same as the probability that two alleles are identical by descent (IBD) in a pedigree (see Section 2.6). So, we have:

$$P(\text{IBD}) = \frac{1}{2N_e} \qquad (3.53)$$

which can just as easily be restated as:

$$\hat{N}_e = \frac{1}{2P(\text{IBD})} \qquad (3.54)$$

The first equation is used to predict what to expect for the packaging of alleles into genotypes. The expected value of the probability of identity by descent depends on the size of the population. The flip side of this same relationship predicts what to expect for the effective population size based on the probability of identity by descent in a population, as shown in the second equation. In big populations the chance that two alleles are IBD is low, whereas, in smaller populations, the chance of identity by descent is greater. When the effective population size is defined by reference to autozygosity or identity by descent, the result is the inbreeding effective population size.

The change in allele frequencies in many replicate populations over generations was the focus of the Markov and diffusion models of genetic drift (Section 3.2). The range of change in allele frequency in these models among many populations could also be expressed as a variance. Earlier in the chapter (Eq. 3.6), the standard error of the mean allele frequency among replicate populations was derived. This leads to the variance in the change in allele frequencies from one generation to the next

($\Delta p = p_{t-1} - p_t$) taken among independent replicate loci

$$\text{Variance}(\Delta p) = \frac{p_{t-1}q_{t-1}}{2N_e} \qquad (3.55)$$

where p and q are allele frequencies at a diallelic locus and the subscripts indicate the generation. As we did above, this equation can be restated by solving for the effective population size:

$$\hat{N}_e = \frac{pq}{2[\text{variance}(\Delta p)]} \qquad (3.56)$$

Here again, we see that the first equation shows that the variance in allele frequencies among many identical replicates depends in the effective population size. By turning the equation around, we can then define how big the effective size is by quantifying the variance in the change in allele frequencies among a group of replicated populations. This definition provides the variance effective population size.

Why distinguish between the inbreeding and variance effective population sizes? In the beginning of the section, the effective population size was defined by how genetic variation in a population behaves over time. In ideal populations that are not changing much in census size over time, the different effective sizes are usually equivalent. In some situations, however, the different types of effective population size are not equivalent since they measure different aspects of genetic variation (Ewens 1982; Crow and Denniston 1988; Crandall et al. 1999).

Effective population size estimated using allele frequencies observed at two generations or time points, call them t_i and t_j, can be estimated with the standardized variance in allele frequency (Nei and Tajima 1981)

$$\hat{F}_c = \left(\frac{1}{k}\right) \sum_{x=1}^{k} \frac{\left(p_{x,t=i} - p_{x,t=j}\right)^2}{\left(p_{x,t=i} + p_{x,t=j}\right)/2 - p_{x,t=i}p_{x,t=j}} \qquad (3.57)$$

where k is the number of alleles at a locus, $p_{x,\,t=i}$ and $p_{x,\,t=j}$ are the frequencies of one of the k alleles at times t_i and t_j. With more than one locus, $\hat{F}_c = \sum_{L=1}^{loci} k_L \hat{F}_{c(L)} / \sum_{L=1}^{loci} k_L$ is used to obtain a multilocus version where the contribution of each locus to the overall standardized variance is weighted by the number of alleles. The variance effective population size is then estimated using

$$\hat{N}_e = \frac{t_j - t_i}{2\left[\hat{F}_c - \left(\frac{1}{2n_{t=i}} + \frac{1}{2n_{t=j}}\right)\right]} \qquad (3.58)$$

where the numerator is the number of time intervals between the samples, and n is the number of individuals sampled (Krimbas and Tsakas 1971; Pollak 1983; Waples 1989). When stronger genetic drift causes more variation in allele frequencies, the estimate of $\hat{F}_c$ is larger, resulting in a smaller estimate of $\hat{N}_e$. Subtracting $1/(2n)$ from $\hat{F}_c$ in the denominator at each of the time points adjusts for additional variance contributed by the sampling of individuals to estimate allele frequencies, an effect that diminishes as sample sizes increase. The complications of estimating N_e from the temporal variance in allele frequency include bias caused by small sample sizes and highly unequal allele frequencies (Jorde and Ryman 2007), loci experiencing natural selection (Goldringer and Bataillon 2004), and additional sampling effects due to pooling individuals for short-read sequencing (Jónás et al. 2016).

The change in heterozygosity can be used to estimate the inbreeding effective population, since the heterozygosity is one minus the autozygosity. In the previous section (Eq. 3.52), we saw that heterozygosity decreases according to $H_t = \left(1 - \frac{1}{2N_e}\right)^t H_0$ because of genetic drift. This expectation can be rearranged to give a way to estimate the effective

population size from the change in heterozygosity in one generation

$$\hat{N}_e = \frac{H_0}{2(H_0 - H_1)} \qquad (3.59)$$

Assuming the effective population size is not too small and there are many replicate loci that can be averaged, we can use the approximation

$$\frac{H_t}{H_0} \approx e^{-\frac{t}{2N_e}} \qquad (3.60)$$

Taking the natural log of both sides

$$\ln\left(\frac{H_t}{H_0}\right) \approx -\frac{t}{2N_e} \qquad (3.61)$$

and then solving in terms of the effective population size

$$N_e \approx -\frac{t}{2\ln\left(H_t/H_0\right)} \qquad (3.62)$$

gives an expression that can be applied to any number of generations.

A good way to understand these approaches to estimate the variance and inbreeding effective population sizes is to employ them with example allele frequency data taken from a finite population experiencing genetic drift. Figure 3.20 shows allele

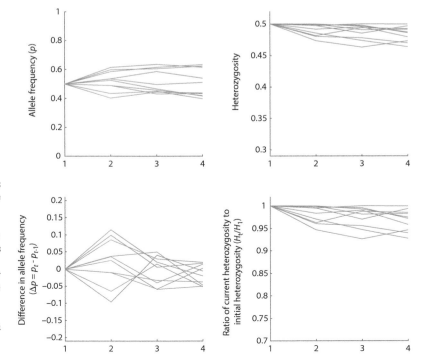

Figure 3.20 Simulated allele frequencies of 10 independent, replicate loci in a single population that experienced effective population sizes of 100, 10, 50, and 100 individuals across four generations. Allele frequencies scatter toward fixation and loss because of genetic drift. The standardized variance in the change in allele frequency (Δp) can be used to estimate the variance effective population size. The inbreeding effective population size can be estimated from the change in heterozygosity through time. The allele frequencies and heterozygosities are available as a spreadsheet on the text website.

frequencies and heterozygosities for 10 independent replicate loci (or one locus observed in 10 independent replicate populations) over four generations in a computer simulation. The figure also shows the change in allele frequencies and heterozygosities between generations. In the simulation, the effective population sizes fluctuated from 100 initially, down to 50 for the second generation, and then back to 100 in the third and fourth generations. There is clearly a greater change in allele frequencies and heterozygosities between the first and second generations than between subsequent generations, consistent with the lower effective population size of 50 in the second generation. Using Eq. 3.42 to obtain the harmonic mean of the effective population sizes, we would expect this simulated population to exhibit the same amount of genetic drift over four generations as a population with a constant effective population size of 50 individuals. This example makes clear that even with small effective population sizes, the change in heterozygosity can be quite small and difficult to estimate with precision unless many individuals and many loci are sampled or a log time period is observed, or both.

The allele frequencies and heterozygosities for the 10 simulated loci seen in Figure 3.20 are given in Tables 3.4 and 3.5. Table 3.4 works through the computations to estimate $\hat{N}_e$ for the 10 replicate loci using the standardized variance in allele frequency over time. In a similar fashion, Table 3.5 shows the steps to estimate $\hat{N}_e$ from change in heterozygosity. Neither estimate of $\hat{N}_e$ is equal to the known effective population size of 50, with the allele frequency estimate at about half of the true value and the heterozygosity estimate about six times larger than the true value. What contributed to this error in both

Table 3.4 Data from simulated allele frequencies in Figure 3.20 used to estimate the effective population size after one generation of genetic drift. Initial allele frequencies were $p = q = 0.5$ at all loci. The change in allele frequency between generation one and generation two is used to estimate the standardized variance in allele frequency ($\hat{F}_c$) and thereby the effective population size ($\hat{N}_e$). Because these are simulated data, any impact of finite sample sizes on $\hat{F}_c$ is ignored. The true N_e is 50.

$p_{t=1}$	$p_{t=2}$	$\hat{F}_c$
0.5	0.4901	0.0004
0.5	0.5857	0.0293
0.5	0.4038	0.0371
0.5	0.5369	0.0054
0.5	0.5254	0.0026
0.5	0.5994	0.0395
0.5	0.4353	0.0167
0.5	0.5357	0.0051
0.5	0.6152	0.0531
0.5	0.4897	0.0004

multilocus average $\hat{F}_c = 0.0190$

$\hat{N}_e = \frac{1}{2\hat{F}_c} = 26.4$

Table 3.5 Simulated allele frequencies in Figure 3.20 used to estimate the effective population size after one generation of genetic drift. Here, the heterozygosity in generation one and generation two is used to estimate the effective population size ($\hat{N}_e$) using the exact and approximate equations for the relationship between heterozygosity and effective population size. Initial allele frequencies were $p = q = 0.5$ at all loci, so $H_{t=1} = 0.5$. The true N_e is 50.

$H_{t=1}$	$H_{t=2}$	$\hat{N}_e = \frac{H_1}{2(H_1 - H_2)}$	$\ln\left(\frac{H_{t=2}}{H_{t=1}}\right)$	$\hat{N}_e \approx -\frac{1}{2}\frac{1}{\ln\left(\frac{H_{t=2}}{H_{t=1}}\right)}$
0.5	0.4998	1250.0	−0.0004	1249.7
0.5	0.4853	17.0	−0.0298	16.8
0.5	0.4815	13.5	−0.0377	13.3
0.5	0.4973	92.6	−0.0054	92.3
0.5	0.4987	192.3	−0.0026	192.1
0.5	0.4803	12.7	−0.0402	12.4
0.5	0.4916	29.8	−0.0169	29.5
0.5	0.4975	100.0	−0.0050	99.7
0.5	0.4735	9.4	−0.0545	9.2
0.5	0.4998	1250.0	−0.0004	1249.7
multilocus average $\hat{N}_e = 292.4$				296.5

estimates? The simulated data used to make the estimates probably does not fit the model assumptions perfectly. Using only 10 replicate loci to estimate the effective population size is clearly not enough given the random variation shown by genetic drift. For example, the average Δp over the 10 loci was not zero, even though as the number of replicate loci increases it is expected to be. The estimates also only utilized a single generation of genetic drift. This serves to remind us that assumptions like "many loci" or "over long time periods" may not be biologically realistic because a data set or organisms themselves may not play by the same rules. The violation of model assumptions often leads to imprecise estimates of population genetic parameters, just as we have seen in this example. It also reminds us again of the difference between a parameter and a parameter estimate with its associated error.

Interact box 3.4 Estimating N_e from allele frequencies and heterozygosity over time

Data for Figure 3.20 are available on the text website as a spreadsheet. Use them to estimate $\hat{N}_e$ from the standardized variance in allele frequency and change in heterozygosity over four generations for each of the 10 replicate loci. These calculations are aided by setting up formulae in a spreadsheet. What is the equivalent $\hat{N}_e$ to a fluctuating population of 100–50–100–100 but in a population that remains constant in size over four generations? Is the estimate of $\hat{N}_e$ close to this expectation? What might contribute to error in the resulting estimate of $\hat{N}_e$?

An R script carries out the genetic drift simulation used to generate Figure 3.20 and produces similar plots. Use the script to examine genetic drift for more loci or more generations. Do estimates of $\hat{N}_e$ improve?

Revisiting the topic of gametic disequilibrium from Chapter 2 will help in understanding an additional method to estimate the effective population size that has found wide application in empirical studies. Chapter 2 explained that finite population size, or genetic drift, can cause departure from two locus

Hardy–Weinberg expected genotype frequencies, or gametic disequilibrium. We can turn this relationship around to make an estimate of the strength of genetic drift based on the observed levels of gametic disequilibrium using the relationship

$$\hat{N}_e = \frac{1}{3\left(\hat{\rho}^2 - \hat{\rho}_{cf}\right)} \tag{3.63}$$

where $\hat{\rho}^2$ is an estimate of the squared two locus disequilibrium correlation and $\hat{\rho}_{cf}$ is an estimated correction factor for the amount of disequilibrium contributed by processes other than genetic drift (see Figure 2.20) combined with contributions caused by some types of data artifacts. The disequilibrium contributed by the finite size of a sample used to estimate $\hat{\rho}^2$ itself is estimated by $\hat{\rho}_{cf} = \frac{1}{S}\left[1 - \frac{1}{(2S-1)^2}\right]$, where S is the sample size of individuals (reviewed in Sved et al. 2013).

Estimates of $\hat{N}_e$ based on gametic disequilibrium observed in samples of multiple unlinked loci have been widely employed in empirical studies, partly because $\hat{\rho}^2$ can be estimated from a single sample of genotypes rather than the two samples required for change in allele frequency or heterozygosity. An additional advantage is that $\hat{N}_e$ based on two locus disequilibrium provides an estimate of the strength of genetic drift from the recent past rather than the deeper past. This is because observed disequilibrium will reflect mostly recent genetic drift (as well as any other ongoing processes) that generate disequilibrium since recombination restores two locus equilibrium rapidly (see Figure 2.19), erasing the impact of processes that caused disequilibrium in the past but are no longer currently acting. Waples (2005) gives a detailed discussion of temporal averaging and interpretation of $\hat{N}_e$ estimates. More details on $\hat{N}_e$ estimation from gametic disequilibrium and the correction factor can be found in Waples (2006) and Hamilton et al. (2018).

The Wright–Fisher model defines drift as occurring in organisms that reproduce only once, so the effective population size is a per generation quantity. Organisms that reproduce repeatedly (or once but asynchronously) have overlapping generations where genetic drift can be thought of as an age- or stage cohort-specific parameter that applies to the temporal subpopulation of individuals in a single age cohort. A version of the effective population size that represents the strength of genetic drift acting on an age cohort population is termed the **effective number of breeders,** or N_b (Waples 1989). With age-structure, each age cohort is the product of genetic drift that took place in the population of parents through random age-specific mortality (the expected probability of surviving into the next age class), and random variation in reproduction (expected fecundity at a given age) (Felsenstein 1971). Overlapping generations can influence estimates of N_e and N_b estimated using temporal changes in allele frequency (Waples and Yakota 2007) or genetic disequilibrium (Waples et al. 2014).

In a widely cited review of 56 species, mostly vertebrates, Frankham (1995) showed that effective population size estimates averaged one-tenth of the census population size when accounting for unequal breeding sex ratio, variance in family size, and fluctuating population size over time. Since that influential review, model development has continued (Wang et al. 2016), and many more empirical estimates have been made using molecular genetic marker data. Empirical studies increasingly include more estimation models, attempts to estimate or control for the impacts of processes such as gene flow and population differentiation, and clearer distinctions between $\hat{N}_e$ and $\hat{N}_b$. Table 3.6 provides some estimates of the effective population size in a range of species. While obtaining precise estimates remains challenging, $\hat{N}_e$ and $\hat{N}_b$ are often on the order of ten or one hundred for many populations and species. There are also additional methods to estimate the effective population size as a composite parameter (N_e multiplied by another parameter) that captures the net action of genetic drift coupled simultaneously with another process such as gene flow, mutation, natural selection, or recombination (Charlesworth 2009). These composite parameters are covered in detail in other chapters.

Breeding effective population size

Up to this point, an implicit assumption has been that a population is an easily recognized and discrete entity. In species where individuals are more or less continuously distributed over large areas, there are no obvious physical or geographic boundaries that define populations. Instead, populations can be defined by average mating and dispersal patterns among individuals that result in limits to the movement of gametes each generation. Based on the size of the breeding and dispersal area, there is the **breeding effective population size**, which is particularly suitable for populations where individuals may occur relatively uniformly over large areas and not form discrete aggregations. Imagine a large, continuous plant population that covers many hectares (plants are a good example since they stay in the same place over time, but the concept applies to all organisms). Now, imagine examining all of the successful mating events for many individuals. The distances of mating events would show a frequency distribution like that in Figure 3.21. The critical feature of the distribution is that the frequency or chance of mating drops off on average as the physical distance between individuals increases. This is a widely observed phenomenon called **isolation by distance** (Wright 1943a). With enough distance

Table 3.6 Empirical estimates of effective population size ($\hat{N}_e$) or effective number of breeders ($\hat{N}_b$) for various species. These estimates are based on a range of methods and assumptions that impact their interpretation.

Species	$\hat{N}_e$	References
Grizzly bear (*Ursus arctos*)	213 – 319	Kamath et al. (2015)
Buzzard (*Buteo buteo*)	25 – 500	Mueller et al. (2016)
Brook trout (*Salvelinus fontinalis*)	40 – 227	Ruzzante et al. (2016)
Brown trout (*Salmo trutta*)	31 – 326	Serbezov et al. (2012)
Review of 305 estimates[a]	104 ± 39	Palstra and Ruzzante (2008)
	267 ± 65	

[a]Mean ± SD from published studies with the larger mean from a subset of 140 considered to have less estimation model bias.

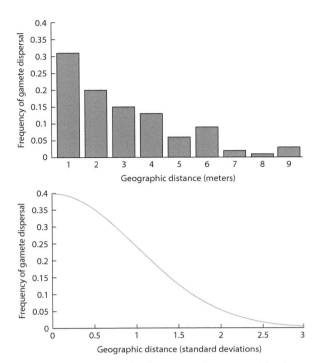

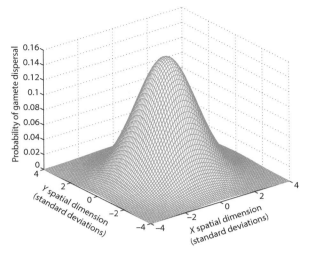

Figure 3.22 An ideal two-dimensional normal distribution used to model the size of genetic neighborhoods and to estimate the breeding effective size (N_e^b) of demes within continuous populations. The radius of the distribution is twice the standard deviation in total gamete dispersal in a generation. The actual physical dimensions of the distribution could range from just a few meters to hundreds or thousands of kilometers.

Figure 3.21 Isolation by distance is characterized by the declining probability of gamete dispersal with increasing geographic distance. The specific shape of the gamete dispersal by distance curve may vary (top), but it is often modeled using a normal distribution (bottom). In a normal distribution, about 95% of observations are expected to fall within ±2 standard deviations from the mean. Empirical estimates of mating and progeny movement in a generation can be used to estimate the variance and thereby the standard deviation of overall dispersal in gametes in order to estimate the area of a genetic neighborhood.

separating them, two individuals have a low probability of mating and can be considered members of distinct genetic populations even if they are not located in geographically distinct populations. The distance required for reproductive isolation by distance may be on the order of meters or thousands of kilometers depending on the species.

Estimating the breeding effective population size depends on the probability distribution of gamete dispersal in space and can be approximated by a normal distribution (other distributions can be used as well). Recall that two standard deviations on either side of the mean contains about 95% of the observations in a normal distribution (see the Appendix). The standard deviation in dispersal, which is the square root of the variance in dispersal, can then be used to describe the probability that a gamete disperses in one dimension (Figure 3.21). Extending this to gamete dispersal into two dimensions, the dispersal

area can be thought of as a circle with the average individual at the center. Since the circle describes a probability distribution, a radius of twice the standard deviation in dispersal in a generation will sweep an area containing about 95% of the observations in a two-dimensional normal distribution (Figure 3.22). This two-dimensional normal distribution model of gamete dispersal probability combined with the average density of individuals is used to quantify the breeding effective population size. Since the area of a circle is πr^2 where r is two standard deviations (or twice the square root of the dispersal distance variance):

$$N_e^b = \pi \left(2\sqrt{\text{dispersal variance}} \right)^2 d \qquad (3.64)$$

which simplifies to

$$N_e^b = 4\pi(\text{dispersal SD})d \qquad (3.65)$$

where d is the average density of individuals. The area described by Eq. 3.61 is known as the **genetic neighborhood** in continuous populations. Multiplying the genetic neighborhood (an area) by the density of individuals (individuals per unit area) gives the breeding effective population size in individuals (see Wright 1943a, Crawford 1984).

Breeding effective population size (N_e^b): The number of individuals found in a genetic neighborhood defined by the variance in gamete dispersal.
Deme: The largest area or collection of individuals where mating is (on average) random.
Genetic neighborhood: An area or a subunit of a population within which mating is random.
Isolation by distance: Decrease in the probability of mating and dispersal of gametes as physical distance increases.

In a classic study, Schaal (1980) estimated the breeding effective population size in Texas bluebonnets (*Lupinus texensis*). This species is pollinated by bumblebees and occurs in large continuous populations that cover many hectares (see color images on the text web page in the section for Chapter 3). An experimental population was constructed using 91 plants with known genotypes at the phosphoglucose isomerase-1 enzyme locus (*Pgi-1*). Mating distances were estimated from seven central plants (homozygous for the *Pgi-1* fast allele) by genotyping progeny from plants without the *Pgi-1* fast allele in the experimental population. Since gametes are also dispersed in seeds every generation, the passive dispersal of seeds was tracked as well. Dispersal distances were very limited, with gametes moving via pollen an average of 1.82 m and seeds moving an average of 0.58 m. The genetic neighborhood size for the experimental population of *L. texensis* was estimated as 6.3 m², containing a breeding effective population size of 95.4 individual plants.

The term **deme**, the largest area or collection of individuals where mating is (on average) random, is often applied to continuous populations and is closely connected with the concepts of breeding effective population size and genetic neighborhoods. The bond orbital in chemistry serves as a metaphor for the deme in population genetics. A bond orbital describes the probability that an electron will be found at some location in space. Similarly, a deme describes the probability that an average individual will move its gametes some distance in space by mating or dispersal in a generation. Members of the same deme are considered able to mate at random, whereas members of two different demes have a low probability (<5%) of mating and are, therefore, members of separate populations. It is important to distinguish demes, or genetically defined population demarcations, from geographically defined populations. From the perspective of effective population size and predicting the behavior of genetic variation, deme is a more useful definition for populations than geographic or spatial definitions.

Effective population sizes of different genomes

Plastid genomes (mitochondria and chloroplasts) are an example where the effective population size is lower than that of the nuclear genome of the same organism. The effective population size of plastid genomes is reduced by two independent factors. First, these genomes are haploid (one copy of the genome per plastid) compared to the two homologous copies of each diploid nuclear chromosome. In addition, most plastids are inherited by offspring from one parent only via the gamete cytoplasm (uniparental inheritance). In species with two equally frequent sexes, uniparental inheritance causes plastid genomes to have half the effective population size of genomes inherited from all possible parents. These two factors combined cause animal mitochondrial genomes and plant mitochondrial and chloroplast genomes to have one-quarter of the effective size of the diploid, biparentally inherited nuclear autosomes. Thus, loci in these genomes experience a higher rate of genetic drift compared to loci in the nuclear genome. Genetic marker studies frequently take advantage of this fact, using plastid genome marker loci to study phenomena such as recent population divergence due to genetic drift where nuclear marker loci would show much less divergence because of a larger effective population size.

3.6 Gene genealogies and the coalescent model

- Genetic drift as a time-backward process.
- Lineages in the present all descent from a most recent common ancestor in the past.
- Exponential distribution approximation for coalescent event waiting times.
- Modeling the branching of lineages to predict the time to the most recent common ancestor.

At this point in the chapter, we will develop an additional approach to model genetic drift based around lineage branching or gene genealogy. Initially, it is necessary to introduce some basic terminology and concepts used in this approach. Although it may not be evident at first, the lineage-branching approach to population genetics has a great deal in common with the material in the first two sections of this chapter. The immediate goal of this section is to establish and motivate the building blocks necessary to model lineage branching events. The next section of this chapter will then show how the concept of effective population size applies in genealogical branching models.

Is it possible to learn something about the past population genetic processes that produced the genetic patterns seen in a sample of individuals? The answer is yes, if we have models of ancestor–descendant relationships, or genealogy, that allow us to predict identity by descent in the past based only on knowledge of the present. With such models, we look at patterns among the individuals available to us in the present and reconstruct versions of events such as genetic drift, gene flow, mutation, or natural selection in the past that could have led to the individuals in the present. These models are referred to collectively as **coalescent theory** since the perspective of the models is to predict the probability of possible patterns of genealogical branching working back in time from the present to the point of a single common ancestor in the past. When two lineages trace back in time to a single ancestral lineage, it is said to be a **coalescent event**, hence the term coalescent theory. This section describes what is called the n-coalescent or sometimes the Kingman coalescent after Sir John Kingman (1982a, 1982b) who proved the existence of the coalescent process and showed that it holds for a wide range of conditions.

A major advantage of coalescent models is that the action of population genetic processes on the branching pattern of lineages is independent of the allelic states of the lineages. More details about the allelic states of lineages and allele frequencies in genealogies are developed in Chapter 5, but, for the current section, bear in mind that each lineage represents an independent copy of a discrete allele or a DNA sequence. Once both the branching processes and the mutation processes that change allelic states are brought together, the coalescent approach serves to make testable predictions for the evolution of DNA sequences under a combination of population genetic processes.

Before moving on with a more formal description, let us consider a metaphor for the coalescence process. Imagine a sealed box full of bugs. Each bug moves around the box at random. Whenever two bugs meet by chance, one of them (picked at random) completely eats the other one in an instant. When a bug is eaten, the population of bugs decreases by one, and the remaining bugs continue to move about the box at random. The time that elapses between bug meetings tends to get longer as the number of bugs in the box gets smaller. This is because chance meetings between bugs depend on the density of bugs in the box. Eventually, the entire box that was full of bugs initially will wind up holding only a single bug after some time has passed. Each bug is analogous to a lineage, and one bug eating another is analogous to a coalescent event. The very last bug is analogous to the lineage that is the most recent common ancestor. If time ran backwards from the last bug to the initial state of many bugs, that would be analogous to the process of genetic drift fixing one allele in a population.

Tracing the pattern of ancestry for allele copies in a finite population experiencing genetic drift provides a means to understand the ancestor–descendant patterns in those allele copies. Figure 3.23 shows an example of random sampling caused by genetic drift in a population of 10 lineages (equivalent to 10 haploid allele copies or 5 diploid individuals) over 10 generations. In the figure, we can trace ancestor–descendant relationships forward in time to see that random variation in reproductive success lead to the single blue lineage becoming the ancestor of all of the descendants in generation 10, an example of fixation caused by genetic drift. In the same population, we can also trace events backward in time to see that the blue lineages in any of the later generations are all descendants of blue lineages in earlier generations. There are a few instances where several blue lineages descend from the same common ancestor in the prior generation.

A central concept in coalescent theory is connecting a group of lineages in the present back through time to a single ancestor in the past. This single ancestor is the first ancestor (going backward in time) of all the lineages in a sample of lineages in the present time and is referred to as the **most recent common ancestor** or **MRCA**. Section 3.2 develops a time-forward model of genetic drift that

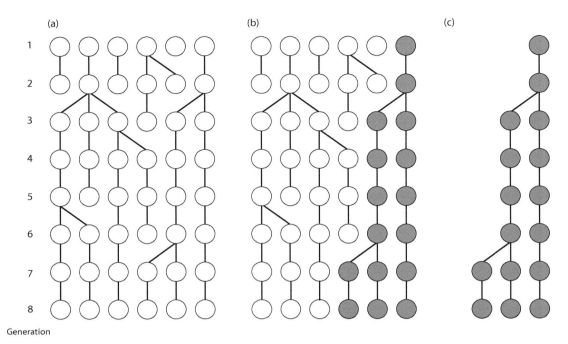

Figure 3.23   An example of sampling of lineages over time in a Wright–Fisher population. Circles show haploid allele copies and lines show the patterns of ancestry. Each generation, sampling with replacement leads to some lineages leaving no descendants and other lineages leaving numerous descendants (A) We can imagine sampling three lineages in the present (blue circles in B). Working backward in time from the present to the past reveals the ancestral links that connects only the sampled lineages, or the genealogy of the sampled lineages (C). In this example, the randomly sampled blue lineages share a most recent common ancestor (MRCA) seven generations in the past. Genealogical probability models describe the distribution of times between sampling events that lead to merging/splitting of lineages as well as the branching patterns (topology) of genealogies.

Interact box 3.5 Sampling lineages in a Wright-Fisher population

Simulating random sampling in a small population under the Wright–Fisher population model helps to see both time-forward and time-backward perspectives. Genetic drift can be viewed as a prospective process where we can predict allele frequency over time when thinking forward in time. We can also view genetic drift retrospectively where a branching process connects those lineages found in the present to fewer and fewer ancestral lineages as time moves backward.

These two aspects of genetic drift can be visualized with several Wright–Fisher simulators, as well as with an R script.

predicts a sample of alleles (or lineages) will eventually arrive at fixation or loss. Fixation is reached by random sampling that expands the numbers of a given lineage or allele in the population. The lineage that reaches fixation can be traced back to a single ancestor at some point in the past. In the process of reaching fixation, a population loses all lineages except one, the one that was fixed by genetic drift. This same genetic drift process can be viewed from a time-backward perspective. A sample of lineages in the present must eventually be the product of a single ancestral lineage at some point back in the past that happened to become more frequent under random sampling. The coalescent model turns the random sampling process around, asking: what is the probability that two lineages in the present can be traced back to a single lineage in the previous generation? Answering this question relies on the same probability tools that were used earlier in the chapter to describe the process of genetic drift.

Coalescence or coalescent event: The point in time where a pair of lineages or genealogies trace back in time to a single common ancestral lineage (to coalesce literally means to grow together or to fuse).
Genealogy: The record of ancestor–descendant relationships for a family or locus.
Lineage A line of descent or ancestry for a homologous DNA sequence or a locus (regardless of whether or not copies of the locus are identical or different).
Most recent common ancestor (MRCA): The first common ancestor of all lineages (or gene copies) at some time in the past for a sample of lineages taken in the present.
Gene copy or **allele copy:** A replicated DNA sequence that has passed from an ancestor to a descendant; used synonymously with the term lineage.
Waiting time: The mean or expected time back in the past until a single coalescence event in a sample of lineages.

A schematic representation of the ancestor–descendant process for two generations can be seen in Figure 3.24A for a set of haploid lineages. Using the rules of random sampling based around the Wright–Fisher model (and its assumptions), we can develop a prediction for the number of generations back in time until two lineages "find" their MRCA or coalesce to a single lineage. Consider a random sample of two of the $2N$ total lineages in the present generation. Given that one of these two sampled lineages finds its ancestor in the previous generation, what is the probability that the other lineage also shares that same common ancestor such that a coalescent event occurs? Given that one of the lineages has a given common ancestor, for coalescence to occur, the other lineage must have the same ancestor among the $2N$ possible ancestors in the previous generation. Thus, the probability of coalescence is $\frac{1}{2N}$ for two lineages, whereas the probability that two lineages do not have a common ancestor in the previous generation is $1 - \frac{1}{2N}$.

The coalescent sampling process just developed for haploid lineages can also be extended to approximate the process of reproduction for diploid lineages. In

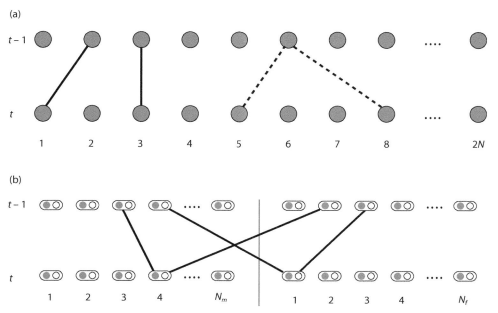

Figure 3.24 Haploid (top, A) and diploid (bottom, B) reproduction in the context of coalescent events. In a haploid population, the probability of coalescence is $\frac{1}{2N}$ (dashed lines), while the probability that two lineages do not have a common ancestor in the previous generation is $1 - \frac{1}{2N}$ (solid lines). In a diploid population, the two gene or allele copies in one individual in the present time have one ancestor in the female population (N_f) and one ancestor in the male population (N_m). Coalescent events in the diploid population arise when the gene copies in males and females are identical by descent. The haploid model with $2N$ lineages is used to approximate the diploid model with $2N = N_f + N_m$ diploid individuals.

diploid reproduction, each offspring is composed of one allele copy inherited from a female parent and another allele copy from a male parent (Figure 3.24B). In a time-backward view, this can be thought of as reproduction where one allele copy finds its ancestor in the male population of the last generation while the other allele copy finds its ancestor in the female population of the last generation. For a given male or female parent, each of their two allele copies has a probability of ½ of being the ancestral copy. As long as the number of males and females in a diploid population is equal and the haploid and diploid population sizes are large, the predictions of the coalescent model are very similar for haploid and diploid populations containing an identical total number of gene copies. The haploid model is more straightforward, and so it is used throughout this section. In practice, the predictions that follow from the haploid coalescent model can be applied to samples of lineages (usually DNA sequences) from diploid organisms.

Like Markov chains, the probability of coalescence displays the Markov property since it is an independent event that depends only on the state of the population at the point of time of interest. Because of this, the basic probabilities of coalescence and noncoalescence between two generations can be used to describe the probability of coalescence over an arbitrary number of generations. If two randomly sampled lineages do not coalesce for $t-1$ generations, then the probability that they do coalesce to their common ancestor in generation t is:

$$\left(1 - \frac{1}{2N}\right)^{t-1} \frac{1}{2N} \qquad (3.66)$$

For example, in a population of $2N = 10$, the chance that two randomly sampled lineages coalesce in four generations is the product of the probability of three generations of not coalescing ($\left(1 - \frac{1}{10}\right)^3 = 0.729$) and the chance of coalescing between any two generations (1/10), which gives a probability of coalescence of 0.0729. The distribution of the probabilities of a coalescent event occurring for two lineages in each of 20 generations for the case of $2N = 4$ is shown in Figure 3.25. It is important to note that only a single coalescent event is possible each generation since we are only considering two lineages.

In practice, the probabilities of coalescence are approximated using an exponential function (see Math box 3.2). As we have seen, the exact probability of coalescence for a pair of lineages is $\frac{1}{2N}$ and the probability of not coalescing is $1 - \frac{1}{2N}$ each generation. The exponential approximation $1 - e^{-\frac{1}{2N}t}$ gives the *cumulative* probability of a pair of lineages coalescing at or before generation t. This probability is symbolized as $P(T_C \leq t)$, where T_C is the generation of coalescence and t is the maximum time to coalescence being considered. Let us use an example of the probability of coalescence at or before four generations have passed in a population of $2N = 10\,000$. The exact probability is the sum of the probabilities of coalescence in each generation, $P(T_C \leq 4) = P(T_C = 1) + P(T_C = 2) + P(T_C = 3) + P(T_C = 4)$. Substituting in

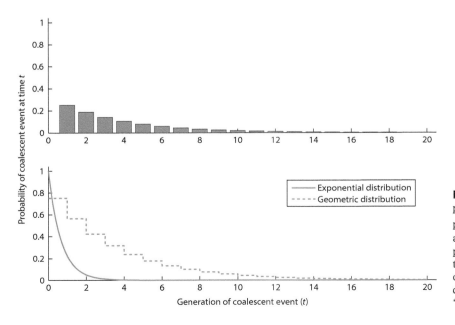

Figure 3.25 The distribution of the probabilities of coalescence over time predicted by $\left(1 - \frac{1}{2N}\right)^{t-1} \frac{1}{2N}$ (top) for a population of $2N = 4$. The bottom panel shows the approximations of these coalescence probabilities based on geometric and exponential distributions with a probability of "success" of 1/4.

expressions for the exact probability of coalescence at each of these four time points gives $P(T_C \le 4) =$

$$\left(1 - \frac{1}{10,000}\right)^0 \frac{1}{10,000} + \left(1 - \frac{1}{10,000}\right)^1 \frac{1}{10,000}$$
$$+ \left(1 - \frac{1}{10,000}\right)^2 \frac{1}{10,000} + \left(1 - \frac{1}{10,000}\right)^3 \frac{1}{10,000}$$

$= 0.0004$. Using the exponential approximation instead, $1 - e^{-\frac{1}{10,000}4} = 0.000\ 3999\ 2$, as the chance that a pair of lineages experiences a coalescence at or before four generations elapses. Thus, the exact probability and the approximation are in close agreement. The exponential approximation of the exact probability improves as the population size increases. See Wakeley (2009) for a more detailed discussion on the relationship between population size and the error of the exponential approximation of coalescence probabilities.

Approximating probabilities of coalescence with the exponential distribution makes computing more practical and also yields several generalizations about the coalescence process. In particular, the geometric and exponential probability distributions can be used to obtain an approximate average and variance for coalescence times. It turns out that, for both types of distribution, the mean time to an event is simply the inverse of the probability of an event occurring. In the coalescence process, the probability of coalescence for a pair of lineages is $\frac{1}{2N}$, so the average time that elapses until coalescence is $2N$ when the coalescence process is approximated with the exponential distribution. The average time to a coalescent event is often called the **waiting time**. Another generalization is that the range of individual coalescence times around that average is quite large. Based on the exponential distribution, the variance in the waiting time is $4N^2$ so that the range of coalescence times around the mean grows rapidly as the size of the population increases. Thus, the length of branches connecting lineages to their ancestors will be highly variable about the mean value. This can be seen in Figure 3.26, which shows six independent realizations of the coalescent tree for six lineages.

It is possible to determine the average time for more than two lineages to find their MRCA. Suppose we want to determine the waiting time for k lineages, where k is less than or equal to the total number of lineages sampled from a population of $2N$. To see the problem in detail, let us consider the case of $k = 3$ lineages. When no coalescence events occur, one lineage finds its ancestor among any of the $2N$ individuals in the previous generation. That means the

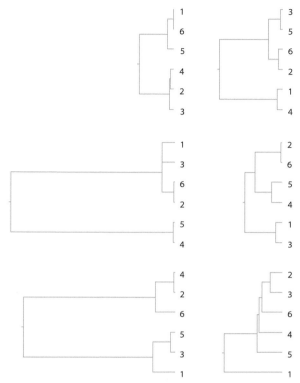

Figure 3.26 Six independent realizations of the coalescent tree for six lineages. All six trees are drawn to the same scale. Each genealogy exhibits coalescent events between random pairs of lineages. The differences in the height of the trees is due to random variation in waiting times. Because of this random variation, average times to coalescence for a given number of lineages are only approached in a large sample of independent trees.

next lineage must find its ancestor among $2N - 1$ individuals in the previous generation and the final lineage must find its ancestor among $2N - 2$ possible parents. Thus, the probability of non-coalescence is

$$\prod_{x=0}^{k-1} \left(1 - \frac{x}{2N}\right) \tag{3.67}$$

If the number of lineages sampled is much smaller than the total number of lineages in the population ($2N$), then the probability of non-coalescence for k lineages can be approximated by

$$1 - \left(\frac{k(k-1)}{2}\right)\left(\frac{1}{2N}\right) \tag{3.68}$$

where $\frac{k(k-1)}{2}$ enumerates the different ways to uniquely sample pairs of lineages from a total of k lineages. (Note that the number of unique pairs of lineages can also be determined with "k draw 2" or $\begin{pmatrix} k \\ 2 \end{pmatrix} = \dfrac{k!}{2!(k-2)!}$.) The probability of a coalescence for all of the possible unique pairs of the k lineages is then the product of the number of unique pairs and the probability of coalescence of each pair

$$\left(\frac{k(k-1)}{2}\right)\left(\frac{1}{2N}\right) \qquad (3.69)$$

Bringing these two probabilities together into an equation like 3.62 gives the probability that k lineages experienced a single coalescent event t generations ago

$$\left(1 - \left(\frac{k(k-1)}{2}\right)\left(\frac{1}{2N}\right)\right)^{t-1}\left(\frac{k(k-1)}{2}\right)\left(\frac{1}{2N}\right) \qquad (3.70)$$

Since this probability also follows an exponential distribution ($e^{-t\left(\frac{k(k-1)}{2}\right)\left(\frac{1}{2N}\right)}$), the average time to coalescence for k lineages in a population of $2N$ is $\dfrac{2N}{\frac{k(k-1)}{2}}$. For example, if $k = 3$ and $2N = 10$, the average time to coalescence is $3\frac{1}{3}$ generations. This is one-third of the average waiting time for two lineages since each of the three unique pairs of lineages (1–2, 1–3, and 2–3) can independently experience coalescence. Figure 3.27 shows the average coalescence times for six lineages based on this same logic. The general pattern is that coalescence times decrease when more lineages are present since there are a larger number of lineage pairs that can independently coalesce.

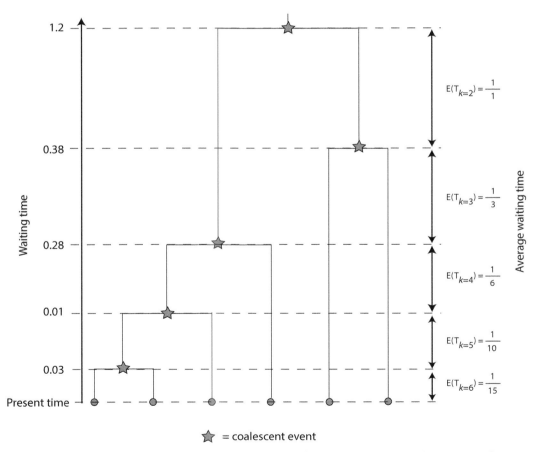

Figure 3.27 A schematic coalescent tree for six lineages shows one realization of waiting times (left y-axis) and average waiting times (right y-axis). With a sample of k lineages, the expected or average time to coalescence is $\dfrac{2N}{\frac{k(k-1)}{2}}$ since each of the independent $\dfrac{k(k-1)}{2}$ pairs of lineages can coalesce. Waiting times can be expressed as continuous time in units of 2N generations. Multiplying the continuous time scale by 2N gives time to coalescence in units of population size. E refers to "expected" and T refers to time to coalescence so that $E(T_k)$ is the expected or average time to coalescence for k lineages. The basic patterns seen in all coalescent trees apply to populations of all sizes, although the absolute time for coalescent events depends on N. Genealogy is not drawn to scale.

Math box 3.2 Approximating the probability of a coalescent event with the exponential distribution

The series of failures (non-coalescence) until a success (coalescence) in the genealogical process can be modeled where time is continuous (a real number) rather than discrete (an integer). The exponential distribution describes situations in which an object initially in one state can change to an alternative state with some probability that remains constant through time. The exponential distribution could be applied to the time until one of the many light bulbs fails, for example, as well as to the time until a coalescent event in a population of lineages (see Figure 3.25). The exponential distribution is described by

$$\text{Probability of change} = ae^{-at} \tag{3.71}$$

where a is the constant probability of changing states in a time interval of one, t is time, and e is the mathematical constant base of the natural logarithm ($e = 2.718\,28 \ldots$). The exponential distribution has a mean of

$$E(x = t) = \frac{1}{a} \tag{3.72}$$

and a variance

$$E(x) = \frac{1}{a^2} \tag{3.73}$$

Under the assumption that the effective population size $2N$ is large (and constant) such that the probability of coalescence ($\frac{1}{2N}$) is small, the probability that a coalescent event occurs

$$\left(1 - \frac{1}{2N}\right)^{t-1} \frac{1}{2N} \tag{3.74}$$

can be approximated using the exponential distribution

$$\frac{1}{2N} e^{-t\frac{1}{2N}} \tag{3.75}$$

for two lineages where t is time in generations. Notice that the constant in the exponential expression (equivalent to the a in Eq. 3.71) is $\frac{1}{2N}$. That means that the average time to coalescence is $\frac{1}{\frac{1}{2N}} = 2N$, and the variance in time to coalescence is $\frac{1}{\left(\frac{1}{2N}\right)^2} = 4N^2$. Setting the exact probability of coalescence approximately equal to the exponential gives an expression

$$\left(1 - \frac{1}{2N}\right)^{t-1} \frac{1}{2N} \approx \frac{1}{2N} e^{-\frac{t}{2N}} \tag{3.76}$$

that can be simplified by canceling the $\frac{1}{2N}$ constant term on both sides

$$\left(1 - \frac{1}{2N}\right)^{t-1} \approx e^{-\frac{t}{2N}} \tag{3.77}$$

Therefore, the exponential distribution approximates the probability of non-coalescence at each time t.

To determine the chances that a coalescence event occurred at or before some time, call it $T_C \leq t$ where T_C is the time to coalescence and t is time scaled in units of $2N$ generations, the cumulative exponential distribution is used. The cumulative distribution

$$P(T_C \leq t) \approx 1 - e^{-t\frac{k(k-1)}{2}} \tag{3.78}$$

effectively adds up the probabilities of coalescence (as one minus the chance of non-coalescence) as time increases to express the probability of a time interval passing without experiencing a coalescent event. The cumulative distribution approaches 1 as time increases since the chance of non-coalescence decreases toward 0 with increasing time. The probability of coalescence increases more rapidly toward 1 for larger numbers of lineages k. This cumulative distribution is used to determine the waiting times needed to construct coalescent trees.

When approximating the probability of coalescent events with the exponential distribution, it is a standard practice to put coalescence times on a continuous scale of units of $2N$ generations. To see how this continuous time scale operates, let j be the time measured as a real number (e.g. 1.0, 1.1, 1.2, 1.3 ... j) in generations. The time to coalescent events t can then be expressed as $t = j/(2N)$. As an example, imagine that a coalescence event occurred at $t = 1.4$ on the continuous time scale. That coalescence event could also be thought of as occurring (1.4) $(2N) = 2.8N$ generations in the past (see Figure 3.27). If the population size was $2N = 100$ lineages, then that coalescent event was (1.4) $(100) = 140$ generations in the past. However, if the population size was $2N = 20$ lineages, then that coalescent event was $(1.4) (20) = 28$ generations in the past. Aside from the practical matter of interpreting a specific time value, the use of a continuous time scale makes an important biological point about the effects of population size on the coalescence process. The basic nature of the coalescent process is identical for all populations no matter what their size. For example, in populations of any size, a single coalescent event will occur faster on average when there is a larger number of lineages sampled than when just a pair of lineages is sampled. Population size just serves to scale the time required for coalescent events to occur. Coalescent events occur more rapidly in small populations compared to bigger populations, a conclusion analogous to that reached for genetic drift earlier in the chapter.

Interact box 3.6 Build your own coalescent genealogies

Building a few coalescent trees can help you to understand how the exponential distribution is put into practice to estimate coalescence times, as well as give you a better sense for the random nature of the coalescence process. You can use a Microsoft Excel spreadsheet to calculate the quantities necessary to build a coalescent genealogy. (As an alternative, you can employ an R script to generate waiting times for a genealogy using the same methods.) The spreadsheet contains the cumulative exponential distributions for a time interval passing without experiencing a coalescent event (see Eq. 3.74) for up to six lineages. To determine a coalescence time for a given number of lineages, k, a random number between 0 and 1 is picked and then compared with the distribution when $\frac{k(k-1)}{2}$ lineages can coalesce. The time interval on the distribution that matches the random number is taken as the coalescence time.

Step 1. Open the spreadsheet and look over all the quantities calculated. Click on cells to view the formulae used, especially the cumulative probability of coalescence for each k. This will help you understand how the equations in this section of the chapter are used in practice. View and compare the cumulative probability distributions graphed for $k = 6$ and $k = 2$.

Step 2. Press the recalculate key(s) to generate new sets of random numbers (see Excel help if necessary). Watch the waiting times until coalescence change. What is the average time to coalescence for each value of k? How variable are the coalescence times you observe in the spreadsheet when changing the random numbers with the recalculate key?

Step 3. Draw a coalescence tree using the coalescence times found in the spreadsheet (do not recalculate until Step 6 is complete). Along the bottom of a blank sheet of paper, draw six evenly spaced dots to represent six lineages. Starting at the top of the random number table, pick two lineages which will experience the first coalescent event. Label the two leftmost dots with these lineage numbers. Then, draw two parallel vertical lines proportional to the waiting time to coalescence (e.g. if the time is 0.5, draw lines that are 0.5 cm). Connect these vertical lines with a horizontal line. Assign the lineage number of one of the coalesced lineages to the pair's single ancestor, at the horizontal line. Record the other lineage number on a list of lineages no longer present in the population (skip over these numbers if they appear again in the random number table). There are now $k - 1$ lineages.

Step 4. Use the random number table to get the next pair of lineages that coalesce. Remember that the single ancestor of lineage pairs that have already coalesced will eventually coalesce with one of the remaining lineages. If one of the lineages of this random pair matches a previous pair's ancestor, begin at the horizontal line indicating that pair's coalescence, and draw a vertical line toward the top of the paper that is the length of the coalescence time for the number of lineages remaining. Draw a vertical line from lineage n to an equal height and connect the two vertical lines with a horizontal line (the line from lineage n will be as long as the sum of all coalescence times to that point). If neither number matches a previous pair's ancestor, draw the branches as in Step 3, beginning at the baseline, but this time add this pair's particular coalescence time to the sum of previous coalescence times to find the vertical branch height.

Step 5. Repeat the process in Step 4 until all lineages have coalesced.

Step 6. Then, add together all of the times to coalescence to obtain the total height of the coalescent tree and sum the height of all of the branches to obtain the total branch length of the tree. How do these compare with the average values for a sample of six lineages?

Step 7. Press the recalculate key combination to obtain another set of coalescence times and repeat Steps 3 through 5 to create another coalescence tree. Draw several coalescence trees to see how each differs from the others. You should obtain coalescence trees like those in Figure 3.25. Your trees will differ from these, because the random coalescence times vary around their average, but the overall shape of your trees will be similar.

This section will conclude by considering several measures of coalescent trees useful to summarize general patterns of the coalescence process. The total time from the present to the point in the past where all k sampled lineages find their MRCA is called the **height of a coalescent tree**. The height of a tree for k sampled lineages is just the sum of the coalescence waiting times as coalescent events reduce the number of lineages from k to $k - 1$ to $k - 2$ down to one. The mean or expected value of the height of a coalescent tree is then

$$E(H_k) = \sum_{i=2}^{k} E(T_i) = 2 \sum_{i=2}^{k} \frac{1}{i(i-1)} = 2\left(1 - \frac{1}{k}\right)$$

(3.79)

where time is in continuous units of multiples of $2N$. Thus, the average height in continuous time of a coalescent tree starts at $2(1-1/2) = 1$ for a pair of lineages and approaches 2 as k grows very large. In time units of $2N$ generations, the average tree height starts at $2N$ and increases to $4N$ as k grows. The average total waiting time for all lineages to coalesce is heavily influenced by the average waiting time for the last pair of lineages to coalesce. Stated another way, average waiting times are shorter when k is larger and increase as lineages coalesce and k gets smaller. This result makes sense since there are fewer and fewer independent pairs of lineages that can coalesce as k decreases (think of $\frac{k(k-1)}{2}$ as k decreases to 2).

The variance in the height of a coalescent tree is the sum of the variances in the coalescence time for each set of coalescent events since all coalescent events are independent:

$$\text{var}(H_k) = \sum_{i=2}^{k} \text{var}(T_i) = 4\sum_{i=2}^{k} \frac{1}{i^2(i-1)^2} \quad (3.80)$$

The variance rapidly approaches the value of about 1.16 as the number of lineages sampled, k, increases to large values. It is important to note that the variance in the height of a genealogy is scaled in time units of $(2N)^2$, rather than the $2N$ time units that apply to the average height of a genealogy

(see Math Box 3.2). The finite maximum value of the variance and the fact that the variance in tree height for $k = 2$ lineages is 1.0 highlight the major impact of the coalescence time of the last pair of lineages on the overall height of a coalescence tree. Figure 3.28 illustrates the variance in the total height of genealogies by displaying the time to MRCA for 1000 replicate genealogies each starting with $k = 6$. The range of time to MRCA is large, and the distribution has a very long tail representing a small proportion of genealogies that take a very long time for all coalescence events to occur.

Another useful measure is the total branch length of a coalescent genealogy of k lineages or L_k. The total branch length is the sum of the waiting times represented along each lineage in the tree. If the branches represented a system of roadways, then the sum of the distances traversed when driving over each segment of road once would be analogous to the total branch length. For example, if a pair of lineages has a waiting time x until coalescence, then the total waiting time is $x + x = 2x$. The average total branch length of a genealogical tree is then just twice the sum of the average waiting times (using continuous time) for each coalescence event:

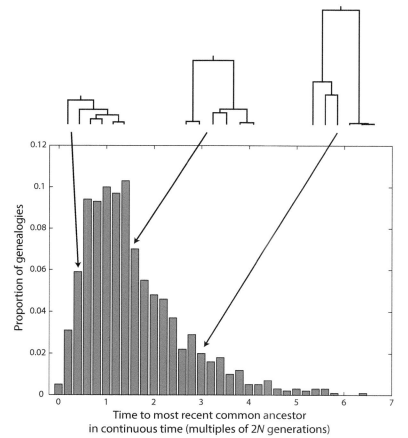

Figure 3.28 The distribution of times to a MRCA (or genealogy heights) for 1000 replicate genealogies starting with six lineages ($k = 6$). The distribution of total coalescence times has a large variance because the range of times is large and also asymmetric with a long tail of a few genealogies that take a very long time to reach the MRCA. The genealogies shown above the distribution are those for the 10th, 50th, and 90th percentile times to MRCA.

$$E(L_k) = \sum_{i=1}^{k-1} 2\frac{1}{i} = 2\sum_{i=1}^{k-1} \frac{1}{i} \qquad (3.81)$$

For example, with $k = 2$ lineages, $i = 1$ in Eq. 3.77 gives an expected total branch length of two. This result makes sense because two lineages are expected to coalesce in $2N_e$ generations or one time unit on the continuous time scale. Multiplying this expected time to coalescence by two for the two independent lineage branches gives a total branch length of two. The expected total branch length starts at two, increasing with greater k and never reaching a maximum. However, the expected total branch length of a genealogy grows more and more slowly as k increases since the expected time to coalescence decreases with increasing k.

This section on the n-coalescent describes the probability model for retrospective lineage branching in a single population. The n-coalescent can be extended to include the influence of numerous population genetic processes in addition to genetic drift. Later sections and chapters will explore how changes in population size (the next section), subdivided populations that experience gene flow (-Chapter 4), mutation (Chapter 5), and natural selection (Chapter 7) influence patterns of coalescence. The application of coalescent theory to DNA sequence data, including examples based on empirical data, will be covered in Chapter 8.

3.7 Effective population size in the coalescent model

- The coalescent model of effective population size.
- Coalescent genealogies with sex ratio and changing population size.
- Coalescent genealogies and population bottlenecks.
- Coalescent genealogies in growing and shrinking populations.

This section will explore the meaning of effective population size in the coalescent model. As in the models of genetic drift developed previously in this chapter, effective population size also plays a critical role in coalescence models. In the context of the coalescent model, the effective population size determines the chance that two gene copies descend from the same ancestor when working back in time from the present to the past.

In the coalescent model, two randomly sampled gene copies have the probability $\frac{1}{2N_e}$ of finding their MRCA in the previous generation. If we call this

the probability of coalescence, P_C, then for a diploid population,

$$P_C = \frac{1}{2N_e} \qquad (3.82)$$

This equation can be rearranged as:

$$N_e = \frac{1}{2P_C} \qquad (3.83)$$

to give the inbreeding effective population size for the coalescent model over one generation. The effective size of a haploid population is defined identically except that the population contains N instead of $2N$ gene copies. (The reasoning used is parallel to that used to arrive at the inbreeding effective population size based on the probability of identity by descent in a finite population.)

Interact box 3.7 Simulating gene genealogies in populations with different effective sizes

Simulating lineage branching events forward in time for populations of different sizes is a direct way to understand how sampling leads to patterns of lineage coalescence. Use the link on the text web page to reach a simulation to model lineage branching events in finite populations.

Start by simulating populations of **N : 4** and **N : 10**. Before pressing the run button (the arrow that points to the right), determine the expected time for all of the gene copies in the present to find a single most recent common ancestor. Use your answer to intelligently set the number of generations (the **G:** entry field) to run the simulation to be able to see most of the lineage coalescence events (note that the maximum number of generations is 30). Run 25 simulations for each population size and tabulate the number of generations for all lineages in the present to find their most recent common ancestor. What does the distribution of coalescence times look like?

Use 100% for a **Speed** value to carry out replicate runs rapidly. The "untangle" button (the leftmost button at the bottom) rearranges the lineages so that the branches do not overlap and is very useful when tracing individual genealogies to visualize common ancestors.

We can apply the coalescent definition of the effective population size to the case of the breeding sex ratio in a population. For the coalescent model, it helps to think of the two sexes as two separate populations where gene copies can find their ancestors (see Figure 3.24). In a diploid population with two sexes and a 1 : 1 breeding sex ratio, half of the gene copies reside in females and half reside in males. A coalescence event requires that two gene copies descend from a single ancestor, either a single male or single female individual. The total probability that two randomly sampled gene copies coalesce in the previous generation, P_C, is the sum of the probability that the coalescence was in the population of females and the probability that the coalescence was in the population of males or

$$P_C = P_C(\text{female population}) + P_C(\text{male population})$$
$$(3.84)$$

Regardless of whether or not the two gene copies come from the male or female population, the probability of coalescence is the chance of sampling the same gene copy twice in the previous generation or $\frac{1}{2N_e^i}$. To take the populations of the two sexes into account, notice that the probability that a gene copy is sampled from either the female or male population is ½. The probability that *both* gene copies are sampled from the population of the same sex is, therefore, (½) (½) = ¼. Putting these probabilities together into Eq. 3.83 gives

$$P_C = \frac{1}{4}\frac{1}{2N_{ef}} + \frac{1}{4}\frac{1}{2N_{em}}$$
$$(3.85)$$

where the effective size of the female and male populations sum to the total effective size ($N_e^i = N_{ef}^i + N_{em}^i$). Based on the definition of the effective population size as the probability of coalescence in Eq. 3.63

$$\frac{1}{2N_e^i} = \frac{1}{4}\frac{1}{2N_{ef}} + \frac{1}{4}\frac{1}{2N_{em}}$$
$$(3.86)$$

which after some rearrangement looks like

$$\frac{1}{2N_e^i} = \frac{1}{8}\left(\frac{N_{em}}{N_{ef}N_{em}} + \frac{N_{ef}}{N_{ef}N_{em}}\right)$$
$$(3.87)$$

$$N_e^i = 4\frac{N_{ef}N_{em}}{N_{em} + N_{ef}}$$
$$(3.88)$$

This leads to an expression for the inbreeding effective population size in terms of the number of males and females. This equation shows us that the effective size of diploid population with two sexes and an equal breeding sex ratio ($N_{ef}^i = N_{em}^i$) is equivalent to an ideal random mating Wright–Fisher population and also gives a means to express the effective population size when the breeding sex ratio is not equal. We also see that the coalescence model allows us to reach exactly the same result that was obtained using the identity by descent approach (Eq. 3.44 and Section 3.4).

The probability that two randomly sampled gene copies do not coalesce, P_{NC}, over some number of generations can be used to show the overall effective population size when the population size is not constant. As the basis of modeling coalescent times with the exponential distribution (the continuous time coalescent), $e^{-\frac{t}{2N_e}}$ is used to approximate $1 - \frac{1}{2N_e}$ as long as N_e does not get too small. This means that P_{NC} can be approximated by an exponential function:

$$P_{NC} = \left(1 - \frac{1}{2N_e}\right)^t \approx e^{-\frac{t}{2N_e}}$$
$$(3.89)$$

where t is the number of generations. Imagine that a population fluctuates in size over three generations as considered in Section 3.3. In such a situation, the probability of two randomly sampled gene copies not experiencing a coalescence event over the three generations could be approximated by

$$P_{NC} \approx e^{-\frac{t}{2N_e}} = e^{\left(-\frac{1}{2N_{e(t=1)}}\right) + \left(-\frac{1}{2N_{e(t=2)}}\right)}$$
$$+ \left(-\frac{1}{2N_{e(t=3)}}\right)$$
$$(3.90)$$

where each term in the exponent of e is distinct since the population sizes fluctuate over time. The exponential terms of e can be solved for the effective population size by taking the natural log of both sizes to eliminate e

$$-\frac{t}{2N_e} = \left(-\frac{1}{2N_{e(t=1)}}\right) + \left(-\frac{1}{2N_{e(t=2)}}\right)$$
$$+ \left(-\frac{1}{2N_{e(t=3)}}\right)$$
$$(3.91)$$

and then multiplying both sides by $1/t$ and then -2 to get

$$\frac{1}{N_e} = \frac{1}{t}\left(\frac{1}{N_{e(t=1)}} + \frac{1}{N_{e(t=2)}} + \frac{1}{N_{e(t=3)}}\right) \quad (3.92)$$

The term on the right side is the harmonic mean of the effective population size as given in Eq. 3.42. Again, the identity by descent and coalescence approaches lead to the same conclusions about effective population size.

Coalescent genealogies and population bottlenecks

Let us examine the distribution of coalescence times in a population that experiences a bottleneck to see how genealogical branching patterns are affected. The situation in Figure 3.29 is analogous to that in Section 3.3 where a population starts out with 100 individuals, is reduced to 10 individuals for one generation, and then returns to a size of 100 individuals in the third generation (see Figure 3.16). For a bottleneck of 100–10–100 over three generations, Eqs. 3.85 and 3.86 can be used to determine the probability of coalescence events. A 100–10–100 population is equivalent to a population of 25 that is constant in size. Therefore, the average probability of two randomly sampled gene copies not finding their common ancestor over the three generations spanning the bottleneck is:

$$P_{NC} \approx e^{-\frac{3}{2(25)}} = 0.9418 \quad (3.93)$$

giving a chance of coalescence of $1 - P_{NC} = 1 - 0.9418 = 0.0582$. This probability is equivalent to $1 - \frac{1}{2N_e}$ for three generations of 100–10–100 or $P_{NC} = (0.995)(0.95)(0.995) = 0.9405$ to give $1 - P_{NC} = 1 - 0.9405 = 0.0595$. (Note the close agreement of the two answers even though a population of 10 is very small and violates the assumption that $e^{-\frac{t}{2N_e}}$ is a good approximation of $1 - \frac{1}{2N_e}$.)

Compare the probability of coalescence $P_C = 1 - 0.995 = 0.005$ for a population of 100 and $P_C = 1 - 0.95 = 0.05$ for a population of 10. During the bottleneck, there is a 10-fold greater chance each generation that two lineages find their ancestor or coalesce than before or after the bottleneck, leading to shorter branch lengths (times to coalescence). Thus, the bottleneck causes increased sampling of lineages and a greater chance that ancestral lineages are lost to sampling. After the bottleneck, the probability of coalescence returns to the same probability as in the first generation. However, the lineages present after the bottleneck are now much more likely to be recently related, so the overall probability of coalescence is like that in a smaller population of constant size.

After a population recovers from a bottleneck, it is possible that the lineages present will all descend from a most recent common ancestor during the bottleneck period. If this occurs, it is equivalent to saying that a single lineage among those present in the pre-bottleneck population becomes the most recent common ancestor of all lineages during the bottleneck. The chance that this occurs increases as the bottleneck

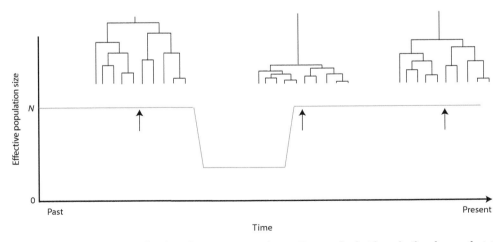

Figure 3.29 The effects of a population bottleneck on gene genealogies. During the bottleneck, the chance that two randomly sampled gene copies are derived from one copy in the previous generation ($\frac{1}{2N_e}$) increases. This can also be thought of as a reduction in the overall height of a genealogical tree caused by the bottleneck since lineages that find their ancestors during the bottleneck lead to short branches. The overall effect of a bottleneck on coalescence among gene copies sample in the present depends on the reduction in the effective population size and the duration. The arrows indicate the point in time when gene copies were sampled from the population.

exhibits a smaller population size or persists for a longer period of time. The expected height of a coalescent tree (the sum of all the time periods between coalescence of pairs of gene copies until there is a single lineage, Eq. 3.75) can be used to show this effect quantitatively. In the continuous coalescent, the expected height in units of $2N$ generations is $2\left(1 - \frac{1}{k}\right)$ where k is the number of gene copies in the present. Figure 3.29 shows $k = 10$ so the expected height of each coalescent tree is $(1.8)(2N_e) = 3.6N_e$ generations. Before the bottleneck, a sample of 10 gene copies from a population of $N_e = 100$ would coalesce to a single lineage in an average of 360 generations. The same sample of 10 gene copies taken from a population of $N_e = 10$ would coalesce to a single lineage in an average of 36 generations. At the time point closest to the present in Figure 3.28, the population has an effective size of about 25 based on the harmonic mean as well as the probability of coalescence over three generations. Therefore, the expected height of the coalescent tree is 90 generations.

Coalescent genealogies in growing and shrinking populations

In a population growing in size over time, the probability of a coalescent event is least near the present because the population is at its largest size. Working back in time from the present to the past, the size of the population is continually shrinking. This means that the probability of a coalescence event $\left(\frac{1}{2N}\right)$ must also be continually increasing as we move back in time toward the MRCA, since N is shrinking. The result is that genealogies in growing populations do not follow the rule that the final coalescence time from two lineages to the MRCA is the longest on average in a genealogy established in the last section for populations of constant size. Instead, genealogies in growing populations tend to have longer times between coalescent events toward the present and shorter times between coalescent events in the past (Figure 3.30). Genealogies from rapidly growing populations, therefore, tend to have longer branches

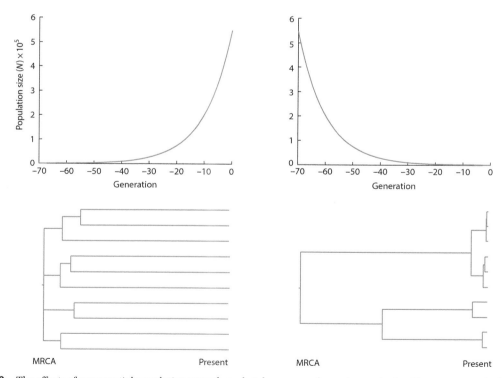

Figure 3.30 The effects of exponential population growth or shrinkage on coalescent genealogies. The upper panels show change in population size over time with exponential growth according to $N(t) = N_0 e^{-rt}$ with $r = \pm 0.1$, yielding relatively slow exponential population growth. The two genealogies illustrate examples of waiting times that might be seen under strong exponential population growth (left) and shrinkage (right). With strong exponential population growth, coalescent times are longest in the present when the population is the largest, leading to genealogies characterized by long branches near the present and very short branches in the past around the time of the most common recent ancestor (MCRA). With exponential population shrinkage, coalescence times are greatest in the past near the MRCA when the population was larger and shortest near the present when the population is at its smallest size. The genealogy on the lower left was obtained using Eq. 3.88 with $r = 100$.

toward the present and shorter branches deep in the tree.

In a population shrinking in size over time, the probability of a coalescent event is greatest near the present because the population size is at its smallest. The effect on genealogies is that coalescent waiting times in the present are even shorter than they would be under constant population size. Similarly, when population size is shrinking, coalescence times in the past are relatively greater than under constant population size because the probability of a coalescence event was greater (Figure 3.30).

A common way to model growing or shrinking populations is to assume that population size is changing exponentially over time. Under exponential growth, the population size at time t in the past is a function of the initial population size in the present N_0 and the rate of population growth r according to

$$N(t) = N_0 e^{-rt} \qquad (3.94)$$

Examples of population size over time under exponential population growth are shown in Figure 3.30. With exponential growth, population size tends to change very rapidly.

The generalizations above regarding coalescent waiting times depend on rapid and sustained changes in population size over time such as under exponential population growth with a constant rate (r). In populations that are changing in size slowly over time, it is possible that coalescence waiting times differ very little from those expected under constant population size. Increases in population size over time cause the probability of coalescence to decrease toward the present. At the same time, the chance of a coalescence event increases toward the present simply due to a larger number of lineages (increasing k) available to coalesce. Only in

populations with rapidly changing population size will the reduction in the probability of coalescence be great enough to overcome the effect of an increasing number of lineages available to coalesce toward the present. In addition, the variance in coalescence waiting times with constant N is large so that only very rapid and sustained change in population size will noticeably impact the distribution of coalescence waiting times.

Changing population size over time complicates finding the distribution of coalescent times. In the previous section, the exponential distribution of coalescence waiting times was obtained by relying on the fact that the probability of non-coalescence, or $\prod_{x=0}^{k-1} \left(1 - \frac{x}{2N}\right)$ where k is the number of lineages available to coalesce, was a constant since N remains the same over time. If N is instead changing rapidly over time, then the probability of non-coalescence also changes rapidly over time. The result is that the distribution of coalescent times cannot be exponentially distributed when population size is changing over time as it is when population size is constant over time. The waiting time between coalescent events in an exponentially growing population is obtained from the equation:

$$t_{k \to k-1} = \ln \left[1 + (N0r)e^{-\tau_i}\left(\frac{-2}{k(k-1)}\right) \ln (U) \right] \qquad (3.95)$$

where U is a uniformly distributed random variable between 0 and 1, N_0 is the initial population size, r is the rate of population growth, and τ_i is the sum of the all past coalescence waiting times up to the current number of lineages k that have not yet coalesced according to

---

Interact box 3.8 Coalescent genealogies in populations with changing size

The coalescent process can be simulated for populations experiencing exponential growth in population size through time. The simulation displays a genealogy based on values for the number of lineages (**n:**) and the growth rate of the population (**exp:**). The growth rate parameter corresponds to the rate of exponential population growth and is, therefore, the key variable in this simulation. Click on the **Recalc** button to begin a new simulation. The resulting genealogy can be viewed as an animation going back in time (click on the **Animation** tab at the top left and use playback controls at bottom) or as a genealogy (click on the **Trees** tab at the top left).

Set the growth rate parameter to zero and click on **Recalc**. View the resulting genealogy. Then, set the growth rate parameter to 128 and click on **Recalc** and view the results. How do the genealogies in growing populations compare with genealogies when population size is constant over time? Rerun the simulation a few times for both values of the growth rate parameter. Also, try other less extreme values of the growth rate parameter, such as 4, 8, and 32.

$$\tau_i = \sum_{k=n}^{i+1} t_k \qquad (3.96)$$

(Slatkin and Hudson 1991). In Eq. 3.95, time is scaled in units of r since $\tau = rt$. As τ_i gets larger, then more time has elapsed meaning that the size of the population has changed more. Note that Eq. 3.95 only applies to populations growing though time. In populations shrinking in size toward the present, there is a chance that a single MRCA will never be reached since the probability of coalescence approaches zero as the population size approaches infinity going back in time from the present to the past.

3.8 Genetic drift and the coalescent with other models of life history

· The Moran model of overlapping generations.
· Multiple merger coalescent models.
· Coalescent effective population size.

To conclude the chapter, it is worth considering alternatives to the Wright–Fisher model that show how genetic drift works in the context of more complex life histories. These additional models can be compared with the Wright–Fisher model to provide insights on how genetic drift could operate in other contexts.

One extension called the Moran model (Moran 1958) allows for a type of overlapping generations where all lineages except one survive from one time period to the next (Figure 3.31A). The population is constant in size over time with $2N$ haploid lineages, and each lineage carries either an A_1 or A_2 allele. Let i represent the number of A_1 alleles in the population with the frequency of A_1 then $p = i/2N$. In each time period, two lineages are sampled at random, the first lineage both survives and also reproduces an identical copy of itself bearing the same allele while the second lineage dies. Because of one reproduction and one death at each time period, the number of copies of one of the alleles in the population can increase by one, decrease by one, or remain the same number. (Sampling is with replacement, and if the same individual is sampled twice, it would reproduce and then die such that the size of the population and the allele frequencies would remain the same.) The number of A_1 alleles increases by one if the first lineage sampled is an A_1 (it survives and reproduces) and the second lineage sampled is an A_2 (it dies) with probability

$$p_{i \to i+1} = \frac{i}{2N} \frac{(2N - i)}{2N} \qquad (3.97)$$

The number of A_1 alleles decreases by one if the first lineage sampled is an A_2 (it survives and

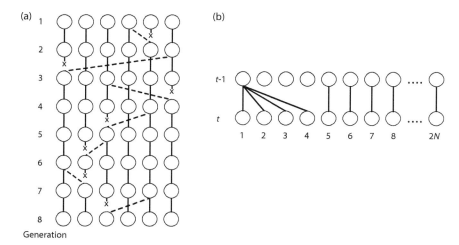

Figure 3.31 Additional models of finite sampling approximate different life history features and serve as alternatives to the Wright–Fisher model. In the Moran model (A), all individuals survive from one generation to the next (solid lines) except one random lineage that dies (x'ed) and is replaced by another random lineage that both survives and also produces one offspring (dashed lines). In the multiple merger coalescent (B), coalescence events can involve three or more lineages simultaneously descending from the same ancestor. Multiple mergers serve to model variance in reproduction success among individuals following distributions other than the Poisson and can include the possibility of several coalescence events occurring simultaneously.

reproduces) and the second lineage sampled is an A_1 (it dies) with probability

$$p_{i \rightarrow i-1} = \frac{(2N-i)}{2N}\frac{i}{2N} \qquad (3.98)$$

The number of allele copies will remain the same if two A1 lineages or two A2 lineages are sampled with probability

$$p_{i \rightarrow i} = \left(\frac{i}{2N}\right)^2 + \left(\frac{(2N-i)}{2N}\right)^2 \qquad (3.99)$$

Because random sampling of two allelic states under the Moran model can be represented as a Markov chain for the number of alleles in a population over time, these transition probabilities can be used to construct a matrix of transition probabilities with eigenvalues that show

$$1 - \frac{2}{(2N)^2} \qquad (3.100)$$

is the rate of change in allele frequencies as they accumulate at the absorbing boundaries of zero and one by genetic drift. In the Moran model, the expected rate of loss of heterozygosity under genetic drift is, therefore, expected to be:

$$H_t = \left(1 - \frac{2}{(2N)^2}\right)^t H_0 \qquad (3.101)$$

In a population of size $2N$ lineages where sampling occurs according to the Moran model, the rate and pattern of genetic drift will be identical to a population of size N lineages that follows Wright–Fisher model sampling (as N grows very large so that $1/N$ and $1/N^2$ do not differ), meaning that drift is twice as strong in the Moran model. This difference is due to changes in the distribution of family sizes rather than any difference in random sampling from overlapping generations (Moran and Watterson 1959; Feldman 1966). We can see this by comparing the variance in reproductive success in the Moran and Wright–Fisher models. For the Moran model, we can find the variance following steps similar to those in Math Box 3.1, using the transition probabilities above with the definitions $p = i/2N$ and $q = (2N-i)/2N$ and multiplying by the squared change in the copies of the A_1 allele for each case

$$\sigma^2 = p(1-p)(+1)^2 + (1-p)p(-1)^2 + (p^2+q^2)(0)^2 \qquad (3.102)$$

which simplifies to

$$\sigma^2 = 2pq \qquad (3.103)$$

or twice the reproductive variance that characterizes the Wright–Fisher model with binomial sampling.

The Moran model also forms the basis of a time-backward genealogy just like the Wright–Fisher coalescent. The conclusions drawn for the Kingman coalescent apply as well to a coalescent with Moran model reproduction but with a different time scale. A Moran model coalescent has exponentially distributed waiting times scaled in units of $\frac{(2N)^2}{2}$ (the Kingman coalescent time scale of $2N$ divided by $2/(2N)$). This difference in time scale is due to the fact that the Moran model requires an average of $2N$ time steps for all individuals to die and be replaced where all individuals are replaced in each time step of the Wright–Fisher model.

The role of variation in family size in the Wright–Fisher model and in the definition of N_e has been covered throughout this chapter, pointing out that the Wright–Fisher model uses Poisson-distributed family size as a standard of reference and showing how family size variation can increase or decrease the strength of genetic drift and thereby alter the effective population size. The coalescent model has been extended in several ways to include the impacts of highly unequal reproduction among individuals, as might be the case for organisms that can produce very large numbers of offspring such as some species of arthropods, fish or plants. This is accomplished through the possibility of **multiple mergers**, or coalescent events where three or more lineages simultaneously descend from the same ancestor in the previous time period (Figure 3.31B). The so-called Ξ-coalescent (the uppercase Greek letter xi) is the most general version where more than one merger of three or more lineages can occur each generation (Möhle and Sagitov 2001). The Λ-coalescent (the uppercase Greek letter lambda) permits only one merger per time interval which can be a multiple merger (Donnelly and Kurtz 1999; Pitman 1999; Sagitov 1999). The ψ-coalescent (ψ is the lowercase Greek letter *psi*) is a variation of the Moran model where there is a chance that the reproductive parent has multiple offspring that replace multiple random lineages that die (Eldon and Wakeley 2006).

Multiple merger models highlight that the Kingman coalescent is a special case of a more general version of the coalescent model, with the constraints that only pairs of lineages can coalesce in single events each time period and that there is relatively low random variation in the production of offspring among individuals. Based on this, Sjödin et al. (2005) suggested a **coalescent effective population size** that exists when the ancestral branching process for a sample of lineages converges to Kingman's coalescent in the limit as population size tends to infinity (see also Wakeley and Sargsyan 2009). Under this definition, if a coalescent effective population size exists, then all aspects of genetic variation in a sample of lineages should conform to the predictions made using the Kingman coalescent scaled for the population size and other parameters of interest. If a coalescent effective population size does not exist, as is the case for some types of multiple merger models, then the predictions of the Kingman coalescent will not describe the ancestral branch pattern accurately even with rescaling. Multiple merger coalescent models can describe types of genetic drift that are not accurately summarized by the Wright–Fisher model, including sweepstakes reproduction in species that are able to produce very large numbers of progeny, a lifecycle of recurrent bottleneck and recolonization events, and strong natural selection on beneficial mutations (Tellier and Lemaire 2014). Some of these applications will be taken up in later chapters.

Chapter 3 review

- In finite populations, allele frequencies can change from generation to generation since the sample of gametes that found the next generation may not contain exactly the same number of each allele as the previous generation. The chance of a large change in the number of alleles decreases as the number of gametes sampled increases. Sampling error in allele frequency causes genetic drift, the random process where all alleles eventually reach fixation or loss.
- The Wright–Fisher model is a simplification of the biological life cycle used to model genetic drift. It makes assumptions identical to Hardy–Weinberg in addition to assuming that each generation is founded by sampling $2N$ gametes from an infinite pool of gametes. The binomial distribution can be used to predict the probability that a Wright–Fisher population goes from some initial number of alleles to any number of alleles in the next generation.

- The action of genetic drift in a very large number of identical replicate populations can be modeled with a Markov chain model (based on the binomial distribution). The model tracks the probabilities of a population with any number of copies of a given allele transitioning to all possible numbers of copies of the same allele in the next generation. Markov models predict that the chances of fixation or loss are equal when $p = q = 1/2$ and that genetic drift reduces genetic variation by $1 - \frac{1}{2N}$ every generation.
- The process of genetic drift in many replicate populations can be thought of as analogous to the diffusion of particles in space. This leads to the diffusion approximation of genetic drift, where the rate at which individual populations reach fixation or loss depends on the diffusion coefficient: a function of population size and allele frequency. The diffusion equation predicts that an allele at an initial frequency of ½ will remain segregating for an average of about $2.8N$ generations.
- The size of a population is defined by the behavior of allele frequencies over time. The effective population size (N_e) is the size of an ideal Wright–Fisher population that shows the same allele frequency behavior over time as an observed biological population regardless of its census population size (N).
- Finite population size and mating among relatives mating are analogous processes since both lead to increasing homozygosity and decreasing heterozygosity over time. The distinction is that genetic drift in finite populations causes changes in both genotype *and* allele frequencies (alleles are lost and fixed) while consanguineous mating changes only genotype frequencies.
- Numerous models predict dynamics of genetic variation based on the effective population size (N_e). As a consequence, there are several definitions of N_e, including the variance effective population size, the inbreeding effective population size, the breeding effective population size, and the coalescent effect population size.
- The effective population size can be estimated from empirical estimates of change in allele frequencies over time, change in heterozygosity over time, or the amount of genetic disequilibrium in a population.
- The genetic effective size (N_e) of a population is often less than the census population size (N) in actual populations because the many ways that finite sampling can occur.
- The average time for a pair of lineages to coalesce is the same as the population size or $2N_e$ with a

large expected range around this average (the variance is $4N_e^2$). In a sample of k lineages, the average time to the first coalescent event is $2N_e$ divided by the number of unique pairs of lineages $(\frac{k(k-1)}{2})$.

- Most coalescent events for a sample of lineages occur in the recent past with only a few lineages having long times to coalescence.
- The effective population size (N_e) can be defined for lineage branching models by reference to the probability of two randomly sampled gene copies descending from the same ancestral copy. This probability decreases as the effective size of populations grows larger. The coalescent model leads to definitions of the inbreeding effective population size that are identical to those obtained using autozygosity.
- Exponential population growth changes the distribution of coalescence times relative to a population with constant population size. When population size is growing, lineages nearest the present tend to have the longest coalescence waiting times because the probability of coalescence grows steadily smaller toward the present. When population size is shrinking rapidly, the probability of coalescence grows steadily greater toward the present causing lineages nearest the present to have the shortest coalescence waiting times.
- Alternatives to the Wright–Fisher model include the Moran model of overlapping generations and multiple merger coalescent models, some versions of which make distinct predictions about the shape and average waiting times in genealogies.

Further reading

For an intriguing account of the role of chance in everyday affairs, see:

Mlodinow, L. (2008). *The Drunkard's Walk: How Randomness Rules our Lives*. New York, NY: Pantheon Books.

For more mathematical details on the binomial and Poisson distributions, Markov chains, and the diffusion equation, see:

Otto, S.P. and Day, T. (2007). *A Biologist's Guide to Mathematical Modeling in Ecology and Evolution*. Princeton, NJ: Princeton University Press.

To learn more about the diffusion equation, its assumptions, and how processes such as mutation, migration, and natural selection can be combined with genetic drift, consult:

Roughgarden, J. (1996). *Theory of Population Genetics and Evolutionary Ecology: An Introduction*. Upper Saddle River, NJ: Prentice Hall.
Ewens, W.J. (2004). *Mathematical Population Genetics. I. Theoretical Introduction*, 2e. New York: Springer-Verlag.

More background on coalescence theory along with many examples and a comprehensive list of references can be found in:

Hein, J., Schierup, M.H., and Wiuf, C. (2005). *Gene Genealogies, Variation and Evolution*. New York: Oxford University Press.
Wakeley, J. (2009). *Coalescent Theory: An Introduction*. Greenwood Village, Colorado: Roberts & Company Publishers.

End of chapter exercises

1. Use the text simulation web site to model genetic drift. (This can also be accomplished with Populus using Model → Mendelian Genetics → Genetic Drift and then the Monte Carlo tab.) Run simulations using the parameter value sets given in the table below and tabulate the results in the spaces provided. Set generation time to 600 generations so all results will be viewed on the same scale (you can reduce generations if the results are difficult to see for some runs).

   Run three sets of simulations for each set of initial conditions to get an idea of how much each run can differ. Fill in the values in the tables below for each set of initial conditions of population size and initial allele frequency.

Population Size	Initial freq(A)	# loci fixed for A	# loci fixed for a	# variable loci	Generation of first and last fix/loss
10	0.5				
10	0.8				
10	0.2				
50	0.5				
50	0.8				
50	0.2				
100	0.5				
100	0.8				
200	0.5				
200	0.8				

What does each line in the graphs represent? What is the relationship between population size and amount of time to fixation or loss of all alleles based on your data? How does initial allele frequency affect time to fixation or loss? Why do the results of each run for a given set of conditions differ slightly?

2   In the simulation program Populus, press the **Model** button and select **Mendelian Genetics** and then **Genetic Drift**. Select the **Markov Model** tab. The model dialog has values for population size and number of A genes. Press the **View** button to see the results graph. Within the results screen, generations are advanced with the **Iterate** button (or the space bar after clicking on **Iterate** once). Run simulations for the values given in the table below and tabulate the results in the spaces provided. Click on the **Iterate** button until almost all of the populations have reached fixation or loss.

Population Size (N)	# A genes	initial freq (A)	~ # generations for 99% fixation	Proportion of populations fixed for A
6	6			
6	3			
12	12			
12	6			
20	20			
20	12			

What do the y-axis values mean? What does each bar in the graphs represent? What is the relationship between population size and amount of time to fixation or loss of all alleles based on your data? How does initial frequency affect time to fixation or loss? What is the ultimate fate of allele frequency in each population? Genetic drift is random – why are the results of each run identical for a given set of conditions?

3   Modify the transition matrix in the R program or Excel spreadsheet of Interact Box 3.2 for a diploid population of three individuals ($2N = 6$). Similar to Table 3.2, set up a matrix where columns represent the initial allelic state and the terms in rows add together to determine the proportion of populations with a given state one generation later. Then, use the binomial formula to calculate the chance that a single population makes each of the allelic state transitions. Think about the problem before carrying out any calculations – it is less work than it may appear at first. Two columns have probabilities of either 0 or 1. Two other columns have the same probabilities but in reversed order. Run the program and examine the graphs.

4   Download the data for Figure 3.20 from the text website. Use the change in allele frequency (Δp) over three generations (time 1 to time 4) for each of the 10 replicate loci to estimate $\hat{N}_e$. The simulated population fluctuated in size over the four generations (100–50–100–100). What is the equivalent $\hat{N}_e$ over three generations in a population that remains constant in size? Is the estimate of $\hat{N}_e$ close to this expectation? What would improve the estimate of $\hat{N}_e$?

5   The heterozygosities for three loci in the three finite populations observed one generation apart are shown in the table below. Use these heterozygosities to estimate the effective population size ($\hat{N}_e$) for each locus and the multilocus average $\hat{N}_e$ for each of the three populations.

Population	Locus	$H(t = 0)$	$H(t = 1)$	$\hat{N}_e$
A	1	0.26	0.246	
	2	0.32	0.33	
	3	0.453	0.43	
B	1	0.28	0.2688	
	2	0.21	0.205	
	3	0.35	0.345	
C	1	0.375	0.384	
	2	0.300	0.292	
	3	0.165	0.161	

How precise are these estimates? Why did the heterozygosity increase in two instances? Does a sample of three loci provide a sound estimate of effective population size? What additional data would improve these estimates?

6   In 1985, Phil Hedrick (Journal of Heredity 76, pp. 127–131) used four coat color phenotypes with a known genetic basis to estimate gametic disequilibrium in cats found in several geographic populations. The loci that contribute to these phenotypes are known to experience high rates of recombination and are therefore not linked. The observed gametic disequilibrium coefficient (D) and correlation ($\hat{\rho}$) for three

locations was 0.0231 and 0.124 in Amsterdam, 0.0249 and 0.110 in Montreal, and 0.0454 and 0.228 in Portsmouth. Sample sizes were 425 in Amsterdam, 325 in Montreal, and 250 in Portsmouth. Estimate the genetic effective population size (N_e) for each location. What processes other than genetic drift and finite sample size might have contributed to observed gametic disequilibrium? How would accounting for these additional processes impact the estimates of N_e?

7   Using the instructions in Interact Box 3.5, build two coalescent genealogies for six lineages ($k = 6$). Draw these genealogies to scale. Provide a time scale in both continuous time and in discrete time units for $2N = 50$. Compare the two resulting genealogies – in what ways are they similar or different?

8   For the two genealogies with k = 6 constructed in the previous question, what is the expected height and the expected variance in time to coalescence in generations on a continuous time scale? What is the height and variance in discrete time if the population size is $2N = 25$? $2N = 500$?

9   Search the literature for a recent research paper that utilizes one or more of the population genetic predictions covered in this chapter. The topic can be any organism, application, or process, but the paper must include a hypothesis test involving a topic related to genetic drift and effective population size. Summarize the main hypothesis, goal, or rationale of the paper. Then, explain how the paper utilized a population genetic prediction from this chapter and summarize the results and conclusions based on the prediction.

10   Construct a simulation model of genetic drift. Instructions to build a spreadsheet model can be found on the text website. These instructions can also be implemented in a programming language such as Python or R.

Problem box answers

Problem box 3.1 answer

In the first generation, both populations have 24 A alleles and 24 a alleles. After one generation, however, there are (0.458) (48) = 22 A alleles in one population and (0.521) (48) = 25 alleles in the other. The chances of observing these allele frequencies can be determined with the binomial formula under the assumptions of the Wright–Fisher model:

$$P_{A=22} = \frac{48!}{22!(26)!}(0.5)^{22}(0.5)^{26}$$
$$= 2.7386 \times 10^{13}(2.3841 \times 10^{-7})$$
$$(1.4901 \times 10^{-8}) = 0.0973$$

$$P_{A=25} = \frac{48!}{25!(23)!}(0.5)^{25}(0.5)^{23}$$
$$= 3.0958 \times 10^{13}(2.9802 \times 10^{-8})$$
$$(1.1921 \times 10^{-7}) = 0.110$$

In both populations, the chance of observing the allele frequency changes under genetic drift is about 10%. The chances of observing these allele frequency changes in both populations is only 0.0107 since the probability of two independent events is the product of their individual probabilities.

Problem box 3.2 answer

Each of the values in the transition matrix is obtained using the binomial formula. The chance that a population at fixation or loss transitions to an allele frequency different than 1 or 0, respectively, is always 0. The chance of transitioning from one to four A alleles is identical to the chance of transitioning from three to no a alleles, since the number of A alleles is four minus the number of a alleles. Using this symmetry permits two columns to be filled out after performing calculations for only one of the columns. The transition probabilities are a function of the sample size only and so are constant each generation. The total frequency of populations in a given allelic state in the next generation depends on initial frequencies of populations in each state ($P_{t=0}(x)$). The expected frequencies of populations in each allelic state, therefore, changes each generation. This Markov chain model is available as a Microsoft Excel spreadsheet on the textbook website under the link to **Problem Box 3.2**.

One generation later ($t = 1$)		Initial state: number of A alleles ($t = 0$)				
State	Expected frequency	0	1	2	3	4
0	$P_{t=1}(0) =$	(1.0000) $P_{t=0}(0)$	+ (0.3164) $P_{t=0}(1)$	+ (0.0625) $P_{t=0}(2)$	+ (0.0039) $P_{t=0}(3)$	+ (0.0000) $P_{t=0}(4)$
1	$P_{t=1}(1) =$	(0.0000) $P_{t=0}(0)$	+ (0.4219) $P_{t=0}(1)$	+ (0.2500) $P_{t=0}(2)$	+ (0.0469) $P_{t=0}(3)$	+ (0.0000) $P_{t=0}(4)$
2	$P_{t=1}(2) =$	(0.0000) $P_{t=0}(0)$	+ (0.2109) $P_{t=0}(1)$	+ (0.3750) $P_{t=0}(2)$	+ (0.2109) $P_{t=0}(3)$	+ (0.0000) $P_{t=0}(4)$
3	$P_{t=1}(3) =$	(0.0000) $P_{t=0}(0)$	+ (0.0469) $P_{t=0}(1)$	+ (0.2500) $P_{t=0}(2)$	+ (0.4219) $P_{t=0}(3)$	+ (0.0000) $P_{t=0}(4)$
4	$P_{t=1}(4) =$	(0.0000) $P_{t=0}(0)$	+ (0.0039) $P_{t=0}(1)$	+ (0.0625) $P_{t=0}(2)$	+ (0.3164) $P_{t=0}(3)$	+ (1.0000) $P_{t=0}(4)$

Problem box 3.3 answer

This captive population has experienced a reduction in population size due to three factors. This case can be thought of as a triple bottleneck. First, the breeding sex ratio became unequal. The effective size based on the number of males (10) and females (15) is:

$$N_e = 4\frac{(10)(15)}{10 + 15} = 24$$

from Eq. 3.44. The mean family size among the 15 females was four, returning the population back to a census size of 60. However, the variance in family size was 6.5 and thus greater than expected for a Poisson distribution. The effective population size based on the variance in family size is:

$$N_e = \frac{4(24)}{6.5 + 16 - 4} = 5.2$$

from Eq. 3.46. Notice that the effective population size used in the numerator is 24, as determined for the unequal sex ratio. In total, the population has fluctuated from $N_e = 60$ before the fire, to census

sizes of 25, and then 60 over three generations. The effective size in generation two was 24 due to the unequal sex ratio. The effective size in generation three was five (after rounding) due to a growing population with high variance in family size. Therefore, the effective population size over three generations is:

$$\frac{1}{N_e} = \frac{1}{3}\left[\frac{1}{60} + \frac{1}{24} + \frac{1}{5}\right]$$

or $N_e = 11.6$. Thus, the fire caused a major reduction in the overall effective population size. The population experienced about the same amount of genetic drift over these three generations as an ideal Wright–Fisher population with an effective size of 12 individuals. One assumption is this analysis is that these animals approximate organisms that breed only once during their lifetimes. This assumption is often unmet for animals, and additional reductions in the effective population size can arise if mating occurs between members of overlapping generations since related individuals can mate (see Nunney 1993 and an application in Ollivier and James 2004).

CHAPTER 4

Population structure and gene flow

4.1 Genetic populations

- Genetic versus geographic organization of populations.
- Isolation by distance and other models of genetic isolation.
- Gene flow and migration.

The expectation that genotypes will be present in Hardy–Weinberg frequencies, covered in detail in Chapter 2, depends on the assumption of random mating throughout a population. Implicit is the view that a population is a single entity where processes such as mating and movement of individuals are uniform throughout, a condition often called **panmixia**. Several processes and features at work in actual populations make this initial perspective of population uniformity unlikely to hold true for many populations. It is often the case that within large populations, the chances of mating are not uniform as assumed by Hardy–Weinberg. Instead, the chance that two individuals mate often depends on their location within the population. This leads to what is called **population structure**, or heterogeneity across a population in the chances that two randomly chosen individuals will mate. The first section of this chapter will introduce biological phenomena that contribute to population structure in mating and migration that can lead to differences in allele and genotypes frequencies in different parts of a population. The goal of the entire chapter is to develop expectations for the impact of population structure on genotype and allele frequencies along with methods to measure patterns of population structure.

To get an initial idea of how a population might be divided into smaller units that behave independently, consider the hypothetical population in Figure 4.1. Initially, all individuals in the population have equal chances of mating regardless of their location. Since mating is random, genotype frequencies in the entire population match Hardy–Weinberg expectations, and allele frequencies are equal on both sides of the creek. Then, imagine that the creek bisecting the population changes permanently into a large river that serves as a barrier to the movement of individuals from one side to the other side. Although some individuals still cross the river on occasion, the rate of genetic mixing or **gene flow** between the two subpopulations bisected by the river is reduced. Lowered levels of gene flow mean that the two subpopulations have allele and genotype frequencies that tend to be independent through time. At later time points in Figure 4.1, the two subpopulations have increasingly different allele frequencies over time due to genetic drift, even though there are Hardy–Weinberg expected genotype frequencies within each subpopulation. In the last time period in Figure 4.1, the allele frequencies in the subpopulations separated by the river are quite different and the genotype frequencies in the total population no longer meet Hardy–Weinberg expectations. In this example, a reduction in gene flow allows the two subpopulations to be acted on independently by genetic drift, ultimately resulting in the population differentiation of allele frequencies. The appearance of geographic barriers restrict gene flow among populations like that in Figure 4.1. Subpopulations – entities recognized with names such as herds, flocks, prides, schools, and even cities – can be formed by a wide range of temporal, behavioral, and geographic barriers that ultimately result in subpopulation allele frequencies that differ from the average allele frequency of the total population.

Another cause of population structure is more subtle, but easy to understand with a thought

Population Genetics, Second Edition. Matthew B. Hamilton.
© 2021 John Wiley & Sons, Inc. Published 2021 by John Wiley & Sons, Inc.
Companion website: www.wiley.com/go/hamilton/populationgenetics

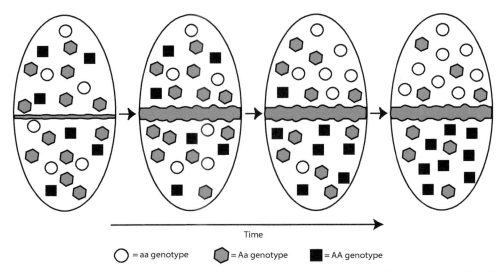

Figure 4.1   An example of population structure and allele frequency divergence produced by limited gene flow. The total population (large ovals) is originally in panmixia and has Hardy–Weinberg expected genotype frequencies. Then, the stream that runs through the population grows into a large river, restricting gene flow between the two sides of the total population. Over time, allele frequencies diverge in the two subpopulations through genetic drift. In this example, you can imagine that the two subpopulations drifted toward fixation for different alleles but neither reaches fixation due to an occasional individual who is able to cross the river and mate. Note that there is random mating (panmixia) within each subpopulation so that Hardy–Weinberg expected genotype frequencies are maintained within subpopulations. However, after the initial time period, genotype frequencies in the total population do not meet Hardy–Weinberg expectations.

experiment. Think of one common species of animal or plant that you encounter regularly at home or work. Think of individuals of this species seeking out mates completely at random. Where would individuals likely find mates? They would probably find mates among the other individuals nearby rather than far away. I thought of the trees that are near my home and also on the university campus where I work. When these trees flower and mate via the movement of pollen, it seems likely to me that trees that are closer are more likely to be mates. I would not expect two trees that are tens or hundreds of kilometers apart to have a good chance of mating. Imagine the species that you thought of and the distances over which mating events might take place. Even if individuals can find mates very far away, there is usually some spatial scale at which the chances of mating are limited. This varies with the species and could be distances as small as a few meters or as large as thousands of kilometers depending on the range of movement of individuals and their gametes.

This phenomenon of decreasing chances of mating with increasing distance separating individuals is termed **isolation by distance** (Wright 1943a, b, 1946; Malécot 1969). Sewall Wright was motivated by data on the spatial frequencies of blue and white flowers of the plant *Linanthus parryae*

(Figure 4.2) to develop expectations for populations experiencing isolation by distance. The patchwork spatial pattern of flower color frequencies in *L. parryae* was considered by Wright as a prime example of the consequences of isolation by distance in continuous populations. Wright (1978) carried out a series of detailed analyses of *L. parryae* flower color frequency data. However, the genetic basis of flower color and the possibility that natural selection in nature shapes the spatial distributions of *L. parryae* flower colors has fueled disagreements for more than 50 years (see Schemske and Bierzychudek 2001; Turelli et al. 2001). Regardless of the processes acting in *L. parryae* specifically, the phenomenon of isolation by distance is ubiquitous in natural populations. Isolation by distance can be considered a null hypothesis in the genetics of natural populations, with the main question being the geographic scale at which it impacts genotype and allele frequencies (see Meirmans 2012).

The patterns of gene flow between and among subpopulations may take many forms, and a range of models capturing this diversity have been described. For example, **isolation by barrier** describes the lower rates of gene flow caused by geographic and habitat obstacles such as the river shown in Figure 4.1 (Vignieri 2005). (In a phylogenetic context, the term **vicariance** is commonly

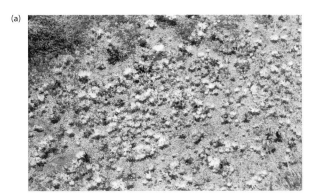

Figure 4.2 The plant *Linanthus parryae* or "desert snow" is found in the Mojave Desert regions of California. *L. parryae* can literally cover thousands of hectares of desert during years with rainfall sufficient to allow the widespread germination of dormant seeds present in the soil. This tiny plant has either blue or white flowers. In some locations, most plants have blue or most have white flowers, while, in other locations, more equal frequencies of the two flower colors are found. Source: Dr. Barbara J. Collins/http://www.clunet.edu/wf.

used to describe similar barriers to migration and gene flow that contribute to speciation.) The **isolation by resistance** (IBR) model allows for variable rates of gene flow, taking its name from the flow of electrons in a circuit as altered by electrical resistors of different strengths (McRae 2006). IBR could be caused by spatial variation in habitats, the different types of matrix (the area surrounding suitable habitat patches over or through which individuals or gametes disperse), as well as by barriers that must be traversed for gene flow to take place. In addition, asymmetric gene flow can be due to **monopolization**, where initial colonizers have a numerical advantage that reduces establishment success of later migrants, which in turn reduces gene flow among subpopulations after their initial establishment (De Meester et al. 2002). Monopolization may result in a pattern of **isolation by colonization** where genetic structure reflects the temporal colonization history and there is not necessarily a relationship between genetic differentiation and geographical distances or environmental differences (Orsini et al. 2013). **Isolation by environment** is the reduction of gene flow and the resulting population genetic differentiation that is independent of geographic distance, as can be caused by a wide range of physical barriers and abiotic variables such as humidity, rainfall, or soil type (Wang and Bradburd 2014). Isolation by environment can also be a product of **isolation by adaptation** where migrants experience reduced establishment or mating success when there is strong natural selection leading to local adaptation causing genetic divergence among populations.

Discriminating among these and other possible causes for rates and patterns of gene flow is a major part of research in **landscape genetics** which employs population genetic predictive models, spatially explicit genotype and allele frequency data, spatial information such as geographic information system (GIS) data around the genetic sampling locations, and spatial statistics to test hypotheses for the geographic processes that shape gene flow and genetic differentiation (Balkenhol et al. 2015). Landscape genetics seeks to identify and test hypotheses for the causes of genetic connections and discontinuities through gene flow as well as to test for correlations between features of the landscape and patterns of genetic differentiation (Manel et al. 2003). One area of focus is the prediction and conservation of dispersal corridors in an effort to preserve gene flow among populations separated by natural habitat variation as well as human impacts such as agriculture, settlements, and structures. For example, Sharma et al. (2013) found that contemporary gene flow was greatest between genetically diverged tiger subpopulations connected by forest corridors.

Computer simulations are a convenient way to explore how isolation by distance influences allele and genotype frequencies. Figure 4.3 shows two simulated populations where each point on a grid represents the geographic location of a diploid individual. In one case, the population exhibits panmixia and individuals find a mate at random from all individuals within a 99×99 individual mating area. In the contrasting case where there is strong isolation by distance, each individual mates at random within a much smaller 3×3 individual area. Both populations start off looking very similar, with

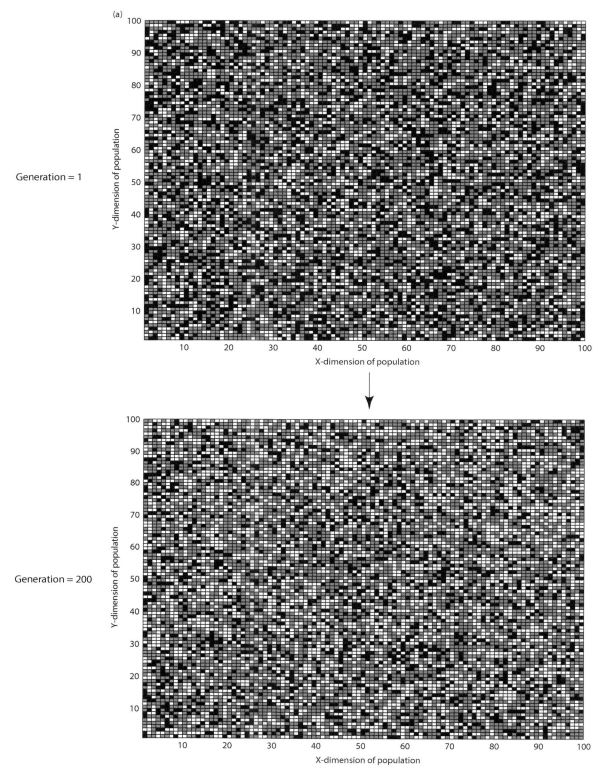

Figure 4.3 Isolation by distance causes spatial structuring of allele and genotype frequencies. In these pictures, a population is represented in two dimensions with each point on a grid representing one diploid individual. The colors represent an individual with a heterozygous (blue) or homozygous (black and white) genotype at each point. In A, there is random mating over the entire population (the mating neighborhood is 99 by 99 squares), while, in B, there is strong isolation by distance (the mating neighborhood is 3 by 3 squares). The population with isolation by distance (B) develops and maintains spatial clumping of genotypes and therefore spatial clumping of allele frequencies. There is no such spatial structure in the population with random mating (A). In the simulation that produced these pictures, the grid is initially populated at random with genotypes at Hardy–Weinberg expected frequencies and $p = q = 1/2$. In each generation, every individual chooses a mate at random within its mating neighborhood and replaces itself with one offspring. The offspring genotypes are determined by Hardy–Weinberg probabilities for each combination of parental genotypes.

(b)

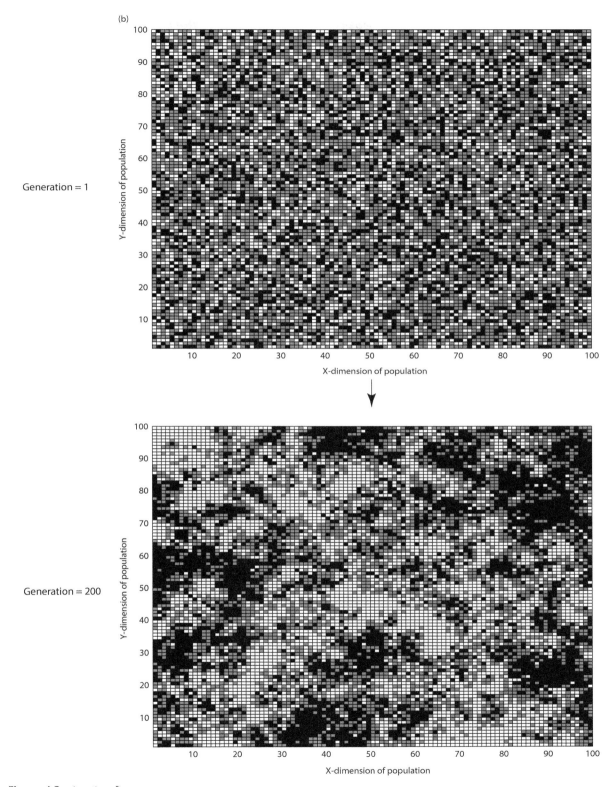

Generation = 1

Generation = 200

Figure 4.3 (*continued*)

Hardy–Weinberg expected genotype frequencies and randomly scattered locations of the three genotypes. After 200 generations, the population with a 99×99 individual mating area (Figure 4.3A) still shows random locations of the three genotypes. However, the population with a 3×3 mating area (Figure 4.3B) has distinct clumps of identical genotypes and fewer heterozygotes. One effect of isolation by distance is clearly local changes in allele frequency in a population, with local regions approaching fixation or loss, akin to the impact of reducing the effective population size (see the breeding effective population size in Chapter 3). Alternatively, isolation by distance can be thought of as a form of mating among relatives, since restricted mating distances cause homozygosity within subpopulations to increase. The patterns of genotypes in the simulated populations bear this out, with an obvious decline in the overall frequency of heterozygotes over time with isolation by distance (Figure 4.3B) but no such decline when there is panmixia (Figure 4.3A).

Isolation by distance: Decreasing chances of mating or gene flow as the geographic distance between individuals or populations increases.
Gene flow: The successful movement of alleles into populations through the movement of individuals (migration) or the movement of gametes.
Panmixia: Random mating, literally meaning "all mixed."
Population structure: Heterogeneity in allele and/or genotype frequencies among parts of a population, usually defined by space or time. Population structure is a pattern that can be caused by many processes such as by the net combination of genetic drift and limited gene flow, as well as by natural selection.
Subpopulation: A portion of the total population that experiences limited gene flow from other parts of the total population so that its allele frequencies evolve independently to some degree; synonymous with **deme**.

Population structure has profound implications for genotype and allele frequencies. Subdivision breaks up a population into smaller units that are each genetically independent to some degree. One consequence is that each subpopulation has a smaller effective population size than the effective size of the entire population if there were random mating. The genetic polymorphism found in a single large panmixic population and a population subdivided into many smaller demes is organized in a different manner. Think of the simple case of a diallelic locus. A single large population may take a very long time to experience fixation or loss due to genetic drift and thus maintain both alleles. In a highly subdivided population, each deme may quickly reach fixation or loss, but both alleles can be maintained in the overall population since half of the subpopulations are expected to reach fixation and half loss for a given allele. Processes that cause population structure can also be thought of as both creative and constraining in evolutionary change (Slatkin 1987a). The genetic isolation among demes caused by subdivision can prevent novel and even advantageous alleles from spreading throughout a population. But, at the same time, genetic isolation allows subpopulations to evolve independent allele frequencies and maintain unique alleles as is required for genetic adaptation to local environments under natural selection, for example.

It is worth noting that there are some important biological distinctions between gene flow and migration or dispersal. Migration (or dispersal) is simply the movement of individuals from one place to another. As such, migration may or may not result in gene flow. Gene flow requires that migrating individuals successfully contribute alleles to the mating pool of populations they join or visit. Thus, migration alone does not necessarily result in gene flow. Similarly, gene flow can also occur without the migration of individual organisms. Plants are a prime example, with gene flow that takes place via the movement of pollen grains (male gametes) but individuals themselves cannot migrate except as seeds. Gene flow can also occur without easily detected migration of individuals, such as cases where individuals move briefly to mate and then return to their original geographic locations. To confuse matters, the variable m (for migration rate) is almost universally used to indicate the rate of gene flow in models of population structure. Even though models do not normally make the distinction, it is wise to remember the biological differences between the processes of migration and gene flow in actual populations.

Box 4.1 Are allele frequencies random or clumped in two dimensions?

How can genetic variation in space be described to look for evidence of isolation by distance or other processes that cause spatial genetic differentiation in populations? The general approach is to compare pairs of individuals or populations, looking at both the similarity of their genotypes and how far apart they are located. Isolation by distance is a form of mating among relatives due to non-random mating, so it causes individuals that are located near to each other to be more related on average.

One classic statistic used to estimate spatial genetic structure is a correlation measure called Moran's I:

$$I_k = \frac{n \sum\limits_{i=1}^{n} \sum\limits_{j=1(i \neq j)}^{n} w_{ij}(y_i - \bar{y})(y_j - \bar{y})}{W_k \sum\limits_{i=1}^{n} (y_i - \bar{y})^2} \quad (4.1)$$

where k represents a distance class (e.g. all populations two distance units apart) so that w_{ij} equals 1 if the distance between location i and j

equals k, and 0 otherwise. Within a distance class k, n is the number of populations, y is the value of a genetic variable such as allele frequency for location i or j, $\bar{y}$ is the mean allele frequency for all populations, and W_k is the sum of the weights w_{ij} or $2nk$. The numerator is larger when pairs of populations have similar allele frequencies that show a large difference from the mean allele frequency.

Like correlations in general, Morans's I takes on values from −1 to +1 when estimated with a large number of samples. A positive value of I means that that allele frequencies between pairs of locations are similar on average, while a negative value means that allele frequencies between pairs of locations tend to differ on average. A value of 0 indicates that differences in subpopulation allele frequencies are not related to the distance between locations or that genetic variation is randomly distributed in space. The spatial locations of genotypes like those shown in Figure 4.3 are perfect situations to use Moran's I (see Figure 4.4).

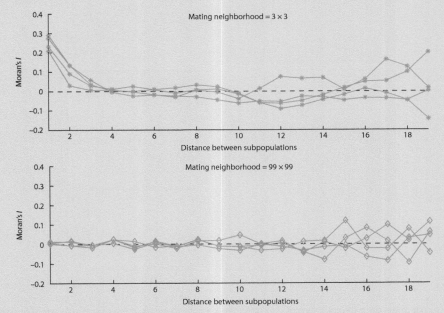

Figure 4.4 Moran's I for simulated populations like those in Figure 4.3. To estimate Moran's I, the 100 by 100 grid was simulated for 200 generations and was then divided into square subpopulations of 10 by 10 individuals. The frequency of the A allele within each subpopulation is y_i, and the mean allele frequency over all subpopulations is $\bar{y}$ in Eq. 4.1. The distance classes are the number of subpopulations that separate pairs of subpopulations. As expected, the simulations with strong isolation by distance (3 by 3 mating neighborhood) show correlated allele frequencies in subpopulations that are close together. However, the simulations with panmixia (99 by 99 mating neighborhood) show no such spatial correlation of allele frequency. The fluctuation of I at the largest distances classes in both figures is random variation due to very small numbers of individuals compared. Each line is based on an independent simulation of the 100 by 100 population.

This first section of the chapter has introduced gene flow and shown how limited gene flow has the potential to shape allele and genotype frequencies and form genetic subpopulations. Section 4.2 will introduce gene flow models and show how gene flow impacts allele frequencies over time. Section 4.3 will introduce an approach to directly measuring gene flow through estimates of parentage determined with genetic markers. Then, in Section 4.4, we will return to the fixation index (or F) from Chapter 2 and extend it for the case of structured populations in order to quantify the pattern of population divergence. Section 4.5 will show how variation in genotype frequencies is found either within populations as heterozygosity or among populations as allele frequency variation. Section 4.6 will present idealized population models that predict the pattern of genetic differentiation using the rates of gene flow and genetic drift, which serve as the means to estimate past gene flow. Section 4.7 introduces maximum likelihood and Bayesian methods to group individuals into genetic populations. The final

section of the chapter incorporates population subdivision into the coalescent model.

4.2 Gene flow and its impact on allele frequencies in multiple subpopulations

- Models of gene flow.
- Continent-island model.
- Two-island model.
- Dispersal kernels.

Gene flow is a mixing process that when acting in isolation eventually homogenizes allele frequencies to be equal in all subpopulations. Gene flow can operate in a wide range of patterns in natural populations, acting to connect subpopulations by the exchange of individuals or gametes. There is a series of gene flow models that have been widely studied in populations genetics and which serve as important points of reference. Some of these gene flow models are illustrated in Figure 4.5. It is likely that gene flow within and among actual subpopulations of

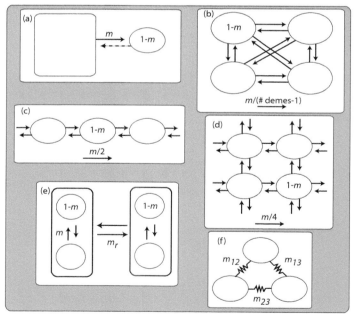

Figure 4.5 Models of population structure make different assumptions about the paths and rates of gene flow among subpopulations. In the "continent-island" model (A), gene flow is unidirectional because the continent population is so large that allele frequencies are not impacted by emigration or drift, while allele frequencies in the small population(s) are strongly influenced by immigration. The "island" model (B) has equal rates of gene flow exchanged by all populations regardless of the number of populations or their physical locations (the island model can also vary the number of populations from two to an infinite number). "Stepping-stone" models restrict gene flow to populations that are either adjacent or nearby in one (C) or two (D) dimensions and thereby incorporate isolation by distance. The "hierarchical island" model (E) has several rates of gene flow for the numerous levels of population organization (Slatkin and Voelm 1991), shown here with one rate of gene flow between demes within the same region and a different rate of gene flow between regions. The "isolation by resistance" model treats the possibility of gene flow rates varying among pairs of subpopulations based on the landscape that separates demes and thereby influences migration (McRae 2006). Gene flow models can also incorporate the extinction and re-colonization of entire subpopulations, a feature commonly added to stepping-stone models. Each panel shows the rate of gene flow indicated by the arrows if m percent of each population is composed of migrants and $1-m$ is composed of non-migrating individuals each generation.

organisms is not as easily categorized nor as invariant as is assumed in these models. Nonetheless, these models remain useful cases that represent possible patterns of gene flow among subpopulations.

The goal of gene flow models is to predict how the process of genetic mixing impacts allele frequencies in subpopulations over time. A useful starting place in such predictions is to assume gene flow is operating alone – there is no genetic drift, no natural selection, and no mutation – and then focus on how and how rapidly gene flow acts in the context of a given gene flow model. This section will present in detail the allele frequency impacts over time seen in two gene flow models.

Continent-island model

Perhaps, the simplest model of gene flow is called the continent-island model (Figure 4.5A). It assumes one very large population where allele frequency changes very little over short periods of time and a smaller population that receives migrants from the large continent population each generation. The island population experiences the replacement of a proportion m of its individuals through migration, with $1 - m$ of the original individuals remaining each generation. (We assume that the proportion m of island individuals replaced by gene flow each generation either die or emigrate to the continent population, which is so large that immigrants do not impact allele frequencies.)

> **Continent-island model:** An idealized model of population subdivision and gene flow that assumes one very large population where allele frequency is constant over time (like a population of many individuals) connected by gene flow with a small population where migrants make up a finite proportion of the individuals present each generation. Gene flow from the island to the continent may occur but is assumed to have a negligible effect on allele frequencies in the continent population.

Based on this situation along with its assumptions, it is possible to predict how gene flow changes allele frequency at a diallelic locus in the island population

over one generation. Allele frequency in the island population one generation in the future (call it p_{t+1}^{island}) is a function of (i) the allele frequency in the proportion of the island population that are not migrants and (ii) the allele frequency in the proportion of the island population that arrives via gene flow from the continent population. This can be stated in an equation as

$$p_{t=1}^{island} = p_{t=0}^{island}(1 - m) + p^{continent}m \qquad (4.2)$$

and used to predict the island population allele frequency after one generation of gene flow from the continent. Expanding the right side of this equation gives

$$p_{t=1}^{island} = p_{t=0}^{island} - p_{t=0}^{island}m + p^{continent}m \qquad (4.3)$$

which can be rearranged to an equation that gives the *change* in allele frequency in the island population over one generation

$$p_{t=1}^{island} - p_{t=0}^{island} = -m\left(p_{t=0}^{island} - p^{continent}\right) \qquad (4.4)$$

in a form readily interpreted in biological terms.

Equation 4.41 predicts that the degree of difference between allele frequencies in the island and continent populations ($p_{t=0}^{island} - p^{continent}$) will determine the direction as well as the rate of change in the island allele frequency as long as the rate of gene flow is not 0 ($m \neq 0$). For example, if $p_{t=0}^{island} > p^{continent}$, then the island allele frequency should decrease. Likewise, the island allele frequency is expected to increase if $p_{t=0}^{island} < p^{continent}$. To use a numerical example, suppose that $p_{t=0}^{island} = 0.1$ and $p^{continent} = 0.9$. The difference between the island and continent allele frequencies is -0.8, so, according to Eq. 4.41, the island allele frequency should increase for any amount of gene flow. If $m = 0.1$, then the island allele frequency will increase by 0.08 to $p_{t=1}^{island} = 0.18$ in one generation.

The expected change in allele frequency due to a single generation of gene flow can also be extended to predict allele frequency in the island population over an arbitrary number of generations. If there is a second generation of gene flow, the allele frequency in the island population is then

$$p_{t=2}^{island} = p_{t=1}^{island}(1 - m) + p^{continent}m \qquad (4.5)$$

Substituting $p_{t=1}^{island}$ as defined in Eq. 4.41 into this equation

$$p_{t=2}^{island} = \left(p_{t=0}^{island}(1-m) + p^{continent}m\right)(1-m) + p^{continent}m$$
$$(4.6)$$

and rearranging terms

$$p_{t=2}^{island} = p_{t=0}^{island}(1-m)^2 + p^{continent}(m(1-m) + m)$$
$$(4.7)$$

to eventually give an expectation for the island allele frequency after two generations of gene flow ($p_{t=2}^{island}$) in terms of the initial island allele frequency ($p_{t=0}^{island}$):

$$p_{t=2}^{island} = p_{t=0}^{island}(1-m)^2 + p^{continent}\left(1 - (1-m)^2\right)$$
$$(4.8)$$

Notice that the exponents are equal to the number of generations that have elapsed. Changing these exponents to an arbitrary number leads to the allele frequency in the island population after t generations have elapsed starting from an initial allele frequency gives

$$p_t^{island} = p_{t=0}^{island}(1-m)^t + p^{continent}\left(1 - (1-m)^t\right)$$
$$(4.9)$$

which can be rearranged to

$$p_t^{island} = p^{continent} + \left(p_{t=0}^{island} - p^{continent}\right)(1-m)^t$$
$$(4.10)$$

The rate of allele frequency change in the island population can also be seen in this equation. The proportion of the island population that made up its initial allele frequency decreases by $(1-m)^t$, approaching zero as time passes due to gene flow. Therefore, the allele frequency difference between the island and continent decreases toward 0 over time and the allele frequency of the island approaches the allele frequency of the continent. Figure 4.6 shows how the island allele frequency approaches the continent allele frequency over time for a range of initial island allele frequencies. Notice the smooth approach to the continent allele frequency: this is a consequence of the fact that the outcome is completely determined by a constant rate of gene flow and random processes such as genetic drift to introduce chance variation are absent. In actual populations, the rate of gene flow could itself vary randomly over time.

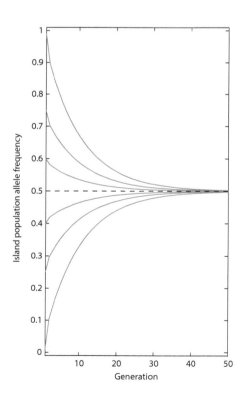
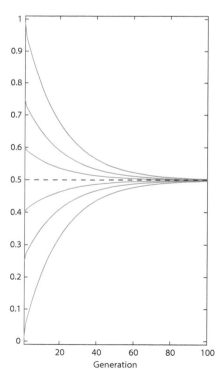

Figure 4.6 Allele frequency in the island population for a diallelic locus under the continent-island model of gene flow. The island population allele frequencies (p^{island}) over time are shown for six different initial values (solid lines). The continent population has an allele frequency of $p^{continent} = 0.5$ shown by the dashed line. In the left panel, $m = 0.1$, and, in the right panel, $m = 0.05$. Equilibrium is reached more slowly when the rate of gene flow is lower. In contrast, the difference in allele frequencies between the island and continent does not affect time to equilibrium for a given rate of gene flow. Note that the time scales in the two graphs differ.

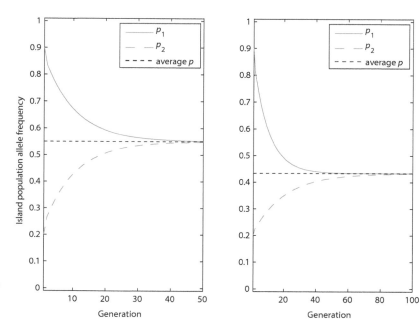

Figure 4.7 Allele frequency in the two-island model of gene flow for a diallelic locus. Starting from allele frequencies of 0.9 and 0.2 and equal rates of gene flow ($m = 0.1$), the subpopulations approach an equilibrium allele frequency of $\bar{p} = 0.5$ (dashed line). With the same initial allele frequencies in the two subpopulations but asymmetric rates of gene flow ($m_1 = 0.1$ and $m_2 = 0.05$), the subpopulations approach an equilibrium allele frequency of $\bar{p} = 0.433$ (dashed line). Equilibrium is reached more slowly in the case of asymmetric rates for one population since the rate of gene flow is lower. Note that the time scales in the two graphs differ.

These predictions of the continent-island model are consistent with intuition. Given that the continent population has a constant allele frequency over time, the island population should eventually reach an identical allele frequency when the two are mixed. How long it takes for the two populations to converge on the same allele frequency depends on the proportion of continent individuals moving to the island each generation. In contrast, the difference in allele frequencies between the island and continent does not alter the time to equilibrium for a given migration rate (see Figure 4.7). This occurs since the rate of change in the island allele frequency is determined by the difference in allele frequencies. Greater differences lead to greater rates of change toward the continent allele frequency. Thus, the continent-island model shows that the process of

gene flow alone is capable of bringing populations to the same allele frequency. Identical allele frequencies between or among populations is really a lack of population structure or panmixia. So, the continent-island can be thought of as a demonstration that gene flow acting in the absence of other processes will eventually result in panmixia.

Two-island model

One simple extension to the continent-island model is to consider the two subpopulations as being equal in size, removing the assumption that one population (the continent) serves as an unchanging source of migrants. The model then represents gene flow between two islands which can each exhibit changes in allele frequency over time. The switch to a

---

Interact box 4.1 Continent-island model of gene flow

Use an R script to explore the continent-island model of gene flow. The script has variables for allele frequencies in the island and continent, the rate at which island alleles are replaced by continent alleles (or the migration rate), and the number of generations to simulate. Start with a continent allele frequency of 0.9, an island allele frequency of 0.1, a rate of gene flow of 0.1, and 100 generations.

Keeping the same values for initial allele frequencies, try a series of values of the migration rate (e.g. $m = 0.1, 0.05, 0.001,$ and 0.001) to see how it affects time to equilibrium; increase the generations if needed.

What is the relationship between the rate of gene flow and time to equilibrium?

two-island model also allows an independent rate of gene flow for each subpopulation, m_1 and m_2. Using reasoning similar to that for the continent-island model, the allele frequency in a subpopulation one generation in the future is the sum of the allele frequency in the proportion of individuals that do not migrate $(1 - m)$ plus the allele frequency in the immigrants. Assuming that $m_1 = m_2 = m$, the allele frequency in either subpopulation is

$$p_{t=1} = p_{t=0}(1 - m) + \bar{p}m \qquad (4.11)$$

where $\bar{p} = \dfrac{p_1 + p_2}{2}$. The allele frequency in the migrants is now the average of the two subpopulations rather than just a constant like the continent allele frequency. This happens since both subpopulations receive immigrants, so the allele frequencies of each subpopulation are approaching the allele frequency in the total population as gene flow mixes the subpopulations. Similar to the result for the continent-island model, the allele frequency in either of the two islands is

$$p_t = \bar{p} + (p_{t=0} - \bar{p})(1 - m)^t \qquad (4.12)$$

after t generations have elapsed. Figure 4.7 shows allele frequencies in the two-island model over time.

When the rates of gene flow are not equal, then the average allele frequency is $\bar{p} = \dfrac{p_1 m_2 + p_2 m_1}{m_1 + m_2}$, or the gene-flow-weighted average of the allele frequencies in the two subpopulations. When $m_1 \neq m_2$, the equilibrium allele frequencies will be closer to the initial allele frequency of the subpopulation with the lower migration rate. This happens because the subpopulation with the lower migration rate experiences less immigration and remains closer to its initial allele frequency, yet it supplies migrants

to the other subpopulation. As seen in Figure 4.14, the time to equilibrium is also longer when the migration rates are asymmetric. Consider the example where migration rates are unequal ($m_1 = 0.01$ and $m_2 = 0.1$) and the initial allele frequencies in the two subpopulations are $p_1 = 0.9$ and $p_2 = 0.1$. The weighted average allele frequency is then $\bar{p} = \dfrac{(0.9)(0.1)}{0.11} + \dfrac{(0.1)(0.01)}{0.11} = 0.827$. This is also the expected allele frequency in both subpopulations at equilibrium.

The main conclusion of the two-island model is that the equilibrium allele frequencies in the two subpopulations are the average allele frequency of the total population when the two migration rates are equal. This conclusion holds when there are a larger number of subpopulations, a result that will be useful to remember when considering the process of gene flow in an island model in combination with another process such as genetic drift.

Neither the continent-island nor the two island models of gene flow account for isolation by distance. The relationship of the rate of gene flow to distance between subpopulations can be modeled by expressing the rate of gene flow as a function of distance. The probability distribution of straight-line (or Euclidian) spatial distances between where an individual is born and where it reproduces is called a **dispersal kernel** (reviewed by Nathan et al. 2012). Dispersal kernel functions can take many forms, and several examples are shown in Figure 4.8. Some dispersal kernels such as the gamma distribution exhibit the highest probability of gene flow at intermediate distances because of density-dependence at short distance (Figure 4.6B). Other dispersal kernels are described as "fat-tailed" or leptokurtic since the probability of gene flow does not decline with distance as rapidly as it does for other distributions. Empirical estimates of dispersal kernels from natural

---

Interact box 4.2 Two-island model of gene flow

Use an R script to explore the two-island model of gene flow. The script has variables for initial allele frequencies for each island, the gene flow rate for each island, and the number of generations to simulate. Start with a continent allele frequency of 0.9, an island allele frequency of 0.1, a rate of gene flow of 0.1, and 100 generations.

Keeping the same values for initial allele frequencies, try a series of values of the migration rate (e.g. $m = 0.1, 0.05, 0.001,$ and 0.001) to see how it affects time to equilibrium; increase the generations if needed.

What is the relationship between the rate of gene flow and time to equilibrium?

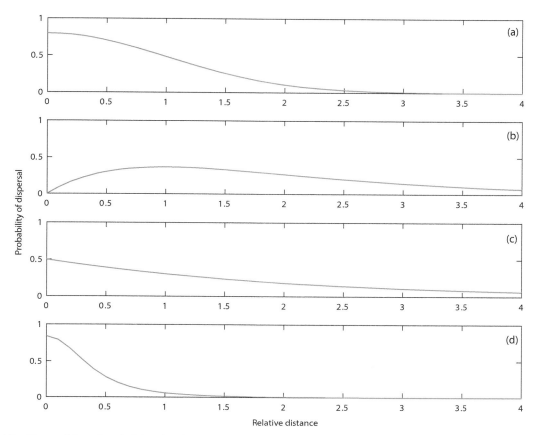

Figure 4.8 Dispersal kernel probability distributions that show how dispersal events can be distributed across geographic distances. The distributions in B and C are fat-tailed compared to the distributions in A and D, and the distribution in D has the thinnest tail. Panel A shows a half-normal distribution with a mean of zero and a variance of one. Panel B is a gamma distribution with a scale parameter of 2 and a shape parameter of 1. Panel C is a Weibull distribution with a scale parameter of 2 and a shape parameter of 1. Panel D is a 2Dt distribution with $a = 0.55$ and $b = 1.8$.

populations suggest that dispersal is often best described by some type of leptokurtic distribution. The overall decline in identity by descent with distance is not strongly dependent on the exact shape of the distribution of gene flow with distance, although it can impact the scale and magnitude of genetic differentiation (Rousset 2008a; Furnstenau and Cartwight 2016).

4.3 Direct measures of gene flow

- Genetic marker-based parentage analysis.

This section of the chapter will introduce and explain the use of molecular genetic markers to identify the unknown parent or parents of a sample of progeny or juveniles and thereby describe the patterns of mating that took place among the parents. Parentage analyses are considered direct measures of gene flow since they reveal and measure the pattern of gamete movement at the scale over which the candidate parents are sampled. Parentage analyses are also commonly used to test hypotheses about what factors influence patterns of mating among individuals. For example, animal parentage studies can test for correlations between mating success and phenotypes or behaviors. Parentage analysis is most often performed in the case where one parent is known and the other parent is unknown and could potentially be any one of a number of individuals or **candidate parents**. Genetic analyses that attempt to identify unknown fathers or unknown mothers from a population of candidate parents are called **paternity analysis** or **maternity analysis**, respectively (see Meagher 1986; Devlin and Ellstrand 1990; Dow and Ashley 1996; reviewed by Jones et al. 2010). Although not detailed here, it is also possible to attempt to infer both unknown parents within a population of candidate parents to estimate the minimum number of parents that contributed

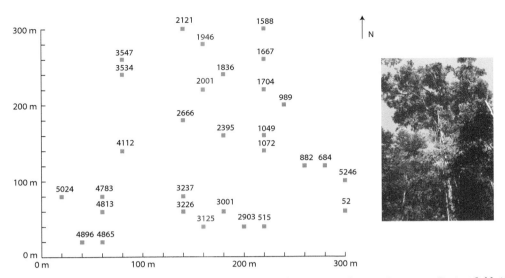

Figure 4.9 An individual *Corythophora alta* tree found at the Biological Dynamics of Forest Fragments Project field sites North of Manaus, Brazil. The map shows the relative locations of the individual trees that make up the candidate parent population within a 9 ha forest plot at Cabo Frio. All trees are can be both maternal parents and candidate paternal parents since the trees are hermaphrodites capable of self-fertilization.

to a group of progeny. This section will review some of the basic concepts required to understand the methods and results of parentage analyses by means of an example paternity analysis. One focus in particular will be the distinction between identifying the true parent of an offspring and identifying a candidate parent that appears to be the true parent due to chance.

To understand the steps carried out in parentage analysis, let us work through an example based on genotype data from the tropical tree *Corythophora alta*, a member of the Brazil nut family (Figure 4.9). All *C. alta* individuals 10 cm in diameter or greater at breast height were sampled from a 9 ha area inside of a large tract of continuous forest. These trees, 10 cm in diameter or greater at breast height, are the candidate parents. A sample of seeds was also collected from some of these trees. The genotypes of both the trees and the seeds were determined for 10 nuclear microsatellite loci (see Box 2.1 for an introduction to this type of genetic marker). A subset of these data is shown in Table 4.1. The goal of the parentage analysis in this case is to determine the fathers of the seeds given the known mothers in order to estimate the proportion of seeds that resulted from pollen transport within the sampled plot compared with pollen transport from outside the sampled plot.

The first step in a parentage analysis is to examine the genotypes of an individual progeny and its known parent for allelic matches. *C. alta* seed genotypes are grouped with their known parent in Table 4.2. For example, in Table 4.2, the genotype

of seed 1-1 from tree 989 is given in the first row and the genotype of the known maternal parent tree (989) is given in the second row. At each locus, one (or sometimes both) of the alleles found in the known parent genotype is observed in the progeny genotype. For seed 1-1 from tree 989, the known parent contributed the 336 allele at locus A, the 106 allele at locus B, the 165 allele at locus C, the 275 allele at locus D, and the 153 allele at locus E. Given those alleles came from the known parent, the true father must have contributed alleles 327, 91, 185, 287, and 153 at loci A through E. This set of single alleles at each diploid locus is called the paternal **haplotype**. We can now scan the genotypes of the candidate parents to see whether there is any individual with a haplotype that contains all of those alleles (this is normally done with the assistance of a computer program). All candidate parents that have a matching haplotype are possible fathers of seed 1-1 from tree 989. In this case, tree 1946 is the only individual with the required haplotype, and, so, 1946 is possibly the father while all of the other candidate parents are **excluded** as fathers due to a genetic mismatch at one or more loci in the paternal haplotype. Repeating this process for the next two seeds also excludes all but a single individual as the father.

The process of excluding potential parents can also accommodate the possibility of methodological errors in genotyping, alleles that do not amplify in polymerase chain reaction (PCR) called null alleles, or changes in allelic state between parent and offspring caused by mutation (Sancristobal and

Table 4.1 Microsatellite genotypes with allelic states given in base pairs for some of the 30 mature individuals of the tropical tree *Corythophora alta* sampled from a 9 ha plot of continuous forest in the Brazilian Amazon. Seed progeny were collected from a known maternal tree. Missing data are indicated by "—".

	Microsatellite locus									
	A		B		C		D		E	
Candidate parents										
684	333	339	97	106	169	177	275	305	135	135
989	330	336	97	106	165	181	275	275	135	153
1072	315	333	103	106	169	179	296	302	138	138
1588	318	327	106	106	165	167	272	293	135	150
1667	324	333	—	—	165	185	275	284	141	159
1704	318	327	103	106	—	—	284	296	144	147
1836	333	339	97	97	181	183	275	296	138	144
1946	327	333	91	106	167	185	284	287	147	153
2001	321	336	—	—	177	181	284	302	138	144
2121	318	333	100	106	179	181	284	302	144	144
2395	327	333	103	103	179	187	275	296	150	159
3001	324	333	91	106	167	183	284	302	147	159
3226	327	327	103	106	163	181	275	275	135	144
3237	324	324	91	103	179	187	284	305	144	159
3547	321	321	103	106	177	179	275	296	—	—
4112	327	327	97	106	169	181	296	302	144	144
4783	321	327	—	—	183	185	290	308	144	156
4813	327	333	106	106	177	179	284	302	135	138
4865	321	327	106	106	167	179	284	296	144	153
4896	315	333	100	106	181	189	275	284	162	162
5024	318	327	100	103	165	167	275	284	147	147
Seed progeny										
989 seed 1-1	327	336	91	106	165	185	275	287	153	153
989 seed 2-1	327	330	103	106	165	181	275	275	135	135
989 seed 3-1	330	336	97	106	165	181	—	—	135	153
989 seed 25-1	321	330	106	106	167	181	275	296	135	153

Table 4.2 Seed progeny genotypes given with the known maternal parent genotype along with the genotype of the most probable paternal parent from the pool of all possible candidate parents. Alleles in the seed progeny that match those in the known maternal parent are underlined. The known maternal parent is also a candidate paternal parent since this species can self-fertilize. Missing data are indicated by "—".

	Microsatellite locus									
	A		B		C		D		E	
989 seed 1-1	327	_336_	91	_106_	_165_	185	_275_	287	_153_	153
989	330	_336_	97	_106_	165	181	_275_	275	_135_	153
1946	327	333	91	106	167	185	284	287	147	153
989 seed 2-1	327	_330_	103	_106_	_165_	_181_	_275_	275	_135_	135
989	330	_336_	97	_106_	165	181	_275_	275	_135_	153
3226	327	327	103	106	163	181	275	275	135	144
989 seed 3-1	_330_	_336_	_97_	_106_	_165_	_181_	—	—	_135_	_153_
989	_330_	_336_	_97_	_106_	165	181	275	275	_135_	_153_
989	330	336	97	106	165	181	275	275	135	153
989 seed 25-1	321	_330_	_106_	106	167	_181_	_275_	296	_135_	_153_
989	330	_336_	_97_	106	165	_181_	_275_	275	_135_	_153_
4865	321	327	106	106	167	179	284	296	144	153

Chevalet 1997). Candidate parents can include those with a perfect match to the inferred parental haplotype as well as individuals that possess genetic mismatches at one (or more) loci. For example, for seed 2-1 from tree 989 in Table 4.2, we see that candidate parent 1588 matches at loci B, C, and E, with missing data at locus D and a mismatch at locus A. If a single genetic mismatch is permitted, then tree 1588 would not be excluded as a possible father. Such error-tolerant parentage analysis requires empirical estimation of genotyping error rates (Adams et al. 2004; Bonin et al. 2004; Hoffman and Amos 2004).

With the exclusion of all but a single candidate parent, it would seem like certain identification of the true parent has been accomplished. Unfortunately, it is always possible that any non-excluded candidate parent is not the actual parent. There is the possibility, by chance alone, that an individual possesses a genotype with the same haplotype as the true parent. Evaluating the chance that a non-excluded candidate parent (sometimes called an **inclusion** or an **included parent**) is not the true parent requires one to determine the chance of such a random match. Let the frequency of an allele in the matching haplotype be p_i where i indicates the locus. At each locus, the chance of matching at random is simply the probability that an individual is either homozygous (p_i^2) or heterozygous ($2p_i[1 - p_i]$) for the allele in question. (In the case of an error-tolerant assignment, p_i is the probability of mistyping an allele

at the ith locus.) Thus, the total probability of a random match for one locus is:

$$P(\text{random match}) = p_i^2 + 2p_i(1 - p_i) \quad (4.13)$$

under the assumptions of random mating and panmixia. Assuming that all of the loci used in a parentage analysis are independent, the probability of a random match for all loci in a given haplotype is the product of the locus by locus frequency of a random match, or:

$$P(\text{multilocus random match}) = \prod_{i=1}^{loci} (p_i^2 + 2p_i(1 - p_i))$$

$$(4.14)$$

where Π indicates chain multiplication over all loci.

Returning to our *C. alta* example, we can calculate the chance of a random match for each of the paternal haplotypes. The haplotypes, allele frequencies (see Table 4.3), probability of a random match at each locus, and probability of a random match at all five loci are given in Table 4.4. Focus first on the haplotype for tree 1946. Given that allele 327 at locus A has an observed frequency of 0.2703 in the population of candidate parents (which is an estimate of the allele frequency in the entire population), the chance of any genotype having one copy of this allele is $(0.2703)^2 + 2(0.2703)(1 - 0.2703) = 0.4675$. We, therefore, expect 46.75%

Table 4.3 Allele frequencies for five *Corythophora alta* microsatellite loci used for paternity analysis.

Microsatellite locus									
A		**B**		**C**		**D**		**E**	
Allele	**Frequency**	**Allele**	**Frequency**	**Allele**	**Frequency**	**Allele**	**Frequency**	**Allele**	**Frequency**
315	0.0405	91	0.0735	163	0.0217	272	0.0238	135	0.2917
318	0.0541	97	0.3088	165	0.2283	275	0.4167	138	0.0625
321	0.1216	100	0.0735	167	0.0761	281	0.0357	141	0.0313
324	0.0541	103	0.1471	169	0.0435	284	0.1429	144	0.2188
327	0.2703	106	0.3971	171	0.0217	287	0.0119	147	0.0625
330	0.1892			177	0.0543	290	0.0119	150	0.0938
333	0.1216			179	0.1304	293	0.0238	153	0.1250
336	0.1216			181	0.2065	296	0.1905	156	0.0208
339	0.0270			183	0.0652	299	0.0119	159	0.0521
				185	0.0435	302	0.0833	162	0.0417
				187	0.0326	305	0.0357		
				189	0.0109	308	0.0119		
				193	0.0109				
				197	0.0543				

Table 4.4 The chance of a random match for the included fathers in Table 4.2. The probability of a random match at each locus is. The combined probability of a random match for all loci in the haplotype is the product of the probabilities of a random match at each independent locus. Paternal haplotype data are treated as missing ("—") for the purposes of probability calculations when progeny genotype data are missing. In the cases where the paternal haplotype has multiple possible alleles at some loci, the highest probability of a chance match is given. The allele frequencies for each locus are given in Table 4.3.

Included father	Microsatellite haplotype					P(multilocus random match)
	A	B	C	D	E	
1946 (seed 1-1)	327	91	185	287	135	
allele frequencies	0.2703	0.0735	0.0435	0.0119	0.2917	
P(random match)	0.4675	0.1416	0.0851	0.0237	0.4983	0.0000665
3226 (seed 2-1)	327	103 / 106	181	275	135	
allele frequencies	0.2703	0.0735 / 0.3971	0.2065	0.4167	0.2917	
P(random match)	0.4675	0.1416 / 0.6365	0.3704	0.6598	0.4983	≤0.03624
989 (seed 3-1)	330 / 336	97 / 106	165 / 181	—	135 / 153	
allele frequencies	0.1892 / 0.1216	0.3088 / 0.3971	0.2283 / 0.2065	1.0	0.2917 / 0.1250	
P(random match)	0.3426 / 0.2284	0.5222 / 0.6365	0.4045 / 0.3704	1.0	0.4983 / 0.2344	≤0.0440

of individuals in the population to have a genotype with either one or two copies of the 327 allele. This is the same as the probability that an individual taken at random from the population (and not necessarily included in the sample of candidate parents) could provide the correct haplotype to be included as a possible father of seed 989 1-1 in Table 4.2. The chances of a random match at each of the five loci are calculated in the same fashion. We see that a genotype that would complement the known parent's haplotype and explain the seed genotype is expected to occur between about 2 and 47% of the time for any single locus. When these probabilities are combined across all five loci, the expected frequency of a random match becomes very small. As shown in Table 4.4, the expected frequency of a random match at all five loci is between 44 in 1000 and 66 in 1 000 000 genotypes under the assumption of random mating. This is a demonstration of the general principle that the ability to distinguish true parentage from apparent parentage due to random matches depends on both the allele frequencies at each locus as well as the total number of loci available. Random matches become less likely as allele frequencies decrease and the number of independent loci increases.

Candidate parent: An individual in the pool of possible parents that shares one or both alleles found in an offspring genotype at all loci.
Cryptic gene flow: Gene flow events incorrectly assigned to candidate parents but actually due to unobserved parents outside the area where candidate parents were sampled, leading to an underestimate of gene flow distances.
Exclusion: Rejection of an individual as a possible parent due to genetic mismatch (neither allele in the individual's genotype is identical to one of the alleles in the progeny genotype).
Exclusion probability: The chance that an individual can be rejected as a candidate parent due to genetic mismatch; depends on allele frequencies and increases with the number of loci and the numbers of alleles per locus employed in a parentage analysis.

We can express the probability that an individual taken at random from a population would be ruled out as a parent due to genetic mismatch. Equation 4.2 gives the probability of a random match at a single locus, or the probability that a genotype has a matching allele by chance alone. If a genotype does *not* match by chance, then it is excluded from possibly being the parent. This means that the **exclusion probability** for a single individual sampled at random from a population is just 1 minus the probability of a random match:

$$P(\text{exclusion}) = 1 - P(\text{random match}) \qquad (4.15)$$

If more than one candidate parent is sampled from a population, the probability of exclusion for each individual is independent (the genotype of each individual represents a random sampling of the alleles present in the population). Therefore, the total probability of ruling out or excluding all candidate parents is the product of the exclusion probabilities for each individual. For a sample of n individuals from a population, the total probability of exclusion is then

$$P(\text{exclusion for } n \text{ individuals}) = (1 - P(\text{random match}))^n$$
$$(4.16)$$

This means that the chance of exclusion decreases as more individuals are sampled from a population. This is the same as saying that the chances of sampling an individual that matches a parental haplotype just by chance increase as more candidate parents are sampled.

Based on the exclusion probability in a population of n candidate parents, we can estimate the chances that a random match *does* occur. Since the exclusion probability is the chance of *not* matching at random, the probability of a haplotype match between a candidate parent and an offspring in a population of n individuals is just 1 minus the probability of exclusion for n individuals, or

$$P(\text{random match in } n \text{ individuals})$$
$$= 1 - P(\text{exclusion for } n \text{ individuals}) \qquad (4.17)$$
$$= 1 - (1 - P(\text{random match}))^n$$

This is the probability that a haplotype matching the true parent will occur at random in a sample of n candidate parents.

The probability of a random match in a sample of n candidate parents (Eq. 4.6) can be thought of as the chance that a candidate parent is mistakenly assigned as the true parent since its genotype provides the matching haplotype by chance, while the true parent is not identified since it is not included in the sample of candidate parents. This phenomenon is referred to as **cryptic gene flow** in paternity analysis since the true gene flow event is not identified, even though a parent has been mistakenly inferred for the progeny. If the true parent was not included in the sample of candidate parents because it was outside the sampling area, incorrectly inferred parentage results in an underestimate of gene flow distances. Equation 4.6 shows that the probability of incorrectly assigning parentage due to random matches increases as the number of candidate parents increases for a given expected genotype frequency.

Returning to the *C. alta* example in Table 4.2, we can determine the chances that one of the candidate parents is incorrectly inferred to be a father, while the true father remains undetected as well as the chances of paternity exclusion with the 30 candidate parents in the study. For seed 3-1, the maternal and paternal parents are the same (Table 4.4), indicating a self-fertilization event. Based on the paternal parent haplotype expected frequency, the chance of paternity exclusion is $(1 - 0.044)^{30} = 0.259$, and the probability of a random match is, therefore, 0.741. Since the four-locus inferred paternal haplotype is expected to occur very frequently (74% of the time) by chance in a sample of 30 candidate parents, there is also a good chance that the seed could appear self-fertilized even though it was actually sired by an individual not in the sample of candidate parents. For seed 989 1-1 where tree 1946 was the only included candidate parent, the chance of paternity exclusion is $(1 - 0.0000665)^{30} = 0.9980$, and the probability of a random match is, therefore, 0.0020. The five-locus inferred paternal haplotype for seed 989 1-1 is expected to appear by chance in only two of 1000 samples of 30 candidate parents given the estimated allele frequencies.

Problem box 4.1
Calculate the probability of a random haplotype match and the exclusion probability

The seed 25-1 from maternal tree 989 shows exact haplotype matches with candidate paternal tree 4865 (see Table 4.2). Using the allele frequencies provided in Table 4.3, calculate the probability of a random match for the paternal haplotype. Then, use this probability of a random match to calculate the exclusion probability for the sample of 30 candidate parents. What loci are the most and least useful in determining paternity for these two seed progenies? Why?

There are four general outcomes for each offspring–known-parent pair in parentage analysis, as follows.

1   A single candidate parent is identified as the parent. Such single parentage assignments need to be interpreted in light of the exclusion probability or likelihood of parentage.
2   Multiple candidate parents are identified for a single progeny. In these cases, one commonly used criterion is to assign as parent the candidate parent with the lowest probability of matching by chance. Additional criteria might also include spatial separation from the known parent, degree of reproductive overlap with the known parent, or reproductive dominance, if such information is available.
3   None of the candidate parents have a genotype that could have combined with the known parent to yield the progeny genotype. In this case, the actual parent may not be present in the sample of candidate parents. Such an outcome is often used to infer that the gene flow event leading to that progeny was from a relatively long distance from a parent outside the sample area of candidate parents (so-called off-plot gene flow). However, it is also possible that the actual parent is in the population of candidate parents but has a genetic mismatch at one or more loci

Interact box 4.3 Average exclusion probability for a locus

In planning a parentage analysis study, it is necessary to determine whether a set of genetic markers will have a sufficiently small probability of exclusion (this is called the power of the genetic markers). As shown in Eq. 4.4, the exclusion probability will depend on the expected genotype frequency for a single parental haplotype. This expected genotype frequency is in turn a function of the number of alleles and the allele frequencies at each locus. Since there are many possible genotypes for a locus with three or more alleles, the *average* probability of exclusion is used to estimate the power of a set of genetic markers to demonstrate nonpaternity (see Chakraborty et al. 1988; Weir 1996).

You can use an Excel spreadsheet that has been set up to calculate the average probability of exclusion (abbreviated as P_E in the spreadsheet) for a case of one locus with six alleles and one locus with 12 alleles. The spreadsheet uses the allele frequencies (that you can modify) to calculate (i) the expected frequencies of each maternal parent–offspring genotype combination and (ii) the exclusion probabilities for the paternal haplotype(s) for each maternal parent–offspring genotype combination. The average exclusion probability is then the average of the exclusion probabilities where each exclusion probability is weighted by the expected frequency of the maternal parent–offspring genotype combination. The spreadsheet follows the derivation for a locus with three alleles given in Table 1 of Chakraborty et al. (1988). The maximum average exclusion probability occurs when all alleles at a locus have identical allele frequencies (e.g. each allele has a frequency of 1/6 when there are six alleles). The maximum average exclusion probability is computed in each spreadsheet according to:

$$\text{Max.prob.exclusion} = \frac{(k-1)\left(k^3 - k^2 - 2k + 3\right)}{k^4} \tag{4.18}$$

where k is the number of alleles at the locus (Selvin 1980).

Compare the average probability of exclusion for cases where the frequencies of each allele are very similar to cases where one or a few alleles are very common and the remaining alleles are rare. How does the evenness of the frequencies for the alleles influence the average exclusion probability? How do you combine the average exclusion probabilities for multiple loci? What is the average exclusion probability of two loci with 12 alleles or two loci with six alleles when allele frequencies are all equal for each locus? How many independent loci with 12 equally frequent alleles would be required for a probability of exclusion of 90% when there are 50 candidate parents?

due to a genotyping error or mutation. An additional alternative is that the actual parent was inside the sampling area of candidate parents when mating took place, but the individual either died or migrated before sampling of the candidate parents was carried out.

4 Parentage is assigned to a candidate parent but that the true parent is an individual not included in the sample of possible parents. When making paternity assignments, the chance of incorrectly assigning paternity within a group of sampled individuals when the father is actually outside the population, or missing a "cryptic gene flow" event, will be related to the expected frequency of a given multilocus genotype.

Parentage analyses measure gene flow by inferring numerous mating events within the population of candidate parents that lead to each sampled progeny or juvenile in a population. This provides estimates of quantities such as the average distance between parents or the number of matings where both parents were within a sample area compared to the number of matings where a parent was outside that area. This means that resulting estimates of gene flow do not rely on any model of population structure or gene flow other than the assumptions that are used to construct the parentage assignments themselves. The resulting estimates of gene flow are, therefore, considered "direct." The clear strength of parentage analyses is that much can be learned about patterns of mating since parental pairings that lead to a specific offspring can often be identified with medium to high confidence.

Parentage analyses have been a critically important tool used to learn about mating and relatedness patterns in wild populations. An example is the numerous studies of parentage among bird nestlings that overturned the long-held idea that birds were usually monogamous breeders. Instead, birds have variable and complex mating patterns where mating outside of nesting pairs by both females and males can be common and juveniles in the nest may not be related to one or both of the nest-attendant "parents" (Westneat and Stewart 2003). Parentage analyses have also been used in a wide variety of plant and animal species to produce detailed descriptions of mating and gene flow patterns.

Although the term "direct" has connotations of precision and ready insight, it is important to recognize that parentage analyses do have limitations when used to infer patterns of gene flow. A major limitation stems from the fact that most parentage studies cover a time scale of only a few generations at most. In all organisms with population sizes that are stable through time, each parent produces just one offspring on average that survives to reproduce successfully. The other progeny die or do not reproduce. This means that many, perhaps even most, of the progeny included in parentage studies ultimately do not reproduce. This problem is particularly acute in long-lived organisms, where parentage studies examine only a very small fraction of progeny produced over a period much less than the average individual lifetime. Gene flow can be thought of as the long-term average of the matings that lead to individuals that survive and contribute progeny to the next generation. How effective parentage studies are at estimating longer-term patterns of gene flow then depends on the sampling duration of parentage studies relative to generation time and how variable parentage patterns are over the short-term compared to their long-term averages.

4.4 Fixation indices to summarize the pattern of population subdivision

- Extending the fixation index to measure the pattern of population structure through F_{IS}, F_{ST}, and F_{IT}.

The first section of the chapter reviewed the processes that contribute to the formation of allele frequency differences among populations. Given that these processes might be acting, it is necessary to develop methods to measure and quantify population structure. The parentage analyses such as those described in the last section can be carried out when genotype data are available for both a sample of candidate parents and a sample of progeny. An alternative situation is where genotype data are determined for individuals sampled within and among a series of geographic locations. This type of sampling is very commonly carried out in empirical studies and requires methods to quantify the pattern of population structure present among the subpopulations and well as the genotype frequencies found within subpopulations. It would be advantageous if such measures could be readily compared to reference situations such as what is expected with no population structure. This was our approach when comparing observed and expected heterozygosity using the fixation index (F) in Chapter 2. We can now extend the fixation index to apply to cases where there are multiple subpopulations. In this more complex case, there can be deviations from Hardy–Weinberg expected frequencies of heterozygotes at two levels: within each subpopulation due to non-random mating and among subpopulations due to population structure. This section of the chapter will develop and explain fixation-index-based measures of departure from expected heterozygosity commonly used to quantify population structure.

Let us look in detail at the case where we can measure the genotypes at a diallelic locus for a sample of individuals located in several different subpopulations. Recall that the heterozygosity in a population is just 1 minus the homozygosity ($H = 1 - F$), so the heterozygosity can be related to the fixation index. With such genotype data, it is possible to compute the observed and expected frequencies of the heterozygote genotype in several ways (Table 4.5). The first way is to simply take the average:

$$H_I = \frac{1}{n} \sum_{i=1}^{n} \hat{H}_i \qquad (4.19)$$

where $\hat{H}$ is the *observed* frequency of heterozygotes in each of the n subpopulations. We could call this $\overline{H}$ since it is the average of the observed heterozygote frequencies in all subpopulations. This is just the probability that a given individual is heterozygous or the average observed heterozygosity. As shown in Chapter 2, the heterozygosity within populations can be increased or decreased relative to Hardy–Weinberg expectations by non-random mating.

Table 4.5 The mathematical and biological definitions of heterozygosity for three levels of population organization. In the summations, *i* refers to each subpopulation 1, 2, 3 ... *n*, and p_i and q_i are the frequencies of the two alleles at a diallelic locus in subpopulation *i*.

$H_I = \dfrac{1}{n}\displaystyle\sum_{i=1}^{n} \hat{H}_I$	The average observed heterozygosity within each subpopulation.
$H_S = \dfrac{1}{n}\displaystyle\sum_{i=1}^{n} 2p_i q_i$	The average expected heterozygosity of subpopulations assuming random mating with each subpopulation or $\overline{2pq}$.
$H_T = 2\overline{p}\,\overline{q}$	The expected heterozygosity of the total population based on the average allele frequencies ($\overline{p}$ and $\overline{q}$) assuming random mating.

Next, we can determine the expected heterozygosity *assuming* each subpopulation is in Hardy–Weinberg equilibrium. This assumption means that the frequency of the heterozygous genotype is expected to be *2pq* for a locus with two alleles. The average expected heterozygosity of subpopulations is then

$$H_S = \frac{1}{n}\sum_{i=1}^{n} 2p_i q_i \qquad (4.20)$$

where p_i and q_i are the allele frequencies in subpopulation *i* and there are *n* subpopulations. We could use the notation $\overline{2pq}$ since the expected heterozygosity is determined for each subpopulation and then averaged. Here, the observed allele frequency is used to estimate the Hardy–Weinberg expected heterozygosity for each subpopulation.

At the most inclusive level in a subdivided population, we can calculate the expected heterozygosity of the total population:

$$H_T = 2\overline{p}\,\overline{q} \qquad (4.21)$$

where $\overline{p}$ and $\overline{q}$ are average allele frequencies for all the subpopulations. The average allele frequency for all subpopulations is equivalent to combining all alleles for all subpopulations into a single population and then simply estimating allele frequencies. In other words, it is the allele frequency for the total population without any divergence among subpopulations taken into account. H_T is, therefore, the Hardy–Weinberg expected frequency of heterozygotes in the entire population if there were no population structure of allele frequencies.

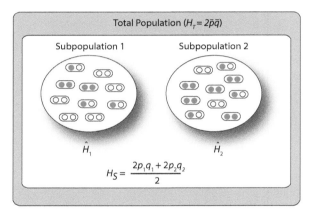

Figure 4.10 Illustration of the hierarchical nature of heterozygosity in a subdivided population. $H_1 = \frac{3}{10}$ and $H_2 = \frac{3}{10}$ to give an average observed heterozygosity of $H_I = \frac{1}{2}\left(\frac{3}{10} + \frac{3}{10}\right) = 0.30$. If *p* is the frequency of one allele (open circles) and *q* the frequency of the alternate allele (filled circles), then $p_1 = 13/20 = 0.65$ and $q_1 = 1 - p_1 = 0.35$ while $p_2 = 7/20 = 0.35$ and $q_2 = 1 - p_2 = 0.65$. The average expected heterozygosity in the two subpopulations is then $H_S = \frac{1}{2}[2(0.65)(0.35) + 2(0.35)(0.65)] = 0.455$. In the total population, the average allele frequencies are $\overline{p} = \frac{1}{2}(0.65 + 0.35) = 0.50$ and $\overline{q} = \frac{1}{2}(0.35 + 0.65) = 0.50$ giving an expected heterozygosity in the total population of $H_T = 2\overline{p}\,\overline{q} = 2(0.5)(0.5) = 0.5$.

These different levels of observed and expected heterozygosity are diagrammed in Figure 4.10 for the case of a total population composed of two subpopulations that each contain 10 diploid individuals. In both subpopulations, three of the 10 individuals are heterozygotes giving observed heterozygote frequencies of $H_1 = \frac{3}{10}$ and $H_2 = \frac{3}{10}$. Together, these yield an average observed heterozygosity of $H_I = \frac{1}{2}\left(\frac{3}{10} + \frac{3}{10}\right) = 0.30$. To determine the average

Table 4.6 The mathematical and biological definitions of fixation indices for two levels of population organization.

$F_{IS} = \dfrac{H_S - H_I}{H_S}$	The difference between average observed heterozygosity and average Hardy–Weinberg expected heterozygosity within each subpopulation due to non-random mating. The correlation between the states of two alleles in a genotype sampled at random from any subpopulation.
$F_{ST} = \dfrac{H_T - H_S}{H_T}$	The reduction in heterozygosity due to subpopulation divergence in allele frequency. The difference between the average expected heterozygosity of subpopulations and the expected heterozygosity of the total population. Alternately, the probability that two alleles sampled at random from a single subpopulation are identical given the probability that two alleles sampled from the total population are identical.
$F_{IT} = \dfrac{H_T - H_I}{H_T}$	The net excess or deficit of heterozygosity caused by the combination of subpopulation divergence of allele frequencies and non-random mating within subpopulations. Alternatively, the correlation between the states of two alleles in a genotype sampled at random from a single subpopulation given the possibility of non-random mating within populations *and* allele frequency divergence among populations.

expected heterozygosity of the subpopulations requires observed allele frequencies for each subpopulation. In subpopulation one, 13 of the 20 alleles are red and seven of the 20 alleles are blue. If p is the frequency of the red alleles and q the frequency of the blue alleles, then $p_1 = 13/20 = 0.65$ and $q_1 = 1 - p_1 = 0.35$. In subpopulation two, the situation is the exact opposite with $p_2 = 7/20 = 0.35$ and $q_2 = 1 - p_2 = 0.65$. The average expected heterozygosity in the two subpopulations is then $H_S = \frac{1}{2}[2(0.65)(0.35) + 2(0.35)(0.65)] = 0.455$. In the total population, average allele frequencies are $\bar{p} = \frac{1}{2}(0.65 + 0.35) = 0.50$ and $\bar{q} = \frac{1}{2}(0.35 + 0.65) = 0.50$. (Notice that obtaining the average of the subpopulation allele frequencies is equivalent to combining all the alleles in the total population and then estimating the allele frequency, as in $\bar{p} = \frac{13 + 7}{40} = 0.50$.) The expected heterozygosity of the total population is then $H_T = 2(0.5)(0.5) = 0.5$.

After calculating the different observed and expected heterozygosities in Figure 4.6, it is apparent that they are not all equivalent. There are differences between the observed and expected heterozygosities at the different hierarchical levels of the population. Recall from Section 2.5 that the difference between observed and Hardy–Weinberg expected genotype frequencies was used to estimate the fixation index or F. In that case, there was only a single population, and we were only concerned with how alleles

combined into diploid genotypes compared with the expectation under random mating. The fixation index can be extended to accommodate multiple levels of population organization, thereby creating measures of deviation from Hardy–Weinberg expected genotype frequencies caused by two distinct processes. With multiple subpopulations, there is a possible excess or deficit of heterozygotes due to non-random mating within subpopulations *and* a possible deficit of heterozygotes among subpopulations compared to panmixia. In the latter case, the fixation index will show how much allele frequencies have diverged among subpopulations due to processes that cause population structure compared with the ideal of uniform allele frequencies among subpopulations expected with panmixia.

Accounting for non-random mating and divergence of subpopulation allele frequency necessitates several new versions of the fixation index. The definitions of these new fixation indices are shown in Table 4.6. Let us employ and interpret each of these versions of the fixation index for the example in Figure 4.6. F_{IS} compares average observed heterozygosity of individuals in each subpopulation and the average Hardy–Weinberg expected heterozygosity for all subpopulations (the I stands for individuals and the S for subpopulations). F_{IS} is identical to the single-population F used in Section 2.5, except that it is now an average for all subpopulations. Using the heterozygosities determined above,

$$F_{IS} = \frac{0.455 - 0.30}{0.455} = 0.341 \qquad (4.22)$$

This is a result that makes biological sense since there are fewer heterozygotes in each subpopulation than would be expected under random mating given the subpopulation allele frequencies. Thus, there is more homozygosity or fixation within the two subpopulations than expected under random mating. The subpopulations on average have a deficit of heterozygosity as expected if there is consanguineous mating taking place.

The next level in the hierarchy is the average expected heterozygosity for subpopulations compared with expected heterozygosity for the total population or F_{ST} (the S stands for subpopulations and the T for the total population). Based on the heterozygosities determined previously,

$$F_{ST} = \frac{0.50 - 0.455}{0.50} = 0.09 \qquad (4.23)$$

This result says that there is somewhat less heterozygosity on average for the two subpopulations compared with the heterozygosity expected in the ideal case where the entire population is panmictic. This is consistent with the fact that the two subpopulations have slightly different allele frequencies, and each has an expected heterozygosity of slightly less than 1/2. However, the total population would have a heterozygosity of 1/2 (the maximum) if there was no allele frequency divergence between the two subpopulations.

The final level in the hierarchy is F_{IT}, the comparison of the average observed heterozygosity for subpopulations with the heterozygosity expected for the total population:

$$F_{IT} = \frac{0.50 - 0.30}{0.50} = 0.40 \qquad (4.24)$$

This gives the combined departure from Hardy–Weinberg expected genotype frequencies due to the combination of non-random mating within subpopulations and divergence of allele frequencies among subpopulations. For this example, homozygosity is 40% greater or heterozygosity 60% less than would be expected in an ideal, randomly mating panmictic population with the same allele frequencies.

Problem box 4.2
Compute F_{IS}, F_{ST}, and F_{IT}

Levin (1978) used allozyme electrophoresis to estimate genotype frequencies for the phosphoglucomutase-2 gene (*Pgm*-2) in *Phlox cuspidata*, a plant capable of self-fertilization. Genetic data were collected from 43 populations across the species range in southeast Texas. Using starch gel electrophoresis, the frequencies of two alleles (fast and slow running) and the frequencies of the heterozygous genotype were recorded for each population. A portion of the data is given in the table below (population numbers match Table 2 in Levin (1978)).

	Subpopulation			
	1	9	43	68
Frequency of *Pgm*-2 fast	0.0	0.93	0.17	0.51
Frequency of *Pgm*-2 slow	1.0	0.07	0.83	0.49
Heterozygote frequency	0.0	0.14	0.34	0.40

Using the heterozygote and allele frequencies, compute the hierarchical heterozygosities H_I, H_S, and H_T and use these to calculate F_{IS}, F_{ST}, and F_{IT}. Is there evidence that *P. cuspidata* individuals engage in selfing? Are the populations panmictic or subdivided?

The individual, subpopulation, and total population heterozygosities are identical in populations after compensating for the degree to which observed and expected heterozygosities are not met at different levels of population organization. The average observed heterozygosity is greater or lesser than the average expected heterozygosity for subpopulations:

$$H_I = H_S(1 - F_{IS}) \qquad (4.25)$$

to the extent that there is non-random mating ($F_{IS} \neq 0$). Similarly, the average expected heterozygosity for subpopulations is less than the expected heterozygosity of the total population under panmixia:

$$H_S = H_T(1 - F_{ST}) \qquad (4.26)$$

to the extent that subpopulations have diverged allele frequencies ($F_{ST} > 0$). The total deviation from expected heterozygosity within and among subpopulations is then

$$H_I = H_T(1 - F_{IT}) \qquad (4.27)$$

Although Eqs. 4.14–4.16 can be considered as rearrangements of Eqs. 4.11–4.13, they also represent a different way to articulate and think of the biological impacts of allele frequency divergence among subpopulations and non-random mating within subpopulations. Each fixation index expresses the degree to which random mating expectations for the frequency of heterozygous genotypes are not met. It is also possible to show how the total reduction in heterozygosity relates to the combined fixation due to non-random mating and subpopulation divergence:

$$1 - F_{IT} = (1 - F_{ST})(1 - F_{IS}) \qquad (4.28)$$

Since using the fixation index to measure allele frequency divergence among subpopulations is the novel concept in this section, let us consider an additional example that focuses exclusively on F_{ST}. Figure 4.11 shows allele frequencies for a diallelic locus in two populations that are composed of six subpopulations. The pattern of allele frequencies among the subpopulations is very different. On the right, all subpopulations have the same allele frequencies, while, on the left, each subpopulation is at either complete fixation or complete loss for one allele. In both sets of populations, $H_T = 2(0.5)(0.5) = 0.5$. The only difference between the two sets of populations is how allele frequencies are organized, or H_S. In the right-hand population, all six subpopulations have allele frequencies of 1/2, giving $H_S = (6(2)(0.5)(0.5))/6 = 0.5$. In the left population, three subpopulations have an allele frequency of zero and three subpopulations have an allele frequency of one. This pattern gives $H_S = (3(2)(1.0)(0) + 3(2)(0)(1.0))/6 = 0.0$. Using these expected heterozygosities for the subpopulations and total population gives $F_{ST} = 0.0$ on the right and $F_{ST} = 1.0$ on the left.

The average allele frequency in the total population is the same in both cases. However, there is a major difference in the way that allele frequencies are organized. On the right, all the subpopulations have identical allele frequencies, as would be expected if the subpopulations were really not subdivided at all. On the left, the subpopulations are highly diverged in allele frequencies as expected with strong population subdivision. Therefore, the different values of F_{ST} reflect the different degrees of allele frequency divergence among the sets of subpopulations. When all of the subpopulations are well mixed and have similar

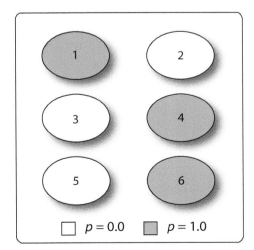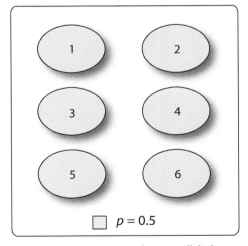

Figure 4.11 Allele frequencies at a diallelic locus for populations that consist of six subpopulations. Allele frequencies within subpopulations are indicated by shading. On the left, individual subpopulations are either fixed or lost for one allele. On the right, all subpopulations have identical allele frequencies of $p = q = 0.5$. In both cases, the total population has an average allele frequency of $\bar{p} = 0.5$ and an expected heterozygosity of $H_T = 2\bar{p}\bar{q} = 0.5$. In contrast, the average expected heterozygosity for subpopulations is $H_S = \overline{2pq} = 0.5$ on the right and $H_S = \overline{2pq} = 0.0$ on the left. $F_{ST} = 1.0$ on the right since the subpopulations have maximally diverged allele frequencies. $F_{ST} = 0.0$ on the left since the subpopulations all have identical allele frequencies. Divergence of allele frequencies among subpopulations produces a deficit of heterozygosity relative to the Hardy–Weinberg expectation based on average allele frequencies for the total population.

allele frequencies, H_S and H_T are identical. Biologically, an F_{ST} value of 0 says that all subpopulations have alleles at the same frequencies as the total population and any single subpopulation has as many heterozygotes as any other subpopulation. As the populations diverge in allele frequency due to whatever process, H_S will decrease and F_{ST} will approach 1. Biologically, an F_{ST} value of 1 says that the genetic variation is partitioned completely as allele frequency differences among the subpopulations with an absence of segregating alleles within subpopulations.

An alternative way to think about the pattern of population differentiation in allele frequency is using the variance in allele frequency relative to the amount of genetic variation in the total population to estimate F_{ST}. The estimate of allele frequency differentiation among subpopulations is then

$$F_{ST} = \frac{\text{var}(p)}{\bar{p}\bar{q}} \qquad (4.29)$$

where the variance in allele frequency among n subpopulations is $\text{var}(p) = \frac{1}{n}\sum_{i=1}^{n}(p_i - \bar{p})^2$ and there are a very large number of subpopulations (Wright 1943a). If there is more variance in allele frequency, then subpopulations differ more in allele frequency and the resulting F_{ST} is larger. For example, in Figure 4.7, the average allele frequency or $\bar{p}$ in both sets of six subpopulations is 0.5. On the right, the variance in p is $\frac{3((0-0.5)^2) + 3((1-0.5)^2)}{6} = 0.25$, while, on the left, the variance in p is $\frac{6((0.5-0.5)^2)}{6} = 0.0$. The leads to $F_{ST} = 1.0$ on the right where there is maximum variance in allele frequencies among the subpopulations given the allele frequencies and $F_{ST} = 0.0$ on the left where there is no variance in allele frequencies among the subpopulations. Measuring the variance in allele frequencies among subpopulations is the basis of several widely employed methods used to estimate $\hat{F}_{ST}$ from the actual genetic marker data.

Estimating fixation indices

Throughout this section, a single locus with two alleles has been used to illustrate hierarchical heterozygosities and fixation indices. These examples are simple, and this section can be thought of as presenting the conceptual derivation of a parameter. In practice, there are numerous details involved in obtaining the fixation index parameter estimates $\hat{F}_{IS}$, $\hat{F}_{ST}$, and $\hat{F}_{IT}$. Each estimator of a fixation index makes a range of assumptions that reflect the type

of genetic data employed (e.g. DNA sequence polymorphisms or variable microsatellite loci) as well as the assumptions about the rates of gene flow, genetic drift, and mutation. Choosing an estimator and then correctly interpreting it requires understanding these assumptions and connections to a model of biological processes (see Rousset 2013).

G_{ST} is a widely employed estimator of the fixation of subpopulations compared to the total population that averages heterozygosities (termed gene diversities by Nei) when loci have more than two alleles and also averages heterozygosities over multiple loci (Nei 1973). Weir and Cockerham (1984) developed the coancestry coefficient for alleles and an estimator of fixation among populations, θ_{ST} (pronounced "theta"), that gives rigorous statistical treatment to features of actual data such as variable sample sizes among loci and subpopulations as well as different numbers of alleles among loci (see also Weir 1996; Weir and Goudet 2017). Excoffier et al. (1992) developed estimators that are analogous to the analysis of variance of allelic state differences within and among subpopulations. This approach, termed the analysis of molecular variance or AMOVA estimates Φ (pronounced "phi") based on a measure of the differences between allelic states, call it d, is sampled from different levels of the population hierarchy. An AMOVA estimate of genetic differentiation among subpopulations is $\Phi_{ST} = \frac{d_T - d_S}{d_T}$ where d_T is the average of allelic state differences between all pairs of alleles in the total population and d_S is the average over all subpopulations of average allelic state differences between all pairs of alleles within each subpopulation. (The exact way that allelic state differences contribute to a genetic distance measure depends on a mutation model, as explained in Chapter 5.) Finally, the estimator ρ_{ST} (pronounced "roe") or R_{ST} is frequently used with microsatellite or simple sequence repeat loci to account for high rates of stepwise mutation that can obscure population structure (Slatkin 1995; see Chapter 5). These estimators are all interpreted in the same way as the biallelic version of the fixation indices.

Wright's original derivation of fixation indices assumed that loci had at most two alleles, and also that mutation had a very low rate and so had a very small impact on allele frequencies or the introduction of new alleles. Relatively recently, microsatellite genetic loci came into widespread use for studies of population differentiation. Microsatellite loci often exhibit many alleles per locus as a consequence of high mutation rates. While a locus with two alleles has a maximum expected heterozygosity of 0.5, a locus

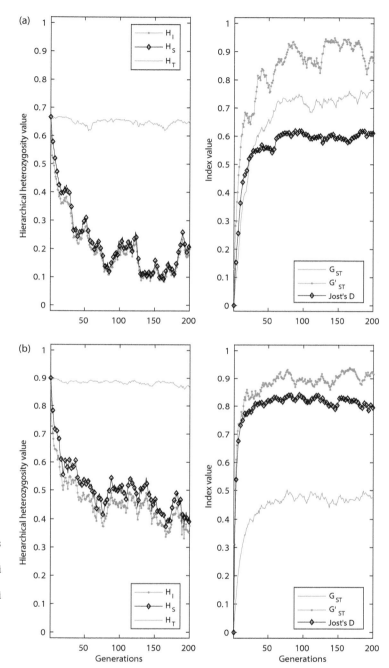

Figure 4.12 Genetic differentiation in a finite island model simulation that illustrates the differences among G_{ST}, G'_{ST}, and D as estimators when loci have more than two alleles. Panel A has loci with three alleles, and panel B has loci with 10 alleles. The simulation was carried out for 200 generations with 10 replicate neutral loci in a finite island model of 20 subpopulations where each subpopulation contained 10 individuals and the rate of gene flow was $m = 0.0005$. G_{ST}, G'_{ST}, and D are multilocus estimates using all loci.

with many alleles can have a maximum expected heterozygosity of nearly one because the frequency of each allele may be only a few percent. A result of such high allelic diversity at each locus is that expected heterozygosity for the total population, H_T, is also high and will result in maximum values of G_{ST} that are less than one. Figure 4.12 gives an example of G_{ST} in a finite island model simulation when H_T is very high. This is an example of the more general phenomenon that the range of F_{ST} is a function of allele frequencies (see Jakobsson et al. 2013; Alcala and Rosenberg 2017 and references therein).

In response to the observation that G_{ST} cannot be greater than the average within-subpopulation homozygosity, Hedrick (2005) proposed a version of G_{ST} that is rescaled by the maximum value given the allele frequencies

$$G'_{ST} = \frac{G_{ST}}{\frac{1 - H_s}{1 + H_s}} \qquad (4.30)$$

(assuming a large number of subpopulations). This results in a measure of genetic differentiation that always ranges between zero and one, analogous to the normalized coefficient of gametic disequilibrium (see Chapter 2). Hedrick suggested that G'_{ST} would allow more accurate comparisons of genetic differentiation when diversity levels were very different as long as mutation rates are much less than the rate of gene flow.

Similarly motivated by cases where heterozygosity is high, Jost (2008, 2009) proposed an estimator that measures deviation from complete differentiation of subpopulations according to $D = \left(\frac{d}{d-1}\right)\frac{H_T - H_S}{1 - H_S}$ where d is the number of subpopulations in a finite island model.

Jost's paper produced some controversy and stimulated a renewed discussion of estimators of genetic differentiation. Responses to Jost's paper helped to clarify the implicit assumptions in D and also in fixation indices (Heller and Siegismund 2009; Ryman and Leimar 2009; Whitlock 2011; Verity and Nichols 2014; Jost et al. 2018). While F_{ST} measures deviations from panmixia among subpopulations, Jost's D measures deviations from complete differentiation among subpopulations.

Based on examining the impacts of both gene flow and mutation on several measures of genetic differentiation in the finite island model, Whitlock (2011) suggested that G'_{ST} and D are not useful metrics of population genetic differentiation for two main reasons. First, it is currently difficult to connect estimates of G'_{ST} and D to evolutionary rate parameters for genetic drift and gene flow that are the key quantities of interest when estimating genetic differentiation (see Section 4.6). Second, the values of G_{ST}, G'_{ST} and D all depend on the mutation rate. When the mutation rate is high relative to the rate of gene flow, genetic differentiation tends to be reduced for G_{ST} because mutation is reducing identity by descent (see more on this in Chapter 5). In contrast, G'_{ST} and D tended to remain close to one, especially with larger numbers of subpopulations, and did not provide a means to distinguish the process or processes causing the high values. See Figure 4.12 to compare G_{ST}, G'_{ST}, and D for loci with different numbers of alleles and different levels of total population heterozygosity.

In Whitlock's study, G'_{ST} did not change much with different rates of gene flow if the mutation rate was high and number of subpopulations was large. For D, its value was also sensitive to mutation rates when the rate of gene flow was relatively high with the consequence that locus to locus variation in the mutation rate would also cause D to have high locus to locus variation. In contrast, G_{ST} was not impacted when mutation rates were much less than gene flow rates. Comparing genetic differentiation in two species, for example, could be assisted by G'_{ST} if mutation rates were similar, but mutation rates are seldom known to facilitate such comparisons. Whitlock's paper highlights how fixation indices are intimately connected with predictions related to rates of genetic drift, gene flow, and mutation and that the selection of an estimator of population genetic differentiation should be motivated by the knowledge of underlying models as well as by goals for hypothesis testing.

4.5 Population subdivision and the Wahlund effect

* Genetic variation can be present as heterozygosity within a panmictic population or as differences in allele frequency among diverged subpopulations.

The previous section of the chapter showed how departures from Hardy–Weinberg expected frequencies of heterozygotes can be used to quantify departures from random mating within demes and allele frequency divergence among demes. This section will further explore heterozygosity within and among several demes with two main goals. The first is to explore the consequences of population subdivision on expected genotype frequencies. The second is to show why F_{ST} functions to estimate allele frequency divergence among demes.

Consider the case of a diallelic locus for two randomly mating demes. The expected heterozygosity for each deme is

$$H_i = 2p_i q_i \qquad (4.31)$$

where i indicates a single subpopulation. The average heterozygosity of the two demes is based on taking the heterozygosity within each subpopulation and then averaging:

$$H_S = \frac{2p_1 q_1 + 2p_2 q_2}{2} \qquad (4.32)$$

In contrast, the heterozygosity in the total population is

$$H_T = 2\bar{p}\bar{q} \qquad (4.33)$$

based on the product of subpopulation average allele frequencies. H_T and H_S both cannot exceed 0.5, the

Figure 4.13 A graphical demonstration of the Wahlund effect for a diallelic locus in two demes. If there is random mating within subpopulations (H_1 and H_2) and in the total population (H_T), the heterozygosity of each falls on the parabola of Hardy–Weinberg expected frequency. The average heterozygosity of subpopulations (H_S) is at the mid-point between the deme heterozygosities. Therefore, H_S can never be greater than H_T based on the average allele frequency (the mid-point between the deme allele frequencies p_1 and p_2). Greater variance in allele frequencies of the demes is the same as a wider spread of deme allele frequencies in the two deme cases.

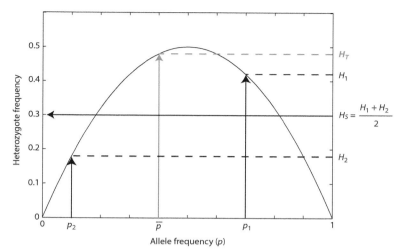

maximum heterozygosity for a diallelic locus. In addition, H_S is an average of H_1 and H_2, so when subdivided populations have different allele frequencies, H_S will always be less than the expected heterozygosity of the total population. These conditions assure that $H_T \geq H_S$ when there is random mating within the subpopulations. This relationship between H_T and H_S is shown graphically in Figure 4.13. This phenomenon is called the **Wahlund effect** after the Swedish geneticist Sten Gosta William Wahlund who first described it in 1928. One result is that F_{ST} will be greater or equal to 0 since the numerator in the expression for F_{ST} is $H_T - H_S$.

Wahlund effect: The decreased expected frequency of heterozygotes in subpopulations with diverged allele frequencies compared with the expected frequency of heterozygotes in a panmictic population of the same total size with the same average allele frequencies.

The Wahlund effect can also be shown in another fashion that more clearly connects it to variation in allele frequencies among subpopulations. The goal will now be to show that the difference between the expected heterozygosity in the total population (H_T) and the average expected heterozygosity of the subpopulations (H_S) depends on the variance in allele frequencies among the subpopulations.

The variance in allele frequencies among a set of subpopulations is

$$\text{Var}(p) = \frac{\sum (p_i - \bar{p})^2}{n} = \frac{\sum p_i^2}{n} - \bar{p}^2 \qquad (4.34)$$

where p_i is the allele frequency in subpopulation i. It also turns out that for a diallelic locus, var(p) equals var(q) since $p = 1 - q$. This result will be used later.

The average expected heterozygosity of the subpopulations

$$H_S = \frac{1}{n} \sum_{i=1}^{n} 2p_i q_i \qquad (4.35)$$

can also be expressed as

$$H_S = \sum 2 \left(\frac{p_i}{n} - \frac{p_i^2}{n} \right) \qquad (4.36)$$

by noticing that $p_i q_i = p_i (1 - p_i) = p_i - p_i^2$ since $p = 1 - q$. This equation can be rearranged to

$$H_S = 2 \left(\frac{\Sigma p_i}{n} - \frac{\Sigma p_i^2}{n} \right) \qquad (4.37)$$

The right-hand term inside the parentheses is identical to a term in the expression for the variance in allele frequency. Rearranging Eq. 4.22 so that $\frac{\sum p_i^2}{n} = \text{var}(p) + \bar{p}^2$ and then substituting gives

$$H_S = 2 \left(\frac{\Sigma p_i}{n} - \text{var}(p) - \bar{p}^2 \right) \qquad (4.38)$$

This too can be simplified by noting that $\frac{\sum p_i}{n}$ is just the average allele frequency, or $\bar{p}$, and then making the substitution so that

$$H_S = 2\left(\bar{p} - \bar{p}^2 - \mathrm{var}(p)\right) \qquad (4.39)$$

The next step is again to use the fact that $p = 1 - q$ to replace $\bar{p} - \bar{p}^2$ with its equivalent expression $\bar{p}\bar{q}$ and multiply the terms inside the parentheses by 2 to finally obtain

$$H_S = 2\bar{p}\bar{q} - 2\mathrm{var}(p) \qquad (4.40)$$

Recall from Eq. 4.21 that $H_T = 2\bar{p}\bar{q}$ and making this substitution then gives

$$H_S = H_T - 2\mathrm{var}(p) \qquad (4.41)$$

Using an equivalent set of substitutions and algebraic rearrangements, it is also possible to show that the expected frequencies of homozygote genotypes in the subpopulations are

$$Freq(AA)_S = \bar{p}^2 + \mathrm{var}(p) \qquad (4.42)$$

and

$$Freq(aa)_S = \bar{q}^2 + \mathrm{var}(p) \qquad (4.43)$$

The changes to homozygosity and heterozygosity caused by allele frequency divergence among populations are exactly analogous to the consequences of consanguineous mating in a single population. In Section 2.6, it was shown that $freq(AA) = p^2 + fpq$ where f is the probability of identity by descent. The Wahlund effect describes a similar phenomenon where allele frequency divergence of populations leads to an increase of homozygosity in subpopulations compared to the heterozygosity expected based on total population allele frequencies.

These equations show that in a subdivided population, the expected genotype frequencies in the subpopulations are a function of the average allele frequencies in the total population as well as the variance in allele frequencies among subpopulations. A set of subpopulations in panmixia is equivalent to a situation where there is no variance in allele frequency ($\mathrm{var}(p) = 0$). In that case, $H_T = H_S$ and F_{ST} is 0 since $H_T - H_S$ is also 0. This result is consistent with the intuitive expectation that extensive gene flow homogenizes allele frequencies among subpopulations. However, when subpopulations have diverged in allele frequencies and $\mathrm{var}.(p) > 0$, then the total population will have a deficit of heterozygotes and

an excess of homozygotes compared to the case of panmixia. This method also provides the prediction that the total deficit of heterozygotes will equal the total excess of homozygotes when $\mathrm{var}.(p) > 0$.

The Wahlund effect is one example of the more general phenomenon that occurs when averaging several values from a nonlinear function and is the basis of Jensen's inequality principle which has numerous applications in ecology and evolution (see Ruel and Ayres 1999).

Interact box 4.4 Simulating the Wahlund effect

The Wahlund effect as a consequence of nonlinear averaging can be seen readily with genotype frequencies for two subpopulations plotted on a de Finetti (or ternary) plot. The website linked on the text web page has an explanation and a snippet of R code that you can use.

Try the first simulation with the default genotype frequency values (where $f = 0$). Vary the allele frequencies for the two subpopulations to make them more or less diverged. What happens to the observed genotype frequencies when you try different levels of non-random mating within subpopulations (a value other than zero for f)?

One consequence of the Wahlund effect is termed **isolate breaking** to describe the increase of heterozygote genotypes that occurs when previously subdivided populations with diverged allele frequencies experience random mating. In human populations, disease phenotypes caused by recessive alleles expressed in homozygotes include cystic fibrosis, albinism, Tay–Sachs disease, and sickle cell anemia. These disorders are more common in relatively insular populations such as Ashkenazi Jews, native American groups, and the Amish but rarer in human populations that have experienced greater amounts of genetic mixing and thereby have less of a heterozygosity deficit due to subdivision. To see the impact of isolate breaking, imagine two randomly mating populations of squirrels that initially do not share any migrants and have allele

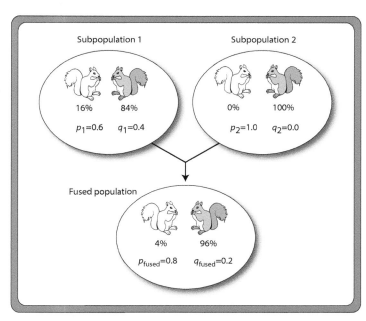

Figure 4.14 A hypothetical example of how the Wahlund effect relates variation in allele frequency between subdivided populations and genotype frequencies in a single panmictic population. Initially, the two subpopulations have different allele frequencies, and, therefore, different frequencies of homozygous recessive albino phenotypes. The average frequency of the albino phenotype is 8% in the subpopulations. When the populations fuse, the allele frequencies become the average of the two subpopulations. However, the genotype frequencies are not the average of the two subpopulations. Rather, homozygotes become less frequent and heterozygotes more frequent than their respective subpopulation averages. In the fused population, the degree to which the frequencies of both homozygotes combined and the heterozygotes differ from their subpopulation averages is the same as the variance in allele frequency between the two subpopulations.

frequencies that have diverged over time (Figure 4.14). Suppose that the population on the left has albino individuals and the basis of the albino phenotype is the completely recessive allele a with a frequency q. The albino allele is completely absent in the population on the right. The average frequency of albino squirrels in the subdivided population is

$$\text{Average frequency } (aa) = \overline{q^2} = \frac{0.16 + 0}{2} = 0.08 \quad (4.44)$$

Relying on Hardy–Weinberg, we can similarly determine the average frequency of dominant homozygotes $\left(\overline{p^2} = \frac{0.36 + 1.0}{2} = 0.68\right)$ and heterozygotes $\left(\overline{2pq} = \frac{0.48 + 0.0}{2} = 0.24\right)$ for the two subpopulations.

Next, imagine that the two populations of squirrels fuse into one randomly mating population. What is the frequency of the recessive allele and expected frequencies of albino squirrels in this fused population after random mating occurs? First, determine the allele frequencies in the fused population:

$$q_{fused} = \frac{0.4 + 0.0}{2} = 0.2 \quad (4.45)$$

and then use that result to determine the expected frequency of homozygous recessive genotypes in the fused population:

$$q^2_{fused} = \left(\frac{0.4 + 0.0}{2}\right)^2 = (0.2)^2 = 0.04 \quad (4.46)$$

There are fewer albino squirrels in the fused population (4%) than there were for the average of the two subdivided populations (8%). You can verify that the other homozygote also decreases in frequency in the fused population. The frequencies of both homozygotes have decreased by 4% in the fused population compared to their average frequencies in the subdivided populations. In contrast, the frequency of heterozygotes in the fused population

$$2pq_{fused} = 2(0.2)(0.8) = 0.32 \quad (4.47)$$

is greater than the average frequency of heterozygotes in the subdivided populations (see Table 4.7).

Now, let us determine the parametric variance in allele frequency among the two populations before and after fusion. Initially, the variance in allele frequency for the two subdivided populations is

$$\text{Var}(q) = \frac{(0.4 - 0.2)^2 + (0.0 - 0.2)^2}{2} = 0.08 \quad (4.48)$$

Table 4.7 Allele and genotype frequencies for the hypothetical example of albino squirrels in Figure 4.9 used to demonstrate Wahlund's principle. Initially, the total population is subdivided into two demes with different allele frequencies. These two populations are then fused and undergo one generation of random mating.

	Initial subpopulations	Fused population
Allele frequency q	0.4 and 0.0	$\dfrac{0.4+0.0}{2}=0.2$
Variance in q	$\dfrac{(0.4-0.2)^2-(0.0-0.2)^2}{2}=0.04$	0
Frequency of aa	$\overline{q^2}=\dfrac{0.16-0.0}{2}=0.08$	$(0.2)^2=0.04$
Frequency of Aa	$\overline{2pq}=\dfrac{0.48-0.0}{2}=0.24$	$2(0.2)(0.8)=0.32$
Frequency of AA	$\overline{p^2}=\dfrac{0.36-1.0}{2}=0.68$	$(0.8)^2=0.64$

whereas var.(q) is 0 after fusion because there is no longer any subdivision for allele frequencies. Take note of the fact that the initial variance in the allele frequencies (0.08) is exactly half the difference between the average frequency of albinos before fusion and the expected frequency of albinos in the fused population. With fusion of the subdivided populations, each homozygote has decreased by 4% and the heterozygote has increased by exactly the same total amount or 8%.

This example shows that removing the allele frequency differences between the two subpopulations by making them into a panmictic population has changed the total population heterozygosity. The result is exactly what is predicted by the Wahlund effect, with more total population heterozygosity under panmixia than under subdivision. Subdivided populations store some genetic variation as differences (variation) in allele frequency among populations at the expense of heterozygosity in the total population. Another way to think of this is that population subdivision is equivalent to mating among relatives that increases the total population homozygosity (or reduces the total population heterozygosity). A fused or panmictic population has a larger effective size than individual subdivided populations with restricted gene flow. In the subpopulations, mating is most probable within the subpopulation rather than with a migrant from the total population. The subpopulations, therefore, have more autozygosity compared to a panmictic population of equivalent size, analogous to the decline in heterozygosity seen in a single finite population due to genetic drift.

An application of Wahlund's principle can be found in forensic DNA profiling. As covered in Section 2.4, the use of DNA markers to determine the expected frequency of a given genotype occurring by chance relies on estimates of allele frequencies in various racially defined human populations. Although allele frequencies at loci used in DNA profiles have been estimated in many populations, there are a limited number of these reference allele frequency databases available. It is, therefore, possible that population-specific allele frequency estimates are not available for some individuals depending on their racial, ethnic, or geographic background. A further complication is that many individuals have ethnically diverse ancestry that may not be represented by any single set of available reference allele frequencies. None of this would be a problem in DNA profiling if human populations exhibited panmixia, since there would then be uniform allele frequencies among all racially defined human populations. However, racially and geographically defined human populations like those used to construct allele frequency reference databases show up to 3–5% population divergence of allele frequencies (Rosenberg et al. 2002).

We can use Wahlund's principle to adjust DNA-profile odds ratios for the effects of population structure. This requires a method to adjust the expected genotype frequency at each locus to account for the increased frequency of homozygotes and the decreased frequency of heterozygotes caused by the divergence of allele frequencies among populations. The adjusted expected frequencies for homozygote genotypes are

$$f(A_iA_i) = p_i^2 + p_i(1 - p_i)F_{IT} \qquad (4.49)$$

and the adjusted expected frequencies for heterozygote genotypes are

$$f(A_iA_j) = 2p_ip_j - (2p_ip_j)F_{IT} = 2p_ip_j(1 - F_{IT}) \qquad (4.50)$$

where i and j represent different alleles at the A locus and F_{IT} measures the total departure of genotype frequencies from frequencies expected under panmixia due to both non-random mating within populations and allele frequency divergence among populations (National Research Council, Commission on DNA Forensic Science 1996). If mating within populations is random ($F_{IS} = 0$), then F_{IT} is equivalent to F_{ST} in these two equations. In that case, applying these corrections increases the frequency of homozygotes and decreases the frequency of heterozygotes in proportion to the degree of allele frequency divergence among populations.

In Section 2.4, the expected frequency of a three-locus DNA profile was determined under the assumptions of Hardy–Weinberg and panmixia. Let us return to that example and adjust the expected genotype frequency and odds ratio to compensate for population structure in human populations. The expected genotype frequencies are given in Table 4.8 based on the upper bound estimate of $F_{ST} = 0.05$ in human populations. The adjustment reduces the expected frequencies of the two heterozygous loci and increases the expected frequency of the homozygous locus. The odds ratio for a chance match of this three-locus genotype was one in 20 408 under the assumption of panmixia and becomes one in 15 152 after adjusting for population structure. Thus, population structure increases the expected frequency of this three-locus genotype by about 35% of its expected frequency under panmixia. A random match for this three-locus genotype is more probable after adjustment for population structure. Compensating for population structure in determining DNA-profile odds ratios is required to obtain an accurate estimate of how often DNA profiles match by chance alone (National Research Council, Commission on DNA Forensic Science 1996). Using Eqs. 4.49 and 4.50 to adjust for population structure is necessary when an appropriate reference allele frequency database is not available, the ethnicity of the individual is not known, or the genotype comes from a person of mixed ancestry, and, therefore, the choice of the appropriate database is not obvious.

Problem box 4.3
Impact of population structure on a DNA-profile match probability

Return to Section 2.4 and Problem Box 2.1 to determine the expected genotype frequency and the probability of a random match after compensation for population structure seen in human populations. Assume that $F_{ST} = 0.05$ for human populations. How does the expected genotype frequency change at individual loci when there is population structure and why? Is the 10-locus genotype still rare enough that the chance of a random match is low?

Table 4.8 Expected frequencies for individual DNA profile loci and the three loci combined with and without adjustment for population structure. Calculations assume that $F_{IS} = 0$ and use the upper bound estimate of $F_{ST} = 0.05$ in human populations. Allele frequencies are given in **Table 2.3**.

Locus	Expected genotype frequency	
	With panmixia	**With population structure**
D3S1358	2(0.2118)(0.1626) = 0.0689	2(0.2118)(0.1626)(1-0.05) = 0.0655
D21S11	2(0.1811)(0.2321) = 0.0841	2(0.1811)(0.2321)(1-0.05) = 0.0799
D18S51	$(0.0918)^2 = 0.0084$	$(0.0918)^2 + 0.0918(1-0.0918)(0.05) = 0.0126$
All loci	(0.0689)(0.0841)(0.0084) = 0.000049	(0.0655)(0.0799)(0.0126) = 0.000066

The next section explores models of population structure that can be used to infer the causes of a given pattern of population structure.

4.6 Evolutionary models that predict patterns of population structure

- The infinite and finite island models.
- Stepping-stone and metapopulation population models.
- Isolation by distance.
- Least cost paths of gene flow and isolation by resistance.
- General expectations and conclusions from the different migration models.

An important goal of estimating genetic differentiation among populations using F_{ST} (or one of its estimators) is to infer the population genetic processes that have caused observed patterns of genetic differentiation. This crucial step requires predictive models that define rates of genetic drift and gene flow in order to predict the pattern and magnitude of F_{ST}. With these predictions, it is then possible to use observations of $\hat{F}_{ST}$ in actual populations to infer quantities related to gene flow and genetic drift.

There are numerous models of population structure that attempt to approximate various gene flow patterns likely to be found in actual populations. However, these models do not necessarily capture the exact mixture of gene flow features in actual populations. It is likely, in fact, that gene flow within and among actual subpopulations of real organisms is not as easily categorized nor as invariant as is assumed in these models. Nonetheless, these models of population structure are useful tools to study the general principles that cause population differentiation. The utility of these different models of population structure is their ability to show the basic and somewhat general features of the impact of rates of gene flow, the size of subpopulations, and the patterns of genetic connectedness among subpopulations on the evolution of genotype and allele frequencies within and among populations.

Infinite island model

One of the oldest and most widely used models of the process of gene flow among a set of subpopulations is Wright's (1931, 1951) infinite island model. Gene flow takes the form of all subpopulations being equally likely of exchanging migrants with any other subpopulation, equivalent to a complete absence of isolation by distance. In addition, the size and migration rate of each subpopulation is most commonly assumed to be equal. The total population is made up of an infinite set of subpopulations each of size N_e with m percent of each subpopulation's gene copies exchanged at random with the rest of the population every generation (see Figure 4.5). Using this model, it is possible to approximately relate the degree of differentiation among subpopulations to a function of the effective population size and the amount of migration.

> **Infinite island model:** An idealized model of population subdivision and gene flow that assumes an infinite number of identical subpopulations (demes) and that each subpopulation experiences an equal probability of gene flow from all other subpopulations.

Let us first consider what will happen in the infinite island model when there is no gene flow among the subpopulations ($m = 0$). Since each subpopulation is a finite island, allele frequencies will vary from one generation to the next because of genetic drift. The expected value of the fixation index for subpopulations compared to the total population is

$$F_{ST} = 1 - e^{-\frac{1}{2N_e}t} \qquad (4.51)$$

where t is time in generations and N_e is the effective size of a single subpopulation (Wright 1943a). In the equation, as time increases, the $e^{-\frac{1}{2N_e}t}$ term gets smaller depending on the effective population size. This serves to approximate what happens to F_{ST} as t increases – the average expected heterozygosity of subpopulations (H_S) decreases and eventually reaches zero with the consequence that F_{ST} reaches one. This results from the process of genetic drift, causing all subpopulations to eventually reach fixation or loss. Note, however, that the total population heterozygosity (H_T) is not impacted by genetic drift since each subpopulation can go to fixation or to loss, but there are an infinite number of subpopulations and so the total population size is infinite.

Next, consider what occurs in the infinite island model when both gene flow and genetic drift are acting at the same time. In Chapter 3, the fixation index

$$F_t = \frac{1}{2N_e} + \left(1 - \frac{1}{2N_e}\right)F_{t-1} \qquad (4.52)$$

was developed as a measure of the probability that two alleles in a genotype are autozygous or identical by descent in a single finite population. We can extend this equation to include the influence of migration on autozygosity when there are numerous subpopulations that experience limited gene flow each generation. The goal is to develop an expression for the fixation index that accounts for both population size and migration. Finite population size causes autozygosity to increase over time in individual subpopulations. Migration counteracts this trend, bringing in alleles from other subpopulations that are not identical by descent, thereby decreasing the autozygosity. Therefore, in general, in subdivided populations, the net autozygosity is the balance of the processes of genetic drift and migration.

When there is gene flow, two modifications need to be made to the probabilities of autozygosity given in the two terms of Eq. 4.52. The first modification involves the probability of autozygosity or $\frac{1}{2N_e}$. With migration, some proportion m of the alleles in a subpopulation arrived via gene flow from other subpopulations, while $1 - m$ of the alleles are contributed by individuals and gametes that did not leave their subpopulation. Therefore, there is some chance that one or both of a pair of alleles was introduced to a subpopulation by migration. A randomly sampled pair of alleles in a subpopulation with zero, one, or two alleles due to gene flow each generation have the probabilities of $(1 - m)^2$, $2m(1 - m)$, and m^2, respectively. Only the $(1 - m)^2$ proportion of genotypes with no alleles introduced by gene flow can contribute to the pool of alleles that may become identical by descent due to finite sampling. We can also see this by noting that $2m(1-m)$ genotypes heterozygous and m^2 genotypes homozygous for alleles entering the subpopulation by gene flow are expected each generation. Together, these two classes of genotypes bearing alleles that entered

the population by gene flow reduce the autozygosity by a factor of $1 - 2m(1 - m) - m^2 = 1 - 2m + 2m^2 - m^2 = (1-m)^2$. This gives $\frac{1}{2N_e}(1 - m)^2$ as the autozygosity adjusted for gene flow. Using the same reasoning, the chance that a randomly sampled pair of alleles in a subpopulation are autozygous due to past mating among relatives (the $\left(1 - \frac{1}{2N_e}\right)F_{t-1}$ term in Eq. 4.52) also needs to be adjusted by a factor of $(1 - m)^2$.

Bring these two changes to the autozygosity together to account for gene flow leads to

$$F_t = \frac{1}{2N_e}(1 - m)^2 + \left(1 - \frac{1}{2N_e}\right)F_{t-1}(1 - m)^2$$
$$(4.53)$$

As seen by examining this equation, when m is between 0 and 1, the effect of gene flow is to reduce the expected value of the fixation index by reducing the probability of identity both in the present (time t) and in the past (time $t - 1$). This makes intuitive sense: if gene flow introduces an allele copy into a subpopulation, it has not been present for sampling events between time $t - 1$ and time t. Therefore, an allele copy introduced by gene flow has not yet had the opportunity to become identical by descent at time t and cannot contribute to the frequency of autozygous genotypes gauged by the fixation index.

Equation 4.53 quantifies the balance of gene flow and genetic drift among multiple subpopulations, so F is identical to F_{ST}. We can make this equation more general by using it to get an expected value of the fixation index among populations (F_{ST}) in the infinite island model when allele frequency differentiation among subpopulations by genetic drift and allele frequency homogenization among subpopulations by gene flow reach a net balance. With the assumption that the migration rate is small and much, much less than the effective population size (see Math Box 4.1), an *approximation* for the expected amount of fixation among subpopulations at equilibrium in an infinite island population is:

$$F_{ST} \approx \frac{1}{4N_em + 1} \qquad (4.54)$$

as shown by Wright (1931, 1951).

Math box 4.1 The expected value of F_{ST} in the infinite island model

At equilibrium when the differentiating effects of genetic drift and the homogenizing effects of gene flow have come into balance, the value of F_{ST} does not change from one generation to the next so that $F_{ST(t)} = F_{ST(t-1)} = F_{ST(equilibrium)}$. If a population is at equilibrium, then we can set both F_t and F_{t-1} equal to F_{eq}. Making this substitution in Eq. 4.53:

$$F_{eq} = \frac{1}{2N_e}(1-m)^2 + \left(1 - \frac{1}{2N_e}\right)F_{eq}(1-m)^2 \tag{4.55}$$

This equation can be solved for F_{eq} most transparently by restating it as

$$F_{eq} = ac + bcF_{eq} \tag{4.56}$$

where $a = \dfrac{1}{2N_e}$, $b = 1 - \dfrac{1}{2N_e}$, and $c = (1-m)^2$. Then, using algebraic manipulation

$$F_{eq} - bcF_{eq} = ac \tag{4.57}$$

$$F_{eq}(1 - bc) = ac \tag{4.58}$$

$$F_{eq} = \frac{ac}{1 - bc} \tag{4.59}$$

Substituting the full expressions for a, b, and c gives

$$F_{eq} = \frac{\dfrac{1}{2N_e}(1-m)^2}{1 - \left(1 - \dfrac{1}{2N_e}\right)(1-m)^2} \tag{4.60}$$

which when multiplied by $\dfrac{2N_e}{2N_e}$ gives the simpler equation

$$F_{ST} = \frac{(1-m)^2}{2N_e - (2N_e - 1)(1-m)^2} \tag{4.61}$$

The terms in the numerator and denominator can be multiplied out to give an expression that is fairly complex (feel free to do the expansion if you are curious). However, if we again invoke the assumption that the migration rate is small and much, much less than the effective population size, then the terms in the expansion of Eq. 4.61 containing m or powers of m can be ignored since they are very small (e.g. if $m = 0.01$ then $2m = 0.02$ and $m^2 = 0.0001$). This then leads to the approximation for the expected value of the fixation index

$$F_{ST} \approx \frac{1}{4N_e m + 1} \tag{4.62}$$

Based on these assumptions, Figure 4.15 shows the expected levels of genetic differentiation among infinitely many island model subpopulations for different levels of the product of the effective population size of each deme and the migration rate among demes ($N_e m$). When $N_e m$ – often called the **effective migration rate** – is large, very little differentiation among demes is expected for a diploid locus since the combination of the effective size of demes and the migration rate is large enough to overcome the genetic differentiation caused by genetic drift. As the effective migration rate declines from a relatively large value such as $N_e m = 10$, genetic differentiation

among demes increases slowly at first and then rapidly once the effective migration rate is less than about 1. An effective migration rate of one individual every other generation ($N_e m = \frac{1}{2N_e}$) is often cited as sufficient to prevent substantial genetic differentiation at a diploid locus in the infinite island model since this rate is enough to mostly counteract the expected rate of loss of heterozygosity by genetic drift ($1 - \frac{1}{2N_e}$) in an isolated population (see Section 3.4).

The relationship between the allele frequencies of subpopulations, the hierarchical measures of heterozygosity, and the fixation indices can be seen in a

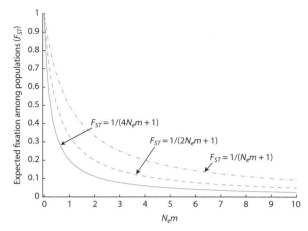

Figure 4.15 Expected levels of fixation among subpopulations depend on the product of the effective population size (N_e) and the amount of gene flow (m) in the infinite island model of population structure. Each line represents expected F_{ST} for loci with different probabilities of autozygosity (from bottom to top $\frac{1}{2N_e}$, $\frac{1}{N_e}$, and $\frac{2}{N_e}$). Marked divergence of allele frequencies among subpopulations ($F_{ST} \geq$ 0.2) are expected when $N_e m$ is below 1 for biparentally inherited nuclear loci with an autozygosity of $\frac{1}{2N_e}$. Y-chromosome or mitochondrial loci (autozygosity = $\frac{2}{N_e}$) are examples where marked divergence among populations is expected at higher levels of $N_e m$.

simulation of gene flow and genetic drift in a subdivided population (Figure 4.16). When gene flow is relatively strong and maintains similar allele frequencies in the subpopulations, the expected heterozygosity of the subpopulations and the total population are also similar and result in low values of the fixation indices (Figure 4.11A). When gene flow is weaker and the subpopulations diverge in allele frequency, the expected heterozygosity of the subpopulations is less than the expected heterozygosity of the total population, resulting in high values of the fixation indices (Figure 4.11B).

An additional point is that F_{ST} can vary considerably among independent replicate loci sampled in an identical fashion from the same subpopulations. Figure 4.17 shows the range of F_{ST} values obtained for 1000 independent loci in a simulation of the finite island model where genetic drift and gene flow among subpopulations in the absence of mutation were acting to change allele frequencies. The range of F_{ST} values for individual loci under the influence of identical population genetic processes is due to the random nature of genetic drift. Each locus has experienced random fluctuations in allele frequencies that has resulted in a range of allele frequency variance among subpopulations. This random variation in F_{ST} due to genetic drift seen in the simulation underscores the need for estimates of F_{ST} to be obtained from the average of multiple loci.

The expected relationship between the fixation index and the effective number of migrants relies on the infinite island model for two reasons. First, in the island model, all subpopulations have an identical rate of migration from all other populations so there is only a single migration rate (m) that applies to all subpopulations. Second, since there are an infinite number of subpopulations, the entire ensemble population will never reach fixation or loss due to genetic drift. In an island model of gene flow where the number of subpopulations is finite, called the **finite island model**, the entire set of populations will eventually reach fixation or loss and F_{ST} will eventually decline to zero since the entire set of subpopulations will eventually reach fixation or loss due to genetic drift in the absence of mutation (Nei et al. 1977; Varvio et al. 1986). The expected amount of genetic differentiation in the finite island model for loci with any number of alleles is

$$G_{ST} \approx \frac{1}{\left(\frac{d}{d-1}\right)^2 4N_e m + 1} \qquad (4.63)$$

where d is the number of subpopulations (Latter 1973; Takahata 1983; Crow and Aoki 1984; Takahata and Nei 1984). This version of G_{ST} corrects the expected amount of differentiation among subpopulations for a finite number of subpopulations. The term $\left(\frac{d}{d-1}\right)^2$ is at a maximum of four with two subpopulations and approaches one as d gets large. For example, when $N_e m = 0.1$ and $d = 10$, the expected value of G_{ST} is about 94% of that expected for an infinite number of demes. This implies that a given level of gene flow is somewhat more effective at homogenizing allele frequencies among fewer subpopulations than among a very large number of subpopulations. For d greater than about 50, the adjustment for a finite number of demes makes little difference and the finite number of subpopulations behaves essentially as an infinite number of subpopulations.

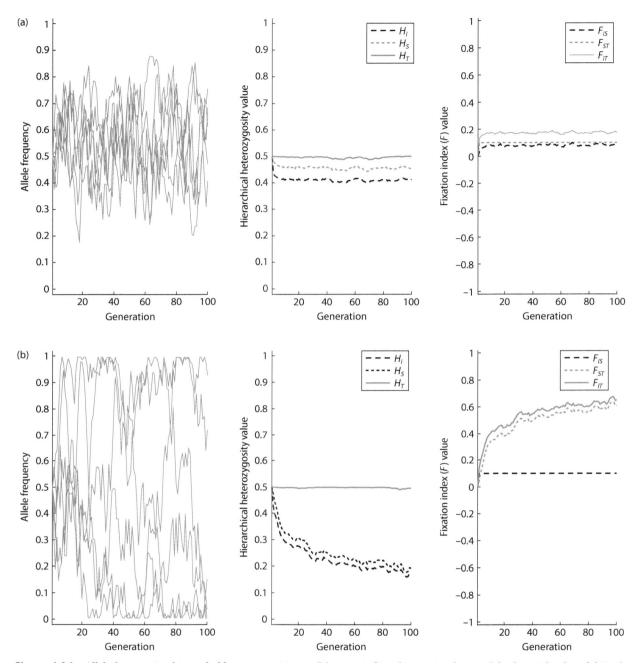

Figure 4.16 Allele frequencies, hierarchal heterozygosities, and fixation indices from a simulation of the finite island model. Each subpopulation contained 10 individuals. The rate of gene flow was $m = 0.2$ in panel A and $m = 0.01$ in panel B. The allele frequencies are shown for 6 randomly chosen subpopulations out of the 200 subpopulations in the simulation. The heterozygosities and fixation indices were calculated from all 200 subpopulations.

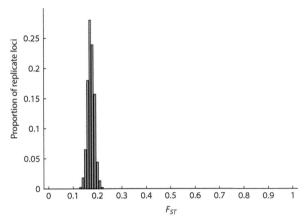

Figure 4.17 The distribution of F_{ST} values for 1000 replicate neutral loci in a finite island model of 200 subpopulations where each subpopulation contains 10 individuals and the rate of gene flow is 10% of each subpopulation ($m = 0.10$). In the distribution, 95% of the replicate loci show F_{ST} values between 0.1459 and 0.2002 while the average of all 1000 replicate loci is 0.1586 (based on the average of H_T and H_S then used to calculate F_{ST}). Replicate loci exhibit a range of F_{ST} values since allele frequencies among subpopulations are partly a product of the stochastic process of genetic drift. In an infinite island model with $N_em = 1.0$, the expected value of F_{ST} is 0.2.

Problem box 4.4
Expected levels of F_{ST} for Y-chromosome and organelle loci

What is the expected value of F_{ST} at equilibrium in the island model for Y-chromosome loci or mitochondrial and chloroplast (organelle) loci? A hint at how to approach the problem is to think about the autozygosity of loci other than diploid autosomes and then make adjustments to Eq. 4.53 that would result in different versions of Eq. 4.54. What level of fixation among populations (F_{ST}) would be expected for these types of loci compared to biparentally inherited diploid loci? What causes the difference in levels of F_{ST} for the different types of loci?

Given that the infinite island models lead to an expected level of genetic differentiation among demes for some level of the effective migration rate, it is natural to reverse the relationship:

$$N_em \approx \frac{1}{4}\left(\frac{1}{F_{ST}} - 1\right) \qquad (4.64)$$

to get an expected effective migration rate given an amount of genetic differentiation among subpopulations in the infinite island model. This equation again reinforces that the level of allele frequency differentiation among subpopulations (F_{ST}) is a function of the balance between the processes of gene flow tending to homogenize allele frequencies among subpopulations and genetic drift causing subpopulations to diverge as they individually approach fixation or loss (N_em) in the context of the infinite island model. This relationship has been used in literally thousands of studies to estimate $\hat{N_e}m$ from empirical estimates of $\hat{F}_{ST}$ in wild populations like the examples in Table 4.9. This equation (or expectations like it but based on different population models) is the basis of the so-called indirect estimates of the number of effective migrants ($\hat{N_e}m$) that cause a given pattern of allele frequency differentiation among populations ($\hat{F}_{ST}$).

It is important to recognize that employing Eq. 4.64 to estimate $\widehat{N_em}$ is really using the infinite island model as an ideal standard rather than actually estimating the long-term effective number of migrants for a specific population. Because of this dependence on the infinite island model, using $\hat{F}_{ST}$ to obtain an estimate of $\widehat{N_em}$ should be interpreted as "the observed level of population differentiation ($\hat{F}_{ST}$) would be *equivalent* to the differentiation expected in an infinite island model with a given number of effective migrants (N_em)." Such a comparison of actual and ideal populations is identical to that used in the definition of effective population size (see Section 3.3). Despite this dependence on a highly idealized model, Slatkin and Barton (1989) concluded that using observed levels of population differentiation to estimate $\widehat{N_em}$ under island model assumptions should be roughly accurate, even when the actual population structure deviates from the island model. In contrast, Whitlock and McCauley (1999) reviewed the many ways in which actual populations will deviate from the infinite island model and the assumptions used to approximate the relationship between $\hat{F}_{ST}$ and N_em, invalidating the indiscriminate use of Eq. 4.64.

Interact box 4.5 Simulate F_{IS}, F_{ST}, and F_{IT} in the finite island model

Use the text simulation website to examine the island model of gene flow among a finite number of subpopulations. Start by examining each of the simulation input parameters. Run the simulation with default parameter values and examine each of the four output graphs, noting the axes and lines plotted on each one.

Run simulations with the following values for the effective population size, the migration rate, and the initial allele frequency. What is $N_e m$ in each case?

N_e	m	Initial allele frequency p
10	0	0.5
10	0.001	0.5
10	0.1	0.5
50	0	0.5
50	0.001	0.5
50	0.1	0.5
100	0	0.5
100	0.001	0.5
100	0.1	0.5

For each simulation run, examine the allele frequencies over time in a sample of subpopulations, the hierarchical heterozygosity measures (H_I, H_S, and H_T), and fixation indices (F_{IS}, F_{ST}, and F_{IT}). What is happening when the allele frequency lines sometimes hit the top or bottom axis (go to fixation or loss) and then reappear? What is the coancestry coefficient and how does it relate to heterozygosity within subpopulations and to F_{IS}?

What are the units of the migration values you entered in the model parameters box? Why does increasing m maintain lower F_{ST} and F_{IT} values? How does migration counteract genetic drift? Is migration always strong enough to overcome the differentiating effect of genetic drift?

Table 4.9 Estimates of the fixation index among subpopulations (F_{ST}) for diverse species based on molecular genetic marker data for nuclear loci. Different estimators were employed depending on the type of genetic marker and study design. Each F_{ST} estimate was used to infer the effective number of migrants ($N_e m$) that would produce an identical level of population structure under the assumptions of the infinite island model.

Species	F_{ST} estimate	$N_e m$ estimate	References
Amphibians			
Alytes muletansis (Mallorcan midwife toad)	0.12–0.53	1.8–0.2	Kraaijeveld-Smit et al. (2005)
Birds			
Gallus gallus (broiler chicken breeds)	0.19	1.0	Emara et al. (2002)
Mammals			
Capreolus capreolus (roe deer)	0.097–0.146	2.2–1.4	Wang and Schreiber (2001)
Homo sapiens (humans)	0.03–0.05	7.8–4.6	Rosenberg et al. (2002)
native Mexican populations	0.136	1.6	Morena-Estrada et al. (2014)
European and Chinese	0.11	2.0	Altshuler et al. (2010)
Microtus arvalis (common vole)	0.17	1.2	Heckel et al. (2005)
Plants			
Arabidopsis thaliana (mouse-ear cress)	0.643	0.1	Bergelson et al. (1998)
Oryza officinalis (wild rice)	0.44	0.3	Gao (2005)
Phlox drummondii (annual phlox)	0.17	1.2	Levin (1977)
Prunus armeniaca (apricot)	0.32	0.5	Romero et al. (2003)
Fish			
Morone saxatilis (striped bass)	0.002	11.8	Brown et al. (2005)
Sparisoma viride (stoplight parrotfish)	0.019	12.4	Geertjes et al. (2004)
Insects			
Drosophila melanogaster (fruit flies)	0.037–0.063	3.7–6.5	Fabian et al. (2012)
Glossina pallidipes (tsetse flies)	0.18	1.1	Ouma et al. (2005)
Heliconius charithonia (butterflies)	0.003	79.8	Kronforst and Flemming (2001)
Corals			
Seriatopora hystrix	0.089–0.136	2.6–1.6	Maier et al. (2005)

The estimate of the effective number of migrants or $\hat{N_e}m$ obtained through the island model is referred to as an indirect estimate of the rate of gene flow. The term indirect is used because the observed pattern of allele frequency differences among subpopulations is used in a model (containing many assumptions) to produce a parameter estimate. This is in contrast to a direct estimate of gene flow from a method like parentage analysis (although Section 4.2 suggests direct methods also depend on assumptions). Such indirect estimates of gene flow have the effect of averaging across all of the past events that led up to the current pattern of allele frequency differentiation among subpopulations. In contrast, direct estimates apply to only those periods of time when parentage or movement is observed. Slatkin (1987a) considers an example where mark–recapture methods suggested the movement of a butterfly among different geographic locations was extremely limited, yet a multilocus estimate of $\hat{F}_{ST}$ suggested almost no allele frequency differentiation among the butterfly populations. One possible explanation is that gene flow was extensive in the past and has very recently decreased but not enough time has elapsed to witness increased population differentiation. Another possibility is that the infrequent gene flow events required to prevent differentiation are not well measured by the mark–recapture technique.

Stepping-stone and metapopulation models

The stepping-stone model, inspired by the flat stones that form a walking path in a Japanese garden, approximates the phenomenon of isolation by distance among discrete subpopulations by allowing most or all gene flow to be only between neighboring subpopulations (Kimura 1953; see Figure 4.5). This gene flow pattern produces an allele frequency clumping effect among the subpopulations qualitatively very similar to that seen in the first section of the chapter for isolation by distance in a continuous population of individuals (Figure 4.3). A classic analysis of the stepping-stone model was carried out by Kimura and Weiss (1964), who showed that the correlation between the states of two alleles sampled at random from two subpopulations depends on (i) the number of subpopulations separating the two sampled subpopulations and (ii) the ratio of gene flow between neighboring colonies and long-distance gene flow where alleles are exchanged among subpopulations at random

distances. As expected for isolation by distance, the correlation between allelic states decreases with increasing distance between subpopulations. Interestingly, the correlation between allelic states drops off more rapidly with distance when subpopulations occupy two dimensions than when they occupy one dimension. In a two-dimensional stepping-stone model, F_{ST} is expected to grow like the logarithm of the number of colonies for fixed values of gene flow parameters (see Slatkin and Barton 1989; Cox and Durrett 2002). Another way of saying this is that increasing levels of gene flow are required to maintain the same level of population structure as the number of colonies increases.

A logical extension of the stepping-stone model is the **metapopulation** model. Metapopulation models approximate the continual extinction and recolonization seen in many natural populations in addition to the process of gene flow. These models are motivated by organisms like pioneer plants and trees that colonize and grow in newly created clearings but eventually disappear from a patch as succession introduces new species and changes the environmental and competitive conditions. Even though each subpopulation of a pioneer species eventually goes extinct, there are other subpopulations in existence at any given time, and new subpopulations are continuously being formed by colonization. A metapopulation is then just a collection of a number of smaller subpopulations or habitat patches (see various definitions of metapopulation and related concepts in Hanski and Simberloff 1997), conceptually similar to the stepping-stone model. However, in metapopulations, the individual subpopulations have some probability of going extinct, and these unoccupied locations that become available can also be colonized to establish a new subpopulation.

Gene flow in a metapopulation can be modeled as two possible types. One type is gene flow among the existing occupied subpopulations like that in island models. The other type is gene flow that occurs when an unoccupied subpopulation is colonized to replace a subpopulation that went extinct. The pattern of gene flow that takes place during colonization may take different forms (Slatkin 1977). The first form is where colonists are sampled at random from all subpopulations, called migrant-pool gene flow. The second form is where colonists are sampled at random from only a single random subpopulation, called propagule-pool gene flow. Migrant-pool gene flow is identical to the pattern of gene flow in the island model where migrants represent the average allele frequencies of all subpopulations. In contrast, propagule-pool gene flow can introduce a genetic bottleneck when a new subpopulation is founded because the colonists are only drawn from a single existing subpopulation.

The impact of the form of colonization on heterozygosity in newly established subpopulations within a metapopulation is described by

$$F_{ST}^{colony} = \frac{1}{2k} + \phi\left(1 - \frac{1}{2k}\right)F_{ST} \qquad (4.65)$$

where F_{ST}^{colony} is the expected allele frequency differentiation in newly established subpopulations, k is the number of diploid colonists, F_{ST} is the degree of allele frequency differentiation among the existing subpopulations, and ϕ (pronounced "phi") is the probability that the two allele copies in a newly established population come from the same subpopulation (Whitlock and McCauley 1990). Colonization corresponds to the propagule pool for $\phi = 1$ (all founding allele copies originate in the same subpopulation) and the migrant pool for $\phi = 0$ (all founding allele copies originate in different subpopulations). All newly founded subpopulations have a chance of being established with alleles that are identical by descent due to sampling from the total population, hence the $\frac{1}{2k}$ term. For those subpopulations that are founded by individuals from a propagule pool (or $\phi = 1$), the chance of alleles being identical by descent and homozygous is greater to the degree that existing subpopulations are differentiated in their allele frequencies. With colonization from the propagule pool, newly founded populations inherit the average level of homozygosity of existing subpopulations plus some additional homozygosity due to sampling from a finite population. With colonization from the migrant pool ($\phi = 0$), founding alleles are always drawn from a different subpopulation so the heterozygosity is the same as the total population heterozygosity ($2\overline{p}\overline{q}$) except for sampling error from a finite number of founders. Using newly established populations of the plant *Silene alba*, McCauley et al. (1995) estimated ϕ between 0.73 and 0.89, suggesting that new populations do experience some additional sampling during their formation that increases population differentiation.

In a metapopulation, extinction and recolonization can be an additional source of gene flow or an additional restriction on gene flow (Maruyama and Kimura 1980; Wade and McCauley 1988). Propagule pool colonization tends to increase the overall population differentiation for all values of the number of diploid colonists (k). In contrast, the change in the overall differentiation with the migrant model depends on the rate of gene flow among existing subpopulations. When the number of diploid colonists (k) exceeds twice the effective number of migrants ($2N_em$), differentiation tends to decrease since colonization accomplishes additional mixing of alleles. While a useful metric of the pattern of differentiation, F_{ST} alone may not be an effective tool to estimate the effective migration rate in metapopulations experiencing extinction and recolonization (Whitlock and Barton 1997; Pannell and Charlesworth 2000).

Isolation by distance and by landscape connectivity

As explained in at the first section of this chapter, isolation by distance is expected to be a basic process that operates in many populations. Rousset (1997) described a relatively simple test for the pattern of isolation by distance that has been widely applied in empirical studies. The test relies on $\hat{F}_{ST}$ estimated between all pairs of subpopulations as well as estimates of the straight-line geographic distance (the so-called Euclidean distance) between all pairs of subpopulations. The regression slope of a plot of $\frac{\hat{F}_{ST}}{1 - \hat{F}_{ST}}$ (often called linearized $\hat{F}_{ST}$) by the logarithm of geographic distance for two-dimensional populations is expected to have a positive slope when there is isolation by distance. This type of relationship is shown in Figure 4.18 based on simulated data for a two-dimensional stepping-stone population model. Rousset (1997) showed that the slope of a regression line for such a plot is a function of the density of reproductive individuals and the variance in gene flow distances. The increase in genetic differentiation with distance also does not depend strongly on the exact shape of the gene flow distribution or dispersal kernel.

Because habitats and landscapes are physically heterogeneous and are occupied by variable sets of species and communities that may impede or facilitate gene flow, simple geographic distance between subpopulations may not be the only cause of genetic

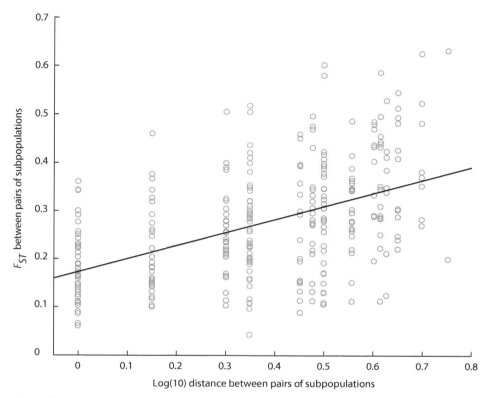

Figure 4.18 Subpopulations in a stepping-stone model exhibit an increase in genetic differentiation between pairs of populations as the geographic distance between them increases as expected under isolation by distance. The line shows a least-squares fit to the points. Simulated data were generated using a generation-by-generation coalescent model for a two-dimensional stepping-stone island model using IBDSim v2 (Leblois et al. 2009). The lattice was 10 x 10 subpopulations of 50 individuals each with absorbing boundaries, a local migration rate of 0.01, a k alleles model with two alleles, and a mutation rate of 0.0005. Ten individuals were sampled from each of the center 25 subpopulations of the lattice for the figure. F_{ST} between all pairs of subpopulations was estimated using Genepop on the web (Rousset 2008b).

differentiation. Recent work has developed a range of ways to quantify landscape variables that vary spatially which could alter the effective migration rate ($N_e m$) and thereby be causally related to genetic differentiation among subpopulations. Landscapes are evaluated for connectivity, where habitat areas are defined and then given a score for the ability of migration or gene flow to take place. This produces a friction surface or **resistance surface** that quantifies the variation in variables hypothesized to cause connectivity variation across the landscape (Spear et al. 2010). Resistance surfaces are used in place of simple Euclidean distances to test for a relationship between the landscape that separates subpopulations and the genetic differentiation of subpopulations.

Two models that make use of landscape information to test for causes of genetic differentiation subpopulation are **least-cost paths** (Adriaensen et al. 2003) and electrical **circuit theory** (McRae 2006; McRae et al. 2008, see comparison between these two methods in Marrotte and Bowman 2017). In these models, the landscape surrounding sampled subpopulations is characterized for variables that are hypothesized to influence genetic connectivity. These variables can include physical attributes such as rainfall or snowfall, elevation, soil type, or average temperature as well as biological variables such as plant cover, community type, predator density, or human occupancy. Least-cost paths are the single track between pairs of populations where rates of gene flow are hypothesized to be the greatest based on attributes of the landscape. A hypothetical landscape is shown in Figure 4.19, divided up into a series of grid squares where the rates of gene flow has been estimated based on

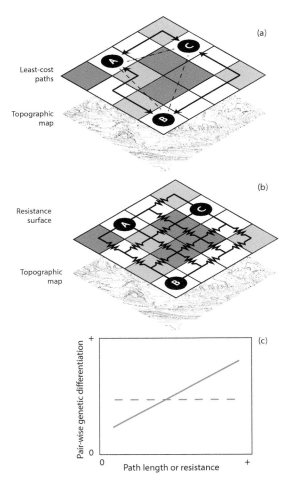

Figure 4.19 Landscapes (represented by the topographical map in the lower planes of A and B) can be sectioned into grid areas (called a raster grid since digital images of landscapes are called raster data) where each pair of cells is scored for gene flow connectivity (upper planes of A and B). In this hypothetical example, the lettered nodes represent subpopulations and darker shading of grid squares represents lower connectivity based on landscape variables. In A, the dashed lines show Euclidean distances between subpopulations and the solid lines are the single least cost paths (greatest connectivity and gene flow) between pairs of subpopulations. In B, the connectivity between all pairs of grid cells is represented by electrical resistors in a mesh circuit. The voltage or current between each pair of subpopulations over all paths in the circuit is analogous to the net effective migration rate between subpopulations. Panel C shows how a graph of linearized F_{ST} by least cost path lengths or resistance for all pairs of subpopulations. A positive slope in the plot (solid line) is evidence that the landscape variable(s) that produced the grid scores are correlated with genetic differentiation between subpopulations.

features of the map in the plane below. Least-cost paths (greatest total rates of gene flow) based on the grid square scores are shown connecting each pair of subpopulations in Figure 4.18A. The least-cost path approach has an implicit assumption that dispersing individuals or gametes are able to evaluate the entire landscape to find the single path with greatest connectivity and does not incorporate multiple possible paths.

An alternative approach is to consider that the landscape provides multiple paths between subpopulations, each path possibly varying in its degree of connectivity. Circuit theory models multiple paths between locations on a landscape using the analogy of the flow of voltage or current in a network of wires and electronic components such as resistors. The most basic circuit that can be used to define resistance among subpopulations is shown in Figure 4.5 where the effective migration rate connections between pairs of subpopulations have been replaced with electrical resistors. The resistance distance is

a function of the probability of the effective migration rate between subpopulations rather than of the geographic distance between subpopulations. Figure 4.18B shows a hypothetical landscape divided up into a series of grid squares. The connectivity between grid cells is represented by electrical resistors for each pair of squares, and the interconnections form an electrical circuit. The voltage or current at any location in the circuit can be found using methods of circuit analysis in electrical engineering. Resistance, voltage, and current in electrical circuits can be interpreted as the connectedness of nodes, analogous to the paths traversed by a large sample of individuals that walk at random along the paths defined by a circuit. McRae et al. (2008) provide an overview of circuit theory along with some examples of the different ways that circuits can be modeled using conductivity, resistance (the inverse of conductance), current, or voltage to predict genetic connectivity between and among subpopulations.

Math box 4.2 Analysis of a circuit to predict gene flow across a landscape

A simple example can illustrate how to analyze a circuit as an analogy of gene flow across a landscape between subpopulations. Electrical circuits are represented as a series of nodes connected by conductors and electrical components. A circuit with several numbered nodes and resistors is shown in Figure 4.20. A and B represent two nodes analogous to subpopulations. The flow of current between A and B will be the greatest where the resistance is lowest, but current will flow over all branches of the circuit. The conductor from nodes 3–5 connects to ground through a resistor and serves to siphon off some of the current, analogous to death of some migrants that traverse node 3 on the way between subpopulations A and B.

The properties of a node, such as its conductance or voltage, can be predicted using Kirchhoff's circuit rules (the sum of currents meeting at a node equals zero; the directed sum of voltages around any closed loop is zero) and Ohm's law (current = voltage/resistance); the current or voltage found at the nodes of a circuit can be expressed as multiple equations with multiple unknowns.

At node 1, there is 1 A of current flowing in from the direction of B with the amount of current flowing out determined by voltage differences across two conductors that both contain resistors

$$1 = \frac{V1 - V3}{R1} + \frac{V1 - V2}{R4} \qquad (4.66)$$

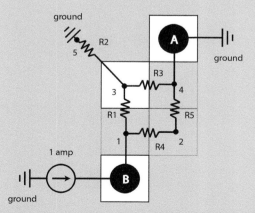

Figure 4.20 An example circuit with resistors representing the genetic connections over a landscape between subpopulations A and B.

At node 2, current flows in from the direction of node 1 and out toward node 4. We can reuse the expression for current across resistor 4 and set it equal to current flowing out across resistor 5 to obtain

$$\frac{V1 - V2}{R4} = \frac{V2 - V4}{R5} \qquad (4.67)$$

At node 3, we reuse the expression for current across resistor 1 for the incoming current and set it equal to current flowing out toward nodes 4 and 5

$$\frac{V1 - V3}{R1} = \frac{V3 - V4}{R3} + \frac{V3 - V5}{R2} \qquad (4.68)$$

As an example, let the resistor values be R1 = R3 = 1, R4 = R5 = 2, and R2 = 20. We also utilize the fact that a conductor connected to ground has a voltage of zero to determine that V4 and V5 are both zero. After substitution and some simplification, this leaves three equations with three unknowns.

$$1 = 1.5V1 - V3 - 0.5V2 \qquad (4.69)$$

$$0 = V2 - 0.5V1 \qquad (4.70)$$

$$0 = 2.05V3 - V1 \qquad (4.71)$$

The unknowns in these linear equations are solved by defining a matrix with the coefficients for each of the variables and then using one of a number of possible methods to solve for the variables. (See the text web page for a link to a website that solves systems of linear equations by methods including matrix inversion and Cramer's rule and shows all the intermediate steps.) The solution for the voltages is V1 = 1.31, V2 = 0.66, and V3 = 0.64. With the voltages known, we can then see that the current flowing to node 4 over resistor R3 is V3/R3 = 0.64, current flowing to node 4 over resistor R5 is V2/R5 = 0.66/2 = 0.33, and current flowing to node 5 over resistor R2 is V3/R2 = 0.64/20 = 0.032.

Returning to the genetic connectedness between A and B, the circuit model shows that 33% of gene flow is along the path 1-2-4, 64% of gene flow along the path 1-3-4, and 3.2% of gene flow across node 3 is unsuccessful.

McRae and Beier (2007) examined the correlations between linearized F_{ST} and geographic distance, least-cost paths, and circuit theory resistances for eight subpopulations of big-leaf mahogany trees located across Central America and for 12 wolverine subpopulations in western North America. In both species, the subpopulations were separated by complex landscapes and distances of hundreds or thousands of kilometers. In both species, the use of circuit theory to represent resistance to gene flow across the landscape was more strongly correlated with genetic differentiation between pairs of subpopulations than was geographic distance alone.

Landscape approaches to testing for causes of genetic differentiation tend to adopt the perspective that gene flow is a strong process and deemphasize the possible contributions of variable rates of genetic drift across a landscape as might be expected with variable effective population sizes and variable rates of gene flow. This is in spite of McRae (2006) having articulated resistance and conductance in terms of the effective migration rate, $N_e m$. An additional challenge is that present-day genetic differentiation is a record of historic processes that have operated over relatively long periods of time. In many species, we might expect that genetic differentiation among subpopulations is a function of the average connectivity on the landscape over time. As such, genetic differentiation and the landscape features and ecological variation used to hypothesize connectedness may not change on similar time scales in general, making inference of causes of genetic differentiation more difficult.

4.7 Population assignment and clustering

- Maximum likelihood assignment.
- Bayesian assignment and clustering.
- Empirical assignment and clustering.

When testing for population genetic differentiation, individuals are often sampled from multiple geographic locations which then serve as the units to compare for genotype and allele frequency differences. However, the geographic units of sampling might not well represent genetic population units. For example, individuals found at several geographic locations might be part of the same genetic population, or individuals found within a single geographic location might be admixed – recently arrived from different origin populations that are genetically diverged. Rather than using sampling locations as evidence of genetic groups a priori, a different approach is to seek evidence from the sample of genotypes itself to establish genetic groups and thereby estimate genetic differentiation. Such **population assignment** or **clustering** methods are now well developed and are widely used with empirical multilocus genotype data to infer genetic subpopulations and to estimate genetic differentiation. These methods have traditionally been divided into those that are model-based and those that are based on statistical estimation. Model-based clustering methods rely on an explicit population genetic model and assumptions about the processes acting on the populations being sampled. Statistical estimation-based methods can be used to determine patterns in genetic data that then place individuals into clusters.

Admixture: In the context of genotype assignment methods, the alleles found in the multilocus genotype of an individual having origins in multiple genetically differentiated populations through hybridization or through past gene flow and mating. Different loci may exhibit different population origins, leading to one multilocus genotype being assigned proportions of population origin.

Likelihood: The support for one model among several possible models given an observed outcome that allows comparison of models. For example, having observed a population where the frequency of heterozygotes equal to $2pq$, a mating model with a fixation index of zero has a higher likelihood than a mating model with a fixation index of one.

Posterior probability: The probability assigned to a hypothesis given its prior probability combined with likelihood evidence through the use of Bayes rule.

Probability: The chance of observing an outcome given a specified model. For example, the probability of a specific genotype given the population allele frequencies and the Hardy–Weinberg model.

Maximum likelihood assignment

One approach to assign a genotype to its source population would be to compute the expected frequency of a genotype given that it was sampled from a population with known allele frequencies under a model of mating. This would provide the likelihood of the genotype (the data) where each of the source populations represents a possible origin (or version of the model with allele frequencies as parameter values). We assign the genotype to the source population (or version of the model) that produces the greatest likelihood of having generated an individual of that genotype. To see a simple version of how this **maximum likelihood** approach to population assignment works, imagine a situation where there are two biallelic loci (call them locus A and locus B) and individuals are found in two subpopulations with known allele frequencies. In population one, allele frequencies are $p_A = 0.3$ and $p_a = 0.7$ and $p_B = 0.4$ and $p_b = 0.6$, while, in population two, $p_A = 0.7$ and $p_a = 0.3$ and $p_B = 0.6$ and $p_b = 0.4$. Now, consider an individual with the genotype AABB. Which population does that individual belong to based on its genotype?

Under random mating, the expected frequency of an AABB genotype is $(p_A^2)(p_B^2)$. So, how likely is it that this individual originated in either of the two populations? To determine this, we need to compute the expected frequency of the genotype using the allele frequencies of each population. In population one, the expected genotype frequencies are $(0.3)^2(0.4)^2 = (0.09)(0.16) = 0.0144$ and $(0.7)^2(0.6)^2 = (0.49)(0.36) = 0.1764$ in population two. Because these expected probabilities can be very small numbers – think of the expected frequency of a multilocus genotype composed of many loci, it is common to express them as a natural logarithm or a base e logarithm. The natural log likelihoods of an AABB genotype are ln $(0.0144) = -4.24$ in population one and ln $(0.1764) = -1.74$ in population two. The statistical approach of maximum likelihood is to choose as the best estimate of the parameter of interest the value or distribution that maximizes the log likelihood of having observed the observed data. In this case, assigning an individual of AABB genotype to membership or ancestry in population two maximizes the log likelihood (the log

likelihood is closest to zero) based on population allele frequencies.

To determine confidence in maximum likelihood population assignments of genotypes requires comparing the likelihood distributions for all possible genotypes in the possible origin populations. These distributions can be generated by simulating a large number of possible genotypes that might be observed under random mating (or some other mating model) given population allele frequencies (see Interact Box 4.6). Simulated likelihood distributions provide confidence intervals for the log likelihood values. If a negative log likelihood falls outside the confidence interval for a population, then it is considered improbable to have originated in that population. In the simple example used above, the AABB genotype could be assigned to either population because of overlapping likelihood distributions, so there is not high confidence in assigning its origin as population two. In general, log likelihood distributions for populations with diverged allele frequencies will have less overlap with more loci and loci that have more alleles per locus.

The maximum likelihood assignment method was originally applied to Canadian polar bear populations (Paetkau et al. 1995) and has been employed in a large number of empirical studies. The method is most effective when (i) allele frequencies have been estimated independently from the set of genotypes being assigned to populations, (ii) allele frequencies are diverged between/among populations, (iii) mating is random and loci are independent (or the fixation index (F) and disequilibrium coefficient (D) are well estimated and are included in genotype frequency expectations) for each possible source population so that the model for expected genotype frequencies is accurate, and (iv) there are many polymorphic loci available such that the expected frequency of any given multilocus genotype is small, making genotypes more unique. See Paetkau et al. (2004) for results of a simulation study examining the power and precision of likelihood assignment.

Bayesian assignment

Bayesian statistical inference is now widely employed to make inferences in population genetics. Bayes theorem is named after the Reverend Thomas Bayes (1701–1761), whose posthumous publication

---

Interact box 4.6 Genotype assignment and clustering

Likelihood and Bayesian genotype assignment and the Bayesian approach to inferring the best supported number of populations are illustrated in spreadsheet models in a file that can be found through the text website.

The **Genotype likelihood** tab has a model where the allele frequencies for two biallelic loci can be set for three populations. A sample of random genotypes are generated for each of the populations based on these allele frequencies. The natural log likelihood is computed for each random genotype, and these log likelihood distributions can be compared among the three populations in a graph. Modify the allele frequencies to see the impact on the log likelihood distributions.

The **Bayesian assignment** tab presents all of the computations required for Bayesian assignment of an AABB genotype to three possible origin populations. Modify the allele frequencies and the distribution of prior probabilities to observe impacts to the posterior probabilities and multilocus and fractional assignments to the three possible origin populations.

The **Bayesian infer _K_** tab of the spreadsheet gives a heuristic version of clustering based on random assignment of six two-locus genotypes to either one or two clusters. This gives some sense of how clustering works but does not include the numerous details of the full method such as prior distributions of allele frequencies or iterative updating of genotype cluster assignments based on allele frequency estimates. Users can resample the genotypes by recalculating the sheet, and also change the pool of six genotypes.

---

in 1763 first introduced the conceptual approach. Independently, Pierre-Simon Laplace developed the same ideas and stated them in an equation in 1774 that defines the probability of an event given observed outcomes of combined events or conditions (see Stigler 1986). Bayes theorem builds on likelihood by placing hypotheses into a context that weighs the evidence of observation by prior beliefs and then normalizes by a total probability (see Math Box 4.2).

Let us work through an example of how Bayes rule can be used to determine the posterior probability distribution that allows an individual of genotype AABB to be assigned one of the several possible origin populations with known allele frequencies. Table 4.10 gives the A and B locus allele frequencies for three possible origin populations. For a genotype and K possible origin populations, Bayes rule can be stated as

$$P(K \mid genotype) = \frac{P(genotype \mid K)P(K)}{P(genotype)} \quad (4.72)$$

Determining the posterior probability that population K is the origin of the genotype requires three quantities that make up the right-hand side of the equation.

$P(K)$ are the prior probabilities that each possible population is the origin of AABB. In this example, the three populations are considered equally likely to be the origin of the AABB individual. Therefore, the prior probabilities for each K are equal at $1/K = 1/3$. There are many alternative prior probability distributions possible. For example, the prior probabilities could be weighted using spatial locations such that origin populations closer to where the genotype was sampled would have higher prior probabilities than those that were at a greater distance.

$P(genotype \mid K)$ are the likelihoods of the genotype given one origin population. Under random mating, the expected frequency of an AABB genotype is $(p_A{}^2)(p_B{}^2)$. This is identical to the computation used for likelihood assignment earlier in this section. For population 1 in Table 4.10, the likelihood of the AABB genotype is $(0.5^2)(0.8^2) = 0.16$. $P(genotype)$ is the total probability of observing an AABB genotype in any of the origin populations. The total probability of observing an AABB genotype is the sum over all populations of the products of genotype likelihood times the population prior probability or $\sum\limits_{K=1}^{pops}$ $P(genotype \mid K)P(K)$. The prior probabilities may be different for each possible source population but

Table 4.10 An example of population assignment of a genotype using Bayes rule. There are two biallelic loci, A and B, and three possible populations. An individual with genotype AABB is being assigned to an origin population based on the posterior probability determined with Bayes rule. The conditional probability of the genotype for a given population is based on Hardy–Weinberg expectations under random mating. The uniform prior probability distribution assumes that an individual was equally likely to have originated in any of the three origin populations. If admixture is possible, two single locus genotypes are assigned to two different origin populations resulting in fractional assignment to populations 1 and 3 ($q_{i1} = q_{i3} = 0.5$).

	Population		
Allele frequencies	**1**	**2**	**3**
$P(A)$	0.5	0.2	0.8
$P(a)$	0.5	0.8	0.2
$P(B)$	0.8	0.7	0.2
$P(b)$	0.2	0.3	0.8
Conditional probabilities of genotype AABB			
$P(\text{G-A locus}|\text{population } K)$	$(0.5)^2 = 0.25$	$(0.2)^2 = 0.04$	$(0.8)^2 = 0.64$
$P(\text{G-B locus}|\text{population } K)$	$(0.8)^2 = 0.64$	$(0.7)^2 = 0.49$	$(0.2)^2 = 0.04$
$P(\text{G-multilocus}|\text{population } K)$	$(0.5)^2(0.8)^2 = 0.16$	$(0.2)^2(0.7)^2 = 0.196$	$(0.8)^2(0.2)^2 = 0.0256$
Prior probabilities			
$P(\text{population } K)$	0.33	0.33	0.33
Probabilities of genotype AABB			
$P(\text{G-A locus})$	$(0.25)(0.33) = 0.083$	$(0.04)(0.33) = 0.013$	$(0.64)(0.33) = 0.213$
$P(\text{G-B locus})$	$(0.64)(0.33) = 0.213$	$(0.49)(0.33) = 0.163$	$(0.04)(0.33) = 0.013$
$P(\text{G-multilocus})$	$(0.16)(0.33) = 0.053$	$(0.0196)(0.33) = 0.007$	$(0.0256)(0.33) = 0.009$
Posterior probabilities			
$P(\text{population } K|\text{G-A locus})$	$\dfrac{0.25*0.33}{0.083 + 0.013 + 0.213} = 0.269$	$\dfrac{0.04*0.33}{0.083 + 0.013 + 0.213} = 0.043$	$\dfrac{0.64*0.33}{0.083 + 0.013 + 0.213} = 0.688$
$P(\text{population } K|\text{G-B locus})$	$\dfrac{0.64*0.33}{0.213 + 0.163 + 0.013} = 0.547$	$\dfrac{0.49*0.33}{0.213 + 0.163 + 0.013} = 0.419$	$\dfrac{0.04*0.33}{0.213 + 0.163 + 0.013} = 0.034$
$P(\text{population } K|\text{G-multilocus})$	$\dfrac{0.16*0.33}{0.053 + 0.007 + 0.009} = 0.780$	$\dfrac{0.0196*0.33}{0.053 + 0.007 + 0.009} = 0.096$	$\dfrac{0.0256*0.33}{0.053 + 0.007 + 0.009} = 0.125$

always sum to one. The total probability of the AABB genotype is $(0.16)(0.333) + (0.0196)(0.333) + (0.0256)(0.333) = 0.0069$, as shown in Table 4.10.

Bringing these three separate quantities together for population 1 as the origin of an AABB individual gives the posterior probability of

$$\frac{(0.16)(0.333)}{0.053 + 0.007 + 0.009} = 0.780 \qquad (4.73)$$

As seen in Table 4.10, the posterior probabilities are lower for assignment to either population 2 or population 3, so the best supported inference is that the AABB individual originated in population 1. Note that the posterior probabilities sum to one.

The simplest version of the Bayesian assignment procedure assumes that all loci in an individual genotype have their origin in a single population. A multilocus assignment of the genotype to population 1 in Table 4.10 is made under the assumption of a single origin population for all loci. Bayesian assignment can also allow for **admixture**, or the possibility that the individual loci in a multilocus genotype of one individual have origins in several different populations because of the past hybridization and mating of individuals between populations. Under the admixture model, q_k^i is the proportion of the multilocus genotype of individual i that originated in population k. Table 4.10 shows population assignment allowing for admixture, and in the example, the A locus genotype is assigned to population 3 while the B locus genotype is assigned to population 1.

A more fully realized example of Bayesian clustering with admixture can be seen in Figure 4.21. Based on the simulated 10 locus genotypes, 20 individuals

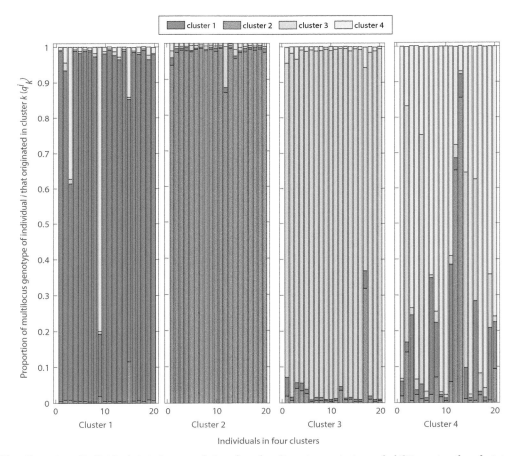

Figure 4.21 Clustering of individuals into four populations based on Bayesian posterior probabilities using the admixture model in STRUCTURE (Pritchard et al. 2000). The bar representing each individual gives the fractional ancestry assigned to one of four clusters. The figure is based on simulated data with 10 biallelic loci of under a finite island model of 20 total populations with $N_e m = 0.1$ and no mutation. Observed $G_{ST} = 0.59$ among the 20 populations after 100 generations when the individuals in four populations were sampled.

sampled from each of four locations have been placed into four clusters. The proportion of the multilocus genotype of each individual that originated within each cluster (q_k^i on the y-axis) reveals that only a few individuals in clusters one through three have a substantial fraction of recent ancestry attributable to other clusters, consistent with a high level of genetic differentiation due to lack of gene flow. In contrast, cluster four shows most individuals with a portion of their loci having recent ancestry in the other clusters as would be expected with recent gene flow and lower genetic isolation.

The Bayesian approach can be extended to test the hypothesis of how many populations the genetic data suggest exist in a sample of genotypes. This allows a hypothesis test for the best supported number of populations based on the observed data rather than assuming that geographic location or other criteria represent genetic entities. The method uses the posterior probability obtained with Bayes rule in a manner similar to genotype assignment above with some modifications. The best clustering is the value of K and assignment of genotypes that maximizes the posterior probability of the observed genotype data.

The first step is to set the range of the number of possible groups or clusters, K, from one to some maximum value. (One possibility is to use the number of geographic locations sampled as maximum K.) The observed set of all N genotypes is then randomly partitioned into each of one to K clusters. For example, for $k = 1$, all N genotypes are assigned to the same cluster, and, for $k = 2$, the genotypes are assigned at random to clusters each of size $N/2$, continuing through each value of K clusters each of size N/K. Then, the allele frequencies for these clusters of random genotypes are computed. With the allele frequencies determined for each cluster, the posterior probability of each genotype is computed given the genotypes assigned to the cluster as well as the posterior probability of the clustering of all genotypes for a given K. Next, a small change to the cluster assignments can be made, with a few genotypes reassigned at random to different clusters for each K (a proposed new assignment of genotypes to clusters). If that reassignment increases the posterior probability, that proposal is retained, otherwise the assignment of genotypes at the prior step is retained. (Markov chain Monte Carlo [MCMC] with the Metropolis-Hastings algorithm or Gibbs sampling are among the methods used to estimate a parameter value that maximizes a posterior probability.) This process of

random changes to genotype assignments to generate new proposals is repeated many times to estimate the maximum posterior probability for each K. The full model to infer K has many more details than are described here, such as prior probability distributions for allele frequencies and an iterative procedure to estimate cluster allele frequencies and then reassign genotypes based on posterior probabilities (Pritchard et al. 2000; Falush et al. 2003).

The number of clusters K with the highest posterior probability of the genotype data is the best supported assignment of genotypes into populations. Table 4.11 gives average posterior probabilities for grouping the individuals into one to six clusters for the simulated genotype data shown in Figure 4.16. Four clusters ($k = 4$) gave the highest probability of the genotype data and is, therefore, the best supported number of clusters. Using simulations, Evanno et al. (2005) showed that inference of cluster size with posterior probabilities is influenced by the type of genetic marker employed and by sample sizes. They suggested using the rate of change in the log probability of the genotype data as K values change to infer the best supported number of clusters (see also Verity and Nichols 2016). An admixture model has been shown to be more robust in population assignment (François and Durand 2010). Kaeuffer et al. (2007) showed how background gametic disequilibrium caused by genetic drift can influence the inference of population clusters. Wang (2017) used simulated data to show that estimated K and individual assignment results in the program Structure (Pritchard et al. 2000) depend on model parameters and that rarely used parameter combinations may be required to accurately assign individuals to source populations. Because clustering approaches lack explicit mutation models, the results may be sensitive to patterns of mutation as well as to data filtering that censors recent mutations at low frequency (Shringapure and Xing 2009; Linck and Battey 2019). Janes et al. (2017) reviewed a large number of studies that estimated K and discussed possible pitfalls in the application and interpretation of Bayesian clustering. Lawson et al. (2018) showed how genetic clustering may lead to an incorrect inference of population history and suggested an approach to evaluate goodness of fit. Alternative models and algorithm variations of Bayesian clustering are implemented in software packages that can be used to analyze genotype data (e.g. Guillot et al. 2005; Corander et al. 2008; Jay et al. 2015).

Math Box 4.3 Bayes Theorem

Imagine that A and B are variables that represent events or conditions that have two levels (e.g. present and absent). The goal is to learn the probability of one level of A using the information about a related variable B. $P(A \mid B)$ is the conditional probability of event A occurring given that B is true. This is equal to

$$P(A \mid B) = \frac{P(A \cap B)}{P(B)} \qquad (4.74)$$

which says that the probability of A given B equals the joint probability of both A and B (the intersection of A and B) divided by the probability of event B. (Note that if A and B are independent then $P(A \cap B) = P(A)P(B)$ and in that case $P(A \mid B)$ is equal to $P(A)$.) Based on rearranging this conditional probability relationship, it is true that

$$P(B \cap A) = P(A \mid B)P(B) \qquad (4.75)$$

Changing the order of A and B, it is also true that

$$P(A \cap B) = P(B \mid A)P(A) \qquad (4.76)$$

Since $P(B \cap A) = P(A \cap B)$, we can substitute the definition of these intersection probabilities to obtain

$$P(A \mid B)P(B) = P(B \mid A)P(A) \qquad (4.77)$$

To determine the conditional probability of A given B, we can then rearrange the equation to obtain Bayes rule

$$P(A \mid B) = \frac{P(B \mid A)P(A)}{P(B)} \qquad (4.78)$$

where $P(A)$ and $P(B)$ are the probabilities of A and B independently of each other (called marginal probabilities). $P(A)$ is the probability of A before any information about B is considered, called the prior probability. $P(B \mid A)$ is the conditional probability or likelihood of B given A. $P(A \mid B)$ is also a conditional probability of event A occurring given that B is true. $P(A \mid B)$ is called the posterior probability, and it is the probability we are seeking to determine. The probability $P(B)$ in the denominator is the sum of the probabilities of all possible outcomes where B is true, sometimes called the normalizing constant. For example, if A has two observed conditions of true (A) and false (A'), then $P(B) = P(A \cap B) + P(A' \cap B) = P(B \mid A)P(A) + P(B \mid A')P(A')$. It is assumed that $P(B)$ is not equal to zero.

Table 4.11 Distribution of natural log posterior probability values for clusters of K from one to six for the simulated data shown in Figure 4.16 estimated using STRUCTURE (Pritchard et al. 2000; Falush et al. 2003). The posterior probabilities of the data given k are the means of 10 independent clustering estimates for each k. The maximum and plateau of the posterior probability both indicate $k = 4$ as the best estimate. Simulated genotype data were 10 biallelic loci for 20 individuals from each of four populations in a finite island model with $N_e m = 0.02$, no mutation, and 20 total populations simulated for 100 generations.

k	Posterior probability
1	−1026.6
2	−863.8
3	−688.7
4	−634.1
5	−656.4
6	−666.8

Empirical assignment methods

Given a set of multilocus genotypes collected from multiple locations, a wide range of statistical methods can be used to summarize the genetic data and carry out statistical tests or provide visualization of the data. A traditional approach to empirical population clustering is carried out by estimating the genetic distances between all genotypes in a sample. Genetic distances quantify allelic similarities between individuals to give a measure of identity by descent. Clustering can be accomplished by estimating genetic distances for pairs of individuals and then grouping genotypes so that the genetic distance within groups is minimized while the genetic distance among groups is maximized. (Genetic distances are covered in more detail in Chapter 5.) This is challenging since both the genetic similarity measure and the clustering procedure must be chosen to recover patterns of interest from the data without full knowledge of the genetic processes operating.

For genetic data with many loci, the high number of dimensions makes interpretation challenging. Principle components analysis (PCA) is a well-established mathematical technique that is applied to a wide variety of data sets with many independent variables which are potentially correlated (Pearson 1901; Hotelling 1933). The technique creates a smaller number of new variables, or axes, that are linear combinations of the original variables. A PCA axis is the sum of the original variables each weighted by a coefficient based on its degree of contribution to the new variable, called loadings. Defining new variables removes the correlations among the original variables. The new variable axes are orthogonal – vectors at right angles – and are independent variables. The new axes are also ranked in the amount of variation they explain in the original data, with the first PCA axis explaining the most variation. (For those readers familiar with linear algebra, the new variables are the eigenvectors of the original data with the magnitude of each eigenvalue quantifying the amount of variance that axis explains. See Otto and Day (2007) for an introduction to the linear algebra concepts.)

Interact box 4.7 Visualizing principle components analysis

PCA can be easier to understand using a visual depiction of relationship between the original and the new axes. There are numerous tutorials on PCA available on the web that utilize graphics or simulations of PCA. Links to several helpful web resources are available on the text web page.

PCA was initially employed as an empirical clustering technique that helped reduce the dimensionality of genetic data sets with many loci and loci with many alleles (Menozzi et al. 1978). Short-read sequencing techniques that generate thousands of single nucleotide polymorphism (SNP) loci per individual have further motivated the use of PCA analysis to reduce dimensionality and provide visual interpretation of genetic variation among groups or geographic locations, aided by the relative computational speed of the method when working with very large data sets.

PCA with genetic data begins with scoring each genotype or haplotype. For an autosomal diploid locus with two alleles, let $G(i,j)$ represent each genotype for individual i at locus j that is scored as 2 for AA, 1 for Aa, or 0 for aa to represent the number of A alleles in the genotype. The genotype scores are re-centered by first determining the mean score for each locus

$$\mu_j = \frac{\sum_{i=1}^{N} G(i,j)}{N} \qquad (4.79)$$

then subtracting μ_j from each individual genotype score to center the genotype scores around the mean. The re-centered genotype scores are then rescaled by the square root of the product of the allele frequencies ($p_j = \mu_j/2$)

$$M(i,j) = \frac{G(i,j) - \mu_j}{\sqrt{p_j(1 - p_j)}} \qquad (4.80)$$

Recalling that the standard deviation in a binomial is $\sqrt{p_j(1 - p_j)}$ shows that this transformation makes each genotype score equal to number of standard deviations from the mean number of A alleles for all genotypes at a locus. (Diploid genotype data for loci with more than two alleles are scored by representing each locus with one column for each of the alleles found at the locus and scoring genotypes of individuals by the number of each allele they contain. With more than two alleles per locus, the rescaling division step is skipped (see Cavalli-Sforza et al. 1994; Patterson et al. 2006). Since genotypes share alleles identical by descent if individuals have common ancestry, PCA on the genotype data estimates the combined coancestry across multiple loci.

The first two axes of a PCA for simulated 10 locus genotype data from 20 individuals from each four sample locations are plotted in Figure 4.22. The first two PCA axes explain about 61% of the variation in the genotypes and the points from each of the populations show a degree of grouping consistent with genetically diverged populations. The points for population 4 are the most spread out, and many are mixed in with points from the other three populations suggesting that population has experienced recent gene flow. For comparison, the PCA in

Figure 4.17 is based on the same genotype data used for the Bayesian clustering with admixture seen in Figure 4.16.

A similar comparison of PCA and Bayesian clustering can be found in Becquet et al. (2007) using genotype data at 310 polymorphic microsatellite loci in 78 common chimpanzees and six bonobos. They used both Bayesian clustering and PCA to test for genetic populations of chimpanzees and for the evidence of any individuals derived from interpopulation mating. The two data analysis methods were in agreement and led the authors to infer three genetic populations of chimpanzees with two wild individuals being recent hybrids.

For PCA with population genetic data, formal statistical tests for genetic differentiation as well as methods to evaluate statistical power to detect genetic differentiation have been described (Patterson et al. 2006). McVean (2009) showed that PCA analysis patterns can be related to the average coalescence times between lineages, thereby providing a critical link between PCA patterns and population genetic predictive models for processes such as population divergence and gene flow and admixture. One

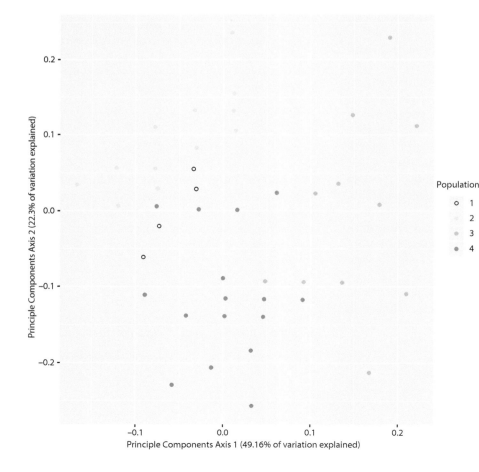

Figure 4.22 Principle components analysis of genotype data exhibits population clustering for 10 diploid loci. The first two principle component axes together explain almost 72% of the genetic variation. Populations show dispersion on the PCA axes consistent with genetic differentiation. The points for population 1 having the least separation from the other populations suggests recent gene flow and ancestry. The figure is based on the same data as Figure 4.16 – 10 biallelic loci of simulated data under a finite island model of 20 total populations with $N_e m = 0.1$ where $G_{ST} = 0.59$ among the 20 populations after 100 generations when 20 individuals were sampled from each four populations.

application is that the total variation explained by a PCA based on sample location allele frequencies is equivalent to an estimate of F_{ST} (McVean 2009). PCA is part of a family of multivariate analysis approaches applied to genetic data to test for population structure, identify hybrids or hybridization, test for evidence of recent mating between and among populations, and identify spatial genetic patterns under a range of models (e.g. Jombart et al. 2008; reviewed by Jombart et al. 2009; Frichot et al. 2012; Francois and Waits 2015).

4.8 The impact of population structure on genealogical branching

- Bugs in many boxes.
- Event times with population subdivision.
- Sample configurations.
- Mean and variance of waiting time in two demes.

In structured populations with gene flow, lineages can move from deme to deme. In a retrospective view, two lineages sampled in the present can experience either coalescence or migration going back in time (Figure 4.23). Determining the mean and variance of time to coalescence in structured populations will show the overall impact of population structure on genealogical trees. In particular, we would like to know whether population structure will alter the average and variance of the height of genealogical trees in comparison with the basic coalescent process in a single panmixic population. We will again utilize the properties of the exponential distribution to approximate the time to an event (see Section 3.6).

Let us begin by thinking about the coalescent process when there is gene flow among several demes in terms of the bugs-in-a-box metaphor used to describe the basic coalescent process. With population subdivision, the bugs are located in multiple boxes with each box representing a deme. Bugs move about within a box at random and eat each other, reducing their numbers. There is also the possibility of migration where a bug is chosen at random and moved to another box. If migration events are very rare, then the individual boxes have a good chance of being reduced to a single bug before a migrant bug enters or leaves the box. It will then take a long time for enough migration events to happen such that the entire group of boxes is reduced to a single bug. When migration events are common, migrant bugs move among the boxes frequently and the boxes are effectively interconnected.

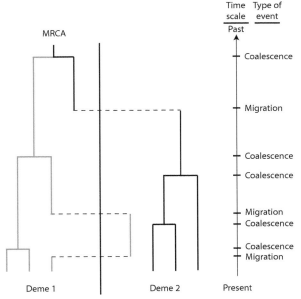

Figure 4.23 A hypothetical genealogy for two demes. Initially, there are three lineages in each deme. The very first event going back in time is the migration of a lineage from deme one into deme two. Immediately after this migration occurs, the chance of coalescence in deme two increases since there are more lineages and the chance of coalescence in deme one decreases since there are fewer lineages. Continuing back in time, a coalescence event occurs in deme one and then a coalescence event occurs in deme two. The lineage that migrated out of deme one migrates back into deme one by chance. Coalescence to the single most recent common ancestor of all lineages cannot occur until the final two lineages are brought together in a single deme by migration.

Therefore, there should be little or no time spent waiting for migration events as the bugs in all the boxes eat their way to a single bug.

Combining coalescent and migration events

Describing genealogies with gene flow can be accomplished by adding another type of possible event that can occur working from the present to a time in the past where all lineages find their most recent common ancestor. We will assume that both coalescent and migration events are rare (or that N_e is large and the rate of migration is small) so that when an event does occur going back in time, it is *either* coalescence *or* migration. In other words, we will assume that migration and coalescence events are mutually exclusive. The fact that events are mutually exclusive is an important assumption. When two independent processes are operating, the coalescence model becomes one of following lineages back in time and waiting for an event to happen. When events are independent but mutually exclusive, the probability of each event is added over all possible events to obtain the total chance that an event occurs. For example, the chance that a diploid genotype for a diallelic locus is a heterozygote under random mating is $2pq$. This is the sum of the independent chance of sampling Aa and the chance of sampling aA since a heterozygote results from one of the two ways of sampling of two different alleles (the probability of a heterozygote under random mating is not $(pq)^2$, which is the chance of sampling Aa and aA *simultaneously*). Therefore, if we can find an exponential approximation for the chance that a lineage migrates to a different deme each generation, we can just add this to the exponential approximation for the chance of coalescence.

In a subdivided population, each generation, there is the chance a lineage in one deme migrates to some other deme. The rate of migration, m, is the chance that a lineage migrates each generation. The chance that a lineage does not migrate is, therefore, $1 - m$ each generation. The chance that t generations pass before a migration event occurs is then the product of the chances of $t - 1$ generations of no migration followed by a migration or

$$P(T_{migration} = t) = (1 - m)^{t-1}m \qquad (4.81)$$

This is in an identical form to the chance that a coalescent event occurs after t generations given in Chapter 3. Like the probability of coalescence, the probability of a migration through time is a geometric series that can be approximated by the exponential distribution (see Math Box 3.2). To obtain the exponent of e (or the intensity of the migration process), we need to determine the rate at which migration is expected to occur in a population.

Now, consider migration events in the context of an island model of gene flow where there are d demes and each deme contains $2N_e$ lineages. The total population size is the sum of the sizes of all demes or $2N_e d$ lineages. When time is measured on a continuous scale with $t = \dfrac{j}{2N_e d}$, one unit of time is equivalent to $2N_e d$ generations. If $2N_e d$ generations elapse and m is the chance of migration per generation, then $2N_e dm$ migration events are expected in the total population during one unit of continuous time. If we define $M = 4N_e m$, then $M/2$ is equivalent to $2N_e m$ or the chance that a lineage in one deme migrates (the per deme migration rate). The chance of migration is independent in all of the demes, so the expected number of migration events in the total

population is the sum of the per deme chances of migration or $\frac{M}{2}d$. This leads to the exponential approximation for the chances that a single lineage in any of the demes migrates at generation t:

$$P(T_{migration} = t) = e^{-t\frac{Md}{2}} \qquad (4.82)$$

on a continuous time scale. When there is more than one lineage, each lineage has an independent chance of migrating but only one lineage will migrate. So, we add the $e^{-t\frac{Md}{2}}$ chance of migration for each lineage over all k lineages to obtain the total chance of migration:

$$P(T_{migration} = t) = e^{-t\frac{Md}{2}k} \qquad (4.83)$$

for k ancestral lineages of the d demes. The chance that one of k lineages migrates at or before a certain time can then be approximated with the cumulative exponential distribution

$$P(T_{migration} \leq t) = 1 - e^{-t\frac{Md}{2}k} \qquad (4.84)$$

in exactly the same fashion that times to coalescent events are approximated.

When two independent processes are operating, the genealogical model becomes one of following lineages back in time and waiting for an event to happen. The possible events in this case are migration or coalescence, so the total chance of any event is the sum of the independent probabilities of each type of mutually exclusive event. Since lineages cannot coalesce unless they are in the same deme, the chance of a coalescent event is

$$P(T_{coal} \leq t) = 1 - e^{-td\sum_{i=1}^{d}\frac{k_i(k_i-1)}{2}} \qquad (4.85)$$

when there are k_i ancestral lineages in deme i, a slightly modified version of the basic coalescent model that takes into account the d demes and time scaling by $2N_e d$. (Note that when $d = 1$, the expected time to coalescence reduces to $\dfrac{k(k-1)}{2}$ on a continuous time scale.) The total chance of *any event*, either coalescence or migration, occurring when going back in time (increasing t) is then

$$P(T_{event} \leq t) = 1 - e^{-t\left[dk\frac{M}{2} + d\sum_{i=1}^{d}\frac{k_i(k_i-1)}{2}\right]} \qquad (4.86)$$

where the exponent is the sum of the intensities of migration and coalescence. In the simplest case of two demes ($d = 2$) with k_1 and k_2 ancestral lineages in each deme, this reduces to

$$P(T_{event} \leq t) = 1 - e^{-t\left[(k_1 + k_2)\frac{M}{2} + \frac{k_1(k_1-1)}{2} + \frac{k_2(k_2-1)}{2}\right]}$$

(4.87)

(the example given in Hudson 1990) where time is scaled in units of the total population size $2N_e d$ or the sum of number of lineages in all of the demes.

When an event does occur at a time given by this exponential distribution in Eq. 4.86, it is then necessary to decide whether the event is a coalescence or a migration. The total chance that the event is either a migration or a coalescence event is $dk\frac{M}{2} + d\sum_{i=1}^{d}\frac{k_i(k_i-1)}{2}$. Therefore, the chance the event is a migration is

$$\frac{dk\frac{M}{2}}{dk\frac{M}{2} + d\sum_{i=1}^{d}\frac{k_i(k_i-1)}{2}} = \frac{kM}{k(M-1) + \sum_{i=1}^{d}k_i^2}$$

(4.88)

whereas the chance that the event is a coalescence is

$$\frac{d\sum_{i=1}^{d}\frac{k(k_i-1)}{2}}{dk\frac{M}{2} + d\sum_{i=1}^{d}\frac{k_i(k_i-1)}{2}} = \frac{\sum_{i=1}^{d}(k_i^2 - k_i)}{k(M-1) + \sum_{i=1}^{d}k_i^2}$$

(4.89)

When the event is a coalescence, the deme is picked at random given that demes with more ancestral lineages have a greater chance of experiencing a coalescent event (the chance that deme j experiences the coalescence is $\frac{\frac{k(k_j-1)}{2}}{\sum_{i=1}^{d}\frac{k_i(k_i-1)}{2}} = \frac{k_j(k_j-1)}{\sum_{i=1}^{d}k_i(k_i-1)}$).

Figure 4.24 shows two realizations of the combined coalescence and migration process when the migration rate is either relatively high or relatively low. The times to each event are determined by the exponential distributions specified by Eq. 4.86.

Interact box 4.8 Gene genealogies with migration between two demes

A coalescent genealogy that includes the possibility of migration among demes can be constructed using the cumulative exponential distribution to determine the waiting time to an event. Once a waiting time is determined, determining whether the event is a migration or a coalescence is accomplished using the probabilities of these events. If the event is coalescence, a random pair of lineages in a random deme is picked to coalesce and the number of ancestral lineages in that deme (k_i) is reduced by 1. If the event is migration, a random lineage is picked and moved into a random deme. Refer to Interact Box 3.4 for step-by-step instructions on how to draw a genealogy using a sample of waiting times.

The text website links to a Microsoft Excel spreadsheet model to calculate the quantities necessary to build a coalescent genealogy with migration between two demes. (As an alternative, there is a simple R script that will provide waiting times to build a genealogy.)

A second way to see coalescence and migration is to use the **Hudson Animator** simulator linked on the text web page. There are three parameters that can be set in the simulation: **n:** sets the number of lineages sampled in *both* demes in the present time (or $k_1 + k_2$ in Eq. 4.86), while **M1:** and **M2:** set the expected number of migrants in deme 1 and deme 2 each time period (or M in Eqs. 4.83 and 4.85). Pressing **Recalc** will calculate the waiting times for a new genealogy. The animation process can be controlled with the buttons below the figure. Waiting times can be seen in at the lower right when the pointer is placed over a circle in the tree. Click on the **Trees** tab at the top left to see how population structure impacts the genealogical tree itself.

Initially, set **n** to 10 and both **M1** and **M2** to the low migration rate of 0.1. Simulate 10 independent genealogies; in each case, record the number of migration events (the light blue circles in the animation) and the total waiting time until coalescence to a single most recent common ancestor. Increase both **M1** and **M2** to a higher migration rate of 1.0 and again simulate 10 independent trees and record the number of migration events and total waiting time to a most recent common ancestor or MRCA. How do the genealogies compare on average when migration rates are lower or higher?

(a)

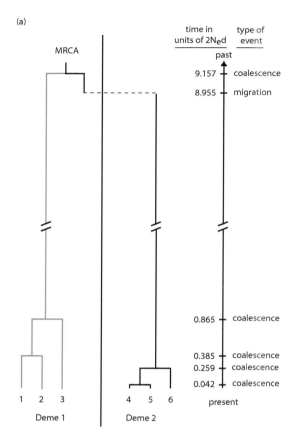

time in units of 2N$_e$d	type of event
past ↑	
9.157	coalescence
8.955	migration
0.865	coalescence
0.385	coalescence
0.259	coalescence
0.042	coalescence
present	

(b)

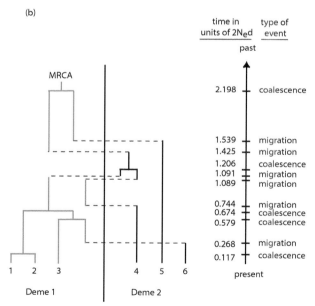

time in units of 2N$_e$d	type of event
past ↑	
2.198	coalescence
1.539	migration
1.425	migration
1.206	coalescence
1.091	migration
1.089	migration
0.744	migration
0.674	coalescence
0.579	coalescence
0.268	migration
0.117	coalescence
present	

Figure 4.24 Genealogies for six lineages initially divided evenly between two demes when the migration rate is low (A) and when the migration rate is high (B). When migration is unlikely, coalescent events within demes tend to result in a single lineage within all demes before any migration event take place. There is then a long wait until a migration event places both demes in one deme where they can coalesce. When migration is likely, lineages regularly move between the demes, and lineages originally in the same deme are as likely to coalesce as lineages initially in different demes. These two genealogies are examples and substantial variation in coalescence times is expected. In panel A, $M = 4N_em = 0.2$, and, in panel B, $M = 4N_em = 2.0$. The two genealogies are not drawn to the same scale.

The average length of a genealogy with migration

Before determining the average time to coalescence for a genealogy in a structured population, it is first necessary to introduce some useful notation to describe the different possible locations of lineages within and among demes. We can define a list (or row vector) that tracks the way lineages are partitioned among all demes as

$$d = (d_1, d_2, d_3, ..., d_n) \qquad (4.90)$$

where each d_i is the number of demes with i lineages and n is the total number of demes. The total number of lineages is then the product of the number of demes containing i lineages and the number of lineages i summed over all possible numbers of lineages per deme or $\sum_{i=1}^{n} id_i$. With a sample of two lineages taken from a total population composed of two demes, there are two possible ways the lineages could be sampled. The two lineages could either be sampled from different demes to give $d = (2,0)$ or sampled from a single deme to give $d = (0,1)$. This notation specifies what is called the **sample configuration** of a number of lineages drawn from some number demes. Figure 4.25 gives several examples of sample configurations for two or three demes. With coalescence to a single ancestral lineage, the sample configuration becomes (1). This sample configuration notation is useful because the mean and variance of coalescence times in a structured population depend on whether lineages are located in the same or different demes.

With that background on sample configurations, let us move on to derive the average coalescence time and expected total length of a genealogy in a structured population. We will focus on the simplest case of two lineages in the context of two demes. We need to determine the chance that two lineages in either of the two possible sample configurations ((2,0) or (0,1)) experience coalescence. Figure 4.26 shows these possible transitions between sample configuration states.

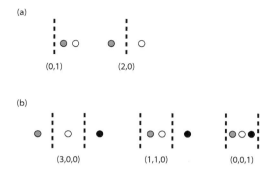

Figure 4.25 Sample configurations for two lineages and two demes (A) and three lineages and three demes (B). Lineages are represented by the dots, and the separation between demes is represented by a dotted vertical line. Only one possibility is given for each sample configuration, even though some configurations can occur in multiple ways. For example, (0,1) can occur when both lineages are in the left-hand deme or when both lineages are in the right-hand deme.

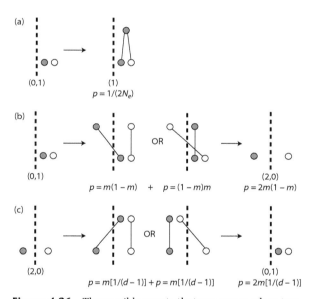

Figure 4.26 The possible events that can occur when two lineages are in the same deme (0,1) or when two lineages are in two different demes (2,0) along with their probabilities of occurring. The separation between demes is represented by a dotted vertical line. Two lineages can coalesce only when they are in the same deme. The probability of coalescence (A), migration of one lineage such that the two lineages are in different demes (B), and migration that places both lineages in the same deme (C) determine the overall chances that two lineages coalesce. The chance that both lineages migrate (with probability m^2) is not shown in B and applies when there are three or more demes.

As in the basic coalescent process, the chance of coalescence is the product of 1 over the population size and the number of unique pairs of lineages that can coalesce. If each deme contains $2N_e$ lineages, the probability of coalescence is $\frac{1}{2N_e}$ for two lineages in

one deme. However, two lineages cannot coalesce unless they are together in the same deme and restricted gene flow will make this less likely to happen.

For two lineages that are together in the same deme or in sample configuration (0,1), there are two possible events that eventually lead to coalescence. The first possible event is simply that the two lineages coalesce with probability $\frac{1}{2N_e}$. The second possible event is that one or both of the two lineages migrates into another deme before they can coalesce. If the proportion of migrants per generation in any deme is m, then the chance that a single lineage is an emigrant is m and the chance that it is not an emigrant is $1 - m$. The chance that one lineage migrates and the other lineage does not is $m(1 - m) + (1 - m)m = 2m(1 - m)$. The chance that both lineages migrate is m^2. The total chance that one or both lineages migrate is then $2m(1 - m) + m^2$, which is approximately $2m$ if m is small and m^2 terms can be ignored. For two lineages in the same deme or (0,1), the total chance that *any* event occurs in the previous generation, either coalescence or migration, is therefore $2m + \frac{1}{2N_e}$.

For two lineages that are in different demes or in sample configuration (2,0), the total chance that one lineage migrates is $2m$ following the same logic as when two lineages are in a single deme. However, to transition from (2,0) to (0,1), the migration event is not into any random deme but must be into the one other deme where the second lineage is found. The chance that migration into a specific deme occurs is $\frac{1}{d - 1}$ where d is the number of demes. The total probability that two lineages initially in separate demes end up in a single deme with the possibility that they can later coalesce is, therefore, $2m\frac{1}{d - 1}$.

To determine the average time to coalescence in two demes, we can use the fact that the average time to an event is one over the probability of each event in a process where waiting times are exponentially distributed. Let $\overline{T}_{(0,1)}$ represent the average time until coalescence for two lineages in the same deme and $\overline{T}_{(2,0)}$ the average time to coalescence for two lineages in different demes. For two lineages in the same deme, the average time to coalescence is the average time to coalescence plus the average time spent in two different demes if there is a migration event. The average time of either coalescence or migration is 1 over the total chance of an event, or $\dfrac{1}{2m + \dfrac{1}{2N_e}}$. When an event occurs, there is a $\dfrac{1}{2N_e}$ chance it is a coalescence and a $2m$ chance that it is a migration event. Bringing this all together leads to an expression for the average time to coalescence for two lineages in the same deme:

$$\overline{T}_{(0,1)} = \frac{\dfrac{1}{2N_e}}{2m + \dfrac{1}{2N_e}} + \frac{2m}{2m + \dfrac{1}{2N_e}}\overline{T}_{(2,0)} \quad (4.91)$$

For two lineages in different demes, the average time to coalescence is the sum of the average time needed to migrate into the same deme and the average time until coalescence once the lineages are in the same deme. Since the chance of migration into the same deme is $2m\dfrac{1}{d-1}$, the average time for two lineages to migrate into the same deme is $\dfrac{d-1}{2m}$. The average time to coalescence for two lineages in different demes this then

$$\overline{T}_{(2,0)} = \frac{d-1}{2m} + \overline{T}_{(0,1)} \quad (4.92)$$

Solving these two equations (see Math Box 4.2) leads to

$$\overline{T}_{(0,1)} = 2N_e d \quad (4.93)$$

and

$$\overline{T}_{(2,0)} = 2N_e d + \frac{d-1}{2m} \quad (4.94)$$

(see Slatkin 1987b; Strobeck 1987; Nordborg 1997; Wakeley 1998).

These average times to coalescence for two lineages in the context of two demes are both simple expressions that are easy to interpret. Equation 4.93 is a bit surprising since it says that the average time to coalescence for two lineages in the same deme is independent of the migration rate and is simply a function of the total population size as in a panmictic population (note that if each of d demes contains $2N_e$ lineages, the total population size is $N_T = 2N_e d$). We can understand why this is the case by imagining what happens as the migration rate changes. If the migration rate decreases, the probability that a lineage migrates into another deme decreases with the effect of shortening the time to coalescence. However, in those cases when a migration event does occur, the lineage would take a longer time to migrate back before coalescence. The average time to coalescence is independent of the migration rate since these two factors exactly balance as the migration rate changes. When two lineages are in different demes, the average time to coalescence increases as the migration rate decreases and as the number of demes increases. The average coalescence time is inversely proportional to the migration rate since migration is required to put two lineages into the same deme by chance. As the number of demes increases, there are an increasing number of places for two lineages to be apart so that more migration events will have to occur until two lineages are together in the same deme.

Average coalescence times within demes and in the total population can also be used to express the degree of population structure. Earlier in the chapter, we used probabilities of autozygosity to express population structure as a difference between the chance that two alleles sampled from the total population are different in state (H_T) and the chance that two alleles sampled from a subpopulation are different in state (H_S) or $F_{ST} = \dfrac{H_T - H_S}{H_T}$. For two lineages drawn at random from the total population of d demes, there is a $\frac{1}{d}$ chance they are from the same deme and a $\dfrac{d-1}{d}$ chance they are from different demes. Therefore, the average coalescence time for two lineages sampled at random from a subdivided population is

$$\overline{T} = \frac{1}{d}\overline{T}_{(0,1)} + \frac{d-1}{d}\overline{T}_{(2,0)} = 2N_e d + \frac{(d-1)^2}{2md} \quad (4.106)$$

and Eq. 4.93 provides the average coalescence time for two lineages sampled from the same deme ($\overline{T}_{(0,1)} = 2N_e d$). Putting these two average coalescence times together

$$F_{ST} = \frac{\overline{T} - \overline{T}_{(0,1)}}{\overline{T}} \quad (4.107)$$

gives an expression for the pattern of population structure from the perspective of coalescence times (Slatkin 1991). Population structure can then be

Math box 4.4 Solving two equations with two unknowns for average coalescence times

We can restate Eqs. 4.91 and 4.92 as

$$\overline{T}_{(0,1)} = x = a + by \qquad (4.95)$$

$$\overline{T}_{(2,0)} = y = c + x \qquad (4.96)$$

where $a = \dfrac{\frac{1}{2N_e}}{2m + \frac{1}{2N_e}}$, $b = \dfrac{2m}{2m + \frac{1}{2N_e}}$, and

$c = \dfrac{d-1}{2m}$. When time is scaled in units of $2N_e$,

then $a = \dfrac{1}{2m + \frac{1}{2N_e}}$.

We can then substitute the equation for x into the equation for y to get

$$y = c + a + by \qquad (4.97)$$

which rearranges to

$$y - by = c + a \qquad (4.98)$$

and then

$$y = \frac{c+a}{1-b} \qquad (4.99)$$

Substituting the values for a, b, and c gives

$$y = \frac{\dfrac{d-1}{2m} + \dfrac{1}{2m + \frac{1}{2N_e}}}{1 - \left(\dfrac{2m}{2m + \frac{1}{N_e}}\right)} \qquad (4.100)$$

The denominator above can be rearranged to
$\frac{2m+\frac{1}{N_e}}{2m+\frac{1}{N_e}} - \frac{2m}{2m+\frac{1}{N_e}} = \frac{\frac{1}{2N_e}}{2m+\frac{1}{2N_e}}$. Next, let $f = 2m + \frac{1}{2N_e}$
and then substitute it into the equation with the rearranged denominator to get

$$y = \frac{\dfrac{d-1}{2m} + \dfrac{1}{f}}{\dfrac{1}{2N_e}} \qquad (4.101)$$

which when the numerator and denominator are multiplied by f gives

$$y = \frac{f\frac{d-1}{2m} + 1}{\frac{1}{2N_e}} \qquad (4.102)$$

Substituting the full expression for f and then expanding gives

$$y = \frac{\left(2m + \frac{1}{2N_e}\right)\frac{d-1}{2m} + 1}{\frac{1}{2N_e}}$$
$$= \frac{\frac{1}{2N_e}\left(\dfrac{2N_e 2m(d-1) + (d-1) + 2N_e 2m}{2m}\right)}{\frac{1}{2N_e}} \qquad (4.103)$$

and then multiplying by $2N_e$ instead of dividing by $\frac{1}{2N_e}$ cancels the $\frac{1}{2N_e}$ term in the numerator and then expanding gives

$$y = \frac{2N_e 2md}{2m} - \frac{2N_e 2m}{2m} + \frac{d-1}{2m} + \frac{2N_e 2m}{2m} \qquad (4.104)$$

which after addition and canceling terms finally gives

$$y = 2N_e d + \frac{d-1}{2m} \qquad (4.105)$$

Equations for $\overline{T}_{(2,0)}$ can then be solved by substituting this expression for y and similar methods of algebraic rearrangement.

thought of as the difference in average coalescence times for a pair of lineages sampled from the total population compared with a pair of lineages sampled within one subpopulation.

In general, population subdivision is expected to increase the time required for lineages to coalesce to a single most recent common ancestor. When gene flow is relatively limited, coalescent events within demes occur much as they would in an isolated panmixic population. However, the single ancestor for each deme must wait for a relatively rare migration event until two lineages in different demes can coalesce to a single ancestor. This tends to produce genealogical trees that have long branches connecting the individual ancestors of different demes. As rates of migration increase, the genealogical tree branch lengths approach the patterns expected in a single panmictic population of the same total size since migration events frequently move lineages among the demes.

Distinguishing coalescent events within and among demes has led to the generalization that genealogies in subdivided populations can be divided into two time scales. One time scale is the recent genealogy of each deme with intra-deme coalescence events and migration events. Another time scale is the older history of ancestral lineages in the total population. Wakeley (1998, 1999) has described the recent time scale as the **scattering phase** and the deeper time scale as the **collecting phase** for a genealogy in a subdivided population. The separation of times scales has also been studied in subdivided populations with extinction and recolonization (Wakeley and Aliacar 2001) and in a continuous population with isolation by distance (Wilkins 2004). A key insight is that under certain conditions, patterns of polymorphism are mostly a product of genealogical sorting in the collecting phase and do not depend much on the scattering phase where migration events occur. Rather, the collecting phase portion of a genealogy is described by the standard coalescent scaled with an appropriate N_e without any impact of migration. The consequence is that it may be impossible to determine if different patterns of genetic polymorphism have been caused by differences in the size of demes, the number of demes, the effective migration rate among demes, or older events such as total population growth since these variables have interchangeable impacts on the overall genealogy.

Chapter 4 review

- Spatial and temporal separation of individuals and subpopulations result in mating that is not random throughout a population. Without enough gene flow to maintain random mating (panmixia), genetic drift causes divergence of allele frequencies among subpopulations.

- Isolation by distance is a very general prediction since gene flow is expected to decrease as the spatial separation of subpopulations increases.

- Numerous models of gene flow, such as the island or stepping stone, describe the patterns of genetic mixing among multiple subpopulations.

- The continent-island and two-island models show that acting over time, gene flow homogenizes allele frequencies to an equilibrium value that depends on the pattern of gene flow rates between and among subpopulations.

- Levels of gene flow can be measured by directly tracking parentage in contemporary populations (a direct estimate). Parentage analyses use genotypes of progeny and one known parent to infer the haplotype of the unknown parent. This unknown parent haplotype is then used to exclude possible parents from the pool of candidate parents. The power of this procedure to identify the true parent depends on the chance that a given haplotype will occur at random in a population.

- F_{IS} measures the average excess or deficit of heterozygous genotypes compared with random mating. F_{ST} measures the deficit of heterozygosity in subpopulations due to population structure compared to heterozygosity expected with panmixia. F_{IT} measures the total excess or deficit of heterozygous genotypes due to both non-random mating within and allele frequency divergence among subpopulations.

- Numerous estimators of the ideal F_{ST} such as G_{ST} account for details of actual genetic data more than two alleles at a locus, averaging over multiple loci, and finite and potentially unequal sample sizes.

- The Wahlund effect demonstrates that genetic variation can be stored as variance in allele frequencies among subpopulations or as heterozygosity within a panmictic population. Fusion of diverged subpopulations or subdivision of a panmictic population converts one type of genetic variation into the other type of genetic variation.

The magnitude and pattern of genetic differentiation among a set of subpopulations can be compared to the predictions from a model of the pattern and magnitude of gene flow and genetic drift to indirectly estimate of rates of gene flow. For example, under the infinite island model, equilibrium F_{ST} is approximately equal to the inverse of four times the effective migration rate plus one ($\frac{1}{4N_em + 1}$). Under isolation by distance in two dimensions, linearized $\hat{F}_{ST}$ is expected to increase with log geographic distance. Metapopulation models include movement of gametes and individuals among subpopulations as well as extinction and re-establishment of subpopulations. Spatially explicit approaches use measured features of the landscape to model least-cost paths of gene flow or circuit theory to model the network of multiple paths with variable rates of gene flow that connect subpopulations.

- Multilocus genotypes can be used to assign individuals to a subpopulation of origin through maximum likelihood or Bayesian posterior probabilities. The Bayesian approach can be extended to determine the best supported number of subpopulations as well as the proportion of the multilocus genotype of each individual that originated within each subpopulation.
- PCA helps reduce the dimensionality of genetic data sets with many loci and loci with many alleles and can be used to identify genetic subpopulations as well as individuals with recent ancestry attributable to several subpopulations.
- Genealogical trees in subdivided populations can be modeled with an exponentially distributed waiting time where the chance of migration and the chance of coalescence are combined.
- In two demes, the average time to coalescence for two lineages in the same deme is the total population size and is independent of the migration rate. For two lineages in different demes, the average time to coalescence gets longer as the number of demes increases and as the migration rate decreases since two lineages can only coalesce when they are in the same deme.
- Population structure and limited gene flow lengthen the average coalescence time of two lineages sampled at random from the population compared to the average coalescence time of two lineages sampled from the same subpopulation.

Further reading

A review of predictions for and estimators of the spatial patterns of genetic variation within and among populations that make up landscape genetics can be found in

Balkenhol, N., Cushman, S.A., Storfer, A.T., and Waits, L.P. (eds.) (2015). *Landscape Genetics*. Chichester, UK: Wiley.

To learn more about the role that the plant *Linanthus parryae* played in the development of the theory of isolation by distance as well as the personalities associated with competing interpretations of the spatial distributions of blue and white flower colors, see

Provine, W.B. (1986). *Sewall Wright and Evolutionary Biology*. Chicago, IL: University of Chicago Press.

For a perspective on parentage analysis in the era of genomics, see

Flanagan, S.P. and Jones, A.G. (2018). The future of parentage analysis: from microsatellites to SNPs and beyond. *Molecular Ecology* 28: 544–567.

A review of concepts and empirical estimates of population structure and indirect estimates of gene flow can be found in

Holsinger, K.E. and Weir, B.S. (2009). Genetics in geographically structured populations: defining, estimating and interpreting F_{ST}. *Nature Reviews Genetics* 10: 639–650.

An overview of approaches to estimating heterozygosity and population differentiation in polyploidy species is given in

Meirmans, P.G., Liu, S., and van Tienderen, P.H. (2018). The analysis of polyploid genetic data. *Journal of Heredity* 109: 283–296.

A review of Bayesian statistical methods in genetics and how the approach has paved the way for new approaches to hypothesis testing in population genetics and genomics can be found in

Beaumont, M.A. and Rannala, B. (2004). The Bayesian revolution in genetics. *Nature Reviews Genetics* 5: 251–261.

A review of the impacts of population structure in the context of the coalescent model can be found in

Charlesworth, B., Charlesworth, D., and Barton, N.H. (2003). The effects of genetic and geographic structure on neutral variation. *Annual Review of Ecology and Systematics* 34: 99–125.

End of chapter exercises

1 In criminal investigations, a multilocus DNA profile obtained from the crime scene as evidence can be compared with millions of multilocus DNA profiles available in database records. Explain why this practice can generate high probabilities of random match of the multilocus DNA profile of the individual who left the evidence DNA sample and a different individual who has a DNA profile in the database.

2 Use the text simulation website **Simulations -> Fixation Indices** for this problem. The simulation dialog has values for the N_e of each deme, the migration rate, and the initial allele frequency in each deme for a diallelic locus. Leave the total number of populations and the coancestry coefficient at default values. Run simulations for the parameter value sets in the table below, and tabulate the results in the spaces provided. Run each simulation **twice** (at least) to get an idea of how much each outcome can differ for the same parameter values. Make note of the generation when each F index initially reaches its approximate maximum. View each set of conditions on a common time scale (up to 500 generations) to aid comparisons.

   This problem can also be carried out using Populus (from the Main Menu, select Mendelian Genetics and then Population Structure; set the number of demes to 10).

Deme size (N_e)	m	Initial freq(A)	F_{IS}	F_{ST}	F_{IT}
10	0	0.5			
10	0.001	0.5			
10	0.1	0.5			
10	0.1	0.8			
50	0	0.5			
50	0.001	0.5			
50	0.1	0.5			
100	0	0.5			
100	0.001	0.5			
100	0.1	0.5			

Define H_I, H_S and H_T both biologically and mathematically. Define F_{IS}, F_{ST}, and F_{IT} both mathematically (using the H notation) and biologically.

In the simulations, the allele frequency lines sometimes hit the top or bottom axis (go to fixation or loss) and then reappear. What is happening in these cases? What are the units of the migration values you specified in the simulation (e.g. 0.001 or 0.1)? What does an increase in this value mean in biological terms? Why does increasing m maintain lower F_{ST} and F_{IT} values? How does migration counteract genetic drift? Is migration always strong enough to do this? Explain using observations from your simulations.

3 Use the text simulation website **Simulations -> Fixation Indices** for this problem. What is the coancestry coefficient (f)? Run simulations for 100 generations with f values of −0.5, 0.0, and 0.5 for $N_e = 20$ and $m = 0.01$. Fill in values observed at 100 generations using the table below.

Coancestry coefficient (f)	H_I	H_S	H_T	F_{IS}	F_{ST}	F_{IT}
−0.5						
0.0						
0.5						

How do H_I and H_S compare for the same value of f? How do F_{IS} and F_{ST} vary with f? Why?

4 In 1965, Sick (Hereditas 54 : 49–69) measured allozyme polymorphism for a hemoglobin gene in codfish. His goal was to document migration and mating patterns of cod for the purposes of stock management and to identify breeding populations. The genotypes he observed in a sample of 1000 fish from each of three geographic areas are given in the table below (F stands for fast and S for slow – the migration rates of each allele in a starch gel). The geographic sampling areas are given on the map (from https://www.google.com/maps/place/Baltic+Sea). Answer each question using **quantitative** reasoning based on the observed allele and genotype frequencies.

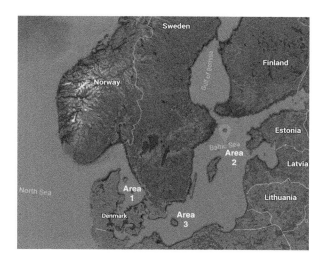

	Genotype		
	FF	FS	SS
Area 1	40	320	640
Area 2	640	320	40
Area 3	340	320	340

What are the observed allele frequencies in the three areas? What can you infer about the mating patterns of cod **within** each area based on genotype frequencies? Are areas 1 and 2 freely experiencing free gene flow? Can the fish in areas 1 and 2 be managed as a single population? What gene flow and population structure processes could explain the observed allele and genotype frequencies within area 3?

5 A set of two subpopulations have non-overlapping alleles at one locus with four alleles. Subpopulation one has allele frequencies of $p_1 = p_2 = 0.5$ and $p_3 = p_4 = 0$, while subpopulation one has allele frequencies of $p_1 = p_2 = 0$ and $p_3 = p_4 = 0.5$. Compute H_S and H_T and then use these to compute G_{ST} and Jost's D. What pattern of genetic variation between the populations does each index capture and why are they different?

6 Personal genotyping services provide information about a client's ancestral population of origin. What types of loci do these services rely on? What models do these services use to estimate the regions of the world where your ancestors lived?

7 In Math Box 4.2, let the resistor values be $R1 = R3 = 1$, $R4 = R5 = 4$, and $R2 = 4$. Solve for the voltages V1, V2, and V3 and then solve for current values at each node. Using the currents, give a biological interpretation for the amount of gene flow along each path between subpopulations A and B and for mortality at node 3.

8 Using the instructions in Interact Box 4.7, construct a coalescent genealogy for six total lineages divided initially in two demes ($k_1 = k_2 = 3$). Use $2N = 50$ and a migration rate of 0.1. (Note that this mutation rate is unrealistically high to be sure there are numerous migration events. Adjust the migration rate if you get too many migration events.) Use graph paper and draw a scale in both continuous and discrete time for the mutation and coalescence events on the genealogy. Repeat the simulation to examine how different a replicate genealogy is based on the same rates of drift and migration. For comparison, simulate and draw a genealogy with a migration rate of 0.01. Explain F_{ST} for the simulated genealogies based on the observed waiting times.

9 Search the literature for a recent research paper that utilizes one or more of the population genetic predictions covered in this chapter. The topic can be any organism, application, or process, but the paper must include a hypothesis test involving a topic such as genetic differentiation among populations, F_{ST} or one of its estimators, the effective migration rate (N_em), or patterns of gene flow. Summarize the main hypothesis, goal, or rationale of the paper. Then, explain how the paper utilized a population genetic prediction from this chapter and then summarize the results and the conclusions based on the prediction.

10 Construct a simulation model of genetic differentiation between two subpopulations experiencing gene flow and genetic drift. Instructions to build a spreadsheet model can be found on the text website. These instructions can also be implemented in a programming language such as python or R.

Problem box answers

Problem box 4.1 answer

The allele frequencies for each of the alleles in the paternal haplotype are obtained in Table 4.3. For tree 4865, there is only one possible paternal allele at each locus. The chance of any genotype having one copy of each paternal allele at each locus is:

A: $(0.1216)^2 + 2(0.1216)(1 - 0.1216) = 0.2284$
B: $(0.3971)^2 + 2(0.3971)(1 - 0.3971) = 0.6365$
C: $(0.0761)^2 + 2(0.0761)(1 - 0.0761) = 0.1464$
D: $(0.1905)^2 + 2(0.1905)(1 - 0.1905) = 0.3447$
E: $(0.1250)^2 + 2(0.1250)(1 - 0.1250) = 0.2344$

The paternal allele is expected to occur in between 14 and 64% of possible genotypes for any individual locus. The probability of a random match at all five loci is $0.2284 \times 0.6365 \times 0.1464 \times 0.3447 \times 0.2344 = 0.0017$ or in 17 out of 10 000 random genotypes. The probability of exclusion is then $1 - 0.0017 = 0.9983$, while the probability of exclusion for a sample of 30 candidate parents is $(0.9983)^{30} = 0.9502$. There is about a 95% chance that there would not be a random match in a sample of 30 candidate parents; therefore, we have high confidence that 4865 is the true father of seed 25-1 from tree 989. For this offspring–maternal parent combination, the B locus is the least useful in resolving paternity since the frequency of the 106 allele is almost 40%. The 167 allele at the C locus is the most useful with a frequency of just over 7%.

Problem box 4.2 answer

$H_I = \frac{1}{n} \sum_{i=1}^{n} \hat{H}_i$ where $\hat{H}$ is the observed frequency of heterozygotes in each of the n subpopulations.

$$H_I = (0.0 + 0.14 + 0.34 + 0.40)/4$$

$$H_I = 0.22$$

The average observed heterozygote frequency is 0.22 or 22%.

$H_S = \frac{1}{n} \sum_{i=1}^{n} 2p_i q_i$ where p_i and q_i are the allele frequencies in subpopulation i.

$$H_S = (2(0.0)(1.0) + 2(0.93)(0.07) + 2(0.17)(0.83) + 2(0.51)(0.49))/4$$

$$H_S = (0.0 + 0.1302 + 0.2822 + 0.4998)/4$$

$$H_S = 0.228$$

$H_T = 2\bar{p}\bar{q}$ where $\bar{p}$ and $\bar{q}$ are average allele frequencies for all the subpopulations. Let f be the frequency of the fast allele and s the frequency of the slow allele so that $f + s = 1$. Then, estimate the average allele frequency for the fast allele in the total population (the slow allele could be averaged as well):

$$\bar{f} = (0.0 + 0.93 + 0.17 + 0.51)/4$$

$$\bar{f} = 0.4025$$

whereas the frequency of the other allele is found by subtraction $\bar{s} = 1 - 0.4025 = 0.5975$.

$$H_T = 2(0.4025)(0.5975)$$

$$H_T = 0.481$$

We can now calculate the F statistics using H_I, H_S, and H_T.

$$F_{IS} = \frac{H_S - H_I}{H_S}$$

$$F_{IS} = (0.228 - 0.220)/0.228$$

$$F_{IS} = 0.035$$

There is no evidence for self-fertilization since these four populations have observed heterozygosities very close to that expected under random mating. Comparing the observed and expected heterozygosity for each population shows that subpopulations 9 and 43 have a slight excess of heterozygotes while subpopulation 68 has about a 10% deficit. These three deviations along with the zero deviation in subpopulation 1 all average out to approximately 0.

$$F_{ST} = \frac{H_T - H_S}{H_T}$$

$$F_{ST} = (0.481 - 0.228)/0.481$$

$$F_{ST} = 0.526$$

There is less heterozygosity within the subpopulations than what we expect under Hardy–Weinberg based on allele frequencies for the total population. This value reflects the substantial differences in subpopulation allele frequencies.

$$F_{IT} = \frac{H_T - H_I}{H_T}$$

$$F_{IT} = (0.481 - 0.220)/0.481$$

$$F_{IT} = 0.543$$

This is the deficit of heterozygosity caused by both non-random mating with populations and allele frequency divergence among subpopulations. In this case, almost all of the deficit in heterozygosity is due to allele frequency divergence among the subpopulations.

The three fixation measures are related by

$$(1 - F_{IT}) = (1 - F_{IS})(1 - F_{ST})$$

Using the values for F_{IS} and F_{ST} to solve F_{IT} gives the same value that was determined by direct computation:

$$(1 - F_{IT}) = (1 - 0.035)(1 - 0.526)$$

$$(1 - F_{IT}) = (0.965)(0.474)$$

$$(1 - F_{IT}) = 0.4574$$

$$F_{IT} = 0.543$$

Based on the data from all 43 subpopulations, Levin (1978) estimated $F_{IS} = 0.70$, $F_{ST} = 0.80$, and $F_{IT} = 0.80$ in *P. cuspidata*.

Problem box 4.3 answer

The Wahlund effect shows that population structure causes heterozygotes to become less frequent and homozygotes to become more frequent by a factor proportional to the amount of allele frequency divergence among populations. Using the allele frequencies in Table 2.3, we can calculate the expected genotype frequencies for each locus with adjustment for population using Eq. 4.36 for homozygous loci and Eq. 4.37 for heterozygous loci:

D3S1358 2(0.2118)(0.1626)(0.95) = 0.0655
D21S11   2(0.1811)(0.2321)(0.95) = 0.0799
D18S51 $(0.0918)^2 + (0.0918)(0.9082)(0.05) = 0.0126$
vWA $(0.2628)^2 + (0.2628)(0.7372)(0.05) = 0.0788$
FGA   2(0.1378)(0.0689)(0.95) = 0.0181
D8S1179 2(0.3393)(0.2015)(0.95) = 0.1299
D5S818   2(0.3538)(0.1462)(0.95) = 0.0942
D13S317 2(0.0765)(0.3087)(0.95) = 0.0448
D7S820   2(0.2020)(0.1404)(0.95) = 0.0539

Assuming that the Amelogenin locus should not be affected by population structure, the expected frequency of the 10-locus genotype after adjustment for population structure is $0.0655 \times 0.0799 \times 0.0126 \times 0.0788 \times 0.0181 \times 0.1299 \times 0.0942 \times 0.0448 \times 0.0539 \times 0.5 = 1.514 \times 10^{-12}$ with an odds ratio of one in 660 501 981 506. Compare that with the expected genotype frequency of 1.160×10^{-12} and odds ratio of one in 862 379 847 814 assuming panmixia. After accounting for population structure, this genotype appears more likely to occur by chance, although its expected frequency is still *extremely* rare.

Problem box 4.4 answer

The joint effects of drift and migration in the fixation index are expressed by

$$F_t = \frac{1}{2N_e}(1 - m)^2 + \left(1 - \frac{1}{2N_e}\right)F_{t-1}(1 - m)^2$$

This can be made more general if x is used to represent the probability of autozygosity and

y is used to represent the probability of allozygosity ($y = 1 - x$):

$$F_t = x(1 - m)^2 + yF_{t-1}(1 - m)^2$$

For the case of a diploid nuclear locus, we used $x = \frac{1}{2N_e}$ and $y = 1 - \frac{1}{2N_e}$ to obtain the relationship between F_{ST} and $N_e m$ at equilibrium. Both Y-chromosome and organelle loci are haploid and uniparentally inherited, giving an effective population size of one-quarter of that for nuclear loci. For example, in humans, the mitochondrial genome is inherited from the maternal parent only or half of the population, and it is also haploid or present in half the number of copies of the nuclear genome. For such loci, we would use $x = \frac{1}{\frac{N_e}{2}} = \frac{2}{N_e}$ and $y = 1 - \frac{2}{N_e}$ to obtain

$$F_{ST} \approx \frac{1}{N_e m + 1}$$

The result is that F_{ST} is expected to be higher for Y-chromosome and organelle loci because their effective population size is smaller (see Fig. 4.15). Compared to diploid nuclear loci, Y-chromosome and mitochondrial loci have levels of F_{ST} fourfold higher when all types of loci share a common migration rate. The greater level of divergence among subpopulations for the Y-chromosome and organelle loci comes about strictly because of differences in the autozygosity for the loci that produced increased rates of fixation or loss due to genetic drift. See Hu and Ennos (1999) and Hamilton and Miller (2002) for more details and references.

CHAPTER 5

Mutation

5.1 The source of all genetic variation

- Types of mutations and rates of mutation.
- How can a low-probability event like mutation account for genetic variation?
- The spectrum of fitness for mutations.
- Estimates of the mutation rate.
- Evolution of the mutation rate.

The previous chapters have discussed in detail genotype frequencies under random and nonrandom mating, the relationship between genetic drift and the effective population size, as well as population subdivision and gene flow. These and all other processes in populations act to shape or change the existing genetic variation in a population. But where does genetic variation come from in the first place? The Hardy–Weinberg expectation shows clearly that particulate inheritance itself does not alter genotype or allele frequencies, so it is not a source of genetic variation. Any form of nonrandom mating alters only genotype frequencies but leaves allele frequencies constant. Genetic drift serves to erode genetic variation, as sampling error leads to allele frequency change and eventually to fixation and loss. Gene flow just serves to partition genetic variation among subpopulations, thereby altering patterns of population structure. The process of **mutation**, the permanent incorporation of random errors in deoxyribonucleic acid or DNA that results in differences between ancestral and descendant copies of DNA sequences, is the ultimate source of all genetic variation.

This chapter will cover the process of mutation starting out with a description of the patterns and rates of mutation. The following sections will present classical population genetic models for the fate of a new mutation, the impact of mutation on allele frequencies in a population, and the predicted balance between the removal of genetic variation by genetic drift and its replacement by mutation. This chapter will also cover several models of the way new alleles are introduced by mutation commonly employed in population genetics, illustrated with applications that highlight the consequences of these models. The final section of the chapter will show how the process of mutation can be incorporated into genealogical branching models.

Mutation is a broad term that encompasses a wide variety of events that lead to alterations in DNA sequences. **Point mutations** lead to the replacement of a single base pair by another nucleotide. Point mutations to chemically similar nucleotides (purine to purine (A↔G) or pyrimidine to pyrimidine (C↔T)) are called **transitions**, while point mutations to chemically dissimilar nucleotides (purine to pyrimidine or pyrimidine to purine) are called **transversions**. Base substitutions that occur within coding genes may or may not alter the protein produced by that gene. **Synonymous** or silent mutations result in the same translation of a DNA sequence into a protein due to the redundant nature of the genetic code, while **nonsynonymous** or missense mutations result in a codon that does change the resulting amino acid sequence.

Mutation can take the form of **insertion** or **deletion** of DNA sequences, often referred to by the short-hand **indels**. Indels within coding regions result in frameshift mutations if the change in sequence length is not an even multiple of three, altering the translation of a DNA sequence and possibly creating premature stop codons. Indels may range in size from a single base pair to segments of chromosomes containing many thousands of base

Population Genetics, Second Edition. Matthew B. Hamilton.
© 2021 John Wiley & Sons, Inc. Published 2021 by John Wiley & Sons, Inc.
Companion website: www.wiley.com/go/hamilton/populationgenetics

pairs. Arrays of multiple copies of homologous genes called **multigene families** are formed by duplication events. Some copies of such duplicated genes may lose functions due to the accumulation of mutations becoming a **pseudogene**. **Gene conversion** may result in the homogenization of the sequences of multiple loci within multigene families. Gene conversion occurs because of an inappropriate mismatch repair that takes place during meiosis. Sections of two homologous chromosomes may anneal when they are single stranded during DNA replication. If these regions differ slightly in sequence, the annealed stretch will contain single nucleotide mismatches. These mismatches will then be repaired to proper Watson–Crick base pairing by enzymes normally involved in proofreading during DNA replication. The process of annealing between two sister chromosomes tends to happen frequently when the same gene has been repeated many times because the gene copies have very similar sequences and the chromosomes can anneal anywhere along the length of the gene array. The result is that all gene copies within multigene regions tend to converge on one random version of DNA sequence without recombination taking place.

Mutation may also take the form of rearrangements where a chromosomal region forms a loop structure that results in a segment breaking and being repaired in reversed orientation, called an **inversion**. **Translocations** are mutations where segments of chromosome break free from one chromosome and are incorporated by repair mechanisms into a nonhomologous chromosome. **Transposable elements**, segments of DNA that are capable of moving and replicating themselves within a genome, are frequent causes of translocation mutations. **Lateral or horizontal gene transfer**, the movement and incorporation of DNA segments between different individuals and even different species, is another possible avenue of mutation that occurs relatively frequently in prokaryotes. For more details on the molecular mechanisms that underlie these different types of mutations, consult a text such as Krebs et al. (2017).

The probability that a locus or base pair will experience a mutation is a critical parameter in population genetics since the rate of mutation describes how rapidly novel genetic variation is added to populations. Although it seems counterintuitive, mutation rates are actually quite difficult to estimate with precision in many types of organisms (see Drake et al. 1998; Fua and Huai 2003; Lynch et al. 2016).

Consider the case of mutation rates at a single **reporter locus** that has a well-understood effect on the phenotype of an organism, like coat color in mice. The data available to estimate mutation rates are the number of progeny that have a different coat color than expected based on the known coat-color genotypes of the parents. It is simple to divide the number of progeny with unexpected coat colors by the total number of progeny examined. However, that calculation estimates the frequency of detectable changes to coat color due to some molecular changes at the coat color locus. That is an estimate of the frequency of all types of mutation anywhere at the locus rather than an estimate of the mutation rate. Such an estimate of mutation frequency could also be incomplete since only mutational changes that caused an obvious change in coat color are included. Not all mutations will be reflected in coat colors, like changes to the third position nucleotide of a codon that are silent, or synonymous, and do not change the resulting amino acid sequence of a gene. Additionally, mutations may vary in their effect on coat color with some mutations having little or no easily observable effect on the phenotype. Therefore, the frequency of observable changes to the phenotype is not equivalent to the mutation rate.

An estimate of the mutation rate requires more information. One critical detail is the number of replications a locus or genome experiences because mutational changes usually occur during the replication process. Different cell types and different species experience different numbers of cell replications during growth and reproduction. For example, in mammals, mutations are more frequent in male gametes than in female gametes because there are many more cell divisions before the production of a sperm than there are before production of an egg. However, the underlying mutation rate could be identical for male and female gametes with the difference due only to the different number of genome replications that occur. Another set of considerations is the size of a locus or genome available to mutate. In the hypothetical mouse coat color, for example, the number of base pairs at the coat color locus is a critical piece of information. The rate of mutation per base pair estimated from the frequency of coat color changes is very different if the locus has 900 or 90 base pairs.

The distinction between mutation frequency and mutation rate highlights the fact that mutation rates in population genetics are expressed in a variety of terms depending on experimental methods and the

Table 5.1 Per-locus mutation rates measured for five loci that influence coat color phenotypes in inbred lines of mice (Schlager and Dickie 1971). Dominant mutations were counted by examining the coat color of F1 progeny from brother–sister matings. Recessive mutations required examining the coat color of F1 progeny from crosses between an inbred line homozygous for a recessive allele and a homozygous "wildtype" dominant allele. The effort to obtain these estimates was truly incredible, involving around seven million mice observed over the course of 6 years.

Locus	Gametes tested	Mutations observed	Mutation rate per locus $\times 10^{-6}$ (95% CI)
Mutations from dominant to recessive alleles			
Albino	150 391	5	33.2 (10.8 – 77.6)
Brown	919 699	3	3.3 (0.7 – 9.5)
Dilute	839 447	10	11.9 (5.2 – 21.9)
Leaden	243 444	4	16.4 (4.5 – 42.1)
Non-agouti	67 395	3	44.5 (9.2 – 130.1)
All loci	2 220 376	25	11.2 (7.3 – 16.6)
Mutations from recessive to dominant alleles			
Albino	3 423 724	0	0 (0.0 – 1.1)
Brown	3 092 806	0	0 (0.0 – 1.2)
Dilute	2 307 692	9	3.9 (1.8 – 11.1)
Leaden	266 122	0	0 (0.0 – 13.9)
Non-agouti	8 167 854	34	4.2 (2.9 – 5.8)
All loci	17 236 978	43	2.5 (1.8 – 3.4)

life cycle of an organism. The target of mutation can be an entire genome, a locus, or a single base pair, while the rate can be expressed in time units of per DNA replication or per sexual generation. Comparisons of mutation rates only make sense when the target size and time period are expressed in identical units. Generally, for population genetic predictions involving sexual eukaryotes, mutation rates per sexual generation are the relevant units. Predictions for prokaryotes such as *Escherichia coli* or yeast would more naturally use mutation rates per cell division.

The most general rule of mutation is that it is a rare event with a low probability of occurrence. In a classic experiment involving literally millions of mice, mutation rates were estimated from five genes with observable effects on the phenotype of coat color (Table 5.1; Schlager and Dickie 1971). Rates of mutation per gene were between 1.8 and 16.6 mutations per 1 million gametes produced. This is equivalent to a mutation rate of $(1.8–16.6) \times 10^{-6}$ per locus per sexual generation. Very similar mutation rates for mice have been reported from more recent irradiation studies as well (Russell and Russell 1996). The rates of mutation from wild type to a novel allele (called **forward** mutations) are nearly a factor of 10 more common than mutations from

a novel allele to wild type (termed **reverse** mutations). This asymmetry of forward and backward mutation rates per locus is a common observation in mutation experiments. It is a product of the fact that there are more ways mutation can cause a normal allele to malfunction than there are ways to exactly restore that function once it is disrupted. In this sense, forward and reverse mutation rates exist only because mutations are detected via their phenotypic effect.

How can such a low-probability event like mutation add more than a trivial amount of genetic variation to populations? Let us calculate an initial answer to that question using humans as our example. Averaging over coding and noncoding parts of the genome, an approximate nuclear genome mutation rate in humans is about 1×10^{-9} mutations per base pair per generation. The haploid genome (one sperm or egg) contains about 3.2×10^9 base pairs (bp). Each genome of each diploid individual will have:

$$\left(1 \times 10^{-9} \text{ mutations bp}^{-1} \text{ generation}^{-1}\right) \atop \left(2 \times 3.2 \times 10^9 \text{ bp}^{-1}\right) = 6.4 \text{ mutations} \quad (5.1)$$

where the factor of 2 is due to a diploid genome. Each of us differs from one of our parents by half this amount, or about three mutations on average. If all mutations are random events evenly distributed throughout the genome, every pair of individuals differs by twice this number of mutations or about 13 mutational differences on average. The overall effect of mutation on available genetic variation depends on the size of a population. The human population in mid-2019 is about 7.571 billion people (see http://www.census.gov/popclock). Based on this population size, there are a total of

$$\left(6.4 \text{ mutations individual}^{-1} \text{ generation}^{-1}\right)$$
$$\left(7.571 \times 10^9 \text{ individuals}\right) = 48.5 \times 10^9 \text{ mutations}$$

$$(5.2)$$

or over 48 billion single nucleotide mutations are expected in the human population. This means that the absolute numbers of mutations per generation are potentially high and their number depends on the rate of mutation, the size of the population, and the size of the genome. We will revisit this topic later in the chapter to make a more formal prediction about the levels of heterozygosity expected when the input of genetic variation due to mutation and the loss of genetic variation caused by genetic drift are at equilibrium.

The impact that a mutant allele (as part of a heterozygous or homozygous genotype) has on the phenotype of an individual can vary greatly. Since natural selection along with genetic drift are critical processes that determine the fate of new mutations, the phenotype is most often considered in the context of its survivorship and reproduction or fitness. The range of the possible fitnesses for an individual mutant allele can be thought of as a **mutation fitness spectrum** like that shown in Figure 5.1. The fitness of all mutation effects to the phenotype is relative to the average fitness of a population (see Chapter 6 for explanations of fitness and average fitness). **Detrimental** or **deleterious** mutations reduce survival and reproduction, while mutations that improve survival and reproduction are **advantageous**. Severely deleterious mutations such as those that result in death (called **lethals**) or failure to reproduce viable offspring are acted strongly against by natural selection and will likely not last for a single generation. Mutations that are strongly deleterious and nearly lethal are sometimes called **sublethals**. Mutations that have small positive or negative effects on fitness (the shaded zone around the mean fitness in Figure 5.1) are called **neutral**

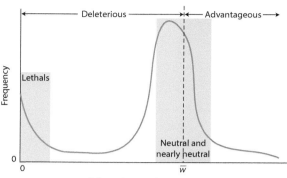

Figure 5.1 A hypothetical distribution of the effects of mutations on phenotypes that ultimately impact the Darwinian fitness of genotypes. Mutations that have a mean fitness less than the mean fitness of the population ($\overline{w}$) are decreased in frequency by natural selection. The shaded area around $\overline{w}$ indicates the zone where mutations have small effects on fitness relative to the effects of genetic drift (the width of the neutral zone depends on the effective population size). The shaded area near zero mean fitness indicates mutations that cause failure to reproduce or are lethal. Lethals are more common since it is a category that includes many degrees of severity resulting from diverse causes. The fitness effects of mutations are inherently difficult to measure because of the rarity of mutation events, the small effect of many mutations, and the dependence of fitness on environmental context.

or **nearly neutral** since their fate will be dictated either totally or mostly by sampling error of genetic drift. The final types are **beneficial** mutations that increase survival and reproduction above the average fitness of the population. It is important to note that the fitness effects of mutations may depend greatly on environmental context (see Fry and Heinsohn 2002) and the genotype at other loci. These different types of mutation will be the subject of models later in this chapter that show how fitness relates to the chance that a new mutation is lost or reaches fixation in a population.

Mutation fitness spectrum: The frequency distribution of the average fitness of new mutations measured relative to the average fitness of a reference population.
Drift barrier hypothesis: The prediction that mutation rates will evolve downward by natural selection until response to natural selection is constrained by genetic drift. A key prediction is that mutation rates and N_e are negatively correlated.

The mutation fitness spectrum plays a central role in a wide range of hypotheses to explain a multitude of phenomena in population genetics and evolution (see Charlesworth and Charlesworth 1998; Orr 2003; Estes et al. 2004; Agrawal and Whitlock 2012; Lynch et al. 2016). Explanations for phenomena as general and diverse as inbreeding depression, the evolution of mating systems, the evolution of sex and recombination, and the rate of adaptation depend in part on the nature of the mutation fitness spectrum. Strongly deleterious or strongly beneficial mutations will be steadily and predictably driven to loss or to fixation, respectively, by natural selection. However, fixation and loss of mutations that have a small impact on fitness (relative to the effective population size) are due in whole or in part to random genetic drift. A consequence is that mildly deleterious mutations may reach fixation by chance and accumulate in a population over time. Similarly, some mildly beneficial mutations may be lost from populations by chance. An accumulation of mildly deleterious mutations reduces individual fitness and may increase the risk of extinction, resulting in natural selection for processes that reduce the load of deleterious mutations in a population. The frequency of beneficial mutations may also place limits on the rate of evolution by positive natural selection. Thus, the specific shape of the frequency distribution shown schematically in Figure 5.1 provides crucial information about the fate of individual mutations as well as the long-term consequences of continual mutation in populations.

Estimating mutation rates

There are three primary approaches used to estimate mutation rates within a single species. The first method was introduced by Luria and Delbruc (1943) in their classic test of two possible explanations for when mutations contribute genetic variation to populations – mutations occur continuously at random or mutations occur in response to stress or stimuli (the latter possibility is sometimes called directed mutation). Luria and Delbruc used a small number of cells to start a liquid culture of *E. coli* susceptible to infection by bacteriophage. They also prepared agar plates with cultured bacteriophage spread over the surface and spread the bacteria on the plates. Because the bacteriophage infected and lysed the bacteria, only a few colonies of bacteria that carried mutations conferring resistance appeared on the plates. Luria and Delbruc predicted that if resistance mutations

occurred at random, the number of mutations would match a Poisson distribution with equal mean and variance. The observed number of bacteriophage-resistant colonies followed a Poisson distribution, rejecting the directed mutation hypothesis. (Meneely (2016) provides an in-depth explanation of the Luria and Delbruc experiment and its expectations based on the Poisson distribution.) This **fluctuation assay** estimates the variance in the number of mutations observed in many replicate lines grown in parallel, at the same time estimating the average number of mutations based on the Poisson distribution (Sarkar et al. 1992). The average number of mutations is divided by the number of cell divisions and by the number of base pairs to estimate the mutation rate. The fluctuation assay has seen contemporary use because it can be coupled with whole genome sequencing as a means to estimate the per base pair mutations rate (e.g. Gou et al. 2019).

Another widely employed method to estimate the mutation rate and the shape of the mutation fitness spectrum relies on founding a series of genetically identical populations and then allowing some to experience mutations for many generations while maintaining a control population that does not experience mutation. Viability and reproduction phenotypes of the mutated populations are then compared with the control population at intervals to estimate the average change in fitness caused by the mutations. Such comparisons are called **mutation-accumulation** experiments since mutations are repeatedly fixed by genetic drift over generations in each of the independent replicate populations (Halligan and Keightley 2009).

If there was absolutely no mutation, the replicate populations in a mutation-accumulation experiment would all maintain identical viability over time since each population started out being genetically identical. Mutation, however, will occur at random and cause independent genetic changes in the different populations, causing the populations to diverge in viability. Imagine that the mutation fitness spectrum is symmetric around the mean fitness so that the frequency of deleterious and beneficial mutations of the same magnitude is equal. That would produce no change in the *average* viability of lines in a mutation-accumulation experiment since there would be equal chances of beneficial or deleterious mutations of the same size that would cancel each other out in a sample of many mutations. However, there would be an increase in the *variance* in viability because the range of viabilities among the populations would

increase with more and more mutations. Next, imagine a mutation fitness spectrum like that shown in
Figure 5.1 where deleterious mutations are more
common than beneficial mutations. As mutations
accumulate, the average viability of lines should
decrease since deleterious mutations are more
common than beneficial mutations. The more
skewed the distribution is toward deleterious
mutations, the faster the average viability should
decrease in the replicate populations.

The results of several classic mutation accumulation studies that estimated the frequency distribution
for mutations that affect viability in *Drosophila melanogaster* have had a major impact on perceptions of
the mutation fitness spectrum (Mukai 1964; Mukai
et al. 1972). Mutation-accumulation experiments in
Drosophila rely on special breeding designs that
maintain the second chromosome without recombination in many replicate homozygous families or
lines over many generations. Mutations of all types
occur on this nonrecombining chromosome and
are fixed by genetic drift within each line of flies
due to a single male founder for each generation.
At intervals of 10 generations, the flies in all of the
different independent lines were assayed for viability
in comparison with a control line that did not experience any mutation due to chromosomal inversions
(again accomplished with special breeding techniques). The change in average viability and variance
in viability found by Mukai et al. (1972) is shown in
Figure 5.2. The variance in viability among the replicate lines increased since the second chromosome
of each line diverged due to the occurrence and fixation of mutations. In addition, the average viability
declined as expected if deleterious mutations were
more common than beneficial mutations. The results
are consistent with deleterious mutations that cause
an average reduction in viability of 5% or less when
homozygous. Therefore, this experiment and others
like it motivated a view of the mutation fitness
spectrum as drawn in Figure 5.1. However, mutation-accumulation experiments have been carried
out in only a relatively few organisms and are inherently unable to detect mutations with very small
effects or mutations that do not affect the phenotype
within the environment where the phenotype is
measured. For example, in a study of chemically
induced mutations in the plant *Arabidopsis thaliana*,
Stearns and Fenster (2016) found an equal frequency of mutations that caused an increase or a
decrease in the value of quantitative traits not
closely related survivorship or reproduction. Of the

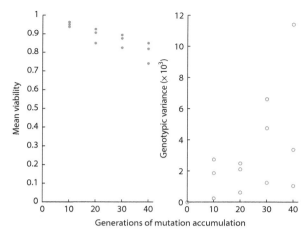

Figure 5.2   The results of the classic *Drosophila melanogaster*
mutation-accumulation experiment carried out by Mukai
et al. (1972). The experiment maintained three distinct sets of
mutation accumulation populations with 25 lines each. The
left panel shows the change in mean viability over time, and
the right panel shows the change in the variance among
replicate independent lines. Each point is the value obtained
from one set of mutation accumulation populations. Mutation
of any type makes the lines diverge genetically and increases
the variance. Mean viability declines over time as expected in
deleterious mutations are more common than advantageous
mutations. Source: Redrawn from Figure 2 in Mukai
et al. (1972).

20 mutation-bearing lines, 19 showed decreased
fitness when plants were grown under natural conditions compared to more favorable growth room
conditions. There is also the possibility that the distribution of mutation fitnesses varies somewhat
among taxa.

Inherent in the mutation fitness spectrum in
Figure 5.1 is that beneficial mutations are rarer than
deleterious mutations. This makes estimating of the
frequency distribution of beneficial mutations even
more difficult than it is for deleterious mutations.
Nonetheless, a number of studies have directly measured the effects of advantageous mutations (see
review by Eyre-Walker and Keightley 2007). Bacterial populations have been used to study mutations
due to their short generation times and ease of constructing and maintaining replicate populations.
Using *E. coli*, several studies have shown that beneficial mutations with small effects on fitness are
much more common than mutations with larger
effects (Imhof and Schlotterer 2001: Rozen et al.
2002). Using a ribonucleic acid or RNA virus, San-
juan et al. (2004) used site-directed mutagenesis
to make numerous single nucleotide mutations. Beneficial mutations were much rarer than deleterious
mutations, but the eight beneficial mutations had

an average of a 7% improvement in fitness and small mutation effects were more common. An important caveat to these studies is that the beneficial mutations detected are biased toward those of larger effects because very small mutation effects cannot be measured, beneficial mutations with larger effects increase in frequency more rapidly under natural selection making them more likely to reach a high-enough frequency to be detected. Further, in asexual organisms without recombination, there is the possibility of competition among lineages bearing different beneficial (relative to their pre-mutation ancestor) mutations such that at equilibrium only the highest fitness lineage reaches fixation while other lineages go to loss. This phenomenon is called **clonal interference**, and its avoidance is one of the possible fitness advantages of recombination.

A third method to estimate mutation rates per sexual generation has become more practical with the availability of rapid and relatively inexpensive whole genome sequencing. Direct comparison of long stretches of DNA sequences in parents and their offspring, or between more distantly related individuals within a pedigree, can be used to identify nucleotide sites that have experienced mutation (Keightley et al. 2014; Narasimhan et al. 2017; Tatsumoto et al. 2017). This method requires an assembled reference genome and deep sequence coverage and so has been used with highly studied or model species.

Older mutation rate estimates relied on phenotypic effects or on reporter loci to estimate the mutation rate. Whole genome sequencing has greatly expanded the ability to observe mutations at the nucleotide level, resulting in improved estimates of the **haploid genome-wide mutation rate per generation (U)** because the majority of the genome can be observed directly. Direct sequencing has also allowed mutation rates to be estimated in an expanding diversity of taxa. Table 5.2 illustrates mutation rate estimates made with direct DNA sequencing. Mutation rates differ across the taxonomic groups surveyed but tend to be more similar within the taxonomic groups. Mutation rates in prokaryotes have a median of 3.28×10^{-10} nucleotide sites per generation with a range of 2.34×10^{-8} to 7.9×10^{-11}, while, in eukaryotes, mutation rates have a median of 2.94×10^{-10} nucleotide sites per generation with a range of 8.15×10^{-10} to 7.61×10^{-12} (Katju and Bergthorsson 2019). As expected, direct estimates of the genome-wide mutation rate with whole-genome sequencing are an average of 125-fold greater than phenotypic estimates.

The mutation rates at microsatellite or simple sequence repeat (SSR) loci are also of interest since such loci are widely employed as selectively neutral genetic markers to study a wide range of population genetic processes (Hodel et al. 2016; Vieira et al. 2016). These repeated DNA regions have very high rates of mutation between 1×10^{-2} and 6×10^{-6} per sexual generation (Ellegren 2000; Steinberg et al. 2002; Beck et al. 2003; Seyfert et al. 2008; Marriage et al. 2009). Mutation rates for SSR loci also vary with the repeat motif (e.g. AT or CA) and the number of base pairs per repeat (e.g. di- or trinucleotide).

Evolution of mutation rates

Mutation rates are ultimately a product of the molecular machinery that maintains and replicates DNA as cells divide and as sexual organisms produce gametes. The loci that influence cellular mechanisms related to DNA errors are called **mutator loci**. A mutator locus, such as a DNA polymerase locus, can exhibit multiple alleles (or haplotypes) with different functional properties that each act to increase or decrease the mutations that occur at other loci. The mutation fitness spectrum (Figure 5.1) shows that the majority of mutations are deleterious, so the addition of new mutations to populations over time will decrease the mean fitness. This leads to the prediction that natural selection, which acts to increase mean fitness, will work to decrease the frequency of mutator alleles over time since they increase the production of mostly deleterious mutations. This prediction is made more complicated by the action of recombination, which will separate a mutator locus from the other loci that bear the mutations caused by the mutator locus. Thus, recombination is expected to reduce the impact of natural selection on mutator loci, leading to higher mutation rates.

The **drift barrier hypothesis** predicts that mutation rates will evolve to the minimum rate achievable by natural selection against mutator alleles and that variation in mutation rates among taxa is a product of variation in the strength of genetic drift. As explained more fully in Chapter 7, genetic drift limits the degree to which genotype frequencies will respond to natural selection. When the effective population size (N_e) is small, genetic drift is strong and natural selection (quantified by the fitness difference s) is limited to decreasing the frequency of only those mutator alleles that have a large effect on the

Table 5.2 Rates of spontaneous mutation expressed per base pair per generation for a range of organisms. These estimates employed comparisons of parents and offspring or other relatives, comparisons of lineages in a mutation-accumulation design, or the fluctuation assay to estimate variance in the number of mutations among replicates. Whole genome sequencing was used to observe a large proportion of the genome in each study.

Organism	Mutation rate	Method	References
Bacteria			
Escherichia coli	2.2×10^{-10}	Mutation accumulation	Lee et al. (2012)
Mycobacterium smegmatis	5.27×10^{-10}	Mutation accumulation	Kucukyildirim et al. (2016)
Eukaryota			
Green alga (*Chlamydomonas reinhardtii*)	9.63×10^{-10}	Mutation accumulation	Ness et al. (2015)
Ciliate (*Tetrahymena thermophila*)	7.61×10^{-12}	Mutation accumulation	Long et al. (2016)
Mouse-ear cress (*Arabidopsis thaliana*)	7.4×10^{-9}	Parent-offspring comparison	Yang et al. (2016)
Animals			
Drosophila melanogaster	2.8×10^{-9}	Parent-offspring comparison	Keightley et al. (2014)
Midge (*Chironomus riparius*)	2.1×10^{-9}	Mutation accumulation	Oppold and Pfenninger (2017)
Atlantic herring (*Clupea harengus*)	2.0×10^{-9}	Parent-offspring comparison	Feng et al. (2017)
Collared flycatcher (*Ficedula albicollis*)	4.6×10^{-9}	Three generation pedigree	Smeds et al. (2016)
Human	1.45×10^{-8}	Related individuals	Narasimhan et al. (2017)
	1.29×10^{-8}	Parent-offspring comparison	Jonsson et al. (2017)
Mouse	5.4×10^{-9}	Mutation accumulation	Uchimura et al. (2015)
	7.0×10^{-9}	Mutation accumulation	Ossowski et al. (2010)
Fungi			
Saccharomyces cerevisiae	1.7×10^{-7}	Fluctuation assay	Gou et al. (2019)
Schizosaccharomyces pombe	2.0×10^{-10}	Mutation accumulation	Farlow et al. (2015)

mutation rate. The point where natural selection and genetic drift are in equal balance can be defined as $4N_e s = 1$. When $4N_e s >> 1$, the fate of mutator alleles is a product of genetic drift alone – often random loss but also occasional random fixation. In contrast, when $4N_e s << 1$, natural selection will drive mutator alleles to loss. With this reasoning, the drift barrier hypothesis predicts that genome-wide mutation rates for different taxa should be correlated with the effective population size (N_e).

The drift barrier hypothesis is supported by empirical patterns of mutation rates estimated for a range of species as well as by population genetic models (Kimura 1967; Lynch 2010, 2011). Mutation rates estimated for a range of taxa are negatively correlated with effective population size as shown in Figure 5.3 (reviewed by Lynch et al. 2016). This is true for both the per base pair and genome-wide mutation rates which are positively correlated. The drift barrier hypothesis is also consistent with the

pattern that mutation rates are 10^2 to 10^3 times lower in single-cell and microorganisms, which have larger N_e compared to vertebrates with smaller N_e. Note that if the mutation rate was a product of natural selection for the production of beneficial mutations, then the mutation rate and the effective population size would be positively correlated since natural selection is more effective as N_e increases.

In an innovative early study, Drake (1991, reviewed in Drake et al. 1998) found that the rate of mutation per effective genome (the portion of the genome that contains coding genes) in DNA-based microbes was 1/300. More recent studies in a larger number of taxa show that multicellular eukaryotes do not exhibit such a relationship, and that the negative relationship in Eubacteria and unicellular eukaryotes is heavily influenced by results for a single taxon (Lynch et al. 2016). The pattern of a negative relationship between the rate of mutation per effective genome observed by Drake can also

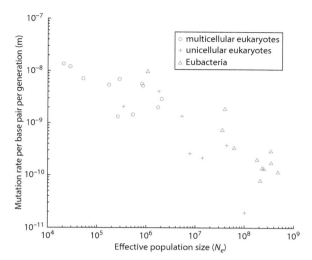

Figure 5.3 Because the majority of mutations are deleterious, the drift barrier hypothesis predicts that natural selection will remove mutator alleles from populations within the constraints of finite population size. Empirical estimates of mutation rates mutation rates per base pair per generation (μ) and estimates of the effective population size (N_e) for a range of taxa show the negative relationship predicted by the drift barrier hypothesis. Source: Data are from Lynch et al. 2016.

be explained by the positive correlation between effective genome size and N_e.

5.2 The fate of a new mutation

- The chance a neutral or beneficial mutation is lost due to Mendelian segregation.
- Mutations fixed by natural selection.
- Frequency of a mutant allele in a finite population.
- Mutations in expanding populations.
- Accumulation of deleterious mutations by Muller's Ratchet without recombination.

How does the frequency of a new mutation change over time after it is introduced into a population? This simple question is central to understanding the chance of fixation and loss for new mutations and, therefore, their ultimate fate in a population. The mutation rate dictates how often a new mutation will appear in a population. But once a mutation has occurred, population genetic processes acting on it will determine whether it increases or decreases in frequency. This section will consider four distinct perspectives on the frequency of a new mutation based on the processes of genetic drift and natural selection. Each of the four perspectives makes different assumptions about the population context in

which a new mutation is found, considering different effective population sizes, levels of recombination, and whether mutations are neutral, advantageous, or deleterious. Naturally, these four perspectives do not cover all possible situations but are meant to explore a range of possibilities and communicate several distinct approaches to determining the fate of a new mutation. Although this section will consider the action of natural selection on mutations, the simple forms of selection assumed should be accessible to most readers. Natural selection and fitness are defined and developed rigorously in Chapter 6.

Chance a mutation is lost due to mendelian segregation

The fate of a new mutation can be tracked by considering its pattern of Mendelian inheritance, as shown by R.A. Fisher in 1930 (see Fisher 1999 variorum edition). Call all the existing alleles at a locus A_x where x is an integer 1, 2, 3 ..., x to index the different alleles, and a new selectively neutral mutation A_m. Any new mutation appears initially as a single-allele copy, and it, therefore, must be found in a heterozygous genotype (A_xA_m). To form the next generation, this A_xA_m heterozygote experiences random mating with the other A_xA_x genotypes in the population. For each progeny produced by the A_xA_m genotype, there is a ½ chance that the mutant allele is inherited and a ½ chance that the mutant allele is not inherited (the A_x allele is transmitted instead). The total chance that an A_xA_m heterozygote passes the mutant allele on to the next generation depends on the number of progeny produced. If k is the number of progeny parented by the A_xA_m heterozygote and there is independent assortment of alleles, then

$$P(\text{mutant lost}) = \left(\frac{1}{2}\right)^k \qquad (5.3)$$

is the probability that the mutant allele is not transmitted to the next generation in any of the progeny. As you would expect, the probability that no mutant alleles are transmitted to the next generation declines as the number of progeny produced increases.

In a population that is constant in size over time, each pair of parents produces two progeny on average that take their places in the next generation. A key phrase here is "on average," meaning that not every pair of parents will produce two progeny: some parents will produce more progeny and some parents will produce fewer. As shown in the context

Table 5.3 The expected frequency of each family size per pair of parents (k) under the Poisson distribution with a mean family size of two ($\bar{k} = 2$). Also given is the expected probability that a mutant allele A_m would not be transmitted to any progeny for a given family size. Note that 0! equals one.

Family size per pair of parents (k)	0	1	2	3	4		K
Expected frequency	e^{-2}	$2e^{-2}$	$2e^{-2}$	$\frac{4}{3}e^{-2}$	$\frac{2}{3}e^{-2}$	...	$\frac{2^k}{k!}e^{-2}$
Chance that A_m is not transmitted	1	$\frac{1}{2}$	$\left(\frac{1}{2}\right)^2$	$\left(\frac{1}{2}\right)^3$	$\left(\frac{1}{2}\right)^4$	...	$\left(\frac{1}{2}\right)^k$

of the variance effective population size in Chapter 3, the Poisson distribution is commonly used to model variation in reproductive success. Here too, we can use a Poisson distribution to determine the expected frequencies of each family size when the average family size is two progeny (Table 5.3). The reason we need to know the expected proportion of the parental pairs that produce a given number of progeny is that each family size has a different probability of not transmitting the mutant allele. For a given family size k, the probability that a mutant allele is not transmitted to the next generation is the product of the expected frequency of parental pairs and the chance of not transmitting the mutant allele:

$$P(\text{mutant lost}) = \left(\frac{2^k}{k!}\right)e^{-2}\left(\frac{1}{2}\right)^k \quad (5.4)$$

or the product of the two terms in each column of Table 5.3. The total probability that the mutant allele is not transmitted to the next generation is the sum over all possible family sizes from 0 to infinity:

$$P(\text{mutant lost}) = \sum_{k=0}^{\infty}\left(\frac{2^k}{k!}\right)e^{-2}\left(\frac{1}{2}\right)^k \quad (5.5)$$

Although this sum looks daunting to calculate, the equation actually simplifies to a very neat result. The e^{-2} term is a constant so that it can be moved in front of the summation

$$P(\text{mutant lost}) = e^{-2}\sum_{k=0}^{\infty}\left(\frac{2^k}{k!}\right)\left(\frac{1}{2}\right)^k \quad (5.6)$$

and the 2^k and $\left(\frac{1}{2}\right)^k$ terms cancel to give

$$P(\text{mutant lost}) = e^{-2}\sum_{k=0}^{\infty}\frac{1}{k!} \quad (5.7)$$

The final trick is to notice that the series determined by the summation $(1 + 1 + \frac{1}{2!} + \frac{1}{3!} + ... + \frac{1}{k!})$ approaches e (e = 2.718 ...) as k goes to infinity. The summation term can then be replaced with e to give

$$P(\text{mutant lost}) = e^{-2}e = e^{-1} \quad (5.8)$$

As promised, the tidy conclusion is that the chance a newly occurring mutant is lost simply due to Mendelian segregation after one generation is $e^{-1} = 0.3679$. Therefore, a new mutation has about a 36% chance of being lost within one generation of its introduction into a population. The world is tough for a new mutation!

This result can be extended to determine the probability that a mutation is lost over multiple generations of Mendelian segregation. A general expression for the *cumulative* probability of a mutation being lost from the population over time is

$$P(\text{mutant lost generation } t) = e^{x-1} \quad (5.9)$$

where x is the probability of loss in the generation before or at time $t-1$. (The series determined by the summation in Eq. 5.7 is really $(1 + x + \frac{x^2}{2!} + \frac{x^3}{3!} + ... + \frac{x^k}{k!})$ and approaches e^{1+x} as k goes to infinity to give $(e^{-2})(e^{1+x}) = e^{x-1}$. When a mutant first appears in a population x = 0.)

Using this result shows that the probability of a new mutation being lost in two generations is

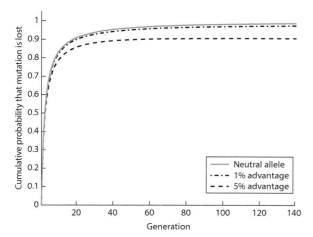

Figure 5.4 The probability that a novel mutation is lost from a population due to Mendelian segregation. A neutral allele is eventually lost from the population while a beneficial mutation has a probability of about twice its selective advantage of fixation. The cumulative probability over time is described by $e^{c(x-1)}$ where x is the probability of loss in the generation before and c is the degree of selective advantage, if any. This expected probability assumes an infinitely large population that has Poisson distributed variance in family size.

$e^{-0.6321} = 0.5315$ or the probability of being lost in three generations is $e^{-0.4685} = 0.6295$. Based on this progression, Figure 5.4 shows the probability that a new mutation is lost over the course of 140 generations. The conclusion from this graph is that a new mutation must eventually be lost from a population given enough time.

We can also ask what impact natural selection might have on this prediction that a new mutation will eventually be lost. Let us imagine that a new mutation is slightly beneficial instead of being neutral. Natural selection will then improve the chances that the new mutation is transmitted to the next generation, giving it a slight advantage over any of the other alleles in the population. Let c be the selective advantage of a new mutation so that a value of 1.0 would indicate neutrality and a value of 1.01 would mean a transmission advantage of 1%. The cumulative probability that an allele is lost at generation t is then

$$P(\text{mutant lost generation } t) = e^{c(x-1)} \quad (5.10)$$

This version of the equation multiplies the exponent for the neutral case by the selective advantage of a beneficial allele. This makes very little difference in the probability that a mutant is lost if only a few generations elapse, but makes a larger difference after more generations have passed (Figure 5.3). In general, the chance that a new beneficial mutation

is not lost is approximately twice its selective advantage, still a very low probability for realistically small values of the selective advantage. However, as Fisher pointed out, if something like 250 independent beneficial mutations occur singly over time, then there is a very small chance ($0.98^{250} = 0.0064$) that all of them would be lost during Mendelian segregation. This suggests that at least some beneficial mutations will be established in populations as mutations continue to be introduced.

The conclusion that a new neutral mutation must always be lost from a population seems at odds with the possibility of random fixation of a new mutation due to genetic drift. Fisher's method of modeling the fate of a new mutation makes the assumption that population size is very large. This assumption allows the use of the expected values for the proportion of parental pairs for each family size under the Poisson distribution and the chance of an allele being lost for each family size, probabilities that should only be met in the limit of many parental pairs that span a wide range of family sizes. Finite numbers of parental pairs would likely not meet these expected values due to chance deviations from the expected value. The assumption of infinite population size is justified because it is used to reveal that particulate inheritance by itself can lead to the loss of new mutations even in the complete absence of genetic drift. Next, we will take up the fate of a new mutation in the context of a finite population.

Fate of a new mutation in a finite population

A second perspective on new mutations is to consider their fate as an allele in a finite population in the absence of natural selection. We can then employ the concepts and models of genetic drift developed in Chapter 3 to predict the frequency of new mutations over time in a population. The first critical observation is to recognize that the initial frequency of any new mutation is simply

$$p_0(\text{new mutation}) = \frac{1}{2N_e} \quad (5.11)$$

because a new mutation is present as a single-allele copy in a population of $2N_e$ allele copies. If the frequency of a new mutation is determined strictly by genetic drift, then each new mutation has a probability of $\frac{1}{2N_e}$ of going to fixation and a probability of $1 - \frac{1}{2N_e}$ of going to loss each generation. This result makes intuitive sense, since a new mutation is very rare and is close to loss but very far from fixation. This result also shows that the chance of fixation or loss of a new mutation depends on the effective population size.

Using the diffusion approximation of genetic drift, it is possible to estimate the average number of generations before a new mutation is either fixed or lost (Kimura and Ohta 1969a). Figure 3.14 and Eq. 3.40 give the average number of generations until fixation or loss for an allele depending on the effective population size and initial allele frequency. Under the assumption that the effective population size is large, those alleles that eventually fix do so in an average of $4N_e$ generations. Those alleles that are lost go to fixation in many fewer generations, approaching zero generations as the population size gets larger and the initial frequency of $\dfrac{1}{2N_e}$ therefore gets smaller. However, since genetic drift is a stochastic process, we expect that the variance around the average time to fixation or loss will be large. In other words, the allele frequency of each new mutation will take a random walk between 0 and 1. Although many mutations may be lost quickly, others may segregate for several or many generations before being lost or fixed.

The fate of new mutations can be seen readily in a simulation. Figure 5.5 shows the frequency of new mutations introduced every 30 generations into a population of $N_e = 10$. Of the seven mutations introduced into the population, six go to loss and only one goes to fixation. This is roughly consistent with the prediction that one in 20 new mutations will fix in a population where $N_e = 10$. Most of the mutations that go to loss do so in fewer than 10 generations;

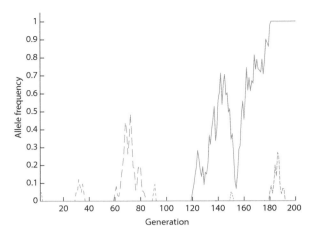

Figure 5.5 The frequencies over time of new mutations that each have an initial frequency of $\frac{1}{2N_e}$. In this example, one new mutation is introduced into the population every 30 generations and $N_e = 10$. All of the mutations except one (solid line) go to loss within a few generations. The one allele that does go to fixation takes a relatively long time to do so compared to the time to loss. At the start of the simulation, the ancestral allele has a frequency of one (not shown). When a new mutation reaches fixation, the original ancestral allele is lost, and the new mutation becomes the ancestral allele.

although in one case, the mutation segregates for about 25 generations. Equation 3.40 predicts that mutations go to loss in an average of about six generations, roughly consistent with the simulation. The mutation that goes to fixation does so in 60 generations, taking a zig-zag trajectory of allele frequency. Equation 3.40 predicts an average of about 39 generations will elapse for those mutations that go to fixation when $N_e = 10$, suggesting that the simulation result is somewhat greater than the expected average time to fixation.

Interact box 5.1 Frequency of neutral mutations in a finite population

The frequency of new mutations acted on by genetic drift can be seen with the text simulation website.

Focus on the top set of graphs labeled Strictly Neutral. Initially, run the simulation once with the default parameter values to get a sense of the outputs. What is shown in the large graph on the right and the two smaller graphs on the left?

Next, run the simulation using the default values of mutations introduced every 20 generations for a total of 200 generations, a population of $N_e = 20$, and a total of 500 independent replicate loci. How many mutations go to fixation or to loss (look at the small histograms)? What is the average time to fixation or to loss? What are the expected number of new mutations that go to fixation or to loss and how do these expectations compare with the results of the simulation? Increase the population size to $N_e = 50$ and view 500 generations. How does the number of new mutations going to fixation and loss change? How does the time that new mutations segregate change?

These predictions for the frequency and fate of new neutral mutations under genetic drift suggest that at least some genetic variation is maintained in populations simply due to the random allele-frequency walk that new mutations take before reaching either fixation or loss. If the population shown in Figure 5.4 were observed at a single point in time, it is possible that it would be polymorphic since a new mutation happened to be somewhere between fixation and loss. Observing many such loci at one point in time, it would be very likely that at least some of them would be polymorphic. This observation forms the basis of the neutral theory of molecular evolution, the hypothesis that genetic variation in populations is caused by genetic drift, covered in Chapter 8.

Mutations in expanding populations

A third perspective on the fate of new mutations will focus on expanding populations. New mutations that occur along the boundary of an expanding population can either go to loss or reach high frequencies and displace existing alleles. Mutations are said to **surf** when they increase in frequency and spread in space from their point of origin along the leading edge of an expanding population, as first shown by Edmonds et al. (2004). Surfing mutations evolve in a manner that differs from the dynamics of mutations that occur within a population of constant size. Neutral, beneficial, or deleterious mutations can all exhibit surfing behavior. Mutation surfing represents a possible explanation for allele frequency clines (geographic gradients) and other spatial patterns of polymorphism in species that have experienced population expansions (reviewed by Excoffier et al. (2009b).

Mutation surfing can be simulated on a two-dimensional grid to represent an expanding population, like that shown in Figure 5.6. Each cell in the grid represents a location that can contain some number of individuals up to a per-cell carrying capacity (K). The haploid individuals within each cell reproduce by generating a number of identical

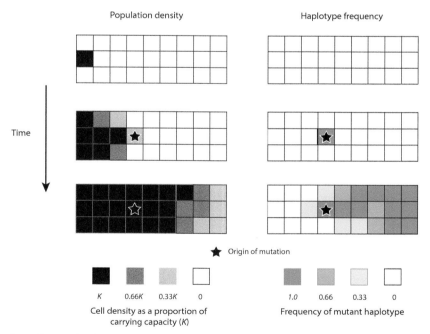

Figure 5.6   A hypothetical example of mutation surfing. A 3 × 9 grid of cells where population density over time is shown in the left column and haplotype frequency is shown in the right column. The population is founded by a single grid cell at carrying capacity (K) located on the left edge of the grid. Individuals in the founding cell reproduce and then their progeny disperse, leading to the establishment of individuals in unoccupied cells. The population expands to the right and eventually most cells reach their carrying capacity. Initially, all individuals possess the original haplotype, but a mutation occurs in a single individual. The case of mutation surfing is shown in this hypothetical example because the mutation persists and reaches high frequencies as the population continues to expand to the right and occupy available cells. The new mutation is acted on by genetic drift at reproduction, during any culling within cells to maintain carrying capacity, and during migration. The haplotype bearing the mutation that surfs to high frequency can also spread to the left of its origin via migration, a pattern more likely to occur when the new mutation has a higher fitness than the original haplotype.

offspring sampled from a Poisson distribution with mean λ. Within each grid cell during each generation, random culling of individuals takes place if needed to maintain the cell at *K* individuals. The last step in each generation is migration with rate *m* where migrants from a given cell move to adjacent cells with equal probability. To simulate neutral mutation, there can be several haplotypes with equal family sizes (λ). To simulate natural selection, less fit haplotypes have smaller mean family sizes (smaller λ) and more fit haplotypes have larger mean family sizes (larger λ). With a simulation model of this type, the population will expand and spread to fill the unoccupied cells of the grid. A haplotype bearing a new mutation will experience genetic drift in several ways – due to random variation in family size, random sampling because of carrying capacity, and random sampling at migration.

Miller (2010) showed that the probability of neutral mutation surfing in one-dimensional populations was positively correlated with mean family size (λ) but negatively correlated with carrying capacity (*K*) and with migration rate (*m*). The first two patterns can be understood in terms of genetic drift – as mean family size increases, drift gets stronger because the variance in family size also increases, and drift gets stronger via culling as the carrying capacity decreases. Increased surfing with lower migration rates occurs because individuals with mutant haplotypes are able to fully occupy a cell with less chance of mixing with nonmutant individuals located behind the expanding front of the population. In two-dimensional grids, the probability of neutral mutation surfing depended weakly on mean family size (λ) and carrying capacity (*K*) but not on migration rate (*m*). Miller (2010) offered two possible explanations for the finding of little relationship between surfing and the migration rate. One explanation is that mutations were modeled only at the expanding front of the population where cells containing the mutation were surrounded by empty or low-density cells irrespective of the mutation rate. A second possible explanation is that when different haplotypes are present along a jagged spatial border at the edge of an expanding population, new mutations lose their spatial "head-start" and are less likely to surf (see also Klopfstein et al. 2006; Hallatschek et al. 2007; Lehe et al. 2012).

Empirical evidence for mutation surfing takes a variety of forms. Bacterial populations grown on agar plates have been used to study mutation surfing directly. Hallatschek et al. (2007) employed bacterial

and yeast strains which produced a visible protein marker to demonstrate how genetic drift in expanding populations often resulted in a single strain spreading across sectors of plates with little mixing of strains. (The text website has a link to the visually striking figures in the Hallatschek et al. 2007 paper.) Bosshard et al. (2017) also employed bacterial populations to show that population expansion leads to higher frequencies of deleterious mutations compared to large, well-mixed populations. That study supports predictions that deleterious mutations can accumulate via mutation surfing (Travis et al. 2007), a phenomenon termed **expansion load** to identify it as a specific cause of genetic load (Peischl et al. 2013; Peischl and Excoffier 2015). Bacterial studies also suggest that surfing can increase the fixation of slightly beneficial alleles and lead to an increased rate of adaptation (Gralka et al. 2016). Frequencies of recessive deleterious alleles increase with geographic distance from Africa consistent with an expansion load, a pattern that supports the possibility that mutation surfing occurred during the expansion of human populations (Hallatschek et al. 2007; Peischel et al. 2016).

Geometric model of mutations fixed by natural selection

A fourth perspective on the fate of new mutations will focus on beneficial mutations, looking first at mutations fixed by natural selection alone and then at mutations fixed by the combined processes of natural selection and genetic drift. In addition to considering how new mutations are lost during segregation, in 1930, Fisher (see Fisher 1999 variorum edition) constructed another model of the fate of mutations that are acted on by the process of natural selection. As discussed earlier in the chapter, mutations may have a range of effects on fitness as well as on any phenotype with variation that has a genetic basis. The model Fisher developed sought to determine the range or distribution of the effect sizes of the beneficial mutations that are fixed by natural selection over time. Are the mutations fixed by natural selection all of large effect or all of small effect, or do they have effect sizes that fall into some type of distribution? It is quite likely that you are aware of the generalizations of this model without being aware of where these conclusions came from or what assumptions are involved. The generalization is that beneficial mutations have small effects: we do not expect to see beneficial changes taking place in single big leaps. This view of evolution is called **micromutationalism**, a concept that has been profoundly influential in evolutionary biology and population genetics (see Orr 1998 and references therein). The model that leads to this conclusion is called the **geometric model of mutation** and is developed in this section.

> **Micromutationalism:** The view that beneficial mutations fixed by the process of natural selection have small effects and, therefore, that the process of adaptation is marked by gradual genetic change.

Fisher imagined a situation where the values of two phenotypes determined the survival and reproduction, or fitness, of an individual organism (fitness is defined rigorously in Chapter 6). An example of two phenotypes might be the number of leaves and the size of leaves to achieve the maximum light capture for photosynthesis for a species of plant. However, the exact nature of the phenotypes is not important in the model as long as they contribute to the fitness of individuals. The critical point to understand is that phenotypic values closer to the maximum fitness value (Fisher called this the "optimum") are favored by natural selection, causing genotypes conferring higher fitness phenotypes to increase in frequency and fix in a population over time. Figure 5.7 shows the model. The values of the two traits are represented by two axes, and the optimum fitness value for the combination of the two traits is at the center, the point labeled O for optimum.

Let us say an individual has values of the two phenotypes that put it at point A on the phenotypic axes, some distance r from the optimum fitness. All the points on a circle of radius r centered at the optimum have the same fitness as the fitness of the individual at point A (the dashed circle in Figure 5.5A). Next, imagine that random mutations can occur to one allele of the genotype of the individual at point A. If the effects of mutations are random, then a mutation could move the phenotype in any direction from point A, and these moves could be of any distance large or small away from point A. Some mutations would move the phenotype a short distance, whereas others cause a long-distance move; some mutations move the phenotype toward the

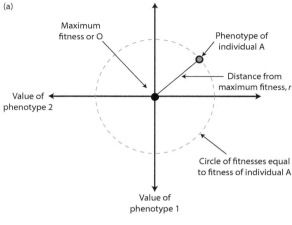

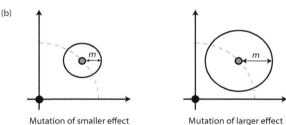

Figure 5.7   R. A. Fisher's geometric model of mutations fixed by natural selection. Panel A shows axes for two hypothetical phenotypes that determine fitness with maximum fitness when both phenotypes have the values at the center point marked with the red dot. An individual (or the mean phenotype of a population) with a phenotypic value is some distance from the maximum fitness. The dashed circle shows a perimeter of equal fitness around the point of maximum fitness. Although only two phenotypes define fitness in this example, the dashed circle of equal fitness would be a sphere with three phenotypes and an n-dimensional hyperspace if n phenotypes contribute to fitness. Panel B shows two mutations with smaller or larger phenotypic effects. The phenotypic effect of the mutations could be in any direction around the current phenotype (solid circles with radius m). Mutations with smaller effects have a better chance of moving the phenotype toward the maximum fitness (more of the area of the mutation effect circle is to the left of the dashed line of equal fitness).

optimum, whereas others move the phenotype away from the optimum.

From this graphical model, can we determine what types of mutational change are likely to be fixed by natural selection and contribute to adaptive change? One conclusion from the model is that mutations with a very large effect on phenotype (change in phenotype greater than $2r$) cannot get the phenotype any closer to the optimum even if they are in the right direction. Since mutations of very large effect always move the phenotype further away from the optimum fitness outside the dashed circle, these will never be fixed by natural selection.

What about mutations with smaller effects? Figure 5.5B shows two situations where the phenotypic effect of a mutation is smaller (change in phenotype less than 2r). On the right, there is a mutation with a larger effect on the phenotype and on the left a mutation with a smaller effect on the phenotype. Both of these mutations could be in any direction, specified by the circle around A with a radius m to indicate the magnitude of the mutational effect. Notice that as the mutation gets larger in effect, less of the circle describing the effect on the phenotype falls inside of the dashed circle that describes the current fitness of the individual at point A. As the phenotypic effect of a mutation approaches zero $(m \to 0)$, its effect circle will approach being half inside the arc of current fitness and half outside the arc of current fitness. Said the other way, as the phenotypic effect of a mutation gets larger and larger, its effect circle encompasses more and more area outside the arc of current fitness. As mutations increase in their phenotypic effect, they have an increasing probability of being in a direction that will make fitness worse rather than better. Therefore, natural selection should fix more mutations of small effect than of large effect since smaller mutations have a greater probability moving the phenotypic value toward the optimum. Mutations with almost zero effect have close to a 1/2 chance of being favorable, while large mutations have a diminishing chance of being favorable.

This can be described in an equation

$$P(\text{mutation improves fitness}) = \frac{1}{2}\left(1 - \frac{m}{2r}\right)$$
$$(5.12)$$

where m is the radius of the phenotypic effect of a mutation and r is the distance to the optimum from the current phenotypic value. As m goes to zero, the probability that a mutation moves the phenotype closer to the optimum approaches 1/2. For mutations of increasing effect, there is a diminishing probability that they improve fitness. When m is equal to twice the value of r, there is no chance that the mutation will improve fitness: a mutation of effect $2r$ could just reposition A on the opposite side of the equal fitness circle around the optimum at best.

Fisher also reasoned that the fitness of organisms depends on many independent traits since the phenotypes of organisms must meet many requirements for successful growth, feeding, avoidance of predation, mating, and so forth. He, therefore, assumed

that the dashed circle of equal fitness shown in Figure 5.5 for illustration was really better represented by a space of many dimensions. In n dimensions, the measure of whether or not a mutation is large or small relative to the distance to the point of maximum fitness (r) is gauged by $\frac{r}{\frac{2r}{\sqrt{n}}} = r\frac{\sqrt{n}}{2r}$ instead of $m/2r$ in Eq. 5.12. The main point is that increasing phenotypic dimensions cause the probability that a mutation improves fitness to decline more rapidly as its phenotypic effect gets larger. The top panel of Figure 5.8 plots a version of Eq. 5.12 which assumes that fitness is determined by many independent phenotypes. This shows the distribution of the probability that a mutation

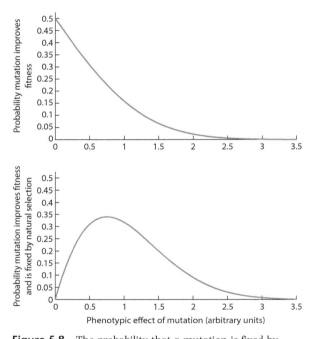

Figure 5.8 The probability that a mutation is fixed by natural selection depends on the magnitude of its effect on fitness. Using the geometric model of mutation and assuming that fitness is determined by many phenotypes, Fisher showed the probability that a mutation improves fitness approaches one half as the effect of a mutation approaches zero (top). This result comes about because smaller mutations have a better chance of moving the phenotype toward the optimum than do larger mutations (see Figure 5.4). Kimura pointed out that mutations with small effects on fitness are also the most likely to be fixed or lost due to genetic drift rather than by natural selection. Combining the chance that a mutation moves the phenotype toward higher fitness, *and* the chance that a mutation has a large enough fitness difference to escape genetic drift suggests that mutations with intermediate effects are most likely to be fixed by natural selection (bottom). Both models assume that mutations of any effect on fitness are equally likely to occur.

improves fitness as the multi-dimensional phenotypic effect of a mutation increases.

The conclusion from the geometric model of mutation is evident in the top panel of Figure 5.6. Mutations with small effects are most likely to bring an organism closer to its fitness optimum and are, therefore, most likely to be fixed by natural selection. Mutations of larger effect have a lower probability of improving fitness and are, therefore, less likely to be fixed by natural selection. Fisher compared the situation to the focus adjustment on a microscope. If a microscope is close to being in focus, then large random changes to the adjustment are likely to make things worse while small random changes are more likely to make the focus better. A logical consequence of Fisher's model is that the mutation fitness spectrum approaches 50% deleterious and 50% beneficial mutations as mutation effects approach 0. This prediction is not consistent with the general picture of the mutation fitness spectrum in Figure 5.1.

Many years later, Kimura (1983) reevaluated the predictions of the geometric model of mutation by relaxing Fisher's implicit assumption of an infinite effective population size. This change allows genetic drift to operate on the frequency of mutations along with natural selection. In a finite population, allele frequency is determined by a combination of sampling error and the effect of natural selection to fix alleles with higher average fitness. Natural selection will only determine the fate of an allele if it is stronger than the randomizing effect of genetic drift. The pressure of natural selection also depends on the phenotypic effect of a mutation – mutations with a larger effect experience a stronger push toward fixation. Thus, the push toward fixation by natural selection is the strongest for those new mutations that have the largest effects. In other words, new mutations with small effects are likely to experience random fixation or loss by genetic drift. The bottom panel of Figure 5.8 shows the probability that a new mutation is fixed by natural selection in a finite population. The mutations with the smallest phenotypic effects are still most likely to move the phenotype toward higher fitness. However, this is now balanced by the effect of genetic drift which has the greatest impact on new mutations with small effects on fitness. The modified result is that new mutations with an intermediate effect on fitness are the most likely to fix under natural selection in a finite population.

Orr (1998) provides an analysis of the effect sizes of mutations that are fixed by natural selection in a finite population that compensates for the fact that the effect of mutations must shrink as a population gets closer to the maximum fitness over time. The net balance of natural selection and genetic drift is considered in detail in later chapters, and the phenotypic effects of loci and alleles are treated in detail in Chapter 9 on quantitative genetics.

Muller's ratchet and the fixation of deleterious mutations

The final perspective on the fate of a new mutation focuses on deleterious mutations that occur within genomes lacking recombination. The combination of mutation, genetic drift, and natural selection result in the progressive loss of the class of genotypes in a population with the fewest mutations in a phenomenon called **Muller's Ratchet** (Muller 1964; Maynard Smith 1978; Charlesworth and Charlesworth 1997). The name is an analogy to a mechanical device like a ratchet wrench that permits rotation in only one direction. Muller's Ratchet results in the accumulation of more and more mutations in a population, which leads to the ever-declining average fitness in populations if most mutations are deleterious. Thus, Muller's Ratchet demonstrates a selective advantage of recombination under some conditions. Muller's Ratchet is closely related to the **Hill-Robertson effect** (Hill and Robertson 1966), the phenomenon where relatively weak natural selection operating on numerous loci with independent fitness values has reduced effectiveness because of disequilibrium caused by reduced recombination and by genetic drift (reviewed by Comeron et al. 2008).

To see how Muller's Ratchet works in detail, consider a finite population of haploid individuals that reproduce clonally. Assume that all mutations at all loci are equally deleterious and acted against by natural selection to the same degree. The selective disadvantage is s at each locus with a mutation, and the total selection coefficient against an individual with n mutated loci is $(1-s)^n$. Further, assume that mutation is irreversible and can only make deleterious alleles from wild-type alleles but not wild-type alleles from deleterious ones. Initially, all individuals in the population start off with no mutations. Mutations that occur decrease the proportion of individuals with no mutations and increase the frequencies of individuals with 1,2,3 ... n mutations. Over time, the frequency of the zero mutation category declines while the frequencies of individuals with

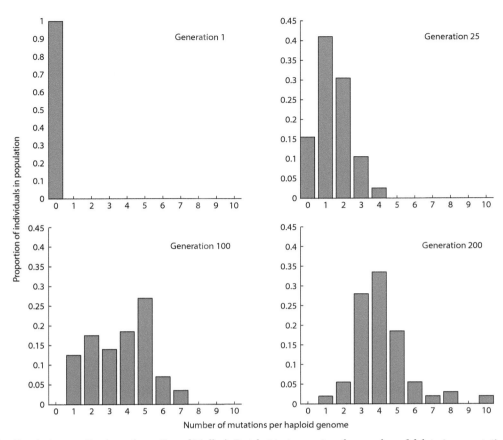

Figure 5.9 Simulation results show the action of Muller's Ratchet in increasing the number of deleterious mutations in the absence of recombination. Initially, all haploid individuals in the population have zero mutations. Mutations occur randomly over time and continually reduce the frequency of individuals with fewer mutations. Genetic drift causes sampling error and the stochastic loss of mutation classes with few individuals. Individuals with more mutations are less likely to reproduce due to natural selection against deleterious alleles. Once the category with fewest mutations (e.g. the zero mutations class) is lost due to genetic drift and mutation, there is no process that can repopulate it. Therefore, the distribution of the number of mutations continually moves to the right but can never move back to the left. The simulation parameters were $N_e = 200$, $\mu = 0.06$, each mutation reduced the chance of reproduction by 1% and each individual had 100 loci.

one or more mutations increases. This process can be seen in the top two panels of Figure 5.9.

Genetic drift and natural selection are also acting along with mutation on the frequencies of individuals with different numbers of mutations. The sampling error of genetic drift can result in the stochastic loss of mutation categories with a low frequency in the population. This effect of genetic drift works regardless of the number of mutations. Any category of mutations lost by drift can be reestablished via mutation from individuals with fewer mutations. However, when all the individuals with the lowest number of mutations are lost from the population, that fewest-mutation category is gone forever. This is because mutation cannot make wild-type alleles that would reduce the number of deleterious mutations. Also, the fewest-mutation category also cannot be reconstituted because there is no recombination. The overall effect of genetic drift is to

push the frequency distribution of the number of mutations toward higher numbers. In contrast, natural selection tends to push the distribution of the number of mutations toward lower numbers since individuals with more mutations are increasingly disfavored by natural selection.

If the effective population size is small, Muller's Ratchet also leads to accelerated rates of fixation to a single allele within the category of individuals with fewest mutations. This occurs since the category of fewest mutations is not renewed by mutation. It is also finite and consists of alleles that have identical fitness, so that genetic drift will eventually cause fixation of a single allele within that mutation category. This effect has implications for genomes with low levels of recombination or in diploid populations with mating systems that lead to high levels of homozygosity that effectively nullify recombination. In these situations, fixation may occur at higher rates

---

Interact box 5.2 Muller's Ratchet

Go to the text website for a simulation of Muller's Ratchet model of new deleterious mutations influenced by genetic drift and natural selection in the absence of recombination.

The simulation starts with a population of haploid, clonal individuals that have no mutations and then lets mutation, genetic drift, and natural selection act. The fitness of each individual determines its chances of contributing progeny to the next generation. The number of progeny produced by each individual is Poisson distributed with a mean of 1 offspring (the mean family size and maximum family size parameters change this distribution). The effective population size, the coefficient of selection against deleterious mutations (the decrease in percent viability for each mutation), and the mutation rate can be set in the simulation. The results are given in terms of the proportion of individuals in the population with a given number of mutations.

Initially, run the simulation using the default values. Then try independently increasing the effective population size (or the population size of haploid chromosomes), the selection coefficient against the deleterious mutations, and the mutation rate. Predict the impact of each simulation parameter on the frequency distribution of the number of mutations per genome before you change each parameter.

---

than for deleterious mutations in genomes with free recombination and the same effective population size (see Charlesworth and Charlesworth 1997).

5.3 Mutation models

- The infinite alleles, k alleles, and stepwise mutation models.
- Understanding the implications of mutation models using the standard genetic distance and R_{ST}.
- The infinite sites and finite sites mutation models for DNA sequences.

Mutation acts in diverse ways and can produce a wide range of changes at the level of alleles and DNA sequences. To study the allele frequency consequences of mutation, it is helpful to construct some simplifying models of the mutation process itself. **Mutation models** attempt to capture the essence of the genetic changes caused by mutation, while, at the same time, simplify the process of mutation into a form that permits generalizations about allele frequency changes. There is no single model of the process of mutation, but rather a series of models that serve to encapsulate different features of the mutation process for different classes of loci and different types of alleles. Often, mutation models are motivated by molecular methods such as allozyme electrophoresis or DNA sequencing used to assay genetic variation in actual populations. This section introduces and describes the major classes

of mutation models. Two types of mutation models for discrete alleles are applied in measures of genetic difference between populations to show the role mutation models play in the interpretation of genetic differences. Mutation models for DNA sequences are applied in the last section of the chapter on mutation in genealogical branching models.

Mutation models for discrete alleles

A repeated theme in earlier chapters was determining expected levels of homozygosity and heterozygosity (autozygosity and allozygosity) under different population genetic processes. A key assumption in many of these expectations is that identity in state can be treated as identity by descent. In other words, alleles identical in state look alike because they descended from a common ancestral allele copy at some point in the past. The **infinite alleles model** of mutation (see Kimura and Crow 1964) is an assumption used to guarantee that identity in state is equivalent to identity by descent. Under the infinite alleles model, each mutational event creates a new allele unlike any other allele currently in the population. Once a given allelic state is made by mutation the first time, it can never be made by mutation ever again. In essence, the allelic state is crossed off the list of possible mutations. The infinite alleles model serves to avoid the possibility that two alleles are identical in state but not identical by descent, as can occur if the same allele can be made by mutation repeatedly over time. Under the infinite alleles model, mutation produces the original copy of each allele but is not an ongoing process influencing the frequency of any allele already in the population. Processes other than mutation are responsible for allele and genotype frequencies after an allele exists in a population. An additional consequence is that the evolutionary "distance" or number of transition events between all alleles is the same since all alleles are produced by a single mutational event and alleles can never accumulate multiple mutations. This means that all alleles can be treated as equivalent when estimating heterozygosity or fixation indices.

The infinite alleles model might roughly approximate the mutational process for molecular markers like allozymes since alleles take discrete states (e.g. fast or slow migration on a gel) and allozyme loci are generally observed to have low mutation rates so it is likely that most alleles in a sample are not recently the product of mutation. A length of DNA sequence might also approximate the infinite alleles model. In a sequence of 500 nucleotides, there are $4^{500} = 1.072 \times 10^{301}$ unique combinations of the nucleotides. If mutation is purely random and mutation changes an existing nucleotide to any other nucleotide with equal probability, many mutations could occur in a population of DNA sequences without producing a duplicate allele since there are a truly staggering number of possible alleles.

Homoplasy: The condition where allelic states are identical without the alleles being identical by descent.
Infinite alleles model: A model where each mutational event creates a new allele unlike any other allele currently in the population so that identity in state for two or more alleles is always a perfect indication of identity by descent.
***k* alleles model:** A mutation model where each allele can mutate to each of the other $k - 1$ possible allelic states with equal probability.
Stepwise mutation model: A mutation model where the allelic states produced by mutation depend on the initial state of an allele. Alleles with a greater difference in state are, therefore, more likely to be separated by a greater number of past mutational events.

There are several features of the mutational process that the infinite alleles model does not account for, and therefore, there are a number of other mutation models. Obviously, there are not an infinite number of alleles possible at actual genetic loci. The **k alleles model** of mutation is an alternative to the infinite alleles model where k refers to a finite integer representing the number of possible alleles. In this model, each allele can mutate with equal probability to each of the other $k - 1$ possible allelic states. With the k alleles model, the same allele can be created by mutation repeatedly, blurring the equivalence of identity in state and identity by descent. As the number of possible alleles or k decreases and as the mutation rate increases, allelic state becomes a poorer and poorer measure of identity by descent since an increasing proportion of alleles with identical states have completely independent histories. The term **homoplasy** refers to allelic states that are identical in state without being identical by descent.

The infinite alleles and k alleles models both assume that the allelic state produced by mutation is independent of the current state of an allele. With these models, each allele has an equal probability of mutating to any of the other allowable allelic states. It is also possible that the state of a new allele produced by mutation is not independent of the initial state of an allele. An example is the common observation that transitions are more common than transversions in diverged DNA sequences. The **stepwise mutation model** accounts for cases where allelic states are somehow ordered and the allelic states produced by mutation depend on the initial state of an allele (Kimura and Ohta 1978). Mutations by slipped-strand mispairing at microsatellite or SSR loci produce new allelic states within one or a few repeats of the initial allelic state much more often than mutations that are many repeats different from the initial allelic state. Microsatellite loci are, therefore, a prime example of ordered, stepwise mutation where alleles closer in state are more likely to be recently identical by descent than alleles that are very different in state.

The role of mutation models is illustrated in summary measures that express the genetic similarity or dissimilarity of individuals or populations called **genetic distances**. The **standard genetic distance** or D measure developed by Nei (1972, 1978a, b) has been widely employed. Given allele frequencies for several populations, D (not to be

Table 5.4 Hypothetical allele frequencies in two subpopulations used to compute the standard genetic distance, D. This example assumes three alleles at one locus, but loci with any number of alleles can be used. D for multiple loci uses the averages of J_{11}, J_{22}, and J_{12} for all loci to compute the genetic identity I.

Allele	Subpopulation 1		Subpopulation 2	
	Frequency	p_{ik}^2	Frequency	p_{ik}^2
1	0.60	$p_{11}^2 = 0.36$	0.40	$p_{21}^2 = 0.16$
2	0.30	$p_{12}^2 = 0.09$	0.60	$p_{22}^2 = 0.36$
3	0.10	$p_{13}^2 = 0.01$	0.00	$p_{23}^2 = 0.00$

confused with the measure of gametic disequilibrium) expresses the probability that two alleles each randomly sampled from two different subpopulations will be identical in state relative to the probability that two alleles randomly sampled from the same subpopulation are identical in state. Table 5.4 gives hypothetical allele frequencies at one locus in two subpopulations that can be used to compute D. With random mating, the total probability that two identical alleles are sampled from subpopulation 1 is

$$J_{11} = \sum_{k=1}^{alleles} p_{1k}^2 = (0.6)^2 + (0.3)^2 + (0.1)^2 = 0.46$$

$$(5.13)$$

and the total probability that two identical alleles are sampled from subpopulation 2 is

$$J_{22} = \sum_{k=1}^{alleles} p_{2k}^2 = (0.4)^2 + (0.6)^2 + (0.0)^2 = 0.52$$

$$(5.14)$$

where p_{ik} indicates the frequency of allele k in population i. The total probability of sampling an identical allele from subpopulation 1 and subpopulation 2 is

$$J_{12} = \sum_{k=1}^{alleles} p_{1k}p_{2k} = (0.6)(0.4) + (0.3)(0.6)$$
$$+ (0.1)(0.0) = 0.42$$

$$(5.15)$$

The normalized genetic identity for this locus is then

$$I = \frac{J_{12}}{\sqrt{J_{11}J_{22}}} = \frac{0.42}{\sqrt{(0.46)(0.52)}} = 0.8589 \quad (5.16)$$

which is used to compute the genetic distance as

$$D = -\ln(I) = -\ln(0.8589) = 0.152 \quad (5.17)$$

When two subpopulations have identical allele frequencies, J_{11} and J_{22} are equal, I is then 1 and the natural log of 1 is 0, giving a genetic distance of 0. D has no upper limit. Although this genetic distance can be calculated for any pair of populations, D for completely isolated populations where divergence is due exclusively to mutation is expected to increase linearly with time under the infinite alleles model. This expectation relies on mutation not causing any homoplasy so that alleles identical in state are always identical by descent. If the infinite alleles model is not met, D underestimates the true genetic distance because the mutational events that record the history of the populations will not be reflected perfectly in the allele frequencies.

A genetic distance is a metric designed to quantify the evolutionary events that separate haplotypes or genotypes. As such, genetic distances are built on the foundation of a model describing the processes acting. For example, there are several genetic distances that implement a stepwise mutation model for microsatellite loci (e.g. Takezaki and Nei 1996) and a comparison of different genetic distance measures for microsatellites found performance depends on mutational patterns (Goldstein et al. 1995). Similarly, a study of numerous genetic distances applied to 500 000 SNP loci for about 900 unrelated individuals in 51 populations also found wide variation in the delineation of populations (Libiger et al. 2009).

These comparative studies highlight the need to be aware of the underlying mutational models when employing genetic distance measures and their fit to observed genetic data.

With an awareness of mutation and the different forms it takes, we can also reflect back on the measures of genetic divergence among populations. Chapter 4 gave the expression for the fixation index among subpopulations relative to the total population ($F_{ST} = \dfrac{H_T - H_S}{H_T}$). It turns out that this assumes the infinite alleles model since it treats all alleles as being an equal number of mutational steps apart with all heterozygotes considered equally distant.

That mutation plays a role in the degree of genetic differentiation among populations was appreciated by Wright (1943b), who stated that "... it requires only a small amount of long range dispersal or mutation to prevent the differentiation of large populations." F_{ST} in the island model where alleles evolve under the k alleles model is a function of drift and gene flow *plus* the impacts of mutation

$$F_{ST} \cong \frac{1}{4N_e(m + \mu + \nu)} \qquad (5.18)$$

where μ and ν are the forward and reverse mutation rates in the case of biallelic loci. Similar forward and reverse mutation rates cause allele frequencies to approach intermediate values such that subpopulations will exhibit lower allele frequency differentiation and decreased F_{ST} when mutation is strong relative to drift. Only when rates of mutation are much less than the rate of gene flow is F_{ST} primarily a function of the net balance of genetic drift and gene flow. Note that mutation under the infinite alleles model will increase genetic differentiation since each new mutation is unique and will only occur within a single subpopulation unless and until copies of the new allele are moved among subpopulations by gene flow.

There is an alternative method to compute the fixation index that relies on the stepwise mutation model instead of the infinite allele model, where the fixation index is measured by

$$\hat{R}_{ST} = \frac{S_T - S_W}{S_T} \qquad (5.19)$$

where S_T is twice the variance in allelic sizes in the total population and S_W is twice the average of the within-subpopulation variance in allelic sizes (Slatkin 1995; Goodman 1997). The states of the alleles then influence the perceived amount of population subdivision. Alleles with states further apart (greater variance in state) are counted more heavily in the estimate of population structure since they are less likely to be recently identical by descent (multiple stepwise mutations would be required to make a large change in state). In contrast, alleles with very similar states (less variance in state) make a smaller contribution to the estimate of population subdivision since they are more likely to be recently identical by descent but changed in state due to mutation. Using the stepwise mutation model and R_{ST} accounts for high rates of mutation that can give the appearance of more or less gene flow than has actually occurred. Table 5.5 gives hypothetical genetic data from two subpopulations, illustrating the degree of population subdivision under the infinite alleles and stepwise mutation models. Whitlock (2011) has shown that various measures of genetic differentiation among populations exhibit different sensitivities to high mutation rates and suggested that measures like R_{ST} with an explicit mutation model are useful when mutation rates are high relative to rates of gene flow.

Interact box 5.3 R_{ST} and F_{ST} as examples of the consequences of different mutation models

Under the infinite alleles model, allelic state is irrelevant in estimating population structure. However, in the stepwise mutation model, allelic states are weighted in the total estimate of population structure. Computing R_{ST} and F_{ST} for two subpopulations in a Microsoft Excel spreadsheet will help you develop a better understanding of how mutational models influence the perception of population structure. Use the Excel spreadsheet to explore how allelic state differences as well as allele frequencies produce different estimates of the amount of population structure.

Table 5.5 A comparison of hypothetical estimates of population subdivision assuming the infinite alleles model using F_{ST} or assuming the stepwise mutation model using R_{ST}. In Case 1, the majority of alleles in both populations are very similar in state. Under the stepwise mutation model, the two alleles are separated by a single change that could be due to mutation. The estimate of R_{ST} is, therefore, less than the estimate of F_{ST}. In Case 2, the two populations have alleles that are very different in state and more than a single mutational change apart under the stepwise mutation model. In contrast, all alleles are a single mutational event apart in the infinite alleles model. The higher estimate of R_{ST} reflects greater weight given to larger allelic state differences.

	Case 1	Case 2
Subpopulation 1	9, 10, 10, 10, 10, 10, 10, 10, 10, 10	9, 10, 10, 10, 10, 10, 10, 10, 10, 10
Subpopulation 2	12, 11, 11, 11, 11, 11, 11, 11, 11, 11	19, 20, 20, 20, 20, 20, 20, 20, 20, 20
Allele size variance in subpopulation 1, S_1	0.10	0.10
Allele size variance in subpopulation 2, S_2	0.10	0.10
Allele size variance in total population, S_T	0.947	52.821
R_{ST}	0.789	0.996
Expected heterozygosity in subpopulation 1, H_1	0.18	0.18
Expected heterozygosity in subpopulation 2, H_2	0.18	0.18
Average subpopulation expected heterozygosity, H_S	0.18	0.18
Expected heterozygosity in total population, H_T	0.59	0.59
F_{ST}	0.695	0.695

Mutation models for DNA sequences

There are two widely used conceptual models of the process of mutation operating on DNA sequences (note that these types of models also apply in principle to amino acid sequences). One approximation for the process of mutation with DNA sequences is the **infinite sites model**. Each allele is an infinite DNA sequence, and each mutation occurs at a different position along the DNA sequence. The infinite sites model can be thought of as an infinite alleles model built specifically for DNA sequences. A key distinction is that the infinite sites model permits the process of mutation to act on each allele in a population any number of times. As a consequence, the evolutionary "distance" between pairs of alleles can vary since a few or many sites differ between pairs of alleles depending on how many mutations have occurred for each allele. Figure 5.10A shows an example of how mutations might occur for DNA sequences under the infinite sites model. For example, after the fourth base-pair position (or site) of sequence 1 mutates from G to C, no more mutations can take place at that site. The sites where mutations took place can, therefore, all be distinguished in alignment of the sequences since each site only experiences a mutation once. Although other processes such as gene drift and natural selection may influence the frequency of the sequences, we can conclude that sequences sharing the same base at a site are identical by descent.

Although no DNA sequence is infinite, the infinite sites model is a reasonable approximation if not too

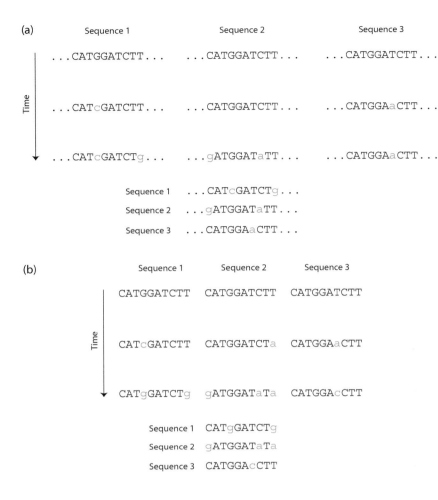

Figure 5.10 Patterns of mutational change in DNA sequences under the infinite sites (a) and finite sites (b) models. Base pair states created by a mutation are in lowercase letters and have been shaded red if mutation occurred a single time or shaded blue if two mutations occurred. In the infinite sites model sequences that are identical in state at the same site are identical by descent because mutations only occur once at each site. In contrast, the finite sites model shows how multiple mutations at the same site act to obscure the history of identity based on comparisons of site differences among DNA sequences. The ellipses (...) that surround the sequences in the top panel indicate that each sequence has infinitely many sites, and only 10 are displayed.

Box 5.1 Single nucleotide polymorphisms

DNA sequencing is now a widely used molecular technique to determine haplotypes and also genotypes. A hypothetical set of DNA sequences from four individuals is shown in Figure 5.11. Each row is a section of a DNA sequence belonging to one individual, and each column is a nucleotide site that is identical by descent, or homologous. Comparing the state(s) of the nucleotide base present at one site in each individual allows the detection of nucleotide variation among individuals.

If the individuals sequenced belong to the same species, then the variable nucleotide sites are **single nucleotide polymorphisms** (SNPs, pronounced "snips"). DNA sequences exhibit at most $k = 4$ alleles per nucleotide site, so patterns and rates of mutation are described with special classes of k alleles models.

If the individuals sequenced belong to different species, then the variable nucleotide sites are described as **sequence divergence**.

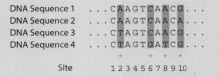

DNA Sequence 1	...CAAGTCAACG...
DNA Sequence 2	...CAAGTCAACA...
DNA Sequence 3	...CTAGTCAACG...
DNA Sequence 4	...CTAGTGATCG...
	* * * *
Site	1 2 3 4 5 6 7 8 9 10

Figure 5.11 Hypothetical DNA sequences at one locus for four individuals and the multiple DNA sequence alignment that identifies polymorphic nucleotide sites (shaded and marked with an asterisk).

much time has passed since sequences shared a common ancestor. If mutation occurs randomly and with equal probability at each site, then any single site has a small chance of experiencing a mutation twice (e.g. the rate of mutation per site squared is small). Over a relatively short period of time, say thousands of generations, only a few mutations are likely to occur, so it is unlikely that one site mutates more than once.

However, actual DNA sequences are finite and the time period for mutations to occur can be very long, so a mutation model taking these facts into account is useful. The **finite sites model** is used for DNA sequences of a finite length. It is similar to the infinite sites model except that now the number of sites is finite and each site can experience a mutation more than once. Multiple mutations have the potential to obscure past mutational events as shown in Figure 5.8B. For example, two sequences are either identical or different at each site even though a site where they differ may have mutated more than once in the past. The fourth site in sequence 1 is such a case. Although there have been two mutations at that site, the second mutation leads to the same nucleotide that was originally in that position. However, in the alignment of all three sequences, the fourth site is identical, and it is not possible to detect the two mutation events that occurred for sequence 1. Consider a similar example of site seven in sequence 3 and what happens when we compare pairs of sequences. Sequences 2 and 3 differ at four sites (1, 7, 8, and 10), but there are actually five mutational events that separate them in the past. Sequences 1 and 3 differ at three sites (4, 7, and 10), but there are actually five mutation events separating them. Thus, multiple mutational changes at the same site work to obscure the complete history of mutational events that distinguish DNA sequences.

Infinite sites model: A model for the process of mutation acting on infinitely long DNA sequences where each mutation occurs at a different position along the DNA sequence and the same position cannot experience a mutation more than once.
Finite sites model: A model for the process of mutation acting on DNA sequences of finite length so that the same site may experience a mutation more than once.

The possibility of multiple mutational changes at the same site, often called **multiple hits**, leads to saturation of mutational changes over time as mutations occur more times at the same sites. Saturation can be "corrected" using nucleotide substitution models that estimate and adjust for multiple mutations at the same site to estimate the "true" number of events that separate two sequences. One such correction called the Jukes–Cantor model is covered in Chapter 8.

One way to understand the impacts of multiple hits is to imagine a situation similar to the beakers containing micro-centrifuge tubes in Chapter 3. Now, the beakers contain a very large number of nucleotides (A, C, T, and G) at equal frequencies. Imagine composing two DNA sequences by drawing nucleotides from the beaker. The chance that a given nucleotide is sampled at random is 25%. Therefore, given one random DNA sequence, there is a 25% chance that another random DNA sequence shares an identical base pair at the same site. Therefore, DNA sequences that have experienced many mutations at the same site are expected to be identical for 25% of their base pairs. Therefore, when there is a possibility of multiple hits, identity in state is not a perfect indicator of identity by descent.

5.4 The influence of mutation on allele frequency and autozygosity

- Irreversible and bidirectional mutation models.
- The parallels between the processes of mutation and gene flow.
- Expected autozygosity at equilibrium under mutation and genetic drift.
- Expected heterozygosity and the biological interpretation of θ.

In developing expectations for allele and genotype frequencies to this point, all processes served only to shape existing genetic variation. To understand the consequences of mutation requires models that predict allele and genotype under the constant input of genetic variation by mutation. This section presents three models for the process of mutation. The first two models are related and ask how recurrent mutation is expected to change allele frequencies over time in a population. The third model predicts genotype frequencies when both genetic drift and mutation are both operating, showing how the combination of these processes influences autozygosity in a population.

Let us develop two simple models to predict the impact of constant mutation on allele frequencies (sometimes called **mutation pressure**) in a single panmictic population that is very large. Both models will focus only on the process of mutation and leave

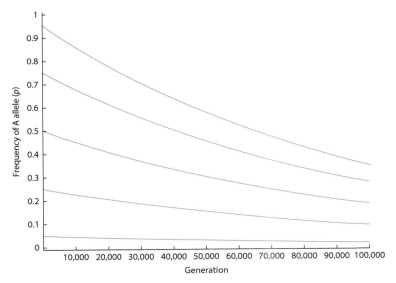

Figure 5.12 Expected change in allele frequency due to irreversible or one-way mutation for a diallelic locus for five initial allele frequencies. Here, the chance that an A allele mutates to an a allele (or the per locus rate of mutation) is 0.00001. This rate of mutation is high compared to estimates of the per locus mutation rate (see Table 5.1). The expected equilibrium allele frequency is $p = 0$ since there is no process acting to replace A alleles in the population. The population has not reached equilibrium even after 100 000 generations have elapsed. Changes in allele frequency due to mutation alone occur over very long time scales.

out other processes such as genetic drift or natural selection. Consider one locus with two alleles, A and a, where the frequency of A is represented by p and the frequency of a is represented by q. For the first model, assume that mutation operates to change A alleles into a alleles but that a alleles cannot mutate into A alleles. This is called the **irreversible** mutation or one-directional mutation model. The chance that mutation changes the state of each A allele every generation is symbolized by μ (pronounced "mu"). The frequency of the A allele after one generation of mutation is then

$$p_{t+1} = p_t(1 - \mu) \tag{5.20}$$

where the $(1 - \mu)$ term represents the proportion of A alleles that does not mutate to a alleles at time t. As long as μ is not 0, then the frequency of A alleles will decline over time because $1 - \mu$ is less than 1. This also must mean that the proportion of the a alleles increases by μ each generation. If the mutation rate is constant over time, then the allele frequency after an arbitrary number of generations is

$$p_t = p_0(1 - \mu)^t \tag{5.21}$$

where p_0 is the initial allele frequency and t is the number of generations that have elapsed.

With irreversible mutation, eventually, all A alleles will be transformed into a alleles by mutation since there is no process that replaces A alleles

in the population. Figure 5.12 shows the expected frequencies of the A allele over time starting at five different initial allele frequencies when the mutation rate is $\mu = 1 \times 10^{-5}$ or 0.000 01. Notice that the time scale to reduce the frequency of the A allele is very long. In this example, the equilibrium allele frequency of $p = 0$ has not been reached even after 100 000 generations. In fact, it takes 69 310 generations to halve the frequency of A with this mutation rate (the halving time is determined by setting $(1-\mu)^t = \frac{1}{2}$) even using a mutation rate at the high end of the observed range (Table 5.1). To generalize from the irreversible mutation model, we can expect that the process of mutation does influence allele frequencies but that substantial changes to allele frequency caused by mutation alone will take thousands or tens of thousands of generations depending on the mutation rate.

Irreversible or one-directional mutation: For a locus with two alleles, a process of mutation that changes A alleles into a alleles but does not change a alleles to A alleles. **Mutation pressure:** The constant occurrence of mutations that add or alter allelic states in a population. **Reversible or bi-directional mutation:** For a locus with two alleles, a process of mutation that changes A alleles into a alleles and also changes a alleles to A alleles.

The assumption of irreversible mutation is not biologically realistic. Mutation can usually change the state of all alleles, resulting in both forward (A → a) and reverse (a → A) mutation for a diallelic locus. The **bidirectional** or **reversible mutation** model takes this possibility into account by using the independent rates of forward mutation (μ) and reverse mutation (ν, pronounced "nu"). With mutation pressure in both directions, we can again ask how mutation will change allele frequencies in a population over time. Each generation, μ of the A alleles mutate to a alleles, while, at the same time, ν of the a alleles mutate to A alleles. The allele frequency after one generation is, therefore,

$$p_{t+1} = p_t(1 - \mu) + (1 - p_t)\nu \qquad (5.22)$$

because the frequency of A alleles will decline due to the proportion of alleles that experience forward mutation ($p_t(1 - \mu)$) but increase due to the proportion of the alleles in the population that experience reverse mutation ($(1 - p_t)\nu$). The general result is that the equilibrium value of the frequency of A is determined by the net balance of the two rates of mutation:

$$p_{equilibrium} = \frac{\nu}{u + v} \qquad (5.23)$$

as derived in Math box 5.1. So, whatever the starting frequency of the A allele, the population will converge to $p_{equilibrium}$ that is closer to one for the allele produced by the higher of the two mutation rates. Figure 5.13 shows the frequency of the A allele over time with bidirectional mutation for five different initial allele frequencies. Because the forward and backward mutation rates used for the figure are not equal but are within a factor of five, both alleles have intermediate frequencies at equilibrium. The number of generations required to reach the equilibrium allele frequency is again very long, just as it is with the irreversible mutation model.

Math box 5.1 Equilibrium allele frequency with two-way mutation

To determine the equilibrium allele frequency for a diallelic locus with the possibility of both backward and forward mutation, we take the basic equation that predicts allele frequency over one generation:

$$p_{t+1} = p_t(1 - \mu) + (1 - p_t)\nu \qquad (5.24)$$

and try to express it as

$$p_{t+1} - a = (p_t - a)b \qquad (5.25)$$

where a and b are constants that depend only on the forward and backward mutation rates μ and ν. Expressing the equation in this way allows us to equate p_{t+1} with a if the $(p_t - a)b$ term goes to 0 under certain limiting conditions. Equation 5.25 can be rearranged by adding a to both sides

$$p_{t+1} = (p_t - a)b + a \qquad (5.26)$$

and then multiplying

$$p_{t+1} = p_t b - ab + a \qquad (5.27)$$

and lastly factoring terms containing a to give

$$p_{t+1} = p_t b + a(1 - b) \qquad (5.28)$$

Equation 5.24 containing the mutation rates can be put into this same form by expanding to give

$$p_{t+1} = p_t - p_t\mu + \nu - p_t\nu \qquad (5.29)$$

which then can be factored to give

$$p_{t+1} = p_t(1 - \mu - \nu) + \nu \qquad (5.30)$$

Comparing Eqs. 5.28 and 5.30 we see that

$$b = (1 - \mu - \nu) \qquad (5.31)$$

and

$$a(1 - b) = \nu \qquad (5.32)$$

Substituting the expression for b into the equation above gives the solution for a

$$a = \frac{\nu}{\mu + \nu} \qquad (5.33)$$

We can then substitute these values of a and b into Eq. 5.25 to get a new expression for the change in allele frequency over one generation:

$$p_{t+1} - \frac{\nu}{\mu + \nu} = \left(p_t - \frac{\nu}{\mu + \nu}\right)(1 - \mu - \nu) \qquad (5.34)$$

Since the expression for change in allele frequency over any one generation interval is identical and the mutation rates are constant over time, we can recast the equation above in terms of the initial allele frequency p_0 and the number of generations that have elapsed

$$p_{t+1} - \frac{\nu}{\mu + \nu} = \left(p_0 - \frac{\nu}{\mu + \nu}\right)(1 - \mu - \nu)^t \qquad (5.35)$$

Notice that as the number of generations grows very large ($t \to \infty$), the $(1 - \mu - \nu)^t$ term approaches 0, making the entire right-hand side of the equation zero. Therefore, when many generations have elapsed, the equilibrium allele frequency is expected to be

$$p_{t \to \infty} = \frac{\nu}{\mu + \nu} \qquad (5.36)$$

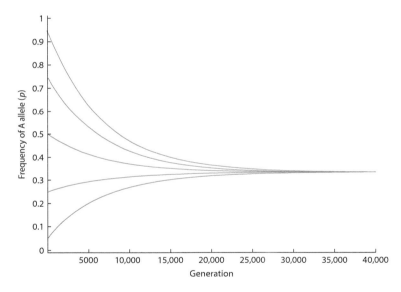

Figure 5.13 Expected change in allele frequency due to reversible or two-way mutation for a diallelic locus for five initial allele frequencies. Here, the chance that an A allele mutates to an a allele (A → a) is 0.0001 and the chance that an a allele mutates to an A allele (a → A) is 0.00005. These mutation rates are toward the high end of the range of estimated mutation rate values (see Table 5.1). The expected equilibrium value is $p = 0.333$ according to Eq. 5.22, an allele frequency that is reached only after tens of thousands of generations. The time to equilibrium is proportional to the absolute magnitudes of the mutation rates, while the equilibrium value depends only on a function of the ratio of the mutation rates.

It turns out that the process of mutation within a population is exactly analogous to the process of gene flow among several subpopulations. Compare allele frequency in Figure 5.12 with irreversible mutation and Figure 4.6 which shows allele frequency under one-way gene flow in the continent-island model. Both processes cause allele frequencies to change toward a state of fixation and loss, and the shape of both curves is identical. Then, compare allele frequency in Figure 5.13 with the process of bidirectional mutation with the process of bidirectional gene flow in the two-island model shown in Figure 4.7. Here, too, the shape of both curves is identical and both processes result in intermediate frequencies of both alleles at equilibrium. The major differences in the mutation and gene-flow graphs are the time scales. In the absence of other processes, gene flow causes allele frequencies to approach equilibrium values in tens or hundreds of generations, whereas mutation requires tens to hundreds of thousands of generations to approach equilibrium allele frequencies. It is important to understand that this difference in time scale is just a product of the vastly different rates for the two processes rather than a

Interact box 5.4 Simulating irreversible and two-way mutation

Go to the text website to simulate the irreversible and two-way mutation models.

Start by running the simulation for 2000 generations, $N_e = 20$, and an initial allele frequency of 0.9 without mutation to see the pattern produced by genetic drift acting alone. Then, click on the mutation checkbox and set the forward mutation rate to 0.01 and the reverse mutation rate to zero (in the irreversible model one of the mutation rates is zero), an initial allele frequency of 0.9, and 2000 generations. Also, simulate a forward mutation rate of 0.001. What is the distribution of allele frequencies among the replicate loci with and without mutation? How rapidly is equilibrium approached with forward mutation rates of 0.01 and 0.001?

For the two-way model, compare the approach to equilibrium over 2000 generations when both backward and forward mutation rates are equal (e.g. both 0.01) and when they are unequal (e.g. 0.01 and 0.005), starting at an initial allele frequency of 0.5. How do the simulation results compare with the expected equilibrium allele frequency for a two-way mutation?

Note that these mutation rates serve as an illustration only and that biologically realistic mutation rates are usually much lower.

fundamental difference in the processes themselves. In the figures, the chance that a gamete migrated was one in 10, while the chance of an allele mutated was between one in 1000 and one in 10 000. While these rates are likely to be on the high end of the range of values found in natural populations, gene flow is expected to occur at much higher rates than mutation as a general rule. The conclusion from this comparison is that gene flow is a much more potent force to change allele frequencies at a single locus over the short term compared to mutation. Mutation does have an effect, but it is long term.

The parallel nature of the processes of gene flow and mutation can be used as an advantage to understand more about the process of mutation. In particular, we can learn more about how mutation will impact autozygosity in finite populations where gene drift is also operating. Recall from Chapter 3 the expression for the level of autozygosity in a finite population caused by genetic drift:

$$F_t = \frac{1}{2N_e} + \left(1 - \frac{1}{2N_e}\right)F_{t-1} \qquad (5.37)$$

Mutation breaks the chain of descent by changing the state of alleles and, therefore, reduces the probability that a genotype is composed of two alleles identical by descent (autozygous). Genotypes with no alleles, one allele, or two alleles impacted by mutation each generation have frequencies of

$(1 - \mu)^2$, $2\mu(1 - \mu)$, and μ^2, respectively. Only the $(1 - \mu)^2$ genotypes with no mutated alleles can contribute to the pool of alleles that may become identical by descent due to finite sampling. From the opposite perspective, note that $2\mu(1-\mu)$ genotypes heterozygous and μ^2 genotypes homozygous for a new mutation are expected each generation. Together, these two classes of genotypes with mutations reduce the autozygosity by a factor of $1 - 2\mu(1-\mu)-\mu^2 = (1 - \mu)^2$. (This is identical to the reasoning used in Chapter 4 for the case of gene flow.)

Mutation will, therefore, reduce the autozygosity caused by finite sampling in the present generation (chance of $\frac{1}{2N_e}$) by a factor of $(1 - \mu)^2$. In addition, mutation will also reduce any autozygosity from past generations (F_{t-1}) because some alleles that are identical by descent may change to new states via mutation, leaving the proportion $(1 - \mu)^2$ of genotypes unaffected by mutation and at the same level of autozygosity. Putting these two separate adjustments for the autozygosity together gives

$$F_t = \frac{1}{2N_e}(1-\mu)^2 + \left(1 - \frac{1}{2N_e}\right)(1-\mu)^2 F_{t-1}$$

$$(5.38)$$

Assuming that the mutation rate is small and much, much less than the effective population size (see derivation in Math box 4.1 for the case of gene flow), an *approximation* for the expected amount of autozygosity at equilibrium in a finite population experiencing mutation is

$$F_{equilibrium} \cong \frac{1}{4N_e u + 1} \qquad (5.39)$$

This result also depends on each mutation giving rise to a new allele that is not present in the population or the infinite alleles model. Since the allozygosity or expected heterozygosity is just 1 minus the autozygosity,

$$
\begin{aligned}
H_{equilibrium} &\cong 1 - F_{equilibrium} \\
&\cong \frac{4N_e u + 1}{4N_e u + 1} - \frac{1}{4N_e u + 1} = \frac{4N_e u}{4N_e u + 1}
\end{aligned}
$$
$$(5.40)$$

This is the expected heterozygosity in a finite population where the "push" on allele frequencies toward fixation and loss by genetic drift and the "push" on allele frequencies away from fixation and loss by mutation have reached a net balance.

The quantity $4N_e\mu$ has a ready biological interpretation when N_e is large and μ is small. In a population of $2N_e$ alleles, the expected number of mutated alleles each generation is $2N_e\mu$. In a sample of two alleles that compose a diploid genotype, the chance that either allele has experienced mutation and are, therefore, not identical by descent is $2(2N_e\mu) = 4N_e\mu$. For example, in a population of $2N_e = 100$, alleles with a mutation rate of one in 10 000 alleles per

generation ($\mu = 0.0001$), the expected number of mutations is 0.01 and the chance that a sample contains two alleles that are not autozygous is 0.02. The quantity $4N_e\mu$ is frequently symbolized by θ (pronounced "theta"). Under the infinite alleles model, θ is the probability that two alleles sampled at random from a population at drift–mutation equilibrium will be allozygous. With $\theta = 0.02$, the expected heterozygosity at drift–mutation equilibrium is 0.0099. It is important to note that equilibrium heterozygosity will be lower than that predicted by θ if the infinite alleles or infinite sites model is not met. This is the case because, with a finite number of allelic states, not all mutation events will produce a novel allele that forms an allozygous pair, or a heterozygote, when sampled with an existing allele in the population. In fact, mutations that make additional copies of existing alleles actually increase the perceived homozygosity due to homoplasy.

Figure 5.14 shows the expected probability of autozygosity and allozygosity at mutation–genetic drift equilibrium. At small values of $4N_e\mu$, there will be an intermediate equilibrium level of autozygosity due to the balance of mutation introducing new alleles and genetic drift moving allele frequencies toward fixation or loss. As $4N_e\mu$ gets large, there is either little drift or lots of mutation, so there will be almost complete heterozygosity (no autozygosity). In the other direction, $4N_e\mu$ near zero indicates very strong genetic drift or very infrequent mutation resulting in high levels of autozygosity and very low heterozygosity. Bear in mind that reaching the expected equilibrium autozygosity or heterozygosity

---

Interact box 5.5 Heterozygosity and homozygosity with two-way mutation

Use the text simulation website to see how a two-way mutation with genetic drift impacts homozygosity and heterozygosity. Focus on the plot at the right that shows the ensemble distribution of allele frequency for many replicate loci.

Start by running the simulation for 500 generations, $N_e = 20$, and an initial allele frequency of 0.5 without mutation to see the pattern produced by genetic drift acting alone. How many loci exhibit complete homozygosity or some heterozygosity? Then, click on the mutation checkbox and set the forward and reverse mutation rates to 0.01. What is the value of $\theta = 4N_e\mu$? How does mutation impact the ensemble distribution of allele frequencies and thereby the number of loci that exhibit some level of heterozygosity? Also, simulate mutation rates of 0.001 and compute the value of $\theta = 4N_e\mu$. How much heterozygosity is seen at the higher mutation rates compared to the lower mutation rates?

How does the simulation mutation model differ from the mutation model assumptions of $\theta = 4N_e\mu$? What are the consequences for the level of heterozygosity maintained by mutation?

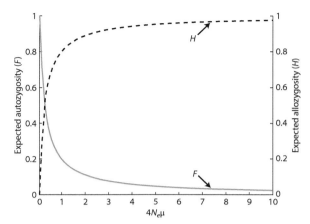

Figure 5.14 Expected homozygosity (*F* or autozygosity, solid line) and heterozygosity (*H* or allozygosity, dashed line) at equilibrium in a population where the processes of both genetic drift and mutation are operating. The chance that two alleles sampled randomly from the population are identical in state depends on the net balance of genetic drift working toward fixation of a single allele in the population and mutation changing existing alleles in the population to new states. A critical assumption is the infinite alleles model, which guarantees that each mutation results in a unique allele and thereby maximizes the allozygosity due to mutations.

The genealogical branching model was introduced in Chapter 3 for a single finite population and then extended in Chapter 4 to account for branching patterns expected with population subdivision. The goal of those sections was to predict genealogical branching patterns without reference to the identity of the lineages represented by the branches. These branching models need to be extended to account for the possibility that mutation occurs. Mutations will alter the genes or DNA sequences that are represented by each lineage or branch in the genealogical tree. Therefore, accounting for mutation will be a critical step in developing a coalescent model that explains differences among a sample of lineages in the present. This section will focus on the action of mutation in the coalescent model along with the state of each lineage in a genealogy. This is accomplished by coupling the process of coalescence and the process of mutation while moving back in time toward the most common recent ancestor. The ultimate goal is to build a genealogical branching model that can be used to predict the numbers and types of alleles that might be expected in a sample of lineages taken from an actual population. For example, one prediction might be the number of alleles expected in a single finite population for a given mutation rate. In this way, the combination of the coalescent process and the mutation process is used to form quantitative expectations about patterns of genetic variation produced by various population genetic processes.

will take many, many generations because mutation rates are low, making mutation a very slow process. If heterozygosity is perturbed from its mutation–drift equilibrium point, a population will take a very long time to return to that equilibrium.

Building a coalescent model with mutation is as simple as adding another type of possible event that can occur between the present and some time in the past (Figure 5.15). We will assume that both coalescent and mutation events are rare (or that N_e is large and the rate of mutation is small) so that when an event does occur going back in time, it is *either*

5.5 The coalescent model with mutation

- Adding the process of mutation to coalescence.
- Longer genealogical branches experience more mutations.
- Genealogies under the infinite alleles and infinite sites models of mutation.

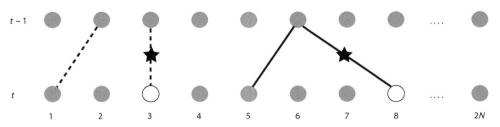

Figure 5.15 Haploid reproduction in the context of coalescent and mutation events. In a haploid population, the probability of coalescence is $\frac{1}{2N}$ (solid lines) while the probability of that the two lineages do not have a common ancestor in the previous generation is $1 - \frac{1}{2N}$ (dashed lines). The process of mutation can also occur simultaneously (stars), changing the state of alleles (filled circles to unfilled circles).

coalescence *or* mutation. In other words, we will assume that mutation and coalescence events are mutually exclusive.

Every generation, there is the chance a mutation occurs. The rate of mutation, μ, can be thought of as the chance that each lineage experiences a mutation each generation. The chance that a lineage does not experience a mutation is, therefore, $1 - \mu$ each generation. The chance that t generations pass before a mutation event occurs is then the product of the chances of $t - 1$ generations of no mutation followed by a mutation, or

$$P(T_{mutation} = t) = (1 - \mu)^{t-1}\mu \qquad (5.41)$$

This equation has an identical form to the chance that a coalescent event occurs after t generations given in Chapter 3. Like the probability of coalescence, the probability of a mutation through time is a geometric series that can be approximated by the exponential distribution (see Math box 3.2).

To obtain the exponential expression or the exponent of e that describes the frequency of mutation events, we need to determine the rate at which mutations are expected to occur. When time is measured on a continuous scale with $t = \dfrac{j}{2N_e}$, one unit of time is equivalent to $2N_e$ generations. If $2N_e$ generations elapse and μ is the rate of mutation per generation, then $2N_e\mu$ mutations are expected during one unit of continuous time. If we define $\theta = 4N_e\mu$, then $\theta/2$ is equal to $2N_e\mu$. This leads to the exponential approximation for the chance that a mutation occurs in a single lineage at generation t:

$$P(T_{mutation} = t) = e^{-t\frac{\theta}{2}} \qquad (5.42)$$

on a continuous time scale. When there is more than one lineage, each lineage has an independent chance of experiencing a mutation but only one lineage can experience a mutation. When events are independent but mutually exclusive, the probability of each event is added over all possible events to obtain the total chance that an event occurs. Adding the $e^{-t\frac{\theta}{2}}$ chance of a mutation for each lineage over all k lineages gives the total chance of mutation:

$$P(T_{mutation} = t) = e^{-t\frac{\theta}{2}k} \qquad (5.43)$$

for k lineages (compare this with the chances of coalescence at time t with k lineages, $e^{-t\frac{k(k-1)}{2}}$, based on

similar logic). The chance that a mutation occurs in one of k lineages at or before a certain time can then be approximated with the cumulative exponential distribution

$$P(T_{mutation} \leq t) = 1 - e^{-t\frac{\theta}{2}k} \qquad (5.44)$$

in exactly the same fashion that times to coalescent events are approximated.

When two independent processes are operating, the coalescence model becomes one of following lineages back in time and waiting for an event to happen. The possible events are mutation or coalescence, so the total chance of *any* event is the sum of the independent probabilities of each type of mutually exclusive event. The total chance of an event occurring while going back in time (increasing t) is then

$$P(T_{event} \leq t) = 1 - e^{-t\frac{k(k-1+\theta)}{2}} \qquad (5.45)$$

where the exponent of e is $-t\left[k\dfrac{\theta}{2} + \dfrac{k(k-1)}{2}\right]$ or the sum of the intensities of mutation and coalescence. When an event does occur at a time given by this exponential distribution, it is then necessary to decide whether the event is a coalescence or a mutation. The total chance that the event is either a mutation event or a coalescence event is $\dfrac{k\theta}{2} + \dfrac{k(k-1)}{2}$. Therefore, the chance the event is a mutation is

$$\frac{\dfrac{k\theta}{2}}{\dfrac{k\theta}{2} + \dfrac{k(k-1)}{2}} = \frac{\theta}{k-1+\theta} \qquad (5.46)$$

while the chance that the event is a coalescence is

$$\frac{\dfrac{k(k-1)}{2}}{\dfrac{k\theta}{2} + \dfrac{k(k-1)}{2}} = \frac{k-1}{k-1+\theta} \qquad (5.47)$$

Using the cumulative exponential distribution specified by Eq. 5.45 and then determining whether each event is a mutation or a coalescence, it is possible to construct a coalescent genealogy that includes the possibility of mutations occurring along each branch (Figure 5.16).

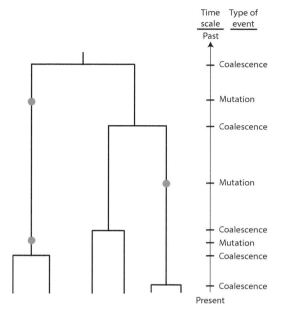

Time scale	Type of event
Past
	— Coalescence
	— Mutation
	— Coalescence
	— Mutation
	— Coalescence
	— Mutation
	— Coalescence
	— Coalescence
Present

Figure 5.16 A genealogy constructed under the simultaneous processes of coalescence in a single finite population and mutation. Working backward in time from the present, both mutation and coalescence events can occur. The blue dots represent mutation events, each assigned at random to a lineage present when the event occurred. Mutation events alter the state of a lineage, causing divergence from the ancestral state of the most recent common ancestor of all the lineages in the present.

The pattern of mutations on genealogical trees has some general features. Since the chance of mutation is assumed to be constant through time, the more time that passes the greater the chance that a mutation occurs. This means that longer branches tend to experience more of the mutations on average in genealogical trees, while shorter branches are less likely to exhibit mutations. Recall the metaphor of branch length as a road that was used in Chapter 3 to describe the total branch length of a genealogy. If mutations are road signs with a constant chance of appearing per distance of roadway, then longer stretches of road are expected to have more signs. Applying this logic to genealogies like that in Figure 5.16 tells us that more mutations are expected during the long average waiting time for coalescence with two lineages ($k = 2$) than are expected to occur when there are six lineages that can coalesce ($k = 6$). Another example would be the pattern of mutations expected for lineages in two demes with different levels of migration (see Figure 4.17). With very limited migration, multiple mutations are expected on the long branches before the single lineages within each deme coalesce. Alternatively, when migration rates are high, then many fewer mutations are expected. In the former case,

---

Interact box 5.6 Build your own coalescent genealogies with mutation

Building a few coalescent trees can help you to better understand the evolution of genealogies when both the processes of mutation and coalescence are operating. You can use an expanded version of the Microsoft Excel spreadsheet used to build coalescent trees in Chapter 3 that now models waiting times for both mutation and coalescence. The spreadsheet contains the cumulative exponential distributions used to determine the time until a coalescent or mutation event (see Eq. 5.45) for up to six lineages. To determine the time that an event occurs for a given number of lineages k and mutation rate, a random number between 0 and 1 is picked and then compared to the cumulative exponential distribution. The time interval on the distribution that matches the random number is taken as the event time. The next step is to determine whether the event was a mutation or a coalescence, again accomplished comparing a random number to the chances of each type of event (Eqs. 5.46 and 5.47).

Step 1: Open the spreadsheet and click on cells to view the formulas used, especially the cumulative probability of coalescence for each k. This will help you understand how the equations in this section of the chapter are put into practice. You can compare the cumulative probability distributions graphed for $k = 6$ and $k = 2$.

Step 2: Look at the section of the spreadsheet under the heading "Event times:" on the right side of the sheet. This section gives the waiting times until an event occurs and then determines if the event was a coalescence or a mutation. Press the recalculate key(s) to generate new sets of random numbers (see Excel help if necessary). Watch the times to an event change.

Step 3: Now, draw a genealogical tree with the possibility of mutations (do not recalculate again until Step 6 is complete). Along the bottom of a blank sheet of paper, draw six evenly spaced dots to represent six lineages.

Step 4: Start at the first "Decide event time:" panel to determine how much time passes (going backward in time) until either a mutation or a coalescence occurs. Then, use the entries under "Decide what type of event:" to determine if the event was a coalescence or a mutation. If the event is a mutation, go to Step 5, otherwise go to Step 6.

Step 5: If the event is a mutation, draw the lines for all lineages back in time by a length equal to the waiting time (e.g. if the time is 0.5, draw lines that are 0.5 cm). Use the random number table to pick one lineage and draw an X on the lineage at the event time to indicate a mutation occurred. If a mutation occurred the number of lineages (k) remains the same. Move down to the next "Decide event time:" panel and obtain the next event time for the same value of k. Repeat Step 5 until the event is a coalescent event.

Step 6: Using the random number table, pick two lineages that will experience coalescence. Label the two left-most dots with these lineage numbers. Then, using a ruler, draw two parallel vertical lines that start at the end of the last event and extend as long as the time to coalescence in continuous time (e.g. if the time is 0.5, draw lines that are 0.5 cm). Connect these vertical lines with a horizontal line. Assign the lineage number of one of the coalesced lineages to the pair's single ancestor at the horizontal line. Record the other lineage number on a list of lineages no longer present in the population (skip over these numbers if they appear in the random number table). There are now $k-1$ lineages.

Step 7: Return to Step 4 until all lineages have coalesced ($k = 1$).

You should obtain a coalescence tree with mutation events on the branches like that in Figure 5.16. Your trees will be different because the random coalescence and mutation times vary around their averages, but the overall shape of your trees (e.g. shorter branches when k is large) and frequencies of mutations (for a given mutation rate) will be similar.

mutations cause lineages to diverge substantially between the two demes, while, in the latter case, the lineages in the different demes have less opportunity to accumulate differences.

A genealogy with generic mutations like that shown in Figure 5.16 is an abstraction until it is joined with a mutation model. Figure 5.17 shows the same genealogy under the assumptions that the most recent common ancestor (MRCA) has an allelic state of A and mutational changes follow the infinite alleles model. Each mutation event is an instance where the current state of a lineage changes to an allelic state not currently present in the population. Because of mutational changes to the ancestral allelic state, the six lineages in the present represent four allelic states. Two of the alleles have a frequency of $2/6 = 33\%$, while the remaining two alleles have a frequency of $1/6 = 16.6\%$. The lineages with the B and C alleles are identical in state, and, therefore, can be considered identical by descent. Figure 5.18 shows a third version of the genealogy with mutations, this time representing each allele as a DNA sequence and employing the infinite sites mutation model. Each mutational event alters a randomly picked site in the DNA sequence under the constraint that a site can only experience mutation a single time. The result is a set of DNA

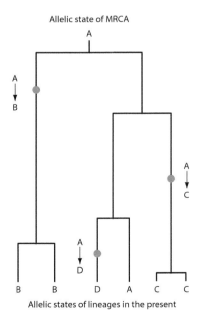

Figure 5.17 A genealogy constructed under the simultaneous processes of coalescence in a single finite population and mutation. Here, the infinite alleles model of mutation is assumed in order to determine the allelic state of each lineage in the genealogy. Arbitrarily assigning allelic state A to the most recent common ancestor, each mutational event then alters the state of the lineage experiencing the mutation. Each mutation changes the allelic state of the lineage to a new allele not present in the population, giving rise to a variety of allelic states among the lineages in the present.

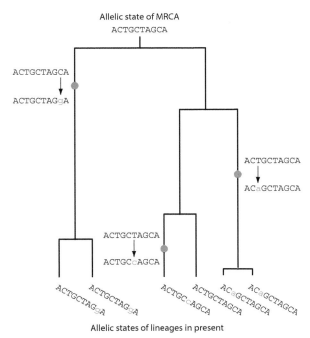

Allelic state of MRCA
ACTGCTAGCA

ACTGCTAGCA
↓
ACTGCTAGgA

ACTGCTAGCA
↓
ACaGCTAGCA

ACTGCTAGCA
↓
ACTGCcAGCA

ACTGCTAGgA ACTGCTAGgA ACTGCcAGCA ACTGCTAGCA ACaGCTAGCA ACaGCTAGCA

Allelic states of lineages in present

Figure 5.18 A genealogy constructed under the simultaneous processes of coalescence in a single finite population and mutation. Here, the infinite sites model of mutation for DNA sequences is assumed in order to determine the allelic state of each lineage in the genealogy. Arbitrarily assigning the DNA sequence ACTGCTAGCA to the most recent common ancestor, each mutational event then alters the DNA sequence of the lineage experiencing the mutation. Each mutation occurs at a random site in the DNA sequence that has not previously experienced a mutation (bases in red, lowercase letters), giving rise to differences in the DNA sequences among the lineages in the present. Here, each base is equally likely to be produced by a mutation, although there are numerous models to specify the pattern of nucleotide changes expected by mutation. Under the finite sites model of mutation, each site in the DNA sequence could experience mutation repeatedly.

sequences that differ at three of 10 nucleotide sites. As with the infinite alleles model, DNA sequences in the present that are identical in state are, therefore, identical by descent.

When a genealogy containing mutations is combined with a mutation model, it results in an explicit prediction of the diversity and types of allelic states expected under the processes that influence the branching patterns. Although the two examples shown here both utilize genealogies resulting from genetic drift in a finite population, mutation could also be combined with processes such as population structure, growing or shrinking population sizes, or natural selection that are used to generate a

coalescent genealogy. Chapter 7 covers genealogies expected with natural selection, and Chapter 8 explains methods to compare the expected patterns of allelic states in genealogies generated by different population genetic processes.

Chapter 5 review

- Mutation is the fundamental process that generates novel genetic polymorphism. Mutation rate estimates can be made from observations of reporter loci, from fluctuation assays, with mutation accumulation lines, and through direct comparisons of DNA sequences in related individuals. While the rate of mutation is challenging to estimate with precision, whole genome sequencing has greatly improved estimates haploid genome-wide mutation rate per generation (U) because a large proportion of the genome can be observed.

- The spectrum of relative fitness for genotypes containing mutations, often estimated with mutation accumulation lines, expresses the distribution of the relative fitness of new mutations and shows that most mutations are deleterious. Mutations that are strongly deleterious or lethal will be purged by natural selection, while strongly advantageous mutations will be fixed by natural selection. Mutations with smaller effects may be neutral or nearly neutral and, therefore, be subject to stochastic fixation or loss due to genetic drift.

- New mutations may be lost simply by Mendelian segregation since there is a $\left(\frac{1}{2}\right)^k$ chance that a given single allele will not be transmitted to k progeny. Neutral mutations are eventually lost while the chance of a new selectively favored mutation escaping loss (becoming fixed) is approximately twice its selective advantage, under the assumption that the population size is very large.

- New selectively neutral mutations are initially at a frequency of $\frac{1}{2N_e}$, so the chance of fixation is $\frac{1}{2N_e}$ and the chance of loss is $1 - \frac{1}{2N_e}$. Most new neutral mutations are expected to be lost from a population very rapidly since their initial frequency is very near 0. For those mutations that do eventually reach fixation, the time to fixation will average $4N_e$ generations.

- Neutral, deleterious, and beneficial mutations that occur along the edge of an expanding populations may reach high frequencies through stochastic mutation surfing.
- Fisher's geometric model of mutation shows that mutations with small effects on phenotype are more likely to be fixed by natural selection since these are the mutations with the greatest chance of improving fitness.
- The combination of mutation, genetic drift, and natural selection in genomes where recombination is absent or restricted leads to a growing accumulation of deleterious mutations in a phenomenon known as Muller's Ratchet.
- The infinite alleles model assumes discrete allelic states where each mutation creates an allele not currently present in the population. The infinite sites model assumes alleles are DNA sequences and that each mutation changes a single nucleotide at a site that has never before experienced mutation. Since mutation cannot form the same allele twice in both of these models, identity in state is always equivalent to identity by descent.
- Genetic distance measures employ a mutation model to quantify the frequency-weighted differences between and among the states of the alleles found within populations. The role of mutation models is illustrated by R_{ST}, an estimator of F_{ST} that utilizes a stepwise mutation model where alleles closer in state are more likely to be more recently related.
- Mutation models for DNA sequences include infinite site to avoid homoplasy and finite sites where homoplasy is generated by multiple mutations at the same nucleotide site.
- Irreversible mutation will eventually lead to loss of the original allele in a population since there is no process to restore the original allele. Bidirectional mutation leads to a net balance of two alleles changing state and an intermediate allele frequency that depends on the forward and reverse mutation rates. Both models show that mutation alone will take thousands or tens of thousands of generations to attain equilibrium allele frequency in a population depending on the mutation rate.
- Under the infinite alleles model, the balance of genetic drift and mutation lead to a prediction for the equilibrium heterozygosity of $\frac{\theta}{\theta + 1}$ where $\theta = 4N_e\mu$ is the effective mutation rate.

- The process of mutation can be added to coalescent genealogies by modeling the waiting time to any event with an appropriate cumulative exponential distribution. When an event does occur, it can be either a coalescence or a mutation that is then reflected in the genealogy. More mutations are likely to occur on longer branches in genealogies since the chance of mutation is constant through time.
- Each mutation event in a genealogy can be interpreted under a specific mutation model such as infinite alleles or infinite sites. The combination of a coalescent genealogy containing mutations and a mutation model yields a prediction for the number and frequency of alleles expected under the process or processes that produced the genealogy.

Further reading

A major figure in evolutionary genetics argues that mutation should be considered the central process of evolution in

Nei, M. (2013). *Mutation-Driven Evolution*. Oxford: Oxford University Press.

For a history of early mutation rate research and the role of a scientist directly involved, see

Peter, D., Keightley, P.D., and Adam, E.-W.A. (1999). Terumi Mukai and the riddle of deleterious mutation rates. *Genetics* 153: 515–523.

For a detailed review of mutation rate estimates based on genome-wide sequencing:

Katju, V. and Bergthorsson, U. (2019). Old trade, new tricks: insights into the spontaneous mutation process from the partnering of classical mutation accumulation experiments with high-throughput genomic approaches. *Genome Biology and Evolution* 11: 136–165.

For a review of the population genetic arguments behind the drift barrier hypotheses along with empirical evidence:

Lynch, M., Ackerman, M.S., Gout, J.-F. et al. (2016). Genetic drift, selection and the evolution of the mutation rate. *Nature Reviews Genetics* 17: 704–714.

End-of-chapter exercises

1 Explain the different methods used to estimate per nucleotide mutation rates. What are the possible advantages and disadvantages of each method?

2 Explain the drift barrier hypothesis for the evolution of the mutation rate. How does it relate to the mutation fitness spectrum? What does it predict?

3 Explain what is being predicted by the effective mutation rate $\theta = 4N_e\mu$ from the points of view of both autozygosity and coalescence. How do the homozygosity and heterozygosity vary with the effective mutation rate and why?

4 How do mutation rates and mutation state changes influence the choice of genetic markers for empirical studies of genetic polymorphism? What makes an effective genetic marker locus?

5 In the Muller's Ratchet model, what would happen if recombination were also acting?

6 Use the text simulation website Simulations -> Drift Selection Mutation for this problem. Check the Mutation box so that both drift and mutation are operating (but natural selection is not operating). Run the simulation, and focus on the bar graph at the right. What does the bar graph show? Vary the simulation parameters for N_e and μ to achieve a range of values of the effective mutation rate $\theta = 4N_e\mu$. For example, run the simulation with N_e values of 5, 20, 100, 500, and 2000. How does the pattern in the graph vary with θ? How do the patterns observed depend on the two-way mutation model?

7 Using the instructions in Interact Box 5.5, construct a coalescent genealogy for six lineages ($k = 6$) that includes mutations. Use $2N = 50$ and a mutation rate of 0.05. (Note that this mutation rate is unrealistically high to be sure that there are numerous mutations to work with. If you generate a genealogy without any mutation events, try again with another set of random numbers.) Use graph paper and draw a scale in both continuous and discrete time for the mutation and coalescence events on the genealogy.

   Assign the MRCA the discrete allelic state of "A" and then assign allelic states to each lineage in the present based on an infinite alleles model of mutation. Under random mating, what are the expected allele and genotype frequencies in the population as well as the expected homozygosity and heterozygosity?

8 Using the instructions in Interact Box 5.5, construct a coalescent genealogy for six lineages ($k = 6$) that includes mutations. Use $2N = 50$ and a mutation rate of 0.05. (Note that this mutation rate is unrealistically high to be sure that there are numerous mutations to work with. If you generate a genealogy without any mutation events, try again with another set of random numbers.) Use graph paper and draw a scale in both continuous and discrete time for the mutation and coalescence events on the genealogy.

   Assign an allelic state of "ATACGTGCAGT-CAGCTGTAC" to the MRCA. Duplicate that sequence to build a multiple sequence alignment for the six lineages in the genealogy. Use a nucleotide substitution model you chose to change nucleotide states for each lineage based on the mutations that occurred on the genealogy. Explain how you modeled the state change caused by each mutation event. (A spreadsheet program can help with this – use a row for each sequence and one column for each nucleotide site.) Based on the resulting multiple sequence alignment after mutation, describe the predicted polymorphism in a sample of six lineages with the number of segregating sites per nucleotide (p_S) and the average number of pairwise differences per nucleotide site (π) as explained in Chapter 8.

9 Search the literature for a recent research paper that utilizes one or more of the population genetic predictions covered in this chapter. The topic can be any organism, application, or process, but the paper must include a hypothesis test involving a topic such as mutation rates, relative fitness of mutations, probability of fixation of a mutation, mutation models, or the effective mutation rate ($4N_e\mu$). Summarize the main hypothesis, goal, or rationale of the paper. Then, explain how the paper utilized a population genetic prediction from this chapter and summarize the results and the conclusions based on the prediction.

CHAPTER 6

Fundamentals of natural selection

6.1 Natural selection

- Translating Darwin's ideas into a model.
- Natural selection as differential population growth.
- Natural selection with clonal reproduction.
- Natural selection with sexual reproduction and its assumptions.

Charles Darwin's (1859) statement of the process of natural selection can be summarized as three basic observations about populations:

- all species have more offspring than can possibly survive and reproduce,
- individual organisms vary in phenotypes that influence their ability to survive and reproduce, and
- within each generation, the individuals possessing phenotypes that confer greater survival and reproduction will contribute more offspring to the next generation.

The result is that phenotypes which cause a predictably greater chance of survival and reproduction will increase in frequency over generations to the extent that such traits have a genetic basis. Darwin's observations initially served as a qualitative model, since an accurate model of genetic inheritance was lacking until Mendel's results were recognized. Once particulate inheritance was understood, the unification of genetics with the principle of natural selection took place in what is now called the **modern synthesis** or **neo-Darwinian synthesis** of evolutionary biology. The major challenge in the modern synthesis for population genetics was to develop expectations for the genetic changes that are caused by natural selection. This section of the chapter develops these basic population genetic expectations for natural selection.

Natural selection with clonal reproduction

At its core, natural selection is actually a process of population growth, so let us start off by examining a simple population growth model. If a population is assumed to have no upper limit in its size, the number of individuals one generation in the future (N_{t+1}) is a product of the number of individuals present now (N_t) multiplied by the **finite rate of increase** of the population λ (pronounced "lambda") to give the expression:

$$N_{t+1} = \lambda N_t \qquad (6.1)$$

In this equation for unbounded population growth, λ is a multiplier that represents the net difference between the number of individuals lost from the population due to death and the number of new individuals recruited to the population by reproduction each generation. If the number of births and deaths are exactly equal, then λ is 1 and the population does not change in size. If there are more births than deaths, then $\lambda > 1.0$ and the population grows, whereas, if there are more deaths than births, then $\lambda < 1.0$ and the population shrinks. The population growth rate can be thought of as the chance that an individual contributes one offspring to the next generation.

Natural selection is really just a special case of this basic population growth model where each genotype has its own growth rate. To see how this works, let us consider a population composed of two genotypes of an asexual organism like a bacterial species that reproduces only by clonal division over discrete generations. Call the two genotypes A and B with

Population Genetics, Second Edition. Matthew B. Hamilton.
© 2021 John Wiley & Sons, Inc. Published 2021 by John Wiley & Sons, Inc.
Companion website: www.wiley.com/go/hamilton/populationgenetics

genotype-specific growth rates or **absolute fitness** values of λ_A and λ_B. The proportions of each genotype in the total population in any generation are:

$$p = \frac{N_A}{N_A + N_B} \qquad (6.2)$$

and

$$q = \frac{N_B}{N_A + N_B} \qquad (6.3)$$

where $N_A + N_B$ is the total population size. Figure 6.1A shows numbers of individuals over time, for example, where genotype A grows faster than genotype B. In absolute numbers of individuals, the population sizes of both genotypes increase over time. However, the proportion of the population made up of individuals with A and B genotypes changes over time in the population (Figure 6.1B). Since the A genotype grows faster, A individuals represent an increasing proportion of the individuals in the total population. This is equivalent to saying that p increases over time while q decreases over time.

Thus, Figure 6.1 shows a case of natural selection favoring the A genotype since it has a higher level of absolute fitness.

An alternative way to represent the changing proportions of the two genotypes in the population is to follow the ratio of the number of A and B individuals, N_A/N_B, over time. The value of the ratio at any point in time will depend on the initial numbers of A and B individuals (call them $N_A(0)$ and $N_B(0)$), the growth rates of the two genotypes, and the number of generations that have elapsed. The ratio N_B/N_A after one generation of population growth is given by

$$\frac{N_B(t = 1)}{N_A(t = 1)} = \left(\frac{\lambda_B}{\lambda_A}\right)\frac{N_B(0)}{N_A(0)} \qquad (6.4)$$

which is akin to dividing a version of Eq. 6.1 for genotype A by a version of Eq. 6.1 for genotype B. In general, we can predict the ratio of N_A/N_B at any time t using

$$\frac{N_B(t)}{N_A(t)} = \left(\frac{\lambda_B}{\lambda_A}\right)^t \frac{N_B(0)}{N_A(0)} \qquad (6.5)$$

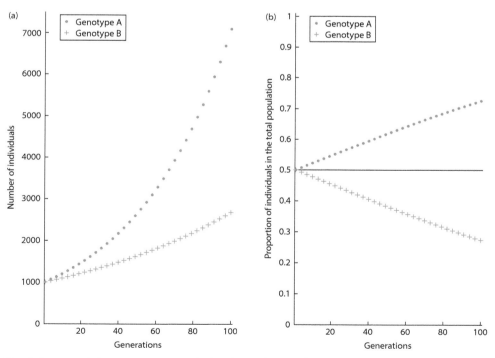

Figure 6.1 Population growth in two genotypes with clonal reproduction that start out with equal numbers of individuals and, therefore, equal proportions in the total population. Genotype A grows 3% per generation ($\lambda = 1.03$), and genotype B grows 1% per generation ($\lambda = 1.01$). Individuals of both genotypes increase in number over time (panel A). Because the genotypes grow at different rates, their relative proportions in the total population change over time (panel B). The solid line shows the initial equal proportions. Eventually, genotype A will approach 100% and genotype B 0% of the total population. Values are plotted for every third generation.

by assuming that genotype-specific growth rates (λ_A and λ_B) remain constant through time.

The ratio of genotype-specific growth rates is called the **relative fitness** and is represented by the symbol w in models of natural selection with discrete generations. Substituting the relative fitness for the ratio of the genotype-specific growth rates in Eq. 6.5 gives

$$\frac{N_B(t)}{N_A(t)} = w^t \frac{N_B(0)}{N_A(0)} \qquad (6.6)$$

Recalling that $N_A + N_B$ gives the total population size N at any point in time and then multiplying both sides of Eq. 6.6 by $\frac{\frac{1}{N}}{\frac{1}{N}}$ gives

$$\frac{\frac{N_B(t)}{N}}{\frac{N_A(t)}{N}} = w^t \frac{\frac{N_B(0)}{N}}{\frac{N_A(0)}{N}} \qquad (6.7)$$

which can be simplified by utilizing Eqs. 6.2 and 6.3:

$$\frac{q_t}{p_t} = w^t \frac{q_0}{p_0} \qquad (6.8)$$

to represent each genotype by its proportion of the total population at any time t. When $w = 1.0$, the two genotypes have identical growth rates and the proportion of each genotype remains constant in the population through time. If $w > 1.0$, then the genotype in the numerator grows faster than the genotype in the denominator, and it will represent a larger proportion of the population over time. Conversely, if $w < 1.0$, then the genotype in the numerator grows less rapidly than the genotype in the denominator, and it will represent a shrinking proportion of the population over time. Using the A genotype as the standard of comparison and the absolute fitness values from Figure 6.1, $w_A = 1.03/1.03 = 1.0$ and $w_B = 1.01/1.03 = 0.981$, and, so, the frequency of the A genotype is expected to increase over time.

The relative fitness can be used to determine the change in frequency of a genotype over time, as shown in Table 6.1. The *change* in genotype

Table 6.1 The expected frequencies of two genotypes after natural selection for the case of clonal reproduction. The top section of the table gives expressions for the general case. The bottom part of the table uses absolute and relative fitness values identical to Figure 6.1 to show the change in genotype proportions for the first generation of natural selection. The absolute fitness of the A genotype is highest and is, therefore, used as the standard of comparison when determining relative fitness.

	Genotype	
	A	**B**
Generation t		
Initial frequency	p_t	q_t
Genotype-specific growth rate (absolute fitness)	λ_A	λ_B
Relative fitness	$w_A = \dfrac{\lambda_A}{\lambda_A}$	$w_B = \dfrac{\lambda_B}{\lambda_A}$
Frequency after natural selection	$p_t w_A$	$q_t w_B$
Generation $t+1$		
Initial frequency p_{t+1}	$\dfrac{p_t w_A}{p_t w_A + q_t w_B}$	$\dfrac{q_t w_B}{p_t w_A + q_t w_B}$
Change in genotype frequency	$\Delta p = p_{t+1} - p_t$	$\Delta q = q_{t+1} - q_t$
Generation t		
Initial frequency	$p_t = 0.5$	$q_t = 0.5$
Genotype-specific growth rate (absolute fitness)	$\lambda_A = 1.03$	$\lambda_B = 1.01$
Relative fitness	$w_A = \dfrac{\lambda_A}{\lambda_A} = \dfrac{1.03}{1.03} = 1.0$	$w_B = \dfrac{\lambda_B}{\lambda_A} = \dfrac{1.01}{1.03} = 0.981$
Frequency after natural selection	$p_t w_A = (0.5)(1.0) = 0.5$	$q_t w_B = (0.5)(0.981) = 0.4905$
Generation $t+1$		
Initial frequency p_{t+1}	$\dfrac{0.5}{0.5 + 0.4905} = 0.5048$	$\dfrac{0.4905}{0.5 + 0.4905} = 0.4952$
Change in genotype frequency	$0.5048 - 0.5 = 0.0048$	$0.4952 - 0.05 = -0.0048$

frequency is the *difference* between frequencies in two generations, $p_{t+1} - p_t$. A difference is commonly symbolized with the Greek capital letter delta (Δ), so we could say that the change in the frequency of the A genotype is given by $\Delta p = p_{t+1} - p_t$. To generate an expression for Δp, we can compare the initial genotype frequency p_t with its frequency a generation later after natural selection has acted via differential growth. We start with the basic expression for the difference in genotype frequency

$$\Delta p = p_{t+1} - p_t \qquad (6.9)$$

If Δp is positive, then the A genotype will increase in proportion in the population while it will decrease in proportion if Δp is negative. Substituting the expression for the expected frequency of the A genotype after natural selection (Table 6.1) gives

$$\Delta p = \frac{p_t w_A}{p_t w_A + q_t w_B} - p_t \qquad (6.10)$$

The $p_t w_A + q_t w_B$ term in the denominator on the right side of Eq. 6.10 is the **average relative fitness** of the population (it is a frequency-weighted average and so depends on the sum of the product of the frequency and relative fitness for each genotype). Positive values of Δp occur when the frequency of the A genotype after natural selection is greater than the average fitness of both genotypes after natural selection. The computations in Table 6.1 show that the frequency of the A genotype multiplied by its relative fitness ($p_t w_A$) is greater than the average fitness so the A genotype will increase in proportion in the population over time. The average fitness will be covered in more detail when considering natural selection in sexual diploid populations.

One advantage of using the relative fitness is that the population growth rates of each genotype do not have to be known to model the proportions of the genotypes over time. Rather, the outcome of the growth process in terms of the relative frequencies of genotypes can be predicted strictly from the ratio of growth rates. This means that Eq. 6.8 potentially applies to organisms with very high absolute growth rates like bacteria as well as to species with absolute growth rates very near one such as elephants. Eq. 6.8 even applies to cases where population sizes are declining through time. If a population is headed

to extinction because it is composed of genotypes that all have growth rates less than 1, the relative fitness will nonetheless accurately express the change in the proportion of genotypes in the population over time. In practice, the relative fitness can be estimated in competition experiments where two or more genotypes are placed in the same environment and their proportions estimated at a later point in time (see Problem Box 6.1).

Absolute fitness: The genotype-specific rate of increase or population growth that predicts the absolute number of individuals of a given genotype in a population over time. Commonly symbolized as W or λ.

Average or mean fitness ($\overline{w}$): The frequency-weighted sum of the relative fitness values of each genotype in the population.

Relative fitness: The growth rate of genotypes relative to one genotype picked as the standard of comparison (often the genotype with the highest absolute fitness). Called Darwinian fitness after Charles Darwin and symbolized as w in models where time is represented in discrete generations (also called Malthusian fitness after Thomas Malthus and symbolized as m in models where time is continuous).

Although simplistic and requiring many assumptions, the model of natural selection among genotypes in organisms with clonal reproduction is nonetheless pertinent to a range of practical situations. One example is the evolution of drug resistance by natural selection in the human immunodeficiency virus (HIV). The genome of HIV (and other retroviruses) is single-stranded RNA. All the proteins inside a virus particle as well as the viral protein envelope itself are encoded by genes in this RNA genome. After infecting a host cell, HIV uses reverse transcriptase produced from its own gene to reverse-transcribe its genome into double-stranded DNA. This DNA version of the retrovirus genome is then integrated into the

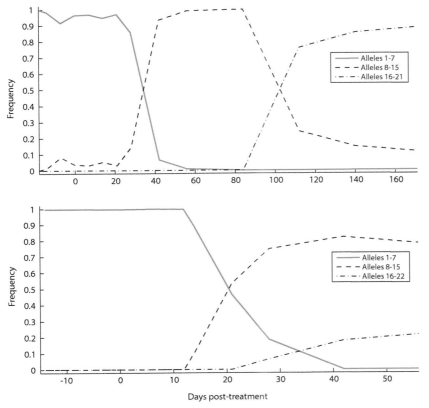

Figure 6.2 Allele frequencies at the protease locus over time in the HIV populations within two patients undergoing protease inhibitor (ritonavir) treatment (Doukhan and Delwart 2001). Alleles found at very low frequencies before drug treatment come to predominate in the HIV population after drug treatment due to natural selection among HIV genotypes for drug resistance. Alleles are bands observed in denaturing-gradient gel electrophoresis (DGGE), a technique that is capable of discriminating single base pair differences among different DNA fragments. DGGE was used to identify the number of different protease locus DNA sequences present in a sample of HIV particles. Protease inhibitor treatment began on day zero. Dr. E. Delwart kindly provided the original data used to draw this figure.

DNA of the host cell, where it is transcribed by the host cell into many new virus RNA genomes. These new viral RNA genomes are packaged into virus particles released from the host cell through a viral protease. One treatment strategy for HIV has utilized drugs that mimic nucleosides (nucleotides without phosphate groups) that interfere with the virus reverse transcriptase but do not interfere with host cell DNA polymerase. Another treatment uses protease inhibitors that interfere with polyprotein cleavage necessary to produce new infectious virus particles. Unfortunately, HIV has shown rapid evolution of drug-resistant genotypes via natural selection. Figure 6.2 shows allele frequencies over time in the population of HIV particles infecting two patients who began protease inhibitor treatment at day 0. Individual HIV particles with protease alleles resistant to the drug have higher replication rates than HIV particles with wild-type protease alleles. This differential growth rate of HIV genotypes, or natural selection at the protease locus, rapidly changed the protease gene allele frequencies in the HIV population found within each patient. The combination of short generation time, high mutation rate, and large effective population size make natural selection a rapid process acting to change allele frequencies in HIV populations.

Problem box 6.1
Relative fitness of HIV genotypes

It is commonly thought that drug-resistant alleles have lower relative fitness than non-resistant alleles in the absence of drug exposure. To test this hypothesis for HIV-1, Goudsmit et al. (1996) monitored the frequency of alleles at codon 215 of the reverse transcriptase gene in an individual newly infected with HIV but who was not undergoing treatment with the nucleoside analog azidothymidine (AZT). Initially, the HIV alleles were all sequences (90% TAC and 10% TCC codons) known to confer AZT resistance. Over time, the non-resistant allele (an TCC codon) increased in frequency to 49% after 20 months.

Use this change in allele frequencies over 601 days and Eq. 6.8 to estimate the relative fitness of the non-resistant allele in the absence of AZT. Assume that the generation time of HIV is 2.6 days and that generations are discrete, and that the wild-type allele was initially present at a frequency of 1.0% and so was not initially detectable. Note that the exponent in an equation like $a = y(x^t)$ where a, y, and x are constants can be removed by taking the log of both sides to get $\log(a) = \log(y) + t\log(x)$.

The haploid model of natural selection makes a noteworthy prediction in the case of multiple beneficial mutations. A phenomenon called **clonal interference** is predicted in the absence of recombination when two or more haplotypes in the population are beneficial (Muller 1932). Multiple beneficial haplotypes will compete with each other, and, eventually, only the single haplotype with the highest fitness will reach fixation while other beneficial haplotypes will go to loss. Clonal interference can be understood by considering an example where both λ_A and λ_B are greater than one, such as the haplotypes shown in Figure 6.1. While the populations of both the A and B haplotype individuals will increase in size over time, their relative fitness values predict the change in their relative frequencies and that only one haplotype will eventually reach fixation. Clonal interference predicts that haploid natural selection will not be able to maintain and fix numerous beneficial mutations if they segregate in a population simultaneously. Empirical evidence for clonal interference among beneficial mutations has been observed in organisms without recombination such as bacteria (Imhof and Schlötterer 2001) and yeast (Lang et al. 2013). This demonstrates that when recombination is absent, the preexisting fitness of a haplotype can influence the fate of new beneficial mutations, as can the relative fitness of other haplotypes found in the population.

It is also worth noting that if both λ_A and λ_B are less than one, the populations of both the A and B haplotype individuals will eventually go to extinction. Yet, the haplotype with the greater relative fitness will indeed increase in frequency over time as the populations of both haplotypes shrink because its rate of decrease is less than that of other haplotypes. Because relative fitness predicts future frequencies of haplotypes and not their absolute population sizes, relative fitness is not an appropriate metric to study the relationship of natural selection and population extinction.

Natural selection with sexual reproduction

The model of natural selection with clonal reproduction leaves out a critical part of the biology of many organisms, namely sexual reproduction. To build a model of natural selection for sexual reproduction, we can combine the Hardy–Weinberg model of genotype frequencies with genotype-specific growth rates to get a general model of natural selection operating on the three genotypes produced by a single locus with two alleles. The blending of these two models leads to a number of assumptions that are listed in Table 6.2 (compare with the assumptions of Hardy–Weinberg alone given in Chapter 2). For now, let us utilize the assumptions that yield the expected genotype frequencies. The consequences of many of the other assumptions are explored throughout the chapter.

Imagine a population of N diploid individuals formed by random mating among the parents and then random fusion of gametes to produce zygotes. When the N zygotes have just formed, before any natural selection, the genotypes are in Hardy–Weinberg expected frequencies. If the total population size at this time is N_t, then the number of zygotes of each genotype is

---

Table 6.2 Assumptions of the basic natural selection model with a diallelic locus.

Genetic
- Diploid individuals
- One locus with two alleles
- Obligate sexual reproduction

Reproduction
- Generations do not overlap
- Mating is random

Natural selection
- Mechanism of natural selection is genotype-specific differences in survivorship (fitness) that lead to variable genotype-specific growth rates, termed viability selection
- Fitness values are constants that do not vary with time, over space or in the two sexes

Population
- Infinite population size so there is no genetic drift
- No population structure
- No gene flow
- No mutation

$$AA : p^2 N_t \quad Aa : 2pqN_t \quad Aa : q^2 N_t \quad (6.11)$$

which defines the initial number of each of the three genotypes analogously to $N_A(0)$ and $N_B(0)$ used in the case of clonal reproduction.

After the initial population of zygotes is formed, natural selection will then operate on the three genotypes. The mechanism of natural selection takes a particular form under the assumptions of the one locus selection model. Each genotype is assumed to experience genotype-specific survival and reproduction during the course of a single generation as diagrammed in Figure 6.3. This leads to a possible reduction in the number of zygotes of each genotype present in the population at the very beginning of the life cycle of a single generation. For the time being, let us assume that any reduction in the numbers of individuals of any genotype comes exclusively from failure to survive to reproductive age but that all adults reproduce equally regardless of genotype. In this situation, the fitness values of each genotype specify the probability of survival to reproduction, termed **viability**. Natural selection then takes the form of **viability selection**.

> **Viability selection:** A form of natural selection where fitness is equivalent to the probability that individuals of given genotype survive to reproductive age but all surviving individuals have equal rates of reproduction.
> **Marginal fitness:** The frequency-weighted and allele-copy-weighted sum of the relative fitness values of genotypes that contain a specific allele; a special case of the average fitness for only those genotypes that contain a certain allele.

As an analog of λ used for clonal reproduction, let ℓ (the cursive letter l) represent the genotype-specific probability of survival to reproductive age. The numbers of individuals of each genotype *after* viability selection at the point of reproduction is then

$$AA : \ell_{AA}p^2 N_t \quad Aa : \ell_{Aa}2pqN_t \quad Aa : \ell_{aa}q^2 N_t$$
$$(6.12)$$

These are the numbers of individuals of each genotype that will engage in random mating to form the next generation. The total number of individuals in the population after selection is then

$$\ell_{AA}p^2 N_t + \ell_{Aa}2pqN_t + \ell_{aa}q^2 N_t \quad (6.13)$$

This is a quantity that can be used to determine the frequency of a genotype or allele in the population after selection. For example, the proportion of the total population made up of individuals with an AA genotype after selection is

frequency of AA genotype
$$= \frac{\ell_{AA}p^2 N_t}{\ell_{AA}p^2 N_t + \ell_{Aa}2pqN_t + \ell_{aa}q^2 N_t} \quad (6.14)$$

Since there are fewer alleles than genotypes, the results of natural selection are often summarized in terms of allele frequencies rather than in terms of genotype frequencies. The allele frequencies in the gametes made by surviving individuals (under the assumptions in Table 6.2) are

Frequency of A allele in gametes

$$= \frac{\ell_{AA}p^2 N_t + \frac{1}{2}(\ell_{Aa}2pqN_t)}{\ell_{AA}p^2 N_t + \ell_{Aa}2pqN_t + \ell_{aa}q^2 N_t} \quad (6.15)$$

and

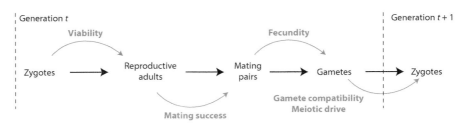

Figure 6.3 A diagram of the life cycle of organisms showing some points where differential survival and reproduction among genotypes can result in natural selection. Viability is the probability of survival from zygote to adult, mating success encompasses those traits influencing the chances of mating and the number of mates, and fecundity is the number of gametes and progeny zygotes produced by each mating pair. Gametic compatibility is the probability that gametes can successfully fuse to form a zygote, while meiotic drive is any mechanism that causes bias in the frequency of alleles found in gametes. Most models of natural selection assume a single fitness component such as viability. In reality, all of these components of fitness can influence genotype frequencies simultaneously.

Frequency of a allele in gametes

$$= \frac{\ell_{aa}q^2N_t + \frac{1}{2}(\ell_{Aa}2pqN_t)}{\ell_{AA}p^2N_t + \ell_{Aa}2pqN_t + \ell_{aa}q^2N_t} \qquad (6.16)$$

In each of these equations, the number of heterozygote individuals after selection is multiplied by one-half since a heterozygote contributes one copy of a given allele to the gamete pool for every two copies of that same allele contributed by a homozygote. These expressions simplify to

Frequency of A allele in gametes

$$= \frac{\ell_{AA}p^2 + \ell_{Aa}pq}{\ell_{AA}p^2 + \ell_{Aa}2pq + \ell_{aa}q^2} \qquad (6.17)$$

and

Frequency of a allele in gametes

$$= \frac{\ell_{aa}q^2 + \ell_{Aa}pq}{\ell_{AA}p^2 + \ell_{Aa}2pq + \ell_{aa}q^2} \qquad (6.18)$$

because N_t can be factored out of each term in the numerator and denominator and then canceled, and the constants of ½ and 2 cancel in the numerator.

As in the case of clonal reproduction, we can utilize relative fitness values for each genotype rather than absolute values of survivorship to reproductive age. We can then replace the ℓ_{AA}, ℓ_{Aa}, and ℓ_{aa} values with the relative fitness values w_{AA}, w_{Aa}, and w_{aa} to give

$$p_{t+1} = \frac{w_{AA}p^2 + w_{Aa}pq}{w_{AA}p^2 + w_{Aa}2pq + w_{aa}q^2} \qquad (6.19)$$

and

$$q_{t+1} = \frac{w_{aa}q^2 + w_{Aa}pq}{w_{AA}p^2 + w_{Aa}2pq + w_{aa}q^2} \qquad (6.20)$$

Notice that the denominator in both expressions is the sum of the fitness-weighted genotype frequencies, or the mean relative fitness $\overline{w}$. Making this substitution gives even more compact expressions for the allele frequencies:

$$p_{t+1} = \frac{w_{AA}p^2 + w_{Aa}pq}{\overline{w}} \qquad (6.21)$$

and

$$q_{t+1} = \frac{w_{aa}q^2 + w_{Aa}pq}{\overline{w}} \qquad (6.22)$$

These expressions show that the increase or decrease in allele frequencies depends on a comparison of the average fitness of genotypes that contain a certain allele (the quantity in the numerator), called the **marginal fitness**, and the average fitness of all genotypes in the population. A larger marginal fitness occurs when individuals with genotypes that make up the marginal fitness have higher viability for a given allele frequency. Table 6.3 summarizes the key quantities used to construct expected genotype and allele frequencies after one generation of viability selection.

As in the case of clonal reproduction, the change in allele frequency over one generation is given by $\Delta p = p_{t+1} - p_t$. For sexual reproduction,

$$\Delta p = \frac{pq[p(w_{AA} - w_{Aa}) + q(w_{Aa} - w_{aa})]}{\overline{w}} \qquad (6.23)$$

and

$$\Delta q = \frac{pq[q(w_{aa} - w_{Aa}) - p(w_{AA} - w_{Aa})]}{\overline{w}} \qquad (6.24)$$

as derived in Math Box 6.1. This equation provides three generalizations that match with our intuitions about natural selection. Allele frequencies do not change when $pq = 0$ or when there is no genetic variation since one allele or the other has reached loss (p or $q = 0$). Allele frequencies do not change when all fitness values are identical − meaning there is no natural selection − so the terms inside the square brackets give a value of zero. Lastly, allele frequencies do not change when the fitness differences weighted by an allele frequency (the $p(w_{AA} - w_{Aa})$ and $q(w_{Aa} - w_{aa})$ terms) cancel each other out to yield a 0 inside the square brackets.

One last point to note is that genotype frequencies (e.g. P_{AA}, P_{Aa}, and P_{aa}) are often used in the expressions above rather than the p^2, $2pq$, and q^2 expected genotype frequencies of Hardy–Weinberg. This is because natural selection over more than one generation can cause genotype frequencies to deviate from those expected by Hardy–Weinberg.

Table 6.3 The expected frequencies of three genotypes after natural selection for a diallelic locus with sexual reproduction and random mating. The absolute fitness of the A genotype is used as the standard of comparison when determining relative fitness.

	Genotype		
	AA	**Aa**	**aa**
Generation t			
Initial frequency	p_t^2	$2p_tq_t$	q_t^2
Genotype-specific survival (absolute fitness)	ℓ_{AA}	ℓ_{Aa}	ℓ_{aa}
Relative fitness	$w_{AA}=\dfrac{\ell_{AA}}{\ell_{AA}}$	$w_{Aa}=\dfrac{\ell_{Aa}}{\ell_{AA}}$	$w_{aa}=\dfrac{\ell_{aa}}{\ell_{AA}}$
Frequency after natural selection	$p_t^2 w_{AA}$	$2p_tq_tw_{Aa}$	$q_t^2 w_{aa}$
Average fitness	$\overline{w}=p_t^2 w_{AA}+2p_tq_tw_{Aa}+q_t^2 w_{aa}$		
Generation $t+1$			
Genotype frequency	$\dfrac{p_t^2 w_{AA}}{\overline{w}}$	$\dfrac{2p_tq_tw_{Aa}}{\overline{w}}$	$\dfrac{q_t^2 w_{aa}}{\overline{w}}$
Allele frequency	$p_{t+1}=\dfrac{p_t\left(p_tw_{AA}+q_tw_{Aa}\right)}{\overline{w}}$	$q_{t+1}=\dfrac{q_t\left(q_tw_{aa}+p_tw_{Aa}\right)}{\overline{w}}$	
Change in allele frequency	$\Delta p=\dfrac{pq[p(w_{AA}-w_{Aa})+q(w_{Aa}-w_{aa})]}{\overline{w}}$	$\Delta q=\dfrac{pq[q(w_{aa}-w_{Aa})+p(w_{Aa}-w_{AA})]}{\overline{w}}$	

Math box 6.1
The change in allele frequency each generation under natural selection

To solve the equation for the change in allele frequency over one generation due to natural selection, start with

$$\Delta p = \frac{p^2 w_{AA} + pqw_{Aa}}{p^2 w_{AA} + 2pqw_{Aa} + q^2 w_{aa}} - p \tag{6.25}$$

where the allele frequencies p and q are all taken in the same generation so the generation subscripts are dropped. First, put both of the terms over a common denominator so they can be subtracted:

$$\Delta p = \frac{p^2 w_{AA} + pqw_{Aa}}{p^2 w_{AA} + 2pqw_{Aa} + q^2 w_{aa}} - \frac{p(p^2 w_{AA} + 2pqw_{Aa} + q^2 w_{aa})}{p^2 w_{AA} + 2pqw_{Aa} + q^2 w_{aa}} \tag{6.26}$$

Then, note that a factor of p can be taken from the numerator of the left-hand term of the difference

$$\Delta p = \frac{p(pw_{AA} + qw_{Aa})}{p^2 w_{AA} + 2pqw_{Aa} + q^2 w_{aa}} - \frac{p(p^2 w_{AA} + 2pqw_{Aa} + q^2 w_{aa})}{p^2 w_{AA} + 2pqw_{Aa} + q^2 w_{aa}} \tag{6.27}$$

which leads to the following (the denominator is hereafter given as $\overline{w}$ for simplicity):

$$\Delta p = \frac{p[pw_{AA} + qw_{Aa} - p^2 w_{AA} - 2pqw_{Aa} - q^2 w_{aa}]}{\overline{w}} \tag{6.28}$$

A trick comes in at this point utilizing the fact that $p = 1 - q$ with a diallelic locus, so that $pq = p(1 - p) = p - p^2$. The first and third terms inside the square brackets of the numerator of Eq. 6.28 ($pw_{AA} - p^2 w_{AA}$) can be expressed alternatively as pqw_{AA}. This then gives

$$\Delta p = \frac{p[pqw_{AA} + qw_{Aa} - 2pqw_{Aa} - q^2 w_{aa}]}{\overline{w}} \tag{6.29}$$

A q can then be factored out of the terms inside the square brackets of the numerator to give

$$\Delta p = \frac{pq[pw_{AA} + w_{Aa} - 2pw_{Aa} - qw_{aa}]}{\overline{w}} \tag{6.30}$$

Then, note the middle two terms inside the square brackets ($w_{Aa} - 2pw_{Aa}$). Since $p + q = 1$, $w_{Aa} - 2pw_{Aa} = (p + q)w_{Aa} - 2pw_{Aa} = qw_{Aa} - pw_{Aa}$. Making this substitution leads to

$$\Delta p = \frac{pq[pw_{AA} + qw_{Aa} - pw_{Aa} - qw_{aa}]}{\overline{w}} \tag{6.31}$$

which can finally be rearranged to

$$\Delta p = \frac{pq[p(w_{AA} - w_{Aa}) + q(w_{Aa} - w_{aa})]}{\overline{w}} \tag{6.32}$$

The same approach can be taken to obtain the expression for Δq.

6.2 General results for natural selection on a diallelic locus

* Selection against a recessive phenotype.
* Selection against a dominant phenotype.
* The general effects of dominance.
* Heterozygote disadvantage and advantage.
* The strength of natural selection.

The previous section presented the basic building blocks of a model for natural selection acting through genotype-specific viability on one locus with two alleles. This section will present the general results of natural selection under this very basic model. This task is simpler than it might seem since all the outcomes of the selection model can be represented by five general categories of fitness values for the three genotypes (Table 6.4). Note that Table 6.4 presents fitness values in terms of **selection coefficients** rather than relative fitness. Selection coefficients are simply the difference between a relative fitness value and 1:

$$s_{xx} = 1 - w_{xx} \text{ or } w_{xx} = 1 - s_{xx} \qquad (6.33)$$

where the subscript xx represents a genotype and the maximum relative fitness is 1. Selection coefficients, therefore, represent the difference in viability for a given genotype and the genotype with the highest viability.

Examining the outcome of selection for each category of fitness values or selection coefficients will illustrate how viability selection is expected to change genotype and allele frequencies in populations. By iterating versions of Eq. 6.14 for all three genotypes as well as Eqs. 6.21 and 6.22, we can visualize the action of natural selection. The behavior of allele frequencies under natural selection can be understood by examining plots of allele frequencies over time to see the direction and rate of allele frequency change. An important general feature of natural selection is the allele frequency reached when allele frequencies eventually stop changing, or the equilibrium allele frequency. The goal of this section is to understand both how and why genotype and allele frequencies change when acted on by a constant force of natural selection over time.

Although it is common to speak of an *allele* favored by natural selection, any change in allele frequencies is really caused by natural selection on *genotypes* due to their different-viability phenotypes. Alleles themselves do not have phenotypes nor fitness values in most types of natural selection (natural selection on gametes or haploid genomes are exceptions). The changing frequency of genotypes is what causes allele frequencies to change. Although two allele frequencies can be displayed with more economy than three genotype frequencies, it is critical not to forget that natural selection directly causes changes in genotype frequency and that change in allele frequencies is an indirect consequence.

The process of natural selection has the special quality that the genotype frequencies reached at equilibrium are always the same as long as the starting frequencies and relative fitness values are constant. Processes that always lead to the same outcome from a given set of initial conditions are called **deterministic** because the end state is completely determined by the initial state. Similar patterns of genotype frequencies in independent

---

Table 6.4 The general categories of relative fitness values for viability selection at a diallelic locus. The variables s and t are used to represent the decrease in viability of a genotype compared to the maximum fitness of one $(1-w_{xx} = s)$. The degree of dominance of the A allele is represented by h with additive gene action (sometimes called "codominance") when $h = 1/2$.

Category	Genotype-specific fitness		
	w_{AA}	w_{Aa}	w_{aa}
Selection against recessive	1	1	$1-s$
Selection against a dominant	$1-s$	$1-s$	1
General dominance (dominance coefficient $0 \leq h \leq 1$)	1	$1-hs$	$1-s$
Heterozygote disadvantage (underdominance for fitness)	1	$1-s$	1
Heterozygote advantage (overdominance for fitness)	$1-s$	1	$1-t$

populations are, therefore, evidence that the process of natural selection is operating. In contrast, the stochastic process of genetic drift would result in random outcomes in each independent population. This also means that there is no need to view replicate outcomes of natural selection for the same set of initial conditions.

Selection against a recessive phenotype

The results of natural selection acting against a completely recessive homozygous genotype (see Table 6.4) are shown in Figure 6.4. The top panel shows the frequencies of the three genotypes over time starting from an initial allele frequency of $p = q = 0.5$. The frequency of the recessive homozygote (aa) declines because that genotype has lower

viability. At the same time, the frequency of the dominant homozygote (AA) increases since it has a higher viability. Even though the heterozygote also has the maximum fitness, its frequency declines from a maximum of 0.5 as A alleles become more frequent and a alleles less frequent over time, reducing the value of $2pq$. The bottom panel summarizes the results of natural selection in terms of allele frequencies over time for five initial allele frequencies. (The one allele frequency trajectory that corresponds to the genotype frequencies in the top panel is given as a colored, dashed line.)

Allele frequencies change more rapidly in the early generations when the initial allele frequency is lower because the selectively favored dominant homozygote and heterozygotes are relatively frequent in the population. Even at an initial dominant allele frequency of 0.05, 9.75% (or $1 - q^2$) of the

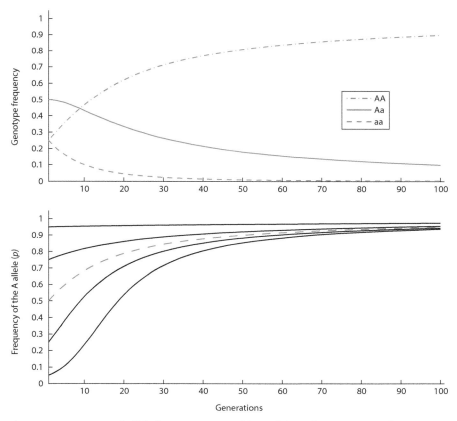

Figure 6.4 The change in genotype and allele frequencies caused by viability selection against the aa genotype exhibiting the recessive phenotype. The top panel shows the change in genotype frequencies over time, while the bottom panel shows the frequency of the dominant allele (A) over time. The colored, dashed line in the bottom panel corresponds to the allele frequencies in the top panel. Because of changes in genotype frequency caused by natural selection, the frequency of the dominant allele rapidly approaches fixation from all five initial allele frequencies. In this illustration, $w_{AA} = w_{Aa} = 1.0$ while $w_{aa} = 0.8$, meaning that 8 individuals with the aa genotype are expected to survive to reproduce for every 10 individuals with the AA or Aa genotype that survive to reproduce each generation. Genotype frequencies assume random mating.

genotypes are AA and Aa. As the frequency of the recessive allele decreases (the dominant allele approaches higher frequencies), the change in allele frequency from one generation to the next steadily declines. For example, starting from an initial allele frequency of 0.05, the allele frequency changes by 0.1 in just a few generations early on but requires many generations when the frequency of the recessive allele is low. This occurs because there are progressively fewer recessive homozygotes and progressively more of the highest fitness genotypes (the dominant homozygote and heterozygotes) in the population, as selection changes the genotype and allele frequencies.

Does the dominant allele go to fixation when there is natural selection against the recessive homozygote? The answer is no, because the heterozygote fitness is equal to the maximum fitness, and every generation heterozygotes will produce gametes that can combine to make the recessive homozygote. In essence, the recessive allele is shielded from natural selection in the heterozygote due to dominance. This is true no matter how large the selection coefficient is against a recessive homozygote.

One way to quantify the sheltering effect of heterozygotes is to examine the proportion of recessive alleles present in heterozygotes compared to recessive alleles present in homozygotes

$$\frac{pq}{q^2} = \frac{p}{q} \qquad (6.34)$$

where the expected frequency of heterozygotes is weighted by one-half since each contains only one recessive allele. When the frequency of the recessive allele is low, $q = 0.05$, for example, the proportion of the genotype frequencies is $0.0475/0.0025 = 19$. This means that there are 19 recessive alleles protected against natural selection in heterozygotes for each recessive allele impacted by natural selection in a homozygous genotype.

Selection against a dominant phenotype

The results of natural selection acting against a completely dominant phenotype shared by the dominant homozygotes and the heterozygotes (see Table 6.4) are shown in Figure 6.5. The top panel shows the

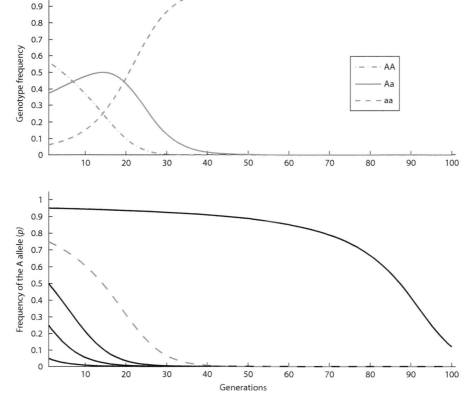

Figure 6.5 The change in the genotype and allele frequency of a completely dominant allele (A) when natural selection acts against the AA and Aa genotypes exhibiting the dominant phenotype. Notice that the frequency of the A allele decreases slowly at first when the A allele is common in the population since the aa genotype is infrequent. The colored, dashed line in the bottom panel corresponds to the allele frequencies in the top panel. In this illustration, $w_{AA} = w_{Aa} = 0.8$ while $w_{aa} = 1.0$. Genotype frequencies assume random mating.

frequencies of the three genotypes over time starting from an initial allele frequency of $p = 0.75$. The frequency of the dominant homozygote (AA) declines due to its lower viability, while the frequency of the recessive homozygote (aa) increases due to its higher viability. Even though the heterozygote has a lower relative fitness than the recessive homozygote, its frequency initially increases since the frequency of the two alleles approaches equality. The frequency of the heterozygote temporarily peaks at the maximum value of $2pq = 0.5$ at the same point that the frequency of the two homozygotes both equal 0.25. The heterozygote frequency then drops again as the frequency of the recessive homozygote continues to increase and the frequency of the dominant homozygote continues to decrease.

The bottom panel of Figure 6.5 shows that the frequency of the dominant allele decreases toward 0 under this type of natural selection. (Again, the one allele frequency trajectory that corresponds to the genotype frequencies in the top panel is given as a colored, dashed line.) At an initial dominant allele frequency of $p = 0.95$, only 0.25% (or q^2) of the genotypes are aa. This makes natural selection slow to change allele frequencies until the frequency of the recessive allele increases enough to make the higher fitness aa genotype more common in the population. The allele frequency trajectories that start at lower initial frequencies for the A allele change more rapidly and bear out this point. Does the recessive allele go to fixation when there is natural selection against the dominant homozygote and heterozygote? In this case, yes, since both the dominant homozygote and the heterozygote have a lower fitness than the favored homozygote, and, therefore, the dominant allele is not shielded from natural selection in the heterozygote.

General dominance

The previous two examples of natural selection against dominant and recessive phenotypes cover the extremes of dominance. The impact of dominant and recessive alleles on the outcome of natural selection on a diallelic locus can be made more general by employing a dominance coefficient, symbolized h. Complete dominance (the heterozygote and a homozygote having identical phenotypes) for one allele is represented by $h = 0$, and complete dominance for the other allele is represented by $h = 1$. When the heterozygote has a phenotype that is the average of the two homozygotes, then $h = \frac{1}{2}$, a situation sometimes called codominance. A dominance

coefficient of $h = \frac{1}{2}$ is more descriptively referred to as **additive gene action** since the phenotype of the heterozygote is the sum of the phenotypic effects of each allele. For example, if phenotypes are AA = 3 spots, Aa = 2 spots, and aa = 1 spot, an A allele contributes 1.5 spots and an a allele contributes 0.5 spots in the heterozygote when gene action is additive. Look at Table 6.4 and verify the fitness of the heterozygote when $h = 0$, 1, and 1/2. This method to specify fitness has the advantage that the results of natural selection can be predicted for any degree of dominance. There is also a strong biological motivation, since alleles commonly show a wide range of dominance or gene action in actual populations, ranging between being completely dominant or completely recessive.

The outcome of selection for three cases of gene action are shown in Figure 6.6. All three cases start at the same initial allele frequency and share the same selection coefficient. However, gene action varies from completely dominant to completely recessive with the additive case in between. The results

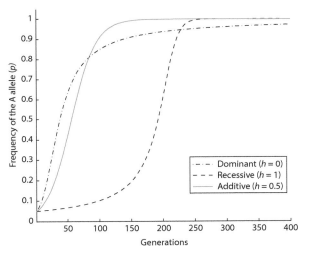

Figure 6.6 Allele frequencies over time for three types of gene action with a low initial allele frequency. In all three cases, the equilibrium allele frequency is fixation or near fixation for the A allele. With complete dominance, natural selection initially increases the allele frequency very rapidly. The approach to fixation for the A allele slows as aa homozygotes become rare since heterozygotes harbor a alleles that are concealed from natural selection by dominance. Natural selection initially changes the frequency of a recessive allele very slowly since homozygote recessive genotypes are very rare. As the recessive homozygotes become more common, allele frequency increases more rapidly. With additive gene action, the phenotype of the heterozygote is intermediate between the two homozygotes so all genotypes differ in their viability. Additive gene action has the most rapid overall approach to equilibrium allele frequency. The degree of dominance is represented by the dominance coefficient, h. In this illustration, the selection coefficient is $s = 0.1$.

of natural selection on a completely dominant allele (rapid change in allele frequency initially but never reaching fixation) and on a completely recessive allele (slow initial change in allele frequency then more rapid change and eventual fixation) are identical to the dynamics seen in earlier examples. The allele frequency trajectory for additive gene action is intermediate. It combines the rapid initial change in allele frequency of complete dominance with the later-stage rapid approach to the equilibrium and fixation of the complete recessive. Equilibrium allele frequency (fixation or near fixation) is reached most quickly with additive gene action.

With completely dominant or recessive alleles, natural selection cannot discriminate between two of the three genotypes since their fitness values are identical. How this lack of difference in fitness values affects natural selection depends on genotype frequencies. In the early generations, the recessive case shows slow change because the heterozygote is selected against and the fittest genotype is rare. In the later generations of the dominance case, heterozygotes shelter the recessive allele from natural selection slowing further change in allele frequency as the recessive homozygote becomes very infrequent. In contrast, the fitness values of all three genotypes are distinct and uniformly different with additive gene action. Additive gene action gives the

maximum difference in marginal and average fitness values across the entire range of possible genotype frequencies under random mating.

Gene action is an important factor in understanding the fate of new mutations acted on by natural selection. Imagine a new mutation in a population that has a high relative fitness when homozygous. As covered in Chapter 5, the initial frequency of any new mutation will be low ($\frac{1}{2N_e}$). A completely or nearly recessive mutation will take a very long time to increase in frequency under natural selection. In contrast, a completely or nearly dominant mutation with the same fitness as a homozygote and starting at the same frequency will increase in frequency very rapidly. The examples in Figure 6.6 where the initial frequency of the A allele is 0.05 are equivalent to a new mutation in a population of $N_e = 10$.

Heterozygote disadvantage

Natural selection when the heterozygote genotype has the lowest relative fitness is known as diversifying selection, or **disruptive selection** (see Table 6.4), are shown in Figure 6.7. Starting from an initial allele frequency of $p = 0.4$, the top panel shows how the aa homozygote eventually reaches

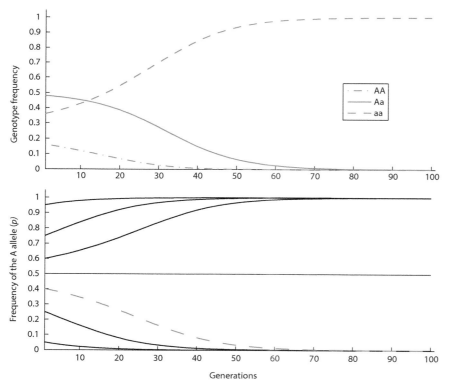

Figure 6.7 The change in the genotype and allele frequency when there is underdominance for fitness and natural selection acts against individuals with Aa genotypes. The equilibrium allele frequency depends on the initial allele frequency. Starting below 0.5 populations head toward loss while starting above 0.5 populations go to fixation. There is an unstable equilibrium at an initial allele frequency of exactly 0.5. From any initial allele frequency, the population converges on a minimum frequency of heterozygotes. The colored, dashed line in the bottom panel corresponds to the allele frequencies in the top panel. In this illustration, $w_{AA} = w_{aa} = 1.0$ and $w_{Aa} = 0.9$. Genotype frequencies assume random mating.

fixation over time. The bottom panel requires close attention in this case, since the equilibrium allele frequency depends strongly on the initial allele frequency in the population. Initial allele frequencies above $p = 0.5$ all lead to the fixation of the AA homozygote, while all initial allele frequencies below $p = 0.5$ lead to fixation of the aa homozygote. When the initial allele frequency in the population is exactly $p = 0.5$, allele frequencies remain constant over time. It turns out that this equilibrium point is not robust to any change in allele frequency, and, so, is called an **unstable** equilibrium. Any slight change in the allele frequencies will result in them changing to alternative stable equilibrium points of fixation or loss. Such an unstable equilibrium is very unlikely to persist in a finite population, since even a slight amount of genetic drift would alter allele frequencies in the population toward one of the stable equilibrium points.

Heterozygote advantage

The results of natural selection acting to increase the frequency of the heterozygous genotype, commonly referred to as heterozygote advantage, overdominance for fitness, or balancing selection (see Table 6.4), are shown in Figure 6.8. The top panel shows the

frequencies of the three genotypes over time starting from an initial allele frequency of $p = 0.05$. The heterozygous genotype increases in frequency due to its higher relative fitness. At the same time, the aa homozygote (initially 90% of the population) declines due to its lower viability. Although the relative fitness of the AA homozygote is lower than that of the heterozygote, its frequency increases toward 25% as allele frequencies approach $p = q = 0.5$ due to the increasing frequency of heterozygotes. The bottom panel shows that for all initial allele frequencies, natural selection causes the population to approach $p = q = 0.5$.

Overdominance for fitness represents a unique exception for the outcome of natural selection on a diallelic locus. Selection against a dominant phenotype results in fixation of the recessive allele and loss of the dominant allele. Similarly, selection against a recessive phenotype results in near fixation of the dominant allele and near loss of the recessive allele. Selection against a heterozygote also results ultimately in fixation of one allele and loss of the other allele. These three forms of natural selection all produce an equilibrium with little or no genetic variation, known as a **monomorphic equilibrium**. In contrast, when heterozygotes have the highest fitness, natural selection maintains both alleles in the population at equilibrium, resulting

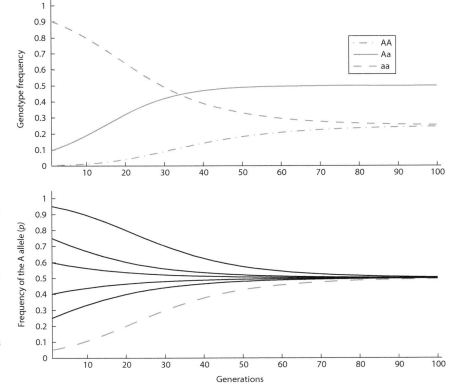

Figure 6.8 The change in the genotype and allele frequency when there is overdominance for fitness and natural selection favors individuals with Aa genotypes. From any initial allele frequency, the population converges on a maximum frequency of heterozygotes. This corresponds to equal allele frequencies with random mating. The colored, dashed line in the bottom panel corresponds to the allele frequencies in the top panel. In this illustration, $w_{AA} = w_{aa} = 0.9$ and $w_{Aa} = 1.0$. Genotype frequencies assume random mating.

in a **polymorphic equilibrium**. Thus, overdominance for fitness is one type of natural selection that is consistent with the permanent maintenance of genetic variation in populations.

The allele frequencies expected at equilibrium with overdominance can be obtained from Eq. 6.23 as shown in Math Box 6.2. The equilibrium allele frequencies are

$$p_{equilibrium} = \frac{t}{s + t} \qquad (6.35)$$

and

$$q_{equilibrium} = \frac{s}{s + t} \qquad (6.36)$$

where s and t are the selection coefficients against the AA and aa homozygotes, respectively (see Table 6.4). The equilibrium allele frequency is higher for the allele in the homozygous genotype that has the smaller selection coefficient (or higher relative fitness).

Math box 6.2 Equilibrium allele frequency with overdominance

By definition, equilibrium allele frequencies are reached when allele frequencies stop changing from one generation to the next. This means that Δp as expressed by

$$\Delta p = \frac{pq[p(w_{AA} - w_{Aa}) + q(w_{Aa} - w_{aa})]}{\overline{w}} \qquad (6.37)$$

which was first shown as Eq. 6.23, should be equal to 0.

Two equilibrium points occur when $p = 0$ or $q = 0$, biologically equivalent to situations where there is no genetic variation in a population. When there is genetic variation (both $p \neq 0$ and $q \neq 0$), the equilibrium point depends on the fitness differences contained in the numerator. Taking the numerator term in square brackets in Eq. 6.37 and setting it equal to 0,

$$p(w_{AA} - w_{Aa}) + q(w_{Aa} - w_{aa}) = 0 \qquad (6.38)$$

and then solving p or q in terms of relative fitness values, will give allele frequencies where Δq is 0. The first step is to substitute $q = 1 - p$:

$$p(w_{AA} - w_{Aa}) + (1 - p)(w_{Aa} - w_{aa}) = 0 \qquad (6.39)$$

and then expand by multiplying the terms

$$pw_{AA} - pw_{Aa} + w_{Aa} - w_{aa} - pw_{Aa} + pw_{aa} = 0 \qquad (6.40)$$

The relative fitness values that are multiplied by p can be brought together

$$p(w_{AA} - 2w_{Aa} + w_{aa}) + w_{Aa} - w_{aa} = 0 \qquad (6.41)$$

and then subtracted

$$w_{Aa} - w_{aa} = -p(w_{AA} - 2w_{Aa} + w_{aa}) \qquad (6.42)$$

Dividing both sides by $-(w_{AA} - 2w_{Aa} + w_{aa})$ gives

$$p = \frac{w_{Aa} - w_{aa}}{2w_{Aa} - w_{AA} - w_{aa}} \qquad (6.43)$$

which expresses p as a function of relative fitness values alone. Substituting the relative fitness values of $w_{AA} = 1 - s$, $w_{Aa} = 1$, and $w_{aa} = 1 - t$ as given in Table 6.4:

$$p = \frac{1 - (1 - t)}{2(1) - (1 - s) - (1 - t)} \tag{6.44}$$

and then carrying out the addition and subtraction then gives the equilibrium allele frequency in terms of selection coefficients for the two homozygous genotypes

$$p = \frac{t}{s + t} \tag{6.45}$$

The strength of natural selection

The strength of selection against a genotype can vary from weak, such as a viability 0.1% less than the most fit genotype, to very strong, such as 50% viability or even zero viability (lethality) of a genotype. Allele frequencies over time (starting from the same initial allele frequency) are plotted in Figure 6.9 for a wide range of selection coefficients in the case of natural selection against a homozygous

recessive genotype. Notice that the shape of the curves in the top and bottom panels of Figure 6.9 are very similar but that the time scale of each plot is very different. Selection coefficients of 10% or greater bring the dominant allele to high frequencies within 100 generations. In contrast, reaching these same allele frequencies takes 10 000 generations when the selection coefficient is between 1.0 and 0.1%. This illustrates the general principle that stronger natural selection (larger selection

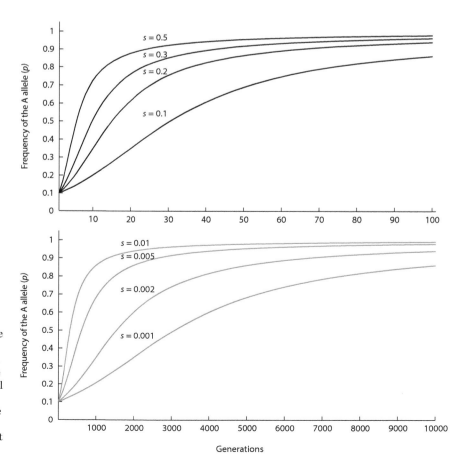

Figure 6.9 The strength of natural selection influences the rate of change in genotype and allele frequencies. In this illustration, selection acts against the recessive homozygote (aa). The top panel shows strong natural selection where viability of the aa genotype is 10–50% less than the other genotypes. The bottom panel shows weak natural selection where viability of the aa genotype is 1–0.1% less than the other genotypes. Note the vastly different time scales in the two plots.

coefficients or larger fitness differences) results in a more rapid approach to equilibrium allele frequencies. This conclusion applies to all of the situations given in Table 6.4 and to the process of natural selection in general.

6.3 How natural selection works to increase average fitness

- Natural selection acts to increase mean fitness.
- The fundamental theorem of natural selection.

In the five relative fitness situations shown in Table 6.4 for natural selection on a diallelic locus, there are always two general outcomes. Directional selection of any type ends in fixation and loss (selection against a dominant phenotype) or nearly fixation and loss (selection against a recessive phenotype). Underdominance too results in fixation

or loss (with one exception unlikely to be realized in finite populations). Overdominance is the exception that maintains both alleles in the population indefinitely. So, the two outcomes are either fixation and loss or intermediate frequencies for both alleles (sometimes called a **balanced polymorphism**). The reason why these two general outcomes occur can be understood by examining the average fitness of a population ($\overline{w}$) as well as the rate of change in allele frequency (Δp) over the entire range of allele frequencies.

Average fitness and rate of change in allele frequency

The mean fitness ($\overline{w}$) over all possible allele frequencies is plotted for each case of natural selection on a diallelic locus in Figures 6.10 and 6.11. For the cases of directional selection in Figure 6.10, note that the

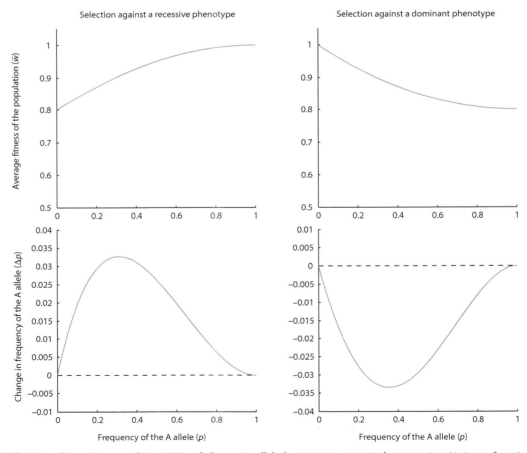

Figure 6.10 Mean fitness in a population ($\overline{w}$) and change in allele frequency over a single generation (Δp) as a function of allele frequency for directional selection. Directional selection reaches allele frequency equilibrium at either fixation or loss, the point of highest mean fitness. Positive values of Δp (above the dashed line) indicate that allele frequency is increasing under selection, while negative values of Δp (below the dashed line) indicate that allele frequency is decreasing under selection. The change in allele frequencies is faster when average fitness changes more rapidly (the slope of $\overline{w}$ is steeper). Here, $w_{AA} = w_{Aa} = 1.0$ and $w_{aa} = 0.8$ for selection against a recessive phenotype and $w_{AA} = w_{Aa} = 0.8$ and $w_{aa} = 1.0$ for selection against a dominant phenotype.

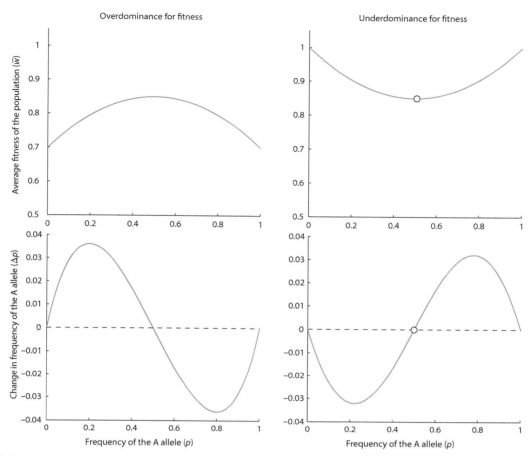

Figure 6.11 Mean fitness in a population ($\overline{w}$) and change in allele frequency over a single generation (Δp) as a function of allele frequency for balancing and disruptive selection. Natural selection changes allele frequencies to increase the average fitness in each generation, eventually reaching an equilibrium when the mean fitness is highest. The change in allele frequencies is faster when average fitness changes more rapidly (the slope of $\overline{w}$ is steeper). The dashed line in the plots of Δp by p shows where allele frequencies stop changing ($\Delta p = 0$) and thus are allele frequency equilibrium points. With underdominance for fitness, Δp is zero when $p = 0.5$, and, so, defines an equilibrium point marked by the circle. However, this equilibrium point is unstable since Δp on either side of $p = 0.5$ changes allele frequencies *away* from the equilibrium point (below $p = 0.5$ Δp is negative leading toward loss and above $p = 0.5$ Δp is positive leading toward fixation). In contrast, with overdominance, Δp on either side of $p = 0.5$ changes allele frequencies *toward* the equilibrium point (below $p = 0.5$ Δp is positive and above $p = 0.5$ Δp is negative), and, thus, $p = 0.5$ is a stable equilibrium point. Here, $w_{AA} = w_{aa} = 1.0$ and $w_{Aa} = 0.7$ for underdominance and $w_{AA} = w_{aa} = 0.7$ and $w_{Aa} = 1.0$ for overdominance.

highest mean fitness corresponds exactly to fixation of the A allele for selection against a recessive phenotype and to loss of the A allele for selection against a dominant phenotype. This same pattern is evident in Figure 6.11 where the highest mean fitness is found at an intermediate allele frequency for overdominance or at fixation or loss for underdominance. These plots of mean fitness by allele frequency show that natural selection acts to increase the mean fitness of the population to its maximum. It is the maximum mean fitness in a population that really defines the equilibrium points for genotype and allele frequencies. The plots of $\overline{w}$ against p reveal the

generalization that the process of natural selection acts to increase the population mean fitness every generation if possible and stops when the mean fitness can no longer increase. In this sense, natural selection can be metaphorically likened to a mountain climber who continually works to find the highest point, but who cannot ever go downhill, and will only rest at the summit. Keeping with this metaphor, plots of $\overline{w}$ against p are called **fitness surfaces**, **adaptive landscapes**, or **adaptive topographies** and represent a topographic map of the mountain at any point where our imaginary mountain climber might venture.

Figures 6.10 and 6.11 also show the change in allele frequency over a single generation (Δp) over all possible allele frequencies for each case of natural selection. Plots of Δp against p reveal when allele frequencies are increasing (Δp is positive) or decreasing (Δp is negative) as well as when allele frequencies are changing rapidly (the absolute value of Δp is large) or slowly (the absolute value of Δp is small). When allele frequencies do not change at all (Δp is zero), then an equilibrium allele frequency has been reached. Notice that fixation or loss of the A allele corresponds to $\Delta p = 0$ for directional selection. For overdominance and underdominance, $\Delta p = 0$ for fixation and loss as well as for the intermediate allele frequency of $p = 0.5$. These allele frequencies are, therefore, equilibrium points because natural selection is not causing any change in allele frequency at these specific allele frequencies.

Also, compare each plot of Δp against p to the corresponding plot of $\overline{w}$ against p. There is a striking relationship between Δp and $\overline{w}$. Both the magnitude and sign of Δp correspond exactly to the slope of the $\overline{w}$ line at any value of p. The slope of $\overline{w}$ is always positive just as Δp is always positive for selection against a recessive, while the slope of $\overline{w}$ is always negative just as Δp is always negative for selection against a dominant (Figure 6.10). This same pattern is seen in Figure 6.11 for overdominance and underdominance, where the slope of $\overline{w}$ is 0 at fixation and loss as well as at $p = 0.5$. The slope of $\overline{w}$ explains why the polymorphic equilibrium point for overdominance is stable while that for underdominance is unstable.

With overdominance, if there is any shift of allele frequencies away from $p = 0.5$, say by genetic drift or mutation, natural selection will return the population back to the equilibrium of $p = 0.5$ (Δp is positive for $p < 0.5$ and negative for $p > 0.5$). In contrast, with underdominance, if there is any shift of allele frequencies away from $p = 0.5$, natural selection will change allele frequencies to the maximum mean fitness values found at fixation and loss (Δp is negative for $p < 0.5$ and positive for $p > 0.5$).

Problem box 6.2
Mean fitness and change in allele frequency

Using the Eqs. 6.35 and 6.36 allows us to predict the allele frequencies at equilibrium for selection with overdominance for fitness. We also need to understand *why* the equilibrium is the point at which genotype frequencies stop changing. Let the fitness values be $w_{AA} = 0.9$, $w_{Aa} = 1.0$, and $w_{aa} = 0.8$. First, calculate expected frequency of the A allele at equilibrium or $p_{equilibrium}$. Then, compute Δp and $\overline{w}$ at $p_{equilibrium}$, $p = 0.9$ and $p = 0.2$. How do the values of Δp and $\overline{w}$ at the three allele frequencies compare? Use Δp and $\overline{w}$ to explain why equilibrium allele frequency is between $p = 0.9$ and $p = 0.2$.

Interact box 6.1 Natural selection on one locus with two alleles

Use the text simulation web site to simulate natural selection (in the Simulations menu select Natural Selection Diallelic Locus). Set the initial genotype frequencies for AA and Aa as well as the relative fitness values for all three genotypes. For each set of fitness values, be sure to simulate *at least* four initial allele frequencies to understand how the outcome might depend on initial conditions (like the bottom panels in Figures 6.4, 6.5, 6.7, and 6.8).
Here are some fitness values to use in simulations:

- Weak selection against recessive: $w_{AA} = 1$; $w_{Aa} = 1$; $w_{aa} = 0.9$ ($h = 0$, $s = 0.1$). Compare with selection against recessive lethal: $w_{AA} = 1$; $w_{Aa} = 1$; $w_{aa} = 0.0$ ($h = 0$, $s = 1.0$)
- Weak selection with additive gene action: $w_{AA} = 1$; $w_{Aa} = 0.95$; $w_{aa} = 0.9$ ($h = 0.5$, $s = 0.1$). Compare with strong selection with additive gene action: $w_{AA} = 1$; $w_{Aa} = 0.7$; $w_{aa} = 0.4$ ($h = 0.5$, $s = 0.6$)
- Weak selection with overdominance: $w_{AA} = 0.98$; $w_{Aa} = 1$; $w_{aa} = 0.95$. Compare with strong selection with overdominance: $w_{AA} = 0.2$; $w_{Aa} = 1$; $w_{aa} = 0.4$.
- Selection against the heterozygote: $w_{AA} = 1$; $w_{Aa} = 0.8$; $w_{aa} = 1$. For this case, be sure to examine the plots for several different initial allele frequencies such as 0.2, 0.5, and 0.8.

The fundamental theorem of natural selection

Sir Ronald Fisher (Figure 6.12) proposed the impressive sounding **fundamental theorem of natural selection** in 1930 (Fisher 1999 variorum edition) as a way to summarize and generalize the process of natural selection on a diallelic locus. In Fisher's words, the fundamental theorem of natural selection was that "the rate of increase in fitness of any organism at any time is equal to its genetic variance in fitness at that time." A modern restatement of the theorem is that "the rate of increase in the mean fitness of any organism at any time ascribable to natural selection acting through changes in gene frequencies is exactly equal to its genic variance in

Figure 6.12 Sir Ronald A. Fisher (1890–1963), photographed in 1943, was a pioneer in the theory and practice of statistics. He invented the techniques of analysis of variance and maximum likelihood as well as numerous other statistical tests and methods of experimental design. Fisher's 1930 book *The Genetical Theory of Natural Selection* established a rigorous mathematical framework that coupled Mendelian inheritance and Darwin's qualitative model of natural selection and is one of the foundation works of modern population genetics. Much of Fisher's work stressed the effectiveness of natural selection in changing gene frequencies in infinite, panmictic populations. Source: Anthony WF Edwards, Master and Fellows of Gonville and Caius College, Cambridge.

fitness at that time" (Edwards 1994). The fundamental theorem has also been interpreted as showing that any change in mean fitness caused by natural selection must always be positive. As Crow (2002) and Edwards (2002) recount, this cryptic yet insightful statement about natural selection has led to a great deal of controversy, misunderstanding, and just plain confusion over many years.

One way to illustrate the idea behind the fundamental theorem is to examine natural selection and the change in the average fitness of a population over time. For simplicity, assume that the organisms are entirely haploid and reproduce asexually or clonally and that generations are discrete (these assumptions are not required by the fundamental theorem itself but make the math much simpler). In the haploid case, the average fitness is fitness of each haplotype weighted by its frequency summed over all haplotypes in the population (recall Eq. 6.10). In an equation, the mean fitness is

$$\overline{w} = \sum_{i=1}^{k} (p_i w_i) \qquad (6.46)$$

where k is the total number of haplotypes in the population. Extending the results in Table 6.1 to an arbitrary number of alleles, the frequency of any single haplotype, call it haplotype i, after natural selection is

$$p'_i = \frac{p_i w_i}{\overline{w}} \qquad (6.47)$$

where the prime symbol is used to represent quantities after one generation of natural selection. Based on this haplotype frequency after selection, the average fitness after one generation of selection is then

$$\overline{w}' = \sum_{i=1}^{k} (p'_i w_i) \qquad (6.48)$$

which when substituting in the expression for p'_i in 6.47 gives

$$\overline{w}' = \frac{1}{\overline{w}} \sum_{i=1}^{k} p_i w_i^2 \qquad (6.49)$$

The change in fitness from one generation to the next standardized by the mean fitness in the initial generation is

$$\Delta\overline{w} = \frac{\overline{w}' - \overline{w}}{\overline{w}} \qquad (6.50)$$

which when substituted in the expression for $\overline{w}'$ from Eq. 6.49 gives

$$\Delta\overline{w} = \frac{\frac{1}{\overline{w}}\sum_{i=1}^{k} p_i w_i^2 - \overline{w}}{\overline{w}} \qquad (6.51)$$

an equation that can be rearranged by multiplying by $\frac{1}{\overline{w}}$ rather than dividing by $\overline{w}$ to give

$$\Delta\overline{w} = \frac{1}{\overline{w}}\left[\frac{1}{\overline{w}}\sum_{i=1}^{k} p_i w_i^2 - \overline{w}\right] = \frac{1}{\overline{w}^2}\sum_{i=1}^{k} p_i w_i^2 - \frac{\overline{w}}{\overline{w}} \qquad (6.52)$$

It turns out that the term $\sum_{i=1}^{k} p_i w_i^2 - 1$ is the variance in fitness when the relative fitness values of all the haplotypes are scaled so that $\overline{w} = 1$ (the variance is $\sum (p_i w_i - \overline{w})^2$ which is equivalent to $\sum p_i w_i^2 - \overline{w}^2$). When the relative fitness values of all the haplotypes is scaled so that $\overline{w} = 1$, this leads to

$$\Delta\overline{w} = \mathrm{var}(w) \qquad (6.53)$$

and the conclusion that the change in mean fitness of the population after one generation of natural selection is equal to the variation in fitness. This variation in fitness is really genetic variation in the case of haploids, due to the frequencies of the different haplotypes in the population as well as to the different fitness values of each haplotype. Therefore, the change in fitness under natural selection is equal to the genetic variation in fitness. Further, since a variance can never be negative, the change in mean fitness by natural selection must then be greater than or equal to zero.

The point of Fisher's fundamental theorem can also be shown graphically for a diploid diallelic locus using a de Finetti diagram (introduced in Chapter 2) that also displays the mean fitness of the population (Figure 6.13). To see this, let $2Q$, P, and R represent the frequencies of the genotypes Aa, AA, and aa, respectively. The ratio of the genotype frequencies can be expressed as the square of half the heterozygote frequency divided by the product of the homozygote frequencies or $\lambda = Q^2/PR$. λ is a measure of departure from Hardy–Weinberg genotype

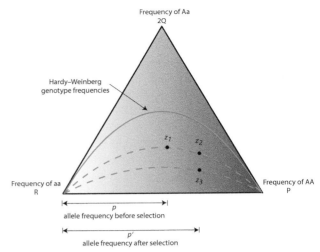

Figure 6.13 A graphical illustration of R.A. Fisher's fundamental theorem of natural selection. The curved lines represent the product of the homozygote frequencies ($P = p^2$ and $R = q^2$) as a constant proportion of the square of the product of the allele frequencies ($Q = pq$) or $\lambda = Q^2/PR$. Hardy–Weinberg genotype frequencies produced by random mating represent the special case of $\lambda = 1$ (solid colored line). Mean fitness is represented by the grayscale gradient with darker tones representing higher mean fitness. In this illustration, genotype frequencies start out at z_1. Suppose that natural selection over one generation changes genotype frequencies to point z_3 (under the conditions that genotype AA has the highest fitness and additive gene action, for example). This change in genotype frequencies can be decomposed into two distinct parts. One part is the change from z_1 to z_2 moving along the curve where λ is constant, but allele frequencies change from p to p'. The other part is the change in the genotype frequencies (changing the value of λ) that occurs by moving vertically on the de Finetti diagram from z_2 to z_3 but keeping allele frequencies constant. The fundamental theorem says that the change in the mean fitness by natural selection is proportional to the change in allele frequency alone. Processes other than natural selection such as mating system dictate the change in genotype frequencies. When natural selection moves the genotype frequencies along a curve of constant λ, then the total change in mean fitness is completely due to changes in allele frequency and genetic variation in fitness is completely additive. Source: modified from Edwards (2002).

frequencies akin to the fixation index F. When genotype frequencies are in Hardy–Weinberg proportions, genotype frequencies are then $2Q = 2pq$, $P = p^2$, $R = q^2$, and $\lambda = 1$. The two dashed lines in Figure 6.13 have values of λ less than 1. Each point of the de Finetti diagram in Figure 6.13 will also represent a value of the mean fitness of the population, depending on the specific values of the relative fitness values of the genotypes. Mean fitness on the de Finetti diagram is represented by the grayscale gradient with darker tones representing higher mean fitness.

The change in mean fitness under natural selection can be thought of as a two-step process on the de Finetti diagram (Edwards 2002). In the first step, allele frequencies change from their current values to some new values while keeping the ratio of genotype frequencies constant. This is equivalent to moving from point z_1 to point z_2 while remaining on the line that defines a constant value of λ. In the second step, the population moves from point z_2 to point z_3 by changing its genotype frequencies but not altering its allele frequencies. The first part of the change in mean fitness caused by selection is due to the change in allele frequencies alone with everything else held constant. This *partial* change in the mean fitness exclusively due to the change in allele frequencies is exactly the same as the **genic variance** or the **additive genetic variance** that is present in the population at point z_1. The second part of the change in mean fitness is due to changes in genotype frequency and is, therefore, caused by factors such as mating patterns or physical linkage resulting in gametic disequilibrium that will change the value of λ as allele frequencies change. The fundamental theorem says that natural selection will change mean fitness by an amount proportional to the additive genetic variance alone. If λ is constant, the total change in mean fitness is just the change due to the variation in allele frequencies. When λ is not constant, changes in genotype frequencies can either increase or decrease mean fitness and can be thought of as causing an average change of zero.

Genetic variation in phenotype due to the substitution of alleles (additive genetic variation) and due to the effects of genotypes is examined from a completely distinct perspective in Chapters 9 and 10 on quantitative genetics. Those chapters also demonstrate the distinction that is made in the fundamental theorem between genetic variation due to changes in allele frequencies and changes in genotype frequencies. Both approaches give the same result that additive genetic variation is the basis of changes in mean phenotype due to natural selection.

6.4 Ramifications of the one locus, two allele model of natural selection

- The classical and balance hypotheses.
- How to explain levels of allozyme polymorphism.

The theoretical work of Fisher, Haldane, and Wright established the core principles of population genetics.

These included the one locus, two allele model of natural selection described in this chapter, along with predictions for how mutation and recombination supply genetic variation, the impacts of mating patterns and gene flow on the hierarchical organization of genetic variation, and how the effective population size regulates the process of genetic drift. Taken collectively, this body of theoretical expectations served to fuse Darwin's concept of natural selection with the principles of Mendelian particulate inheritance. These expectations form the foundation of population genetics and were labeled **neo-Darwinism** by Huxley (1942).

The Classical and Balance Hypotheses

Whereas the neo-Darwinian synthesis achieved by population genetics reached orthodoxy in the 1930s and 1940s, a long-running debate began to take shape. Under the logic of early neo-Darwinism, natural selection was the dominant evolutionary force in almost all aspects of evolutionary change. It was then a matter of debate as to what type of natural selection – directional, stabilizing, or disruptive – was most common in captive and natural populations. The answer to this question gradually turned into two broad points of view based on what one assumed and how one interpreted available data on genetic variation. Dobzhansky (1955) labeled these schools of thought the **classical hypothesis** and the **balance hypothesis**. Both hypotheses rely on natural selection as the principle process operating in populations, although they differ greatly in the predicted consequences.

Classical hypothesis: The point of view that directional natural selection is the dominant process in populations, predicting relatively little genetic variation except when selection pressures are heterogeneous in time or space or are frequency-dependent.
Balance hypothesis: The point of view that balancing natural selection is the dominant process in populations, predicting extensive genetic variation caused by overdominance for fitness.

The classical hypothesis was that directional selection was the predominant process in populations, and from this, two major predictions arise as a consequence. The first prediction was that, under random mating, populations contained individuals homozygous at most loci. The second prediction was that populations harbored relatively little genetic variation since the equilibrium points for any sort of directional selection on a diallelic locus are fixation and loss or near fixation and loss. The classical school recognized the existence of "wild-type" alleles, or alleles at high frequency in a population because such alleles were of higher fitness and were brought to high frequency by directional selection. Alternative "mutant" alleles that appeared in populations were most often deleterious but on very rare occasions would have a higher fitness than the current wild-type allele and would then become the new wild-type allele.

The classical school predictions were supported by a range of empirical observations, especially from laboratory populations of organisms such as *Drosophila*. In such populations, phenotypes are of the wild type (within some range of variation) and mutations with visible phenotypic effects appear rarely but are almost universally deleterious and do not reach high frequencies.

The classical hypothesis predicted that genetic variation in populations was produced in four ways (Dobzhansky 1955). First, deleterious mutations continually occur and segregate for a short time before they are eliminated by directional natural selection. Most of these deleterious mutations are likely to be recessive and thus exist mostly in heterozygote genotypes where they are sheltered from natural selection. (Dobzhansky pointed out that these are the sorts of mutations that cause hereditary diseases when homozygous.) Second, some proportion of mutations are selectively neutral because they have marginal fitness values very near the mean fitness. A third possibility is that rare beneficial mutations are found in a population before they have become established as the new wild-type allele. The fourth possibility is that some mutations are slightly beneficial in one environment but slightly deleterious in another environment. Alleles will then persist in a population exposed to environments that are heterogeneous in time or space. This last category motivated numerous population genetics models where directional selection occurs but fitness varies in time or space, or individual fitness is frequency-dependent (see Chapter 7).

The balance hypothesis took the alternate perspective that overdominance for fitness was the general rule in most populations so that balancing selection was the dominant process that regulated genetic variation. (Note that frequency-dependent selection with a rare-allele advantage is also considered a form of balancing selection. See Chapter 7.) Under balancing selection, heterozygotes would have higher frequencies than in the absence of selection or under directional selection. The balance hypothesis, therefore, predicted that loci would maintain two or more alleles indefinitely. Owing to balancing selection, heterozygotes would be more frequent and homozygotes much less frequent than expected by Hardy–Weinberg or under directional selection. Of the new mutations that enter a population, only those that exhibited overdominance as heterozygotes were expected to be retained in the population.

As Dobzhansky (1955) explained, the balance hypothesis was also associated with predictions about the inter-relationship among loci. Natural selection on multiple loci can lead to the accumulation of gametic disequilibrium (see Chapter 2). High levels of gametic disequilibrium are expected under balancing selection in populations with all loci at intermediate allele frequencies since only that subset of gametes that produce multilocus heterozygote zygotes would have high fitness. (Note that, under the classical hypothesis, there is also strong natural selection, but relatively little gametic disequilibrium in absolute terms is expected because populations would be close to fixation for the wild-type allele.) Using this expectation for gametic disequilibrium and then assuming that most loci experience balancing selection leads to the prediction of **coadapted gene complexes** or **supergenes** within species (Hedrick et al. 1978; Thompson and Jiggins 2014). A supergene is a haplotype or genotype at numerous loci that is held together and frequently inherited as an intact unit because gametic disequilibrium is very strong. A coadapted gene complex is a supergene where natural selection acts or has acted so that the alleles or genotypes at each locus have high fitness in the context of the alleles or genotypes at all other loci. Stated another way, selection will increase the frequency of new mutations that interact well with the alleles and heterozygous genotypes at other loci. In contrast, any mutations that have reduced relative fitness caused by interactions among loci will be reduced in frequency by natural selection. Thus, the notion of a coadapted gene complex assumes that epistasis for fitness is common.

Supergenes and coadapted gene complexes presented both a research agenda and a conceptual challenge to biologists of the classical/balance hypothesis era. A great deal of research was devoted to the study of multilocus genetic variation in laboratory and natural populations. At the same time, numerous models of multilocus natural selection and recombination were developed and studied. Very high levels of gametic disequilibrium caused by balancing selection and epistasis for fitness served to negate the Mendelian process of independent assortment. What then was the source of genetic variation required for evolutionary change? The answer was often sought in population genetic mechanisms that had the potential to recombine or break up supergenes.

The notion that supergenes held together by gametic disequilibrium are common in natural populations is now recognized as one end of a continuum. Contemporary population genetics has internalized an appreciation of the processes that cause gametic disequilibrium and evidence that multiple loci are not necessarily independent. There are now well-characterized examples of genome regions with high levels of gametic disequilibrium such as the major histocompatibility complex (*Mhc*) loci in mammals. These loci experience balancing selection because of their functional role in recognizing non-self-peptide fragments and compose a large chromosomal region that has relatively high levels of gametic disequilibrium. The supergene prediction has now been refined into a series of more specific hypotheses tailored to diverse situations. Non-independence of loci is central to models of molecular evolution that seek to explain polymorphism within populations, as embodied by concepts such as hitch-hiking, background selection, and genetic draft (see Chapter 8). Quantitative genetics recognizes non-independence of traits caused by phenotypic and genetic correlations. The idea that selection favors alleles that interact well across multiple loci is now called the **Dobzhansky–Muller model**, and it serves as an explanation of how isolated populations might develop reproductive isolation that leads to speciation (reviewed by Coyne and Orr 2004).

The study of **ecological genetics** can be traced to efforts to test the classical and balance hypotheses with empirical data. Today, ecological genetics is defined as the study of genetic variation within species in the context of environmental variation and organismal interactions. Ecological genetics seeks to identify the causes of patterns of genetic polymorphism, often with reference to the assumed or demonstrated pressures of natural selection imposed by ecological context. Early ecological genetics was focused on testing the classical and balance hypotheses for genetic variation. On the one hand was the classical school prediction that relative fitness of genotypes varied in time and space. On the other hand, the balance school predicted that overdominance for fitness was very common. Both of these possibilities were testable to some extent in natural populations, by measuring the relative fitness of phenotypes with a known genetic basis or by observing the frequency of genetic polymorphisms. Dobzhansky was among the first to study "laboratory" organisms in the wild. He pioneered field research in *Drosophila* and established a tradition of empirical research that is now the norm in population genetics. Edmund B. Ford was also instrumental in the establishment of the field of ecological genetics. Ford studied wild butterflies and moths and wrote the influential book *Ecological Genetics* (1975) first published in 1964.

Many of the widely known empirical studies in ecological genetics take on new meaning when viewed through the lens of the classical hypothesis/balance hypothesis debate. For example, industrial melanism in peppered moth (*Biston betularia*) populations in England was evidence for the classical school position since it shows that directional selection pressures vary among populations based on proximity to industrial centers whose soot stained tree trunks black (reviewed by Majerus 1998). (It is no coincidence that Bernard Kettlewell, who performed much of the original peppered moth work as a research scientist at the University of Oxford, was supervised by E.B. Ford.) Widely known human population examples are human leukocyte antigen (HLA) loci found within the major histocompatibility complex (MHC) of immune system genes. Balancing selection acts on HLA loci because variable HLA cell-surface antigen proteins and ligands (binding targets) for other immune cells provide better immune function. As a consequence, HLA loci are among the most polymorphic in the human genome, and each locus exhibits many alleles (Solberg et al. 2008). Another widely known example – blood-group protein genotypes and sickle cell anemia in malarial areas of Africa – is considered in detail in Chapter 7.

How to explain levels of allozyme polymorphism

Another long-running controversy in population genetics grew out of the classical hypothesis/balance hypothesis debate. The new controversy revolved around how to explain genetic polymorphism within natural populations observed with a then radically new technique. The technique was gel electrophoresis of enzyme polymorphisms, or allozymes (see Box 2.2). Two papers published in 1966 ushered in the new controversy. Hubby and Lewontin (1966) presented allozyme estimates of heterozygosity for 21 loci estimated from multiple populations of 15–20 *Drosophila pseudoobscura* individuals. Nine of these 21 loci exhibited between two and six alleles segregating within populations. The Hubby and Lewontin paper showed a technique that could be used to determine both the proportion of loci that possessed more than one allele and the level of heterozygosity for each polymorphic locus.

The controversy over the causes of allozyme polymorphism changed the focus of much of population genetics within the span of only a few years starting in the mid-1960s. Initially, the classical and balance hypotheses were considered as primary explanations. In fact, in a paper published along with the allozyme data themselves, Lewontin and Hubby (1966) argued that the level of heterozygosity observed (averaged over populations 30% of loci were polymorphic) were inconsistent with the balance hypothesis because of the segregation load that would have been required (see the Genetic load section in Chapter 7). The remaining explanation for the allozyme polymorphism within the context of the time was directional natural selection consistent with the classical hypothesis.

The balance hypotheses experienced some setbacks from empirical data around the same time. Apparent overdominance in maize was shown to decline over multiple generations (Moll et al. 1964; reviewed by Crow 1993b). True overdominance should persist indefinitely as a function only of heterozygosity. These maize results, however, were consistent with the prediction that overdominance was actually caused by gametic disequilibrium between loci bearing beneficial dominant alleles and other loci bearing deleterious recessive alleles. When two individuals homozygous for different alleles at two such loci are crossed, the progeny will experience a great increase in fitness because the recessive deleterious phenotype will be masked by dominance. The maize results demonstrated that apparent overdominance phenomena were caused by the combination of simple dominance and linkage rather than by true overdominance.

The classical hypothesis/balance hypothesis debate soon receded. Selective neutrality (see Chapter 8) was an explanation under the classical hypothesis that predicted a low level of genetic variation in populations. This idea of selectively neutral alleles, which was developed and mathematically formalized starting in the 1950s and 1960s, emerged as a primary null hypothesis for genetic polymorphism. The neutral theory hypothesized that many loci have selectively neutral alleles and that polymorphism was a product of the non-equilibrium random walk that new neutral mutations experience because of genetic drift (Kimura 1983).

The waning of the balance hypothesis and the ascension of the two components of the classical hypothesis produced what was labeled the **neo-classical theory** of population genetics by Lewontin (1974). This label came about because both elements of the classical hypothesis explanation for polymorphism – selectively neutral mutations and mutations under directional or purifying selection – are drawn from the early classical hypothesis. Under the neo-classical hypothesis, the debate became one about the relative contributions of neutral mutations or mutations acted on by natural directional selection to levels of genetic polymorphism. There was also continuing work on the selection element of the classical hypothesis, which was updated and bolstered with empirical support from more elaborate theoretical models and ecological genetic studies. The balance hypothesis remains relevant even today and empirical evidence suggests that balancing selection does operate in natural populations (Charlesworth 2006; Fijarczyk and Babik 2015).

Chapter 6 review

- The synthesis of Darwin's concept of natural selection with Mendelian particulate inheritance that forms that basis of population genetics is termed neo-Darwinism.
- For haploid organisms, natural selection is a population growth process where different genotypes vary in genotype-specific population growth rates. The ratio of genotype-specific growth rates is the relative fitness, and it predicts the genotype that will approach fixation in an infinitely expanding population over time.
- When recombination is absent, natural selection can result in clonal interference where beneficial haplotypes compete and only the most fit reaches fixation and other haplotypes are lost.
- Natural selection in diploid organisms also relies on the relative fitness to express genotype-specific growth rates with the addition of sexual reproduction such that pairs of parents can produce a predictable frequency of genotypes in their progeny under random mating.

- The outcomes of natural selection on viability for a diallelic locus can be generalized into directional selection (a homozygote most fit) that results in fixation and loss (or very nearly fixation and loss), balancing selection (heterozygote advantage) that maintains both alleles forever, and disruptive selection (heterozygote disadvantage) that results in fixation or loss depending on initial genotype frequencies.
- The fundamental theorem of natural selection shows us that the change in mean fitness by natural selection is proportional to the additive genetic variation in fitness.
- The degree of dominance and recessivity for viability phenotypes impacts the rate of change of genotype frequencies under natural selection because there is not a perfect relationship between genotype and phenotype. Natural selection changes genotype frequencies fastest when gene action is additive.
- The classical hypothesis predicts that directional natural selection is common in natural populations, with the consequences that genetic polymorphism is limited at most loci. What genetic variation exists is explained by deleterious mutations, along with some neutral mutations and very few beneficial mutations.
- The balance hypothesis predicts that balancing natural selection due to overdominance for fitness is common in natural populations, and predicts that there should be ample genetic polymorphism maintained by selection. The balance hypothesis also predicted selection would cause gametic disequilibrium over relatively large regions of the genome.

Further reading

Arguably, the first comprehensive treatment of natural selection that came out of the modern synthesis and still a worthwhile read today is:

Fisher, R.A. (1999). *The Genetical Theory of Natural Selection: A Complete Variorum Edition*. Oxford: Oxford University Press (originally published in 1930).

Another early classic of the modern synthesis that established the mathematical connections between Mendelian genetics and natural selection is:

Haldane, J.B.S. (1990). *The Causes of Evolution*. Princeton, NJ: Princeton University Press (originally published in 1932).

For a history of early population genetics beginning with Darwin and Mendel and ending with Fisher, Haldane, and Wright, see:

Provine, W.B. (1971). *The Origins of Theoretical Population Genetics*. Chicago, IL: University of Chicago Press (this book was originally published in 1971 while the 2001 edition has an afterword by Provine).

For a wide-ranging consideration and critique of aspects of the classical/balance hypothesis debate written in the midst of the allozyme era in population genetics, see:

Lewontin, R.C. (1974). *The Genetic Basis of Evolutionary Change*. New York: Columbia University Press.

End-of-chapter exercises

1. Imagine a population containing two haplotypes with absolute fitness values of $\lambda_A = 0.8$ and $\lambda_B = 0.9$ and initial populations of $N_A = 1000$ and $N_B = 1000$. What are the expected population sizes in the next generation? What are the expected haplotype frequencies after one generation of natural selection?
2. A population initially consisted of 4000 AA, 5100 Aa, and 3000 aa individuals. After a severe winter, 3000 AA, 3400 Aa, and 1500 aa individuals survived to reproduce.

   (A) Determine the absolute fitness, relative fitness, and the selection coefficients for each genotype.
   (B) Based on these relative fitness values, predict the genotype and allele frequencies after another generation of natural selection. Use the genotype and allele frequencies for the reproducing individuals in part A as initial frequencies and assume that mating is random.

3. One population has absolute fitness values of $W_{AA} = 0.7$, $W_{Aa} = 0.8$, and $W_{aa} = 0.9$. Another independent population has absolute fitness values of $W_{AA} = 0.85558$, $W_{Aa} = 0.97779$, and $W_{aa} = 1.1$. What are the genotype-specific relative fitness values in each population? What genotype(s) and allele will increase in frequency in each of the populations? How will the mean fitness ($\overline{w}$) change over time in each population under natural selection? How will total population size (N) of each population change over time? What does this example

illustrate about the difference between absolute fitness and relative fitness?

4 Can natural selection ever purge a population of a strongly deleterious allele that is completely recessive when there is random mating? Why or why not? What if there is mating among relatives?

5 What is the relationship between $\overline{w}$ and equilibrium genotype and allele frequencies?

6 If overdominance for relative fitness were common in populations, what would you expect for levels of genetic polymorphism? In comparison, what levels of polymorphism would you expect if directional selection were common? Use results from the text simulation website to support your answer.

7 Search the literature for a recent research paper that utilizes one or more of the population genetic predictions covered in this chapter.

The topic can be any organism, application, or process, but the paper must include a hypothesis test involving a topic such as directional selection, relative fitness, overdominance, or the population average fitness. Summarize the main hypothesis, goal, or rationale of the paper. Then, explain how the paper utilized a population genetic prediction from this chapter and then summarize the results and the conclusions based on the prediction.

8 Construct a simulation model of natural selection on one locus with two alleles. Instructions to build a spreadsheet model can be found on the text website. These instructions can also be implemented in a programming language such as Python or R.

Problem box answers

Problem box 6.1 answer

To solve for relative fitness given initial and final allele frequencies and time elapsed, we need to rearrange Eq. 6.8 by taking the logarithm of both sides:

$$\log\left(\frac{q_t}{p_t}\right) = t\log(w) + \log\left(\frac{q_0}{p_0}\right)$$

P > to remove the exponent. We will let p represent the frequency of the wild-type allele and q the combined frequency of the drug-resistant alleles. Based on 601 days between allele frequency estimates, $t = 231$ generations elapsed. Substituting these values gives

$$\log\left(\frac{0.51}{0.49}\right) = (231)\log(w) + \log\left(\frac{0.99}{0.01}\right)$$

$$0.01737 = (231)\log(w) + 1.9956$$

$$-1.9782 = (231)\log(w)$$

$$-1.9782/231 = \log(w)$$

$$-0.008564 = \log(w)$$

$$10^{-0.008564} = w$$

$$w = 0.9805$$

The relative fitness of the drug-resistant alleles is 98% of the wild-type allele, and, therefore, the wild-type allele increases in frequency over time when AZT is not present.

Problem box 6.2 answer

$$p_{equilibrium} = t/(s+t) = 0.2/(0.1+0.2) = 2/3$$

For equilibrium allele frequencies:

$$\overline{w} = 0.9(0.667)^2 + (1)2(0.667)(0.333)$$
$$+ 0.8(0.333)^2 = 0.9333$$

$$\Delta p = \frac{(0.667)(0.333)[0.667(0.9-1)+0.333(1-0.8)]}{0.9333} = 0$$

or calculated using the marginal fitness

$$p_{t+1} = \frac{0.9(0.667)^2 + 1(0.667)(0.333)}{0.9333} = 0.667$$

$$\Delta p = 0.667 - 0.667 = 0$$

At $p = 0.9$ $\left(p > p_{equilibrium}\right)$

$$\overline{w} = 0.9(0.9)^2 + (1)2(0.9)(0.1) + 0.8(0.1)^2$$

$$= 0.917$$

$$\Delta p = \frac{(0.9)(0.)[0.9(0.9-1) + 0.1(1-0.8)]}{0.917}$$

$$= -0.0069$$

or calculated using the marginal fitness

$$p_{t+1} = \frac{0.9(0.9)^2 + 1(0.9)(0.1)}{0.917} = 0.8931$$

$$\Delta p = 0.8931 - 0.9 = -0.0069$$

At $p = 0.2$ $\left(p < p_{equilibrium} \right)$

$$\overline{w} = 0.9(0.2)^2 + (1)2(0.2)(0.8) + 0.8(0.8)^2 = 0.868$$

$$\Delta p = \frac{(0.2)(0.8)[0.2(0.9-1) + 0.8(1-0.8)]}{0.868} = 0.0258$$

or calculated using the marginal fitness

$$p_{t+1} = \frac{0.9(0.2)^2 + 1(0.2)(0.8)}{0.868} = 0.2258$$

$$\Delta p = 0.2258 - 0.2 = 0.0258$$

At $p = 0.9$, $\overline{w}$ is lower than at $p_{equilibrium}$. Therefore, Δp is negative, meaning that natural selection is causing allele frequencies to decrease. At $p = 0.2$, $\overline{w}$ is also lower than at $p_{equilibrium}$. Therefore, increasing allele frequencies (positive Δp) by natural selection causes an increase in mean fitness. At $p_{equilibrium}$, $\overline{w}$ is at its maximum for these relative fitness values; so, Δp is 0 because selection will no longer change the allele frequencies.

CHAPTER 7

Further models of natural selection

7.1 Viability selection with three alleles or two loci

- Mean fitness surfaces.
- Natural selection on one locus with three alleles.
- Natural selection on two diallelic loci.

Chapter 6 established a series of general predictions about the action of natural selection when fitness is equivalent to genotype-specific viability determined by a single locus with two alleles. The conditions required for the basic diallelic locus model of natural selection are quite restrictive and are probably not met often in biological populations. The goal of this chapter is to extend our understanding of the model of natural selection to increasingly complex and general genetic situations. In a sense then, this chapter explores the process of natural selection under assumptions that might better approximate conditions found in some natural populations. In the first section, we will retain the viability natural selection model and its assumptions but modify the numbers of alleles at a locus and the number of loci. The goal is to examine the outcomes of viability selection when fitness is determined by either a single locus with three alleles or two loci each with two alleles.

A useful tool that we will employ to understand the dynamics of genotype frequencies, allele frequencies, and mean fitness under natural selection is called a **fitness surface**. A fitness surface is a graph that shows genotype frequencies of a population on some axes along with the mean fitness of the population at each possible point in the range of genotype frequencies. For one locus with two or three alleles, a de Finetti diagram can be used as a fitness surface, as shown in Figure 7.1. The three axes represent genotype frequencies of a population on the plot. Each

point inside the triangle defines three genotype frequencies that are then used to compute the mean fitness of the population. The mean fitness is represented by shading as well as contour lines that connect points of equal mean fitness. Since contour plots of mean fitness are interpreted exactly like topographic maps where contour lines are used to represent elevation, they are also called fitness landscapes or **adaptive landscapes**.

The highest point on a fitness surface represents equilibrium genotype frequencies under natural selection. A fitness surface also shows how natural selection will change genotype frequencies over time if the process of natural selection operates like a hiker who can only travel uphill. For any point on a fitness surface, natural selection will act to increase mean fitness of the population and shift genotype frequencies in a direction that increases the mean fitness. Once the population is at a point where mean fitness cannot increase, natural selection has reached an equilibrium and stops changing genotype frequencies. In Figure 7.1, the entire surface is a tilted plane; that is, it has its highest point at the left vertex or where the AA genotype is fixed in the population. Therefore, natural selection will change genotype frequencies such that the population climbs in mean fitness until reaching fixation for AA.

Natural selection on one locus with three alleles

With an understanding of fitness surfaces, let us now turn to the classic case of natural selection on three alleles at the human hemoglobin β gene (see Allison 1956; Modiano et al. 2001). The hemoglobin protein is found in red blood cells and is responsible for binding and then carrying oxygen from the lungs to the entire body. Adult hemoglobin is formed from four

Population Genetics, Second Edition. Matthew B. Hamilton.
© 2021 John Wiley & Sons, Inc. Published 2021 by John Wiley & Sons, Inc.
Companion website: www.wiley.com/go/hamilton/populationgenetics

Figure 7.1 A fitness surface made by including mean fitness on a de Finetti plot of the three genotype frequencies for a diallelic locus. The colored lines indicate the possible trajectories of genotype frequencies as natural selection increases the mean fitness of the population. The fitness values are $w_{AA} = 1.0$, $w_{Aa} = 0.6$, and $w_{aa} = 0.2$, so the highest mean fitness is found in the lower left apex when the population is fixed for the AA genotype. This highest fitness point can be reached by continually increasing mean fitness from any initial point on the surface. Gene action is additive because alleles have a constant impact fitness regardless of the allele they are paired with in a genotype. An A allele always contributes 0.5 and an a allele 0.1 toward the fitness of a genotype.

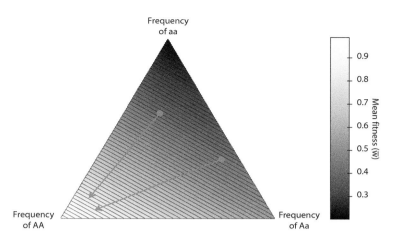

Table 7.1 Relative fitness estimates for the six genotypes of the hemoglobin β gene estimated in Western Africa where malaria is common. Values from Cavalli-Sforza and Bodmer (1971) are based by deviation from Hardy–Weinberg expected genotype frequencies. Values from Hedrick (2004) are estimated from relative risk of mortality for individuals with AA, AC, AS, and CC genotypes and assume 20% overall mortality from malaria.

	Genotype					
	AA	**AS**	**SS**	**AC**	**SC**	**CC**
Relative fitness (*w*) from Cavalli-Sforza and Bodmer (1971)						
Relative to w_{CC}	0.679	0.763	0.153	0.679	0.534	1.0
Relative to w_{AS}	0.89	1.0	0.20	0.89	0.70	1.31
Relative fitness (*w*) from Hedrick (2004)						
Relative to w_{CC}	0.730	0.954	0.109	0.865	0.498	1.0
Relative to w_{AS}	0.623	1.0	0.109	0.906	0.498	1.048

separate proteins, two α (or "alpha") proteins and two β (or "beta") proteins. The hemoglobin β gene encodes the β-protein, which is often referred to as β-globin or *Hb*. The *Hb* A allele is the most common allele in human populations. Although several hundred *Hb* alleles have been identified in human populations, the *Hb* S allele is a common low-frequency allele. The S allele is characterized by a nucleotide change that results in the substitution of the hydrophobic amino acid valine in place of the hydrophilic glutamic acid at the sixth amino acid position of the β-globin protein. Individuals homozygous for the S allele exhibit changes in red blood cell morphology ("sickling") and impaired oxygen transport that lead to chronic anemia (Ashley-Koch et al. 2000). The *Hb* C allele is also present at low frequencies in West African and southeast Asian populations. Individuals who are CC homozygotes have mild to moderate anemia and enlargement of the spleen

that is often asymptomatic (e.g. Fairhurst and Casella 2004).

The fitness of *Hb* genotypes depends on the environment where people live. In areas of the world without the malarial parasite *Plasmodium falciparum*, genotypes that result in anemia and related conditions have lower fitness. However, in regions where malarial infection is common, certain *Hb* genotypes confer resistance to infection by *P. falciparum* that may partly or completely compensate for any disadvantage due to anemia. Two estimates of the relative fitnesses of the six *Hb* genotypes in Western Africa where malaria is common are shown in Table 7.1.

A seemingly obvious prediction from Table 7.1 is that natural selection in populations where malaria is common would increase the frequency of the CC genotype and eventually fix the C allele. But, is this really what will happen? The answer comes from examining fitness surfaces for the six *Hb* genotypes.

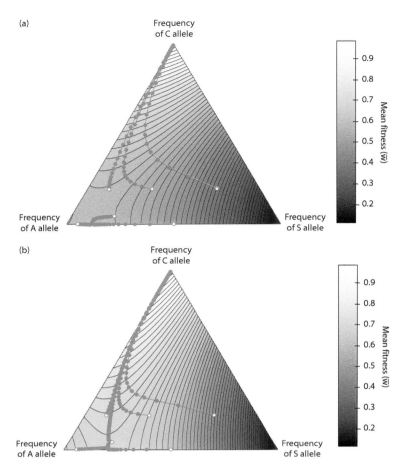

Figure 7.2 Fitness surfaces for the A, S, and C alleles at the human hemoglobin β gene when malaria is common. The surface in panel A corresponds to the top set of fitness values in Table 7.1, while panel B shows the surface for the bottom set of values. The tracks of circles represent generation-by-generation allele frequency trajectories due to natural selection over 50 generations calculated with Eq. 7.3. In panel A, when the initial frequency of the C allele is relatively high, the equilibrium of natural selection is the fixation of the CC genotype. In contrast, when the C allele is initially rare (a frequency of less than about 7%), selection reaches an equilibrium with only the A and S alleles segregating and the C allele going to loss. In panel B, selection will eventually fix the CC genotype from any initial frequency of the C allele. However, when the C allele is at low frequencies, the increase in the C allele each generation is extremely small so that selection would take hundreds of generations to fix the CC genotype. The six initial allele frequency points, shown as open circles, are identical for the two surfaces.

With three alleles, there are six genotype frequencies, which is too many to represent in a de Finetti plot like Figure 7.1. But since the allele frequencies must sum to 1, we can represent the fitness surface on a ternary graph where each axis represents one of the three allele frequencies. Fitness surfaces drawn in this way are shown Figure 7.2 for the two sets of fitness values given in Table 7.1. These three allele fitness surfaces are now rippled or hilly compared to the fitness surface in Figure 7.1.

Understanding how genotype frequencies will change on a fitness surface requires calculating the change in allele frequencies due to selection for a series of points on the surface. The sign and magnitude of the change in allele frequency will be a function of the slope of the fitness surface at any point we examine. To do this for the fitness surfaces in Figure 7.2, we need to extend the viability model of natural selection to three alleles at one locus. We can compute the mean fitness of the population

$$\overline{w} = w_{AA}p^2 + w_{BB}q^2 + w_{CC}r^2 + w_{AB}2pq \\ + w_{AC}2pr + w_{BC}2qr \tag{7.1}$$

where p, q, and r represent the frequencies of the three alleles A, B, and C. We can also use the marginal fitness of the genotypes that contain each of the alleles to compute whether an allele will increase or decrease in frequency due to the average fitness of all the genotypes that carry the allele. When there is random mating, the marginal fitness for the A allele is

$$\overline{w}_A = \frac{w_{AA}p^2 + w_{AB}pq + w_{AC}pr}{p} = w_{AA}p + w_{AB}q + w_{AC}r \tag{7.2}$$

where the frequencies of the heterozygous genotypes are multiplied by ½ since they carry one copy of the A allele. The marginal fitness is a way to compare the ratio of p in the current generation with the frequency of p in the next generation that will result from natural selection changing genotype frequencies. Allele frequencies change each generation due to differences between the marginal fitness of each allele and the average fitness of the entire population. The change in the frequency of the A allele is

$$\Delta p = p \frac{(\overline{w}_A - \overline{w})}{\overline{w}} \qquad (7.3)$$

Allele frequency after one generation of selection is then simply $p_{t+1} = p + \Delta p$. Similar expressions are obtained easily for the B and C alleles. Also, note that this approach can be extended to an arbitrary number of alleles at one locus as long as genotypes are in Hardy–Weinberg frequencies at the start of each generation before the action of selection.

Returning to the fitness surfaces, Figure 7.2A is an interesting case because it has two stable equilibrium points. One of the equilibrium points matches our intuition after inspecting Table 7.1 that the CC genotype should be fixed by selection. When the initial frequency of the C allele is relatively high, all three trajectories of allele frequencies over 10 generations of selection calculated with Eq. 7.3 are clearly headed for fixation of CC. In contrast, when the initial frequency of the C allele is low, the trajectories of allele frequencies show that the C allele will be lost from the population. This is counterintuitive given that the CC genotype has the highest relative fitness. This result is a consequence of the fitness surface. When C is at low frequency, its marginal fitness is actually less than the mean fitness. In other words, the fitness surface is going down in elevation toward higher frequencies of the C allele. Since natural selection only works to increase mean fitness, the C allele is reduced in frequency to loss.

To see one possible consequence of this fitness surface, imagine that the A and S alleles are older in human populations than the C allele and the A and S alleles have reached equilibrium frequencies. Using Eq. 6.35 and Table 7.1, the equilibrium frequency of the A allele would be $t/(s + t) = 0.8/(0.11 + 0.8) = 0.88$, and, therefore, the equilibrium frequency of S would be $1 - 0.88 = 0.11$. Next, imagine that the C allele occurs in the population at a later time due to mutation. Since mutation rates are low, the resulting frequency of the C allele will also be low, and most C alleles would occur in AC and SC heterozygotes. All heterozygotes have overdominance (AS) or underdominance (SC and AC) for fitness. In particular, the SC heterozygote has a lower relative fitness than AA and AS genotypes, so its marginal fitness would be negative when C is at a low frequency. Thus, to get from a mean fitness state where C is infrequent, we would have to go through a dip of lower mean fitness as C initially increases. At the higher initial frequencies of C, however, mean fitness increases steadily until CC fixes.

So, if A and S alleles were ancestral, natural selection alone would drive a newly introduced C allele to loss despite the high relative fitness of the CC homozygote.

Problem box 7.1
Marginal fitness and Δp for the *Hb* C allele

Compute the mean fitness, marginal fitness of the C allele, and the change in the C allele using the two sets of initial allele frequencies given below and relative fitness values from the top of Table 7.1. Use Δp along with Figure 7.2A to predict the equilibrium that will be reached by natural selection for both initial allele frequencies.

Initial allele frequencies set 1: $p = 0.75$, $q = 0.20$, $r = 0.05$
Initial allele frequencies set 2: $p = 0.70$, $q = 0.20$, $r = 0.10$

For the fitness surface in Figure 7.2B, natural selection will eventually fix the CC genotype from any initial frequency of the C allele. However, when the C allele is at low frequency, selection increases the frequency of the C allele very slowly. This is because the marginal fitness of the C allele is only very slightly greater than the mean fitness below a frequency of about 15% when the frequency of the A allele is also high. This can be seen on the fitness surface by noting the wide spacing between contour lines toward the left vertex. Widely spaced contour lines indicate areas with little slope. These are areas where the mean fitness of the population is either constant or nearly constant for a range of genotype frequencies. Such flat areas on fitness surfaces can be stable or unstable equilibrium points and are regions where selection is a weak process because the marginal fitness values are very close in value to the mean fitness.

Determining which of the different hemoglobin β genotype fitness values best describe actual populations is not the main point of this example. Rather, the hemoglobin β gene serves to illustrate that dominance for fitness, the order of appearance of alleles in a population, and the relative fitness values may all interact to determining the outcome of natural selection with three alleles.

Interact box 7.1
Natural selection on one locus with three or more alleles

Direct simulation of selection on one locus with three alleles is an easy way to see that equilibrium points depend strongly on over- and underdominance for fitness. Populus has the ability to simulate selection on a locus with three or more alleles. Launch Populus, and, in the **Model** menu, choose **Natural Selection** and then **Selection on a Multi-Allelic Locus**. Click on each of the radio buttons for the display options to see how the results can be displayed. Note that when using the default fitness values, the P_3 allele goes to fixation. The options dialog box can be made larger by dragging the tab at the bottom right, making it easier to see the parameter fields.

Then, try some different fitness values:

Additivity: $w_{11} = 0.6$, $w_{12} = 0.7$, $w_{13} = 0.8$; $w_{21} = 0.7$, $w_{22} = 0.8$, $w_{23} = 0.9$; $w_{31} = 0.8$, $w_{32} = 0.9$, $w_{33} = 1.0$
Overdominance: $w_{11} = w_{22} = w_{33} = 0.3$; $w_{12} = w_{13} = w_{21} = w_{23} = w_{13} = w_{31} = 1.0$
Underdominance: $w_{11} = w_{22} = w_{33} = 1.0$; $w_{12} = w_{13} = w_{21} = w_{23} = w_{13} = w_{31} = 0.3$

Also be sure to vary the allele frequencies for each set of fitness values. You might try frequencies of all alleles equal at 1/3, and then one allele more common, with 0.67, 0.12, and 0.21 (the default values).

Natural selection on two diallelic loci

Since phenotypes, and therefore fitness, may be caused by more than one locus, a logical step is to extend the model of natural selection to two loci. Biologically, there is strong motivation to consider selection at more than one locus since many phenotypes are known to show variation caused by multiple loci (see Chapter 9). The fate of two mutations could also be considered as two-locus selection. Natural selection on two loci is inherently more complicated than at a single locus because of gametic disequilibrium. As covered in Chapter 2, both linkage and natural selection itself produce gametic disequilibrium that must be accounted for in a two-locus model of natural selection. Because natural selection on two loci is considerably more complex than on just one locus, the goal of this section is to provide a general introduction to two-locus models. It is important to recognize at the outset that there is no easily summarized set of equilibria for two-locus selection as there are for selection on a diallelic locus. The outcome of two-locus selection depends on the balance between natural selection and recombination between loci as well as the initial genotype frequencies in the population.

Two-locus natural selection is commonly approached from the perspective of gametes because gametic disequilibrium is expressed in terms of gamete frequencies. With two diallelic loci, there are 16 possible genotypes that result from the union of four possible gametes. Let the frequencies of the gametes AB, Ab, aB, and ab be x_1, x_2, x_3, and x_4. Table 7.2 shows the relative fitness values for all possible combinations of four gametes. There are only 10 unique fitness values if the same gamete inherited from either parent has the same fitness in a progeny genotype. For example, if an Ab gamete from either a male or female parent has the same fitness in an AB/Ab progeny genotype, then $w_{12} = w_{21}$ in the fitness matrix.

The expected frequencies of each gamete from the 10 possible parental matings are shown in Table 7.3 under the assumption of random mating and a recombination rate of c between the two loci (compare with Table 2.12). The frequencies of each gamete in the next generation can be obtained by summing each of the columns in Table 7.3 while also weighting each expected frequency by the relative fitness of each genotype. For example, the expected frequency of AB gametes after one generation of natural selection and recombination is

Table 7.2 Matrix of fitness values for all combinations of the four gametes formed at two diallelic loci (top). If the same gamete inherited from either parent has the same fitness in a progeny genotype (e.g. $w_{12} = w_{21}$), then there are 10 gamete fitness values shown outside the shaded triangle. These 10 fitness values can be summarized by a genotype fitness matrix (bottom) under the assumption that double heterozygotes have equal fitness ($w_{14} = w_{23}$) and representing their fitness value by w_H. The double heterozygote genotypes are of special interest since they can produce recombinant gametes.

	AB	**Ab**	**aB**	**Ab**
AB	w_{11}	w_{12}	w_{13}	w_{14}
Ab	w_{21}	w_{22}	w_{23}	w_{24}
aB	w_{31}	w_{32}	w_{33}	w_{34}
Ab	w_{41}	w_{42}	w_{43}	w_{44}

	BB	**Bb**	**Bb**
AA	w_{11}	w_{12}	w_{22}
Aa	w_{13}	w_H	w_{24}
Aa	w_{33}	w_{34}	w_{44}

$$x_{1(t+1)} = \frac{\begin{array}{l} w_{11}\,x_1^2 + w_{12}x_1x_2 + w_{13}x_1x_3 \\ + (1-c)w_{14}x_1x_4 + rw_{23}x_2x_3 \end{array}}{\bar{w}} \quad (7.4)$$

which is directly comparable with Eq. 6.21 for one allele at a diallelic locus. This can be simplified by expanding the $(1-c)w_{14}x_1x_4$ term:

$$x_{1(t+1)} = \frac{\begin{array}{l} w_{11}\,x_1^2 + w_{12}x_1x_2 + w_{13}x_1x_3 \\ + w_{14}x_1x_4 - cw_{14}x_1x_4 + cw_{23}x_2x_3 \end{array}}{\bar{w}}$$

$$(7.5)$$

and then factoring an x_1 out of the first four terms and c out of the last two terms

$$x_{1(t+1)} = \frac{\begin{array}{l} x_1(w_{11}x_1 + w_{12}x_2 + w_{13}x_3 + w_{14}x_4) \\ - c(w_{14}x_1x_4 - w_{23}x_2x_3) \end{array}}{\bar{w}}$$

$$(7.6)$$

An additional simplification is possible if we assume that the fitness of genotypes with the same number of A and B alleles is equal. For example, the double heterozygotes AB/ab and Ab/aB have the same number of A and B alleles, so we can reasonably

Table 7.3 Expected frequencies of gametes under viability selection for two diallelic loci in a randomly mating population with a recombination rate of c between the loci. The expected gamete frequencies assume that the same gamete coming from either parent will have the same fitness in a progeny genotype (e.g. $w_{12} = w_{21}$). Eight genotypes have non-recombinant and recombinant gametes that are identical and so do not require a term for the recombination rate. Two genotypes produce novel recombinant gametes, requiring the inclusion of the recombination rate to predict gamete frequencies. Summing down each column of the table gives the total frequency of each gamete in the next generation due to mating and recombination.

Genotype	Fitness	Total frequency	Frequency of gametes in next generation			
			AB	**Ab**	**aB**	**ab**
AB/AB	w_{11}	x_1^2	x_1^2			
AB/Ab	w_{12}	$2x_1x_2$	x_1x_2	x_1x_2		
AB/aB	w_{13}	$2x_1x_3$	x_1x_3		x_1x_3	
AB/ab	w_{14}	$2x_1x_4$	$(1-c)\,x_1x_4$	$(c)\,x_1x_4$	$(c)\,x_1x_4$	$(1-c)\,x_1x_4$
Ab/Ab	w_{22}	x_2^2		x_2^2		
Ab/aB	w_{23}	$2x_2x_3$	$(c)\,x_2x_3$	$(1-c)\,x_2x_3$	$(1-c)\,x_2x_3$	$(c)\,x_2x_3$
Ab/ab	w_{24}	$2\,x_2x_4$		x_2x_3	x_2x_3	
aB/aB	w_{33}	x_3^2			x_3^2	
aB/ab	w_{34}	$2x_3x_4$			x_3x_4	x_3x_4
ab/ab	w_{44}	x_4^2				x_4^2

assume they have equal fitness values (Table 7.2). This assumption allows us to equate the fitness values of those double heterozygotes where recombination plays a role in the gametes that are produced. Applying this assumption to Eq. 7.6, we can set $w_{14} = w_{23}$ and then the c $(w_{14}x_1x_4 - w_{23}x_2x_3)$ term becomes $cw_{14}(x_1x_4 - x_2x_3)$ to give

$$x_{1(t+1)} = \frac{\begin{array}{c} x_1(w_{11}x_1 + w_{12}x_2 + w_{13}x_3 + w_{14}x_4) \\ -cw_{14}(x_1x_4 - x_2x_3) \end{array}}{\overline{w}}$$
$$(7.7)$$

This helps because the gametic disequilibrium parameter D is the difference between the product of the coupling gametes and the product of the repulsion gametes (see Eq. 2.27). In the notation of this section, $D = x_1x_4 - x_2x_3$. We can then substitute D for $x_1x_4 - x_2x_3$ in Eq. 7.7 to give

$$x_{1(t+1)} = \frac{\begin{array}{c} x_1(w_{11}x_1 + w_{12}x_2 + w_{13}x_3 + w_{14}x_4) \\ -cw_{14}D \end{array}}{\overline{w}} \quad (7.8)$$

Eq. 7.8 shows that the frequency of AB gametes after one generation of natural selection is a function of three things. First, the viabilities of the three genotypes that produce AB gametes can change genotypes frequencies and thereby impact the frequency of AB gametes in the next generation (recombination does *not* alter the frequency of AB gametes produced by the AB/AB, AB/Ab, and AB/aB genotypes). Then, an additional part of the frequency of AB gametes is determined by the combination of recombination, fitness values of the double heterozygotes, and initial gametic disequilibrium in the population. Double heterozygotes could be more or less frequent than expected by random mating as measured by D. Also, the frequency of recombination and the relative fitness of the genotypes will determine how many AB gametes are produced. If D and r could be ignored, the frequency of AB gametes would be analogous to the frequency of one of the four possible gametes for a single locus with four alleles.

Expanding on the idea that the four gamete frequencies can be treated like the frequencies of four alleles at one locus, we can utilize some of the expressions developed earlier in the chapter for a single locus. The marginal fitness for each of the two-locus gametes ($\overline{w}_i$) is obtained by summing the frequency-weighted fitness value of each of the gametes that a given gamete could pair with to make a genotype

$$\overline{w}_i = \sum_{j=1}^{4} x_j w_{ij} \quad (7.9)$$

Similarly, the average fitness of the population is the frequency-weighted average of the fitness values for all of the possible gamete combinations

$$\overline{w} = \sum_{i=1}^{4} \sum_{j=i}^{4} x_i x_j w_{ij} \quad (7.10)$$

The marginal fitness and mean fitness can be combined with Eq. 7.8 to give an expression for the change in a gamete frequency under selection and recombination.

To continue with the AB gamete as an example, notice that the marginal fitness $\overline{w}_1$ is equal to $x_1w_{11} + x_2w_{12} + x_3w_{13} + x_4w_{14}$. Making this substitution in Eq. 7.8, dividing by the mean fitness and substituting w_H for w_{14} or w_{23}, gives the change in the AB gamete frequency over one generation of natural selection

$$\Delta x_1 = \frac{x_1\overline{w}_1 - cw_HD}{\overline{w}} \quad (7.11)$$

This is exactly like the expression for Δp for a diallelic locus (compare with Eq. 6.23). Using analogous steps for the other three gametes gives the recursion equations for change in gamete frequency after one generation of natural selection and recombination:

$$\Delta x_2 = \frac{x_2\overline{w}_2 + cw_HD}{\overline{w}} \quad (7.12)$$

$$\Delta x_3 = \frac{x_3\overline{w}_3 + cw_HD}{\overline{w}} \quad (7.13)$$

$$\Delta x_4 = \frac{x_4\overline{w}_4 - cw_HD}{\overline{w}} \quad (7.14)$$

Equations 7.11–7.14 show that the change in gamete frequency under natural selection is due to both fitness values and recombination. If there is no recombination ($c = 0$), then each gamete is analogous to a single allele. The outcome of selection is then like four alleles at a single locus as dictated by the gamete fitness values. The process of recombination may either reinforce or oppose the changes in gamete frequencies due to natural selection. For example, if gametes Ab and aB have the highest fitness values and there is no recombination, then Δx_2 and Δx_3 would be positive while Δx_1 and Δx_4 would be negative (when not at equilibrium). The gamete frequency changes caused by recombination would amplify the effect of natural selection on gamete frequencies since the $cw_H D$ term would increase Δx_2

and Δx_3 but decrease Δx_1 and Δx_4. In contrast, if gametes AB and ab have the highest fitness and there is recombination, then the $cw_H D$ term would decrease Δx_1 and Δx_4 but increase Δx_2 and Δx_3 in opposition to natural selection.

Examining natural selection on two loci with and without recombination demonstrates that selection and recombination working in opposition can produce counterintuitive equilibrium gamete frequencies. Figure 7.3 shows a fitness surface where gene action is completely additive. Since the fitness surface is a tilted plane, our earlier experience with one locus selection suggests that the equilibrium

(a)

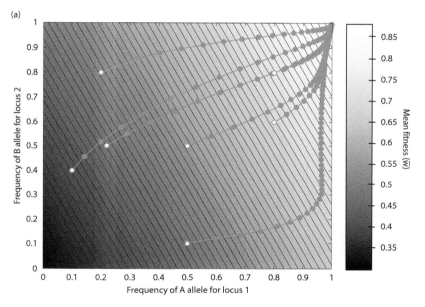

(b)

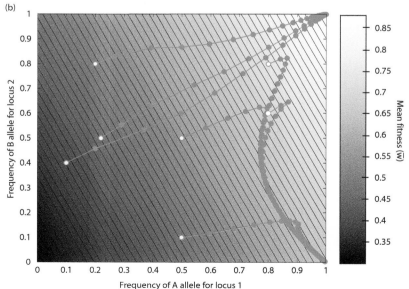

Figure 7.3   A fitness surface for two loci that each have two alleles where gene action is additive. The blue dots show generation-by-generation allele frequencies based on equations for change in gamete frequency (Δx_1 through Δx_4) for seven different initial sets of the four gamete frequencies. When recombination is a weak force ($c = 0.05$), equilibrium allele frequencies are dictated by natural selection, and all initial gamete frequencies eventually reach the highest mean fitness point (panel A). In contrast, when recombination is a strong force ($c = 0.5$), then equilibrium allele frequencies depend on initial gamete frequencies (panel B). When recombination is strong, equilibrium allele frequencies may not correspond to the highest mean fitness. Relative fitness values are $w_{AABB} = 0.9$, $w_{AABb} = 0.8$, $w_{AAbb} = 0.7$, $w_{AaBB} = 0.7$, $w_{AaBb} = 0.6$, $w_{Aabb} = 0.5$, $w_{aaBB} = 0.5$, $w_{aaBb} = 0.4$, $w_{aabb} = 0.3$. The seven initial allele frequency points, shown as open circles, are identical for the two surfaces.

under natural selection should be the highest fitness point. When recombination is a weak force relative to selection (Figure 7.3A), then the change in gamete frequencies follows the slope of the fitness surface and the equilibrium point reached from all initial gamete frequencies is the highest mean fitness. However, when recombination is strong relative to selection (Figure 7.3B), then equilibrium gamete frequencies depend strongly on initial gamete frequencies.

Figure 7.4 shows another example of two-locus selection on a fitness surface with two peaks shaped like a saddle due to dominance and epistasis at the two loci. When recombination is weak (Figure 7.4A), then the equilibrium points depend on the slope of the fitness surface at the initial gamete frequencies since populations move uphill due to the strong force of selection. However, when recombination is strong relative to selection (Figure 7.4B), then gamete frequencies will change in opposition to selection and change in directions that decrease mean fitness. When recombination is strong, as in Figures 7.3 and 7.4, the gamete frequency trajectories take sharp turns and move downhill on the fitness surfaces due to the force of recombination. This happens because recombination works toward

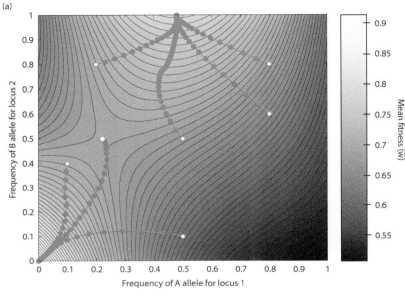

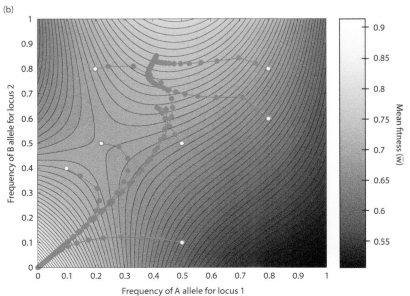

Figure 7.4 A fitness surface for two loci that each have two alleles where gene action exhibits epistasis. When recombination is a weak force ($c = 0.05$), equilibrium allele frequencies are dictated by natural selection. Equilibrium allele frequencies depend on initial gamete frequencies since the two highest mean fitness points are separated by a fitness valley (panel A). When recombination is strong ($c = 0.5$), allele frequencies change such that mean fitness actually decreases for a time before increasing again to eventually reach the lower of the two mean fitness peaks (panel B). The two initial gamete frequencies in the upper right of the surface reach an equilibrium point where fitness is not maximized and there is gametic disequilibrium ($D = 0.041$). Relative fitness values are $w_{AABB} = 0.61$, $w_{AABb} = 0.58$, $w_{AAbb} = 0.50$, $w_{AaBB} = 1.0$, $w_{AaBb} = 0.77$, $w_{Aabb} = 0.50$, $w_{aaBB} = 0.64$, $w_{aaBb} = 0.62$, $w_{aabb} = 0.92$. The seven initial allele frequency points, shown as open circles, are identical for the two surfaces.

gametic equilibrium ($D = 0$), whereas selection works toward the highest mean fitness. When one process is much stronger, then it will win out over the other process to determine the equilibrium. When the two processes are of approximately equal strength, then the result is a compromise that may produce an equilibrium that is neither gametic equilibrium nor maximum mean fitness.

The fitness surface in Figure 7.4 demonstrates an additional point about the action of natural selection on two loci. Gene action is a key variable in determining the equilibrium reached by natural selection. With additive gene action for two loci, the genotype at one locus has the same fitness value regardless of the genotype at the other locus. This means continual small changes in genotype frequencies that each increase mean fitness will eventually reach the highest mean fitness. In contrast, with non-additive gene action (dominance and epistasis), those same small, generation-by-generation changes in allele frequencies may lead to local maxima because the fitness surface is not a plane. Such peaks and valleys of mean fitness occur when the genotype at one locus has an impact on fitness values at another locus. Fitness surfaces, therefore, have increasingly complex topography as dominance and epistasis increase and additive gene action decreases. For this reason, natural selection is sometimes described as shortsighted or myopic because it operates based on mean fitness each generation rather than on some plan that accounts for the entire mean fitness surface. The result is that the equilibrium produced by natural selection when mean fitness surfaces possess multiple maxima depends strongly on initial genotype frequencies.

Although there is no general set of equilibrium gamete frequencies for two-locus selection with an arbitrary set of fitness values, many special cases have been examined that have produced some general conclusions (see Hastings 1981, 1986; reviewed by Ewens 2004). By itself, low frequencies of recombination (small c) make it more likely that selection will result in gametic disequilibrium at equilibrium gamete frequencies even with random mating. The combination of non-additive gene action and infrequent recombination also make gametic disequilibrium at equilibrium gamete frequencies more likely. High rates of self-fertilization (strong departure from random mating) add an additional force on gamete frequencies that can either compliment or act in opposition to selection and recombination (Hastings 1985; Holsinger and Feldman 1985).

Since mean fitness may decrease with selection and recombination, Fisher's fundamental theorem does not hold for two-locus selection (see Turner 1981; Hastings 1987). A critical conclusion from examining two-locus natural selection is that generalizing from the results of one-locus selection models to multiple loci may be biologically misleading except in limiting cases such as when there is very little recombination and there is no epistasis. As a final note, keep in mind that the two-locus model discussed here features the two loci contributing to a single fitness function. Therefore, the two-locus model here is distinct from the models of two linked loci where each locus has an independent fitness function such as the Hill–Robertson or Muller's Ratchet models.

7.2 Alternative models of natural selection

- Moving beyond the assumptions of fitness as constant viability in an infinitely growing population.
- Natural selection via different levels of fecundity.
- Natural selection with frequency-dependent fitness.
- Natural selection with density-dependent fitness.

The model of natural selection considered thus far equates fitness with the viability of genotypes. This is equivalent to assuming that while individuals of different genotypes vary in survival to adulthood, all genotypes are equal in terms of any other phenotypes that may impact numbers of progeny an individual contributes to the next generation. Looking again at Figure 6.3, you can see that there are numerous points in the reproductive life cycle where genotypes may have differential success or performance. Genotypes may differ in phenotypes such as the production and survival of gametes, mating success, gamete genetic compatibility with other gametes, and parental care. It is even possible that some alleles at a locus have an advantage during segregation of homologous chromosomes and are more likely to be found in gametes, a phenomenon called **meiotic drive** (see the historical background on this process in Birchler et al. 2003). Each of these points in the life cycle is a situation where genotypes will potentially have different levels of performance, eventually leading to different frequencies of genotypes in the progeny. The basic viability model of natural selection also assumes that fitness values are constant through time and space. Instead, it may be that fitness actually changes in response to the conditions found in different populations or in response to the changes in genotype frequency brought on by natural selection. In order to accommodate these potential biological situations, modifications to the model of natural selection are required. This section is devoted to extending the basic viability model of natural selection in a variety of ways to predict how natural selection works for different components of fitness and for changing fitness values. It is not possible to cover all possible models of natural selection exhaustively since there are many. Instead, each of the three models detailed in this section gives some insight into the dynamics of natural selection when one of the major assumptions of the one-locus, two-allele viability model is changed.

Natural selection via different levels of fecundity

Natural selection due to differences in genotype viability is sometimes called **hard selection** since genotype frequency changes come about from the death of individuals and their complete failure to reproduce. In contrast, natural selection due to differences in the fecundity (production of offspring) of individuals with different genotypes causes changes in the frequency of genotypes within the progeny of each generation. Fecundity selection is called **soft selection** because all individuals in the parental generation reproduce, although by differing amounts.

A fecundity model for natural selection on a diallelic locus requires a different approach than was taken for viability selection. A major difference is that fitness depends on the *pair* of genotypes that mate. This means that there are nine different fitness values in a fecundity selection model, as shown in Table 7.4. Another difference is that predicting the genotype frequencies of the progeny is going to be slightly more complicated than for simple random mating. Variation in fecundity may alter the number of progeny produced by each mating pair from the frequency expected by random mating alone. This requires accounting for the expected progeny genotype frequencies that arise from each mating pair weighted by the fecundity of that mating pair as

Table 7.4 Fitness values based on the fecundities of mating pairs of male and female genotypes for a diallelic locus along with the expected genotype frequencies in the progeny of each possible male and female mating pair weighted by the fecundity of each mating pair. The frequencies of the AA, Aa, and aa genotypes are represented by X, Y, and Z, respectively.

Fitness values:

Male genotype	Female genotype		
	AA	Aa	aa
AA	f_{11}	f_{12}	f_{13}
Aa	f_{21}	f_{23}	f_{23}
aa	f_{31}	f_{32}	f_{33}

Expected progeny genotype frequencies:

Parental mating	Fecundity	Total frequency	Offspring genotype frequencies		
			AA	Aa	aa
AA × AA	f_{11}	X^2	X^2	0	0
AA × Aa	f_{12}	XY	$\frac{1}{2}XY$	$\frac{1}{2}XY$	0
AA × aa	f_{13}	XZ	0	XZ	0
Aa × AA	f_{21}	YX	$\frac{1}{2}YX$	$\frac{1}{2}YX$	0
Aa × Aa	f_{22}	Y^2	$Y^2/4$	$(2Y^2)/4$	$Y^2/4$
Aa × aa	f_{23}	YZ	0	$\frac{1}{2}YZ$	$\frac{1}{2}YZ$
aa × AA	f_{31}	ZX	0	ZX	0
aa × Aa	f_{32}	ZY	0	$\frac{1}{2}ZY$	$\frac{1}{2}ZY$
aa × aa	f_{33}	Z^2	0	0	Z^2

shown in Table 7.4. X, Y, and Z are the frequencies of the genotypes AA, Aa, and aa, respectively. This is the same notation used for the proof of Hardy–Weinberg in Chapter 2.

This fecundity selection model is more complex than a viability model because the equations used to solve for genotype frequencies after one generation are functions of genotype frequencies in the parental generation. The mean number of offspring of each genotype after one generation of fecundity selection is found by summing the offspring frequencies (columns in Table 7.4) each weighted by its fecundity. For the progeny with the AA genotype, the average fecundity, or $\bar{f}$, is

$$\bar{f}X_{t+1} = f_{11}X^2 + f_{12}\frac{1}{2}XY + f_{21}\frac{1}{2}YX + f_{22}\frac{Y^2}{4}$$
$$(7.15)$$

which simplifies to

$$\bar{f}X_{t+1} = f_{11}X^2 + (f_{12} + f_{21})\frac{1}{2}XY + f_{22}\frac{1}{4}Y^2$$
$$(7.16)$$

Using similar steps, the equations for the average fecundities of the Aa and aa genotypes are

$$\bar{f}Y_{t+1} = (f_{12} + f_{21})\frac{1}{2}XY + (f_{13} + f_{31})XZ$$
$$+ \frac{1}{2}f_{22}Y^2 + (f_{23} + f_{32})\frac{1}{2}YZ$$
$$(7.17)$$

and

$$\bar{f}Z_{t+1} = f_{33}Z^2 + (f_{32} + f_{23})\frac{1}{2}YZ + f_{22}\frac{1}{4}Y^2$$
$$(7.18)$$

The total average fecundity ($\bar{f}$) is the sum of the average fecundity for each genotype, so that $\bar{f}X_{t+1}$, $\bar{f}Y_{t+1}$, and $\bar{f}Z_{t+1}$ give the proportion of the total number of offspring composed of any genotype after one bout of reproduction. Compare these equations for the average fecundity as functions of genotype frequencies with Eqs. 6.21 and 6.22 that are in terms of allele frequencies. Since random mating does not occur by definition when there is fecundity selection, general equilibrium points cannot be found for arbitrary sets of the nine fecundity values

shown in Table 7.4. Rather, the change in genotype frequencies caused by fecundity selection model must be understood by considering special cases of fecundity values.

One special case of fecundity selection occurs when the total fecundity of a mating is always the sum of the fecundity value of each genotype for each sex, a situation called additive fecundities, analogous to additive gene action (Penrose 1949). Let the fecundity values of females be f_{AA}, f_{Aa}, and f_{aa} and the fecundity values of males be m_{AA}, m_{Aa}, and m_{aa}. With an additive fecundity model, the fecundity values given in Table 7.4 would be $f_{11} = f_{AA} + m_{AA}$ and $f_{12} = f_{AA} + m_{Aa}$ as two examples. With additive fecundities, higher fecundities for heterozygotes result in both alleles being maintained in a population at equilibrium, as is true for overdominance in the viability selection model. A second special case is when fecundities are multiplicative (Bodmer 1965). For example, $f_{11} = f_{AA}m_{AA}$ and $f_{12} = f_{AA}m_{Aa}$. Depending on fecundity values for the three genotypes, there can be equilibrium points where both alleles are maintained in the population. A third special case that has been examined extensively is when there are four fecundity parameters that correspond to the degree of heterozygosity of each mating pair (Hadeler and Liberman 1975; Feldman et al. 1983). In these cases, depending on the specific fecundity values used, it is also possible that fecundity selection can maintain both alleles in the population since equilibrium points are reached when all three genotypes have non-zero frequencies. Nevertheless, the fecundity selection model does not result in the maintenance of genetic variation for arbitrary fitness values more often than the basic viability model of selection (Clark and Feldman 1986). This means that fecundity models predict that natural selection frequently results in fixation or loss of alleles at equilibrium, just as the viability model does for directional selection.

Pollak (1978) has shown that mean fecundity does not necessarily increase with fecundity selection. This means that the mean fecundity is not necessarily maximized at equilibrium genotype frequencies for fecundity selection, in contrast to the way natural selection maximizes mean fitness in the viability model for one locus with two alleles.

Hybridization between genetically modified and wild sunflowers provides an example of how a simple fecundity selection model can be used to understand changes in allele frequencies. Using transgenic biotechnology, it is now routine practice to permanently incorporate foreign genes into crop plants. There is the possibility that such transgenes can escape into the wild through the hybridization that occurs between some crop plants and wild relatives that are often weeds (reviewed by Snow and Palma 1997). In the case of sunflowers, mating between pure crop genotypes and wild plants produces hybrids with seed production that is only 2% of that shown by wild plants, but hybrids and wild plants have identical survival rates. Cummings et al. (2002) established three experimental populations with half crop–wild-plant F1 hybrids and half wild plants. In these populations, the initial frequency of crop-specific alleles was 25%. The frequencies of the crop-specific alleles at three allozyme loci dropped to about 5% in the next generation. The crop-specific allele frequency in the next generation best matched the allele frequency predicted by a fecundity selection model with additive fecundities.

Natural selection with frequency-dependent fitness

In the basic viability model, we considered fitness values as invariant properties of the genotypes. Another way to say this is that a fitness value w_{xx} is constant regardless of conditions or the frequencies of any of the genotypes. It seems intuitive to expect that the fitness of a genotype may depend on its frequency in a population, and there is direct evidence for frequency-dependent fitness in natural populations. For example, mating success of males with different chromosomal inversion genotypes in *Drosophila* depends on chromosome inversion frequencies in the population (see Álvarez-Castro and Alvarez 2005). In plants, the frequency of different flower colors in a population may impact the frequency of visits by pollinators and thereby cause frequency-dependent mating success (e.g. Gigord et al. 2001; Jones and Reithel 2001). A whimsical example of frequency-dependent fitness values comes from the left and right curvature of the mouths of Lake Tanganyikan cichlid fish *Perissodus microlepis* that pluck and eat scales from other fish. There appears to be an advantage to the rarer phenotype, presumably since the fish that are attacked anticipate approach of the cichlid from the side of the more frequent mouth phenotype (Hori 1993). Frequency-dependent selection is the basis of one form of **balancing selection** (along with heterozygote advantage) and was originally described by E. B. Poulton in 1884 (see Allen and Clarke 1984). Negative

frequency-dependent selection has been widely invoked as an explanation of how natural selection can maintain genetic polymorphism, and many examples have been documented in a range of organisms. Brisson (2018) reviews situations where negative frequency-dependent selection may be difficult to distinguish from other forms of natural selection.

We can construct a simple selection model where fitness (as genotype-specific viability) depends on genotype frequency and, therefore, changes as genotype frequencies change. The key concept in frequency-dependent selection models is creating a measure of fitness that changes. Suppose that the fitness of a genotype decreases as that genotype becomes more common in the population, called **negative frequency dependence**. The relative fitness values are:

$$w_{AA} = 1 - s_{AA}p^2$$

$$w_{Aa} = 1 - s_{Aa}2pq \qquad (7.19)$$

$$w_{aa} = 1 - s_{aa}q^2$$

where s_{xx} represents the genotype-specific selection coefficient. Genotypes have higher fitness when they are rare since relative fitness decreases as the product of the selection coefficient and the genotype's frequency increases. Note that the selection coefficient itself is a constant and can be thought of as a *per-capita* decrease in relative fitness.

As with other models of selection, the equilibrium points in this model of natural selection can be found by determining when the change in allele frequency (Δp) is equal to 0. The expression for change in allele frequency over one generation of fecundity selection is

$$\Delta p = \frac{pqs(q-p)(p^2 - pq + q^2)}{\overline{w}} \qquad (7.20)$$

for the special case of the selection coefficient being equal for all genotypes as derived in Math Box 7.1. Two equilibrium points occur at fixation and loss ($p = 1.0$ and $p = 0.0$) since the pq term in the numerator is 0. There is also an equilibrium point at $p = \frac{1}{2}$ since the $q - p$ term is 0.

Figure 7.5 shows the relative fitness values in Eq. 7.19 when all selection coefficients are equal to 1. It is interesting to note that, at $p = \frac{1}{2}$, the fitness of the heterozygote is less than that of the two homozygotes, so this model of natural selection does not

Math box 7.1 The change in allele frequency with frequency-dependent selection

Start with the expression for change in allele frequency under viability selection given in Eq. 6.23:

$$\Delta p = \frac{pq[p(w_{AA} - w_{Aa}) + q(w_{Aa} - w_{aa})]}{\overline{w}} \tag{7.21}$$

and then substitute the definitions of the frequency-dependent fitness values given in Eq. 7.19. If the selection coefficients for all genotypes are equal so that they can be represented by s without any subscripts, this gives

$$\Delta p = \frac{pq[p(1 - sp^2 - (1 - s2pq)) + q(1 - s2pq - (1 - sq^2))]}{\overline{w}} \tag{7.22}$$

Expanding the terms inside the square brackets gives

$$\Delta p = \frac{pq[p - sp^3 - p + s2p^2q + q - s2pq^2 - q + sq^3]}{\overline{w}} \tag{7.23}$$

which simplifies by canceling the positive and negative p and q terms and factoring out an s to give

$$\Delta p = \frac{pqs[-p^3 + 2p^2q - 2pq^2 + q^3]}{\overline{w}} \tag{7.24}$$

The term in square brackets can then be factored to give

$$\Delta p = \frac{pqs(q - p)(p^2 - pq + q^2)}{\overline{w}} \tag{7.25}$$

Interact box 7.2 Frequency-dependent natural selection

Populus can be used to simulate frequency-dependent natural selection. In the **Model** menu, choose **Natural Selection** and then **Frequency-Dependent Selection (Diploid Model)**. In the model dialog, you can set the frequency-sensitive relative fitness values for the three genotypes produced by one locus with two alleles. The $s1$, $s2$, and $s3$ values correspond to s_{AA}, s_{Aa}, and s_{aa} in Eq. 7.19. First, try the values of $s1 = 0.3$, $s2 = 1.0$, and $s3 = 0.3$. Explain why the graph of genotype-specific fitness by allele frequency (the middle panel) looks the way it does.

Next, enter the selection coefficients of $s1 = 0.7$, $s2 = 1.0$, and $s3 = 0.2$. Then, compare the p by delta p graph in the top panel with the p by mean fitness graph in the bottom panel to see the mean fitness at equilibrium allele frequency. Does natural selection always cause allele frequencies to reach an equilibrium that corresponds to the maximum value of the mean fitness? Through educated guesses, try to find selection coefficients where equilibrium allele frequencies do and do not correspond to the maximum mean fitness value.

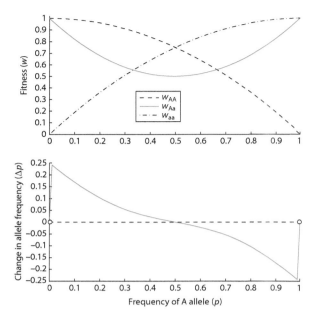

Figure 7.5 The relative fitness of each genotype (w_{xx}) and the change in allele frequency (Δp) across all frequencies of the A allele under frequency-dependent natural selection. There is a stable equilibrium point at $p = 0.5$ in this particular case, even though the heterozygote has the lowest fitness. Two unstable equilibria at fixation and loss are marked with open circles. Here, the relative fitness values are $w_{AA} = 1 - s_{AA}p^2$, $w_{Aa} = 1 - s_{Aa}2pq$ and $w_{aa} = 1 - s_{aa}q^2$ with $s_{AA} = s_{Aa} = s_{aa}$.

require overdominance for fitness to maintain genetic variation at equilibrium. However, with independent selection coefficients for each genotype, there is no stable polymorphism in general. With arbitrary fitness values for the three genotypes, there are many possible outcomes, many without stable polymorphism.

Natural selection with density-dependent fitness

An assumption in most models of natural selection is that populations are able to grow without any limits. The first section of this chapter developed a natural selection model where the size of the population of any genotype one generation in the future was its population size currently multiplied by a constant (refer back to Eq. 6.1). This model is obviously unrealistic because no organism can grow without some eventual limits on population size. Organisms are limited by the space and resources available to them, limitations that lead to changes in the rate of growth as the density of individuals changes over time. To

incorporate such limits in a model of natural selection, we can alter our basic genotype-specific population growth equations to incorporate an upper bound on the population size as well as a rate of population growth that changes with population size.

A simple model where population growth has an upper bound is called **logistic growth** and the upper limit is called the **carrying capacity** (symbolized by K). Logistic growth depends on feedback between the growth rate and the size of the population according to

$$\lambda = 1 + r - \frac{r}{K}N \qquad (7.26)$$

where N is the population size and r is the rate of increase (λ used as the growth rate earlier in the chapter can be equated to r by $\lambda = 1 + r$). Biologically, r represents the rate of growth in excess of the replacement rate of 1. When $N = 0$, then $\frac{r}{K}N$ is 0 and the growth rate is at its maximum. But when $N = K$, then $\frac{r}{K}N$ equals r and the population replaces itself but does not change in size.

Logistic population growth can be applied to the three genotypes at a diallelic locus by defining genotype-specific carrying capacities and rates of increase to obtain absolute fitness values for each genotype

$$\lambda_{AA} = 1 + r_{AA} - \frac{r_{AA}}{K_{AA}}N \qquad (7.27)$$

$$\lambda_{Aa} = 1 + r_{Aa} - \frac{r_{Aa}}{K_{Aa}}N \qquad (7.28)$$

and

$$\lambda_{aa} = 1 + r_{aa} - \frac{r_{aa}}{K_{aa}}N \qquad (7.29)$$

Here, the sum of the numbers of individuals with each genotype is equal to the total size of the population ($N_{AA} + N_{Aa} + N_{aa} = N$). The average absolute fitness in the population is the average of the r and r/K values for each genotype:

$$\bar{\lambda} = 1 + \bar{r} - \frac{\bar{r}}{K}N \qquad (7.30)$$

where $\bar{r} = p_t^2 r_{AA} + 2p_t q_t r_{Aa} + q_t^2 r_{aa}$ and $\frac{\bar{r}}{K} = \frac{p_t^2 r_{AA}}{K_{AA}} + \frac{2p_t q_t r_{Aa}}{K_{Aa}} + \frac{q_t^2 r_{aa}}{K_{aa}}$.

The final step is to use these results to modify our previous results for unbounded population growth to take logistic growth into account. First, we can modify the growth in the total size of the population (Eq. 6.1 for unbounded growth) so that

$$N_{t+1} = \bar{\lambda} N_t \qquad (7.31)$$

This equation predicts that the total population size cannot exceed the largest carrying capacity. We can also follow allele frequencies over time by modifying the expression for allele frequency after one generation of selection:

$$p_{t+1} = \frac{\lambda_{AA} p_t^2 + \lambda_{Aa} p_t q_t}{\bar{\lambda}} \qquad (7.32)$$

which is the logistic growth version of Eq. 6.19.

Both the numbers of individuals of each genotype and the allele frequency under density-dependent natural selection can be seen in Figure 7.6 when the AA genotype has the highest carrying capacity. Starting out with a very small N, the numbers of

individuals of all genotypes increase over time. However, once N approaches the lowest carrying capacity, K_{aa} in this illustration, the number of aa individuals peaks and then declines. This happens because the absolute fitness of aa approaches 1 in the fewest generations while the other two genotypes continue to add individuals to their populations because their carrying capacities are higher. The same phenomenon also occurs to the Aa genotype because it has the next lowest carrying capacity. The AA genotype has the highest carrying capacity, and the number of individuals of that genotype eventually grow to the point where it makes up the entire population.

The general result for density-dependent selection is that the carrying capacities of the three genotypes will determine the eventual genotype and allele frequencies when populations approach their carrying capacities. When K_{AA} is the highest, then the A allele goes to fixation, and when K_{aa} is the largest, then the a allele goes to fixation. Alternatively, when K_{Aa} is the largest, then there is an equilibrium with both alleles segregating, and K_{Aa} is the smallest, then either A or a will reach fixation depending on initial allele frequencies. These results are qualitatively identical to those for unbounded growth.

In contrast, the results of density-dependent and density-independent natural selection do not agree when population size is restricted to low numbers. The total population size N may be much less than the carrying capacity in highly disturbed or inhospitable environments where individuals have low reproductive output or high turnover. To see this, consider Eqs. 7.27–7.29 for genotype absolute fitness values. As N for each genotype approaches 0, the $\frac{r}{K}N$ term will also approach 0, leaving each absolute fitness increasingly determined by the genotype-specific growth rate. The genotype with the highest growth rate should then increase the fastest and dictate genotype and allele frequencies at equilibrium (if r_{AA} is the highest, A fixes; if r_{aa} is the highest, a fixes; if r_{Aa} is the highest, both alleles segregate; and if r_{Aa} is the lowest, then initial allele frequencies dictate fixation of either A or a). This effect can be seen in Figure 7.6 where allele frequencies change toward a lower frequency of p when the population is small. Despite the fact that the population is expected to fix for p at carrying capacity, the growth rate of the aa genotype is the greatest, so it has the greatest impact on allele frequencies when the population is small.

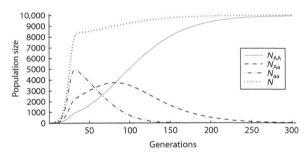

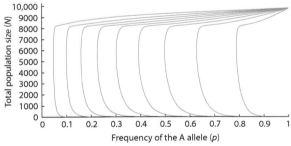

Figure 7.6 The results of density-dependent natural selection on the numbers of individuals of different genotypes (N_{AA}, N_{Aa} and N_{aa}) and allele frequencies in a population of total size N. At the upper limit of N, the equilibrium allele and genotype frequency is determined by the genotype with the highest carrying capacity (K). In contrast, the genotype with the highest growth rate (r) has the greatest impact on allele frequency when the population is small. In this example, $K_{AA} = 10\,000$, $K_{Aa} = 9000$, and $K_{aa} = 8000$ with $r_{AA} = 0.2$, $r_{Aa} = 0.25$, and $r_{aa} = 0.3$.

Interact box 7.3 Density-dependent natural selection

Populus can be used to simulate density-dependent natural selection. In the **Model** menu, choose **Natural Selection** and then **Density-Dependent Selection w/ Genetic Variation**. In the options dialog, you can set the genotype-specific carrying capacity and growth rates. Click on the radio button for **Nine-Frequency** to display results for nine initial allele frequencies (the **Single Frequency** button shows the results in terms of the total population size N). The **N** text box sets the initial population size. Press the **View** button to see the simulation results.
    Parameter values to simulate:

K_{AA} = 8000, K_{Aa} = 8000, and K_{aa} = 10 000; r_{AA} = 0.4, r_{Aa} = 0.4, and r_{aa} = 0.3; generations = 100
K_{AA} = 8000, K_{Aa} = 10 000, and K_{aa} = 8000; r_{AA} = 0.4, r_{Aa} = 0.3, and r_{aa} = 0.35; generations = 100
K_{AA} = 8000, K_{Aa} = 6000, and K_{aa} = 9000; r_{AA} = 0.5, r_{Aa} = 0.3, and r_{aa} = 0.4; generations = 100

7.3 Combining natural selection with other processes

- Natural selection and genetic drift acting simultaneously.
- Genetic differentiation under neutral gene flow or natural selection.
- The balance between natural selection and mutation.
- Genetic load.

Natural selection takes place at the same time that other processes are also operating and having an impact on allele frequencies. These other processes may work in concert with natural selection and work toward the same equilibrium allele frequencies, or they may work against natural selection toward alternative equilibrium allele frequencies. Since many population genetic processes are likely to be acting simultaneously in actual biological populations, it is important to put natural selection into the context of other processes that impact allele frequencies. This section first considers allele frequencies when natural selection and genetic drift are in opposition and then natural selection and mutation acting in opposition. When natural selection and other processes act in concert, this simply shortens the number of generations required to reach equilibrium and does not alter equilibrium allele frequencies.

Natural selection and genetic drift acting simultaneously

Wright (1931) showed the probability that a population has a given allele frequency when exposed to the simultaneous processes of natural selection, genetic drift, and mutation as given by

$$\varphi(p) = Cp^{(4N_e\mu - 1)}q^{(4N_e\nu - 1)}e^{(4N_espq)} \qquad (7.33)$$

where φ (pronounced "phi") means a probability density, p and q are the allele frequencies, N_e is the effective population size, μ and ν are the forward and backward mutation rates, s is the selection coefficient, and C is a constant used to adjust the total probability across all allele frequencies to sum to 1.0 for each value of N_es. This equation is a probability density function for an ensemble population like those discussed in Chapter 3 for genetic drift. It describes the chance that one of many replicate populations will reach any allele frequency between zero and one at equilibrium given the values of the effective population size and the selection coefficient.

This equation is most easily understood by examining its predictions in graphical form (Figure 7.7). When N_es is near 0, either genetic drift is very strong because the population is tiny or the selection coefficient is extremely small and so the populations are evolving in a neutral fashion under drift alone. In either of these cases, genetic drift is the dominant process and will eventually result in either fixation or loss in all populations. At the lowest values of N_es in Figure 7.7, populations are most likely to have allele frequencies near 0 or near 1 as expected under genetic drift alone. Alternatively, N_es can take on large values in two general situations. One is when genetic drift is very weak because the effective population size is very large and there is some natural selection favoring heterozygotes (s can occupy a wide range as long as it is not extremely small). The other is when the selection coefficient is large and the effective population size is at least 10 or so individuals and, therefore, genetic drift is not extreme. At the largest values of N_es in Figure 7.7,

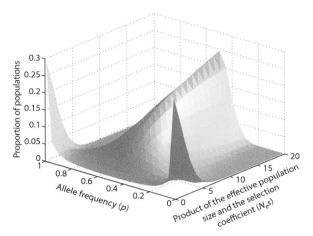

Figure 7.7 The expected distribution of allele frequencies for a very large number of replicate finite populations under natural selection where there is overdominance for fitness ($w_{AA} = w_{aa} = 1 - s$ and $w_{Aa} = 1$). In an infinite population, the expected allele frequency at equilibrium is 0.5. However, in finite populations, the equilibrium allele frequency will depend on the balance of natural selection and genetic drift. This balance is determined by the product of the effective population size and the selection coefficient ($N_e s$). Low values of $N_e s$ mean that selection is very weak compared to drift and each population reaches fixation or loss. High values of $N_e s$ mean that selection is strong compared to drift and most populations reach an equilibrium allele frequency near 0.5. Here, forward and backward mutation rates are equal ($\mu = \nu = 0.00001$).

the probability is greatest that the allele frequency in a population will be near 0.5. Equation 6.35 shows that 0.5 is the expected equilibrium allele frequency under balancing selection when $w_{AA} = w_{aa}$ and a population is infinite. Therefore, when $N_e s$ is large, selection is stronger than drift and the equilibrium allele frequency is determined mostly by natural selection. At intermediate values of $N_e s$, the equilibrium allele frequency for many populations is between the equilibrium allele frequencies expected under genetic drift acting alone or natural selection acting alone.

To summarize the balance of natural selection and genetic drift, Motoo Kimura suggested a simple rule of thumb for a diploid locus (Kimura 1983). If four times the product of the effective population size and the selection coefficient is much less than one ($4N_e s << 1$), then selection is weak relative to sampling and genetic drift will dictate allele frequencies. Alternatively, if four times the product of the effective population size and the selection coefficient is much greater than one ($4N_e s >> 1$), then selection is strong relative to sampling and natural selection will dictate allele frequencies. When four times the

product of the effective population size and the selection coefficient is approximately one ($4N_e s \approx 1$), then allele frequencies are determined about equally by drift and selection.

Figure 7.8 shows an example of the balance between genetic drift and natural selection in replicated laboratory populations of the fruit fly *Drosophila melanogaster* (Wright and Kerr 1954). The plot shows allele frequencies at the *Bar* locus for 108 replicate populations that were each founded from four males and four females every generation (since *Bar* is hemizygous in males, the effective population size is equivalent to six rather than eight diploid individuals). Although the vast majority of populations fix for the wild-type allele due to strong natural selection favoring the recessive phenotype of wild-type homozygotes, three populations fix for the *Bar* allele by the end of the experiment. The selection coefficient against the *Bar* homozygote was estimated at 0.63, giving an upper bound estimate of $4N_e s \approx 15$ in the experiment (N_e may have been less than six in actuality). Even with this value of $4N_e s$, natural selection is not sufficiently strong to dictate equilibrium allele frequency in all populations.

It is also possible to gain biological insight into $N_e s$ by recognizing that it is analogous to the quantity $N_e m$ that dictates the balance between genetic drift and gene flow discussed in Chapter 4. Both $N_e s$ and $N_e m$ represent the net balance of the pressure on allele frequencies toward eventual fixation or loss due to genetic drift and the countervailing force driving allele frequencies toward a specific allele frequency caused by either natural selection or by gene flow. In the case of natural selection, the specific allele frequency is dictated by the relative fitness values of genotypes, while, in the case of gene flow, the specific allele frequency is the average allele frequency for all demes.

Genetic differentiation among populations by natural selection

Natural selection is expected to act at the same time as the neutral processes of genetic drift and gene flow, with potentially contrasting impacts on the genetic differentiation of loci. When relative fitness is a function of subpopulation or geographic location, selection is expected to shape the population differentiation of loci. Natural selection can impact population differentiation in two main ways. Directional selection where relative fitness values differ among populations, or **local adaptation**, is

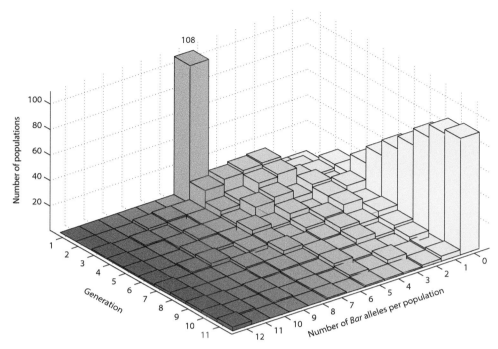

Figure 7.8 Frequency of the *Bar* allele in 108 replicate *Drosophila melanogaster* populations over 10 generations (Wright and Kerr 1954). Each population was founded from four males and four females. The *Bar* locus is found on the X chromosome and so is hemizygous in males, making the effective population size equivalent to about six individuals. The eyes of *Drosophila melanogaster* individuals homozygous for the wild-type allele are oval, but heterozygotes and homozygotes for the partially dominant *Bar* allele have bar-shaped eyes with a reduced number of facets. Females homozygous for the Bar allele produced 37% of the progeny compared to females homozygous or heterozygous for the wild-type allele. Despite this strong natural selection against *Bar*, three populations fixed for *Bar* by the end of the experiment. Compare with the similar example in **Figure 3.11** where the locus is selectively neutral.

Interact box 7.4 The balance of natural selection and genetic drift at a diallelic locus

You can use the text simulation website to simulate the simultaneous action of genetic drift and natural selection in many identical finite populations. Use the **Simulations** menu to choose the **Drift Selection Mutation** model. In the dialog, check the natural selection box and be sure to leave the mutation box unchecked.

Run the model a few times using the default parameters to understand the output. The graph on the left shows allele frequency trajectories of a sample of replicate loci/population over time. The histogram on the right shows the distribution of allele frequencies for all populations (analogous to a two-dimensional slice through Figure 7.7 for a single value of N_es). Note that mutation will have almost no effect on the outcome of allele frequencies as long as backward and forward mutation rates are equal and very small (you can test this by setting one very high mutation rate such as 0.1 to see the impact).

Simulate drift and selection, one keeping the selection coefficient constant and varying N_e. Before each of these runs, you should compute the value of $4N_es$ and make a prediction about the distribution of allele frequencies among the populations. Set $w_{AA} = w_{aa} = 0.9$ and $w_{Aa} = 1.0$ (or $s = 0.1$) and then run separate simulations for $N_e = 5$, 20, and 200.

Describe the ensemble distribution of alleles frequencies for each level of N_es shown in the histogram plot and explain what the plot shows about the relative balance of drift and selection.

expected to result in different genotype frequencies among populations and, therefore, result in population differentiation. Alternatively, directional selection or balancing selection where fitness values are uniform among populations is expected to maintain similar genotype frequencies among multiple populations in opposition to genetic drift, leading to reduced genetic differentiation. In all of these cases, selection coefficients need to be large enough so that natural selection is strong enough to overcome the randomization of allele frequencies among populations by genetic drift and the simultaneous homogenization of allele frequencies caused by gene flow.

One approach to identify loci that have experienced natural selection is to observe some locus-by-locus pattern in a sample of many loci and identify those loci that have outlier values based on a model of neutral evolution and predictions for the impact of natural selection. The basic principle behind this approach is due to a paper by Lewontin and Krakauer (1973), who pointed out that natural selection will operate differently for each locus based on the relative fitness values of genotypes and the strength of selection but that neutral processes of genetic drift and gene flow act uniformly over all loci with random variation. The logic is that most loci are evolving neutrally or nearly neutrally and that the relatively few loci evolving under strong natural selection can be recognized because they fall outside the range values observed for the neutral loci.

Lewontin and Krakauer (1973) proposed using the evolutionary variance of $\hat{F}_{ST}$ for a sample of loci to identify outlier loci under the assumptions of an island model of gene flow. Their predictions for natural selection were twofold. First, loci experiencing similar selection pressures among subpopulations would exhibit less variance than neutral loci diverging by drift because selection would counteract random divergence of allele frequencies. Second, loci experiencing heterogeneous selection coefficients among subpopulations due to local adaptation would exhibit stronger differentiation than neutral loci diverging by drift alone. Several aspects of the Lewontin and Krakauer (1973) test were soon identified as weaknesses of the approach (Nei and Maruyama 1975; Robertson 1975a, 1975b; Nei and Chakravarti 1977; Nei et al. 1977). One potential problem is if rates of gene flow are greater among some subpopulations than among others contrary to the island model, then the allele frequencies of some demes would be correlated (not independent), which could violate the variance of $\hat{F}_{ST}$ assumptions

(a positive covariance in $\hat{F}_{ST}$ between some demes would increase the total variance in $\hat{F}_{ST}$). A second potential problem is the pattern of relatedness among subpopulations, which could range from uniformly related to hierarchically related, caused by the random timing of mutations, again with potential consequences for the variance of $\hat{F}_{ST}$. A third problem was the empirical difficulty of obtaining a precise estimate of the variance in $\hat{F}_{ST}$ since genetic data were then very difficult to collect. A final problem was that expectations relied on analytical approximations because stochastic computer simulations of $\hat{F}_{ST}$ distributions were very difficult to carry out.

With ongoing innovation in methods to observe SNP genetic polymorphism at a genomic scale, it is now possible to estimate the variance in $\hat{F}_{ST}$ using a very large sample of loci. (Such studies have taken the ambiguous moniker of "genome scans.") This led to renewed efforts to use the distribution of $\hat{F}_{ST}$ to identify loci that have experienced natural selection as well as additional effort to develop appropriate null expectations for neutral loci. For example, Akey et al. (2002) used a set of about 26 000 single nucleotide polymorphisms (SNPs) in human genomes sampled from three populations (African American, East Asian, and European American) and assumed an island model of gene flow. Akey showed that SNPs in coding loci had an average $\hat{F}_{ST} = 0.107$ compared with average $\hat{F}_{ST} = 0.123$ in coding regions, consistent with purifying selection common over all populations. In total, 174 loci were identified as $\hat{F}_{ST}$ distribution outliers and, therefore, having experienced strong natural selection.

There are different perspectives on how sensitive the distribution of $\hat{F}_{ST}$ values among neutral loci is to the patterns of gene flow among subpopulations and to the history of population size and growth rate. The separation of time scales into shallow and deep time in the coalescent for a subdivided population (see Chapter 4) suggests that the distribution of $\hat{F}_{ST}$ may be robust to variation in population demographic history and gene flow patterns since the deeper time portion of the genealogy has a strong influence on the patterns of polymorphism compared to the shallower recent time epoch where migration events mostly occur (see Beaumont 2005). A Bayesian statistical estimation approach (Foll and Gaggiotti 2008) has addressed a range of potential shortcomings that can lead to false-positive identification of loci as outliers based on posterior probabilities of a model of natural selection and a model of neutral genetic differentiation for each locus.

In contrast, simulation studies suggest that $\hat{F}_{ST}$ distributions can differ between different models of gene flow, making inference of outlier loci contingent on the model of gene flow employed to establish a null distribution. Excoffier et al. (2009a) used computer simulations of genetic loci under infinite allele and stepwise mutation models to compare the $\hat{F}_{ST}$ distributions for an island model and a hierarchical island model. They showed that using finite island model expectations for the distribution of $\hat{F}_{ST}$ when populations actually experienced hierarchical island model gene flow patterns lead to many false-positive outlier loci. In addition, they reanalyzed empirical SNP data sets for humans and stickleback fish, showing many fewer outlier loci under hierarchical island expectations than under an island model. Figure 7.9 illustrates this pattern with $\hat{F}_{ST}$ distributions from simulations of both the classical island model and the hierarchical island model. While the average $\hat{F}_{ST}$ of both distributions is nearly identical, the shape of the distributions is somewhat different, especially

the proportion of loci in the tails of the distributions. Variable rates of recombination among neutral loci may impact the shape of $\hat{F}_{ST}$ distributions as well (Booker et al. 2020).

Empirical validation studies for the $\hat{F}_{ST}$ outlier approach have also been carried out by examining $\hat{F}_{ST}$ for loci known to be experiencing strong natural selection, frequently using human genetic data. For example, Lohmueller et al. (2006) examined $\hat{F}_{ST}$ between European and African human populations for SNPs at 48 loci known to be associated with disease phenotypes but found that these loci did not have outlier $\hat{F}_{ST}$ values (see similar studies by Myles et al. 2008 and Adeyemo and Rotimi 2010). In contrast, Brandt et al. (2018) showed that human leukocyte antigen or *HLA* genes in the major histocompatibility complex (*MHC*) had lower $\hat{F}_{ST}$ than other loci consistent with balancing natural selection because of overdominance for fitness (see also Marigorta et al. 2011).

While the empirical genomic data to estimate $\hat{F}_{ST}$ for large numbers of loci in more diverse species are becoming increasingly common, the fundamental challenges of the Lewontin and Krakauer test

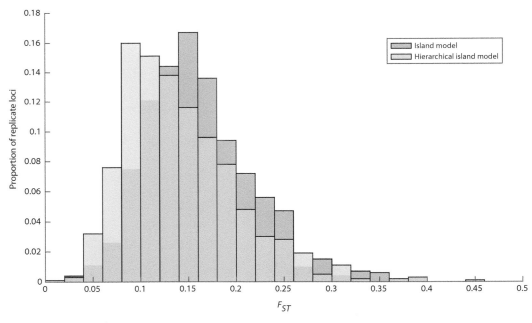

Figure 7.9 Distributions of $\hat{F}_{ST}$ for 1000 independent biallelic loci under the island model and the hierarchical island models of gene flow after 110 generations. The central tendency of both distributions is nearly identical with F_{ST}(island) = 0.2626 and F_{ST}(hierarchical) = 0.2548, but the shape and tails of the distributions differ. The differences in $\hat{F}_{ST}$ distributions impact the perception that loci are either more or less differentiated than expected under the null model of neutral evolution. Simulation parameters were 20 total subpopulations each with N_e = 20, island migration rate of m = 0.04, hierarchical migration rates of m_d = 0.05 among demes within groups and m_g = 0.02 among groups with four groups and five demes per group, and no mutation.

remain. Tests for loci exhibiting more or less genetic differentiation than expected under a neutral null model will require careful consideration of the numerous aspects of the null model for the distribution of $\hat{F}_{ST}$, including a range of possible models of gene flow.

The balance between natural selection and mutation

Natural selection takes place at the same time that mutation is working to alter allele frequencies and reintroduce alleles that selection may be driving to loss. Therefore, the process of natural selection may be counteracted to some degree by mutation. If a completely recessive allele is both deleterious when homozygous and also produced by spontaneous mutation, there are contrasting forces acting on its frequency. Mutation pressure will continually reintroduce the allele into a population, while natural selection will continually work to drive the allele to loss. What equilibrium allele frequency is expected when the opposing processes of mutation and natural selection balance out?

Let us assume that there are two alleles at one locus and that the a allele is completely recessive and has a frequency of q. Also, assume the case of selection against a recessive as given in Table 6.4. An expression for the change in allele frequency per generation for the specific case of natural selection against a recessive homozygote can be obtained by substituting the fitness values of $w_{AA} = w_{Aa} = 1$ and $w_{aa} = 1 - s$ into Eq. 6.24

$$\Delta q_{selection} = \frac{pq[q((1-s)-1) + p(1-1)]}{(1)p^2 + (1)2pq + (1-s)q^2} \quad (7.34)$$

which then rearranges to

$$\Delta q_{selection} = \frac{-spq^2}{1 - sq^2} \quad (7.35)$$

Let us further assume that mutation is irreversible and that the probability that an A allele mutates to an a allele is μ. The change in the frequency of the allele due to mutation each generation is then

$$\Delta q_{mutation} = \mu p \quad (7.36)$$

At equilibrium, the action of natural selection pushing the allele toward fixation and the pressure of mutation increasing the allele frequency exactly

balance so that allele frequency does not change. This means that, at equilibrium,

$$\Delta q_{mutation} + \Delta q_{selection} = 0 \quad (7.37)$$

Substituting in the expressions for $\Delta q_{mutation}$ and $\Delta q_{selection}$ into this equation yields

$$\mu p = \frac{spq^2}{1 - sq^2} \quad (7.38)$$

If we assume that the frequency q of the a allele is low, then q^2 is very small and the quantity $1 - sq^2$ is approximately 1. This approximation leads to

$$\mu p = spq^2 \quad (7.39)$$

which can be solved in terms of genotype frequency:

$$q^2 = \frac{\mu}{s} \quad (7.40)$$

or in terms of allele frequency:

$$q_{equilibrium} = \left(\frac{\mu}{s}\right)^{\frac{1}{2}} \quad (7.41)$$

Thus, the expected frequency of a deleterious recessive allele at an equilibrium between natural selection and mutation depends on the ratio of the mutation rate and the selection coefficient. Equation 7.41 shows that even if a recessive homozygous genotype is lethal ($s = 1$), the expected frequency of the allele is $\mu^{0.5}$ and the expected frequency of the lethal genotype is μ due to recurrent mutation.

The balance between selection and drift is illustrated in Figure 7.10. The Δq due to mutation is always positive, while the Δq due to selection in this case is negative. Figure 7.9 uses the absolute value of each Δq to show where the change in allele frequency for the two processes intersects. This intersection is the equilibrium point. The expected equilibrium allele frequency is

$$q_{equilibrium} = \left(\frac{1 \times 10^{-6}}{0.1}\right)^{\frac{1}{2}} = 0.0032 \text{ , in agreement}$$

with the figure.

Consanguineous mating results in an excess of homozygosity and a deficit of heterozygosity compared to random mating. If f is the degree of departure from Hardy–Weinberg expectations (see Eq. 2.20), then the equilibrium allele frequency from Eq. 7.41 can be restated as

$$q^2 + fpq = \frac{\mu}{s} \quad (7.42)$$

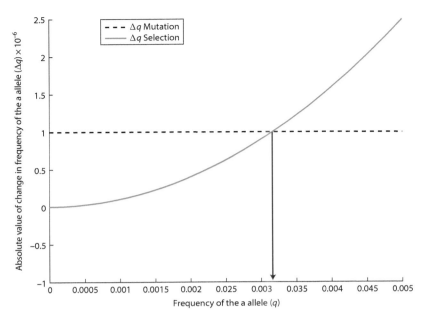

Figure 7.10 The absolute value of the change in allele frequency due to mutation ($\Delta q_{\text{mutation}}$) and due to natural selection ($\Delta q_{\text{selection}}$) when there is selection against a recessive homozygote. Mutation continually makes new copies of the recessive allele while selection continually works toward loss of the recessive allele. The equilibrium allele frequency occurs when the processes of mutation and selection exactly counteract each other. Here, $s = 0.1$ and $\mu = 1 \times 10^{-6}$, so the expected equilibrium is $q_{\text{equilibrium}} = 0.0032$, as shown by the vertical arrow.

Under the assumption that q is small compared to f, the approximate equilibrium allele frequency when mutation and selection reach a balance with consanguineous mating is

$$q_{equilibrium} = \frac{\mu}{fs} \qquad (7.43)$$

(see Haldane 1940; Morton 1971). Since recessive deleterious mutations are only perceived by natural selection when homozygous, consanguineous mating increases the effectiveness of selection by increasing the proportion of homozygous genotypes in the population. This means that selection is more effective at eliminating the recessive homozygote (there are fewer heterozygotes that shelter the allele), and the equilibrium allele frequency for mutation–selection balance occurs at a lower allele frequency. It is counterintuitive that populations which cease consanguineous mating and engage in random mating may temporarily experience an increase in deleterious allele frequencies and a decrease in average fitness due to less effective natural selection.

Genetic load

For natural selection to change allele frequencies in a population, individuals of some genotypes must experience higher rates of death (either actual for viability selection or reproductive for fecundity selection) than other genotypes. Natural selection works

by culling individuals of some genotypes in favor of individuals of other genotypes, increasing the mean fitness in the process. The amount of death or failed reproduction associated with natural selection was

---

Interact box 7.5 Natural selection and mutation

You can use the text simulation website to simulate the simultaneous action of natural selection and mutation. Use the **Simulations** menu to choose the **Drift Selection Mutation** model. In the dialog set $N_e = 20$, check the natural selection box and set $w_{AA} = w_{Aa} = 1.0$, and $w_{aa} = 0.9$ (or $s = 0.1$ with complete dominance for the A allele), and check the mutation box and set the forward mutation rate to $\mu = 1 \times 10^{-3}$ and the reverse mutation rate to $\nu = 0$. Compute the expected equilibrium for these parameter values using Eq. 7.41 and then run the simulation.

Genetic drift will have a minimal effect on the outcome of allele frequencies as long as the effective population size is large. You can test this prediction by trying simulations with N_e values such as 10, 100, 500, and 1000 and comparing the equilibrium allele frequencies for each case.

first called the **load** by Muller (1950). Genetic load comes in two forms. **Substitutional load** refers to the reduction in mean fitness caused during the fixation of beneficial mutations or purging of deleterious mutations. Distinctly, the production of individuals with lower fitness genotypes by Mendelian segregation during reproduction is called the **segregational load**. Sexual reproduction leads to segregational load because both recombination and independent assortment produce novel progeny genotypes that possess a range of fitness values. Those progeny genotypes with a lower fitness perish (or do not reproduce) under viability selection. In principle, the genetic load places an upper limit on the ability of natural selection to change genotype frequencies in a population. Deaths due to viability selection cannot greatly exceed the total demographic excess of a population – the number of individuals produced each generation beyond those needed for demographic replacement – for long or the population will go extinct.

The genetic load has been a tool used to estimate the upper limits to the processes of mutation and of natural selection. One approach has been to determine how strong natural selection can be before an unrealistic genetic load occurs. The substitutional load has been used to estimate mutation parameters in populations, such as the rate of fixation of beneficial mutations, the rate of deleterious mutations, or the decline in fitness associated with deleterious mutations. Substitutional load also played a role in attempts to set acceptable thresholds for human radiation exposure and to understand mutation impacts on human health (Muller 1950; Crow 1997; Lynch 2016). The segregational load was used as a counterargument against the balance hypothesis during the early days of the neutral theory of molecular evolution. In these roles, the genetic load has been controversial (see reviews by Wallace 1991; Crow 1993b). The genetic load continues to be employed in population genetics and evolutionary biology. For example, the concept has been invoked in the accumulation of deleterious alleles that might cause mutational "meltdowns" in small and endangered populations (e.g. Lynch and Gabriel 1990), and in arguments for the fitness advantage of sexual reproduction (reviewed by de Visser and Elena 2007).

The genetic load concept (although not the term) originated in Haldane's work (1937). Haldane's result can be seen with the general dominance model of selection on a single diallelic locus where relative fitness values are $w_{AA} = 1$, $w_{Aa} = 1 - hs$, and $w_{aa} = 1 - s$ where s is the selection coefficient and h is the dominance coefficient (see Table 6.4). Assume a population at Hardy–Weinberg equilibrium, with complete dominance ($h = 0$), and a maximum fitness of one. In such a population, the fitness-weighted genotype frequencies are p^2, $2pq$, and $q^2 - sq^2$. The mean fitness in the population is then the sum of the frequency-weighted fitness values or $\overline{w} = p^2 + 2pq + q^2 - sq^2$. The mean fitness is, therefore, $\overline{w} = 1 - sq^2$ because $p^2 + 2pq + q^2 = 1$. The process of forward mutation (A to a) will make new recessive alleles in the population each generation, transforming some number of Aa genotypes into aa genotypes. Assuming that natural selection and mutation are at equilibrium means that selection removes aa genotypes as fast as they are made by mutation. The rate at which new aa genotypes are made from Aa genotypes is the forward mutation rate or μ. The mean fitness of the population is then $\overline{w} = 1 - \mu$ (also assuming that reverse mutation can be ignored). With incomplete dominance ($h \neq 0$), $\overline{w} = 1 - 2pqhs - sq^2$ and the mean fitness can be shown to be approximately $\overline{w} \approx 1 - 2\mu$.

The genetic load is defined as

$$L = \frac{w_{max} - \overline{w}}{w_{max}} \qquad (7.44)$$

and expresses the difference between the maximum fitness (w_{max}), which corresponds to the most fit genotype, and the average fitness ($\overline{w}$) in a population at a given point in time (Crow 1958). If w_{max} is defined as one, then the genetic load is given more simply by

$$L = \frac{1 - \overline{w}}{1} = 1 - \overline{w} \qquad (7.45)$$

Substituting the mean fitness from Haldane's result above and recalling that $w_{max} = 1$, the genetic load under complete dominance is $L = 1 - (1 - \mu) = \mu$ and under incomplete dominance $L = 1 - (1 - 2\mu) \approx 2\mu$. This predicts that the load is a function only of the mutation rate but not the selection coefficient.

If we assume, as Haldane did, that loci are completely independent (no linkage disequilibrium and no epistasis), then the mean fitness in a population of individuals with multilocus genotypes is the product of the relative fitness values of all loci multiplied together

$$\overline{w} = \prod_{i=1}^{loci} \overline{w}_i = \prod_{i=1}^{loci} (1 - 2\mu_i) \qquad (7.46)$$

$$\cong \prod_{i=1}^{loci} e^{-2\mu_i} = \prod_{i=1}^{loci} e^{-U} \qquad (7.47)$$

where U is the genome-wide deleterious mutation rate. The genome-wide genetic load is then expected to be $L = 1 - e^{-U}$.

The straightforward conclusion is that higher mutation rates lead to higher genetic loads because a larger number of deleterious mutations must be purged by natural selection to remain at selection/mutation equilibrium. This translates into the conclusion some proportion of heterozygous (without complete dominance) and homozygous individuals that must die or fail to reproduce each generation. The load never disappears because while selection removes deleterious alleles from a population, mutation continually supplies new ones.

Later, Haldane (1957) used a mutational-load argument to estimate the rate of substitutions. His goal in this analysis was to understand the rate of *phenotypic* evolution, perceived at that time to be relatively slow consistent with Darwinian emphasis on gradual change. Haldane concluded that natural selection could accomplish no more than about one beneficial substitution every 300 generations and that this would require the deaths of 30 times the population size present in one generation. This result made a major impact on some researchers in population genetics. Haldane's conclusion was often applied very generally as a fundamental limit on the rate of natural selection. As show later by Ewens (2004), Haldane's result implicitly assumed that demographic excess in humans is limited to 10% of the population size, and, so, was not as general as some had originally thought.

Segregational load: The decrease in the mean fitness in a population caused by individuals with low fitness genotypes that are introduced in a population by each generation of Mendelian segregation.
Substitutional load: The decrease in the mean fitness in a population caused by the introduction of deleterious mutations or eventual substitution of beneficial mutations. Also called mutation load.

To understand the segregational load, assume a standard diallelic locus model of overdominance for fitness where relative fitness values are $1 - s$ for the AA genotype, 1 for the Aa genotype, and $1 - t$ for the aa genotype where s and t are selection coefficients. As shown in Chapter 6, the expected equilibrium allele frequencies for this model of natural selection are

$$p_{eq} = \frac{t}{s + t} \qquad (7.48)$$

and

$$q_{eq} = \frac{s}{s + t} \qquad (7.49)$$

Using these equilibrium allele frequencies, it is possible to express the equilibrium frequency of heterozygotes expected under balancing selection as

$$H_{eq} = 2p_{eq}q_{eq} = 2\left(\frac{t}{s + t}\right)\left(\frac{s}{s + t}\right) = \frac{2st}{(s + t)^2} \qquad (7.50)$$

This shows that the frequency of heterozygotes in a population depends on the magnitude of the selection coefficients against the homozygous genotypes. For example, if both homozygotes have 10% lower viability than the heterozygote, $s = t = 0.1$ and at equilibrium half of the population is composed of heterozygotes $\left(H_{eq} = \frac{2(0.1)(0.1)}{(0.1 + 0.1)^2} = 0.5\right)$. Weaker selection results in less heterozygosity at equilibrium, while stronger selection results in more equilibrium heterozygosity.

We can combine the expected frequency of heterozygotes at equilibrium and the mean fitness to get an expression for the genetic load at equilibrium under balancing selection in terms of the homozygote selection coefficients:

$$\overline{w} = 1 - \frac{st}{s + t} \qquad (7.51)$$

as worked out in Math box 7.2. This population mean fitness can also be expressed in terms of the heterozygosity maintained by balancing selection by utilizing the equilibrium frequency of heterozygotes given in Eq. 7.50. Notice that $\frac{st}{s + t} = \left(\frac{2st}{(s + t)^2}\right)\left(\frac{s + t}{2}\right)$ so that the mean fitness for one locus can be written as

$$\overline{w} = 1 - H_{eq}\left(\frac{s + t}{2}\right) = 1 - H_{eq}\overline{s} \qquad (7.52)$$

This equation has the biological interpretation that the mean fitness at equilibrium in a population under balancing selection is one minus the product of the equilibrium heterozygosity and the mean selection coefficient. A population experiencing balancing selection, therefore, always has at least some genetic load since the mean fitness will never reach the maximum fitness of one (mating among heterozygotes will generate additional lower fitness homozygotes every generation).

Let us use some data on heterozygosity and selection coefficients similar to those that were available in the 1960s to compute the segregational load caused by balancing selection. Allozyme surveys in *Drosophila* of the time estimated that average heterozygosity was around 0.3. A homozygote disadvantage of 10% was considered reasonable under the balance hypothesis, giving a selection coefficient of $s = 0.10$. Putting these two values into Eq. 7.52 gives the mean fitness as

$$\overline{w} = 1 - (0.3)(0.1) = 1 - 0.03 = 0.97 \qquad (7.60)$$

Math box 7.2 Mean fitness in a population at equilibrium for balancing selection

To solve the equation for the mean fitness in terms of the selection coefficients *s* and *t* on the homozygous genotypes, start with the standard expression for the mean fitness:

$$\overline{w} = p^2 w_{AA} + 2pq w_{Aa} + q^2 w_{aa} \qquad (7.53)$$

and then substitute the fitness values for each genotype:

$$\overline{w} = p^2(1 - s) + 2pq(1) + q^2(1 - t) \qquad (7.54)$$

Then, express the genotype frequencies as equilibrium allele frequencies given in terms of the selection coefficients (Eqs. 7.48 and 7.49):

$$\overline{w} = \left(\frac{t}{s + t}\right)^2(1 - s) + 2\left(\frac{t}{s + t}\right)\left(\frac{s}{s + t}\right)(1) + \left(\frac{s}{s + t}\right)^2(1 - t) \qquad (7.55)$$

and then multiply through to give

$$\overline{w} = \frac{t^2(1 - s)}{(s + t)^2} + \frac{2st}{(s + t)^2} + \frac{s^2(1 - t)}{(s + t)^2} \qquad (7.56)$$

Expanding the numerators of the first and last terms gives

$$\overline{w} = \frac{t^2 - t^2 s + 2st + s^2 - s^2 t}{(s + t)^2} \qquad (7.57)$$

Notice that $s^2 t + t^2 s = st(s + t)$ and $t^2 + 2st + s^2 = (t + s)(t + s)$ in the numerator; so, making these substitutions

$$\overline{w} = \frac{(s + t)^2}{(s + t)^2} - \frac{st(s + t)}{(s + t)^2} \qquad (7.58)$$

and then simplifying gives

$$\overline{w} = 1 - \frac{st}{s + t} \qquad (7.59)$$

so that the genetic load is 0.03. Allozyme surveys of the time also suggested that about one-third of all loci in *Drosophila* had more than one allele segregating. Extrapolated to the entire genome, thought to be composed of around 8000–10 000 loci, it was believed that perhaps 2000–3000 loci were variable. If each locus is completely independent, then the mean fitness for the entire genome is

$$\bar{w} = [1 - (0.3)(0.1)]^{3000} = (0.97)^{3000} = 2.07 \times 10^{-40}$$
$$(7.61)$$

and the segregational load is $L = 1 - 2.07 \times 10^{-40}$, which is nearly the maximum value. This produces a conclusion that the genetic load would be enormous. Interpreted in biological terms, this genetic load means that an individual heterozygous at 3000 loci would have to produce 10^{40} progeny (the inverse of the load) for each progeny produced by an individual homozygous at all loci. Using lower average heterozygosity per locus and selection coefficients still leads to very high genetic loads.

Segregational and substitutional loads played an important role in attempts to explain the proportion of segregating loci and levels of heterozygosity in the 1960s. Balancing selection was considered and rejected by Lewontin and Hubby (1966) as a hypothesis to explain the first allozyme polymorphism data in *Drosophila*. Expectations for the substitutional and segregational loads were explored and developed in a series of papers authored and coauthored by Kimura (Kimura 1960, 1967; Kimura et al. 1963; Kimura and Maruyama 1966). The genetic load ultimately played a key part in the argument given by Kimura in his proposition of the neutral theory of molecular evolution (Kimura 1968). Kimura argued that there was too much genetic polymorphism or too fast a rate of divergence for all genetic changes to be caused by natural selection because the resulting genetic load would be too large. Kimura's alternative hypothesis was that many polymorphisms are selectively neutral. The neutral explanation greatly reduces the genetic load since a smaller portion of polymorphisms would be caused by selection and accrue a genetic load.

There are a series of counterarguments to the idea that natural selection is limited by the segregation load (reviewed in Wallace 1991; Crow 1993b; section 2.11 in Ewens 2004). One criticism revolves around the point of reference used for the maximum fitness in a population (w_{max} above). Haldane and

Kimura defined load relative to the most fit genotype in the population. In the case of the balancing selection, the most fit genotype would be one that is heterozygous at all loci. However, a genotype that is heterozygous at *all* loci would be very, very infrequent. For example, imagine that all allele frequencies are equal to 0.5 and fitness is determined by 100 loci that contribute equally to fitness. The expected frequency of a 100-locus heterozygote is $(0.5)^{100} = 7.89 \times 10^{-31}$ in a randomly mating population. Ewens (reviewed in 2004) showed that if genotypes with fitness values four standard deviations greater than the population mean of one are used as a reference point (still a rare genotype), then w_{max} equals 1.98. This implies that the most fit individuals need to produce about two progeny for every single progeny produced by the average fitness individuals. This cost of selection seems tolerable for many populations and species.

Another counterargument focuses on the form that natural selection takes while it works to cull individuals with less-fit genotypes from a population. Estimates of genetic load commonly assume that every locus is independent so that natural selection must act against homozygous genotypes at each and every locus independently. This is equivalent to assuming multiplicative fitness across multiple loci, seen as an exponent in Eq. 7.61. This assumption has the consequence of maximizing the perceived genetic load. It is possible that a selective death of one individual culls numerous deleterious alleles simultaneously (see Kondrashov and Crow 1988). For example, if selection results in the death of an individual because it bears a homozygous genotype at one locus, it also has the effect of culling any other deleterious alleles at other loci in that individual from the population at the same time. In a similar vein, both epistasis and pleiotropy would reduce the genetic load compared to the multiplicative model of Haldane. With epistasis, fitness is a function of interactions among numerous loci, and multiple deleterious mutations in one genotype might cause its fitness value to decrease faster than under a multiplicative model resulting in more efficient removal by selection. With pleiotropy, the genotype at a given locus impacts several phenotypes, and selective death of a single individual with a deleterious allele could cause increased mean fitness at several phenotypes related to fitness.

Another category of approaches to load focuses on the ways in which natural selection operates, leading to alternative predictions for the degree of load in populations. For completely recessive deleterious alleles ($h = 0$), the load is half that for partial dominance because each selective death of a recessive homozygote will remove two deleterious mutations while no heterozygotes will die. Similarly, any increased homozygosity because of mating among relatives (fixation index $F > 0$) would also serve to reduce genetic load for partly recessive deleterious alleles ($h > 0$) since heterozygote selective deaths will decrease (Barrett and Charlesworth 1991). The truncation form of directional selection (see Figure 9.10) was modeled as an alternative to the assumption of independent loci with multiplicative fitness. In truncation selection, individuals bearing less than some threshold number of deleterious mutations produces an epistasis-like effect where individuals culled by selection have a greater average number of deleterious mutations than surviving individuals (Crow and Kimura 1979, see example in Crow 1997). In this model, truncation selection has the potential to reduce the average number of mutations each generation of selection by enough to compensate for the expected number of deleterious mutations that occur anew each generation so that load does not increase.

With viability selection, fitness is binary and individuals selected against die before reproduction. An alternative is that selection instead acts as continuous fitness variation among reproducing adults such as fecundity variation so that fitness is a matter of degree among individuals that all survive. Some degree of fecundity selection (in lieu of some smaller amount of viability selection) could reduce the number of selective deaths required for substitution or segregation. Instead, genetic load would take the form of variation in fitness among individuals of different genotypes, such as fecundity variation. This is sometimes described as hard (fitness is binary) and soft (fitness is continuous) forms of natural selection. Weak purifying viability selection against weakly deleterious mutations occurring at many unlinked nucleotide sites can produce a large load (Kondrashov 1995). Charlesworth (2013) showed two alternative forms of selection that generate smaller loads. One was weak stabilizing selection on large numbers of loci where genetic load is produced by modest variation in fitness values among individuals due to segregating nucleotide sites. The other was a form of soft purifying selection where individuals compete for limiting resource.

A final category of models for genetic load focus on population ecology and its interaction with evolutionary change. The standard model of natural selection assumes unbounded population growth (see Sections 6.1 and 7.2). In contrast, most populations are expected to exhibit density-dependence with the consequence that many individuals perish each generation because of intraspecific competition for resources. The deaths caused by density-dependence have the potential to remove deleterious alleles and lessen genetic load. The degree to which load is removed by density-dependence will depend on the degree to which selection culls individuals before or after they consume resources that could be used by surviving individuals to reproduce. Deleterious mutations that reduce intraspecific competitive ability can actually increase population size if individuals of lower fitness are culled by selection before they consume resources since that leaves more resources for surviving individuals to increase their reproduction (Clarke 1973a, b). Similarly, genetic load has the potential to influence two-species interspecific competition and can lead to the extinction of one species because any resources made available by selective deaths can be utilized by the species that is the stronger competitor (Agrawal and Whitlock 2012).

The concept of genetic load continues to motivate research into models of mutation and selection that seek to understand the fate and possible accumulation of deleterious mutations, especially as empirical data provide a growing number of estimates of the genome-wide deleterious mutation rate (U).

7.4 Natural selection in genealogical branching models

- The problem with selection in genealogical branching models.
- Directional selection and the ancestral selection graph.
- Genealogies and balancing selection.

The final topic in this chapter is the process of natural selection in the context of genealogical branching models. Representing natural selection in genealogical branching models will require a change in perspective about how selection works and also an expansion of the ways that events are represented on genealogies. The major goal of this section is to introduce ways of modeling selection on genealogies to understand how the operation of natural selection might change the height and total branch length of genealogical trees compared with the case of coalescence patterns due to genetic drift alone.

Adding natural selection to the genealogical branching model introduces a serious complication to the Wright–Fisher model of sampling that the basic genealogical branching model is built on. Recall from Chapter 3 that the basic coalescent model assumes that when going one generation back in time, the chance that any two lineages coalesce is $\frac{1}{2N}$. This probability results from the assumption that all lineages in any given generation have an equal chance of being chosen as a common ancestor working back in time from the present. In the basic coalescent model where alleles are selectively neutral, each lineage within a generation has an *equal and constant* probability of becoming an ancestral lineage when working back in time to find the most recent common ancestor (MRCA). From the time-forward perspective, with neutral evolution each lineage has an equal probability of being sampled and represented in the next generation. In general, with selective neutrality, the haplotype of a lineage does not influence its sampling properties.

Natural selection violates the basic assumption that all lineages have equal and constant probabilities of coalescence. When natural selection operates, some lineages tend to increase in frequency over time, whereas other lineages tend to decrease in frequency over time due to fitness differences among the lineages caused by differences in haplotype relative fitnesses. These changes in the frequencies of lineage copies translate into probabilities of coalescence that change over time as well. A lineage bearing a haplotype favored by selection will increase over time and, therefore, have a *decreasing* probability of coalescence working back in time from the present. Similarly, a lineage bearing a lower fitness haplotype will decrease in frequency over time and, therefore, have an *increasing* probability

of coalescence moving back in time. Thus, natural selection presents a fundamental contradiction to the sampling process built into the genealogical branching model.

Directional selection and the ancestral selection graph

Fortunately, there is a clever and relatively simple way to modify the genealogical branching model to accommodate directional natural selection (Neuhauser and Krone 1997; Neuhauser 1999). This modification relies on treating coalescence and natural selection as distinct processes that can both possibly occur working back in time from the present toward the MRCA in the past exactly as was done to combine coalescence with migration or mutation. The first step in including natural selection in the genealogical branching model is to alter slightly our view of the sampling process. Figure 7.11 shows five lineages of $2N$ total across the span of one generation. If there is selective neutrality, then moving forward in time each lineage is sampled once to found the next generation. This is equivalent to the absence of a coalescence event moving backward in time. Alternatively, if there is natural selection operating, then lineages with one haplotype are favored and will increase in frequency over time. In Figure 7.11, the possible action of natural selection is shown by the dotted line. If the lineage represented by the open circles has a higher fitness haplotype, then it will displace the lineage of the lower fitness haplotype (closed circles). This displacement event is analogous to growth in the population size of the fitter haplotype given that the total population size is constant.

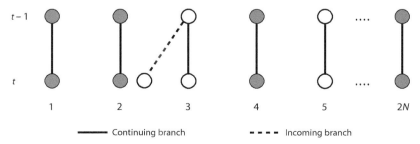

Figure 7.11 Haploid reproduction with the possibility of coalescence and natural selection events. Each haploid lineage replicates itself and is included in the next generation if there are no coalescence events (solid lines). A lineage makes an extra copy of itself (dashed line) and has the potential to displace one copy of another lineage. If the lineage making the extra copy of itself (open circle) has a higher fitness haplotype than a randomly chosen lineage (closed circle), then it will displace the lineage of the lower fitness haplotype. Therefore, the outcome of a lineage duplication event that may result in natural selection depends on the haplotype states of the specific lineages involved. The solid lines are continuing branches and the dashed line is an incoming branch.

We can treat the dual continuing/incoming branching process as two independent parts of the overall coalescence process. When two independent processes are operating, the coalescence model is based on waiting for *any* event to occur and then deciding which type of event happened. When events are independent but mutually exclusive, the probability of each event is added over all possible events to obtain the total chance that an event occurs. As was done for both migration and mutation, we assume that coalescent and natural selection events are rare, or that N_e is large and the selection coefficient is small. This assumption makes sure that natural selection and coalescence events are mutually exclusive and that when an event does occur going back in time it is *either* coalescence *or* natural selection.

Natural selection depends on the chance that the fitter haplotype makes an incoming branch that displaces a lineage bearing a less-fit haplotype. Twice the rate of natural selection events is

$$\sigma = 4N_e s \qquad (7.62)$$

(σ is pronounced "sigma") where $2N_e s$ is the expected number of natural selection events that will occur for a single lineage during one unit of continuous time (see Section 5.5 for a fuller explanation of such a rate in the context of mutation rates), and the fitter haplotype has a relative fitness of $1 + s$ compared to a relative fitness of 1 for the less-fit haplotype. When s is of the order of $\frac{1}{2N}$, then the rate of natural selection events would be comparable to the rate of coalescent events due to neutral sampling. The exponential approximation for the chance that a natural selection event occurs at generation t is then

$$P\left(T_{\text{incoming branch}} = t\right) = e^{-t\frac{\sigma}{2}} \qquad (7.63)$$

for a single lineage and

$$P\left(T_{\text{incoming branch}} = t\right) = e^{-t\frac{\sigma}{2}k} \qquad (7.64)$$

for k lineages on a continuous time scale. The chance that an incoming branch due to natural selection displaces one of k lineages at or before a certain time can then be approximated with the cumulative exponential distribution

$$P\left(T_{\text{incoming branch}} \leq t\right) = 1 - e^{-t\frac{\sigma}{2}k} \qquad (7.65)$$

in exactly the same fashion that times to coalescent events are approximated.

Combining natural selection with coalescence and mutation into a three-process waiting-time distribution is now simple. Since the three types of events are mutually exclusive, we add the chance that a natural selection event occurs to the chance that a coalescence or mutation event occurs to get the expected waiting time until an event:

$$P(T_{event} \leq t) = 1 - e^{-t\left(\frac{k(k-1)}{2} + \frac{\sigma}{2}k + \frac{\theta}{2}k\right)} \qquad (7.66)$$

When an event does occur according to the waiting time in Eq. 7.66, it is then necessary to determine the type of event. Since the total chance that an event occurs is $\dfrac{k(k-1)}{2} + \dfrac{k\sigma}{2} + \dfrac{k\theta}{2}$, the chance an event is a coalescence is

$$\frac{\dfrac{k(k-1)}{2}}{\dfrac{k\sigma}{2} + \dfrac{k(k-1)}{2} + \dfrac{k\theta}{2}} = \frac{k-1}{k-1+\sigma+\theta} \qquad (7.67)$$

the chance an event is due to natural selection is

$$\frac{\dfrac{k\sigma}{2}}{\dfrac{k\sigma}{2} + \dfrac{k(k-1)}{2} + \dfrac{k\theta}{2}} = \frac{\sigma}{k-1+\sigma+\theta} \qquad (7.68)$$

and the chance that the event is a mutation is

$$\frac{\dfrac{k\theta}{2}}{\dfrac{k\sigma}{2} + \dfrac{k(k-1)}{2} + \dfrac{k\theta}{2}} = \frac{\theta}{k-1+\sigma+\theta} \qquad (7.69)$$

Using Eq. 7.66 and then determining whether each event is selection, mutation, or coalescence, it is possible to construct what is known as an **ancestral selection graph** (Figure 7.12). The term ancestral selection graph is used to describe the outcome of the three processes since it explicitly shows possible natural selection events. Natural selection events result in an *addition* of branches and thereby serve to visualize selection events that are not apparent on a genealogy alone. When branching occurs due to a natural selection event (going back in time), the resulting branch is called the **incoming branch** to represent a possible lineage displacement. The lineage that the incoming branch splits off from is called the **continuing branch**. The incoming branch coalesces with a randomly chosen lineage at a later time determined by the waiting-time distribution and

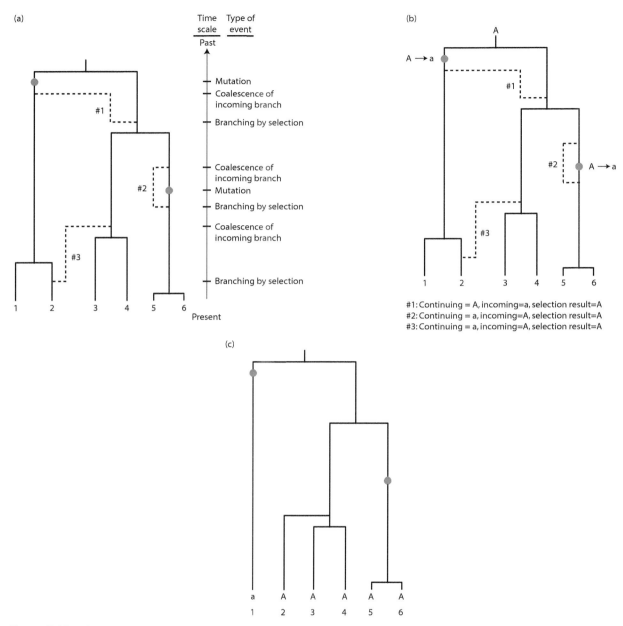

Figure 7.12 The ancestral selection graph used to include natural selection in the genealogical branching model. In A, the waiting times between events and the types of events are determined until the MRCA is reached by working backward in time from six lineages in the present. Branching and coalescence events due to natural selection (dashed lines) and mutation events are identified. Natural selection causes the addition of one "incoming" branch to the number of lineages that can coalesce and this incoming branch can then coalesce with any lineage. In B, a haplotype state is assigned to the ultimate ancestor and allelic states are traced forward in time to determine the outcome of mutation and natural selection events. At each of the selection events, the state of the continuing branch and the incoming branch are compared. In this example, A is the fitter haplotype and it displaces the a haplotype when continuing and incoming branches coalesce. When the haplotypes of continuing and incoming branches are identical, there is no change in haplotype state. In C, the haplotype states of the lineages in the present are assigned once all of the selection events have been resolved. In this example, selection causes a slight increase in the total branch length because the selection event at #3 displaces a shorter branch.

assumes the state of the branch where it coalesces. Refer again to Figure 7.12 to see the distinction between incoming and continuing branches. Even though natural selection events make more branches, the coalescence process is faster and will eventually result in coalescence to the MRCA (the coalescence rate is proportional to k^2 whereas the selection rate is proportional to k).

Figure 7.12 shows one outcome of the coalescence–natural-selection–mutation process. Figure 7.12A shows the events that occurred working back in time from six lineages in the present. The first potential natural selection event (labeled #3 because it is the last event when working forward in time) occurred on lineage 2, causing a branching event. The incoming branch eventually coalesced with the lineage that is ancestral to lineages 3 and 4. The second potential natural selection event caused a branching event on the lineage that is ancestral to lineages 5 and 6. That incoming branch coalesced again with the same lineage after a short time. The final potential natural selection event caused a branch from one of the two internal lineages near the MRCA that coalesced with the other lineage present near the MRCA.

The actual outcome of these three selection events can only be determined once the state of the MRCA and the fitness of the two haplotypes have been assigned. In Figure 7.12B, the ancestor is assigned a state of A which is also assumed to be the higher fitness haplotype. Then, moving forward in time on the ancestral selection graph, the outcome of each instance of natural selection can be determined. For selection event #1, the incoming branch has a haplotype of a due to the mutation while the continuing branch has the ancestral A haplotype. Since A is more fit, the state of the continuing branch is kept and the state of the incoming branch discarded. (This is exactly like the open circles being less fit and the closed circles more fit in Figure 7.11.) At the second selection event (#2), the incoming branch has a state of A and the continuing branch has a state of a due to a mutation. Here, the incoming branch displaces the continuing branch and the lineage has a state of A thereafter. The incoming branch has a state of A and the continuing branch a state of a at the final selection event (#3). This results in the displacement of the continuing branch.

The genealogy that results after resolving the potential natural selection events is shown in Figure 7.12C. Given the ancestral state and high-fitness haplotype, selection events #1 and #2 had no impact on the branching pattern of the tree. In

contrast, selection event #3 caused a change in the branching pattern that moved the coalescence point of lineage 2 from the continuing branch to the coalescence point of the incoming branch. This reflects the fact that after natural selection acted, lineage 2 was identical by descent to a different lineage. This change in the branching pattern causes the total branch length of the tree to be slightly longer than it was without natural selection. In this case, the height of the tree is not changed by natural selection.

Problem box 7.2
Resolving possible selection events on an ancestral selection graph

Use Figure 7.12B and trace the lineage states forward in time, assuming that the state of the MRCA is an a haplotype and that A is the fitter haplotype. Are the resulting lineage states in the present the same as originally given in the figure? Has the height of the genealogy changed?

As another exercise to test your knowledge of the ancestral selection graph, resolve the genealogy in Figure 7.12B using the a allele as the state of the MRCA, alternatively assuming that the A and then the a allele is the fitter haplotype.

The conclusions to be drawn from the ancestral selection graph with two alleles are straightforward. Weak to moderate directional natural selection tends to have only a minor impact on average times to coalescence. Stated another way, the action of directional natural selection does not greatly alter the average times to coalescence compared with a strictly neutral genealogy with the same number of lineages. When the selection coefficient and the mutation rate are approximately equal, the mean time to the MRCA is shortened slightly (Neuhauser and Krone 1997; see also Przeworski et al. 1999). However, the difference in the average coalescence times with directional selection and with strict neutrality is slight given the wide variation in coalescence times due to finite sampling. Strong natural selection for advantageous alleles or selection against deleterious mutations is expected to reduce the total height of genealogical trees because of lineages bearing states that are strongly disadvantageous (see Charlesworth et al. 1993, 1995).

Interact box 7.6 Build an ancestral selection graph

Building an ancestral selection graph can help you to better understand the evolution of genealogies when both the processes of natural selection, genetic drift, and mutation are all operating. You can use an expanded version of the Microsoft Excel spreadsheet from Chapter 5 and modify it to include natural selection events. (Alternatively, there is an R script that can be used for this exercise.) The spreadsheet contains the cumulative exponential distributions used to determine the time until addition of an incoming branch for selection, a coalescence event, or a mutation event (see Eq. 7.48). To determine the time that an event occurs for a given number of lineages k and mutation rate, a random number between 0 and 1 is picked and then compared to the cumulative exponential distribution. The time interval on the distribution that matches the random number is taken as the event time. The next step is to determine whether the event was a selection, coalescence, or mutation, again accomplished by comparing a random number to the chances of each type of event (Eqs. 7.49–7.51).

Step 1: Open the spreadsheet and notice the fields highlighted in yellow where the values of key parameters are set. You can also click on cells to view the formulas used, especially the cumulative probability of coalescence for each k. The cumulative probability distributions are graphed for $k = 6$ and $k = 2$.

Step 2: Look at the section of the spreadsheet under the heading "Event times:" on the right side of the sheet. This section gives the waiting times until an event occurs and then determines if the event was selection coalescence, or mutation. Press the recalculate key(s) to generate new sets of random numbers (F9, command =, or control =). Watch the times to an event change.

Step 3: Start with six initial lineages ($k = 6$) and draw an ancestral selection graph with the possibility of mutations. Along the bottom of a blank sheet of paper, draw six evenly spaced dots to represent six lineages.

Step 4: Recalculate the sheet to draw new random numbers. Start at the first "Decide event time:" panel to determine how much time passes (going backward in time) until an event occurs. Then, use the entries under "Decide what type of event:" to determine if the event was an incoming branch, a coalescence, or a mutation. If the event is a mutation event go to Step 5, if it is an incoming branch go to Step 6, otherwise go to Step 7.

Step 5: For a mutation event, first draw the lines for all lineages back in time by a length proportional to the waiting time (e.g. if the time is 0.5, draw lines that are 5 cm). Use the random number table to sample one lineage and draw an X on the lineage at the event time to indicate a mutation occurred. If a mutation occurred the number of lineages (k) remains the same. Go to Step 4.

Step 6: For a selection event that adds an incoming branch, first draw the lines for all lineages back in time by a length proportional to the waiting time (e.g. if the time is 0.5, draw lines that are 5 cm). Use the random number table to sample one lineage that will experience a branch addition and draw a dashed horizontal line to the left or right to show a new lineage branch. (That branch will be extended as more time is added in later iterations.) Assign the lineage the number $k + 1$. Edit the number of lineages parameter to increase k by one. Go to Step 4.

Step 7: For a coalescence event, first draw the lines for all lineages back in time by a length proportional to the waiting time (e.g. if the time is 0.5, draw lines of 5 cm). Use the random number table to sample two lineages that will experience coalescence. Connect the vertical lines of the coalescing lineages with a horizontal line. Assign the lineage number of one of the coalesced lineages to the pair's single ancestor at the horizontal line. Record the other lineage number on a list of lineages no longer present in the population (skip over these numbers if they appear in the random number table). Edit the number of lineages parameter to decrease k by one. Go to Step 8.

Step 8: Return to Step 4 until all lineages have coalesced ($k = 1$).

Step 9: Assign an allelic state to the MRCA and define the mutation model (e.g. k alleles model). Also define the which allelic state(s) have high fitness. Then, starting with the MRCA and moving forward in time, resolve each instance of a mutation. Go back and start at the MRCA again and move forward in time to resolve each incoming branch due to natural selection, removing lineages that carry low fitness haplotypes or retaining lineages that carry high-fitness haplotypes.

You should obtain an ancestral selection graph like that in Figure 7.12. Your tree will be different because the random coalescence and mutation times vary around their averages, but the overall shape of your tree (e.g. shorter branches when k is large) and frequencies of mutations (for a given mutation rate) will be similar.

Genealogies and balancing selection

Natural selection where heterozygotes have the highest fitness, a type of **balancing selection**, can also be incorporated in genealogical branching models (Hudson and Kaplan 1988; Kaplan et al. 1988; Nordborg 1997; see Hudson 1990). Earlier in the chapter, we saw that balancing selection is expected to maintain both alleles at a diallelic locus segregating in the population at equilibrium. The two allele frequencies at equilibrium will depend on the selection coefficients against the two homozygous genotypes. Since the haploid genealogical model does not have diploid genotypes nor sexual reproduction, we will need to take an alternative approach rather than specify multiple genotype fitness values.

Balancing selection is a special case of natural selection because it works counter to the fixation and loss due to genetic drift. In a genealogical branching model, genetic drift is represented by the process of coalescence. So, to approximate the overall effect of balancing selection, we need a process that will delay coalescence to the same degree that selection favors heterozygotes in the diploid selection model. This same overall result can be obtained by modeling balancing selection along the lines of population structure with two demes. Although it sounds like an odd approach, population structure and balancing selection have similar effects for different reasons. In structured populations, two lineages cannot coalesce unless they are in different demes (refer to Figure 4.23). Gene flow events that move lineages into different demes, therefore, tend to delay coalescence events. Using this same logic, we can model balancing selection in a single panmictic population as a process where there are two lineage types. A switching process akin to gene flow (or mutation) changes lineage types at random, while the coalescence process operates at the same time. If two lineages must be of the same type to coalesce, then the switching process will prevent coalescence among the lineages that are of different types.

Let the two lineage types be A and B and their respective frequencies in the population be p and q so that $p + q = 1$ (see Figure 7.13). Every generation lineages of one type may switch to the other type with rate μ. Twice the expected number of the $2N$ total lineages in the population that switch types each generation is then $\nu = 4N\mu$. The expected number of lineages switching each generation serves as a surrogate for the strength of balancing selection

since frequency switching will let lineages escape coalescence. Using this switching rate, the expected waiting time until an A lineage switches to a B lineage is

$$P(T_{A \to B} \leq t) = 1 - e^{k_A \frac{t}{2}\left(\frac{q}{p}\right)} \qquad (7.70)$$

and the expected waiting time until a B lineage switches to an A lineage is

$$P(T_{A \to B} \leq t) = 1 - e^{k_B \frac{t}{2}\left(\frac{p}{q}\right)} \qquad (7.71)$$

The ratio of the frequencies (q/p and p/q) serves to adjust the exponent for the relative frequencies of the two lineage types.

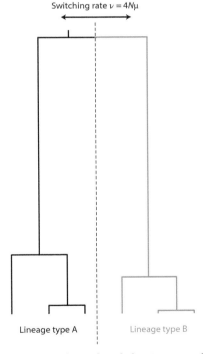

Figure 7.13 A genealogy where balancing natural selection is modeled by type switching. Every generation, lineages of one type (here A and B) may switch to the other type with rate μ. Twice the expected number of the $2N$ total lineages in the population that switch types each generation is then $\nu = 4N\mu$. Since lineages can only coalesce when they are of the same type, type switching increases the average time to coalescence. This is analogous to natural selection favoring heterozygotes because overdominance also extends the segregation times of alleles. Genealogical trees that result from balancing selection modeled as type switching tend to have longer branches compared to genealogies that result from genetic drift or directional natural selection.

Next, we need to express the waiting time until a coalescence event, keeping in mind that lineages can only coalesce if they are of the same type. Given that there are $2Np$ lineages of type A and $2Nq$ lineages of type B and coalescence events are mutually exclusive, the expected waiting time until a coalescence event is

$$P(T_{coalescence} \leq t) = 1 - e^{\frac{k_A(k_A-1)}{2}\left(\frac{1}{p}\right) + \frac{k_B(k_B-1)}{2}\left(\frac{1}{q}\right)} \quad (7.72)$$

If switching events and coalescence events are all mutually exclusive, the individual exponents can be added together to obtain the total waiting time to any event

$$P(T_{coalescence} \leq t) = 1 - e^{\frac{k_A(k_A-1)}{2}\left(\frac{1}{p}\right) + \frac{k_B(k_B-1)}{2}\left(\frac{1}{q}\right) + k_A\frac{k}{2}\left(\frac{q}{p}\right) + k_B\frac{k}{2}\left(\frac{p}{q}\right)}$$

$$(7.73)$$

Given that an event has occurred with a known waiting time, the type of event can be determined by drawing a random number between 0 and 1 and comparing it with the cumulative total of the chance of each event divided by the total probability of all events.

Genealogical trees that result from balancing selection modeled as type switching tend to have longer branches compared with genealogies that result from genetic drift alone (Figure 7.13). This is due to the increase in average waiting times between coalescence events caused by lineage type switching. The final two lineages in particular are expected to take a long time to coalesce since they must switch to the same type. The results of two allele balancing selection are qualitatively similar to genealogies with long waiting times for the last two lineages expected with subdivided populations. If mutation is also operating along with balancing selection, then genealogies with longer branches would also accumulate more mutations since the number of mutation events is proportional to the total branch length of a genealogy. Lineages of the two different types, in particular, are expected to have more mutational changes between them than would be expected in a genealogy under the basic neutral coalescent model.

An additional model of balancing selection exists for populations with more than two alleles and equivalent (overdominant) fitness values for the possible heterozygous genotypes (Vekemans and Slatkin 1994; Uyenoyama 1997). Such multi-

allelic balancing selection is distinct from balancing selection with only two alleles, resulting in genealogies that have long coalescence times for the lineages near the present (or shorter times to coalescence further back in time nearer the MRCA) compared to neutral genealogies. Classic examples of multi-allelic balancing selection are the many alleles found at single self-incompatibility loci in some plants (e.g. Schierup et al. 1998).

Chapter 8 further explores the consequences of natural selection on the shape of genealogies in the context of tests of the null model that patterns of genetic polymorphism differ from that expected by drift and mutation alone.

7.5 Shifting balance theory

- Sewall Wright's classic model of natural selection, genetic drift, gene flow and mutation on an adaptive landscape.

If forced to choose a single model in all of population genetics that has had the longest running impact, Sewall Wright's shifting balance model and its associated fitness surfaces would certainly be among the top picks. Wright first described the fitness surface in a 1932 presentation at the 6th International Congress of Genetics. For that presentation, Wright was asked to make a non-mathematical presentation of his theoretical work in population genetics. Wright attempted to distill his very long and mathematically sophisticated 1931 paper that developed fundamental expectations for numerous population genetic processes. In a biography of Wright, Provine (1986) argued that fitness surfaces as a heuristic aid to understand allele and genotype frequency dynamics "was one of his single most influential contributions to modern evolutionary biology." Indeed, the fitness surface is a commonly used metaphor in population genetics even today. The goal of this section is to introduce Wright's adaptive landscape metaphor and then explain Wright's interpretation of how genetic drift, gene flow, natural selection, and mutation might interact in the context of genetically subdivided populations known as the **shifting balance process**.

Allele combinations and the fitness surface

At the beginning of his 1932 paper, Wright starts with the observation that there are a very large number of possible genotypes (what he called "allelomorph combinations") in any given species. He gives the example that for 1000 loci with 10 alleles each, there are 10^{1000} possible combinations of alleles that might make up a 1000 locus gamete haplotype. (Wright points out that this is a staggeringly large number, which he compares to the estimated number of electrons and protons in the visible universe.) He then assumes that wild-type alleles have a frequency of 0.99 at these 1000 loci, while the infrequent alternate alleles at these loci confer phenotypes only slightly different than wild type. Wright reasoned that there will be only a very small proportion of individuals in a population exhibiting phenotypes caused by any combination of more than 20 non–wild-type alleles in a 1000 locus haplotype. For example, the expected frequency of a single haplotype containing 20 non–wild-type and 980 wild-type alleles is $(0.01^{20})(0.99^{980}) = 5.3 \times 10^{-45}$, while the expected frequency of a haplotype containing all 1000 wild-type alleles is $0.99^{1000} = 4.3 \times 10^{-5}$. A haplotype composed of all wild-type alleles is 40 orders of magnitude more frequent than a haplotype with just 20 non–wild-type alleles.

Under these assumptions, the range of phenotypes exhibited in a single population will represent only a very small part of the possible range of phenotypes because haplotypes composed mostly of wild-type alleles are very common and haplotypes with numerous non–wild-type alleles are very infrequent. Wright draws the conclusion that "The population is thus confined to an infinitesimal portion of the field of gene combinations" even though there is only a very small chance that two individuals possess identical haplotypes. In other words, the individuals in any population represent only a small portion of the very large range of possible phenotypic variations because haplotypes composed mostly of wild-type alleles are very common while haplotypes composed of numerous non–wild-type alleles are very rare.

With this perspective on the large number of allele combinations in a haplotype or gamete, Wright goes on to describe what a surface or plot would be like "If the entire field of possible gene combinations be graded with respect to adaptive value under a particular set of conditions ...." The **fitness surface** presented by Wright (1932) is shown in Figure 7.14 as a two-dimensional representation of at least three dimensions. The x and y axes, although unlabeled by Wright in the original, represented the range of possible allelic combinations in a haplotype or genotype. The fitness of each genotype is represented by the

Figure 7.14 Sewall Wright's original adaptive landscape diagram. The high fitness points on the surface are indicated by +, while the low fitness points are indicated by −. Original caption: "Diagrammatic representation of the field of gene combinations in two dimensions instead of many thousands. Doted lines represent contours with respect to adaptiveness." Source: From Wright (1932).

height of each point (the dimension perpendicular to the page when the surface is drawn in two dimensions). In contemporary usage, the x and y axes represent allele frequencies that vary between zero and one, while the height of each point on the surface is the mean fitness of a population at those allele frequencies. (The distinction between Wright's surface and the contemporary interpretation of an adaptive landscape is taken up later in this section.) The dashed lines represent contour lines of constant fitness. The high points on the surface are indicated by a "+" symbol, while the low points are indicated by a "−" symbol. Since Wright's fitness surface is exactly analogous to a topographic map with peaks and valleys, the term **adaptive landscape** is frequently used to describe three dimensional graphs where allele or genotype frequencies are represented in two dimensions and a measure of fitness is represented in a third dimension.

Wright then considered how evolution by natural selection would move a population on such an adaptive landscape. He considered the possibility that "a particular combination [of alleles] gives maximum adaptation and that the adaptiveness of other combinations falls off more or less regularly according to the number of removes." In contemporary terminology, this is equivalent to additive gene action where the genotypic value is a linear function of the effects of the alleles in the genotype (see Chapters 9 and 10). Under additive gene action, a fitness surface is simply a plane where the highest point can be reached by moving through intermediate steps of increasing population average fitness. Alternatively, Wright also imagined fitness surfaces where "... it is possible that there may be two peaks ...." Fitness surfaces with multiple peaks result from dominance and epistasis, types of gene action that can produce large changes in average fitness with only minor changes in allelic combinations because the genotypic value is a nonlinear function of the effects of the alleles in the genotype. Wright imagined that "In a rugged field of this character, selection will easily carry the species to the nearest peak, but there may be innumerable other peaks which are higher but which will be separated by 'valleys'." (There are examples of fitness surfaces under strict additivity, dominance, and epistasis throughout Chapter 7.)

One of the basic challenges of proposing that evolutionary change occurs by natural selection exclusively can be visualized with the aid of Wright's adaptive landscape diagram. Natural selection acts to increase the mean fitness of a population based on the slope of the adaptive landscape immediately around the current allele frequency position of a population. This leads to changes in allele frequency that, if unopposed by other processes, will eventually lead to the maximum mean fitness that can be achieved by continuously moving uphill on the adaptive landscape. The possible problem is that selection does not take a broad view of the adaptive landscape while causing generation-to-generation changes in allele frequencies. The process of natural selection "sees" only the slope of the adaptive landscape immediately around a population's current position in allele frequency space. If the adaptive landscape has multiple peaks, the process of selection alone can "strand" populations at points on the surface that are local mean fitness maxima but not among the highest levels of mean fitness that are possible. In terms of the landscape metaphor, natural selection is like a climber who must always go up in elevation and can never descend (even temporarily) nor cross a valley. Because of this increasing fitness requirement, natural selection may not be capable of reaching the highest peaks on a fitness surface.

Wright proposed his shifting balance model as a possible mechanism that would prevent populations from getting stuck forever on a limited number of fitness peaks. Wright (1932) states clearly that "The problem of evolution as I see it is that of a mechanism by which the species may continually find its way from lower to higher peaks" on a fitness surface. He goes on to say that "there must be some trial and error mechanism on a grand scale by which the species may explore the region surrounding the small portion of the [fitness surface] field it occupies. To evolve, the species must not be under strict control of natural selection. Is there such a trial and error mechanism?" These sentences capture what seems to have been the main question in Wright's mind when he wrote his 1932 paper.

Wright's view of allele frequency distributions

After describing the fitness surface metaphor and asking what mechanisms might overcome the limitations of natural selection, Wright then attempted to summarize his work at that time on the shape of allele frequency distributions that would be expected under the action of the basic processes of population genetics. He summarized these ideas very succinctly without reference to any equations in a total of three paragraphs that refer to one figure (see Figure 7.15). The three panels of Figure 7.15A refer to the expected equilibrium distributions of allele frequencies for one locus in many independent replicate finite populations in an entire species. The processes of natural selection, genetic drift, and mutation act simultaneously to shape these allele frequency distributions. A narrow distribution of allele frequencies among independent populations is expected when $4NU$ (equivalent to θ) and $4NS$ (four times the product of the effective population size and the selection coefficient) are large (left panel in Figure 7.15A). This distribution results from a large selection coefficient in favor of the wild-type allele (S with solid arrow) with genetic drift that is relatively weak (dashed arrows labeled $\frac{1}{4N}$ where N is the effective population size). Frequent forward mutation (U with solid arrow) prevents complete fixation of the wild-type allele, shifting the distribution to the left. A lower rate of reverse mutation (V with dashed arrow) increases the frequency of the wild-type allele a small amount, shifting the distribution to the right. A wider distribution of allele frequencies among independent populations is expected when $4NU$ and $4NS$ take intermediate values as shown in the middle panel of Figure 7.15A. In that case, the forces of selection, mutation, and genetic drift (all solid arrows) are approximately equal to each other so that replicate populations exhibit a range of wild-type allele frequencies. In the right panel of Figure 7.15A, the distribution of allele frequencies is horseshoe shaped (most replicate populations near

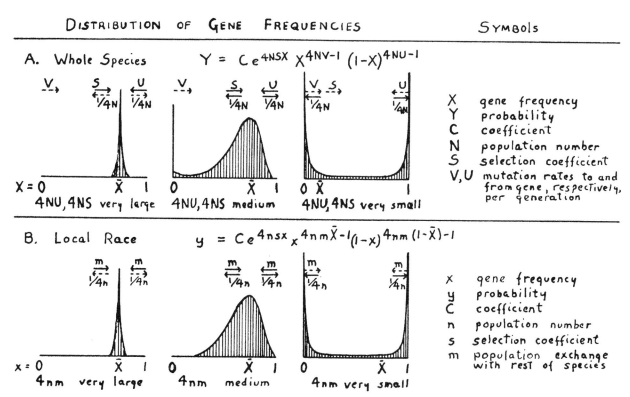

Figure 7.15 Wright's schematic representation of the simultaneous action of multiple population genetic processes leading to equilibrium distributions of allele frequencies. Each distribution represents the allele frequencies of many replicate populations or an ensemble distribution (in A, numerous populations that make up a species are independent, while, in B, subpopulations are interdependent in an island model). The magnitude and direction of the effects of a process on the distribution of allele frequencies is indicated by arrows bearing letters. Solid arrows indicate stronger processes, and dashed arrows indicate weaker processes. For example, in the left-most distribution of panel B, strong migration relative to drift maintains subpopulation allele frequencies with little divergence, while weak genetic leads to a modest spread of subpopulation allele frequencies around the average allele frequency of the total population. In all panels, the frequency of the wild-type allele (*x*) is given on the *x* axis and the frequency of populations with a given allele frequency is given on the *y* axis. The probability on the *y* axis is given by the equation at the top of A and B for the allele frequency on the *x* axis given values for the population parameters. Original caption: "Random variability of a gene frequency under various specified conditions." Source: From Wright (1932).

fixation or loss) when $4NU$ and $4NS$ are small because genetic drift is strong (solid arrows for $\frac{1}{4N}$) relative to natural selection and mutation (dashed arrows for *S*, *U*, and *V*).

The three panels of Figure 7.15B refer to the equilibrium distributions of allele frequencies at one locus in many finite subpopulations connected by some degree of gene flow. The forces that dictate the distribution of allele frequencies are now genetic drift (arrows labeled $\frac{1}{4n}$ where *n* is the effective population size of demes) and migration (arrows labeled *m* for the rate of migration in an island model). Here, Wright assumed that migration was much stronger than mutation and selection so these later two processes could be ignored. At one extreme shown in the left panel of Figure 7.15B, a narrow distribution of allele frequencies among demes is expected when $4nm$ is large because migration rates are high (solid

arrows) and and/or genetic drift is relatively weak due to large effective population size (dashed arrows). The middle panel of Figure 7.15B shows the case when genetic drift and migration approximately balance, leading to a wide distribution of intermediate allele frequencies in demes. At the other extreme, shown in the right panel of Figure 7.15B, the distribution of allele frequencies is horseshoe shaped (most populations near fixation or loss) when $4nm$ is small because migration rates are low and and/or genetic drift is strong due to small effective population size.

Evolutionary scenarios imagined by wright

With the allele frequency distributions established, Wright then returned to the movement of populations on fitness surfaces given different parameters for the processes of natural selection, mutation, genetic drift and migration. Wright saw the position of populations on the landscape, the area occupied by populations on the landscape, and the very topography of the landscape itself as subject to change over time. His goal was to illustrate scenarios with sufficient trial and error (or genetic drift) that the tendency of natural selection to strand a population on a single fitness peak could be overcome. Figure 7.16 illustrates the six possibilities that Wright considered. In each of these six cases, the allele frequency distributions in Figure 7.15 define the range of variation in fitness expected among individuals in a population or deme. The connection arises if each locus in the allele frequency distributions of Figure 7.15 is interpreted as having a phenotypic effect and genotypes homozygous for the wild-type allele have the highest fitness. Broader allele

frequency distributions then lead to a larger number of possible allele combinations that would produce a wider range of fitness values.

Panels A and B in Figure 7.16 show Wright's ideas about the area of the adaptive landscape that would be occupied by a species in the context of large $4NS$ and $4NU$ values. Imagine a population initially occupies some area around a fitness peak, as shown by the dashed circle inside the shaded circle in Figure 7.16A. If the selection coefficient against genotypes with non–wild-type alleles were to decrease or the forward mutation rate were to increase, the area a population occupies around the adaptive peak would spread out (larger shaded circle). This would correspond to the allele frequency distribution changing from one like that in the left panel of Figure 7.15A to one like that in the middle panel of Figure 7.15A. In contrast, when selection against

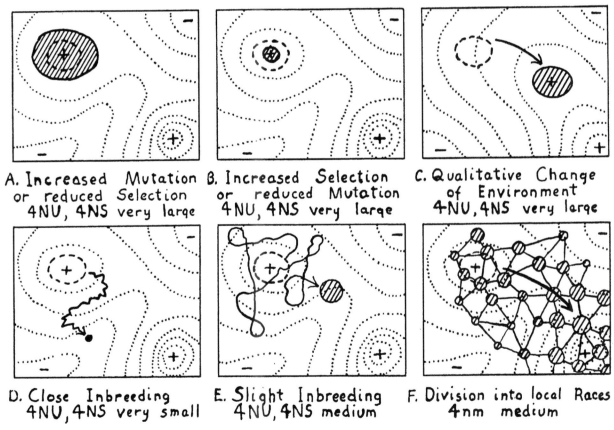

A. Increased Mutation or reduced Selection 4NU, 4NS very large

B. Increased Selection or reduced Mutation 4NU, 4NS very large

C. Qualitative Change of Environment 4NU, 4NS very large

D. Close Inbreeding 4NU, 4NS very small

E. Slight Inbreeding 4NU, 4NS medium

F. Division into local Races 4nm medium

Figure 7.16 Wright's representation of the action of drift-mutation balance (dictated by the magnitude of $4NU$), drift selection balance (dictated by the magnitude of $4NS$), and drift-migration balance in the island model (dictated by the magnitude of $4\,nm$). Wright's parameters are N for effective population size, U for mutation rate, S for the selection coefficient in directional selection, and $4\,nm$ for the effective migration rate in the infinite island model. The word "inbreeding" is used in the population sense where finite population size leads to genetic drift. Original caption: "Field of gene combinations occupied by a population within the general field of possible combinations. Type of history under specified conditions indicated by relation to initial field (heavy broken contour) and arrow." Source: From Wright (1932).

genotypes with alleles other than the wild type becomes stronger or forward mutation rates decrease in the context of large 4NS and 4NU, the area on the fitness surface occupied by a population shrinks because populations take on a narrower range of allele frequencies (compare the circle made by a dashed line to the smaller shaded circle in Figure 7.16B). This case corresponds to the allele frequency distribution changing in the opposite direction, from one like the middle panel of Figure 7.15A to one like that in the left panel of Figure 7.15A. In the case of Figure 7.16A, the average fitness of the species decreases, making it possible that "... the spreading of the [fitness surface] field occupied may go so far as to include another and higher peak ..." or that a fitness valley is crossed. It is also possible that the fitness peak itself could become taller if beneficial mutations occurred in a population and were fixed by selection. However, Wright pointed out that rates of mutation were very low so that evolutionary change of this sort would be very slow.

Wright also considered how the shape of the adaptive landscape itself might change. Figure 7.16C shows Wright's concept of how the adaptive landscape might change over time due to changes in the environmental context of a population (still in the context of large values of 4NS and 4NU). Because genotypic fitness values are defined by the physical and biological environment a species experiences, the fitness values of allele combinations may very well change over time. This would lead to a reshaping of the fitness surface itself, with peaks and valleys changing elevation or the position of peaks on the adaptive landscape shifting over time. While such change in the fitness surface would cause populations to track high fitness peaks, Wright saw this as "change without advance in adaptation" because populations were not necessarily occupying a number of peaks in the fitness landscape nor were populations necessarily evolving to higher levels of mean fitness.

This theme of constant environmental change driving a continual redefinition of the genotypes having highest fitness in a population was emphasized by Fisher (1999 variorum edition). Under this view, rugged adaptive landscapes are less problematic since a population can be thought of as occupying a region of allele frequency space where the topography of the fitness elevations changes over time. If the fitness landscape is continually remodeling itself, a population will not be stranded on a fitness peak since eventually the peak itself will move or change position. This view is also the foundation of the "Red Queen" or "arms race" model of Van Valen (1973), where a species must constantly experience adaptive change to keep pace with a continually changing genotypic fitness values ultimately caused by a perpetually changing environmental context defined by other species that themselves are constantly changing.

Another set of possibilities that Wright considered, diagrammed in Figure 7.16D and E, focused on effective population size. Wright pointed out that if the effective population size was very small relative to the selection coefficient and the mutation rate (Figure 7.16D), a population would likely experience fixation or loss at all loci due to genetic drift (see the allele frequency distribution in the right panel of Figure 7.15A). As a consequence, a population would cease attraction to fitness peaks, would wander at random around the fitness landscape, and would also experience the fixation of deleterious alleles leading to inbreeding depression. If the effective population size became small rapidly, then, after fixation and loss at most loci, movement on the landscape would be very slow since most new mutations would be unlikely to segregate for long. In contrast, a finite population with a medium effective population size relative to the selection coefficient and the mutation rate (Figure 7.16E) would occupy a fairly large area on the surface and would experience some random movement around a fitness peak but would not stray too far from the peak. This would occur because the population would experience an approximate balance between natural selection and genetic drift and would also have the input of new mutations over time (see the allele frequency distribution in the center panel of Figure 7.15A). Wright saw populations experiencing a balance of genetic drift, natural selection and mutation (medium values of 4NU and 4NS) as being able to shift fitness peaks and a means by which "the species may work its way to the highest peaks in the general field." The limitation is that peak shifting by one such population was expected by Wright to be a very slow process that would only occur if the mutation rate was approximately equal to the reciprocal of the effective population size.

The final case Wright considered, shown in Figure 7.16F, was the situation where a species was subdivided into a number of finite demes (or "small local races") that were nearly genetically isolated but did experience some gene flow. Here, Wright was thinking of the allele frequency distribution in the center panel of Figure 7.15B, where there is a balance between genetic drift leading to population differentiation and gene flow leading to the homogenization of allele frequencies that produces a broad distribution of allele frequencies among demes. Wright's idea was that many semi-independent finite demes would move positions on the fitness landscape more rapidly than a single panmictic population. Wright also conjectured that those demes that did reach higher fitness peaks would produce more migrants. The effect of more migrants would be for a deme on a higher fitness peak to shift the allele frequencies of the demes that received those migrants toward the positions of higher fitness peaks. This process of higher rates of gene flow from those demes on higher fitness peaks is often called **interdemic selection** since it is equivalent to natural selection acting at the level of demes with different levels of demographic productivity. Thus, Wright envisioned that a species made up of many subdivided demes experiencing approximately equal pressures of natural selection and genetic drift could explore more of the fitness surface and would be more likely to find more of the higher fitness peaks than natural selection alone. Wright concluded that "subdivision of a species into local races provides the most effective mechanism for trial and error in the field of gene combinations."

The shifting balance process is often summarized by the simultaneous operation of three "phases" of population genetic change in a subdivided population. Phase I involves genetic drift within demes that causes the allele frequency position of each deme to shift randomly with respect to the position of fitness peaks. Phase II is the operation of natural selection on demes such that the allele frequency position of demes is shifted toward and higher up fitness peaks, with taller peaks exerting a stronger influence on allele frequencies. Phase III is interdemic selection such that the rates of emigration from demes are proportional to the mean fitness of a population. Thus, demes at higher fitness peak elevations export more migrants and compose a larger proportion of the immigrant pool of other demes, shifting allele frequencies of all demes toward the allele frequency locations of taller fitness peaks.

Critique and controversy over shifting balance

While Wright's metaphor of the adaptive landscape and proposal of the shifting balance theory has stimulated the thinking of biologists for decades, his

ideas have also generated sustained controversy. The fitness surface metaphor itself has been one focus of critique because the original fitness surface described by Wright is problematic in some respects. As Provine (1986) describes, Wright employed two distinct versions of the fitness surface. One version of the fitness surface illustrates the fitness of *each genotype* based on an ordering of the allele combinations in genotypes. In this version of the fitness surface, each combination of alleles has a relative fitness and defines one point on the landscape. This type of surface has been compared to the pixels that make up a photographic print or digital image (Ruse 1996). In the genotype version of the fitness surface, what is represented biologically by the dimensions other than that representing fitness is not clear since the genotype axes do not relate to the frequency of genotypes or alleles in a population. Another version of the fitness surface plots the *mean fitness of a population* for all possible allele frequencies. The contemporary usage of the fitness surface metaphor is often in the population mean fitness sense, with axes representing allele frequencies and one dimension representing the mean fitness of a population at those allele frequencies, although there are exceptions (e.g. Weinreich et al. 2005). Wright often switched back and forth between these two types of fitness surfaces in his writing, leading to ambiguity and confusion (Provine 1986). Fitness surfaces have been constructed and interpreted in a wide variety of ways since Wright's work (Gavrilets 2004; Skipper 2004).

Coyne et al. (1997) presented a detailed and vigorous critique of Wright's shifting balance theory that examined evidence for and against operation of the three phases in actual populations. They reexamined the theoretical basis of the shifting balance theory with the benefit of more than 60 years of work in theoretical population genetics and considered empirical evidence for the shape of fitness surfaces and operation of the stages of the shifting balance process. They concluded that "although there is some evidence for the individual phases of the shifting balance process, there are few empirical observations explained better by Wright's three phase mechanism than by simple mass selection." Other authors responded in defense of shifting balance theory or offered alternative points of view (e.g. Wade and Goodnight 1998; Peck et al. 1998), generating a cascade of replies and counter-replies (Coyne et al. 2000; Goodnight and Wade 2000; Peck et al. 2000). While it is not possible here to consider in detail all of the points raised in that debate, disagreements over elements of the shifting balance process serve to highlight difficulties that arise when attempting to predict the outcome of multiple population genetic processes operating simultaneously.

The third phase of the shifting balance process, production of migrants in proportion to the population mean fitness and shifting of demes via differential contributions to the immigrant pool, is particularly problematic (see Crow et al. 1990). The difficulty is that the migration rate must be low enough to permit population subdivision into semi-isolated demes but at the same time high enough to permit the exchange of individuals (or gametes) among subpopulations that leads to interdemic selection. A general objection is that interdemic selection is a form of group selection, a process where there is greater survival or reproduction of a population of individuals compared to other populations such that some populations go extinct while others persist and expand. Williams (1966, 1992) has presented the classical arguments that natural selection on additive genetic variation among individual genotypes is expected to act more rapidly than selection on groups because the frequencies of individuals can change more rapidly than the frequencies of populations. Nonetheless, possible evidence for group selection in the context of differential migration has been shown in experiments with the flour beetle *Tribolum castaneum* (Wade and Goodnight 1991; Wade 2013). Large changes in the number of individuals per population were observed over nine generations by selecting individuals to be founders of the next generation based on the total number of individuals in a population. In contrast, the size of populations did not change over time when founding individuals were selected at random with respect to population size. The interpretation of numerous *Tribolum* experiments of this type (reviewed by Goodnight and Stevens 1997) has been controversial since there is disagreement over what exactly constitutes group and individual selection and what type of selection is imposed by the experimental procedures (see Coyne et al. 1997; Getty 1999; Wade et al. 1999). In support of the third phase, Bitbol and Schwab (2014) used a haploid model and simulations to show that population subdivision can accelerate the crossing of fitness valleys and plateaus.

Another aspect of the disagreement over shifting balance theory involves the dual nature of the concept of epistasis (see Cheverud and Routman 1995; Whitlock et al. 1995; Fenster et al. 1997; Brodie 2000; Cordell 2002). Epistasis exists when genotypes at two or more loci result in a genotypic value that is greater or less than the sum of the genotypic effects of the loci when taken individually (see Chapter 9). The existence of an interaction between two or more loci indicates the existence of **physiological epistasis** (also known as functional or mechanistic epistasis). The term physiological epistasis simply recognizes that certain genotypes at two or more loci interact in the production of a phenotype. The contribution, if any, of such physiological epistasis to population level quantities is a function of the frequencies of interacting genotypes in a population. The term **statistical epistasis** is used to refer to the amount of standing population variation in genotypic values caused by interactions among loci. In the symbols and concepts of Chapters 9 and 10, statistical epistasis is V_I. The amount of statistical epistasis present in a population is a function of the frequencies of interacting multilocus genotypes and, therefore, a function of population allele frequencies, as it is for additive and dominance variance (V_A and V_D), as well as a function of mating system and the rate of recombination.

Wright implicitly assumed that statistical epistasis was abundant in natural populations. While there is evidence that statistical epistasis exists in natural and laboratory populations (MacKay 2001; Cordell 2002; Carlborg and Haley 2004; see chapters in Wolf et al. 2000), statistical epistasis is not widespread in populations, although it remains difficult to estimate. There is currently no consensus over the relative contribution of epistasis to overall quantitative trait variation, although there is recognition that empirical detection of epistasis is limited by experimental designs and statistical power (see Whitlock et al. 1995). Some conclude that there is a lack of evidence for strong or frequent statistical epistasis in natural populations. Others suggest that there is some evidence for epistasis in natural populations, and since epistasis is difficult to detect, it is premature to draw a conclusion about the prevalence of epistasis. These disparate views translate into difficulty summarizing the nature of adaptive landscapes in populations that are genetically variable.

When a population is at fixation and loss for all loci, there can be no statistical epistasis since there is no variation in genotypic value, even though physiological epistasis may exist and could even be strong. An alternative definition of epistasis is useful for populations that may exhibit little statistical epistasis and yet have abundant physiological epistasis. **Sign epistasis** is a special case of physiological epistasis in populations with little or no genetic variation (Weinreich et al. 2005). A locus exhibits sign epistasis when a new mutation exhibits higher than average fitness on some genetic backgrounds defined by other loci but lower than average fitness on other genetic backgrounds. The sign of the fitness value is, therefore, a function of the other loci that make up the genetic background of the allele. Weinreich et al. (2006) examined five single nucleotide mutations in the β-lactamase gene of bacteria. Four of these mutations result in missense versions of the β-lactamase gene so that they are deleterious individually and selected against in antibiotic environments. The fifth mutation is a noncoding change 5′ to the gene. However, when all five mutations occur simultaneously, they lead to a version of the β-lactamase gene that confers resistance to β-lactam antibiotic drugs such as penicillin. When each mutation occurs on a genetic background with the other four mutations present, its fitness is positive since the five mutation version of the gene has high fitness in antibiotic environments. In general, sign epistasis predicts that some allele combinations may have a limited number of mutational combinations that lead to increased fitness.

The fitness landscape concept continues to be relevant to a variety of questions in evolutionary biology such as the degree to which phenotypes produced by numerous evolutionary intermediate steps are repeatable (see reviews by de Visser and Krug 2014, Fragata et al. 2019). Yi and Dean (2019) suggested that the emerging ability to experimentally alter and understand the proximate details of how and why nucleotide and amino acid changes alter molecular phenotypes will lead to empirically defined fitness landscapes that explain evolutionary trajectories. They reviewed seven illustrative studies where the observations of genetic variation were combined with experimental work to understand the pathway of genotype to phenotype to fitness brought on by individual genotype changes. Advances in the ability to make highly specific

experimental genotype alterations that lead to phenotypic and fitness variation (e.g. Karageorgi et al. 2019) is a breakthrough that seems likely to lead to much more empirical research on the causal pathways that define genotype–phenotype-fitness and a deeper understanding of fitness landscapes.

Chapter 7 review

- Mean fitness in a population can be viewed as a graph of average fitness by genotype or allele frequencies called a fitness surface. Natural selection acts as an uphill climber on a fitness surface, moving genotype frequencies uphill based on the slope at the current genotype frequencies.
- Fitness surfaces will have multiple peaks and valleys if there is dominance or epistasis.
- When fitness depends on three alleles at one locus, the results of natural selection depend on initial genotype frequencies in the population if there is strong over- and underdominance.
- The net balance of recombination and natural selection may result in equilibria that do not correspond to mean fitness maxima. Recombination works toward gametic equilibrium irrespective of mean fitness and may act in opposition to natural selection.
- Although the viability model of fitness is used as a standard, changes in genotype frequencies and their equilibria are often distinct when fitness is defined as differential fecundity, carrying capacity, or when fitness values vary in time and space.
- When both natural selection and genetic drift are acting, selection is strong relative to genetic drift when $4N_es$ is much greater than 1, selection is weak relative to genetic drift when $4N_es$ is much less than 1, and selection and drift are about the same strength when $4N_es$ is approximately 1.
- When both natural selection and mutation are acting, deleterious alleles will be maintained in a population at a level that increases with the mutation rate and with consanguineous mating but decreases with the selection coefficient against the allele when homozygous.
- Genetic load results from the selective deaths (either reproductive or actual) that must occur as frequencies of mutations or genotypes change under natural selection. In principle, the amount of natural selection is limited by the genetic load a

population can tolerate without going extinct. Genetic load arguments were used as indirect evidence for the neutral theory of molecular evolution.

- Compared with the predictions of the Haldane–Muller model, the genetic load depends strongly on the genotype used as a reference point for w_{max} and may be reduced if deleterious mutations are recessive; homozygosity is increased by identity by descent; or truncation, balancing, or soft selection operates.
- In genealogical branching models, directional selection can be modeled as an ancestral selection graph. Weak directional selection does not greatly alter the total branch length nor the total height of genealogical trees on average.
- In genealogical branching models, balancing selection can be modeled as a lineage type switching process similar to gene flow or mutation. With two alleles, balancing selection lengthens the average time to coalescence for the final two lineages since they have to switch to the same type to coalesce. With three or more alleles, balancing selection will tend to increase coalescence times and lengthen terminal branches in a genealogical tree.
- Sewall Wright's metaphor of the adaptive landscape is a heuristic device designed to articulate how the process of natural selection alone is constrained since it can only increase population mean fitness.
- Shifting balance theory was a hypothesis about how the simultaneous action of natural selection, genetic drift, mutation, and population subdivision might lead to the exploration of a larger portion of the adaptive landscape than would be possible under selection alone.
- Adaptive landscapes are an enduring metaphor in population genetics and have renewed relevance in experimental studies that examine the phenotypic and fitness impacts of precision genotype editing.

Further reading

A classic and approachable treatment of two-locus selection that features fitness surfaces can be found in:

Lewontin, R.C. and White, M.J.D. (1960). Interaction between inversion polymorphisms of two chromosome pairs in the grasshopper, *Moraba scurra. Evolution* 14: 116–129.

For case studies, perspective, and basic theory relating to genotype interactions at two or more loci that influence fitness, see chapters in:

Wolf, J.B., Brodie, E.D. III, and Wade, M.J. (eds.) (2000). *Epistasis and the Evolutionary Process*. Oxford: Oxford University Press.

Extensions of the basic viability natural selection model that account for biological variations such as spatial and temporal variation in fitness, fitness trade-offs, competition, and predation can be found in:

Roff, D.A. (2001). *Life History Evolution*. Sunderland, MA: Sinauer Associates.

A review of empirical studies that employed the F_{ST} outlier approach to test for local experiencing local adaptation is provided by:

Haasl, R.J. and Payseur, B.A. (2016). Fifteen years of genomewide scans for selection: trends, lessons and unaddressed genetic sources of complication. *Molecular Ecology* 25: 5–23.

A review of the classic model of genetic load and numerous more recent models that expand our understanding of the relationship between deleterious alleles and natural selection:

Agrawal, A.F. and Whitlock, M.C. (2012). Mutation load: the fitness of individuals in populations where deleterious alleles are abundant. *Annual Review of Ecology, Evolution, and Systematics* 43: 115–135.

For background and explanation of the original fitness surface along with some response to criticisms, see:

Wright, S. (1988). Surfaces of selective value revisited. *American Naturalist* 131: 115–123.

End-of-chapter exercises

1 Refer to the top set of relative fitness values in Table 7.1 and the fitness landscape in

Figure 7.2. If only the A and S alleles are present in the population, what are the equilibrium allele frequencies expected under natural selection in an infinite population?

2 Refer to the top set of relative fitness values in Table 7.1. Assume that the frequency of the C allele was initially 0.001 (a very large frequency under mutation pressure alone). Could one generation of genetic drift to cause the C allele to attain a high enough frequency such that it then continues to increase in frequency by natural selection? First, compute the change in the frequency of the C allele (or Δr using r as the symbol for the frequency of the C allele) under natural selection for a range of low frequencies of C. Try values of $r = 0.001$–0.1. Based on Δr, what minimum frequency does the C allele need to attain before it is increased by natural selection ($\Delta r >$ zero)?

Would genetic drift be sufficient to cause the frequency of the C allele to cross the fitness landscape valley such that $\Delta r >$ zero? Estimate the expected variance in the C allele frequency under genetic drift for effective population sizes

of 100, 10, and 2. To do this, use the binomial distribution to estimate the expected variance in the change of allele frequency by finite sampling.

Using a spreadsheet or computer code to compute Δr will make it easier to compare several sets of parameter values.

3 Search the literature for a recent research paper that utilizes one or more of the population genetic predictions covered in this chapter. The topic can be any organism, application, or process, but the paper must include a hypothesis test involving a topic such as adaptive landscapes, selection and gametic disequilibrium, selection in finite populations ($4N_e s$), density- or frequency-dependent selection, tests for selection using F_{ST} outlier loci, or adaptive landscapes. Summarize the main hypothesis, goal, or rationale of the paper. Then, explain how the paper utilized a population genetic prediction from this chapter and summarize the results and the conclusions based on the prediction.

Problem box answers

Problem box 7.1 answer

$w_{AA} = 0.679 \; w_{AS} = 0.763 \; w_{SS} = 0.153 \; w_{AC}$
$= 0.679 \; w_{SC} \; 0.534 \; w_{CC} = 1.0$

Initial allele frequencies set 1:
$p = 0.75, q = 0.20, r = 0.05$
$\overline{w} = 0.679(0.75)^2 + 0.153(0.2)^2$
$\quad + 1(0.05)^2 + (0.763)2(0.75)(0.2)$
$\quad + (0.679)2(0.75)(0.05) + (0.534)2(0.2)(0.05)$
$\overline{w} = 0.382 + 0.00612 + 0.0025 + 0.2289 + 0.051$
$\quad + 0.0107 = 0.6812$
$\overline{w}_C = 1(0.05) + (0.679)(0.75) + (0.534)(0.2)$
$\quad = 0.6661$
$$\Delta p = \frac{0.05(0.6661 - 0.6812)}{0.6812} = -0.0011$$
$$p_{t+1} = 0.05 - 0.0011 = 0.0489$$

The C allele goes to loss because its marginal fitness is lower than the mean fitness.

Initial allele frequencies set 2 :
$p = 0.7, q = 0.20, r = 0.1$
$\overline{w} = 0.679(0.7)^2 + 0.153(0.2)^2 + 1(0.1)^2$
$\quad + (0.763)2(0.7)(0.2) + (0.679)2(0.7)(0.1)$
$\quad + (0.534)2(0.)(0.1)$
$\overline{w} = 0.333 + 0.0061 + 0.01 + 0.2136 + 0.0951$
$\quad + 0.0214 = 0.6792$
$\overline{w}_C = 1(0.1) + (0.679)(0.7) + (0.534)(0.2) = 0.6821$
$$\Delta p = \frac{0.1(0.6821 - 0.6792)}{0.6792} = 0.05$$
$$p_{t+1} = 0.1 + 0.05 = 0.15$$

The C allele goes to fixation because its marginal fitness is greater than the mean fitness, a condition that will hold until the C allele is fixed in the population.

Problem box 7.2 answer

At selection events #1 and #2, the incoming branch displaces the continuing branch. The incoming and continuing branches have identical states at selection event #3. The result is

that all lineages have the state A in the present. In this case, the height of the genealogy is shorter because of the outcome of selection event #1. See Figure 7.17.

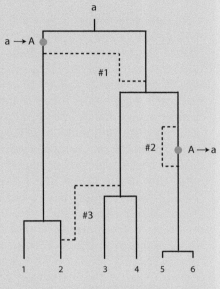

#1: continuing = a, incoming=A, selection result=A

#2: continuing = a, incoming=A, selection result=A

#3: continuing = A, incoming=A, selection result=A

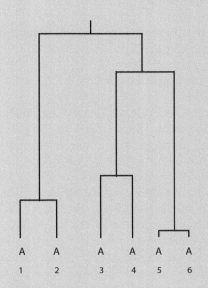

Figure 7.17   An ancestral selection graph where the MRCA has the haplotype state a and where A is the fitter haplotype. The A haplotype displaces the a haplotype when continuing and incoming branches coalesce.

CHAPTER 8

Molecular evolution

8.1 The neutral theory

- The neutral theory and its predictions for levels of polymorphism and rates of divergence.
- The nearly neutral theory.
- The selectionist–neutralist debates.

The field of molecular evolution involves the study of DNA, RNA, and protein sequences with the goal of elucidating the processes that cause both change and constancy among sequences over time. One approach to molecular evolution is to focus on a specific gene, seeking to test hypotheses about what parts of that specific sequence are most likely involved in some function or in the regulation of transcription. Another type of inquiry in molecular evolution involves testing hypotheses about the population genetic processes that have operated on sequences in the past using DNA sequence data. This latter type of research often seeks to distinguish whether a pattern of variation in a sample of DNA sequences is consistent with genetic drift or with certain forms of natural selection. The common feature of all hypothesis tests in studies of molecular evolution is the use of null and alternative hypotheses for the patterns and rates of sequence change. This chapter will introduce the conceptual foundations behind many of the most commonly used null and alternative hypotheses in molecular evolution. Although this chapter focuses exclusively on DNA sequences, the concepts presented are sometimes applicable to protein sequences as well.

The **neutral theory** now forms the basis of the most widely employed null model in molecular evolution. The neutral theory adopts the perspective that most mutations have little or no fitness advantage or disadvantage and are therefore **selectively neutral**. Genetic drift is therefore the primary evolutionary process that dictates the fate (fixation or loss) of newly occurring mutations. When it was originally proposed, the neutral theory was a major departure from orthodox population genetic theory of the time. In the 1950s and 1960s, it was widely thought that most mutations would have substantial fitness differences, and therefore, the fate of most mutations was dictated by natural selection. Motoo Kimura (shown in Figure 8.1) argued instead that the interplay of mutation and genetic drift could explain many of the patterns of genetic variation and the evolution of protein and DNA sequences seen in biological populations (Kimura 1968, reviewed in Kimura 1983). King and Jukes (1969) also proposed a similar idea at around the same time. (The debate over the neutral theory as well as some of the logic behind Kimura's proposal of the neutral theory is covered at the end of this section of the chapter.) The neutral theory null model makes two major predictions under the assumption that genetic drift alone determines the fate of new mutations. One prediction is the amount of **polymorphism** for sequences sampled within a population of one species. The other prediction is the degree and rate of **divergence** among sequences sampled from separate species.

Figure 8.1 Motoo Kimura (left) and James Crow (right) in 1986 on the occasion of Kimura being awarded an honorary doctoral degree at the University of Wisconsin, Madison. Kimura pioneered the use of diffusion equations to determine quantities such as the average time until fixation or until loss for neutral mutations. Based on these foundations, he proposed the neutral theory of molecular evolution in 1968. Kimura and Crow collaborated to develop some of the basic expectations for neutral genetic variation. Crow mentored and collaborated with many influential contributors to population genetics, including Kimura. Source: James F. Crow.

> **Divergence:** Fixed genetic differences that accumulate between two completely isolated lineages that were originally identical when they separated from a common ancestor.
> **Polymorphism:** The existence in a population of two or more alleles at one locus. Populations with genetic polymorphisms have heterozygosity, gene diversity, or nucleotide diversity measures that are greater than 0.

Polymorphism

The balance of genetic drift and mutation that determines polymorphism in the neutral theory is diagrammed in Figure 8.2. Each line indicates the frequency of an allele over time. New mutations enter the population (lines at the bottom edge), and their frequency in the population is a random walk between fixation and loss caused by genetic drift. To see how this random walk results in polymorphism, hold a straight edge such as a ruler to form a vertical line at any single time point. If the vertical line intersects any allele frequency lines, then the population has genetic polymorphism at that time point since there are multiple alleles segregating in the population. More alleles segregating in the population indicate more polymorphism. Segregating alleles, and therefore polymorphism, results from the random walk in frequency that each mutation takes under genetic drift. Most mutations segregate for short periods of time and then are lost from the population. However, since their frequency is dictated by random sampling, some alleles may reach high frequencies before eventually being lost. A small proportion of mutations will eventually be fixed in the population after a random walk in allele frequency. Under neutral theory, polymorphism results from the transient dynamics of allele frequencies before they reach fixation or loss end points. The process that underlies Figure 8.2 can be approximately simulated in Interact Box 5.1.

The neutral theory's prediction for levels of polymorphism in a population follows directly from the predicted dynamics of allele frequency under genetic drift (see Chapters 3 and 5). Chapter 3 showed that the initial frequency of an allele is also its chance of eventual fixation. For new mutations that start out as one copy in the entire population of $2N$ allele copies, the chance of eventual fixation is $\frac{1}{2N}$ while the chance of eventual loss is $1 - \frac{1}{2N}$. The diffusion approximation of genetic drift in Chapter 3 also showed that the average time to fixation of a new mutation approaches $4N$ generations while the average time to loss approaches just $2\left(\frac{N_e}{N}\right)\ln\left(2N\right)$ generations as N gets large. (To see these results, set $p = \frac{1}{2N}$ in Eq. 3.40 and evaluate the expressions as N increases toward infinity.) The average time to fixation also has a large variance, so the standard deviation of the number of generations to fixation is expected to be about half of the mean or $2.15N_e$ generations (Kimura and Ohta 1969b; Narain 1970; Kimura 1970, 1983).

For example, in a population of $N = 1000$, the average time to loss is about 15 generations (assuming $\frac{N_e}{N} = 1$), but the average time to fixation for those alleles that eventually fix is 4000 generations. In addition, the time to fixation is highly variable since genetic drift is a stochastic process or a random walk. The standard deviation of the time to fixation is large, at approximately 2150 generations, consistent with

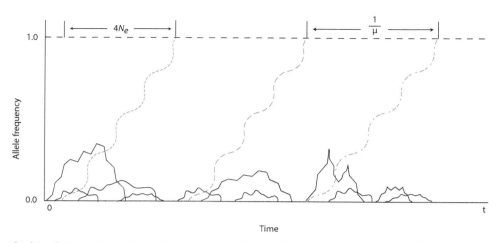

Figure 8.2 The fate of selectively neutral mutations in a population. New mutations enter the population at rate μ and an initial frequency of $\frac{1}{2N}$. Allele frequency is a random walk determined by genetic drift. The time that a new mutation segregates in the population, or the segregation time of a mutation, depends on the effective population size. However, the chance that a new mutation goes to fixation (equal to its initial frequency) is also directly related to the effective population size. These two effects of the effective population size cancel each other out for neutral alleles. The neutral theory then predicts that the rate of fixation is μ and therefore the expected time between fixations is 1/μ generations. For that subset of mutations that eventually fix, the expected time from introduction to fixation is $4N_e$ generations. Source: After Figure 3.1 in Kimura (1983).

a broad range in times to fixation. Relatively few mutations are expected to fix, but those mutations that do go to fixation segregate for a much longer time on average than the large proportion of mutations that go to loss. While mutations are segregating before their eventual end point of fixation or loss, there is polymorphism in the population.

An additional way to understand the expected polymorphism of neutral alleles in a population is to examine the equilibrium balance between genetic drift causing alleles to go to fixation and the input of new alleles in a population by mutation. Chapter 5 showed that for the infinite alleles model of mutation, the combined processes of mutation and genetic drift produce equilibrium heterozygosity:

$$H_{equilibrium} = \frac{4N_e\mu}{4N_e\mu + 1} \qquad (8.1)$$

that depends on the effective population size N_e and the mutation rate μ (see Eq. 5.39). In this view of neutral mutations, polymorphism results from either a high rate of input of mutations even if drift is strong, a long dwell time for each mutation due to a large effective population size even if mutations are infrequent, or intermediate levels of mutation and genetic drift.

The neutral theory prediction for polymorphism can be readily compared with polymorphism expected under positive (higher than average genotype fitness) and negative (lower than average

genotype fitness) natural selections (Figure 8.3). New mutations that are deleterious will go to loss faster than neutral mutations since natural selection will deterministically reduce their frequencies, and there will be little or no random walk in allele frequency. In contrast, new mutations that are advantageous will increase in frequency to fixation, again deterministically under natural selection without a random walk in frequency. A locus with new mutations that are influenced by directional natural selection should show less polymorphism than a locus with neutral mutations. The other possibility is that some advantageous mutations are influenced by balancing selection due to overdominance for fitness. In that case, two or more alleles will have very long times of segregation since natural selection will maintain several alleles at intermediate frequencies between fixation and loss with the result of increased levels of polymorphism in the population.

As a whimsical metaphor, compare the average times to fixation under directional selection, neutral evolution, and balancing selection with the time it takes a population genetics student to go from his or her lab (initial mutation) to the coffee shop and back (fixation) at different career stages (different processes). A new, overworked student goes directly to get a cup of coffee and returns immediately without stopping to talk to anyone, so the trip is short and direct like directional selection. With more experience, a student has a bit more free time and will pause more often to greet friends, like a random

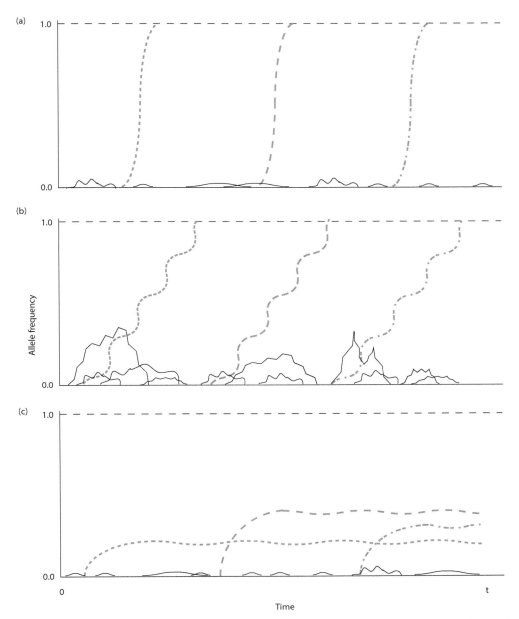

Figure 8.3 The dwell time for new mutations is different if fixation and loss is due to genetic drift or natural selection. With neutral mutations (B), most mutations go to loss fairly rapidly and a few mutations eventually go to fixation. For both eventual fixation or loss of neutral mutations, the path is a random walk implying that the time to fixation or loss has a high variance. For mutations that fix because they are advantageous (A), directional selection fixes them rapidly in the population. Therefore, under directional selection, alleles segregate for a shorter time and there is less polymorphism than with neutrality. For mutations that show overdominance for fitness, natural selection favoring heterozygote genotypes maintains several alleles in the population indefinitely. Therefore, balancing selection greatly increases the segregation time of alleles and increases polymorphism compared to neutrality. Both cases of natural selection (A and C) are drawn to show negative selection acting against most new mutations. If new mutations are deleterious, then the time to loss is very short and there is very little random walk in allele frequency since selection is nearly deterministic.

walk. As they approach graduation, a student takes a roundabout path to the shop and stops to talk frequently such that the coffee break takes a very long time, like balancing selection.

Divergence

The neutral theory also predicts the rate of divergence between sequences. Genetic divergence

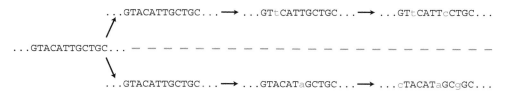

Figure 8.4   The process of divergence for two DNA sequences that descended as identical copies of an ancestral sequence. Each sequence experiences neutral mutations, some of which are eventually fixed by genetic drift. These fixed mutations replace all other alleles and are, therefore, substitutions (indicated by lower case letters). As substitutions accumulate, the two sequences diverge from the ancestral sequence as well as from each other. In this example, the two sequences are eventually divergent at 5 of 12 nucleotide sites due to substitutions. The dashed line is meant to indicate complete isolation of the two populations containing the derived sequences.

occurs by **substitutions** that accumulate in two DNA sequences over time. Think of two DNA sequences that are copies of the same ancestral sequence (Figure 8.4). The two sequences were originally identical before any substitutions occurred. Over time, mutations occurred in each population and some were fixed by chance due to genetic drift (see Figure 8.2). Each fixed mutation causes a change in the base pairs at random nucleotide positions in the sequence, causing each sequence to diverge slightly from its ancestor as well as from its sibling sequence. A biological example is two species that recently diverged from an ancestral species with no genetic variation. Both of the new species would be founded with identical DNA sequences and thereafter be reproductively isolated from each other. DNA sequences compared between the two species would each experience DNA sequence divergence due to mutations occurring over time.

> **Substitution:** The complete replacement of one allele previously most frequent in the population with another allele that originally arose by mutation.

The neutral theory predicts the rate at which allelic substitutions occur and thereby the rate at which divergence occurs. Predicting the substitution rate for neutral alleles requires knowing the probability that an allele becomes fixed in a population and the number of new mutations that occur each generation. A new mutation in a population of diploid individuals is initially present as just a single copy out of a total of $2N$ copies of the locus. Therefore, the initial frequency of a new mutation is $\frac{1}{2N}$. Under genetic drift, the chance of fixation of any neutral

allele is simply its initial frequency (see Chapter 3). Each generation, the chance that an allele copy mutates is μ and there are a total of $2N$ allele copies. Therefore, the expected number of new mutations in a population each generation is $2N\mu$. Multiplying the probability of fixation by the expected number of mutations per generation

$$k = (2N\mu)\frac{1}{2N} \qquad (8.2)$$

gives the rate at which alleles that originally entered the population as mutations go to fixation per generation, symbolized by the substitution rate k. Notice that this equation simplifies to

$$k = \mu \qquad (8.3)$$

The necessary assumption is that the substitution process is viewed on a time scale that is long relative to the average time to fixation for an individual mutation. If more than $4N_e$ generations have elapsed, then it is likely that all the alleles in a population will have descended from one allele due to genetic drift. The probability that the lucky allele that is fixed in the population was a new mutation is μ.

This result is remarkable because it says that the probability that a neutral mutation goes to fixation each generation, or the rate of substitution, is simply equal to the mutation rate. Notice that the predicted substitution rate does not depend on the effective population size. This is because a mutation in a smaller population has a greater chance of fixation but there are fewer new mutations each generation, while a mutation in a larger population has a smaller chance of reaching fixation but there are more mutations introduced each generation. The rate of input of new mutations in a population and the chance of fixation due to genetic drift exactly balance out when N changes. Note that this same result holds in the case of haploid loci since there are a total of N alleles and the probability of fixation of a new mutation is $\frac{1}{N}$.

Based on the rate of substitution, the neutral theory also predicts that the substitutions that ultimately cause divergence should occur at a regular average rate. For waiting time processes, the time between events is the reciprocal of the rate of events. Using a clock that chimes on the hour as an example, the rate of chiming is 24 per day (or 24/day). Therefore, the expected time between chiming events is 1/24 of a day or 1 hour. Since the rate of neutral substitution is μ, the *expected* time between neutral substitutions is $1/\mu$ generations (see Figure 8.2). For example, if the mutation rate of a locus is 1×10^{-6} (one nucleotide change per 10^{6} gametes per generation), then the expected time between neutral substitutions is 10^{6} generations *on average*. This offers one explanation of why different loci diverge at different rates: the different loci simply have distinct mutation rates that lead to variable neutral substitution rates.

Nearly neutral theory

The **nearly neutral theory** considers the fate of new mutations if some portion of new mutations is acted on by natural selection of different strengths (Ohta and Kimura 1971, Ohta 1972, reviewed in Ohta 1992 and Gillespie 1995). The nearly neutral theory recognizes three categories of new mutation: neutral mutations, mutations acted on strongly by either positive or negative natural selection, and mutations acted on weakly by natural selection relative to the strength of genetic drift. This latter category contains mutations that are nearly neutral since neither natural selection nor genetic drift will determine their fate exclusively.

For a new mutation in a finite population that experiences natural selection, the forces of directional selection and genetic drift oppose each other. Recall from Chapter 3 that genetic drift causes heterozygosity to decrease at a rate of $\frac{1}{2N_e}$ per generation. Thus, $\frac{1}{2N_e}$ quantifies the "push" on a new mutation toward fixation caused by genetic drift. The selection coefficient (s) on a genotype describes the "push" on alleles toward fixation or loss due to natural selection. The force of selection on a new mutation can be quantified using the result from Chapter 5 that

the chance of fixation is approximately $2s$ (see the Mendelian segregation model in 5.2). Setting these forces equal to each other

$$2s = \frac{1}{2N_e} \qquad (8.4)$$

gives the conditions where genetic drift and natural selection have approximately equal influence on the fate of allele frequencies. When $2s$ is within an order of magnitude of the reciprocal of effective population size, an allele can be described as **net neutral** or **nearly neutral** since natural selection and genetic drift are approximately equal forces dictating the probability of fixation of an allele. Next, notice that multiplying both sides of Eq. 8.4 by $2N_e$ gives $4N_es = 1$ as the condition where the processes of genetic drift and natural selection are equal. When $4N_es$ is much greater than 1, natural selection is the stronger process, whereas when $4N_es$ is much less than 1, genetic drift is the stronger process.

Using more sophisticated mathematical techniques, Kimura (1962) showed that the probability of fixation for a new mutation in a finite population is

$$P_{fixation} = \frac{1 - e^{-4N_esp}}{1 - e^{-4N_es}} \qquad (8.5)$$

where p is the allele frequency ($p = \frac{1}{2N_e}$ for a single mutation, but, in general, p is assumed to be much less than 1), N_e is the effective population size, and s is the selection coefficient assuming codominance. This equation is plotted in Figure 8.5 along with the constant probability of fixation for a new mutation expected under the neutral theory.

The nearly neutral theory predicts that the rate of substitution will depend on the effective population size for the proportion of mutations in a population that are nearly neutral ($4N_es \approx 1$). The nearly neutral theory, therefore, predicts that the amount of polymorphism in populations depends for some mutations on the effective population size. A consequence is that subdivided populations and different species can exhibit different levels of polymorphism based on their effective population size. Similarly, rates of divergence can also vary between species due to differences in N_e. This is in contrast to the neutral theory, which predicts that the rate of substitution is independent of the effective population size.

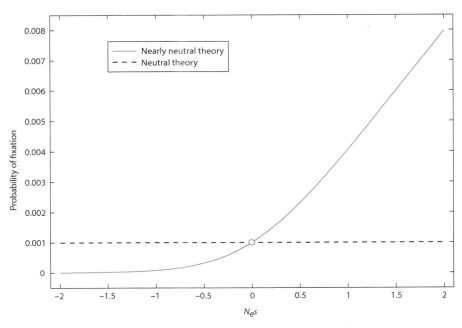

Figure 8.5 The probability of eventual fixation for a new mutation under the neutral and nearly neutral theories. Under the nearly neutral theory, the probability of fixation depends of the balance between natural selection and genetic drift, expressed in the product of the effective population size and the selection coefficient ($N_e s$). When negative selection operates against a deleterious allele, the selection coefficient and $N_e s$ are negative. Values of $N_e s$ near zero yield a fixation probability close to that predicted by the neutral theory. Only when the absolute value of $N_e s$ is large does natural selection exclusively determine the probability of fixation. The neutral theory assumes that neutral mutations are not influenced by selection and have a constant probability of fixation dictated by the effective population size. In this example, the initial allele frequency is 0.001, or the frequency of a new mutation at a diploid locus in a population of 500. Source: After Ohta (1992).

Interact box 8.1 Compare the neutral theory and nearly neutral theory

Use the text simulation website to compare the predictions of the neutral theory and nearly neutral theory. In the **Simulations** menu, choose the **Neutral Theory** model.

Run the model using the default parameter values. Look at the larger plots on the left under **Strictly neutral** and in **Nearly neutral** to understand what they show. Also, look at the histogram plots on the right to understand that each displays the distribution of times to fixation for those alleles that fix as well as time to loss for those alleles that go to loss. The nearly neutral model also shows the distributions of relative fitness coefficients for alleles that are fixed and alleles that are lost.

Run the simulation with a series of increasing values of N_e. In each of these runs, compute the value of $N_e s$. How does the outcome of the nearly neutral model change as $N_e s$ varies?

The selectionist–neutralist debates

The neutral theory of molecular evolution was proposed by Kimura in 1968 (reviewed in Kimura 1983, 1989), coupling models of genetic drift with the then novel (and scarce) data on rates of amino acid divergence. Kimura used the amino acid divergence data in mammals to estimate that the rate of nucleotide substitution genome-wide was about one site every other year. This implied a very large genetic load (see Chapter 7) if natural selection was the principle process that governed the eventual substitution of new mutations. Kimura showed, instead, that if new mutations are neutral (meeting the condition that $|2N_e s|$ $< <1$), then the genetic load was low enough to seem reasonable. (In the process of making the load calculation, he also showed that the rate of substitution of a neutral mutation was approximately the mutation rate.) In the same paper, Kimura considered the level of polymorphism in the allozyme data of Hubby and Lewontin (1966), estimating that an effective population size of between 2300 and 9000 would produce the levels of heterozygosity observed in *Drosophila* under neutral evolution.

The proposal of the neutral theory ushered in a new era in population genetics that saw the development of numerous models constructed with the goal of explaining levels of polymorphism or rates of divergence (reviewed by Nei 2005). At the same time, there was an increasing volume of genetic data available to test population genetic models. New data often revealed patterns of polymorphism or divergence that were not strictly compatible with the neutral theory, motivating continual extensions to the neutral theory. At the same time, numerous advances in predictions for how natural selection at the molecular level (e.g. hitch-hiking, codon bias, and background selection) were developed as alternative hypotheses to the neutral theory. After Kimura's (1968) initial proposal of the neutral theory, genetic load faded in importance while levels of polymorphism and rates of divergence became the primary issues.

Pan-neutralism and **pan-selectionism** are caricatures that illustrate some of the exaggerated stances taken in selectionist–neutralist debates. Figure 8.6 shows versions of the mutation fitness spectrum that schematically represent the extreme

positions of pan-neutralism and pan-selectionism (compare with Figure 5.1). Both of these points of view are extreme because they rely on a picture of the fitness of mutations that does not match most observations. The neutral theory was often misunderstood as a proposal that *all* nucleotide or amino changes were selectively neutral (Figure 8.6A). Pan-neutralism cannot explain patterns of molecular evolution such as differences in divergence rates at synonymous and nonsynonymous sites within a gene (which share a common mutation rate and effective population size). Functional constraint due to natural selection (first suggested by King and Jukes 1969) can readily explain such observations. At the same time, the neutral theory helped illustrate implications of the classical and balance hypothesis perspectives. Both of the classical and balance hypotheses emphasize natural selection and tend toward pan-selectionism. As Figure 8.6B illustrates, arguing for a complete absence of neutral mutations is an unreasonable position because it requires a mutation fitness spectrum with a discontinuity around a selection coefficient of zero.

One part of the selectionist–neutralist debate involved the level of allozyme polymorphism (estimated by heterozygosity or gene diversity) in populations. Kimura and Crow (1964) had developed an expectation for the level of heterozygosity at drift-mutation equilibrium in the infinite alleles model (see Chapter 5). After the proposal of the neutral theory, this expectation was applied to the growing body of protein polymorphism data. Observed polymorphism should match this prediction if the neutral theory was correct. The problem at the time was that observed heterozygosities fell into too small of a range (see Lewontin 1974). Based on mutation rate estimates, the species sampled should have differed a great deal in their effective population size if mutations were neutral. Thus, there was perceived to be too little polymorphism to be consistent with neutral theory predictions. But at the same time, there was perceived to be too much polymorphism to be consistent with the classical hypothesis. Time has shown that estimates of key population parameters like mutation rates, assumed effective population sizes, and the estimates of polymorphism were sometimes imprecise enough to cause fairly large deviations from expectations. There was also the limitation that empirical data were not general since the sample of genes, population, and species was limited. It was during this period that Lewontin (1974) lamented that neutral theory expectations depended on the

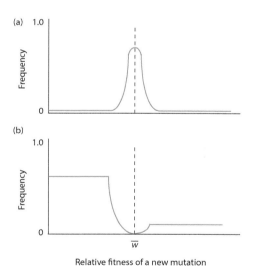

Figure 8.6 The caricatures of the mutation fitness spectrum drawn to illustrate the extreme views of pan-selectionism and pan-neutralism. Under a pan-neutralist view (A), almost all mutations have little or no impact on fitness and so are selectively neutral. Under a pan-selectionist view, (B) there would be almost no mutations that are selectively neutral, many mutations that are selected against and a relatively high frequency of beneficial mutations. Neither of these extremes is supported by most observations, except perhaps in isolated cases. This illustration is inspired by figures in Turner (1992) and Crow (1972).

product of a large unknown number (the effective population size) and a small unknown number (the mutation rate), a comment he frequently repeated.

Numerous natural selection-based explanations of polymorphism were developed as alternatives to or extensions of the neutral theory. One innovative hypothesis put forth to explain the deficit of polymorphism relative to neutral theory expectations was that of genetic hitch-hiking around positively selected sites (Maynard Smith and Haig 1974). A reduction in polymorphism was predicted for neutral loci in gametic disequilibrium with positively selected loci, explaining a possible cause of reduced polymorphism. A decade later, Kreitman's (1983) DNA sequence data set showing polymorphism for the alcohol dehydrogenase or *Adh* locus in *Drosophila melanogaster* gave new life to the balancing selection hypothesis. The *Adh* locus had originally been studied with allozyme techniques and was known to exhibit high levels of polymorphism. The *Adh* DNA sequences identified the basis of the allozyme polymorphism as a threnine/lysine difference and also exhibited synonymous variation around these non-synonymous sequence changes. Overall, the *Adh* sequences exhibited too much polymorphism to be consistent with strict neutrality but were consistent with a model of balancing selection. These data contributed to a renewal of the balance hypothesis and sparked continued debate about the processes regulating polymorphism at the DNA level.

Another part of the selectionist–neutralist debate revolved around rates of divergence. The neutral theory predicted that rates of divergence should be constant through time since the expected time between neutral substitutions is the inverse of the mutation rate. Amino acid sequence data collected in the 1970s and DNA sequence data collected in the 1980s showed that the variance in the number of substitutions was too large to be compatible with the neutral theory. One explanation for the variance was that the rates of fixation of new mutations were largely a consequence of natural selection. Since selection strength could vary for different mutations and also be variable through time and space, natural selection would cause variation in substitution rates. In particular, natural selection could cause periods of little or no substitution because deleterious mutations would be selected against while beneficial mutations would be fixed rapidly. The high variation in rates of substitution, called the overdispersed molecular clock, stimulated many extensions of the simple neutral model in an attempt to explain the degree of variation in substitution rates (see Gillespie 1991; Culter 2000a).

Originally, the neutral theory divided mutations into two discrete categories, those that were neutral and those that were either deleterious or advantageous and, therefore, acted on by natural selection. Kimura suggested that most mutations were in the neutral category. An alternative view was proposed by Ohta in an attempt to explain phenomena inconsistent with strict neutral theory (reviewed in Ohta 1992; Ohta and Gillespie 1996), who suggested a third category of mutations that were weakly deleterious and, therefore, nearly neutral. A mutation is strictly neutral if there is no selection coefficient on it so that its fate is always dictated only by genetic drift. In contrast, the fate of a nearly neutral mutation depends on its effective population size context. With a selection coefficient that is small relative to the effective population size, the fate of a mutation is determined by genetic drift. Alternatively, the same mutation can be acted on by natural selection if the selection coefficient is large compared to the effective population size.

The phenomenon of **codon bias**, where observed frequencies of synonymous codons do not match random predictions, is an example where both neutral processes and natural selection are required to explain observed patterns (reviewed by Plotkin and Kudla 2011). Natural selection helps to explain why substitution rates are lower at nonsynonymous sites than at synonymous sites – a greater proportion of changes at nonsynonymous sites are deleterious and, therefore, fixation is prevented by natural selection. There is also evidence that codon usage varies based on gene functions such as translation speed (e.g. LaBella et al. 2019). The nearly neutral theory explains why selection for codon usage is stronger in species with larges effective population sizes. At the same time, the frequency of fixation at synonymous sites is also determined by mutational biases and by genetic drift.

The selectionist–neutralist debates produced a rich set of expectations for polymorphism and divergence that encompasses the full range of population genetic processes and is able to explain many observed phenomena. Most contemporary population genetic models have the common feature that the strength of natural selection is expressed relative to the strength of genetic drift as measured by the effective population size. Natural selection and genetic drift are now seen as inseparably linked processes that lie on a continuum between the extremes of pure selection (large N_es) and pure drift (small N_es). While the debate over the relative roles of natural selection and genetic drift in producing patterns of polymorphism or rates of divergence continues, genetic drift plays a central role in predictions. The neutral theory is now universally utilized as a null hypothesis in all of population genetics and its varied empirical applications.

8.2 Natural selection

- Positive selection, negative selection, and balancing selection.
- Disequilibrium makes natural selection at two loci less efficient.
- Genetic hitch-hiking and selective sweeps.
- Background selection.
- Hitch-hiking and rates of divergence.

Natural selection is a fundamental process that shapes the variation observed in DNA, RNA, and protein sequences. The models of natural selection employed in molecular evolution follow those described in Chapters 6 and 7 and are named with reference to the various ways that natural selection can act on alleles, especially new mutations. It is common to employ the predictions of the haploid viability model of natural selection, but keep in mind that the diploid viability model of natural selection is used as well (with a reminder that most often only diploid genotypes have fitness, not alleles). When considering DNA sequences, alleles are usually defined as the different base pairs that can be present at a single nucleotide site, termed single nucleotide polymorphisms (SNPs) when two or more nucleotides are present in a population at one nucleotide site.

Two general types of directional natural selection acting on nucleotide sites are recognized given that new mutations have fitness values either above or below the population mean fitness. Haplotypes with relative fitness above mean fitness experience **positive selection**, increasing in frequency and eventually reaching fixation. In contrast, **purifying selection** or **negative selection** occurs when haplotypes have relative fitness below the mean fitness and decrease in frequency and decrease in frequency until they are lost from the population. Both positive and purifying selections are expected to result in relatively little polymorphism at the loci under selection since high fitness alleles will change in frequency

deterministically (not randomly) as they approach equilibria of fixation or loss. **Balancing selection** is also possible, caused by negative frequency-dependent selection (see Chapter 7), higher relative fitness of a heterozygous genotype (see Chapter 6), or by relative fitness values that vary in space or time. The action of balancing selection will maintain two (or more) alleles in a population at intermediate frequencies, greatly extending the time to fixation or loss for haplotypes that contribute to high fitness heterozygous genotypes. Therefore, loci experiencing balancing selection are expected to maintain polymorphism. As with all forms of natural selection, the strength of selection will be relative to the effective population size as explained in the previous section of the chapter.

Purifying or negative selection: The deterministic decrease in frequency and eventual loss of alleles found in genotypes that possess less than average relative fitness.
Positive selection: The deterministic increase in frequency and eventual fixation of alleles found in genotypes that possess greater than average relative fitness.
Balancing selection: Heterozygote advantage (or relative fitness values that are negatively correlated with frequency) that maintains intermediate frequencies for two or more alleles at one locus.

It is common practice in population genetics to treat loci or nucleotide sites as independent entities, but there is the possibility that evolutionary processes both depend on and impact neighboring loci or nucleotide sites. The possibility that natural selection operating on a new beneficial mutation at one locus would depend on beneficial or deleterious mutations at other loci was recognized early (Fisher 1999, first published in 1930, Muller 1932). However, observed patterns of nucleotide polymorphism in a wide array of organisms suggest that linkage plays a large role in patterns of molecular evolution. (Recall from Chapter 2 that a low recombination rate is among numerous causes gametic disequilibrium between loci in any population.) A diverse set of models has been developed to predict what might happen in populations when two loci experience natural selection and the loci are not independent.

These models adopt the perspective of fitness defined by individual loci without epistasis and that disequilibrium between loci is a product of the combination of mutation, linkage, and drift. This is in contrast to models of natural selection when fitness has a multilocus basis with the potential for epistasis, and disequilibrium among those loci is a product of response to natural selection (see Chapter 7).

The impacts of gametic disequilibrium on new mutations are of particular interest in molecular evolution because the fate of new mutations dictates levels of polymorphism and rates of divergence. When new mutations enter a population as a single copy, they initially experience very high levels of gametic disequilibrium. A new mutation that is present in only one copy will be uniquely associated with the other alleles that just by chance occur on the same chromosome where the mutation occurred. The gametic disequilibrium experienced by new mutations has substantial consequences for neighboring sites in the genome if new mutations are acted on by natural selection. First, let's explore changes in polymorphism caused by gametic disequilibrium between neutral nucleotide sites and nucleotide sites where mutations are acted on by natural selection.

A seminal study of the probability of fixation of a beneficial allele in a finite population was conducted by Hill and Robertson (1966, reviewed by Comeron et al. 2008). They considered two loci, focusing on the chance of fixation of a beneficial allele one locus experiencing positive selection while a second locus also bears a beneficial allele, in the context of limited recombination between the loci as well as genetic drift. Imagine the first locus has a beneficial allele A and the second locus a beneficial allele B. Selection can rapidly change allele frequencies when AB gametes are in excess because of gametic disequilibrium caused by drift and mutation. But when there is an excess of aB and Ab gametes due to gametic disequilibrium caused by drift and mutation, selection is less likely to fix the A allele compared to when there is a free recombination between the loci. Over many replicate simulations, they found that the AB combination reached fixation less often when the product of the effective population size and recombination rate (Nc) was small than when Nc was large. The magnitude of the effect also depended on the initial frequency of the beneficial allele, the strength of natural selection, and was most pronounced when the two alleles had similar relative fitness values. This phenomenon is termed the **Hill–Robertson effect**

(Felsenstein 1974) or sometimes Hill–Robertson interference.

Maynard Smith and Haigh (1974) coined the term **genetic hitch-hiking** to describe the consequences of gametic disequilibrium between a locus experiencing strong positive selection and neighboring loci. In that model, as natural selection drives a beneficial mutation to high frequency in a population, the neutral alleles in gametic disequilibrium with the selected mutation also reach high frequency because they happened to be part of the haplotype where the advantageous mutation initially occurred. Hitch-hiking can result in an increase in frequency and fixation of deleterious alleles found at linked sites. Another consequence of hitch-hiking is a loss of polymorphism in the population for neutral alleles, since only one set of neutral alleles in disequilibrium with the advantageous mutation remains in the population once the advantageous mutation approaches fixation by natural selection. The original hitch-hiking model was extended by Kaplan et al. (1989) to repeated beneficial mutations and linked regions where recombination rates vary.

To see the consequences of gametic disequilibrium for new mutations, consider what happens when a favorable mutation arises in a population. Assume for now that the population is composed of haploid individuals that reproduce clonally so that there is no recombination. Figure 8.7 illustrates the changes to allele frequencies over time when a favorable mutation enters a population. Initially, the population contains five different haploid sequences. Each of these haplotypes bears a number of neutral mutations, and each haplotype also has an intermediate frequency in the population that is the product of genetic drift. An advantageous mutation, indicated by a star in the figure, occurs by chance on one of the haplotypes. Over time, the chromosome bearing the favorable mutation will increase in frequency since it has a higher fitness and the other haplotypes will decrease in frequency. Eventually, depending on the relative fitness of the mutation, the haplotype bearing the advantageous mutation will approach fixation in the population. Because there is no recombination in this example, the advantageous mutation is only found on one haplotype. Thus, the advantageous mutation is in complete gametic disequilibrium with two neighboring neutral mutations that happened to be part of that haplotype.

The reduction in polymorphism caused by genetic hitch-hiking is called a **selective sweep** because while an advantageous mutation and the neutral

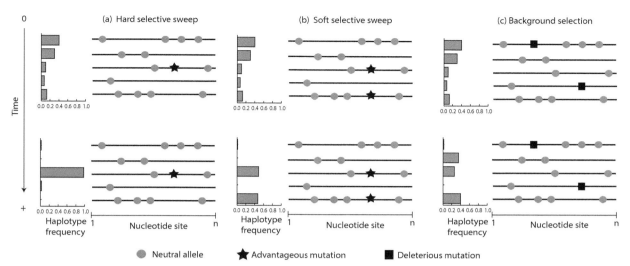

Figure 8.7 The impact of natural selection on new mutations as well as on associated nucleotide sites. Imagine a single population that contains five distinct DNA haplotypes. Each DNA sequence is distinguished by a number of neutral mutations and has a frequency given by the histogram on the left. Initially, the population has polymorphism since there are intermediate frequencies of each DNA sequence. In A, the third DNA sequence experiences a mutation that is strongly advantageous and natural selection acts to increase its frequency rapidly. The hard sweep caused by the advantageous mutation results in very little polymorphism at linked sites because only those original neutral mutations in disequilibrium with the advantageous haplotype remain in the population. In B, the population initially has polymorphism as well as a particular allele found on several haplotypes that was initially neutral but becomes advantageous. The soft sweep brings the haplotypes bearing the advantageous mutation to high frequencies, along with neutral polymorphisms at linked sites. In C, deleterious mutations occur on several haplotypes that leads negative selection and those haplotypes being lost from the population. Such background selection reduces polymorphism in a population to the alleles found on haplotypes that do not experience deleterious mutations.

polymorphisms that are linked to it are swept to high frequency by natural selection, other neutral polymorphisms not in gametic disequilibrium with the selected site are swept out of the population at the same time. It is important to emphasize that the reduction in polymorphism seen in selective sweeps is an indirect consequence of natural selection since only the advantageous mutation itself has a selection coefficient that is not effectively zero. A frequent (although imperfect) analogy for the impacts of selective sweeps on linked neutral sites is that association with a selected site accelerates time to fixation and loss in a manner akin to a reduced effective population size, leading to reduced polymorphism. Another way to think of the impact of selective sweeps is that a site acted on by positive selection and its linked sites are associated because they share the same coalescent genealogy (see the final section of the chapter for the coalescent with recombination).

Selective sweeps are also distinguished based on initial conditions when positive selection beings to act on a haplotype. **Hard sweeps** are the result of

strong positive selection acting on a new mutation (Figure 8.7A). Because new mutations are initially in strong disequilibrium with other loci, linked site polymorphism is expected to be minimal. Because strong selection will result in fixation after a short time, there is little time for new neutral mutations to occur at nearby sites. Because time is short, there should be few recombination events which maintain extensive linkage disequilibrium. With a hard sweep, linked sites are expected to exhibit low levels of polymorphism as if they have a greatly reduced effective population size compared to strictly neutral loci. Hard sweeps are also expected to reduce polymorphism in a wider region around the selected site because fixation occurs rapidly and leaves little time for recombination events (Figure 8.8).

The term **soft sweep** characterizes positive selection acting on a haplotype that is already present in a population and is at an intermediate frequency as part of the standing genetic variation in the population (Figure 8.7B). Imagine that all of the alleles at one locus are initially evolving neutrally and are in moderate gametic disequilibrium with adjacent loci. Then, a change in the environment or a sharp increase in the effective population size alters the average fitness of one variant making it advantageous. With such a change in fitness, positive selection will increase the frequency of the advantageous allele and linked sites. But since the advantageous site was found initially on a background of more polymorphic sites, that more variable pool of haplotypes will increase in frequency along with the advantageous allele. Soft sweeps are, therefore, expected to reduce polymorphism in a narrower region around the selected site since recombination events may have already occurred in the haplotypes bearing the beneficial allele (Figure 8.8). With a soft sweep, linked sites are also expected to exhibit a more modest reduction in polymorphism as if they have a moderately reduced effective population size compared to strictly neutral loci. A related model is the multiple

origin soft sweep where the same advantageous mutation (or one that is very similar in location and fitness) occurs repeatedly (Hermisson and Pennings 2005, 2017).

An alternative prediction for genetic polymorphism at linked sites is motivated by the fact that many more mutations are deleterious than are beneficial (see Chapter 5). This suggests that negative selection on deleterious mutations should be common and could influence polymorphism at other loci. In fact, negative selection against deleterious mutations is expected to reduce polymorphism through a process called **background selection** (Charlesworth et al. 1993, 1995, reviewed by Charlesworth 2012). Deleterious mutations occur on some haplotypes in a population, causing those haplotypes to have a higher rate of loss (Figure 8.7C). Negative selection causes the deleterious allele to

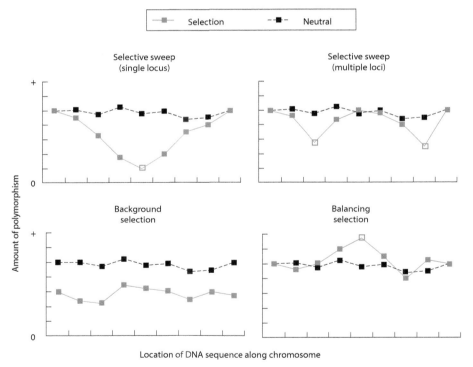

Figure 8.8 Consequences of natural selection on polymorphism at nucleotide sites experiencing selection (open boxes) as well as on associated nucleotide sites (filled boxes) with recombination. Polymorphism at strictly neutral reference loci is provided for comparison and would be a function of the effective population size and the mutation rate. Strong positive selection on one site leads to a selective sweep that both removes polymorphism at even relatively distant linked sites because selection is strong relative to recombination. Positive selection on multiple loci (or on standing genetic variation) leads to a selective sweep with a smaller, more localized reduction of linked site variation as selection is less strong compared to recombination. Background selection is expected to cause a relatively even reduction in polymorphism across all loci because selection drives haplotypes bearing deleterious mutations to loss, leaving other haplotypes that possess some neutral polymorphism. Balancing selection leads to polymorphism at the site under selection as well as to increased polymorphism at linked sites because they will have longer segregation times to accumulate neutral polymorphisms.

go to loss, bringing its associated neutral mutations to loss as well. The haplotypes that do not experience a deleterious mutation remain in the population and contribute to polymorphism. Under the background selection model, recombination takes place among those haplotypes without deleterious mutations because they remain segregating for a longer time in the population, leading to relatively uniform polymorphism across loci (Figure 8.8). This is in contrast to the decreased polymorphism expected near those loci having experienced selective sweeps. Like hitch-hiking, the impact of background selection can be thought of as a lower N_e at neutral sites because linkage to deleterious mutations accelerates time to loss.

Background selection: A reduction in polymorphism caused by the combination of purifying selection driving strongly deleterious mutations to loss and gametic disequilibrium that leads to the loss of neutral alleles associated with deleterious mutations.

Genetic hitch-hiking: The process by which selectively neutral alleles increase or decrease in frequency due to their association with alleles that are under the influence of natural selection.

Selective sweep: The reduction or elimination of polymorphism in a region of DNA sequence surrounding a site where a beneficial mutation has increased in frequency due to positive natural selection. The reduction of polymorphism is a result of gametic disequilibrium between a beneficial mutation and neighboring neutral sites that has not been broken down by recombination.

Gillespie (2000) studied the **pseudo-hitch-hiking model** to predict the impacts on polymorphism of recurrent positive selection producing a series of selective sweeps in large populations. In that model, linked site polymorphism depends strongly on the rate of recombination and the rate and relative fitness values of new beneficial mutations. Gillespie (2001) termed the consequences of recurrent selective sweeps on liked nucleotide site polymorphism **genetic draft** because neutral mutations in

disequilibrium are like an unpowered glider, floating their way to fixation by riding along on the up-currents of increased probability of fixation caused by positive selection. Gillespie has shown that genetic draft is a stochastic process because the neutral mutations that do reach fixation do so by random association with selected mutations. Thus, positive selection on beneficial mutations causes finite random sampling from the pool of available neutral mutations even if the effective population size is infinite.

Another possibility is that new mutations are acted on by **balancing selection**, which would eventually bring new beneficial mutations at the same site to intermediate frequencies and maintain them in the population for very long periods of time. Balancing selection is also expected to impact polymorphism at neutral sites that are in gametic disequilibrium with the selected site. When a new beneficial mutation appears at a site under balancing selection, the initial increase in its frequency has an effect like a selective sweep. However, long-term balancing selection leads to an increase in polymorphism because balancing selection maintains multiple alleles that persist in the population for long periods of time (Figure 8.8). These selected alleles can then accumulate mutations at neutral sites that are in gametic disequilibrium, gathering polymorphism over time. Compared to independent neutral sites, neutral sites in gametic disequilibrium with sites under balancing selection have greatly increased segregation times and so have a greater opportunity to experience mutation that leads to the accumulation of polymorphism (reviewed by Charlesworth 2006).

The impact of these modes of natural selection acting individually, or in combination, depends on a number of population genetic parameters that are likely to vary across genome regions as well as among species. The strength of the association between sites under selection and neutral sites will vary with the recombination rate. Mating among relatives and self-fertilization is expected to reduce the realized recombination rate because crossing-over between pairs of homozygous loci has no effect (Glémin et al. 2006, Andersen et al. 2012). The frequency of selection will depend on the absolute rate of mutation rate while the mode of selection will depend on the relative rates of beneficial and deleterious mutation. The strength of selection relative to other processes will vary with the distribution of selection coefficients. Response to selection is also a function of the effect size of mutations and the degree of dominance or recessive gene action, often called the genetic architecture (see Chapter 9). Population differentiation has the potential to impact the action of these types of natural selection (Bierne 2010). For example, when selective sweeps contribute to local adaptation, then their signatures might be limited to only one a number of subpopulations and the impacts of selection could be further modulated by the history of gene flow among subpopulations. Effective population size and variations in it over time due to demographic changes will also influence the action of selection since the expected number of mutations per generation is a function of the effective mutation rate ($\theta = 4N_e\mu$), and the effective population size will determine the relative strength of selection and genetic drift.

Hitch-hiking and rates of divergence

Genetic divergence between and among different species is a product of the substitutions that have occurred within each species. When the probability of fixation of a new mutation is dictated by natural selection, then divergence rates change compared to rates expected by genetic drift. Positive natural selection speeds up divergence because advantageous mutations fix faster on average than they would under genetic drift alone. Alternatively, negative natural selection slows divergence rates since deleterious mutations go to loss rapidly, and fewer neutral mutations remain that might fix and generate substitutions. The question that remains, then, is whether or not the rate of divergence at neutral sites will be sped up or slowed down if they are in linkage disequilibrium with nucleotide sites that are acted on by natural selection.

The expected rate of substitution within a species is determined by the scaled mutation rate ($2N_e\mu = \frac{\theta}{2}$) and the probability of fixation for mutations (P_F), which can be stated in an equation as

$$k = \frac{\theta}{2}P_F \qquad (8.6)$$

As shown earlier in the chapter, for independent neutral mutations, $P_F = \frac{1}{2N}$, as this is also the initial frequency of each new mutation. Here, we use a new symbol to express the fixation probability because the probability of fixation for neutral mutations (P_F) may well be different when a mutation is in gametic disequilibrium with a selected site than when a mutation is independent.

Assume that we have a neutral locus with two alleles A and a. The frequency of the A allele is x, and the frequency of the a allele is, therefore, $1 - x$. Further assume that this neutral locus is completely linked to another locus where all alleles are under the influence of very strong positive natural selection. Let's imagine that a new mutation occurs at the selected locus that is infinitely advantageous and so goes to fixation instantly. What is the probability that the A allele at the neutral locus is also fixed due to hitch-hiking? The A allele at the neutral locus has a frequency of x, and therefore, there is also the probability x that the new advantageous mutation at the selected locus is linked to the A allele. Thus, there is the probability x that the A allele will sweep to fixation with the new mutation at the selected locus. However, the probability that the A allele is fixed by genetic drift is also x since that is the initial frequency of the A allele. Therefore, even complete linkage to the selected allele does not alter the fixation probability for the A allele.

Birky and Walsh (1988) presented this idea and a more general analytical case along with the results of simulations, all showing that neither positive nor negative natural selection will change substitution rates at neutral sites. This occurs because the increased probability of substitution of the copies of a neutral allele linked to a selected site is counterbalanced by a decrease of exactly the same amount in the probability of fixation of all the neutral allele copies that are not linked to a selected site.

Empirical studies

With background on the numerous natural selection models and their indirect impacts on linked site polymorphism, along with an explanation of why divergence between species is not impacted by linkage to selected sites, we can fully appreciate the results of some empirical studies that set out to test for different possible forms of natural selection acting along with limited recombination.

A common result in studies of genetic variation at numerous loci in *Drosophila* populations is that levels of polymorphism are positively correlated with rates of recombination (reviewed by Hudson 1994; Cutter and Payseur 2013). There are several hypotheses for the cause of a relationship between polymorphism and rates of recombination. One hypothesis consistent with strict neutrality is that the recombination rate at a locus is somehow related to its mutation rate such that the scaled mutation rate is greater in regions with greater recombination (Hellmann et al. 2003, 2005). For example, the molecular processes that cause recombination might also cause point mutations. Recall that the neutral theory predicts that levels of polymorphism and rates of divergence are correlated since both are ultimately products of the mutation rate. This leads to the prediction that both levels of polymorphism and divergence rates should correlate with the recombination rate if neutral processes explain the relationship between recombination and polymorphism at the *Drosophila* loci (Begun and Aquadro 1992). Data

to test this neutral hypothesis for the correlation between polymorphism and recombination rate in *Drosophila* are shown in Figure 8.9. While polymorphism in *D. melanogaster* clearly increases with the recombination rate, levels of divergence between *D. melanogaster* and *Drosophila simulans* for a subset of the same loci are independent of the recombination rate. These data, therefore, reject the strict neutral hypothesis. An alternative explanation is the operation of natural selection on beneficial mutations that has resulted in selective sweeps. The strength of genetic hitch-hiking and the amount of

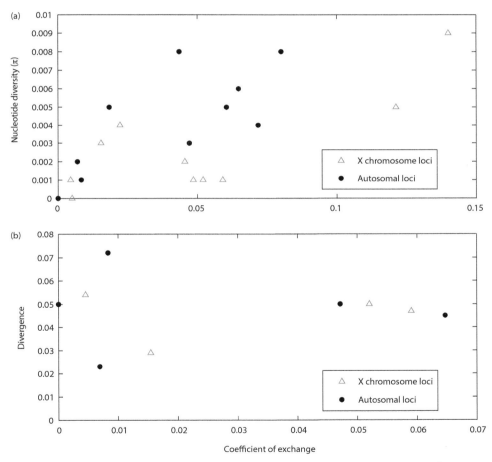

Figure 8.9 Plots of nucleotide diversity within *Drosophila melanogaster* populations (A) and divergence between *D. melanogaster* and *Drosophila simulans* (B) by the coefficient of exchange, a measure of the recombination rate, for numerous loci. The nucleotide diversity at a locus decreases along with the recombination rate declines (A). The correlation of polymorphism and the recombination rate at a locus could be explained by the neutral theory if loci with lower recombination rates also happen to have lower mutation rates. Under this neutral hypothesis, divergence rates would also be correlated with recombination rates since both polymorphism and divergence increase as the mutation rate increases. An alternative explanation for the correlation of recombination and polymorphism is the action of natural selection on beneficial mutations that has caused hitch-hiking and a reduction of polymorphism due to selective sweeps. Divergence rates at a subset of the loci (B) suggest that divergence and recombination rates are independent, rejecting the neutral hypothesis for these data. Source: Data from Begun and Aquadro (1992).

polymorphism lost to selective sweeps decrease with increasing recombination because recombination reduces gametic disequilibrium between a selected mutation and neighboring sites. However, selective sweeps have no impact on the rate of divergence so that divergence will show no relationship to the recombination rate. Therefore, the selective sweep hypothesis is consistent with the pattern that polymorphism increases with the recombination rate in the *Drosophila* DNA sequence data.

An ongoing challenge is to distinguish selective sweeps from background selection (Innan and Stephan 2003; Stephan 2010). Genomic-scale studies of polymorphism are providing more opportunities to test hypotheses for the impacts of both types of natural selection on polymorphism. One widely cited study by Lohmueller et al. (2011) observed that polymorphism was reduced in regions of lower recombination in humans and that low frequency alleles were less common in regions near genes. To establish predictions tailored for the parameters of human populations, they carried out a number of simulations designed to generate patterns of polymorphism expected under background selection alone, or background selection acting in combination with positive selection leading to selective sweeps. Their simulation results suggested that the patterns observed in their human polymorphism data could be explained by background selection alone or background selection in conjunction with a small number of selective sweeps. However, simulated selective sweeps alone were not consistent with the correlation between polymorphism and recombination observed in the data.

Mitochondrial DNA has been used very widely as a genetic marker in studies of polymorphism in animals (Avise et al. 1987; Harrison 1989). It has been widely assumed that mitochondrial polymorphism is the product of neutral evolution with populations being near drift-mutation equilibrium and that mitochondrial diversity is positively correlated with effective population size. Yet, the absence of recombination in mitochondrial genomes provides the necessary conditions for selective sweeps and for background selection. This has led many to hypothesize that the mitochondrial genome polymorphism should be lowered by natural selection and its impact on linked sites (reviewed by Dowling et al. 2008; Galtier et al. 2009).

A test for nonneutral patterns of polymorphism in animal mitochondrial genomes was carried out using comparisons of polymorphism measured for nuclear allozyme loci, nuclear DNA sequences, and mitochondrial DNA sequences for a large and taxonomically broad sample of animal species (912, 417, and 1683 species, respectively, for the three data types). Bazin et al. (2006) used this comparative approach to test the neutral hypothesis that polymorphism within species for all three types of data should be correlated since each class of loci shares a similar effective population size. (Because mitochondrial genomes are haploid and uniparentally inherited, their effective population size is four times less than that of biparentally inherited diploid nuclear loci.) Based on census population sizes, insects, echinoderms, and mollusks are expected to have larger effective population sizes than mammals, fish, reptiles, and birds. The neutral theory predicts that the taxa with larger effective population sizes should also have higher levels of polymorphism for the same loci. The neutral prediction was met for nuclear allozyme and DNA sequence data because polymorphism was higher for insects, echinoderms, and mollusks than for mammals, fish, reptiles, and birds. In contrast, mitochondrial polymorphism was both low and nearly uniform across all the animal groups and did not show a correlation with levels of nuclear allozyme and DNA sequence polymorphism. This result can be explained by genetic hitch-hiking that has caused selective sweeps in the nonrecombining mitochondrial genome.

The Bazin et al. (2006) study had several limitations. No adjustment was made for the phylogenetic relatedness of the species that were compared, which can lead to an inflation of correlations since some species are similar due to recent ancestry (Felsenstein 1985). Also, highly diverged species were compared that may have differed in their mitochondrial genome mutation rates which were unmeasured, in addition to other measured variables like census population size. Motivated by these limitations, Nabholtz et al. (2008) carried out a follow-up study that carried out the same tests, but in a smaller, more recently diverged taxonomic group of 277 mammal species, and with adjustments for population differentiation, substitution rates as a proxy for lineage mutation rates, and phylogenetic relationships. The refined study found a correlation between allozyme heterozygosity and mitochondrial nucleotide diversity consistent with neutral predictions. There was also strong evidence for taxonomic variation in mitochondrial nucleotide diversity, suggesting that mitochondrial mutation rates vary widely among species. The authors concluded that mitochondrial polymorphism did not show evidence of selective sweeps or strong background selection, but that mitochondrial polymorphism still may only weakly reflect recent effective population size as commonly assumed.

Predictions for natural selection and its impact on associated sites, which are not mutually exclusive, together serve as alternatives to a strict neutral null model. These forms of natural selection help explain why levels of DNA sequence polymorphism in some regions of the genomes and in some species might depend only weakly on the effective population size, a conjecture sometimes called Lewontin's paradox (Lewontin 1974, 1985a). The question of what relative balance of evolutionary processes best explains patterns of genetic polymorphism persists today, updated with continuing refinement of model predictions and greatly expanded empirical data (e.g. Leffler et al. 2012; Hague and Routman 2016).

8.3 Measures of divergence and polymorphism

- Measuring divergence of DNA sequences.
- Nucleotide substitution models correct divergence estimates for saturation.

- DNA polymorphism measured by the number of segregating sites and nucleotide diversity.

Most natural and laboratory populations contain at least some, and often a large amount, of genetic variation represented by different alleles found at the many loci in the genome. The smallest possible unit of the genome is a homologous **nucleotide site**, or single base-pair position in the exact same genome location, that could be compared among individuals. Genetic variation within species at such nucleotide sites is characterized by the existence of DNA sequences that have different nucleotides (e.g. some individuals have an A and other individuals have a T at the 37th base pair from the start codon in the same gene) and is called **nucleotide polymorphism**. Nucleotide polymorphisms are sometimes referred to as **single nucleotide polymorphisms** or **SNPs** (pronounced "snips"). Genetic variation between and among different species is called **divergence**. It is a product of the substitutions that have occurred within each species that together contribute to DNA sequence differences. This section of the chapter covers the commonly used measures of divergence and polymorphism estimated from DNA sequence data.

Box 8.1 DNA sequencing

Steady advances in DNA sequencing techniques (Heather and Chain 2016) have ushered in a flood of DNA sequence data used to observe SNPs and sequence divergence. Continued progress in DNA sequencing promises even faster, less expensive data collection. The first generation DNA sequencing technique, or Sanger sequencing, utilizes DNA sequencing-by-synthesis using DNA polymerase to copy a sequence template and occasionally incorporate chain-terminating dideoxynucleotides (dideoxynucleotides lack a 3′ hydroxyl –OH group necessary for continued 5′-to-3′ DNA synthesis) into the many copies. Electrophoresis is used to separate the resulting populations of different length sequences at one base pair resolution, and a radioactive or florescent label is used to detect the state of the dideoxynucleotide that terminated each sequence (Figure 8.10). Sanger sequencing generates single sequences of up to about 750 nucleotides.

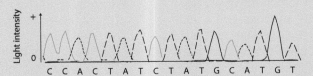

Figure 8.10 An electrophorogram resulting from electrophoresis of single base-pair fragments produced by dideoxynucleotide-terminated Sanger sequencing where each nucleotide has a different molecular label.

(continued)

Box 8.1 (continued)

The primary second-generation sequencing technique, called pyrosequencing, also relies on a DNA polymerase to copy a sequence template. When a complementary deoxynucleotide is added by the DNA polymerase to the copied DNA strand, a pyrophosphate is released. An enzyme (ATP sulfurylase) converts the released pyrophosphate to ATP, causing a chemoluminescent enzyme to release light. A detector measures the light emitted when A, T, C, or G nucleotide solutions are washed over the DNA template and the next nucleotide is added to the copied strand. Pyrosequencing can be carried out with many independent template strands in parallel using microfluidic devices, and there are now a wide range of methods that utilize the general approach. Pyrosequencing generates single sequences of up to about 300 nucleotides, substantially shorter than those produced by Sanger sequencing. Manipulation and assembly of these short-read sequences present substantial computational challenges.

A third generation (sometimes called the fourth generation) of sequencing techniques promise much longer single sequence reads with potential advantages of fewer template preparation steps and improved assembly of sequences. One of these techniques is called single molecule real-time (SMRT) sequencing where a DNA polymerase is attached to the bottom of a well called a zero-mode waveguide (ZMW). The ZMWs are narrower than the wavelength of the light used to excite florescence from a labeled nucleotide, resulting in illumination only at the bottom of the well where the DNA polymerase is building a copy of the template. The ZMW well greatly reduces background florescence and permits all labeled nucleotides to be in solution at all times, removing the need for time-consuming nucleotide solution changes. A second approach is nanopore sequencing (Feng et al. 2015) where either individual nucleotides or a single DNA strand are electrophoresed through a very small orifice. The pore is either a biological protein such as alpha hemolysin or a solid-state device. The nucleotide states can be detected by measuring changes in electrical current as the different nucleotides pass through the pore, or with fluorescent probes.

DNA divergence between species

The most fundamental method to quantify molecular evolution is by comparing two DNA sequences. This comparison is a two-step procedure. First, the two DNA sequences must be **aligned** such that homologous nucleotide sites for each sequence are all lined up in the same columns. For example, if two coding genes were sequenced, then one way to align them would be to match up the first three nucleotides that make up the start codon. (Methods of sequence alignment are beyond the scope of this text, but readers can refer to text such as Page and Holmes (1998) for more details.) The second step is to determine the number of sites that have different nucleotides. The number of nucleotide sites that differ between two DNA sequences divided by the total number of nucleotide sites compared gives the proportion of nucleotide sites that differ, often called a ***p*-distance** as a shorthand for proportion distance. This is a basic measure of the evolutionary events that have occurred since two DNA sequences descended from a common ancestor, when they were each identical copies of the same sequence.

> ***p*-distance:** The number of nucleotide sites that differ between two DNA sequences divided by the total number of nucleotide sites, a shorthand for proportion – distance, sometimes symbolized as *d* for distance.

An example of divergence between a pair of sequences is shown in Figure 8.4. Consider the two sequences at the far right in the present time after some divergence that has introduced substitutions. There are five nucleotide sites that have different nucleotides out of a total of 12 nucleotide sites. Therefore, the *p*-distance is $5/12 = 0.3125$ or 31.25% of the nucleotide sites have diverged.

The *p*-distance between two DNA sequences sampled from completely independent populations should increase over time as substitutions within each population replace the nucleotide that was originally shared at each site due to identity by descent. If the two DNA sequences represent two distinct species or completely isolated populations, then the *p-distance* is a measure of divergence between the two species.

DNA sequence divergence and saturation

Saturation is the phenomenon where DNA sequence divergence appears to slow and eventually reaches a plateau even as time since divergence continues to increase. Saturation in nucleotide changes over time is caused by substitution occurring multiple times at the same nucleotide site, a phenomenon called **multiple hit** substitution (see the related topic of multiple hit mutation in Chapter 5). Substitutions that occur repeatedly at the same site have the effect of covering up information about past substitutions, since only the most recent substitution can be observed and measured as divergence between two DNA sequences. Computing the *p-distance* between two sequences leads to under-estimates of number of substitutions that have occurred and, therefore, an under-estimate of the degree of divergence. The top panel of Figure 8.11 shows divergence that increases linearly with time since divergence (dashed line) and that exhibits saturation (solid line). Actual DNA sequence data routinely exhibit some degree of saturation, as shown in the bottom panel of Figure 8.11 for the mitochondrial cytochrome *c* oxidase subunit II gene sequenced for several bovine species (ungulates including domestic cattle, bison, water buffalo, and yak) that diverged between 2 and 20 million years ago.

Saturation can be understood by imagining the process of assembling a DNA sequence at random and comparing it with another existing DNA sequence. Think of drawing individual nucleotides out of a bucket containing equal numbers of A, C, G, and T base pairs. If a nucleotide site in the existing sequence is an A, there is a 25% chance that a randomly drawn nucleotide will be an A and the sites will match. On the other hand, there is a 75% chance it will not be an A (it will be a T, C, or G) and the two sites will be diverged. Thus, a DNA sequence assembled from random draws of nucleotides at equal frequency should be 75% divergent or 25% identical to another sequence on average. The consequence is that two sequences which

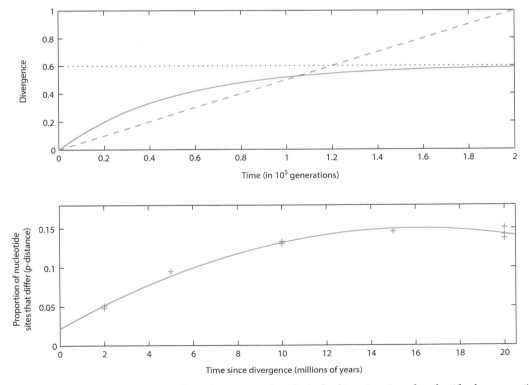

Figure 8.11 Substitutions that occur repeatedly at the same nucleotide site lead to saturation of nucleotide changes as time since divergence from a common ancestor increases. The rate of substitutions does not change and the total number of substitutions continues to increase over time, as shown by the dashed line in the top panel representing the true number of substitutions. In contrast, multiple substitutions at the same sites leads to a slowing and leveling off in the estimate of divergence (solid line, top panel). Therefore, the amount of divergence leads to the perception that the rate of divergence decreases over time. The bottom panel shows divergence and saturation at the mitochondrial cytochrome c oxidase subunit II gene among bovine species (ungulates including domestic cattle, bison, water buffalo, and the yak) that diverged between 2 and 20 million years ago. In the top panel, $\alpha = 1 \times 10^{-6}$. The bottom panel data are from Janecek et al. (1996) and the line is a quadratic regression fit.

originated from a common ancestor do not continue to get increasingly divergent over time. Eventually, the maximum divergence will plateau at 75% as continued mutation essentially randomizes the shared sites between the two sequences.

There are a wide variety of methods to correct the observed divergence between two DNA sequences to obtain a better estimate of the true divergence after accounting for multiple hits. These correction methods are called **nucleotide substitution models** and use parameters for DNA base frequencies and substitution rates to obtain a modified estimate of the divergence between two DNA sequences. The simplest of these is the Jukes and Cantor (1969) nucleotide-substitution model, named after its authors. Working through the derivation of the Jukes–Cantor model is worthwhile to gain some insight into how nucleotide-substitution models operate.

The Jukes–Cantor model starts out by assuming that any nucleotide in a DNA sequence is equally likely to be substituted with any of the other three nucleotides. For example, if a site currently has a C, then substitution of an A, T, or G all have the same chance of occurring. Figure 8.12 shows three possible events for a nucleotide site. The site may (i) experience one and only one substitution, (ii) not experience any substitutions over time, and (iii) experience a substitution that changes the nucleotide at the site and then another independent substitution that restores the original nucleotide at the site. In the first situation, perceived

divergence and actual divergence are the same and no correction is required. The perceived divergence in the second and third cases is the same, but very different events have occurred. In the third case, substitutions have occurred for some portion of nucleotide sites that appear to have no divergence. Nucleotide substitution models such as Jukes–Cantor serve to estimate the frequency of nucleotide sites that appear to have not diverged but actually have diverged.

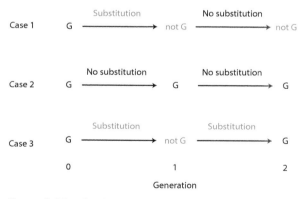

Figure 8.12 The three types of events that a single nucleotide site may experience over two generations. A nucleotide site may initially have a G, for example. In case 1, a single substitution event in generation 1 changes the G to an A, C, or T nucleotide, the nucleotide also present at generation 2. A *p*-distance measure of divergence will accurately count the number of substitutions in case 1. In cases 2 and 3, the nucleotide site still retains the same nucleotide it had initially, giving the impression that there have been no substitutions. In case 2, this impression is accurate. However, in case 3, there have been two substitution events that are not accounted for in a simple *p*-distance measure of divergence.

In the Jukes–Cantor model, the probability of a nucleotide substitution is customarily represented by α (pronounced "alpha"). Since there are three nucleotides that can each be substituted for a nucleotide currently at a site and all three are equally likely to be substituted, the probability of any substitution is 3α. So if the nucleotide is initially a G at generation zero, the probability that it is also a G one generation later is

$$P_{G(t=1)} = 1 - 3\alpha \qquad (8.7)$$

Since the chance of substitution is independent in each generation, the probability of no substitutions over two generations is

$$P_{G(t=2)} = (1 - 3\alpha)^2 \qquad (8.8)$$

This gives the probability that a nucleotide does not change over two generations as shown in case 2 of Figure 8.12.

We also need to determine the probability that a nucleotide changes twice, as shown in case 3 of Figure 8.12. From generation zero to generation one, the probability of a substitution is 3α. This probability can also be written as 1 minus the probability of no substitution, or $1 - P_{G(t=1)}$. From generation one to generation two, there is only one base that can be substituted to make the site match its initial state. The chance that this occurs is the probability of a substitution or α. Bringing these two independent probabilities together gives the probability that a multiple hit nucleotide substitution occurs, which restores the nucleotide initially present

$$P_{G(t=2)} = \alpha\left(1 - P_{G(t=1)}\right) \qquad (8.9)$$

This is the probability that two substitutions occur, neither of which would not be detectable by comparing two DNA sequences at time two.

Combining these two results gives the probability that a nucleotide site has the same base pair after two generations:

$$P_{G(t=2)} = (1 - 3\alpha)P_{G(t=1)} + \alpha\left(1 - P_{G(t=1)}\right) \qquad (8.10)$$

regardless of the number of substitutions that have occurred over two generations. Since the probabilities of substitution and no substitution are independent each generation, this equation can be written in a more general form as

$$P_{G(t+1)} = (1 - 3\alpha)P_{G(t)} + \alpha\left(1 - P_{G(t)}\right) \qquad (8.11)$$

or as a recurrence equation that applies to any two time periods one generation apart. This recurrence equation can also be expressed as the change in the probability that an initial nucleotide site remains unchanged over time. Since the change in a quantity is the difference between what it is now and what it was one time step t in the past, the change in the probability that a given nucleotide is found at a site over one generation is then

$$\Delta P_{G(t)} = (1 - 3\alpha)P_{G(t)} + \alpha\left(1 - P_{G(t)}\right) - P_{G(t)} \qquad (8.12)$$

Expanding the terms on the right side of this equation gives

$$\Delta P_{G(t)} = P_{G(t)} - 3\alpha P_{G(t)} + \alpha - \alpha P_{G(t)} - P_{G(t)} \qquad (8.13)$$

which then simplifies to

$$\Delta P_{G(t)} = \alpha - 4\alpha P_{G(t)} \qquad (8.14)$$

The model we have considered to this point treats time as discrete steps, as shown in Figure 8.12. If we consider the rate of change at any time t, then the change in the probability that a nucleotide site appears the same with changes in time is a differential equation, $\frac{dP_{G(t)}}{dt} = \alpha - 4\alpha P_{G(t)}$. The solution to this equation is

$$P_{G(t)} = \frac{1}{4} + \left(P_{G(t=0)} - \frac{1}{4}\right)e^{-4\alpha t} \qquad (8.15)$$

which is analogous to an exponential growth equation for a population with a carrying capacity. As t gets large, the $e^{-4\alpha t}$ term approaches 0 so that the probability and the $P_{G(t)}$ approaches ¼.

Using the continuous time equation, the probability that a nucleotide site is a G over time depending on its initial state and the rate of substitution is shown in Figure 8.13. If the nucleotide at a site is initially a G, then $P_{G(t)} = 1$ and the probability the site remains a G over time is

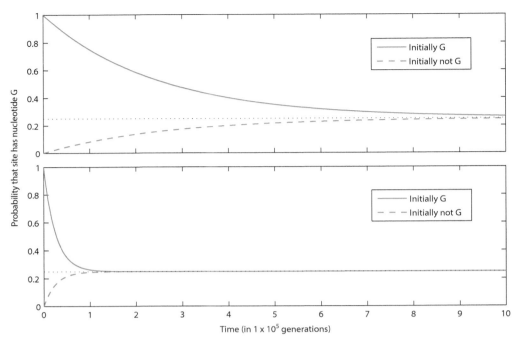

Figure 8.13 The probability that a nucleotide site retains its original base pair under the Jukes-Cantor model of nucleotide substitution. If a nucleotide site originally has a G base, for example, the probability of the same base being present decline steadily over time. If a nucleotide site was initially not a G (it was an A, C, or T), the probability that a G is present at the site increase over time. The probability that a given base is present always converges to 25% because that is the probability of sampling a given base at random if the probability of substitution to each nucleotide is equal. In the top panel, $\alpha = 1 \times 10^{-6}$ while in the bottom panel $\alpha = 1 \times 10^{-5}$.

$$P_{G(t)} = \frac{1}{4} + \frac{3}{4}e^{-4\alpha t} \qquad (8.16)$$

This is a probability that declines exponentially toward ¼ as t increases. In contrast, if the nucleotide at a site is not initially a G, then $P_{G(t)} = 0$. The probability the site remains a G over time is

$$P_{G(t)} = \frac{1}{4} - \frac{1}{4}e^{-4\alpha t} \qquad (8.17)$$

which increases from 0 to ¼ as time increases. Recall that ¼ is the probability that a site in a sequence assembled from randomly drawn nucleotides (that are equally frequent) matches the same site in an existing sequence. Also, notice that the approach to ¼ will be faster as the substitution rate α increases due to the $-4\alpha t$ term in the exponent.

Let's not forget that the original goal was to correct observed divergence between sequences or *p-distances* for multiple hit mutations. The model of sequence change we have so far is the foundation

of a correction, but we need to do some more work to obtain an actual correction method. If we think of two DNA sequences originally identical by descent at every nucleotide site at time 0, at some later time t, the probability that any site will possess the same nucleotide is

$$P_{I(t)} = \frac{1}{4} + \frac{3}{4}e^{-8\alpha t} \qquad (8.18)$$

where $P_{I(t)}P_{I(t)}$ indicates the probability that the nucleotides at a given site are identical. The exponential term is now $e^{-8\alpha t}$ because there are two DNA sequences that can change independently so that the chance of the nucleotide remaining the same decreases twice as fast with time. The probability that two sites are different or divergent – call it d – over time is 1 minus the probability that the sites are identical so that

$$d = \frac{3}{4}\left(1 - e^{-8\alpha t}\right) \qquad (8.19)$$

The exponent can be removed from this equation by taking the natural logarithm of the right side and rearranged to give

$$8\alpha t = -\ln\left(1 - \frac{4d}{3}\right) \qquad (8.20)$$

This equation states that eight times the substitution rate multiplied by time is related to the amount of divergence we expect to see between two DNA sequences. For two DNA sequences that were originally identical by descent, we expect that each site has a $3\alpha t$ chance of substitution. Since there are two sequences, there is a $6\alpha t$ chance of a site being divergent between the two sequences. If we set expected divergence $K = 6\alpha t$, then we notice K is close to the $8\alpha t$ above. In fact, K is ¾ of the expression for $8\alpha t$, so that

$$K = -\frac{3}{4}\ln\left(1 - \frac{4d}{3}\right) \qquad (8.21)$$

where d is the observed proportion of sites that differ between two DNA sequences, or the *p-distance*. K is then the estimate of the actual number of sites that have experienced divergence events corrected for multiple hits with the Jukes–Cantor nucleotide-substitution model.

A few examples will help show how the Jukes–Cantor model correction works in practice. Imagine two DNA sequences that differ at 1 site in 10 so that the *p-distance* is 10% or $d = 0.10$. This level of observed divergence is an under-estimate because it does not account for multiple hits. To adjust for multiple hits, we compute corrected divergence as

$$K = -\frac{3}{4}\ln\left(1 - \frac{4}{3}(0.10)\right) = 0.1073 \qquad (8.22)$$

which shows that at the low apparent divergence of 10%, there are expected to be 0.7% of sites that had experienced multiple hits. The true divergence is then estimated as 10.73% or slightly greater than the apparent divergence. If apparent divergence was greater, say $d = 0.40$, then we should expect a larger correction. In that case,

$$K = -\frac{3}{4}\ln\left(1 - \frac{4}{3}(0.40)\right) = 0.5813 \qquad (8.23)$$

so the correction for multiple hits is much larger at a bit over 18% for a total corrected divergence of 58.13% of nucleotide sites. A frequently used convention is that a capital K refers to a saturation-corrected estimate of divergence while lower case k or d is an uncorrected estimate of divergence.

The Jukes–Cantor is the simplest possible nucleotide substitution model because it assumes that all nucleotides are equally frequent in DNA sequences and that all sites experience the same substitution rate. Many DNA sequences, however, exhibit variation in these parameters, which is not accounted for in the Jukes–Cantor model. There are numerous models of nucleotide substitution of increasing complexity that take these factors into account by using an increasing number of parameters to represent the different types of substitution rates (Posada and Crandall 2001). Figure 8.14 illustrates a hierarchy of some of the nucleotide-substitution models available. These different models can be distinguished by examining DNA sequence data to test each model assumption. For example, the Jukes–Cantor model assumes that all nucleotides have equal frequencies. If a sample of DNA sequences shows base frequencies that deviate significantly from equal frequencies of 25%, then the F81 model is a better choice because it assumes arbitrary base frequencies. Both the JC and F81 nucleotide-substitution models assume that transition and transversion rates are equal and that substitution rates are constant among sites. It is now common practice to estimate the substitution model that best approximates the patterns of nucleotide change in a DNA sequence data set (Posada and Crandall 1998).

DNA polymorphism measured by segregating sites and nucleotide diversity

Variable DNA sequences at one locus within a species represent different alleles that are present in the population. Since DNA sequences are composed of many nucleotide sites, defining alleles is somewhat more complex than if alleles are discrete. Imagine obtaining a sample of n individuals from a population and determining the DNA sequence of L nucleotides for one gene or genomic region for each individual (see Tajima 1993b). For simplicity, consider each individual as haploid or homozygous. The first step would be to construct a **multiple sequence**

Substitution model assumptions   Substitution models compared

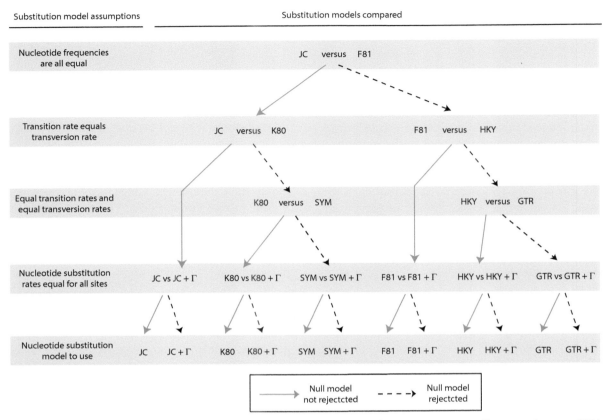

Figure 8.14 The hierarchy of nucleotide substitution models that can be used to correct apparent divergence between DNA sequences to better estimate the actual number of substitutions that have occurred. The Jukes-Cantor model is the simplest and assumes that there is just one rate of substitution that applies to all nucleotide changes and is constant among nucleotide sites. Other nucleotide substitution models include an increasing number of parameters to represent more features of DNA sequence evolution, in particular variable rates of substitution among various categories of nucleotides. If nucleotide substitution rates are variable among different sites, this variation can be modeled by a gamma distribution indicated by the Greek letter Γ. Nucleotide substitution models are JC = Jukes-Cantor (Jukes and Cantor 1969), F81 = Felsenstein 81 (Felsenstein 1981), K80 = Kimura 80 (Kimura 1980), HKY = Hasegawa-Kishino-Yano (Hasegawa et al. 1985), SYM = symmetrical model (Zharkikh 1994) and GTR = general time reversible (Rodriguez et al. 1990). Source: Figure after Posada and Crandall (1998).

alignment so that the homologous nucleotide sites for each sequence are all lined up in the same columns (Figure 8.15). With such a multiple sequence alignment, there are two commonly used measures that characterize the pattern of DNA polymorphism in a sample of DNA sequences from a single species.

One measure of DNA polymorphism is the **number of segregating sites**, S. A segregating site is any of the L nucleotide sites that maintain two or more nucleotides within the populations, sites 2, 6, and 8 in Figure 8.15. The total number of segregating sites is S and can be expressed as the number of segregating sites per nucleotide site, p_S, by dividing the number of segregating sites by the total number of sites:

$$p_S = \frac{S}{L} \tag{8.24}$$

The frequency of DNA sequences with a given nucleotide at a site does not influence S (compare sites 2 and 6 in Figure 8.15), but S will increase as the number of individuals sampled increases since DNA sequences with additional polymorphisms will be added to the sample.

The number of segregating sites (S) under neutrality is a function of the scaled mutation rate $4N_e\mu$. Watterson (1975) first developed a way to estimate θ from the number of segregating sites observed in a sample of DNA sequences. The expected number of segregating sites at drift–mutation equilibrium can

Sequence 1   A A T G T C A A C G
Sequence 2   A A T G T C A A C G
Sequence 3   A T T G T C A A C G
Sequence 4   A T T G T G A T C G
                 *   *   *

Site number   1 2 3 4 5 6 7 8 9 10

Figure 8.15 A hypothetical sample of four DNA sequences that are each 10 nucleotides long. There a total of three segregating sites ($S = 3$) or 3/10 segregating sites per nucleotide ($p_S = 0.3$). The nucleotide diversity is calculated by summing the nucleotide sites that differ among all unordered pairs of DNA sequences. In this example, there is an average of 1.33 pairwise nucleotide differences per sequence pair or an average of 0.133 pairwise nucleotide differences per nucleotide site.

Segregating sites (S and p_S):

Sites 2, 6, and 8 have variable base pairs among the four sequences (columns marked with *).
These are segretating sites. Therefore, for these sequences, $S = 3$ segrating sites and $p_S = 3/10 = 0.3$ segregating sites per nucleotide site examined.

Nucleotide diversity (π):

1   A A T G T C A A C G $d_{12} = 0$
2   A A T G T C A A C G

1   A A T G T C A A C G $d_{13} = 1$ 2   A A T G T C A A C G $d_{23} = 1$
3   A T T G T C A A C G   3   A T T G T C A A C G

1   A A T G T C A A C G $d_{14} = 3$ 2   A A T G T C A A C G $d_{24} = 3$ 3   A T T G T C A A C G $d_{34} = 2$
4   A T T G T G A T C G   4   A T T G T G A T C G   4   A T T G T G A T C G

$$\Sigma d_{ij} = 0 + 1 + 3 + 1 + 3 + 2 = 10$$

number of pairs of sequences compared = $[n(n\text{-}1)]/2 = [4(3)]/2 = 6$

$\hat{\pi} = 10$ differences/6 pairs = 1.67 average pairwise differences

$\hat{\pi} = 1.67$ avg. differences/10 sites = 0.167 pairwise differences per site

be determined using the coalescent model (Watterson used a different approach). Under the infinite sites model of mutation, each mutation that occurs increases the number of segregating sites by 1. The expected number of segregating sites is, therefore, just the expected number of mutations for a given genealogy. If each lineage has the probability μ of mutating each generation and there are k lineages, then the expected number of mutations in one generation is $k\mu$. If the expected time to coalescence for k lineages is T_k, then $k\mu T_k$ mutations are expected for each value of k. The expected number of mutations (E indicates an expectation or average) is obtained by summing over all k between the present and the most recent common ancestor (MRCA)

$$E[S] = E\left[\sum_{k=2}^{n} \mu k T_k\right] = \mu \sum_{k=2}^{n} k E[T_k] \qquad (8.25)$$

where n is the total number of lineages in the present. To see an illustration of this equation, refer to Figure 3.25 where $n = 6$ in the summation of Eq. 8.24 and imagine summing up the probability of a mutation in each time interval between coalescent events.

A fundamental result of the coalescent model is that the probability of k lineages coalescing is

$\frac{k(k-1)}{2}\left(\frac{1}{2N_e}\right)$. Therefore, the expected time to coalescence is the inverse of the probability of coalescence or $\frac{2(2N_e)}{k(k-1)}$. This expected time to coalescence can then be substituted into Eq. 8.24 to give

$$E[S] = \mu \sum_{k=2}^{n} k \frac{2(2N_e)}{k(k-1)} \qquad (8.26)$$

This equation simplifies by canceling each k, taking the constant $4N_e$ outside the summation, and adjusting the range of the summation to remove the -1 after k in the denominator:

$$E[S] = 4N_e\mu \sum_{k=1}^{n-1} \frac{1}{k} \qquad (8.27)$$

to give the expected number of segregating sites in a sample of n DNA sequences. Notice that $\theta = 4N_e\mu$ can substituted in Eq. 8.27 to give

$$E[S] = \theta \sum_{k=1}^{n-1} \frac{1}{k} \qquad (8.28)$$

The variance in the number of segregating sites is the sum of variances due to the mutation process (Poisson so the variance equals the mean) and to the coalescence process (exponential so the variance is the expected value squared) with zero covariance since the processes are independent

$$\text{var}(S) = \theta \sum_{k=1}^{n-1} \frac{1}{k} + \theta^2 \sum_{k=1}^{n-1} \frac{1}{k^2} \qquad (8.29)$$

The variance in the number of segregating sites in the neutral model is considerable because of the combined random variation in coalescence times and mutation events.

Once the expected number of segregating sites $E[S]$ is known, it can be solved for θ as a way to estimate the scaled mutation rate using the observed number of segregating sites, $\hat{S}$. Start by rearranging Eq. 8.28

$$\theta = \frac{E[S]}{\sum_{k=1}^{n-1} \frac{1}{k}} \qquad (8.30)$$

to solve for θ in terms of the number of segregating sites divided by the total branch length of the genealogy. If we define a new variable, $a_1 = \sum_{k=1}^{n-1} \frac{1}{k}$, and use $\hat{S}$ instead of $E[S]$, then

$$\hat{\theta}_S = \frac{\hat{S}}{a_1} \qquad (8.31)$$

using the absolute number of segregating sites or

$$\hat{\theta}_S = \frac{\hat{p}_S}{a_1} \qquad (8.32)$$

using the observed number of segregating sites per nucleotide site sampled. It is also possible to obtain an expression for the variance in $\hat{\theta}_S$ by substituting the definition of a_1 into Eq. 8.29 to give

$$\text{var}(\hat{\theta}_S) = \hat{\theta}_S a_1 + \hat{\theta}_S^2 \sum_{k=1}^{n-1} \frac{1}{k^2} \qquad (8.33)$$

An estimate of the scaled mutation rate determined from the number of segregating sites in a sample of DNA sequences is sometimes symbolized as $\hat{\theta}_W$ (W for Watterson). The importance of these two final quantities is that $4N_e\mu$ and its variance under the standard neutral model can be estimated from the number of segregating sites observed in a sample.

A second measure of DNA polymorphism is the **nucleotide diversity** in a sample of DNA sequences, symbolized by π (pronounced "pie") or sometimes θ_π, and also known as the **average pairwise differences** in a sample of DNA sequences (Nei and Li 1979; Nei and Kumar 2000). The nucleotide diversity is equivalent to the heterozygosity measured using alleles represented by DNA sequences (assuming random mating and the infinite sites model of mutation). The nucleotide diversity summarizes nucleotide polymorphism by averaging the number of nucleotide site differences found when each unique pair of DNA sequences in a sample is compared. In contrast with the proportion of segregating sites, the nucleotide diversity is sensitive to the frequency of each DNA sequence allele in a sample since more frequent sequences appear in more of the pairwise comparisons. The nucleotide diversity is the sum of the number of nucleotide differences seen for each pair of DNA sequences

$$\hat{\pi} = \frac{1}{\frac{n(n-1)}{2}} \Sigma_{i=1}^n \Sigma_{j>i}^n d_{ij} \qquad (8.34)$$

where i and j are indices that refer to individual DNA sequences, d_{ij} is the number of nucleotide sites that differ between sequences i and j, and n is the total number of DNA sequences in the sample. The number of unique pairwise comparisons in a sample of n sequences is $(n[n-1])/2$, and so dividing the sum of d_{ij} by this number gives the average number of differences per pair of sequences. The average number of pairwise differences can also be divided by the number of nucleotide sites examined (L) to express $\hat{\pi}$ per nucleotide site. Figure 8.15 shows an example computation of $\hat{\pi}$ for a hypothetical sample of four DNA sequences.

In larger samples that may include multiple identical DNA sequences, the nucleotide diversity can be estimated by

$$\hat{\pi} = \frac{k}{k-1} \Sigma_{i=1}^k \Sigma_{j=i}^k p_i p_j d_{ij} = 2 \frac{k}{k-1} \Sigma_{i=2}^k \Sigma_{j=i}^{i-1} p_i p_j d_{ij} \qquad (8.35)$$

where p_i and p_j are the frequencies of alleles i and j, respectively, in a sample of k different sequences that each represent one allele. The first version of the equation sums all elements in the matrix of the number of differences per pair of alleles, including the diagonal elements. The second version is twice the

Table 8.1 Nucleotide diversity (π) estimates reported from comparative studies of DNA sequence polymorphism from a variety of organisms and loci. All estimates are the average pairwise nucleotide differences per nucleotide site. For example, a value of π = 0.02 means that two in 100 sites vary between all pairs of DNA sequences in a sample.

Species	Locus	π	References
Drosophila melanogaster	*anon*1A3	0.0044	Andolfatto (2001)
	Boss	0.0170	
	transformer	0.0051	
Drosophila simulans	*anon*1A3	0.0062	
	Boss	0.0510	
	transformer	0.0252	
Caenorhabditis elegans[a]	*tra-2*	0.0	Graustein et al. (2002)
	glp-1	0.0009	
	COII	0.0102	
Caenorhabditis remanei[b]	*tra-2*	0.0112	
	glp-1	0.0188	
	COII	0.0228	
Arabidopsis thaliana[a]	*CAUL*	0.0042	Wright et al. (2003)
	ETR1	0.0192	
	RbcL	0.0012	
Arabidopsis lyrata ssp. *petraea*[b]	*CAUL*	0.0135	
	ETR1	0.0276	
	RbcL	0.0013	

[a] mates by self-fertilization.
[b] mates by outcrossing.

sum of the lower diagonal elements. These are equal since the d_{ii} elements are all equal zero. These versions of the formula provide an average of d_{ij} because each elements of the sum is weighted by the frequency of each type of DNA sequence found in a sample. The nucleotide diversity can be underestimated if there are rare sequence polymorphisms in a population that are unlikely to be sampled (see Renwick et al. 2003). Information on the sampling variance of π can be found in Nei and Kumar (2000).

Some values of π from different organisms and loci are shown in Table 8.1. Estimates of nucleotide diversity are useful because π is a measure of heterozygosity for DNA sequences. As such, the value of π is a function of $4N_e\mu$ under an equilibrium between genetic drift and mutation. With an estimate of π and the mutation rate at a locus (μ), it is then possible to estimate the effective population size. Because π is an estimator of the scaled mutation rate θ, it is sometimes referred to as $\hat{\theta}_\pi$.

Interact box 8.3 Estimating π and S from DNA sequence data

It is a worthwhile exercise to estimate the number of segregating sites (*S*) and the nucleotide diversity or average pairwise differences (π) from a sample of multiple DNA sequences.

The first step is to obtain DNA sequence data from Genbank. The text web page gives step-by-step instructions to obtain DNA sequences for the mitochondrial cytochrome *b* gene in a sample of 30 African sable antelope (Pitra et al. 2002). These 30 sequences must first be aligned into a multiple sequence alignment. Then, the aligned sequences can be used to estimate the number of segregating sites (*S*) and the average pairwise differences (π).

The θ_π and θ_S measures are based on the general approach of quantifying the **site frequency spectrum** of sequence polymorphism. The segregating sites in a sample of n sequences can be categorized according to the number of sequences that share each allelic type (Fu 1995, 1997). This serves to describe the $2(n\text{-}1)$ branches in a genealogy since segregating sites with a common allelic state share the same branch of a genealogy. Segregating sites can be classified into haplotypes where an allelic state occurs i or $n\text{-}i$ times. The variable ξ_i (lowercase Greek xi) is the number of segregating sites where a haplotype (or mutant type) occurs i or $n\text{-}i$ times. Fu (1995) showed that the number of segregating sites is a function of the scaled mutation rate according to $E[\xi_i] = \frac{1}{i}\theta$ for i between 1 and $n\text{-}1$ in a sample of n sequences. Using the $n = 4$ sequences in Figure 8.15 as an example and assuming that sequence 1 is ancestral in order to count derived allelic states, both sites 6 and 8 have one haplotype with a derived nucleotide state for $\xi_1 = 2$, site 2 has two haplotypes with derived nucleotide states for $\xi_2 = 1$, and $\xi_3 = 0$ since there are no sites where three haplotypes have a derived nucleotide state.

The two measures of sequence polymorphism seen earlier in the section can be written as functions of ξ

$$\hat{\theta}_S = \frac{1}{a_1}\sum_{i=1}^{n-1}\xi_i \qquad (8.36)$$

and

$$\hat{\theta}_\pi = \frac{2}{n(n-1)}\sum_{i=1}^{n-1}i(n-i)\xi_i \qquad (8.37)$$

Two other sequence polymorphism measures can be constructed that give more weight to either high frequency haplotypes

$$\hat{\theta}_H = \frac{2}{n(n-1)}\sum_{i=1}^{n-1}i^2\xi_i \qquad (8.38)$$

or more weight to low-frequency haplotypes

$$\hat{\theta}_L = \frac{1}{n-1}\sum_{i=1}^{n-1}i\xi_i \qquad (8.39)$$

These measures capture different aspects of sequence polymorphism and are utilized in hypothesis tests for processes acting on sequence polymorphism that are described later in the chapter.

8.4 DNA sequence divergence and the molecular clock

- The molecular clock hypothesis.
- Dating divergence events with a molecular clock.

One key result of the neutral theory is the prediction that the rate of substitution is equal to the mutation rate. A corollary of this prediction is that the expected number of generations between substitutions is the reciprocal of the mutation rate. For example, if the mutation rate is 1×10^{-5} base pairs replicated in error per generation, then we expect to wait an average of 10^5 generations to see one mutation in a single gene copy. Thus, the neutral theory provides a null model for the rate of divergence of homologous genes or genome regions between isolated populations or species called the molecular clock hypothesis. This section will first present data that demonstrate the molecular clock and then show how the molecular clock hypothesis can be used to date evolutionary events based on DNA divergence. The section will conclude by showing why divergence may appear to decrease over time as many substitutions accumulate and how divergence estimates can be corrected using models of the mutation process.

As the clock metaphor suggests, the molecular clock hypothesis predicts that divergence accumulates with uniform regularity over time, just like the ticking of a clock. This means that the divergence between two species should increase as the time since they shared a common ancestor recedes further into the past. Such a pattern was originally observed for hemoglobin proteins by Zuckerkandl and Pauling (1962, 1965), who first hypothesized a molecular clock. A classic example of the molecular clock is the increase of divergence with increasing time seen in the NS gene of the human influenza A virus (Figure 8.16). Buonagurio et al. (1986) used influenza virus isolated from samples originally taken between 1933 and 1986. They then estimated the number of nucleotide substitutions, or the p-distance, between each sequence and the inferred ancestral sequence. The linear increase in divergence with time is the pattern expected by the molecular clock hypothesis.

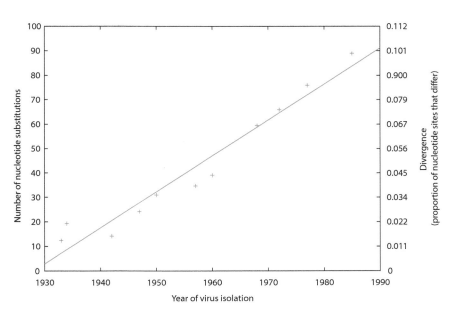

Figure 8.16 Rates of nucleotide change in the NS gene that codes for "nonstructural" proteins based on 11 human influenza A virus samples isolated between 1933 and 1985. The number of years since isolation and DNA sequence divergence from an inferred common ancestor are positively correlated. The pattern of increasing substitutions as time since divergence increases is expected under the molecular clock hypothesis. The observed rate of substitution was approximately 1.9×10^{-3} substitutions per nucleotide site per year, a very high rate compared to most genes in eukaryotes. The line is a least-squares fit. Source: Data from Buonagurio et al. (1986).

Molecular clock hypothesis: The neutral theory prediction that divergence should occur at a constant rate over time so that the degree of molecular divergence between species is proportional to their time of separation, synonymous with rate constancy or rate homogeneity.

Another important early advance for the molecular clock came when Richard Dickerson (1971) compared rates of substitution in proteins from cytochrome *c*, hemoglobin, and fibrinopeptide genes and observed that the average rates of change were very different for the three proteins (Figure 8.17). Based on knowledge of the function of the proteins at the time, Dickerson argued that the rate of molecular evolution was faster when fewer sites were subject to functional constraints on amino acid changes. That is, faster molecular evolution occurred when more sites were neutral and free to evolve by genetic drift. Slower molecular evolution occurred when a larger proportion of amino acid changes were eliminated by natural selection because they decreased or eliminated protein function. Thus, those sites that have not diverged in sequences compared among species may be constant due to selective constraint for function. Under this view, novel at the time, the portions of protein or DNA sequences that are invariant over time and shared among species indicate regions of functional importance. The neutral

theory served as a key concept to explain why different loci might have molecular clocks that tick at different rates.

Dating events with the molecular clock

A useful application of the molecular clock is to date divergence events between species. For some organisms, the fossil record and geological context provide a means to date when species originated, went extinct, or exhibited evolutionary transitions. However, many types of organisms have no fossil record and not all phenotypes fossilize, presenting a problem for dating biological events. If the amount of DNA sequence divergence between two species is known, then this information can be utilized to date their divergence. The molecular clock hypothesis asserts that for neutral alleles the rate of substitution is simply the mutation rate, or $k = \mu$. If an absolute substitution rate expressed in fixations per time interval is available, multiplying that rate by a time gives an expected number of substitutions. The number of diverged nucleotide sites between two species also increases at twice the rate of substitutions since each lineage will experience substitutions independently. Bringing these two observations together gives the expected amount of divergence

$$k = 2T\mu \qquad (8.40)$$

between two species that diverged *T* time units ago. If divergence between two species as well as the rate of

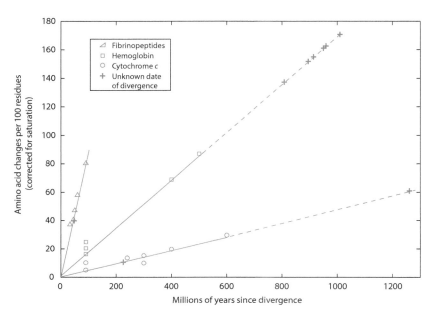

Figure 8.17 Rates of protein evolution as amino acid changes per 100 residues in fibrinopeptides, hemoglobin and cytochrome *c* over very long periods of time. Rates of divergence are linear over time for each protein as expected for a molecular clock. Different proteins have different clock rate due to different mutations rates and degrees of functional constraint imposed by natural selection. Amino acid changes between pairs of taxa with unknown divergence times are plotted on dashed lines with the same slope as lines through points for taxa with estimated divergence times. The six points with unknown divergence times for hemoglobin represent divergence of ancestral globins into hemoglobins and myoglobins in the earliest animals, events that the molecular clock estimates to have happened between 1.1 billion and 800 million years ago. Source: Data from Dickerson (1971). See also Robinson et al. (2016).

divergence is known, this relationship can be rearranged to solve for the unknown of time instead:

$$T = \frac{k}{2\mu} \qquad (8.41)$$

> **Absolute substitution rate:** A rate of molecular change estimated in substitutions per year based on the combination of a sequence divergence estimate from two taxa and an estimate of the time that has elapsed since those taxa diverged.

Figure 8.18 illustrates a situation with three taxa and two divergence times. Dating events with the molecular clock requires nucleotide divergences (adjusted for saturation) for each pair of species are estimated (K_{AB}, K_{AC}, and K_{BC}) and one divergence time is known. Imagine that our goal is to determine the time of divergence for species A and B (T_2) given that divergence time T_1 is known. The absolute rate of substitution that occurred over the known divergence time can be estimated:

$$\mu = \frac{1}{2}\left(\frac{K_{AC}}{2T_1} + \frac{K_{BC}}{2T_1}\right) \qquad (8.42)$$

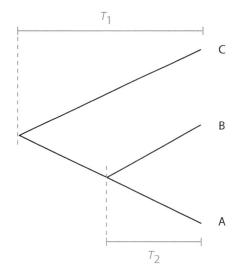

Figure 8.18 A schematic phylogenetic tree that can be used with to date divergence events under the assumption of a constant rate of divergence over time or a molecular clock. T_1 is the time in the past when species C and the ancestor of species A and B diverged. T_2 is the time in the past when species A and B diverged. If either T_1 or T_2 are known, the rate of molecular evolution per unit of time can be estimated from observed sequence divergences. This rate of divergence can then be used to estimate the unknown amount of time that elapsed during other divergences.

based on the average of the divergences observed between species pair A and C (K_{AC}) and species pair B and C (K_{BC}) and the time of divergence T_1. This rate can then be used to solve for the unknown divergence time T_2:

$$T_2 = \frac{K_{AB}}{2\mu} \qquad (8.43)$$

By substituting the definition of μ in Eqs. 8.42 and 8.43 becomes

$$T_2 = \frac{2T_1 K_{AB}}{K_{AC} + K_{BC}} \qquad (8.44)$$

As an example, imagine that DNA sequence divergences are estimated as $K_{AB} = 0.10$ $K_{AC} = 0.31$, and $K_{BC} = 0.36$ substitutions per site and the divergence time T_1 is estimated with fossil and geologic data to be 10 million years. The interval T_2 is then estimated as

$$T_2 = \frac{2(10 \text{ million years})(0.10 \text{ substitutions per site})}{(0.31 \text{ substitutions per site}) + (0.36 \text{ substitutions per site})}$$

$$= 2.985 \text{ million years}$$

$$(8.45)$$

The answer makes intuitive sense. The DNA divergence between species A and B is about one-third of the average DNA sequence divergence between species pairs A–B and A–C. Since A, B, and C diverged from a common ancestor 10 million years ago, then T_2 when A and B diverged is about one-third of that time. Assumptions in these time estimates are that rates of substitution are constant over time, among lineages, and over loci. These assumptions will be explored critically later in the chapter.

The molecular clock has been widely used to date major evolutionary transitions, establish times when the ancestors of many different organisms first evolved, and test hypotheses related to divergence times (reviewed by Hedges et al. 2015). One example is testing the hypothesis that the early mammal evolution was facilitated by ecological niches that opened up when the dinosaurs went extinct. The molecular clock suggests that the earliest mammal lineages had appeared well before the extinction of the dinosaurs was complete (Bromham et al. 1999; Bininda-Emonds et al. 2007). Thus, the estimated divergence time does not support the

Problem box 8.1 Estimating divergence times with the molecular clock

In the present day, dicotyledonous plants represent the majority of land plants. The divergence of ancestral seed plants into monocotyledonous and dicotyledonous plants was, therefore, a major evolutionary transition. Based on DNA divergence data for synonymous sites at nine mitochondrial genes in a range of plants (Laroche et al. 1995), a molecular clock can be used to date this event.

Table 8.2 gives DNA divergence data for comparisons of maize and wheat, both monocotyledons, with an estimated divergence time of approximately 60 million years ago. First, use the maize–wheat DNA divergence data to calibrate the absolute rate of substitution per million years for each locus. Then, use this rate of change to estimate the time when monocots and dicots split given their degree of DNA sequence divergence. In terms of Figure 8.18, the maize–wheat split is T_2 and the monocot–dicot split is T_1.

Table 8.2 DNA divergence estimates for three loci in maize and wheat, both monocotyledons.

Locus	Nucleotide sites	Synonymous sites	Substitutions per site Wheat-maize	Substitutions per site Monocot-dicot
coxI	1461	495	0.0504	0.2060
atp9	195	67	0.1374	0.4439
nad4	1272	456	0.0381	0.1101

hypothesis that mammals evolved at the time that habitats were left empty as dinosaurs disappeared.

A second illustration is the classic question of when humans and their close ancestors diverged. Using calibration times of 13 million years for the divergence between orangutans and humans and 90 million years between artiodactyls (hoofed mammals with an even number of digits, such as cattle, deer, and pigs) and primates as well as numerous loci, Glazko and Nei (2003) estimated that the divergence of humans and chimpanzees occurred 5–7 million years ago. Additional molecular clock studies in primates have shown evidence for variation in the rates of substitution among lineages and across the different types of genomic changes that can impact dates inferred using a molecular clock. Moorjani et al. (2016) found that the whole-genome molecular clock for 10 primate lineages showed the greatest uniformity for substitutions at CpG sites (where a cytosine nucleotide is followed by a guanine nucleotide). Using a molecular clock for only CpG sites produced an estimate of 12.1 million years ago for human and chimpanzee divergence. There is evidence that the human lineage has recently experienced a slower molecular clock compared to gorillas and chimpanzees, which increases molecular clock divergence time estimates and places them in better agreement with fossil evidence (Scally and Durbin 2012; Besenbacher et al. 2019).

A third example is the employment of the molecular clock to date the origin of the human immunodeficiency virus or HIV. The date that HIV was transmitted from primates to humans has been a critical question in understanding the origins of the virus and its patterns of evolutionary change in an effort to prevent possible future transmissions of disease from animals to humans. The main hypothesis for the origin of HIV is that haplotypes of simian immunodeficiency virus (SIV) infected humans who hunted primates and processed the carcasses. The rate of substitution in DNA sequences from HIV-1 exhibits a clock-like pattern of substitution rates even with recombination events, natural selection on some nucleotide changes, and after adjustment for sampling date since some virus samples were collected at different times (Leitner and Albert 1999; Salemi et al. 2001; Liu et al. 2004; Park et al. 2016). The application of the molecular clock suggests that HIV-1 subtype M, the haplotype that experienced pandemic transmission in humans, was introduced to humans in the 1920s. The molecular clock for HIV has now been used for a detailed

reconstruction of the geographic origins and early dissemination of the virus in Kinshasa, Democratic Republic of Congo (then Zaire) and then its spread in Africa and around the world (see Faria et al. 2014). (The controversial hypothesis that HIV was first spread to humans through contaminated polio vaccine made from cultured chimpanzee cells that was administered in the former Belgian Congo between 1957 and 1960 is inconsistent with the molecular clock divergence estimate for HIV-1 and is also refuted by other direct evidence (Cohen 2001; Worobey et al. 2004)). Phylogenetic and molecular clock evidence also suggests two transmission events where SIV lineages circulating in primates jumped to humans – the HIV-1 lineage is most closely related to SIV in chimpanzees and the HIV-2 lineage is most closely related to SIV in sooty mangabeys (Gao et al. 1992; Lemey et al. 2003; Keele et al. 2006).

The use of the molecular clock to estimate times of divergence is complicated by numerous issues in practice (reviewed by Arbogast et al. 2002), contributing to the statistical uncertainty of date estimates. First, calibration times usually have considerable ranges, leading to uncertainty in any divergence time estimated from the molecular clock. In addition, variation in rates of substitution over time,

Interact box 8.4 Molecular clock estimates of evolutionary events

TimeTree (Kumar et al. 2017) is a searchable tree of life that brings together results from studies that employed the molecular clock to date evolutionary events. The project allows users to date divergence times for a pair of taxa, to produce a time-scaled phylogenetic tree, or view phylogenetic patterns displayed on a geologic time line. The project has collected information for nearly 100 000 species and counting with the goal of representing the entire tree of life.

Visit the TimeTree website and view divergence time estimates for a pair of species (**Node Time**), such as human and chimpanzee. Use the **Timetree** feature for a group of taxa such as "primates". Then, explore times for other taxa of your choosing.

among lineages, among different loci and among different classes of mutational changes are now considered the rule rather than the exception, complicating the methods needed to estimate divergence times. Recent advances in the molecular clock have been based on Bayesian approaches to estimation of coalescence times (reviewed by dos Reis et al. 2016). These approaches have advantages such as a range of ways to model the variation in substitution rates, the use of prior probabilities for node dates based on fossil evidence, and estimation from combined sequence and phenotype data. Variation in substitution rates is explored further in the next section.

8.5 Testing the molecular clock hypothesis and explanations for rate variation in molecular evolution

- Rate heterogeneity in the molecular clock.
- The Poisson process model of the molecular clock.
- Ancestral polymorphism and the molecular clock.
- Relative rate tests of the molecular clock.
- Possible causes of rate heterogeneity.

The molecular clock predicts that selectively neutral homologous sequences (meaning sequences that were once identical by descent) with equal mutation rates should experience a similar number of substitutions per unit time as divergence increases. Therefore, the molecular clock hypothesis provides a null model to examine the processes that operate during molecular evolution. It is possible to directly test the molecular clock hypothesis and thereby test this null model. Rejecting the molecular clock hypothesis suggests that the sequences compared evolve at unequal rates, a situation referred to as **rate heterogeneity**. Rejecting the molecular clock hypothesis is a way to identify processes that influence the chance of substitution such that rates of fixation are either higher or lower than expected by genetic drift alone. For example, the previous section showed how natural selection changes the probability of fixation and, therefore, the rate of substitution. So, for example, one sequence taken from a population where most mutations are deleterious and selected against and another sequence taken from a population where most mutations are neutral would have different rates of substitution and show different numbers of substitutions over a fixed time interval (see Figure 8.3). Thus, testing for equal rates of substitution, or **rate homogeneity**, is a useful

step in identifying the processes that may be operating in molecular evolution.

> **Rate heterogeneity:** Variation in the rate of substitution over time or among different lineages for homologous genome regions.

The molecular clock and rate variation

Since the neutral theory leads to the molecular clock hypothesis, evidence for rate heterogeneity would appear to be evidence that genetic drift is not the main process leading to the ultimate substitution of most mutations. Rejecting the hypothesis of rate homogeneity would suggest that natural selection is operating on mutations such that their rates of substitution are either sped up or slowed down relative to substitution rates under genetic drift. The probability that a new mutation is fixed by natural selection depends on the selection coefficient, s, and the effective population size rather than just $\frac{1}{2N_e}$ as it does for genetic drift. Natural selection is, therefore, very unlikely to produce a molecular clock because s, N_e, and μ are not likely to be constant through time or among different lineages. Before reaching the conclusion that natural selection explains all rate heterogeneity, however, it is necessary to dig deeper into the molecular clock hypothesis. The molecular clock is potentially more complex than was revealed at the beginning of the chapter. Understanding these complications is a necessary prerequisite to understanding the range of alternative hypotheses that may explain heterogeneity in rates of molecular evolution.

The molecular clock was originally proposed by Zuckerkandl and Pauling (1962, 1965, Zuckerkandl 1987) to model amino acid substitutions. It was based on a simple statistical method used to describe events that happen at random times given some rate at which events occur (such models are called point processes). The simplest point process for a molecular clock is a Poisson process, a stochastic process which is defined in terms of the count of events, $N(t)$, since time was equal to 0. In a Poisson process, the expected number of events between two times follows a Poisson distribution. Assuming that all substitutions are independent events, the probability of a substitution is very small, and that the number of time intervals is very large, a Poisson clock gives the probability of observing some number of substitutions after a time period has elapsed as

$$\text{Probability}(N(t) \text{ substitutions at time } t) = \frac{e^{-\lambda t}(\lambda t)^{N(t)}}{N(t)!}$$

(8.46)

where $N(t)$ is the total number of substitutions (an integer), t is the time in years, and λ is the rate of substitutions per year. Under this model, the expected number of substitutions at time t is λt or the product of the substitution rate and the number of time steps that have elapsed. A critical thing to notice in this model is that the rate of substitutions, λ, is constant and does not change with time nor with the total number of substitutions, $N(t)$. The Poisson molecular clock is illustrated in Figure 8.19. The top panel shows the probability that $N(t)$ is between 0 and 14 for one time step when the rate of substitution is $\lambda = 4$. The bottom panel of Figure 8.15 shows variation in the number of substitutions among five replicate sequences that all evolved for the same period

of time at the same constant rate of substitution ($\lambda = 4$).

The Poisson model for a molecular clock implies that the time intervals between substitutions are random in length (Figure 8.20). Thus, substitutions that follow a Poisson molecular clock will be separated by variable lengths of time. This stands in contrast to our everyday notion of a clock or watch, which has uniform lengths of time separating each event (events are seconds, minutes, and hours). Therefore, a molecular clock that is based on a random process has inherent variation in the number of substitutions that occur over a given time interval even though the rate of substitution remains constant. This means that independent lineages that each diverged from a common ancestor at the same time can display variation in the number of substitutions that have occurred. In other words, if substitution is a random process, then we expect some variation in the number of substitutions among lineages and loci even if divergence time and the substitution rate are constant. This explanation for variation in the numbers of substitutions is often referred to by saying that rates of molecular evolution follow a **Poisson clock**.

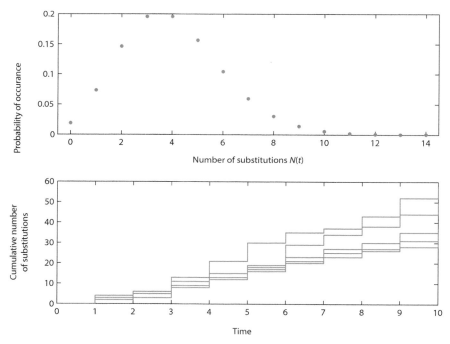

Figure 8.19 Substitution patterns under a Poisson process. The top panel shows the probability distribution for the number of substitutions that might occur during one time interval. $N(t)$ between zero and nine all have probabilities of greater than 0.01. The bottom panel shows the cumulative number of substitutions under a Poisson process for five independent trails. Each trail is akin to an independent lineage experiencing substitutions. The average number of substitutions is approximately 40 (4 multiplied by the number of time intervals) but there is variation among the lineages. In both panels the rate of substitution is that same at $\lambda = 4$.

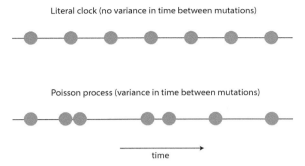

Literal clock (no variance in time between mutations)

Poisson process (variance in time between mutations)

time

Figure 8.20 Two representations of rate at which substitution events (circles) occur over time. Mutations might occur with metronome-like regularity, showing little variation in the time that elapses between each mutation event. If substitution is a stochastic process, an alternative view is that the time that elapses between substitutions is a random variable. The Poisson distribution is commonly used distribution to model the number of events that occur in a given time interval, so the bottom view is often called the Poisson molecular clock. Note that, in both cases, the number of substitutions and time elapsed is the same so that the average substitution rate is identical.

The Poisson process model of the molecular clock leads to a specific prediction about the variation in numbers of substitutions that should be observed if the rate of molecular evolution follows a Poisson process. The Poisson distribution has the special property that the mean is equal to the variance. Therefore, the mean number of substitutions and the variance in the number of substitutions should be equal for independent DNA sequences evolving at the same rate according to a Poisson process. A ratio to compare the mean and variance in the number of substitutions is called the **index of dispersion**, defined as

$$R(t) = \frac{\text{variance } N(t)}{E(N(t))} \qquad (8.47)$$

where E indicates an expected or mean value. The index of dispersion defines the degree of spread among divergence estimates that should be seen under a Poisson process, just as a Markov chain defines the spread in allele frequencies that is expected in an ensemble of finite populations. If the numbers of substitutions in a sample of pairwise sequence divergences follow a Poisson molecular clock, then the variance and mean number of substitutions should be equal and therefore $R(t)$ should equal 1. If the variance is larger than the mean, then $R(t)$ is greater than 1, a situation referred to as an **overdispersed molecular clock** since substitution

rates have a wider range of values than predicted by the Poisson process model (reviewed by Cutter 2000a).

> **Overdispersed molecular clock:** Absolute divergence rates from many independent pairs of species that overall exhibit more variance in divergence rate than expected by the Poisson process molecular clock model; a dispersion index value that is greater than 1.

Ancestral polymorphism and poisson process molecular clock

The molecular clock modeled as a Poisson process assumes it is possible to compare pairs of DNA sequences that were derived from a single DNA sequence in the past and then diverged instantly into two completely isolated species. Actual DNA sequences usually have a more complex history that involves processes that operated in the ancestral species followed by the process of divergence in two separate species (Figure 8.21). In the ancestral species, the number and frequency of neutral alleles per locus in the population were caused by a population process such as genetic drift and mutation (assuming that the ancestral species was panmictic). This zone of **ancestral polymorphism** is the period of time when genetic variation in the ancestral species was dictated by drift–mutation equilibrium. Within this ancestral population, two lineages split at some point and eventually became lineages within the two separate species (Figure 8.17). DNA sequences from these two lineages were sampled in the present to estimate substitution rates. Recognizing this more complex history of diverged DNA sequences shows two things. First, it points out that lineages and species may have diverged at different points in time, with lineages often diverging earlier than species. Second, it shows that two distinct processes can contribute to the nucleotide differences between sequences seen as substitutions when observed in the present. Referring to Figure 8.17, during the time period *T*, polymorphisms among sequences were caused by the population processes dictating polymorphism in the ancestral species. Later, during the time period *t*, substitutions were the product of the divergence process between species.

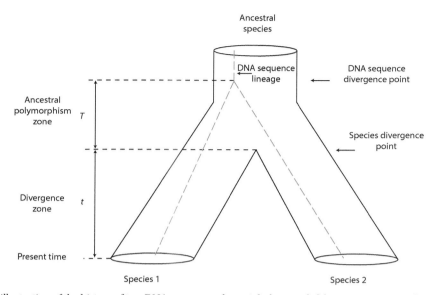

Figure 8.21   AN illustration of the history of two DNA sequences that might be sampled from two species in the present time in order to estimate the rate of substitutions. The history is like a water pipe in an upside-down "Y" shape. The tube at the top contains the total population of lineages in the ancestral species, eventually splitting into populations of lineages that compose two species. The time when two lineages diverged from a common ancestor is not necessarily the same as the time of speciation. Therefore, a population process governing polymorphism operates for T generations in the ancestral species while a divergence process operates for t generations in the diverged species. The polymorphism process initially dictates the number of nucleotide changes between two sequences. Later, the divergence process dictates the number of nucleotide changes between two sequences. In two DNA sequences sampled in the present, it is impossible to distinguish which process has caused the nucleotide changes observed.

The existence of both ancestral polymorphism and divergence processes complicates testing for overdispersion of the molecular clock (Gillespie 1989, 1994). To make this point, Gillespie articulated the distinction between **origination processes** and **fixation processes**. An origination process describes the times at which the subset of new mutations that will ultimately fix first enter the population. A fixation process, in contrast, describes the times at which the subset of new mutations that will ultimately fix reach a frequency of 1 in the population. At a conceptual level, it is clear that the two processes are not identical. The distribution of origination times is a product of the causes of mutation. Times until fixation depend on both the causes of mutation (the origination process) and on the causes of fixation, such as genetic drift in a finite population of neutral alleles or natural selection. Measuring times until the fixation of new mutations would require that we are able to follow the populations of diverging species over time and watch as new mutations segregate and eventually go to fixation and loss, recording the times for those that fix and then calling these the substitution times. In Figure 8.2, originations are the events at the bottom

of the y axis and fixations are events at the top of the y axis. In practice, we have only the accumulated amino acid or DNA differences between pairs of species observed at one point in time. Such sequence differences are a product of the origination process because they are a sample of mutations that came into the population some time ago and have fixed by the time we observe them. This is not the same thing as having observed the "tick" of fixations over a long period of time.

Gillespie and Langley (1979) showed that a molecular clock combining polymorphism and divergence does not necessarily comprise a Poisson process where the index of dispersion is expected to equal one. To see this, it is necessary to develop expectations for the mean and variance in the number of nucleotide differences between two sequences when both polymorphism and divergence processes are operating over the history of two DNA sequences.

As shown earlier in the chapter for the infinite sites model of mutation, the expected number of segregating sites (S) for a sample of two DNA sequences (so $a_1 = 1$) is with. This is a prediction for the amount of polymorphism expected under neutrality in a finite population. This result tells us that the mean

Table 8.3 Mean and variance in the number of substitutions at a neutral locus for the cases of divergence between to species and polymorphism within a single panmictic population. The rate of divergence is modeled as a Poisson process, so the mean is identical to the variance. The mutation rate is μ and the $\theta = 4N_e\mu$. Refer to Figure 8.17 for an illustration of divergence and ancestral polymorphism.

	Expected value or mean	Variance
Ancestral polymorphism	θ	$\theta + \theta^2$
Divergence	$2t\mu$	$2t\mu$
Sum	$2t\mu + \theta$	$2t\mu + \theta + \theta^2$

and variance of the number of nucleotide sites are expected to be different in a sample of two DNA sequences from the ancestral polymorphism zone in Figure 8.17.

Now shift focus to the divergence zone of Figure 8.17. For one DNA sequence from species 1 and another from species 2, the divergence time is $2t$ because each species has diverged independently for t generations. Based on the Poisson process in Eq. 8.38, both the expected number of diverged sites and the variance in the number of diverged sites are $2\mu t$. The means and variances in the number of nucleotide sites between two sequences are shown in Table 8.3 as they apply to polymorphism and divergence in Figure 8.17.

Given the means and variances of the number of changes between two DNA sequences for both polymorphism and divergence processes, we can then combine these expectations into a new expression for the index of dispersion. The index of dispersion for the number of differences between a sequence from species 1 and a sequence from species 2 is then

$$R(t) = \frac{\text{variance } N(t)}{E[N(t)]} = \frac{2\mu t + \theta + \theta^2}{2\mu t + \theta} \quad (8.48)$$

where $\theta = 4N_e\mu$. (This requires the assumptions that the time of lineage divergence T is a random variable with a geometric distribution as in a genealogical branching model, and that N_e is large and μ is small.) As shown in Math Box 8.1, this new version of the dispersion index can be rewritten as

$$R(t) = 1 + \frac{\theta^2}{E[N(t)]} \quad (8.49)$$

If we assume that there is no ancestral polymorphism or that $T = 0$ in Figure 8.17, then θ is 0 and $R(t)$ is identical to what is expected under the Poisson process molecular clock dictated by divergence only.

The major conclusion is that the two-process version of $R(t)$ is expected to be greater than one when there is any ancestral polymorphism even when DNA changes follow a constant molecular clock. Stated another way, ancestral polymorphism increases the variance in the number of substitutions

---

Math box 8.1 The dispersion index with ancestral polymorphism and divergence

Define a new variable $\alpha = \dfrac{t}{2N_e}$ to scale time by $2N_e$ generations. Then, notice that

$$\theta\alpha = 4N_e\mu\frac{t}{2N_e} = 2\mu t \quad (8.50)$$

The numerator and denominator of

$$R(t) = \frac{2\mu t + \theta + \theta^2}{2\mu t + \theta} \quad (8.51)$$

can then be rewritten as functions of α

$$R(t) = \frac{\theta(1 + \alpha) + \theta^2}{\theta(1 + \alpha)} \quad (8.52)$$

and rearranged to

$$R(t) = \frac{\theta(1 + \alpha)}{\theta(1 + \alpha)} + \frac{\theta^2}{\theta(1 + \alpha)} \quad (8.53)$$

The expected number of DNA changes at time t is $2\mu t + \theta$ (see Table 8.3), which is equivalent to $\theta(1 + \alpha)$. The index of dispersion can then be written

$$R(t) = 1 + \frac{\theta^2}{\theta(1 + \alpha)} = 1 + \frac{\theta^2}{E[N(t)]} \quad (8.54)$$

seen for pairs of sequences evolving under a constant rate compared with a pure divergence process. Unfortunately, the index of dispersion in Eq. 8.42 seems impossible to estimate in practice because θ cannot be estimated in the ancestral species as it does not exist any longer. However, the main point of this model is not to provide a practical test of the molecular clock. Instead, the model shows how $R(t) > 1$ is not necessarily strong evidence to reject a constant rate of substitution. One cause of $R(t) > 1$ is that the Poisson process accurately describes the substitution process but that substitution rates are not constant. Alternatively, the Poisson process of model of divergence itself may not be accurate even though the rate of DNA change is constant. The latter possibility suggests that the index of dispersion may be a poor way to test the neutral molecular clock hypothesis.

Ancestral polymorphism also presents difficulties for dating divergences using the molecular clock (Maddison 1997; Arbogast et al. 2002). The problem arises because sequence lineage history (genealogy) and species divergence history (species phylogeny) are not identical. Two sequences sampled in the present from two different species have been accumulating substitutions since the MRCA of the two sequences gave rise to the lineages (Figure 8.17). The total sequence divergence between two species that would be used to date a speciation event has occurred during two distinct time intervals. One time interval T is the period when the two lineages accumulated changes in the ancestral species. The second time interval t is the period when substitutions accumulated after the current species split. Estimates of time since divergence estimate the total elapsed time since the divergence of the two *lineages* rather than just the time since divergence of the two species. Thus, the use of the molecular clock to date divergence time yields over-estimates of the species divergence time. As the divergence time t increases relative to the polymorphism time T, the degree of over-estimation shrinks. However, it is usually impossible to determine t relative to T in practice and so the degree of over-estimation of the species divergence time is usually unknown.

Relative rate tests of the molecular clock

One method to circumvent some of the limitations inherent in comparing absolute rates of divergence is to compare relative rates instead. The **relative rate test** compares the number of nucleotide or amino acid changes since divergence from an ancestor represented by a DNA sequence from closely related species (Sarich and Wilson 1967; Fitch 1976). Rates of nucleotide substitution in two different species can be estimated by comparing the number of DNA or amino acid changes that have occurred independently in each of two species using a third outgroup species to assign sequence changes to each lineage. If rates of substitution are equal in the two species, then the number of sequence changes should be equal in the two species within a statistical confidence interval. Unequal numbers of sequence changes lead to rejection of the null hypothesis that the two species have an equal rate of substitution. Relative rate tests avoid the need for a date of divergence that is often imprecise and also do not rely on the dispersion index and its underlying assumption that the molecular clock is a simple Poisson process.

Tajima's (1993a) 1D test of the molecular clock is a relative rate test that uses the number of nucleotide substitutions that occurred along two lineages being compared as well as an outgroup lineage. The basis of the test is shown in Figure 8.22. In the figure, the letters i, j, and k are used to represent the identity of the nucleotide found at the same nucleotide site in each of the three sequences. The outgroup is used to identify the point in time that nucleotide changes

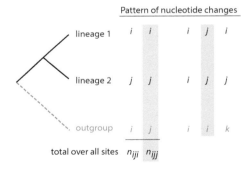

Figure 8.22 Patterns of nucleotide changes that are possible when comparing DNA (or amino acid) sequences from two lineages and an outgroup. The letters i, j, and k are used to represent the identity of the nucleotide found at the same nucleotide site in each of the three sequences. For example, *iij* indicates that the first two lineages have an identical base pair and the third lineage has a different base pair. Tajima's 1D relative rate test utilizes substitutions that can be unambiguously assigned to one lineage (*iji* and *ijj*). If rates of substitution are identical for lineages 1 and 2, then $E(n_{iji}) = E(n_{ijj})$. For the patterns *jji* and *ijk*, the lineage where the substitution took place cannot be determined unambingously. The pattern *iii* indicates identical nucleotides in all three sequences and, therefore, no substitution events.

took place since lineages 1 and 2 should share the same base pair as the outgroup due to identity by descent if no substitution has occurred. Only changes that can be assigned unambiguously to a lineage are useful when comparing rates between lineages 1 and 2. Nucleotide substitutions of the pattern *iji* indicate the change occurred on lineage 2, whereas the pattern *ijj* indicates the change occurred on lineage 1. These two instances allow unambiguous assignment of a substitution to a lineage to estimate the numbers of substitutions. The other three possible nucleotide patterns cannot be used to estimate rates of substitution for one lineage. Nucleotide sites with the pattern *iii* are not useful because no substitution occurred and there is no information available to estimate the rate of change. For nucleotide sites with the pattern *jji*, the substitution to *j* could have occurred in the ancestor to lineages 1 and 2 or both lineages 1 and 2 could have experienced a substitution but it is not clear which event occurred. The pattern *ijk* for a nucleotide site indicates that no two lineages share a nucleotide, so again it is unclear at what point in the past these substitutions occurred and they cannot be used to estimate the rates of substitution for lineages 1 and 2.

Under the molecular clock hypothesis, the number of substitutions that occurred on lineage 1 should be identical to the number of substitutions that occurred on lineage 2. Since the divergence time is identical for lineages 1 and 2, identical substitution rates for the two lineages would give the same number of substitutions observed on each lineage. Therefore, the number of substitutions observed for sequence 1 that occurred on lineage 1 (*ijj*) should be equal to the number of substitutions observed for sequence 2 that occurred on lineage 2 (*iji*):

$$E\left(n_{ijj}\right) = E\left(n_{jij}\right) \qquad (8.55)$$

where E means expected or average value, n_{ijj} is the total number of nucleotide substitutions that occurred on lineage 1, and n_{jij} is the total number of nucleotide substitutions that occurred on lineage 2. This expectation can be tested with the chi-squared statistic

$$\chi^2 = \frac{\left(n_{ijj} - n_{iji}\right)^2}{n_{ijj} + n_{iji}} \qquad (8.56)$$

where there is one degree of freedom. A chi-squared value greater than 3.84 indicates that it is unlikely that the difference in the number of substitutions between the two lineages is due to chance. In other words, a large chi-squared value is evidence to reject the molecular clock hypothesis that substitution rates are equal for the two lineages and is evidence of rate heterogeneity. The chi-square approximation is accurate as long as n_{ijj} and n_{jij} are both 6 or greater.

Tajima's 1D test for equal divergence rates in two taxa is simple to employ because it does not require an explicit nucleotide substitution model. Hamilton et al. (2003) took advantage of this aspect of the 1D test when they compared rates of divergence using both nucleotide and insertion/deletion (indel) variation between species of Brazil nut trees. (Because a range of molecular mechanisms leads to the formation of indels, there are no generally employed models of sequence change by indels and many relative rate tests cannot be used with indel variation.) Comparing substitution rates among eight species with the 1D test, they found that two tree species consistently failed to support a molecular clock for both nucleotide and indel changes. One species (*Lecythis zabucajo*) had an accelerated rate of substitution, whereas the other species (*Eschweilera romeucardosoi*) had a slowed rate of substitution.

Relative rate tests provide no information about rates of molecular evolution in the outgroup taxon nor any information about absolute rates of DNA sequence change. The outcome of relative rate tests depend critically on the outgroup used (Bromham et al. 2000). As the time since divergence of the common ancestor of both taxa and the outgroup increases, so does the time over which the evolutionary rates are averaged. If rate heterogeneity is a short-term or recent phenomenon, then averaging from a distant outgroup may obscure it. Conversely, if rate heterogeneity is only apparent over long time periods, the rate of substitution may appear homogeneous if a recently diverged outgroup is employed. Finally, since natural selection depends on population-specific fitness values, it is considered unlikely that selection acting simultaneously on both lineages subject to a relative rate test would result in rate homogeneity.

Three-taxon relative rate tests that incorporate nucleotide-substitution models and use a maximum likelihood framework are described in Gu and Li (1992) and Muse and Weir (1992). A variety of relative rate tests that utilize phylogenetic trees are also available that test the molecular clock hypothesis using sequences from many taxa simultaneously (see Nei and Kumar 2000; Page and Holmes 1998).

Patterns and causes of rate heterogeneity

Ohta and Kimura (1971) were the first to carry out a test of the Poisson process molecular clock with rigorous statistical comparisons. They used protein sequences from three loci (β globin, α globin, and cytochrome *c*) sampled from a range of species. Based on the observed divergences between pairs of sequences and estimates of the time that has elapsed since those species diverged, they estimated a series of absolute rates of divergences. These absolute rates varied widely (the dispersion index for their data falls between 1.37 and 2.05), leading them to reject the hypothesis of a constant molecular clock (see Gillespie 1991). A few years later, Langley and Fitch (1974) published a larger analysis of absolute substitution rates for the same three loci as well as fibrinopeptide A and used phylogenies to better estimate the number of substitutions for each species. They too found that the dispersion index was greater than one for all loci. These papers attracted a great deal of attention because the variation in rates of sequence change required explanation. Since these early results, a great deal of data on both absolute and relative rates of molecular evolution show

clearly that rates of molecular evolution are commonly more variable than expected by a Poisson process model. In fact, rate heterogeneity is now considered the norm and a constant rate of molecular evolution the exception. This section focuses on hypotheses to explain variation in rates of molecular evolution.

Under neutrality, variation in the rate of divergence at different loci can be explained by differences in rates of mutation. Similarly, variable rates of divergence at the same locus in different species can be explained by different mutation rates among species. Such variation in rates of molecular evolution for the same locus in different species is called a **lineage effect** on the molecular clock (Gillespie 1989, 1991 see Carruthers et al. 2019). There may be rate heterogeneity evident at a locus even after accounting for variation among lineages, called **residual effects**. Residual effects are the variation in the rate of divergence or unevenness in the tick rate of the molecular clock *within* lineages over time (see Figure 8.16). Residual effects are sometimes described as a pattern where substitutions occur in bursts or clusters with periods of no change in between. Another way to think of residual effects is as substitution rate differences for some loci in some lineages, or as lineage by locus interaction variance. The substitution rate variation from lineage variation, locus variation, and lineage by locus interaction is illustrated in Figure 8.23.

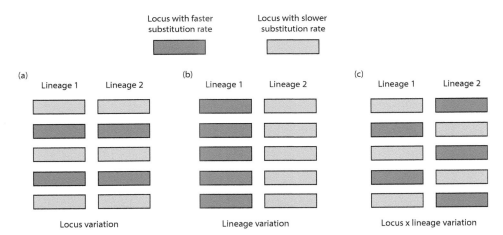

Figure 8.23 Three types of variation in substitution rate, each associated with a different mechanism. Variation among loci in substitution rates (A) can be explained under the neutral theory by loci having different mutation rates. Variation in substitution rates among lineages (B) can be explained by any mechanism common to all loci within a lineage such as uniformly faster mutation rates, or a generation time effect. Locus by lineage interaction variation (C), or residual variation, can be explained by natural selection where loci experience variable selection pressures in different lineages. These three types of substitution rate variation are not mutually exclusive.

Lineage effect: Variation in the rate of divergence among multiple species that could be explained by the different lineages having variable neutral mutation rates.
Replication-independent causes of mutation: Causes of mutation that can occur at any time and are, therefore, independent of the rate of cell division. Examples include environmental mutagens such as ultraviolet radiation, γ particles, and chemicals.
Residual effect: Variation or unevenness in the rate of divergence within a lineage that cannot be explained by rate heterogeneity among lineages or among loci.

Kimura (1983) argued that mutation rates in different species are roughly constant per year. This could be true if the processes that caused mutations were constant over time units like years. Examples are **replication-independent** causes of mutation such as exposure to ultraviolet radiation, γ particles, or chemical mutagens. The free radical ions constantly produced within cells are another example of a replication-independent cause of mutation. It seems likely that exposure to these extrinsic causes of mutation is constant over calendar time and so a portion of mutations due to replication-independent causes have a rate that is also set in calendar time.

Returning to the basis of the neutral theory shows why different species might not experience substitutions at the same rates. As shown at the beginning of the chapter, the neutral theory predicts that the substitution rate is equal to the mutation rate. But since the mutation rate is measured in nucleotide changes per generation, then the substitution rate is also expressed in per generation terms. This leads to the difficulty that a constant molecular clock might not exist if species differ in their generation times. As an example, imagine two species with identical mutation rates of $\mu = 1 \times 10^{-5}$ errors per base pair per generation. Now, imagine the species have generation times of 10 and 100 years. The species with the shorter generation time has

$$\mu = \frac{1 \times 10^{-5} \text{ mutations generation}^{-1}}{10 \text{ years generation}^{-1}} \quad (8.57)$$
$$= 1 \times 10^{-6} \text{ mutations per year}$$

whereas the species with the longer generation time has

$$\mu = \frac{1 \times 10^{-5} \text{ mutations generation}^{-1}}{100 \text{ years generation}^{-1}} \quad (8.58)$$
$$= 1 \times 10^{-7} \text{ mutations per year}$$

Thus, the constant molecular clock per generation predicted by the neutral theory can produce variable rates of substitution per year when comparing species with different generation times.

The observation that neutral mutation rates that are constant per generation may simultaneously be variable per year leads to the **generation-time hypothesis**, a neutral explanation for variation in rates of substitution as caused by differences in generation times of species that have constant rates of substitution per generation. Numerous studies have shown evidence for a generation time effect in rates of substitution (Li et al. 1987, 1996; Ohta 1993, 1995). Substitution rates observed over many nuclear genes in different groups of mammals are shown in Table 8.4.

Rodents have shorter generation times than primates and artiodactyls. Substitution rates are also negatively correlated with generation times. In contrast, comparisons within these groups, such as comparing rates of substitution between mice and rats, show nearly equal substitution rates. The rate's speeding up in rodents compared to primates, and artiodactyls is a classic example of the generation

Table 8.4 Number of substitutions per nucleotide site observed over 49 nuclear genes for different orders of mammals. Divergences are divided into those observed at synonymous and nonsynonymous sites. Primates and artiodactyls (hoofed mammals such as cattle, deer, and pigs with an even number of digits) have longer generation times than do rodents. There were a total of 16 747 synonymous sites and 40 212 nonsynonymous sites.

Mammal group	Synonymous sites	Nonsynonymous sites
Primates	0.137	0.037
Artiodactyls	0.184	0.047
Rodents	0.355	0.062

Source: Data from Ohta (1995).

time effect and is consistent with a neutral explanation for heterogeneity in the rate of molecular evolution.

> **Generation-time hypothesis:** The hypothesis that variation in rates of substitution is due to differences in generation times among species that have constant rates of substitution per generation. This explanation for rate heterogeneity is consistent with neutral molecular evolution.
> **Replication-dependent causes of mutation:** Causes of mutation that occur during replication of DNA, such as replication errors, so that the rate of mutation depends on the rate of cell division.

A generation-time effect can be explained by **replication-dependent** causes of mutation. If mutations occur mostly during the process of cell division when chromosomes are replicated, then more cell replications per generation leads to a higher rate of neutral divergence per generation. In animals, variation in replication-dependent mutation rates per generation may be explained by the fixed number of cell divisions leading to germ-line cells (cells that produce gametes). This explains the observation that mutations occur more frequently in male gametes than in female gametes since more germ-line cell divisions occur in males than in females. The generation-time effect in animals could then be explained if generation times are correlated with the number of germ-line cell divisions (e.g. animals with longer generation times have more germ-line cell divisions). Yet, plants with shorter time intervals to first flowering have been shown to have higher rates of substitution (Gaut 1998; Kay et al. 2006). Variation in rates of molecular evolution in plants suggests that germ-line cell divisions is not the only explanation for rate heterogeneity because plants do not have separate germ and somatic cell lines.

The **metabolic rate hypothesis** proposed by Martin and Palumbi (1993; reviewed by Rand 1994) was based on the observation that sharks have rates of synonymous substitution five to seven times lower than that observed in primates and artiodactyls despite the fact that all taxa examined have relatively similar generation times. Mutation rates may be correlated with metabolic rate of organisms for several reasons. Organisms with high metabolic rates have rapidly operating cellular functions and one of these cellular functions is DNA replication. Therefore, high rates of metabolism cause high rates of DNA replication and high rates of replication-dependent mutation. Alternatively, aerobic respiration within cells produces free oxygen radicals that cause oxidative damage of DNA. Therefore, high metabolism increases the rate of exposure to mutagenic agents and increases replication-independent rates of mutation. These two mechanisms that couple metabolic rate and mutation rates are not mutually exclusive, and both can occur at the same time. Gillooly et al. (2005) proposed a model of the substitution rate that explicitly includes effects of body size and temperature and suggested that the molecular clock may indeed be constant after accounting for rate variation from these causes.

Heterogeneity in the rate of divergence can also be explained by the nearly neutral theory (reviewed by Ohta 1992). To see this, let f_0 stand for the proportion of mutations that are selectively neutral because the pressure of negative selection acting on them is weak relative to the effective population size. (All mutations are assumed to be deleterious and advantageous mutations so rare they can be ignored, an assumption of the nearly neutral theory that is problematic (see Gillespie 1995)) The remaining $(1 - f_0)$ mutations have a large enough deleterious effect that they are acted on by negative selection and are not neutral. The rate of substitution of neutral mutations under the nearly neutral theory is then

$$k = f_0\mu \qquad (8.59)$$

analogous to Eq. 8.3 for the neutral theory. This equation says that the rate of divergence will be higher when more mutations are effectively neutral (f_0 is larger) and lower when fewer mutations are effectively neutral (f_0 is smaller).

Because the proportion of mutations that are effectively neutral depends on the effective population size, the rate of divergence also varies with the effective population size under the nearly neutral theory. In the nearly neutral theory, all substitutions are the result of genetic drift, as in the neutral theory. But a larger effective population size leads to fewer mutations that are effectively neutral and, therefore, a smaller pool of neutral mutations that can ultimately reach fixation. In contrast, a smaller effective population size leads to more mutations being effectively neutral and, therefore, a larger pool of neutral mutations that can ultimately experience substitution. Thus, the nearly neutral theory predicts that the rate of divergence is negatively correlated with the effective population size because changes in N_e result in changes in f_0. Under the nearly neutral theory, rate heterogeneity can then be explained by different effective population sizes among lineages or loci that cause f_0 to vary.

Generation time may also influence perceived variation in substitution rates among species under the nearly neutral theory. In the nearly neutral theory, both mutation and substitution rates are expressed in per-generation terms as they are in the neutral theory. This should lead to generation-time effects on the substitution rate just as in the neutral theory. However, generation time effects may be canceled out under the nearly neutral theory because of a negative correlation between generation time and effective population size. In the nearly neutral theory, the proportion of mutations that are effectively neutral depends on the effective population size. Independently, longer generation times lead to fewer substitutions per year, whereas shorter generation times result in more substitutions per year if the mutation rate is constant per generation. The effective population size and the generation time should act independently on substitution rates. However, it turns out that generation time and effective population size are not generally independent (Chao and Carr 1993). For example, mice have short generation times and a large effective population size, whereas elephants have long generation times and a small effective population size. Therefore, the impacts of generation time and effective population size tend to cancel each other out, resulting in a nearly neutral theory prediction that substitution rates do not show a generation time effect.

It is possible to test the nearly neutral theory prediction that the impacts of generation time and effective population size on rates of molecular evolution tend to cancel each other out. Ohta (1995) carried out such a test by comparing rates of substitution at synonymous and nonsynonymous sites for 49 genes in primates, artiodactyls, and rodents (recall from earlier in this section that divergence rates for these same animals support the generation time hypothesis). Ohta divided the DNA sequence data into divergence observed at synonymous or nonsynonymous sites within exons. Mutations at nonsynonymous sites are exposed to natural

selection since they alter the amino acid sequence of a protein and, therefore, have a phenotypic effect. In contrast, synonymous site mutations are not perceived by natural selection (or selection is much weaker) since they do not alter the amino acid sequence. The nearly neutral theory predicts that nonsynonymous substitution rates should be lower than synonymous substitution rates because of the negative selection on the pool of nonsynonymous mutations (nonsynonymous f_0 is smaller). In addition, divergence rates at nonsynonymous sites should not exhibit a generation time effect because of the negative correlation between generation time and effective population size for mutations that are nearly neutral. Table 8.4 shows Ohta's divergence data for primates, artiodactyls, and rodents at synonymous and nonsynonymous sites. Synonymous substitution rates are an order of magnitude greater than nonsynonymous rates, as expected if nonsynonymous sites experience frequent negative selection against mutations. The synonymous rate is 2.59 times faster for rodents than for primates. In contrast, the nonsynonymous rate is 1.68 times faster for rodents than for primates. Therefore, divergence at nonsynonymous sites shows less of a generation-time effect, also consistent with the nearly neutral theory.

8.6 Testing the neutral theory null model of DNA sequence polymorphism

- The Hudson–Kreitman–Aguadé (HKA) test.
- The McDonald–Kreitman (MK) test.
- Mismatch distributions.
- Tajima's D

This section provides the opportunity to apply the conceptual results of the neutral theory developed earlier in the chapter to test the neutral null model for the causes of molecular evolution. Some tests take advantage of the neutral theory predictions for levels of polymorphism and divergence, while others rely on coalescent model results that were developed in earlier chapters. These tests have been widely employed in empirical studies of DNA sequences sampled from a wide array of loci, genomes, and species (see review by Ford 2002). The tests described in this section have contributed much to our knowledge of how natural selection has acted on DNA sequences as well as our understanding of how multiple population genetic processes (mating, gene flow, genetic drift, mutation, changes in N_e, and natural selection) interact in natural populations.

HKA test of neutral theory expectations for DNA sequence evolution

The HKA test, so named after its authors Hudson, Kreitman, and Aguade (Hudson et al. 1987), is a test that compares neutral theory predictions for DNA sequence evolution with empirically estimated polymorphism and divergence. The test utilizes the expectation that under neutrality both polymorphism within species and divergence between species are a product of the mutation rate. In fact, under neutral evolution levels of polymorphism and divergence at a locus should be correlated because they are both products of the very same mutations. If a locus has a high mutation rate, for example, then the population should be highly polymorphic (see Eq. 8.1). At the same time, divergence at that locus when compared with another species should also be substantial since the rate of substitutions that contribute to divergence is also equal to the mutation rate (see Eqs. 8.2 and 8.3). Alternatively, a combination of both low polymorphism and low levels of divergence should be apparent if a neutral locus has a low mutation rate. In this way, expected levels of polymorphism and divergence are not independent under neutrality. Evidence that divergence and polymorphism are not correlated would be at odds with neutral expectations and, therefore, evidence to reject the neutral null model for the locus under study.

The HKA test requires DNA sequence data from two loci. One locus is chosen because it is selectively neutral and serves as a reference or control locus. Examples of a neutral reference locus include noncoding regions of the genome or duplicate copies of genes that are not functional (pseudo-genes), both of which are expected to be relatively free of functional constraints on nucleotide substitutions. The other locus used is the focus of the test and the locus for which the neutral null model of evolution is being tested.

The HKA test also requires that DNA sequence data for two loci be collected in a particular manner. First, DNA sequences for two loci must be obtained from two species to estimate divergence between the species for both the neutral reference and the test loci. In addition, DNA sequences from multiple individuals within one of the species need to be obtained to estimate levels of polymorphism present at both loci. Polymorphism is measured by nucleotide diversity (π) for each locus. Divergence is estimated by comparing the DNA sequences for both loci between an individual of each species, employing a nucleotide substitution model to correct for homoplasy.

Once the estimates of polymorphism and divergence are made from DNA sequence data, they can be compared in a format like that shown in Table 8.5. Panel a in Table 8.5 shows the neutral theory expectations for polymorphism and divergence at the two loci. Under neutrality and the infinite sites model, DNA sequence polymorphism is expected to be $\theta = 4N_e\mu$ and divergence is expected to be $k = 2T\mu$. The test and reference locus may have different mutation rates. But, note that the effective population size is constant when polymorphism is estimated for the two loci since the loci are sampled from the same species. The divergence times are also equal for the two loci since they are estimated from the same species pair. The ratio of the two divergence estimates at the test and reference loci is expected to equal the ratio of the test locus mutation rate over the reference locus mutation rate ($\frac{\mu_T}{\mu_R}$) since the factor of $4N_e$ cancels out. The ratio of the two divergence estimates at the test and reference loci is also expected to equal $\frac{\mu_T}{\mu_R}$. Therefore, under neutrality, the ratio of polymorphism estimates at the two loci as well as the ratio of the divergence estimates at the two loci should be equal since they both represent ratios of the mutation rates at the two loci. Similarly, the ratios of polymorphism over divergence for each locus are expected to be equal under neutrality. The ratios can be tested for equality using a chi-square test.

Table 8.5b shows an idealized illustration of polymorphism and divergence estimates that would be consistent with the neutral null model of DNA sequence evolution. In this idealization, the two loci do have different mutation rates that lead to different amount of polymorphism and divergences. However, the ratios are equal as expected if the fate of mutations is due to genetic drift only.

Table 8.5c shows a classic example of divergence and polymorphism estimated in fruit flies to carry out the HKA test (Hudson et al. 1987). The locus tested for neutral evolution is the gene for alcohol dehydrogenase (*Adh*) and the reference locus is sequence upstream (5′) to the coding region that does not possess an open reading frame. Polymorphism was estimated for the two genes from a sample of *D. melanogaster* individuals and divergence for the two loci

Table 8.5 Estimates of polymorphism and divergence for two loci sampled from two species that form the basis of the HKA test. The correlation of polymorphism and divergence under neutrality results in a constant ratio of divergence and polymorphism between loci independent of their mutation rate as well as a constant ratio of polymorphism or divergence between loci (A). Case B shows an illustration of ideal polymorphism and divergence estimates that would be consistent with the neutral null model. Data for the *Adh* gene and flanking region (Hudson et al. 1987) in case C is not consistent with the neutral model of sequence evolution because there is more *Adh* polymorphism within *Drosophila melanogaster* than expected relative to flanking region divergence between *D. melanogaster* and *D. sechellia*.

A. Neutral case expectations

	Test locus	Neutral reference locus	Ratio (test/reference)
Focal species polymorphism (π)	$4N_e\mu_T$	$4N_e\mu_R$	$\dfrac{4N_e\mu_T}{4N_e\mu_R} = \dfrac{\mu_T}{\mu_R}$
Divergence between species (K)	$2T\mu_T$	$2T\mu_R$	$\dfrac{2T\mu_T}{2T\mu_R} = \dfrac{\mu_T}{\mu_R}$
Ratio (π/K)	$\dfrac{4N_e\mu_T}{2T\mu_T} = \dfrac{4N_e}{2T}$	$\dfrac{4N_e\mu_R}{2T\mu_R} = \dfrac{4N_e}{2T}$	

B. Neutral case illustration

	Test locus	Neutral reference locus	Ratio (test/reference)
Focal species polymorphism (π)	0.10	0.25	0.40
Divergence between species (K)	0.05	0.125	0.40
Ratio (π/K)	2.0	2.0	

C. Empirical data from *Drosophila melanogaster* and *D. sechellia*

	Adh	5' *Adh* flanking region	Ratio (*Adh*/flank)
D. melanogaster polymorphism (π)	0.101	0.022	4.59
Between species divergence (K)	0.056	0.052	1.08
Ratio (π/K)	1.80	0.42	

was determined with sequences from *Drosophila sechellia*. If the 5' flanking region is truly neutral, then the *Adh* data show too much polymorphism within *D. melanogaster*. An excess of *Adh* polymorphism is also indicated by the large ratio of polymorphism for the two loci within *D. melanogaster* compared with the ratio of divergences for the two loci. It is now widely accepted that the *Adh* locus in *D. melanogaster* exhibits an excess of polymorphism consistent with balancing selection.

Although the HKA test is ingenious, it does have some limitations and assumptions. One difficulty in practice is the ability to identify an unambiguously neutral reference locus. For example, the 5' flanking region used by Hudson et al. (1987) as a neutral reference locus very likely contains promoter sequences that are functionally constrained by natural selection. Innan (2006) described a modification to the HKA test to use the average of multiple reference loci.

Implicit in the HKA test is the assumption that each of the two species used is panmictic. Population

subdivision has the potential to alter levels and patterns of nucleotide polymorphism and divergence (see review by Charlesworth et al. 2003) depending on how individuals are sampled. Consider levels of polymorphism in a subdivided species where F_{ST} is greater than 0. Population subdivision causes lower polymorphism for individuals sampled within subpopulations due to both reduced effective population size that increases drift within demes and increased autozygosity due to a higher probability of mating within demes. In contrast, there will be larger genetic differences for individuals sampled from two different demes due to differentiation among demes that would result in high perceived levels of polymorphism. If the HKA test is carried out for a species with population structure, sampling needs to be conducted to avoid taking sequences from only one or a few demes that could lead to an erroneous conclusion of too little polymorphism compared to the neutral expectation. Ingvarsson (2004) showed how the HKA test can lead to incorrect rejection of the neutral null hypothesis when there is population subdivision. That paper also gives an example of a population-structure-corrected HKA test applied to organelle DNA sequence data from the plant species *Silene vulgaris* and *Silene latifolia* which both exhibit strong population structure.

The McDonald–Kreitman (MK) test

The MK test is a test of the neutral model of DNA sequence divergence between two species (McDonald and Kreitman 1991). Like the HKA test, it is named after its authors, McDonald and Kreitman. The MK test is also conceptually similar to the HKA test because it too establishes expected ratios of two classes of DNA changes at a single locus under neutrality. The MK test requires DNA sequence data from a single coding gene. The sample of DNA sequences is taken from multiple individuals of a focal species to estimate polymorphism. The test

also requires a DNA sequence at the same locus from another species to estimate divergence.

The neutral expectations for the MK test are given in Table 8.6. The two classes of DNA change used in the MK test are **synonymous** and **nonsynonymous** (or replacement) changes. Nonsynonymous mutations within coding regions may alter the amino acid specified by a codon. Due to the redundancy of the genetic code, some mutations within coding regions will not change the amino acid specified by the codon and are, therefore, synonymous changes.

If genetic drift is the only process influencing the fate of a new mutation, levels of polymorphism and divergence within each category of DNA change should be correlated because they are both determined in part by the mutation rate. Fixed differences between species are caused by mutations that have gone to fixation, with expected divergence under the

Table 8.6 Estimates of polymorphism and divergence (fixed sites) for nonsynonymous and synonymous sites at a coding locus form the basis of the MK test. Under neutrality, the number of nonsynonymous sites divided by the number of synonymous sites is equal to the ratio of the nonsynonymous and synonymous mutation rates. This ratio should be constant for both nucleotide sites with fixed differences between species and polymorphic sites within the species of interest (A). Case B shows an illustration of ideal nonsynonymous and synonymous site changes that would be consistent with the neutral null model. Data for the *Adh* locus in *Drosophila melanogaster* (McDonald and Kreitman 1991) in C show an excess of *Adh* nonsynonymous polymorphism compared to that expected based on divergence. Data for the *Hla*-B locus for humans show an excess of polymorphism and more nonsynonymous than synonymous changes consistent with balancing selection (Garrigan and Hedrick 2003).

A. Neutral case expectations

	Fixed differences	Polymorphic sites
Nonsynonymous sites (*N*)	$N_F = 2T\mu_N$	$N_P = 4N_e\mu_N$
Synonymous sites (*S*)	$S_F = 2T\mu_S$	$S_P = 4N_e\mu_S$
Ratio (*N/S*)	$\dfrac{N_F}{S_F} = \dfrac{2T\mu_N}{2T\mu_S} = \dfrac{\mu_N}{\mu_S}$	$\dfrac{N_P}{S_P} = \dfrac{4N_e\mu_N}{4N_e\mu_S} = \dfrac{\mu_N}{\mu_S}$

B. Neutral case illustration

	Fixed differences	Polymorphic sites
Nonsynonymous changes	4	15
Synonymous changes	12	45
Ratio	0.33	0.33

C. Empirical data from *Adh* locus for *Drosophila melanogaster* (McDonald and Kreitman 1991)

	Fixed differences	Polymorphic sites
Nonsynonymous changes	7	2
Synonymous changes	17	42
Ratio	0.412	0.048

D. Empirical data for the *Hla*-B locus for humans (Garrigan and Hedrick 2003)

	Fixed differences	Polymorphic sites
Nonsynonymous changes	0	76
Synonymous changes	0	49
Ratio	–	1.61

neutral theory of $2T\mu$. Nucleotide sites that have two or more nucleotides within the focal species exhibit polymorphism, with an expected level of $4N_e\mu$ under neutral theory. Since synonymous and nonsynonymous mutations may occur at different rates, we can assign each category of DNA change a different rate (μ_N and μ_S). Both the ratio of nonsynonymous and synonymous fixed differences and the ratio of nonsynonymous and synonymous polymorphic sites are expected to be equal to μ_N/μ_S under neutral theory. The MK test, therefore, compares these two ratios for equality as a test of neutral theory. The neutral case illustration in Table 8.6b gives an example where $\mu_N < \mu_S$ and there is a higher level of polymorphism than divergence. Nonetheless, the ratios of the number of nonsynonymous over synonymous changes are constant for fixed differences and polymorphic sites as expected if both classes of mutations are neutral.

An MK test based on numbers of synonymous and nonsynonymous changes at the *Adh* locus that were fixed between *Drosophila* species or polymorphic within *D. melanogaster* (McDonald and Kreitman 1991) is given in Table 8.6c. Using fixed sequence differences as a reference point, fewer substitutions between species are nonsynonymous than synonymous. The rate of nonsynonymous substitutions is 41.2% of the substitution rate for synonymous substitutions. Under neutrality, we expect about 41% of polymorphic sites within *D. melanogaster* to be at nonsynonymous sites. In contrast, the observed data show that only about 4.5% of polymorphic sites are nonsynonymous. Thus, polymorphic sites have too many synonymous changes or too few nonsynonymous changes to be consistent with neutral levels of polymorphism. (Note that if using polymorphic sites as the frame of reference, then in this case there is an elevated rate of divergence at nonsynonymous sites compared to neutral expectations.)

A common observation in studies of coding DNA sequences is that the numbers of nonsynonymous and synonymous DNA changes are not equal. A neutral explanation for this pattern is that these two types of DNA change have different underlying mutation rates. It is expected that nonsynonymous changes will be more frequent than synonymous changes if mutations occur at random nucleotide sites. In fact, 96% of nucleotide changes in the first nucleotide position of a codon, all changes in the second position and 30% of changes at the third position are nonsynonymous. Overall, if mutation occurs at random within coding sequences, 75.3% of all mutations will be nonsynonymous and 24.7% will be synonymous.

An alternative explanation is that rates of synonymous and nonsynonymous mutation are roughly equal, but that nonsynonymous mutations commonly alter proteins in ways that impair their function. Nonsynonymous mutations that disrupt function also reduce fitness and are, therefore, acted against by purifying natural selection. It is also possible that some nonsynonymous mutations result in enhancement of function and are fixed rapidly by positive selection. A third, nonneutral alternative is that nonsynonymous mutations are maintained at intermediate frequencies in the population by balancing selection. An example of strong balancing selection is the human leukocyte antigen (*Hla*) B gene in humans as detected with an MK test by Garrigan and Hedrick (2003) using divergence between humans and chimpanzees (Table 8.6d). There are no fixed DNA differences between humans and chimpanzees for this locus, suggesting low rates of mutation since divergence of these two species. In contrast, the human populations show high levels of polymorphism and 1.6 nonsynonymous changes for every synonymous change, inconsistent with the neutral hypothesis that polymorphism and divergence are correlated. *Hla* genes form the major histocompatibility complex (MHC) region that encodes cell-surface antigen-presenting proteins important in immune system function. Heterozygotes for these loci have higher fitness since they present more diverse cell-surface antigens.

Mismatch distributions

The previous section explored how a sample of DNA sequences from a population can be used to test neutral expectations by comparing estimates of θ based on polymorphism measured with nucleotide diversity (π) and the number of segregating sites (S). Both π and S summarize the patterns of sequence variation into a single number. Specifically, the nucleotide diversity π is really an average of the differences between all pairs of sequences in a sample. Instead of using an average to measure polymorphism, we can directly examine the distribution of all individual pairwise sequence comparisons. This is commonly called the **mismatch distribution**, and it is the frequency distribution of the number of nucleotide sites that differ between all unique pairs of DNA sequences in a sample. The mismatch distribution is a tool that can be used to infer the history of the population that gave rise to a sample of DNA sequences. It can be used to infer past changes in the effective size of a population using selectively neutral DNA sequences. Alternatively, in populations that have maintained a constant size over time, these distributions can be used to identify the action of natural selection.

> **Mismatch distribution:** The frequency distribution of the number of nucleotide sites that differ between all unique pairs of DNA sequences in a sample from a single species. It is also known as the distribution of pairwise differences.
>
> **Haplotype frequency distribution:** The distribution of the frequency of each sequence haplotype in a population assuming that individuals are haploid or homozygous. It is also known as the site frequency spectrum.

Let's assume complete neutrality of mutations to focus on using mismatch distributions to develop expectations for patterns of DNA sequence differences in stable, growing, and shrinking populations. The properties of the mismatch distribution arise directly from expected patterns that characterize neutral genealogies. Chapter 3 shows that the last pair of lineages ($k = 2$) takes the longest average time to coalesce in standard neutral genealogies for populations with constant N_e. When there is mutation, the two oldest lineages in the population also differ by the largest number of mutations since the expected number of mutations is proportional to the length of time a lineage exists. In populations that maintain constant N_e, these oldest two lineages experience numerous mutations and, therefore, have a high degree of mismatch. This pattern of long lineages having multiple mutations can be seen in the genealogy in Figure 8.24.

Working from the past to the present, the two oldest lineages in any genealogy give rise to additional lineages. The younger progeny lineages inherit all

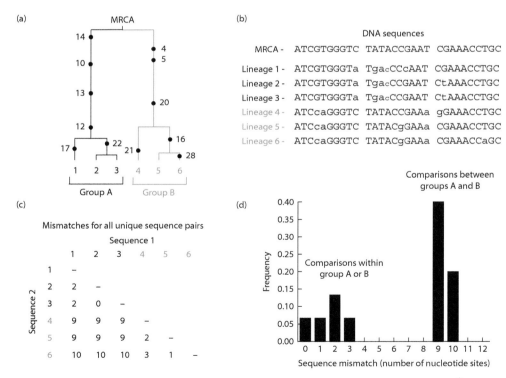

Figure 8.24 Estimates of the scaled mutation rate θ are estimated differently using nucleotide diversity ($\hat{\theta}_\pi$) and the number of segregating sites ($\hat{\theta}_S$) depending on the location of mutations in a genealogy. Each mutation makes a single segregating site under the assumptions of the infinite alleles model no matter where it occurs. However, mutations on internal branches will appear in multiple pairwise comparisons and cause π to be larger (A). In contrast, mutations that occur on external branches (B) that cause a nucleotide change in only a single lineage contribute less to π. Each mutation is counted four times (d_{13}, d_{23}, d_{14}, and d_{24}) in A but three times (d_{12}, d_{23}, d_{24}) in B when computing π.

the mutations that have occurred on the progenitor lineages and may also experience additional new mutations. Since the lineages closer to the present tend to have shorter times to coalescence (the probability of coalescence increases with larger k), they also tend to accumulate fewer mutations. Looking at Figure 8.23, the three lineages within group A would each inherit the four mutations that occurred on the internal branch that was their ancestor. Because lineages 1, 2, and 3 within group A share the mutations of their ancestral lineage, they also tend to have fewer nucleotide sites that mismatch. For example, lineages 1 and 2 differ by only the two mutations that occurred near the present (mutations at nucleotide sites 17 and 22). Lineages 4, 5, and 6 within group B also have low levels of mismatch by the same logic.

In contrast, the level of sequence mismatch is high when lineages are compared between groups A and B, as shown in Figure 8.24. For example, lineages 1 and 4 differ by nine mutations. This high level of mismatch occurs because sequences from distantly related lineages are separated by much more time since they shared a common ancestral lineage, leading to many more mutational changes that independently altered each DNA sequence. Another way to think of the situation is that closely related lineages differ only by a few young mutations, while distantly related lineages differ by more mutations, many of which are old and have been resident in the population for a long time.

The mismatch distribution has distinct patterns depending on the demographic history of the population (Slatkin and Hudson 1991; Rogers and Harpending 1992). Mismatch distributions from populations that have experienced a constant N_e over time tend to have two clusters of values in the mismatch distribution. Such a bimodal distribution is the characteristic signature of genealogies in populations with a relatively constant N_e in the past. The bimodal pattern is caused by roughly equal times to coalescence of all internal and external branches. In contrast, populations that had rapidly growing or shrinking N_e in the past tend to have distinct mismatch distributions. In populations that have rapidly growing N_e, most coalescence events happen early in the genealogy near the MRCA since the probability of coalescence decreases toward the present (see the left-hand genealogy in Figure 8.26). This leads to long external branches that each experience many unique mutations. The mismatch distribution then has a high frequency

of sequence pairs with a high degree of mismatch and few sequence pairs with a low degree of mismatch. Alternatively, populations that experienced continual declines in N_e have genealogies where most coalescence events happen near the present because the probability of coalescence increases toward the present (see the right-hand genealogy in Figure 8.26). In a shrinking population, the mismatch distribution tends to have a high frequency of sequence pairs with low mismatch counts.

A related way to view polymorphism is by examining the distribution of haplotype frequencies in a sample of sequences. Such **haplotype frequency distributions** show the proportion of sequences in a population that represent each of the observed sequence alleles (assuming that individuals are haploid or homozygous). Under neutrality and constant effective population size (see Figure 8.26), a range of haplotype frequencies are expected from very frequent to rare. When populations are growing rapidly or there is balancing selection, there is expected to be an excess of rare haplotypes produced by the excess length of external branches in the genealogy. When populations are shrinking rapidly or there is strong directional selection, there is expected to be an excess of high-frequency haplotypes and very few rare haplotypes because most of the branch length lie in the internal branches of the genealogy.

Mismatch and haplotype frequency distributions can help identify instances of expansion or contraction in the effective population size if sequences are neutral. Alternatively, if sequences are known to come from a population with a constant effective size, then these distributions can be used to identify the action of natural selection. Several tests are available that use the haplotype frequency or mismatch distributions to evaluate the null hypothesis of constant effective population size through time using DNA sequences (Fu and Li 1993; Fu 1996, 1997; Schneider and Excoffier 1999; Mousset et al. 2004; Innan et al. 2005). It is important to note several limitations of these tests. First, recombination has the potential to impact the mismatch distribution along with population demography. Recombination events assemble novel sequence haplotypes from existing haplotypes and in doing so break up mutations that are associated due to identity by descent. Therefore, recombination obscures the history of mutations and in the extreme would lead to a uniform mismatch distribution. Second, coalescence is a stochastic process, and there is an inherently large variance in times to coalescence (see

Chapter 3). This leads to a large variance in the shape of mismatch distributions even when N_e is constant. Therefore, tests that utilize the mismatch distribution can only be expected to detect very large and sustained shrinkage or expansion of N_e.

Tajima's D

Tajima's D is a test of the standard coalescent model (neutral alleles in a population of constant size) that is commonly applied to DNA polymorphism data sampled from a single species (Tajima 1989a, b). The test uses the nucleotide diversity and the number of segregating sites observed in a sample of DNA sequences to make two estimates of the scaled mutation rate $\theta = 4N_e\mu$. This section will refer to an estimate of θ based on the nucleotide diversity as $\hat{\theta}_\pi$ and an estimate of θ based on the number of segregating sites as $\hat{\theta}_S$. Tajima's D test relies on the fact that $\hat{\theta}_\pi$ and $\hat{\theta}_S$ are expected to be approximately equal under the standard coalescence model where all mutations are selectively neutral and the population remains a constant size through time. The null hypothesis of the test is that the sample of DNA sequences was taken from a population with constant effective population size and selective neutrality of all mutations. Natural selection operating on DNA sequences as well as changes in effective population size through time lead to rejection of this null hypothesis.

Tajima's D takes advantage of the fact that mutations that occurred further back in time in a genealogy are counted more times when computing the nucleotide diversity (π) from all unique pairs of sequences. In contrast, the position of a mutation on a genealogy does not influence the number of segregating sites (S) since any number of sequences bearing a given nucleotide always represents just one segregating site (Figure 8.25). The coalescent process with genetic drift acting alone when there is constant effective population size results in approximately the same total length along interior and exterior branches in a genealogy (with substantial stochastic variation). In contrast, processes that alter the probability of coalescence also change the ratio of interior and exterior branch length, changing π and S and thereby changing Tajima's D (Figure 8.26). A common interpretation of Tajima's $D < 0$ is that directional selection or a selective sweep has acted (see Braverman et al. 1995), and when Tajima's $D > 0$, then balancing selection on a locus with

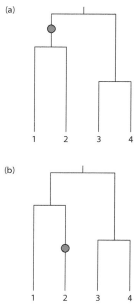

Figure 8.25 Differences in the shape of genealogies are the basis of Tajima's D test. In the standard coalescent model of genealogical branching, the probability of coalescence is constant per lineage over time. The standard coalescent therefore gives expected branch lengths when all alleles are selectively neutral and the effective population size is constant (center). Changes in the effective population size over time (population growth, population bottlenecks) change the probability of coalescence over time as well. Natural selection also alters the probability of coalescence based on the fitness of alleles each lineage bears. Changes in the effective population size and natural selection alter the expected time to coalescence and therefore the expected branch lengths in a genealogical tree. If the chance of coalescence is greater in the present than in the past (right), most coalescent events occur near the present and internal branches are long in comparison to external branches. If the chance of coalescence is smaller in the present than in the past (left), most coalescent events occurred in the past and external branches are long in comparison to internal branches. Since the chance of a mutation is constant over time, lineages with longer branches are expected to experience more mutations.

two alleles has acted. However, both sustained increases in the effective population size over time and balancing selection for three or more alleles can lead to decreasing probabilities of coalescence toward the present time and result in longer external branches. Both strongly shrinking effective population size over time and population bottlenecks are expected to cause increasing probabilities of coalescence toward the present time and, therefore, shorter external branches. In addition, population divergence can impact Tajima's D, with longer external branches in a genealogy if lineages are sampled from numerous differentiated demes, or longer

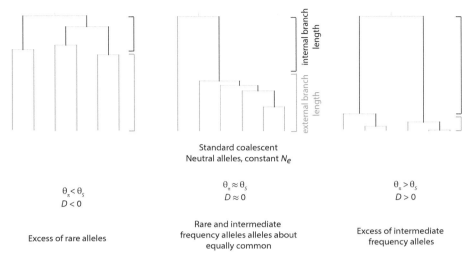

internal branch length

external branch length

Standard coalescent
Neutral alleles, constant N_e

$\theta_\pi < \theta_S$
$D < 0$

$\theta_\pi \approx \theta_S$
$D \approx 0$

$\theta_\pi > \theta_S$
$D > 0$

Excess of rare alleles

Rare and intermediate
frequency alleles alleles about
equally common

Excess of intermediate
frequency alleles

Figure 8.26 The basis of the mismatch distribution. Panel A shows a neutral genealogy that bears multiple mutation events. Each mutation event is represented by a circle and the number of the random nucleotide site that mutated assuming the infinite sites mutation model. The six lineages in the present can be separated into two groups (called A and B) based on their ancestral lineage when there were only two lineages in the population. The DNA sequences for each lineage are shown in B based on the 30 base pair sequences assigned to the most recent common ancestor (MRCA) with mutations shown in lowercase letters. Panel C shows the number of nucleotide sites that are different or mismatch between pairs of DNA sequences. The mismatch distribution shown in panel D is a histogram of the mismatches for the 15 pairs of DNA sequences compared. Neutral genealogies from populations with constant N_e through time tend to show bimodal mismatch distributions. The cluster of observations with few mismatches results from sequence comparisons between recently related lineages (comparisons within group A or group B). In contrast, sequences from distantly related lineages that do not share the same ancestor when $k = 2$ (comparisons between groups A and B) tend to have more mismatches.

interior branches if lineages are sampled from two differentiated demes.

An alternative way to think about how Tajima's D works is to consider the frequency distribution of alleles under different types of natural selection or population demography. Mutations that happen to occur on internal branches in a genealogy have an intermediate frequency because they are inherited by lineages that arise later in time. In contrast, mutations that happen to occur on external branches have a low frequency since they are unique to a single lineage. Since total internal and total external branch length are expected to be about equal under the standard coalescent model, intermediate and rare alleles are also expected to be about equal in frequency. Both strong population growth and multiallelic balancing selection can lead to an excess of rare mutations since these processes increase the external branch length. In contrast, strong purifying selection, strongly shrinking population size, or a population bottleneck can lead to an excess of intermediate frequency mutations because these processes increase the amount of internal branch length.

Tajima's D statistic is computed from the difference between $\hat{\theta}_\pi$ and $\hat{\theta}_S$ divided by the standard deviation of

$$D = \frac{\hat{\theta}_\pi - \hat{\theta}_S}{\sqrt{\mathrm{var}(\hat{\theta}_\pi - \hat{\theta}_S)}} = \frac{\hat{\theta}_\pi - \frac{p_S}{a_1}}{\sqrt{e_1 p_S + e_2 p_S (p_S - 1)}}$$

(8.60)

where p_S is the number of segregating sites per nucleotide site. Recall that the standard deviation is the square root of the variance, so that dividing by the standard deviation puts D in units of standard deviations away from the mean of 0 expected for standard coalescent genealogies. Only when the observed result is about two standard deviations away from the mean do we reject the null hypothesis of $D = 0$ and thereby reject the null model of a neutral genealogy with constant effective population size (see confidence limits in Table 2 of Tajima 1989a).

The quantities used to compute the variance are

$$e_1 = \frac{n+1}{3a_1(n-1)} - \frac{1}{a_1^2}$$

(8.61)

and

$$e_2 = \frac{c}{a_1^2 + a_2} \tag{8.62}$$

where

$$a_1 = \sum_{k=1}^{n-1} \frac{1}{k} \tag{8.63}$$

$$a_2 = \sum_{k=1}^{n-1} \frac{1}{k^2} \tag{8.64}$$

and

$$c = \frac{2(n^2 + n + 3)}{9n(n-1)} - \frac{n+2}{a_1 n} + \frac{a_2}{a_1^2} \tag{8.65}$$

where n is the number of sequences sampled and assuming that there is no recombination.

Although the variance of D is a complex expression, it can still be understood intuitively. The formula includes both **sampling** and **evolutionary variance**. Sampling variance comes from taking a sample of DNA sequences and using them to estimate π and S. As with any finite sample of data from a larger underlying population, repeating the sampling procedure would result in a slightly different estimate of the parameters of interest because the sample is not perfectly representative of the full population. Sampling variance decreases as sample sizes increase since estimates are based on a larger and larger proportion of the underlying population. In contrast, evolutionary variance is caused by the variable outcomes of the random evolutionary processes of genetic drift and mutation. Evolutionary variance can only be estimated by sampling multiple independent realizations of the same random process, for example, taking samples of DNA sequences from multiple populations that independently experienced genetic drift after being isolated from the same ancestral population. The coalescence and

---

Problem box 8.2 Computing Tajima's *D* from DNA sequence data

To study the population history of *D. simulans*, Baudry et al. (2006) sampled flies from multiple populations in Africa, Europe, and the Antilles. From these flies, they sequenced four genes located on the X chromosome. Using part of their DNA sequence data, test the hypothesis that *D. simulans* meets the assumptions of the standard neutral coalescent model via Tajima's *D*.

DNA sequences from the *runt* locus for flies sampled in Europe and Mayotte (an overseas collectivity of France composed of several islands in the Indian Ocean, between northern Madagascar and northern Mozambique) exhibited the following patterns:

Population	*n* sequences	Nucleotide sites	*S*	π
Europe	15	556	17	0.012 436
Mayotte	15	538	34	0.013 525

Use the number of segregating sites (*S*) to calculate the number of segregating sites per nucleotide site (p_S) and then estimate $\hat{\theta}_S$ per site according to Eq. 8.32. Then compute Tajima's *D* according to Eq. 8.60.

What do you conclude about the history of these two *D. simulans* populations? Note that your estimates of Tajima's *D* will differ from those in Baudry et al. (2006) because they used only synonymous site polymorphisms whereas you have used polymorphisms at all sites. Why did Baudry et al. (2006) use only synonymous site polymorphisms? The DNA sequence data files are available on the text website.

mutation processes each have a great deal of evolutionary variance. For example, under the standard coalescent model, there is a wide range of coalescence times for k lineages around an average with variance in coalescence time that is the largest for two lineages (see Section 3.6).

The value of Tajima's D is influenced by changes over time in the size of populations, population structure, and the action of natural selection. Therefore, Tajima's D is not a simple test for the action of natural selection alone as is sometimes assumed (see Fu 1997; Li 2011). The null model is based on a constant mutation rate through time (a molecular clock), the infinite sites model of mutation, the Wright–Fisher model with nonoverlapping generations, and a panmictic population at drift–mutation equilibrium (see Tajima 1996 on the first two points). Although a large value of D serves to reject the standard coalescent model for a given set of DNA polymorphism data, distinguishing among changes in effective size through time, population structure, and natural selection is a challenge. Demographic changes over time that impact effective population size, or population differentiation due to the effective migration rate, are expected to impact all loci. In contrast, natural selection is expected to be a locus-by-locus phenomenon based on the fitness impacts of locus polymorphisms. Therefore, comparing estimates of D from numerous loci is one approach to infer the processes that acted to shape patterns of genetic polymorphism. DNA polymorphism patterns at numerous genes in humans, for example, often show negative values of Tajima's D. These results are now generally considered to be caused by a low level of population structure as well as a history of very rapid population growth in the recent past that characterizes human populations rather than balancing selection operating independently on many human loci (Ptak and Przeworski 2002; Tishkoff and Verrelli 2003).

Tajima's D test serves to compare two measures of sequence polymorphism, with θ_π being most sensitive to intermediate frequency alleles, and θ_S being most sensitive to low-frequency alleles (see Zeng et al. 2006). Several other tests have been described that compare related measures of sequence polymorphism but that capture different patterns in the distribution of allele frequencies. These tests utilize the number of sites at which a sequence occurring i times in the sample differs from the ancestral sequence (ξ_i described earlier in the chapter). These

tests rely on several predictions about the impact of natural selection in allele frequency distributions. One prediction is that a selective sweep will cause one haplotype to reach high frequency – the site that experienced a beneficial mutation and linked site polymorphisms. Any recent mutations that occurred on that high-frequency haplotype will be present and generate a pool of low frequency alleles. A second prediction aids in determining if positive selection or background selection could have been acting to alter haplotype frequencies. A new advantageous mutation that is the basis of a selective sweep is a derived haplotype state. In contrast, background selection acts to remove deleterious mutation from a population with the result that the ancestral haplotype is preserved. Therefore, using an outgroup sequence can show which haplotypes are ancestral and which are derived, thereby helping to distinguish positive selection and background selection.

Fay and Wu (2000) suggested using

$$H = \hat{\theta}_\pi - \hat{\theta}_H \tag{8.66}$$

to compare the measures of sequence polymorphism most sensitive to intermediate frequency alleles ($\hat{\theta}_\pi$) and most sensitive to high-frequency alleles ($\hat{\theta}_H$). Small values of H indicate an excess of high-frequency derived alleles compared to a neutral model, consistent with positive selection. Fay and Wu's H is not as sensitive as Tajima's D to strong population size increases or decreases since such demographic changes are expected to impact low frequency alleles most strongly and have less impact on more frequent alleles that have their mutational origins deeper in a genealogy.

Zeng et al. (2006) proposed a way to standardize Fay and Wu's H by dividing by the standard deviation of $\hat{\theta}_\pi - \hat{\theta}_H$. Since both Tajima's D and Fay and Wu's H use intermediate-frequency alleles as their reference for comparison, they also suggested an additional statistic comparing measures of low and high-frequency alleles using

$$E = \frac{\hat{\theta}_L - \hat{\theta}_S}{\sqrt{\text{var}\left(\hat{\theta}_L - \hat{\theta}_S\right)}} \tag{8.67}$$

Negative values of E provide evidence of the recovery of low frequency alleles expected to occur after a selective sweep at the locus being studied. Zeng et al. (2006) also described a way to combine D and H into one joint test. Their simulations showed how this related suite of tests can be used to detect deviations from the null model of neutrality and constant population size. For example, the power of Tajima's D, H, and joint DH increases rapidly as the frequency of the advantageous mutation increases, but that E becomes a more powerful test for positive selection after a beneficial allele has reached fixation. Also, D and E were sensitive to population growth, while H was most sensitive to reduction in population size.

As a test for positive selection, Li (2011) proposed using the observed maximum frequency of derived mutations (ξ_{max}) alone to determine the probability of observing a given level of ξ_{max} under a neutral model. The ξ_{max} test was designed to be insensitive to the confounding effects of changes in population size, background selection, or population differentiation.

This all serves to reinforce that the multiple measures of the site frequency spectrum are a function of the manner in which samples are collected, the evolutionary history of the sampled sequences, and that observed polymorphism is impacted by all of the population genetic process that shape genealogies.

8.7 Recombination in the genealogical branching model

- Genealogies and recombination.
- Ancestral recombination graph.
- Consequences of recombination.

Recombination plays an important role in generating genetic polymorphism by generating haplotypes that are new combinations of existing haplotypes and by changing the frequencies of existing haplotypes. Describing genealogies with recombination can be accomplished by starting with the basic coalescent and adding another type of possible event that can occur working from the present to a time in the past where all lineages find their MRCA (Hudson 1983). We will again utilize the properties of the exponential distribution to approximate the waiting time to an event (see Section 3.6).

Let us begin with recombination in a population of diploid, sexual organisms with equal population sizes of females and males ($N_m = N_f$). Figure 8.27A shows inheritance over one generation where the two chromosomes in an individual at time t are sampled from one male and one female parent at time t-1. In

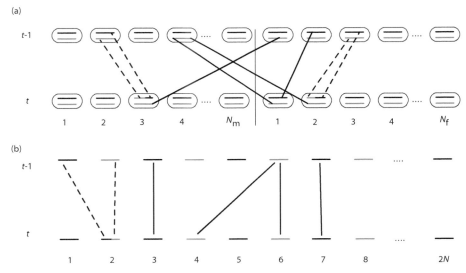

Figure 8.27 Diploid (A) and haploid (B) reproduction in the context of coalescent events with the possibility of recombination. In a diploid population, the two chromosomes in one individual in the present have one ancestor in the female population (N_f) and one ancestor in the male population (N_m). In addition, chromosomes in an ancestor may have recombined such that a single chromosome in the present descends from segments of two chromosomes in the past. In a haploid population, the probability of coalescence is $\frac{1}{2N}$ (solid lines) while the probability that two lineages do not have a common ancestor in the previous generation is $1 - \frac{1}{2N}$ (solid lines). Recombination events (dashed lines) lead to bifurcation of lineages when moving back in time because one chromosome in the present descends from chromosomal segments in two ancestors in the past. As with the basic coalescent, a haploid population with $2N$ lineages is used to approximate a diploid population with $2N = N_f + N_m$ individuals.

the parents, there is a chance that recombination has occurred such that the chromosome transmitted to the next generation is a new combination of the two parental chromosomes. For example, male lineage 3 at time t inherits a recombined chromosome from its paternal ancestor, lineage 2. Because the population is finite, there is the possibility that one of the parental lineages at time t-1 is the ancestor of two of the progeny chromosomes at time t such as female lineages 1 and 2 that inherit the same chromosome from their paternal ancestor (male lineage 4 at time t-1).

The process of coalescence with recombination can be approximated by using a haploid population of size $2N$ (Figure 8.27B) as it is for the basic coalescent model. Working backward in time, lineages sample an ancestral chromosome at random, with

the possibility that two lineages share the same ancestor resulting in a coalescent event. At the same time, there is the possibility that recombination occurs so that the two segments of a chromosome on either side of a recombination point have different ancestors. Recombination is shown in Figure 8.27B for lineage 2 which has a chromosome with one section inherited from lineage 1 and another section inherited from lineage 2.

For coalescence and recombination, it is possible to construct an **ancestral recombination graph** (Griffiths 1991; Griffiths and Marjoram 1997) or ARG (Figure 8.28). Recombination events result in an *addition* of lineages going back in time to represent recombination events. (Because they share a process causing the addition of lineages going back in time, the ancestral recombination graph and the ancestral

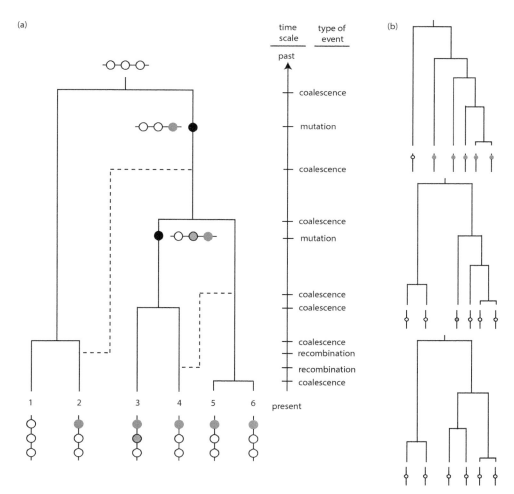

Figure 8.28 An ancestral recombination graph with mutation for three loci in six lineages (A). The event types and relative times between events are given on the scale at the right. Panel B shows the three genealogies for each of the three loci after resolving the recombination events on the genealogy in panel A. The genealogies for the three loci are correlated but not identical due to the action of recombination.

selection graph described in Chapter 7 work in a similar fashion.) Any lineage generated by recombination can then coalesce with a randomly sampled lineage at a time farther back in the past. As a consequence of recombination, different loci (or chromosomal segments) can have different genealogies and different MRCAs. Even though recombination events add branches moving backward in time, the coalescence process is faster and will eventually result in coalescence to a single ancestor (the rate of coalescence is proportional to k^2, whereas the recombination rate is proportional to k). This single ancestor of all potentially recombining loci is called the grandmost recent common ancestor or GMRCA.

An assumption of the haploid model is that coalescent and recombination events are independent and are infrequent enough that they do not occur simultaneously (or that N is large and the rate of recombination is small) so that when an event does occur going back in time it is *either* coalescence *or* recombination. When events are independent, the probability of each event is added over all possible events to obtain the total chance that an event occurs. Therefore, if we obtain an exponential approximation for the chance of recombination each generation, we can just add this to the exponential approximation for the chance of coalescence.

A lineage will experience recombination, with the rate c, each generation. The chance that a lineage does not experience a recombination event is, therefore, $1 - c$ each generation. The chance that t generations pass before a recombination event occurs for a single lineage is then the probability of $t - 1$ generations of no migration followed by a recombination event or

$$P(T_{recombination} = t) = (1 - c)^{t-1}c \qquad (8.68)$$

This is in an identical form to the chance that a coalescent, migration, or mutation event occurs after t generations given in previous chapters. Like the probability of coalescence, the probability of recombination through time is a geometric series that can be approximated by an exponential distribution (see Math Box 3.2).

To obtain the exponential distribution for the recombination process, we need to determine the rate at which recombination is expected to occur in a population. If we define $\rho = 4Nc$, then $\rho/2$ is equivalent to $2Nc$ or the expected number of recombination events in a population of $2N$ lineages per generation, also called the **population recombination rate**. When

time is measured on a continuous scale, one unit of time is equivalent to $2Nc$ discrete generations, or $2Nc$ recombination events are expected in the population during one unit of continuous time.

The definition of the population recombination rate leads to the exponential approximation for the probability that a lineage experiences recombination at exactly generation t

$$P(T_{recombination} = t) = e^{-t\frac{\rho}{2}} \qquad (8.69)$$

When there is more than one lineage, each lineage has an independent chance of recombination, so the $e^{-t\frac{\rho}{2}}$ chance of recombination for each lineage is summed over all k lineages to obtain the total chance of recombination

$$P(T_{recombination} = t) = e^{-t\frac{\rho}{2}k} \qquad (8.70)$$

for k lineages. The chance that one of k lineages has a recombination event at or before a certain time can then be approximated with the cumulative exponential distribution

$$P(T_{recombination} \leq t) = 1 - e^{-t\frac{\rho}{2}k} \qquad (8.71)$$

in exactly the same fashion that times to coalescent events are approximated.

When two independent processes are operating, the genealogical model becomes one of following lineages back in time and waiting for an event to happen. The possible events in this case are recombination or coalescence, so the total chance of any event is the sum of the probabilities of each type of independent event. The total chance of *any event*, either coalescence or recombination, occurring when going back in time (increasing t) is then

$$P(T_{event} \leq t) = 1 - e^{-t\left[k\frac{\rho}{2} + \frac{k(k-1)}{2}\right]} \qquad (8.72)$$

where the exponent is the sum of the intensities of recombination and coalescence.

When an event does occur at a time given by this exponential distribution in Eq. 8.72, it is then necessary to decide whether the event is a coalescence or a recombination. The total chance that the event is either a recombination or a coalescence event is $k\frac{\rho}{2} + \frac{k(k-1)}{2}$. The chance an event is recombination is, therefore,

Interact Box 8.5 Build an Ancestral Recombination Graph

A coalescent genealogy that includes the possibility of recombination can be constructed using the cumulative exponential distribution to determine the waiting time to an event. Once a waiting time is obtained, determining whether the event is a coalescence or a recombination is accomplished using the probabilities of these two types of events. If the event is a coalescence, a random pair of lineages is picked to coalesce and the number of lineages (k) is reduced by 1. If the event is recombination, a random lineage is picked and it bifurcates to generate two new lineages.

Step 1: With k lineages, draw a time to the next event using the exponential distribution.

Step 2: Determine if it is a coalescence or recombination event.

Step 3A: If it is a coalescent event, sample two lineages at random to coalesce after the waiting time has elapsed. Label these lineages with a solid line. Decrease k by one.

Step 3B: If it is a recombination event, sample one lineage at random to generate a new recombining branch after the waiting time has elapsed. Label the recombination lineage with a dotted line. Increase k by one.

Step 4: If $k \geq 2$, go to step 1. Otherwise, the GMRCA has been reached.

Step 5: Assign an ancestral chromosome (DNA sequence or multilocus haplotype) to the GMRCA. Starting with the GMRCA, trace the branches forward in time with each lineage carrying its ancestral haplotype state. The dashed lineages that were generated by recombination carry a segment of their ancestral chromosome determined by randomly assigning a recombination point along the chromosome and dividing the ancestral chromosome into two sections. One of the two chromosomal sections is carried along the dashed line and paired with the complementary haplotype section carried by the lineage it connects to generate a recombinant chromosome. Note that mutations are required to see an impact of recombination on the genealogy (otherwise, all lineages carry the ancestral haplotype), so feel free to add a few hypothetical mutations.

The text website links to an R script or a Microsoft Excel spreadsheet to calculate the quantities necessary to build a coalescent genealogy with recombination.

$$\frac{k\frac{\rho}{2}}{k\frac{\rho}{2} + \frac{k(k-1)}{2}} = \frac{\rho}{\rho + k - 1} \quad (8.73)$$

while the chance that the event is a coalescence is

$$\frac{\frac{k(k-1)}{2}}{k\frac{\rho}{2} + \frac{k(k-1)}{2}} = \frac{k-1}{\rho + k - 1} \quad (8.74)$$

after multiplying the left-hand expressions by $\frac{2/k}{2/k}$. When the event is a coalescence, two lineages are picked at random to coalesce and k decreases by one. When the event is a recombination, one haplotype is sampled at random and one point along its chromosome is also sampled at random. Then, two ancestors are sampled at random, one for each of the chromosome segments separated by the

recombination point. Therefore, a recombination event causes the number of ancestors to *increase* working backward in time and the number of lineages goes to $k + 1$.

Consequences of recombination

The ancestral recombination graph shows how in the absence of recombination multiple positions along a haplotype are associated and share the same genealogy. In contrast, multiple positions along a haplotype separated by a recombination break point are independent and may have distinct genealogies. The degree of correlation among genealogies for the sites along a chromosome that is inversely proportional to the recombination rate is an alternative way to understand linkage disequilibrium (McVean 2002). One consequence of recombination involves the number of segregating sites. While recombination does not change the mean number of segregating sites (S), the variance in S decreases as the population recombination rate ($4Nc$) increases (Hudson 1983). This occurs because as an increasing number of sites have an independent genealogy, the variance in S contributed by coalescence time decreases, leaving a greater fraction of the total variance contributed by the process of mutation (see Eq. 8.29).

The ancestral recombination graph is a useful conceptual tool to predict the possible impacts of recombination on patterns of neutral DNA sequence polymorphism. However, the ARG is unwieldly as a framework for estimating population genetic parameters such as the population recombination rate from empirical DNA sequence data (Rasmussen et al. 2014; Hubisz and Siepel 2020). One challenge is that, in the ARG, some of the chromosome segments in the history of a sample may not be part of the chromosomes in the sample of lineages in the present. Such "trapped" sections of chromosomes can make it difficult to estimate the history of an ARG for observed data. In an alternative approach, the coalescent with recombination was recast as a process that operates sequentially along the length of a chromosome (Wiuf and Hein 1999). In what is called the **sequential coalescent**, a chromosome can be divided into a series of nucleotide segments each with a genealogy generated by drift and mutation, and these site genealogies are then associated by the pattern of past recombination events. A difficulty of the sequential coalescent is that it does not exhibit the Markov property since the genealogy of the next locus along a chromosome depends on the genealogy of all prior loci. Approximations of the sequential coalescent model that restore the memorylessness feature are called the **sequentially Markovian coalescent** or SMC (e.g. McVean and Cardin 2005; Wilton et al. 2015). Different versions of the SMC model a reduced set of possible coalescence events within linked segments for tractability, such as only those coalescence events for a pair of haplotypes. Then, population parameters related to the coalescence process are estimated from observed DNA sequence data using hidden Markov model methods to infer past population demography (Spence et al. 2018). The result is a distribution of estimated coalescence times for the many segments of the genome, with peaks at times when larger proportions of the genome had a common ancestor such as during population bottlenecks.

An early application of these concepts was to estimate past population demography given recombination in human populations based on 12 whole-genome sequences of individuals from a range of populations (Li and Durbin 2011). That study used the two alleles found at loci along homologous pairs of diploid chromosomes to employ a special case of the model called the pairwise sequentially Markov coalescent (PSMC). The study used simulated genomic data to test the capacity of PSMC to estimate population size parameters in a range of demographic scenarios. Estimates from the PSCM include the effective mutation rate, a parameter that can be scaled to calendar time with an estimate of the mutation rate and the generation time. In that way, the study inferred that the ancestors of different contemporary human populations experienced numerous bottlenecks and that the differentiation of modern humans started as early as 120 000 years ago.

Chapter 8 review

- The neutral theory is a widely used null hypothesis in molecular evolution, predicting patterns and rates of DNA sequence change under the assumption that all mutations have no fitness advantage or disadvantage. Even though genetic drift leads to fixation or loss, neutral alleles experience a random walk to these end points that results in transient genetic variation.

- Neutral theory predicts that polymorphism, or genetic variation within populations, is a function of the effective population size and the mutation rate. Larger effective population sizes or higher mutation rates result in higher levels of equilibrium polymorphism.

- Neutral theory predicts that the rate of divergence, the accumulation of fixed nucleotide differences between two species, is a function of only the mutation rate.

- Nearly neutral theory uses the assumption that many mutations are effectively neutral because their selection coefficients are less than the pressure of genetic drift. When $4N_es = 1$, genetic drift and natural selection are equally likely to dictate the fate of a new mutation.

- The proposal of the neutral theory of molecular evolution sparked a controversy between neutralists and those who favored selection-based explanations of rates of divergence and levels of polymorphism. The neutralist–selectionist debate led to many innovations in the expectations for both genetic drift and natural selection.

- Nucleotide sites are acted on directional natural selection, causing purifying (or negative) selection on deleterious alleles and positive selection on deleterious alleles. Natural selection also indirectly alters levels of polymorphism at associated neutral nucleotide sites depending on rates of recombination.
- Balancing selection can increase polymorphism at sites experiencing selection as well as at linked neutral sites that segregate longer than expected by drift and have time to accumulate numerous neutral mutations over time.
- Genetic hitch-hiking reduces polymorphism at neighboring sites because positive selection will bring a beneficial mutation and linked neutral mutations to fixation. This results in a selective sweep where linked site polymorphism is lost.
- Hard selective sweeps are the results of strong selection acting on a single beneficial mutation with the greatest loss of linked site polymorphism. Soft selective sweeps are the result of selection on a standing polymorphism or on beneficial alleles at multiple loci, with a lesser reduction of polymorphism.
- Negative selection drives deleterious mutations to loss along with neutral mutations present at linked sites, reducing polymorphism in a process called background selection.
- Divergence rates of neutral nucleotide sites are not impacted by natural selection at linked nucleotide sites.
- Apparent divergence between two DNA sequences may be underestimated because of multiple hit mutations or homoplasy. Nucleotide substitution models serve to correct observed divergence for multiple hits, giving a better estimate of actual divergence.
- Nucleotide diversity (π) and the number of segregating sites (S) are two measures of DNA sequence polymorphism that can be used to estimate $\theta = 4N_e\mu$.
- The molecular clock hypothesis uses the neutral theory prediction that divergence occurs at a constant rate over time to estimate the time that has elapsed since the two sequences shared a common ancestor.
- Heterogeneity in the rate of divergence over time is common and leads to difficulty equating divergence with time since divergence.
- Under a Poisson process model, the variance in substitution rates should equal the mean substitution rate to give an index of dispersion of one. The index of dispersion is often not equal to one,

suggesting either that the substitution rate is not dictated by purely neutral processes or that the Poisson model is not an apt description of the neutral substitution process.
- Rate heterogeneity can be consistent with neutral evolution such as when mutation rates are constant per generation but generations span different lengths of time. Alternatively, rate heterogeneity may be caused by natural selection that changes the probability of substitution for mutations depending on their fitness.
- The HKA and MK tests examine the neutral prediction that polymorphism and divergence should be proportional since both are functions of the mutation rate.
- Mismatch distributions are a product of the branching patterns of genealogies and are expected to be bimodal under the standard neutral model with constant population size and panmixia.
- Tajima's D compares θ estimated from the average pairwise differences and from the number of segregating sites, quantities which are expected to be equal under the null model of neutrality, constant population size, and panmixia.
- DNA sequence polymorphism can be characterized using numerous measures that give greater weight to low, medium, or high-frequency alleles which are impacted differently by processes such as population growth or natural selection.
- The ancestral recombination graph combines the coalescent with recombination, showing how nucleotide sites have genealogies that are correlated or independent depending on the pattern of past recombination events.

Further reading

Motoo Kimura provided an accessible overview of neutral theory in:

Kimura, M. (1989). The neutral theory of molecular evolution and the world view of neutralists. *Genome* 31: 24–31.

For a review of the nearly neutral theory, see:

Ohta, T. (1992). The nearly neutral theory of molecular evolution. *Annual Reviews of Ecology and Systematics* 23: 263–286.

A concise summary of why neutral theory is a central tenant of explanations for molecular polymorphism and divergence is provided by:

Jensen, J.D., Payseur, B.A., Stephan, W. et al. (2019). The importance of the neutral theory in 1968 and 50 years on: a response to Kern and Hahn 2018. *Evolution* 73: 111–114.

A review of models of natural selection in the context of linkage and mutation with a synthesis of empirical studies

Cutter, A.D. and Payseur, B.A. (2013). Genomic signatures of selection at linked sites: unifying the disparity among species. *Nature Reviews Genetics* 14: 262–274.

A review and synthesis of the numerous models of selective sweeps can be found in:

Stephan, W. (2019). Selective sweeps. *Genetics* 211: 5–13.

An overview of molecular clock calibration and estimation approaches, as well as estimation software can be found in:

Ho, S.Y.W. and Duchêne, S. (2014). Molecular-clock methods for estimating evolutionary rates and time-scales. *Molecular Ecology* 23: 5947–5965.

For a review of empirical studies where directional or balancing natural selection has been invoked to explain observed polymorphism, see:

Hedrick, P.W. (2006). Genetic polymorphism in heterogeneous environments: the age of genomics. *Annual Review of Ecology Evolution and Systematics* 37: 67–93.

For a review of the wide range of hypothesis tests for natural selection using DNA sequence polymorphism and divergence, see

Vitti, J.J., Grossman, S.R., and Sabeti, P.C. (2013). Detecting natural selection in genomic data. *Annual Review of Genetics* 47: 97–120.

An accessible introduction to the sequentially Markovian coalescent and population genomic inference methods based on it can be found in

Mather, N., Traves, S.M., and Ho, S.Y.W. (2020). A practical introduction to sequentially Markovian coalescent methods for estimating demographic history from genomic data. *Ecology and Evolution* 10: 579–589.

End-of-chapter exercises

1   Imagine a set of $k = 6$ lineages sampled from a population and locus where $\theta = 5$. What is the expected number of segregating sites per locus? What is the variance in the expected number of segregating sites? What mutation model is being assumed such that it is unnecessary to know the number of nucleotides being sampled to determine the expected number of segregating sites?

2   Using the example data in Figure 8.15, compute $\hat{\theta}_S$, $\hat{\theta}_\pi$, $\hat{\theta}_L$, and $\hat{\theta}_H$. Explain how these four different estimators weight the different classes of sequence differences. Add more segregating sites or more sequences to that hypothetical data set to illustrate how low, medium, and high frequency alleles impact these four measures.

3   Imagine differentiated allele frequencies among numerous subpopulations with local adaptation experience by a few loci acted on by selection and most loci evolving under drift and gene flow. Based on your knowledge of selective sweeps and background selection, what would you predict for the relationship between F_{ST} and the recombination rate for many SNPs sampled across the genome? (See Keinan and Reich 2010 for a discussion of predictions and a test in human populations.)

4   Search the literature for a recent research paper that utilizes one or more of the population genetic predictions covered in this chapter. The topic can be any organism, application, or process but the paper must include a hypothesis test involving a topic such as neutral and nearly neutral theory, purifying or background selection, selective sweeps, Tajima's D, the HKA test, genetic hitch-hiking, or the molecular clock. Summarize the main hypothesis, goal, or rationale of the paper. Then, explain how the paper utilized a population genetic prediction from this chapter and summarize the results and the conclusions based on the prediction.

Problem box answers

Problem box 8.1 answer

For the wheat–maize divergence 60 million years ago, divergence rates are

$$coxI \quad \frac{0.0504 \text{ substitutions per site}}{2(60 \text{ million years})} = 0.00042 \text{ substitutions per site per million years}$$

$$atp9 \quad \frac{0.1374 \text{ substitutions per site}}{2(60 \text{ million years})} = 0.001145 \text{ substitutions per site per million years}$$

$$nad4 \quad \frac{0.0381 \text{ substitutions per site}}{2(60 \text{ million years})} = 0.000318 \text{ substitutions per site per million years}$$

Using these absolute divergence rates, the divergence time of the monocot–dicot split is estimated as

$$coxI \quad \frac{0.2060 \text{ substitutions per site}}{(2)0.00042 \text{ substitutions per site per million years}} = 245.2 \text{ million years}$$

$$atp9 \quad \frac{0.4439 \text{ substitutions per site}}{(2)0.001145 \text{ substitutions per site per million years}} = 193.8 \text{ million years}$$

$$nad4 \quad \frac{0.1101 \text{ substitutions per site}}{(2)0.000318 \text{ substitutions per site per million years}} = 173.1 \text{ million years}$$

In both cases, there is a factor of 2 in the denominator because there are two lineages accumulating substitutions independently during divergence. The estimated divergence time clearly depends on the locus used since the molecular clock ticks at slightly different rates per million years for each locus. The average divergence time of 204 million years ago for these three loci matches the average of about 200 million years based on all available data (Laroche and Bousquet 1995).

Problem box 8.2 answer

The Mayotte and Europe populations both contain a sample of 15 sequences for the *runt* locus so that $n = 15$

$$a_1 = \sum_{k=1}^{n-1} \frac{1}{k} = \sum \left(\frac{1}{1} + \frac{1}{2} + \frac{1}{3} + \frac{1}{4} + \frac{1}{5} + \frac{1}{6} + \frac{1}{7} + \frac{1}{8} + \frac{1}{9} + \frac{1}{10} + \frac{1}{11} + \frac{1}{12} + \frac{1}{13} + \frac{1}{14} \right)$$

$$= 3.2516$$

and

$$e_1 = \frac{n+1}{3a_1(n-1)} - \frac{1}{a_1^2} = \frac{15+1}{3(3.2516)(15-1)} - \frac{1}{3.2516^2} = -0.02258$$

Next,

$$a_2 = \sum_{k=1}^{n-1} \frac{1}{k^2} = \sum \left(\frac{1}{1^2} + \frac{1}{2^2} + \frac{1}{3^2} + \frac{1}{4^2} + \frac{1}{5^2} + \frac{1}{6^2} + \frac{1}{7^2} + \frac{1}{8^2} + \frac{1}{9^2} + \frac{1}{10^2} + \frac{1}{11^2} + \frac{1}{12^2} + \frac{1}{13^2} + \frac{1}{14^2} \right)$$

$$= 1.576$$

and

$$c = \frac{2(n^2 + n + 3)}{9n(n-1)} - \frac{n+2}{a_1 n} + \frac{a_2}{a_1^2} = \frac{2(15^2 + 15 + 3)}{9(15)(15-1)} - \frac{15+2}{(3.2516)(15)} + \frac{0.576}{(3.2516)^2}$$

$$= 0.04813$$

so that

$$e_2 = \frac{c}{a_1^2 + a_2} = \frac{0.04813}{3.2516^2 + 1.576} = 0.00396$$

In the European population, there are 17 segregating sites out of a total of 556 sites so that

$$\hat{\theta}_S = \frac{p_S}{a_1} = \frac{17/556}{3.2516} = 0.0094$$

whereas, in the Mayotte population, there are 34 segregating sites out of a total of 538 sites so that

$$\hat{\theta}_S = \frac{S}{a_1} = \frac{34/538}{3.2516} = 0.0194$$

Tajima's D for the European population is

$$D = \frac{\hat{\theta}_\pi - \hat{\theta}_S}{\sqrt{e_1 p_S + e_2 p_S (p_S - 1)}} = \frac{0.0124 - 0.0094}{\sqrt{(0.0226)0.0306 + (0.00396)(0.0306)(0.0306 - 1)}}$$

$$= 0.0030/0.02393 = 0.1254$$

whereas Tajima's D for the Mayotte population is

$$D = \frac{\hat{\theta}_\pi - \hat{\theta}_S}{\sqrt{e_1 p_S + e_2 p_S (p_S - 1)}} = \frac{0.0135 - 0.0194}{\sqrt{(0.0226)(0.0632) + (0.00396)(0.0632)(0.0632 - 1)}}$$

$$= -0.0059/0.0345 = -0.0171.$$

Neither population has patterns of DNA sequence polymorphism that deviate from those expected under the standard neutral model.

CHAPTER 9

Quantitative trait variation and evolution

9.1 Quantitative traits

- Components of phenotypic variation.
- Components of genotypic variation (V_G).
- Inheritance of additive (V_A), dominance (V_D), and epistasis (V_I) components of genotypic variation.
- Genotype-by-environment interaction ($V_{G \times E}$).
- Additional sources of phenotypic variation.

In the other chapters of this book, the concept of phenotype employed is somewhat simplistic. This is out of necessity because the emphasis in other chapters is on expectations for genotype and allele frequencies rather than on understanding the causes of variation in phenotype. Phenotypes were assumed to be completely determined by the genotype and to have two or three discrete classes that correspond exactly to the three genotypes of a single locus with two alleles. (A minor exception is the two-locus model of natural selection where the phenotype is fitness.) While there certainly are examples of phenotypes in natural populations that fit this description, the majority of phenotypes are probably not well characterized by these assumptions. This chapter will expand the concept of phenotype and develop the concepts needed to understand the relationship between various types of genetic variation and phenotypic variation. The chapter will introduce the various components of quantitative trait variation, show how these components can be used to describe inheritance of phenotypes, and also explore the action of natural selection and genetic drift on complex phenotypes. The chapter will wrap up with a section devoted to genetic mapping methods used to identify and characterize individual loci that cause quantitative trait variation.

Think of variable phenotypes such as human height, the number of ears on a corn plant, daily milk production in domestic cows, wood density in a tree species, or the probability of onset of a disease such as diabetes or hypertension. Think next of complex behavioral phenotypes such as sexual preference or propensity to substance addiction in humans, success in male–male contests for mates, or the quality of mates chosen by females. Also, think of phenotypes related to Darwinian fitness such as individual size, the number of gametes, the number of progeny, or number of days an individual survives. While each of these classes of phenotypes seems unrelated, they all share features in common as **quantitative traits** (also called *metric traits*). Quantitative traits are sometimes called *multifactorial traits* because the variation in a phenotype among individuals has multiple causes. Quantitative trait variation among individuals is a product of differences in genotype produced by multiple genes as well as differences in environmental conditions experienced by each individual.

The hallmark of quantitative traits is a broad range of variation characterized by a continuous distribution of individual phenotypes in a population (Figure 9.1). **Continuous traits** have a scale of measurement that is naturally continuous, such as quantifying height in centimeters or weight in kilograms. **Meristic traits** exhibit a large number of discrete classes, such as the number of bristles on a fruit fly or the number of leaves on a tree that forms a distribution of phenotypic values. **Threshold** or **liability traits** are continuously distributed phenotypes with some trait value that defines an upper or lower limit. Trait values above or below the threshold define qualitatively distinct categories such as "normal" and "symptomatic." The production of insulin is one example, where human populations show a continuous distribution of insulin production and

Population Genetics, Second Edition. Matthew B. Hamilton.
© 2021 John Wiley & Sons, Inc. Published 2021 by John Wiley & Sons, Inc.
Companion website: www.wiley.com/go/hamilton/populationgenetics

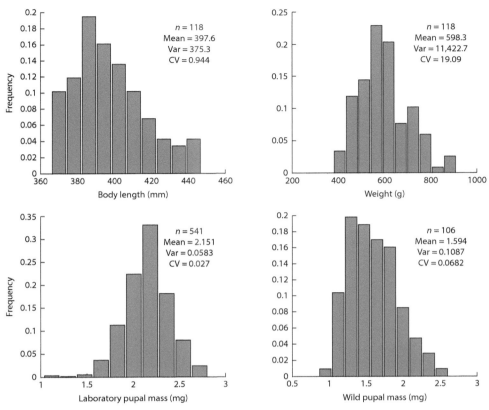

Figure 9.1 Examples of continuous quantitative trait distributions. The top panels show the distributions of body length and weight in a sample of 3-year-old striped bass (*Morone saxatilis*). The bottom panels are pupal mass distributions for mosquitoes raised in laboratory or field conditions. Each panel gives the sample size (*n*), mean, variance, and coefficient of variation (CV) that quantify the phenotypic distribution. Striped bass data are from L. Pieper (unpublished). Source: Mosquito pupal mass data from Armbruster and Conn (2006) and P. Armbruster (unpublished).

individuals are clinically recognized as diabetic when insulin production drops below a threshold level.

In quantitative genetics, the words **phenotype**, **trait**, and **character** are all considered synonymous. The term **value** is used to refer to the phenotype in the same units that it is measured in. **Phenotypic value** refers to the observed phenotype of an individual, for example, observing that an individual fish has a value of 400 mm for the phenotype of body length.

> **Quantitative trait or character:**
> A phenotype where values for numerous individuals in a population are continuously distributed and that variation has both genetic and environmental causes.
> **Value:** The phenotypic measurement of an individual in the units of trait measurement; the mean phenotypic measurement of a population.

Biologists have been aware of continuously distributed phenotypes since Mendel's time. After the

recognition of Mendel's work in the early twentieth century, there was a major controversy involving Mendelian genetics and quantitative genetics. The biometric school was a branch of genetics devoted to understanding the inheritance of continuously distributed phenotypes. Members of the biometric school pioneered the application of statistical methods to quantify and compare continuous phenotypic variation. Francis Galton (half-cousin of Charles Darwin) founded the biometric school through his study of human phenotypes and was an innovator in math and statistics as well as the founder of the eugenics movement (see Gillham 2001). Adherents of the biometric school argued that continuous traits were due to a distinct set of biological causes and could not be explained by Mendel's theory of particulate inheritance. Galton tried unsuccessfully to develop a model that explained the inheritance of quantitative traits without reference to Mendelian genetics (see Provine 1971; Bulmer 1998). In 1918, Ronald A. Fisher (the same R.A. Fisher who contributed the fundamental theorem of natural selection) published a seminal paper showing definitively how single Mendelian loci that individually produced discrete genotypes could combine to result in continuously distributed phenotypes.

The continuous distribution of phenotypes under Mendelian inheritance is due to **polygenic variation**. The continuous variation in phenotype results from the simultaneous segregation of several to many independent Mendelian loci. Figure 9.2 shows the phenotypic distribution for a trait determined by two independent Mendelian loci. In this two-locus illustration, the expected frequency distribution of phenotypes in the population is stepped and not smooth. However, the distribution clearly resembles a normal distribution with symmetry about a single central mode. This hypothetical two locus pigment trait has five classes between 10 and 90% pigment that are evenly spaced as expected for a quantitative trait. One assumption in Figure 9.2 is that the population of individuals is large. This ensures that there is not a lot of chance sampling variation that would cause an observed distribution of

phenotypes to differ greatly from its expected frequencies (e.g. due to chance no individuals with 10% pigment phenotypes are observed). This assumption is implicit in all of the expectations in quantitative genetics.

As the number of loci determining a trait increases, the phenotypic differences between adjacent genotype classes also decreases, resulting in a phenotypic distribution that is smoother. For a phenotype where there is codominance (also known as semi-dominance) and allele frequencies are all ½, the expected frequencies of each class of phenotypic values can be found by taking the expected frequencies of the three genotypes at a single locus and raising it to the power of the number of loci:

$$\left(\frac{1}{4} + \frac{1}{2} + \frac{1}{4}\right)^{n} \qquad (9.1)$$

where n is the number of loci that contribute equally to the quantitative trait. From this equation, it is also apparent that a smoother distribution of phenotypic values would result from loci with more than two alleles because each locus would produce more than three genotypes under random mating. For the two locus phenotype shown in Figure 9.2, the expected frequencies of the phenotypes are found by multiplying the frequencies of the Hardy–Weinberg genotype frequencies for each locus:

$$\left(\frac{1}{4}AA + \frac{1}{2}Aa + \frac{1}{4}aa\right)\left(\frac{1}{4}BB + \frac{1}{2}Bb + \frac{1}{4}bb\right) \qquad (9.2)$$

and then summing the frequencies of those genotypes that have identical phenotypes.

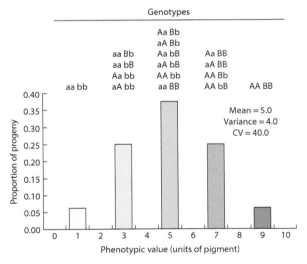

Figure 9.2 The phenotypic distribution for a trait determined by two Mendelian loci. This hypothetical phenotype might be something like flower color, ranging between nearly white and deep blue. The genotypes are those expected in a large number of progeny from a cross between two doubly heterozygous (AaBb) parents. Alternatively, the genotype frequencies are those expected in a large number of progeny from a parental population with Hardy–Weinberg genotype frequencies where mating is random and all allele frequencies are ½. Each a or b allele in a genotype causes ¼ unit of pigment in the phenotype, while each A or B allele in a genotype causes 2¼ units of pigment in the phenotype. As the number of loci contributing to the trait increases, the expected frequency of any individual genotype decreases and the phenotypic distribution will become smoother. For this phenotypic distribution, the mean = 5, the variance = 4, and the CV = 40.

Problem box 9.1
Phenotypic distribution produced by Mendelian inheritance of three diallelic loci

Calculate the expected genotype frequencies for three locus genotypes in a population where mating is random, all loci have two alleles, and allele frequencies at all loci are ½. Then, construct a histogram of phenotypic values using the minimum and maximum phenotypic values used in Figure 9.2 by assuming that alleles have phenotypic values of 1/6 (lower-case letter alleles) and 1½ (capital-letter alleles).

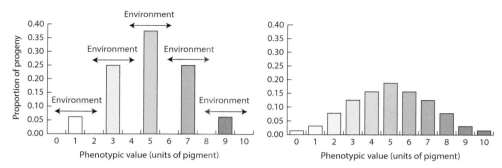

Figure 9.3 The effect of environmental variation on phenotypic variation. The phenotypic distribution on the left is produced by two Mendelian loci with all allele frequencies equal ½ as in Figure 9.2. If the environment causes some variation in the phenotype expressed by each genotype, then the distribution of phenotypes produced by polygenic variation becomes both smoother and wider. In this illustration, environmental variation causes 50% of the individuals of each genotype to randomly increase or decrease 1 unit in phenotypic value. While the average effect of environmental variation here is a zero change in phenotypic value, the phenotypic variance increases.

Another primary cause of variation among individuals in quantitative traits is the environment. Even if there is only a single genotype, the phenotype expressed by each individual in a population will vary somewhat depending on the environmental conditions each individual experiences. For example, the biomass and fruit production of plants is impacted by the amount of sunlight and nitrogen each individual receives. Another example of environmental variation in quantitative traits is the role of diet and exercise in human disease, where better conditions tend to lessen the frequency or severity of disease phenotypes. Figure 9.3 shows how environmental differences among individuals contribute to the continuous distribution of phenotypic variation. The left-hand panel of Figure 9.3 shows the five phenotypes produced by a trait due to two diallelic loci. If each phenotypic class expressed by a genotype is modified by environmental variation, the distribution of phenotypes becomes both wider and smoother as shown in the right-hand panel of Figure 9.3. The environmental variation experienced by individuals is also likely to be a truly continuous variable, unlike the discrete categories of genotypes produced by multiple loci, which then causes continuous variation in phenotypes.

Components of phenotypic variation

Now that we have seen how discrete Mendelian genetic variation for multilocus genotypes combined with continuous environmental variation produces continuous phenotypic distributions, let's represent the genetic and environmental causes of phenotypic variation in notation. In quantitative genetics, it is

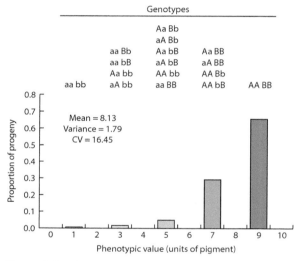

Figure 9.4 Phenotypic distributions for a trait determined by two loci in populations with low additive genetic variation (V_A). Genetic variation in phenotypes is additive since each a or b allele in a genotype contributes ¼ unit of pigment while each A or B allele in a genotype contributes 2¼ units of pigment irrespective of genotype. However, total genetic variation is relatively low since allele frequencies are near fixation and loss (the frequency of a and b alleles is 0.1 while the frequency of A and B alleles 0.9). Compare with Figure 9.2 where allele frequencies are all equal to ½ and allelic effects are additive.

customary to symbolize expected quantitative trait variation with a V. The V variable always bears a subscript to indicate a specific cause of phenotypic variation. The total variation in phenotype is represented by V_P and examples of total phenotypic variance are shown in Figures 9.1 and 9.4. This total phenotypic variation has both genetic and environmental causes. The phenotypic variation caused by variation in genotypes in the population is represented by V_G. Independently, the variation in phenotype caused by the environment is represented by V_E. The equation

$$V_P = V_G + V_E \qquad (9.3)$$

is used to represent the principle that the total variation in phenotype in a population is the sum of phenotypic variation caused by genotype differences among individuals and phenotypic variation caused by the different environments that individuals experience.

Quantifying and comparing the variance of quantitative traits utilize a set of summary statistics. Since quantitative trait distributions usually approximate hump-shaped normal distributions, statistics that describe normal distributions are useful (see the Appendix for a primer on basic statistics used in quantitative genetics). The middle or central tendency of a quantitative trait distribution is described by its average or mean. The spread of the observations around this central tendency is described by the variance of the distribution. The coefficient of variation or CV (CV $= \dfrac{\sqrt{\text{var}(x)}}{\overline{x}}(100)$ where $\overline{x}$ is the trait mean and *var[x]* is the trait variance) is used to compare the variances of distributions after correcting for differences due only to the value of the mean. The CV expresses the magnitude of the standard deviation as a percentage of the mean. As an illustration, look at the quantitative trait distributions in Figure 9.1. Body length and weight for fish have much greater mean values than pupal weights for mosquitoes. But, using the CV, we can compare the variation in these traits to see that the standard deviation is about 5% of the mean for body length in striped bass and almost 21% of the mean for pupal mass in wild-reared mosquitoes. The CV also allows us to properly compare the distributions of pupal mass for laboratory-reared mosquitoes and mosquitoes grown in the wild to see that the spread of pupal mass values around the mean is about two times wider in the wild mosquitoes (CV = 20.68) than in laboratory-reared mosquitoes (CV = 11.23).

Biologically, Eq. 9.3 helps us to recognize and quantify the determinants of phenotypic variation. Equation 9.3 divides the causes of quantitative trait variation into those due to heredity and those due to the environment, or into the phenotypic variation caused by nature and that caused by nurture. Look again at the bottom panels of Figure 9.1 that show pupal mass distributions for mosquitoes. Imagine that the laboratory and wild populations of mosquitoes have roughly the same genotype frequencies or V_G.

The greater pupal mass variance or V_P in the wild population could then be explained by greater environmental variance or V_E, a common observation since laboratory conditions tend to be more uniform and benign. As an alternative, imagine that the laboratory population has less V_G since it was founded from a small sample of individuals and it also has less V_E since the conditions individuals experience are more uniform. Then, the greater pupal mass variance (V_P) in the wild could be caused by both more genetic variation (V_G) and more environmental variation (V_E) than experienced by individuals in the laboratory. In this fashion, it is possible to quantify and compare the relative causes of phenotypic variation.

The V notation compactly expresses the multiple causes of total phenotypic variation, V_P, in quantitative traits. Genotypic variation (V_G) and environmental variation (V_E) are not the only causes of total phenotypic variation. Table 9.1 summarizes additional causes of total phenotypic variation. Notice that V_G is actually broken down into three distinct components due to the effects of alleles, the effects of dominance, and the effects of gene interaction or epistasis (epistasis is symbolized V_I since V_E is already taken to represent environmental variance). In Chapter 10, the Mendelian basis of quantitative traits is used to derive quantitative expectations for V_G and its components. For now, let's continue to develop an intuitive understanding of the causes of total phenotypic variation in addition to V_G and V_E.

Components of genotypic variation (V_G)

Thus far, we have distinguished between the genetic and environmental causes of phenotypic variation. In quantitative genetics, a primary goal is to explain the hereditary causes of phenotypic variation. To fully understand the genetic contribution to total phenotypic variation, it is necessary to recognize that V_G is itself made up of three separate causal components. The components of the total genotypic variation are

$$V_G = V_A + V_D + V_I \qquad (9.4)$$

where V_A is the additive genetic variation, V_D is the dominance genetic variation, and V_I is the interaction or epistasis genetic variation.

The **additive genetic variance** is the genotypic variance caused by the cumulative phenotypic effects of alleles when they are assembled into genotypes. **Additive** simply means that the phenotypic effect of each allele can be added together to determine the phenotypic value of any genotype. When

Table 9.1 Symbols commonly used to refer to categories or causes of variation in quantitative traits. Variation is indicated by V while the specific cause of that variation is indicated by a capital letter subscript. Total genetic variation (V_G) in phenotype can be divided into three subcategories.

Symbol	Definition
V_P	Total variance in a quantitative trait or phenotype
V_G	Variance in phenotype due to all genetic causes
V_A	Variance in phenotype caused by additive genetic variance or the effects of alleles
V_D	Variance in phenotype caused by dominance genetic variance or deviations from additive values due to dominance
V_I	Variance in phenotype caused by interaction genetic variance (epistasis between and among loci)
V_E	Variance in phenotype caused by environmental variation
V_{GxE}	Variance in phenotype caused by genotype-by-environment interaction
V_{Ec}	Variance in phenotype caused by environmental variation shared in common by parents and offspring or by relatives

alleles have additive effects, then the specific pairing of alleles in a genotype, be it a homozygote or a heterozygote, has no impact on the way alleles combine to produce a phenotype. In other words, additivity describes the situation when each allele has the same effect on the phenotypic value regardless of the context where it is found. Figure 9.2 shows an illustration of additive genetic variation in a quantitative trait. Each a or b allele in a genotype contributes ¼ unit of pigment in the phenotype, whereas each A or B allele in a genotype contributes 2¼ units of pigment in the phenotype. The phenotype of any genotype can be determined simply by adding together the phenotypic effects of the two alleles.

When gene action is additive, the amount of additive genetic variance (V_A) in a population depends on allele frequencies. There is more additive genetic variance in phenotype when alleles are at intermediate allele frequencies than when they are near fixation and loss. This is because intermediate allele frequencies result in all possible genotypes (assuming random mating) being represented in the population, which in turn produces a wide range of phenotypes. Figure 9.2 shows the wide range of phenotypes found in a population where genetic variation is additive and all alleles are at a frequency of ½.

To see how genetic variation causes phenotypic variation when allelic effects are strictly additive, compare Figures 9.2 and 9.4. In both figures, phenotypes for each genotype are determined by adding together the phenotypic effects of the alleles in a genotype (each a or b contributes ¼ unit of pigment and each A or B contributes 2¼ units of pigment). However, the allele frequencies in Figure 9.4 are closer to fixation and loss so there is a much less even genotype frequency distribution in the population. The change in allele frequencies leads to two very common genotypes and three genotypes that are very rare. This reduction in genetic variation causes a reduction in phenotypic variation. With allele frequencies nearer fixation and loss, the frequency distribution of phenotypes is now narrower and clumped around values at the upper end of the range.

Additive genetic variance (V_A): The proportion of the total genotypic variance (V_G) caused by the sum of phenotypic effects of alleles when they are assembled into genotypes.

In addition to the additive effects of alleles, quantitative trait variation is also caused by the effect of genotypes. Dominance and epistasis are properties of genotypes that can be thought of in two conceptually distinct ways (see Wade 1992; Cheverud and Routman 1995; Phillips 1998). In a physiological or functional sense, dominance and epistasis describe the way phenotypes map to genotypes. Both are forms of interaction. With dominance, the phenotype depends on the combination of alleles within a locus that compose a genotype (allele interactions). With epistasis, the phenotype depends on the combination of genotypes at two or more loci (genotype interactions). Two diallelic

loci can produce eight distinct ways by which genotypic values are determined by combinations of two-locus genotypes. These eight genetic effects are illustrated in Table 9.2. In general, there are a total of $3^n - 1$ distinct genetic effects for n diallelic loci (Cockerham 1954; see also Goodnight 2000).

At the same time, both dominance and epistasis have a population-level meaning that is statistical.

Table 9.2 The eight uncorrelated (or orthogonal) types of genetic effects that can occur between two diallelic loci. Four of the eight types of genetic effects are interactions that give rise to V_I. With additive-by-dominance interaction, the effect of an allele at the additive locus depends on the genotype it is paired with at the locus with dominance. The effect of one A_1 allele is $a = +0.5$ when an A genotype is paired with B_1B_1, $a = +2$ with B_1B_2, and $a = +1$ with B_2B_2 and there is a dominance deviation of $d = 0$ at the A locus since all A_1A_2 genotypes have the mid-point value of zero. The additive-by-dominance-by-additive interaction example uses A allele effects of $a = \pm0.5$ with B_1B_1 and B_2B_2, and $a = \pm1$ with B_1B_2. For dominance-by-dominance, the dominance deviations for A_1A_2 and B_1B_2 are $d = +2$, and the $A_1A_2B_1B_2$ genotype has a dominance deviation of twice that of the single loci for a value of $-1 + 2 + 2 = 3$. Note that the allelic effects and dominance deviations are arbitrary in these examples. The genotypic values assume all allele frequencies are ½.

Genetic Effect		Genotypes and Phenotypes		
		A_1A_1	A_1A_2	A_2A_2
Additive A locus ($a = \pm0.5$)	B_1B_1	1	0	−1
	B_1B_2	1	0	−1
	B_2B_2	1	0	−1
Additive B locus ($a = \pm0.5$)	B_1B_1	1	1	1
	B_1B_2	0	0	0
	B_2B_2	−1	−1	−1
A locus dominance ($d = +1$)	B_1B_1	0	1	0
	B_1B_2	0	1	0
	B_2B_2	0	1	0
B locus dominance ($d = +1$)	B_1B_1	0	0	0
	B_1B_2	1	1	1
	B_2B_2	0	0	0
Additive-by-additive Interaction ($a = \pm0.5$ for both loci)	B_1B_1	1	0	−1
	B_1B_2	0	0	0
	B_2B_2	-1	0	1
Additive (A locus)-by-dominance (B locus) interaction	B_1B_1	1	0	-1
	B_1B_2	4	0	−4
	B_2B_2	−2	0	2
Dominance (A locus)-by-additive (B locus) interaction	B_1B_1	−1	0	1
	B_1B_2	2	0	−2
	B_2B_2	−1	0	1
Dominance-by-dominance interaction	B_1B_1	−1	1	−1
	B_1B_2	1	3	1
	B_2B_2	−1	1	−1

Source: Table after Cheverud (2000) and Goodnight (2000).

Both dominance and epistasis can be thought of as the "leftover" part of genotypic variance in the population that is not explained by the additive genetic variance due to differences in alleles among individuals (see Huang and Mackay 2016). The magnitudes of V_D and V_I in the statistical sense are a function of both the genotype frequencies in the population and the relationship between genotype and phenotype. These non-additive parts of the phenotypic variance are sometimes called dominance and epistasis *deviations* since they are measured as differences from the variance that would be expected in a population if all genetic effects were additive.

The distinction between additive and dominance genetic variance in quantitative traits can be seen with a modification of the rules that specify the relationship between genotypes and phenotypes. Under additivity, each genotype's phenotypic value is determined by the sum of the phenotypic effects of the alleles that compose it. An alternative relationship between the genotype and phenotype is seen with complete dominance, where heterozygotes have the same phenotype as one of the homozygotes. With dominance, the phenotypic value of the heterozygote is no longer determined by adding together the phenotypic effects of the two alleles that compose it. When gene action shows dominance, the phenotypic contribution of an allele depends on pairing of alleles in the genotype where it resides. Dominance causes the phenotype to be defined by the genotype context rather than being independent of the pairing of alleles in a genotype as it is under additivity.

The additivity rule that applies in Figure 9.2 is changed to complete dominance in Figure 9.5A. Dominance markedly changes the phenotypic frequency distribution, even though allele frequencies remain constant in the two panels. In Figure 9.2, the phenotypic distribution is symmetric and exhibits every possible value of the phenotype. In contrast, Figure 9.5A shows that about 56% of the individuals in the population have phenotypic values of 9. There are no longer any individuals with phenotypic values of 3 or 7 units of pigment. In this example, dominance increases V_P somewhat compared to additivity (see Figure 9.2) because more genotypes (heterozygotes and dominant homozygotes) exhibit genotypic values at the upper extreme of the distribution. In general, dominance can either increase or decrease V_P depending on

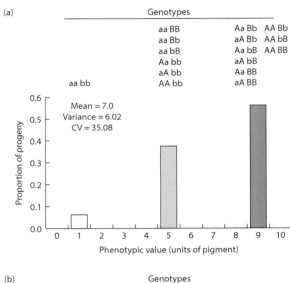

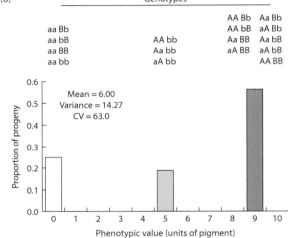

Figure 9.5 Phenotypic distributions for a trait determined by two loci with either complete dominance (panel A) or epistasis (panel B). Due to dominance, the majority of phenotypes are 9 units of pigment and no individuals display 3 or 7 units of pigment phenotypes. In this example of two diallelic loci, the frequency of heterozygotes is at a maximum because all allele frequencies are ½. In general, total phenotypic variation in the population can be increased or decreased by dominance. The bottom phenotypic distribution shows an example of epistasis identical to the two locus system that determines coat color in Labrador retriever dogs. One dominant locus controls pigment color (BB and Bb genotypes have black coats and bb genotypes have brown coats), while a second completely dominant locus controls the presence or absence of pigment in hairs (AA and Aa genotypes have pigmented hair and aa genotypes have unpigmented hair). In this example of dominance by dominance epistasis, phenotypic expression of the genotype at the coat color locus depends on the genotype at the pigmentation locus. In both graphs, the genotype frequencies are those expected under Hardy–Weinberg assumptions where all allele frequencies are ½. The mean and variance are based on a population of 1000 individuals.

allele frequencies in the population. The Mendelian basis of dominance and the dominance variance are explained in detail in Chapter 10.

> **Dominance genetic variance (V_D):** The proportion of the total genotypic variance (V_G) caused by the deviation of genotypic values from their values under additive gene action caused by the combination of alleles assembled into a single-locus genotype.
> **Epistasis or interaction genetic variance (V_I):** The proportion of the total genotypic variance (V_G) due to the deviation of genotypic values from their values under additive gene action caused by interactions between and among loci.

Epistasis literally means "standing on" and denotes interactions between two or more loci that dictate the phenotypic value of a multilocus genotype. When epistasis is absent, the phenotypic value of a multilocus genotype is the sum of the phenotypic value of all single locus genotypes. With epistasis, the phenotypic contribution of the genotype at one locus depends on the genotypes of other loci that it is paired with. When there is epistasis, then the *combination* of genotypes at two or more loci dictates the phenotypic value of a genotype. In the most extreme case, epistasis produces a phenotypic value for each multilocus genotype that is unique to that combination of genotypes. For example, with epistasis, the phenotypic value of the AA genotype paired with the BB genotype (AABB) cannot be predicted from the phenotypic values of the AABb or AAbb genotypes. The interaction genetic variance (V_I) is caused by deviations of genotypic values from the values that would occur if each locus had additive effects.

Since the interactions between two loci can take many forms, a range of terminology has been used to describe the impact of locus interactions on phenotypic values (see Phillips et al. 2000). For example, the term **synergistic epistasis** describes the situation where a genotype has a larger effect on the phenotypic value in the presence of certain other genotypes than would be expected under additivity. In contrast, **antagonistic epistasis** describes an interaction where a genotype has a smaller effect on the phenotypic value in the presence of certain other genotypes than would be expected under additivity.

A classic example of two-locus epistasis is the coat-color phenotypes of Labrador retriever dogs (see Figure 9.5B). For the sake of illustration, assume that one completely dominant locus controls pigment color, with BB and Bb genotypes having black coats and bb genotypes having brown coats (see Kerns et al. 2007 for a more complete description of the genetic basis of coat color in dogs). A second completely dominant locus controls the presence or absence of pigment in hairs. At the pigmentation locus, AA and Aa genotypes have pigmented hair and, therefore, exhibit the black or brown coat color determined by the B locus. (The pigmentation locus is often symbolized as the E locus, but A is used here for consistency across examples.) However, if the pigmentation locus genotype is aa, then the coat color locus has no effect and the coat is yellow because hair pigment is not produced. In a population of Labrador retriever dogs, interaction between the coat color and pigmentation loci will alter the mean and variance of coat color phenotypes relative to two loci that combine additively. The exact impact of the interaction on the phenotypic mean and variance will depend on the genotype frequencies at the two loci.

The distinction between phenotypic variation produced by additive gene action and dominance or epistasis helps clarify a subtle point of terminology in quantitative genetics. The problem is what exactly to call V_G. It is sometimes called genetic variation because multiple alleles in a population lead to variation that under additive gene action produces multiple phenotypes. But with dominance and epistasis, genetic variation is a product of multiple genotypes that then produce a range of phenotypes in a population. Calling V_G *genotypic variation* is probably best since it encompasses phenotypic variation due to both alleles and genotypes.

Inheritance of additive (V_A), dominance (V_D), and epistasis (V_I) genotypic variation

Another way to appreciate the differences among the V_A, V_D, and V_I components of the total genotypic variation is to consider an example of inheritance across a generation. Additive genetic variation (V_A) has a distinct pattern of inheritance compared to dominance and epistasis genetic variation (V_D and V_I). A critical distinction among the three components of the total genotypic variance is that V_A is caused by the average phenotypic effects of *alleles*, while V_D and V_I are caused by the average phenotypic effects of *genotypes*.

Additive genetic variation is the component of the genotypic variance that causes the phenotypic resemblance between relatives. For example, parents and their offspring or siblings (brothers and sisters) have a higher degree of phenotypic resemblance than two randomly sampled individuals in the same population. This average phenotypic resemblance comes about because relatives share alleles that are identical by descent. When alleles combine additively, then shared alleles translate into shared phenotypic values. Only when alleles have additive effects does genetic variation contribute to average resemblance between parents and offspring or among related individuals.

Examples of additive gene action across one generation are shown in Table 9.3. In the top half of the table, alleles at one locus are assumed to act additively to determine phenotypic values. When crossing BB × bb parents or Bb × Bb parents, the parental mean phenotype and the progeny mean phenotype are always identical. The equality of mean phenotypes across one generation is remarkable, given that the genotypes of the parents and progeny are not identical. In fact, in the BB × bb cross, none of the progeny share their genotype with the parents since all progeny possess Bb genotypes. What is common between the parents and progeny in each case are allele frequencies. As long as alleles combine additively to determine phenotypic values, identical allele frequencies in two separate populations will produce identical mean phenotypes. This can be seen well in the second cross (B) for additive gene action, where a population of all Bb heterozygotes has the same allele frequencies and same mean phenotype as a population of ¼ BB, ½ Bb, and ¼ bb individuals.

In contrast to the additive effects of alleles are the genotype effects of dominance and epistasis. Dominance can be thought of as an interaction between the two alleles that make up a single-locus diploid genotype. With dominance, the genotypic value of the heterozygote is not just the sum of the two allelic effects but is some other value depending on how the two alleles interact when packaged into one genotype. While dominance is a continuous variable, complete dominance, where the heterozygote and the dominant homozygote have identical phenotypes, is a useful point of reference. In complete

Table 9.3 Examples of parental and progeny mean phenotypes that illustrate the impacts of additive gene action (top) or complete dominance gene action (bottom). For both types of gene action, the phenotypic value of each genotype is given and the genotypes of two possible parental crosses are shown along with the genotypes in the progeny from each cross. Under additive gene action, the mean phenotypic values are identical in the parents and progeny because phenotypic values are a function of allele frequencies and allele are identical in parents and progeny. In contrast, under complete dominance, parent and progeny mean phenotypic values differ because phenotypic values are a function of the genotype and genotype frequencies differ between parents and progeny.

Additive Gene Action

Genotypes	BB Bb bb			
Phenotypes	3 2 1			
	Cross			Mean phenotype
A) Parents	BB x bb			(3 + 1)/2 = 2
Progeny	Bb			2
B) Parents	Bb x Bb			2
Progeny	¼ BB, ½ Bb, ¼ bb			¼(3) + ½(2) + ¼(1) = 2

Complete Dominance

Genotypes	BB Bb bb			
Phenotypes	3 3 1			
	Cross			Mean phenotype
A) Parents	BB x bb			(3 + 1)/2 = 2
Progeny	Bb			3
B) Parents	Bb x Bb			3
Progeny	¼ BB, ½ Bb, ¼ bb			¼(3) + ½(3) + ¼(1) = 2.5

dominance, alleles behave differently in how they contribute to phenotype depending on whether a genotype is composed of two identical alleles (homozygous genotypes) or two dissimilar alleles (heterozygous genotypes).

Two examples of the average phenotypic resemblance between parents and offspring under complete dominance are shown in the bottom half of Table 9.3. When crossing BB × bb parents (a) to yield a population of Bb progeny, the parental mean phenotype and the progeny mean phenotype are not identical. This lack of parent–offspring phenotypic resemblance occurs because the parental population and the progeny population do not share any genotypes in common. The parents are both homozygotes, while all progeny are heterozygotes. The mean phenotype for parents and progeny is closer for the Bb × Bb parental cross (b). This occurs because 50% of the progeny have an identical genotype to the parents. Thus, with dominance, shared genotype frequencies will lead to phenotypic resemblance. Because particulate inheritance breaks up genotypes, alleles are inherited but diploid genotypes are not. Genotype frequencies are a consequence of how gametes combine to make progeny genotypes. Thus, the genotype effects of dominance (V_D) do not contribute to average phenotypic resemblance between parents and offspring. Like dominance, epistasis (V_I) is also a property of the genotype that does not contribute to average phenotypic resemblance between parents and offspring because genotypes are not inherited. An exception is that additive by additive epistasis does contribute to resemblance between parents and offspring.

Whereas V_D does not contribute to the resemblance of parents and offspring, dominance variance can cause phenotypic resemblance between other types of relatives. In particular, dominance variance contributes to the phenotypic resemblance among full siblings (brothers and sisters). This occurs because full siblings can inherit identical genotypes since they share the same two parents in common. Thus, a shared heterozygote genotype would cause two full siblings to share a genotypic value caused by dominance. Resemblance among relatives is explored more fully in Section 10.6.

Genotype-by-environment interaction ($V_{G\times E}$)

Phenotypic variation can be caused by the combination of genotypes and environments in a population. Up to this point, we have assumed that genotypes are

all equally sensitive to their environments, meaning that a change of environment would impact the phenotype of all genotypes to the same extent. In fact, genotypes very often have different degrees of sensitivity to environmental conditions. This cause of phenotypic variance is called **genotype-by-environment interaction** and is symbolized by $V_{G\times E}$. This adds another term to the expression for the independent causes of total phenotypic variation in a population:

$$V_P = V_G + V_E + V_{G\times E} \qquad (9.5)$$

In one form of genotype-by-environment interaction, genotypes are extremely sensitive to changes in the environment such that the total phenotypic variance changes markedly between two or more environments. In another form of genotype-by-environment interaction, genotypes change phenotypic rank in different environments. For example, genotype AA has a larger phenotypic value than genotype Aa in environment one, but, in environment two, the order is reversed with genotype Aa having the larger phenotypic value.

> **Genotype-by-environment interaction ($V_{G\times E}$):** The contribution to total phenotypic variation caused by genotypes that vary in their sensitivity to different environments; also known as phenotypic plasticity.

Hypothetical genotypic (V_G), environmental (V_E), and genotype-by-environment interaction ($V_{G\times E}$) contributions to total phenotypic variation are illustrated in Figure 9.6. The figure illustrates the results of an imaginary experiment where individuals of four different genotypes are subjected to two environments. Lines connect the phenotype measured for the same genotype in the two environments. Genotypic variation only (V_G; Figure 9.6A) means that the four genotypes have different phenotypes but that no genotype changes its phenotype between the environments (notice that the spread of phenotypes does not change between the environments when there is only V_G). Environmental variation only (V_E; Figure 9.6B) means that the four genotypes have identical phenotypes but the phenotype changes between the two environments (the four genotype lines are not drawn exactly on top of each other so each can be seen). A combination of both

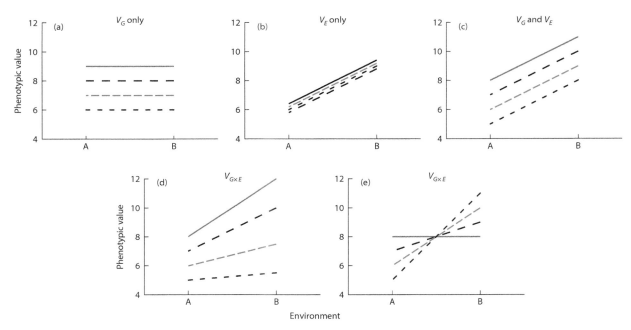

Figure 9.6 Examples of phenotypic variation due to genetic (V_G), environmental (V_E), and genotype-by-environment (V_{GxE}) causes shown in norm of reaction plots. In all graphs, the phenotypic values of four genotypes within each of two environments (here called A and B) are plotted. Lines connect the phenotypic values of one genotype measured in the two environments. Panel A shows genotypic variation, where the four genotypes have different phenotypic values, but the phenotypic value of each genotype does not change between environments. Environmental variation is shown in panel B since all genotypes have identical phenotypes (lines are staggered so each can be seen) but the phenotype changes between environments. Both genotypic and environmental variations are shown in panel C since genotypes differ in phenotype and genotypes have different phenotypes in the different environments. Genotype-by-environment interaction, illustrated in panels D and E, means that genotypes differ in the phenotypic value expressed in two or more environments. One type of genotype-by-environment interaction is characterized by lines connecting the genotypes that are not parallel (panel D), leading to changes in the phenotypic variance. In a second type of genotype-by-environment interaction, the rank order of phenotypic values exhibited by genotypes changes across environments and leads to crossing lines in norm of reaction plots (panel E).

genotypic (V_G) and environmental (V_E) variations in phenotype means that the genotypes have different phenotypes and the phenotypes also change between environments (Figure 9.6C). Notice that, with V_G and V_E, the lines are parallel, showing that each genotype has an identical change in phenotype caused by the change in environment.

Two types of genotype-by-environment interaction are shown in Figure 9.6. Both examples illustrate the hallmark of genotype-by-environment interaction: different genotypes vary in their response to changes in their environment. One example shows that the range of genotypic values of the four genotypes depends on the environment (Figure 9.6D). In environment A, there is less variation in genotypic values, and, in environment B, there is more variation in genotypic values. When the lines connecting genotypes are not parallel in a norm-of-reaction plot, then genotypic variance changes with environment. Another example shows a different environmental sensitivity of genotypes that causes a change of phenotypic ranks between

the two environments (Figure 9.6E). Three of the genotypes respond to the change in environment and demonstrate increased genotypic values in environment B. The genotype with a value of 5 in environment A experiences the largest change in phenotype, showing a value of 11 in environment B. In contrast, the genotype indicated by the solid line is completely insensitive to the environmental change. Crossing lines indicate genotype-by-environment interactions as changes in rank order of genotypic values in norm-of-reaction plots.

An early and still classic example of a genotype-by-environment interaction comes from Clausen et al. (1948; see Nunez-Farfan and Schlichting 2001). These researchers sampled a group plants in the genus *Achillea* at a single location in Aspen Valley, California (elevation 1950 m). They then sprouted vegetative cuttings of each Aspen Valley plant to create multiple individuals of each genotype for the numerous sampled genotypes. The newly sprouted cuttings were planted at three sites with increasing elevations between sea level (Stanford)

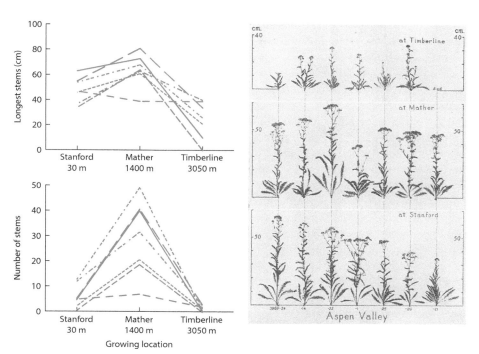

Figure 9.7 The longest stem and number of stems phenotypes for seven *Achillea* genotypes originally sampled at Aspen Valley, California, cloned from vegetative clippings, and then transplanted at three elevations. *Achillea* phenotypes show that genetic (V_G), environmental (V_E), and genotype-by-environment interaction ($V_{G\times E}$) contribute to total phenotypic variation (V_P). Source: Data are from Table 11 and the photograph from Figure 17 of Clausen et al. (1948). Used with permission of Carnegie Institution of Washington.

and high in the mountains above treeline (Timberline). The sprouted cuttings of each of the original Aspen Valley plants were genetically identical, so plants with identical genotypes could be transplanted at each elevation. The transplanted cuttings were monitored, and their phenotypes were measured over several years. Figure 9.7 shows the values of two phenotypes (longest stem and number of stems) measured for seven different genotypes grown at each of the three elevations.

Genotypic, environmental, and genotype-by-environment interaction all contributed to phenotypic variation in *Achillea*. V_G is evident because the phenotypes for the different genotypes within an environment were clearly not identical. V_E was evident because the average phenotype varied across the environments, being highest at Mather, intermediate at Stanford, and lowest at Timberline. $V_{G\times E}$ was also evident because the ranks of the phenotypic values for each genotype clearly changed across the three environments. In other words, genotypes that demonstrated phenotypic values above the mean in one environment had a below average phenotypic value in another environment and vice versa. Therefore, in *Achillea*, the phenotypic variation seen across these three environments had three causes. It was due to a combination of differences in phenotypic values among genotypes, differences in phenotypic values among environments, and differences in the change in phenotypic value of genotypes across the three environments.

Genotype-by-environment interaction has been observed for diverse phenotypes and a wide range of organisms. An example related to human health involves the risk of colorectal adenoma, a disease characterized by pre-cancerous tumors of the colon, genotypes at the *UGT1A6* (UDP glycosyltransferase 1 family, polypeptide A6) locus, and the use of aspirin. Regular aspirin intake reduces the risk of colorectal adenoma for individuals with all *UGT1A6* genotypes compared to no use of aspirin. However, individuals homozygous or heterozygous for the "slow" *UGT1A6* allele, a variant that leads to slower aspirin metabolism, show significantly reduced risk of colorectal adenoma when individuals take aspirin compared with other *UGT1A6* genotypes (Bigler et al. 2001; Chan et al. 2005; Hubner et al. 2006). Thus, individual risk of colorectal adenoma depends on the combinations of *UGT1A6* genotype and aspirin "environments." Additional examples of genotype-by-environment interactions in human disease are reviewed by Hunter (2005). See many more examples of genotype-by-environment interactions in Pigliucci (2001) and DeWitt and Scheiner (2004).

The genotype-by-environment interaction can also be thought of as a correlation between genotypic value and environmental impact on the phenotype. The decomposition of $V_P = V_G + V_E$ without any contribution of $V_{G\times E}$ makes the assumption that the phenotypic value of a genotype and any environmental impact on the phenotype of that genotype are completely independent. The absence of a correlation between genotypes and effects of the environment is illustrated in Figure 9.3, where phenotypic values are as likely to increase as to decrease due to their environment. In contrast, a correlation between genotypic value and environment is common in agricultural contexts. For example, domestic animals are often fed in proportion to their individual size or productivity, such as adjusting the amount of feed given to individual cows according to their production of milk. This feeding practice introduces a correlation between genotypic value and environmental variation that then has an impact on the total phenotypic variation. In the case of cows being fed based on milk production, total phenotypic variation should increase since individuals producing more milk (because of their genotypic value) are fed more (a non-random environment), which in turn makes their milk production even greater due to the impact of having more to eat.

Additional sources of phenotypic variance

In many organisms, progeny share an environment with their mother, their father, or both parents for some period of time as embryos or during their development and growth. This common environment, sometimes symbolized V_{Ec} for *common* environmental variance, can cause parents and offspring to resemble each other to a greater degree than they would if they each inhabited randomly sampled environments. **Maternal effects**, the correlation between environment and genotype for mothers and their offspring, are common in mammals since mothers supply the pre-natal environment and often provide extensive postnatal care for young. As an example, consider mammals that nurse their progeny. A mother living in an environment rich with food is likely to have a greater body mass since she is well fed. Because the mother receives ample nutrition, she will also be able to produce ample milk for her offspring. If the offspring grow faster and larger due to their mother's ample milk supply, then there will be positive covariance between the mother's body size and their own body size. This positive covariance in body size is caused by the rich environment shared by the mother and her offspring. A shared common environment can also lead to an increase in phenotypic resemblance among full or half siblings since they too share the same environment. Thinking again of a mother providing an ample supply of milk due to a rich environment, the progeny of such a mother could exhibit positive

covariance among their phenotypes (a similar large body size or fast growth rate) due to their common milk supply environment compared to siblings raised in different random environments.

The total phenotypic variation is modified for nonindependence of genotypic value and environmental variation by

$$V_P = V_G + V_E + 2(\text{cov}_{GE}) \qquad (9.6)$$

where cov_{GE} is the covariance between genotypic value and environmental variation in phenotype. Since a covariance can be either positive or negative, the total phenotypic variance can be either increased or decreased by a correlation between the genotypic value and the impact of the environment. The cov_{GE} is almost never quantified in practice and is assumed to be zero. The contribution of cov_{GE} (if any) to V_P is, therefore, lumped in with nonadditive causes of V_G. Despite this, it is important to remember that any covariance between genotypes and environments is a potential cause of increased or decreased variance in phenotype.

Math box 9.1 Summing two variances

In quantitative genetics, it is common to sum variance components to obtain a total variance. For example, the total phenotypic variance is the sum of the genotypic variance and the environmental variance according to $V_P = V_G + V_E$. When summing variances of two variables, it is possible that they are not completely independent of each other. Therefore, it is necessary to account for the possibility of covariance between the variables and to adjust the total variance. When summing the variance of two variables X and Y to obtain the total variance,

$$\text{var}(X + Y) = \text{var}(X) + \text{var}(Y) + 2\text{covar}(X, Y) \qquad (9.7)$$

A covariance can either be positive (e.g. when the value of X is large, the value of Y tends to be large) or negative (when the value of X is large, the value of Y tends to be small). Unless X and Y are independent (for any value of X the value of Y is random), the sum of two variances may be increased or decreased by any covariance.

Phenotypic variation among individuals may also be caused by **epigenetic variation**, changes in gene expression that are not caused by DNA sequence polymorphisms but by DNA modifications that are inherited. Activation, reduction, or blocking of gene expression in eukaryotes can be caused by covalent bonding of methyl groups to histones or to DNA sequences or by short non-coding RNAs in the cytoplasm that can bind to DNA sequences (Angers et al. 2010; Springer and Schmitz 2017). These processes generate **epigenetic marks**, and different patterns of marks generate variable **epialleles** that impact phenotypic variation cause phenotypic variation among individuals (Richards 2008). Most epigenetic marks are cleared during meiosis or after fusion of gametes and are not transmitted to offspring. However, some epialleles exhibit transgenerational epigenetic inheritance for a few to many generations termed **soft inheritance** (Richards 2006; Angers et al. 2010). One example is a study where the plants *Arabidopsis thaliana* and *Solanum lycopersicum* (tomato) were experimentally exposed to caterpillar or simulated mechanical herbivory treatment, and resistance to caterpillar herbivory was then measured in subsequent generations (Rassman et al. 2012). In both species, caterpillars were smaller on herbivory-treated plants than on control plants, and this induced defense to herbivory was epigenetically inherited for two generations in *Arabidopsis*. Further, some *A. thaliana* mutant lineages deficient in certain molecular pathways did not exhibit induced resistance, suggesting that genetic variation among individuals could contribute to variation in epigenetic resistance to herbivory.

The process of epigenetic marking can have environmental and genetic causes. This means that any phenotypic variation among individuals (V_P) caused by epialleles can be partitioned into genotypic variation (V_G) and environmental variation (V_E) among individuals, as well as genotype-by-environment interaction (V_{GxE}) and genotype–environment covariance (cov$_{GE}$) (Banta and Richards 2018). Further, the genotypic component of epigenetic variation (V_G) can be composed into phenotypic variance caused by additive (V_A), dominance (V_D), and interaction (V_I) components depending on genetic causes of epialleles.

Another possible source of genotypic variance (V_G) comes from gametic disequilibrium between loci as well as autozygosity within loci. As detailed in Chapter 2, physical linkage, finite population size,

admixture of genetically diverged populations, mutation, and natural selection can all produce gametic disequilibrium. Additionally, consanguineous mating also causes an increase in gametic disequilibrium. When consanguineous mating occurs, one result is increased homozygosity for individual loci. An increase in homozygosity leads to gametic disequilibrium because genotypes at two or more loci within an individual will not be a random combination of all possible genotypes but will tend to be combinations of homozygous genotypes.

All forms of gametic disequilibrium contribute to the **disequilibrium covariance** of genotypic values that increases or decreases the total phenotypic variance (Cockerham 1956; Weir et al. 1980). The degree of gametic disequilibrium can be measured as a covariance *between individuals* of the correlation between genotypic values at two loci *within* an individual. To visualize this covariance, first imagine the possible correlations between the genotypic values at two loci within a single individual. For example, consider the two locus genotype AABB in a population. If the A and B alleles both contribute to larger phenotypic values, the AA genotypic value and the BB genotypic value are positively correlated in an individual with the AABB genotype. Under random mating and free recombination, the AA genotypic value and the BB genotypic value are found together in the same individual only occasionally since the two loci are independent. Free recombination and random mating mean that AA and BB single locus genotypes co-occur at random. In contrast, imagine the AABB genotype is common in the population due to consanguineous mating or strong gametic disequilibrium that causes high frequencies of AB gametes. In that case, there will be a positive correlation of genotypic values between the A and B loci within individuals because the AA and BB single locus genotypes tend to occur together more frequently than expected under independent assortment.

Under random mating in a large population, there should be little gametic disequilibrium for a trait not under selection and the fixation index (F) should be approximately zero. This leads to a disequilibrium covariance of zero. Since the genotypic disequilibrium covariance is not estimated in practice, these conditions become implicit assumptions in the decomposition of $V_P = V_G + V_E$. As with correlations between genotypic value and environment, the contribution (if any) of genotypic covariance to V_P is lumped in with nonadditive causes of V_G.

CHAPTER 9

9.2 Evolutionary change in quantitative traits

- Heritability and the breeder's equation.
- Changes in quantitative trait mean and variance due to natural selection.
- Estimating heritability by parent–offspring regression.
- Response to selection on correlated traits.
- Long-term response to selection.
- Neutral evolution of quantitative traits.

Evolutionary change in quantitative traits is caused by the processes that reduce, shape, or increase variation in the genotypes that underlie the genotypic portion of the total phenotypic variance. Just like the individual loci considered in earlier chapters, the multiple loci that compose a quantitative trait will experience genetic drift and mutation and will also be subject to natural selection. Change in quantitative trait means and variances will occur to the extent that these processes change the allele and genotypes frequencies at those loci that contribute to a quantitative trait. Because it is usually not possible to track the individual loci that underlie a quantitative trait, the genetic basis of quantitative traits is tracked by summary measures that describe a population. A critical distinction that we need to bear in mind is the difference between additive genetic variation (V_A) that leads to resemblance between relatives, and dominance and interaction genetic variation (V_D and V_I) that is not inherited. Because additive genetic variation does lead to parent–offspring resemblance, it is the basis of the action of natural selection on quantitative traits.

Heritability and the Breeder's equation

The **heritability** is used to express the proportion of the total phenotypic variance (V_P) that is caused by either all types of genotypic variance (V_G) or by only the additive genetic variance (V_A). Utilizing the equation for the components of the total phenotypic variance $V_P = V_G + V_E$, both sides of the equation can be multiplied by $1/V_P$ to obtain

$$\frac{V_P}{V_P} = \frac{V_G}{V_P} + \frac{V_E}{V_P} \tag{9.8}$$

This divides the total phenotypic variance, or 100% of V_P, into the proportion caused by genotypic variation and the proportion caused by environmental variation. The proportion of the total phenotypic

variance caused by genotypic variance defines the **broad-sense heritability**

$$h_{BS}^2 = \frac{V_G}{V_P} \tag{9.9}$$

Since genotypic variance is composed of the separate components $V_A + V_D + V_I$, the proportion of the total phenotypic variance caused by only the additive genetic variance defines the **narrow-sense heritability**:

$$h_{NS}^2 = \frac{V_A}{V_P} \tag{9.10}$$

The narrow sense heritability expresses the proportion of the total phenotypic variance made up of genotypic variance that contributes to the parent–offspring resemblance. The remaining proportion of V_P is caused by genotype components of the total genotypic variance ($V_D + V_I$) and the environmental variance, none of which contribute to phenotypic resemblance between parents and offspring (with the exception of additive by additive epistasis).

Idealized heritabilities are proportions and, therefore, range between 0.0 and 1.0 (although *estimates* may fall outside this range due to estimation error). The symbol for the heritability is always h^2, and there is no biological meaning in the square root of h^2. This is a convention that has held since Wright used the symbol in a 1921 paper (Wright 1921). Unless explicitly noted otherwise, the word heritability and h^2 without a subscript will hereafter refer to the narrow-sense heritability, as is common practice.

As an illustration, broad-sense and narrow-sense heritabilities for five human phenotypes related to blood pressure are shown in Figure 9.8. In these examples, only the additive and dominance components of genotypic variance were estimated but the epistasis component was not. Each pie chart divides the total phenotypic variation (V_P) into the dominance (V_D) and additive (V_A) components of the total genotypic variance as well as environmental variance (V_E). The broad-sense heritability for fat-free body mass shows that 76% of V_P was caused by genotypic variation. The remaining 24% of variation in fat-free body mass was caused by environmental differences among individuals. The genotypic variation can be further divided by comparing the broad-narrow and narrow-sense heritabilities. Additive genetic variation explained 45% of V_P, while 31%

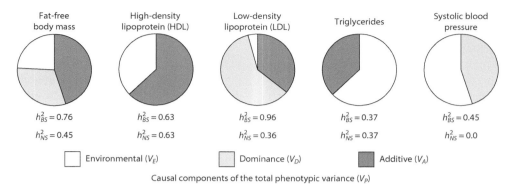

Figure 9.8 Broad-sense and narrow-sense heritabilities for five blood pressure related quantitative traits in humans. Each pie chart divides the total phenotypic variation (V_P) into its causal components of dominance (V_D) and additive (V_A) genotypic variance as well as environmental variance (V_E). These heritabilities were estimated in a small population of Hutterites, a self-reliant, communal group of Anabaptists that traces its origin to followers of Jakob Hutter who fled Austria in the sixteenth century to escape religious persecution. The 806 individuals in this study are descendants of 64 ancestors so that many individuals have a nonzero probability of sharing a genotype that is identical by descent because since their parents are distantly related. This improves the precision of estimates of dominance variance. Source: Estimates from Abney et al. (2001).

(the difference between h_{BS}^2 and h_{NS}^2) of V_P was caused by dominance. Notice that the proportion of V_P explained by each of V_A, V_D, and V_E varies considerably among the five traits in this population. While variation in low-density lipoprotein (LDL) levels among individuals is caused by a combination of V_A, V_D, and V_E – as was the case for fat-free body mass – the relative contributions of each changed greatly, with dominance explaining 60% of V_P and environment only 5% of V_P. Variation in the high-density lipoprotein (HDL) and triglyceride level phenotypes were explained by V_A and V_E with no measurable contribution of V_D. Variation in systolic blood pressure was explained by about equal contributions of V_D and V_E with no measurable contribution of V_A. While variation in systolic blood pressure does have a genetic basis, there was very little phenotypic resemblance between relatives due to the additive effects of alleles.

It has long been observed that there is a correlation between the type of trait and the magnitude of its narrow-sense heritability. Life-history traits such as survival and number of progeny generally exhibit the lowest narrow-sense heritabilities, whereas physiological (e.g. efficiency of food conversion, percentage of milk fat, and serum cholesterol concentration), behavioral (e.g. parental care, mating, and phototaxis), and morphological (e.g. body size, height, and bristle number) phenotypes show progressively greater narrow-sense heritabilities (Mousseau and Roff 1987; Roff and Mousseau 1987; Falconer and Mackay 1996). One hypothesis is that traits more closely related to fitness are likely

to be under stronger and more consistent natural selection pressures than arbitrary morphological phenotypes so that long-term response to selection has depleted additive genotypic variation for life-history traits but not for morphological traits. An alternative possible explanation for this pattern is that the relative magnitudes of V_A, V_D, and V_E are different for morphological, behavioral, and life-history traits. In an analysis of 182 quantitative traits, Houle (1992) used a measure of additive genotypic variation standardized by the trait mean to show that life-history phenotypes actually have higher additive genotypic variance (V_A) as well as residual variance (all remaining variance not due to additive genotypic variance) than traits presumably experiencing weaker natural selection. In the traits analyzed by Houle (1992), lower heritability of fitness-related traits was caused by higher levels of nonadditive genotypic variance (V_D and V_I) and environmental variance rather than by a lack of additive genotypic variation.

A persistent debate has been over the proportion of total genotypic variation that is additive (V_A) compared to the proportions that are due to dominance (V_D) or epistasis (V_I). It has been argued that most quantitative genetic variation is additive, based on empirical estimates of variance components and also on predictions for components of V_G when most alleles are close to fixation and loss rather than at intermediate allele frequencies (Hill et al. 2008). In contrast, Huang and Mackay (2016) compared the traditional additive partitioning of genetic variance (see Chapter 10) to several alternative versions where genetic variance was parameterized as either dominance variance or by additive by additive variance. Using simulated and empirical data, they showed that the traditional additive model explained the greatest portion of variance even when it did not match the genetic model that generated the simulated phenotypic data. Based on these findings, they concluded that the relative contributions of V_A, V_D, and V_I cannot be determined only from variance components estimated using the traditional additive model.

It is important to recognize that all heritability estimates are highly contextual. A heritability estimate made in one population may not be representative of heritability in another population of the same species because both the genetic and environmental causes of phenotypic and genotypic variation may differ among populations. Clearly, allele frequencies may differ among populations, causing V_A to differ. Genotype frequencies may also differ among populations, causing V_G to differ. Even if allele frequencies are identical, differences in mating system can impact V_D and V_I because homozygosity is increased or decreased by mating patterns. The range of environments or their quality may also differ among populations, so that even if genotype and allele frequencies are constant, V_P can increase or decrease and thereby cause the heritability to change as well.

Changes in quantitative trait mean and variance due to natural selection

To begin, it is important to understand that natural selection on quantitative traits operates on differences in phenotypic value. The phenotypic values of individuals dictate individual fitness, whereas the individual multilocus genotypes that underlie a quantitative trait are not directly perceived by natural selection and are unknown to an observer. In contrast, in the models of natural selection in Chapters 6 and 7, natural selection acted on different fitness values possessed by known genotypes. This distinction is biologically meaningful. In selection on phenotypic values, the manner in which phenotypic values are determined by alleles and genotypes (i.e. additivity, dominance, or epistasis) has a direct impact on how effective selection is in changing the allele frequencies that underlie the mean phenotype in a population. As we will see in this section, selection is most effective when variation in phenotypic values is due to additive variation and least

effective when phenotypic variation is due to dominance and epistasis.

There are three general types of natural selection experienced by quantitative traits, as illustrated in Figure 9.9. Under **directional selection,** individuals with phenotypic values at one edge of the phenotypic distribution have higher relative fitness, so response to selection results in an increasing or decreasing mean phenotype. When **stabilizing selection** operates, phenotypic values around the population mean have the highest fitness. In contrast, under **disruptive selection,** phenotypic values at the outer edges of the phenotypic distribution have higher relative fitness.

Math box 9.2 Selection differential with truncation selection

When truncation selection acts on a quantitative trait, determining the selection differential requires determining the mean of the parental phenotypic distribution above the truncation point (μ_S). This can be done using the partial moments of a normal distribution. In general, the partial moment of a normal distribution is

$$\int_0^{+\infty} \frac{1}{\sqrt{2\pi}} e^{-\frac{x^2}{2}} \qquad (9.11)$$

With a normal distribution that has a mean of zero, a variance of σ^2 and a truncation point of t, the mean of the portion of the distribution between t and infinity is

$$\int_0^{+\infty} \frac{\sigma}{\sqrt{2\pi}} e^{-\frac{t^2}{2\sigma^2}} \qquad (9.12)$$

In the special case where the truncation point is at the 50th percentile of the distribution, the truncation point is the mean of the distribution so that $t = 0$. With $\sigma = 1$, the mean of the distribution above the truncation point is then

$$\frac{1}{\sqrt{2\pi}} = 0.3989 \qquad (9.13)$$

which is the phenotypic mean of the selected parents, μ_S.

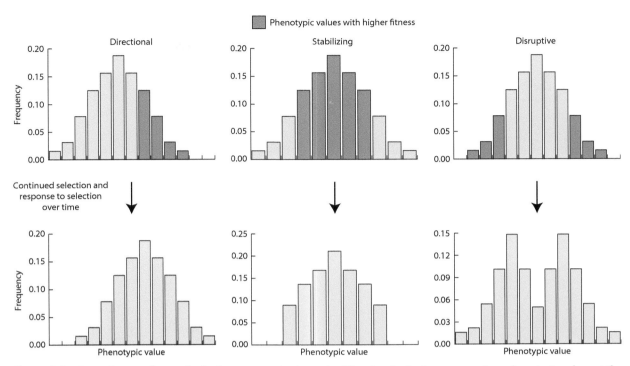

Figure 9.9 General types of natural selection on quantitative traits. Directional selection occurs when phenotypic values at the upper or lower end of the distribution have the highest fitness. Stabilizing selection occurs when intermediate trait values have the highest fitness. Disruptive selection occurs when traits values at the edges of the phenotypic distribution have the highest fitness. Response to directional selection increases or decreases the population mean phenotype. Response to stabilizing or disruptive selection does not change the mean but decreases or increases the variance of the phenotypic distribution.

The response to **stabilizing** or **disruptive selection** on quantitative traits is distinct from the response to directional selection. When a trait responds to stabilizing selection, the mean phenotypic value in a population does not change. Instead, the variance in phenotypic value decreases over time since individuals with phenotypic values at the upper and lower extremes have lower fitness values and do not contribute to future generations. Similarly, the mean phenotypic value in a population does not change with response to disruptive selection. Disruptive selection causes the variance in phenotypic values to increase since individuals at the upper and lower extremes of the distribution have higher fitness than individuals with phenotypic values near the mean. Disruptive selection results in the phenotypic distribution widening over time, and the distribution of phenotypic values can eventually become bimodal.

The heritability provides the basis for predicting the outcome of natural selection on quantitative traits according to

$$R = h^2 s \qquad (9.14)$$

where R is the response to selection, h^2 is the narrow-sense heritability, and s is the selection differential that measures the strength of natural selection. R and s are measured in the same units used to measure phenotypic value (e.g. kilograms and centimeters). Equation 9.11 is commonly called the **breeder's equation** because it predicts the change in mean phenotype in a population that will occur due to one generation of artificial selection as often employed by animal and plant breeders. The response to selection predicted by the breeder's equation is intuitive. Stronger natural selection or phenotypic variation that has a greater basis in additive genetic variance will result in a greater change in the mean phenotype in a population.

To better understand how natural selection changes the mean phenotypic value, let us work through an example of directional selection based on phenotypic value. The change in mean phenotype in a population caused by natural or artificial selection depends on both the force and amount of selection that are applied to the population as well as genetic variation in the trait. The **selection differential** is one way to measure the strength of directional natural selection on quantitative traits. The selection differential is the difference between the phenotypic mean of the entire population and the phenotypic mean of that subset of individuals selected on the basis of their phenotypic value to be parents of the next generation. The **selection threshold** (sometimes called the truncation point) seen in Figure 9.10 is the lower bound of phenotypic values in the group of parents selected to mate. The selection differential is computed as the difference in the phenotypic mean of the selected parents (μ_S) and the phenotypic mean of the entire P1 population (μ)

$$s = \mu_S - \mu \qquad (9.15)$$

As shown in Figure 9.10, the selection differential for this case is

$$s = 12.5 - 10.0 = 2.5 \qquad (9.16)$$

which expresses that the selected parents have a 2.5-unit greater average phenotypic value than the full population that they were sampled from. Larger selection differential values indicate stronger

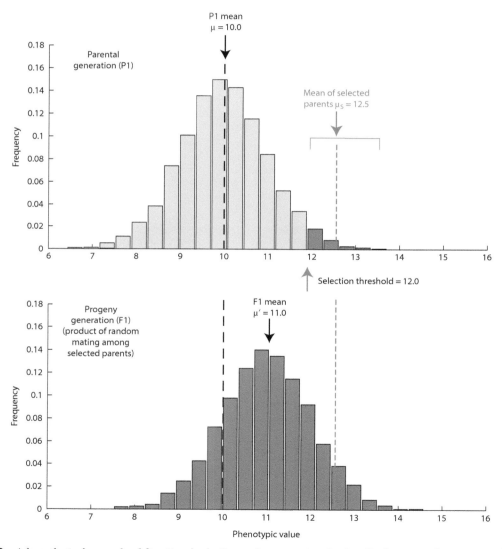

Figure 9.10 A hypothetical example of directional selection and response to selection. In the parental generation, those individuals with a phenotypic value of 12.0 or greater are allowed to mate. The selection differential is $s = 12.5$–$10.0 = 2.5$. Random mating among this subset of the parental population produces a distribution of progeny phenotypic values with a mean value of 11.0. The response to selection is $R = 11.0$–$10.0 = 1.0$. The realized heritability is, therefore, $h^2 = 1.0/2.5 = 0.40$. Response to selection is proportionate to the degree to which parental phenotypic values are caused by the phenotypic effects of alleles. Parental phenotypic values being caused by genotypes or the environment do not lead to response to selection since these causes of phenotypic value are not inherited by offspring.

selection. Although not true in this example, selection differentials are often expressed in units of standard deviations by standardizing phenotypic values in the parental population to have a mean of 0 and a variance of 1.

Imagine now that those individuals with phenotypic values above the selection threshold in the P1 population mate at random and then produce a large population of progeny that are reared in the same environment as the parents. The phenotypic distribution of the progeny is shown in the lower panel of Figure 9.10. The progeny have a mean phenotypic value of $\mu' = 11.0$. The difference between the F1 population phenotypic mean and the phenotypic mean of the entire P1 population expresses how much the phenotypic mean was changed by selection for larger trait values in the parents. The response to selection is

$$R = \mu' - \mu = 11.0 - 10.0 = 1.0 \qquad (9.17)$$

The phenotypic mean in this example was increased one unit by directional selection in the P1 population.

The response to selection is a function of the amount of phenotypic variation that is due to additive genetic variation according to the breeder's equation. Since both the response to selection and the selection differential are known, it is then possible to estimate the heritability by rearranging the breeder's equation

$$\hat{h}^2 = R/s \qquad (9.18)$$

When estimated in this way, $\hat{h}^2$ is called the **realized heritability** since it is estimated from the observed response to selection rather than predicted by resemblance of parents and progeny in the absence of selection. Using the selection differential and response to selection calculated for Figure 9.10, the realized heritability is then

$$\hat{h}^2 = 1.0/2.5 = 0.40 \qquad (9.19)$$

This tells us that 40% of the variance in trait values in the parental population was caused by additive genetic variation based on the definition of heritability.

Why was there a response to selection? Why did μ' increase in value compared to μ? The phenotypic value in the selected group of parents was greater than the rest of the parental population partly due

to the alleles that they possessed in their multilocus genotypes for this trait. When they bred, these alleles were passed down to their offspring. Selection changed the frequency of alleles that confer larger trait values because allele frequencies in the P1 individuals above the selection threshold were different than in the P1 population as a whole. Alleles that contributed to larger trait values became more frequent in the progeny population than they were initially in the parental population. Therefore, the mean phenotypic value in the progeny population increased relative to the mean phenotypic value of the parental population.

But why is the progeny mean phenotypic value (μ') not equal to mean value of the selected parents (μ_S)? Parents in the selected group had phenotypic values above the truncation point partly due to causes other than the effects of the alleles in their multilocus genotypes for the trait. Part of the phenotypic variance in the parental generation (V_P) was caused by factors that do not contribute to the resemblance of parents and offspring. The genotypic values of the selected parents were due to the combination of alleles in their genotypes. Such genotype effects cause dominance variance (V_D) and interaction variance (V_I) in quantitative trait values. However, these components of the genotypic variance are not inherited and do not contribute to resemblance between parents and offspring on average. In addition, some of the phenotypic variance in the parental population could have been caused by environmental variance (V_E) that would also not contribute to resemblance between the selected parents and the offspring. In the example of Figure 9.10, 60% (or 1 $- \hat{h}^2$) of the variation in the P1 phenotypic values was caused by the combination of nonadditive genetic variation ($V_D + V_I$) and environmental variation (V_E). This 60% is the percentage of the selection differential that did *not* produce a response to selection.

Estimating heritability by parent–offspring regression

Another method used to estimate the heritability based on the resemblance between parents and their offspring is parent–offspring regression. Parent–offspring regressions predict V_A without actually carrying out a response-to-selection experiment. This method to estimate heritability is, therefore, applicable to populations where selection experiments cannot be carried out. The estimation of heritability by parent–offspring regression can even be carried out in natural populations if a reliable method to identify the parents of offspring, such as paternity analysis, is available. This method takes advantage of the fact that the phenotypes of offspring resemble the phenotypes of their parents to the extent that phenotypic values are caused by shared alleles. Neither dominance nor epistasis components of the genotypic variance are inherited by progeny so they do not cause progeny to resemble their parents. This phenomenon was explained in the first section of the chapter and illustrated in Table 9.3. Now, we will revisit the resemblance between parents and offspring in greater depth.

By comparing the phenotypes of parents and their offspring, it is possible to determine the degree of phenotypic resemblance. The necessary data come from the measures of phenotypic values for pairs of parents as well as the phenotypic values of all progeny produced by each pair of parents. Figure 9.11 illustrates plots of parent–offspring regressions using one-locus genotypes assuming random mating among the parents. In both panels, the slope of the regression line between the phenotypic values of offspring and the average phenotypic values of their parents (called the **mid-parent value**) provides an estimate of phenotypic resemblance between parents and offspring. Because resemblance between parents and offspring must be due to inherited alleles rather than the effects of genotypes, the degree of resemblance is a function of the heritability. Therefore, the slope of the regression line estimates the heritability.

The parent–offspring regression for additive gene action is shown in Figure 9.11A. With strictly additive effects of alleles, the phenotypic values of all progeny are determined by counting up the number of a and A alleles in their genotypes. Under additivity, the a allele contributes 2.5 and the A allele contributes 7.5 to the genotypic value. The mean phenotype of all progeny from one pair of parents corresponds exactly to the average number of A and a alleles transmitted to their progeny. For example, mating aa and Aa parents results in 75% a and 25% A alleles transmitted to progeny. The average progeny phenotype is then $0.75(2.5) + 0.25(7.5) = 3.75$ for each allele in its genotype. Since there are two alleles, the average value among all progeny is $2(3.75) = 7.5$. This corresponds exactly to the aa × Aa mid-parent value. In the additive case, this same logic can be applied to all parental matings. The result is always that the mid-parent phenotypic value equals the average progeny phenotypic value.

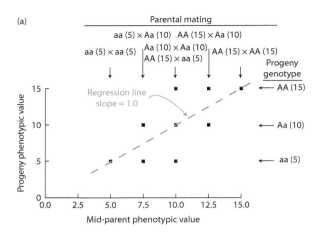

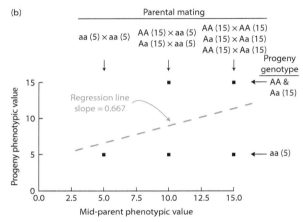

Figure 9.11   Parent–offspring regressions used to estimate heritability (h^2) under the assumption of strict additivity (a) or complete dominance (b). The resemblance or covariance between the mid-parent phenotypic value and the mean progeny phenotypic value is greater on average with additivity (slope = 1.0) than with dominance (slope = 0.667). The slope of the regression line is equal to the heritability since it measures resemblance of parents and offspring. The mid-parent phenotypic value is the average phenotypic value of two parents that mate and produce progeny. Phenotypic values are given in parentheses next to each genotype. Each dot represents the mean phenotypic value of many progeny that result from a given parental mating. Both panels assume Hardy–Weinberg expected genotype frequencies in the parental populations, allele frequencies of $p = q = 0.5$, and that $V_E = 0$.

The parent–offspring regression for complete dominance is shown in Figure 9.11B. We again see that when the combination of alleles in a genotype defines the phenotypic value of a heterozygote, the resemblance between parents and their offspring is not as large. In Figure 9.11B, dominance causes the AA and Aa genotypes to have identical phenotypes. This means, for example, that while two Aa genotypes have an average phenotypic value of

15, their progeny have an average phenotypic value of $0.25(5) + 0.5(15) + 0.25(15) = 12.5$. Dominance masks the presence of a alleles in the heterozygotes. These a alleles result in the production of a portion of aa progeny in some parental matings with a phenotypic value of 5 that causes the progeny average to be less than 15. With dominance, the progeny mean phenotype no longer corresponds to a weighted average of the number of a and A alleles in their genotypes.

Interact box 9.1 Estimating heritability with parent–offspring regression

Populus can be used to simulate parent–offspring regressions while varying the population allele frequencies, the degree of dominance, and the sample size of families.
Go to the text web page to see a step-by-step guide as well as questions to answer based on simulation results.

It is possible to prove that the slope of the regression line for offspring values and mid-parent values estimates the heritability. The slope of a regression line, b, between variables x and y is defined by

$$b = \frac{\text{cov}(y, x)}{\text{var}(x)} \qquad (9.20)$$

Based on the probability that half-sibling progeny share alleles identical by descent (see Section 10.6), the expected covariance between the mid-parent values and offspring values is

$$\text{cov}(\text{offspring}, \text{mid} - \text{parent}) = \tfrac{1}{2}\, V_A \qquad (9.21)$$

It is also the case that the expected variance of the mid-parent values is equal to $\tfrac{1}{2}V_P$. This comes about since the variance of the phenotypic values of a large sample of *pairs* of parents (say n pairs where n is large) is expected to be half the variance of a large sample of individual parents (or $2n$ individual parents) because of the smaller value of n used to compute the variance. In other words, if the phenotypic variance of a large number of individuals is x, then the expected phenotypic variance of one-half of these

individuals is $\tfrac{1}{2}x$. Bringing these two points together, we can restate the regression coefficient as

$$
\begin{aligned}
b &= \frac{\text{cov}(\text{offspring}, \text{mid} - \text{parent})}{\text{var}(\text{mid} - \text{parent})} \\[2mm]
&= \frac{\dfrac{1}{2} V_A}{\dfrac{1}{2} V_{P(mid-parent)}} = \frac{V_A}{V_P}
\end{aligned}
\qquad (9.22)
$$

to prove that the slope of the regression line estimates the heritability.

Response to selection on correlated traits

Our exploration of the heritability and the response to selection thus far has focused on single traits in isolation. While this approach helped to introduce these concepts, it is an unrealistic perspective on how quantitative traits evolve under natural selection in actual organisms. Phenotypes are not actually isolated, completely compartmentalized units that are unrelated. In contrast, many phenotypes are highly inter-related in the sense that they are not independent. Correlations among trait values can be manifest in two forms. The values for two traits may simply be correlated within individuals, a phenomenon called **phenotypic correlation**. The degree to which trait values are inherited in common is measured by the **genetic correlation**. As a simple example of these correlations, think of domestic cats and their traits of body weight and tail length. If in a population of individuals, heavier cats tend to have longer tails, their weight and tail length show a phenotypic correlation. Imagine that artificial selection was applied to the cat population for greater body weight and that body weight responded to this selection because of additive genetic variation. An associated increase in tail length at the same time, in the absence of any selection on tail length, would demonstrate a genetic correlation between tail length and body weight.

Genetic correlation: Non-independence of inherited values for two traits. A correlation between the breeding values of two quantitative traits.
Phenotypic correlation: Non-independence of the values of two or more quantitative traits within individuals.

Phenotypic correlation between traits within the same individual may be caused by the environment. For example, environmental conditions may tend to have the same effect on several traits resulting in a positive correlation or have opposite effects resulting in a negative correlation. A genetic component of a phenotypic correlation, or a genetic correlation, can have two causes that are not mutually exclusive. One cause is pleiotropy, or the phenomenon where a single locus contributes to variation in two or more phenotypes. Another cause is genetic linkage such that two distinct loci, one locus affecting one trait only, are in strong gametic disequilibrium. Both situations lead to correlated changes in phenotypes for two or more traits when genotype frequencies change in a population.

Both phenotypic and genetic correlations are represented mathematically in matrices that have as many rows and columns as there are phenotypes being considered. The values that represent the phenotypic variance and covariance are contained in a **P matrix**. The diagonal elements in **P** represent trait variances, whereas the off-diagonal elements represent the covariance for each pair of traits. Similarly, the **G matrix** contains individual values that measure both the additive genetic variance and covariance. The diagonal elements in a **G** matrix are simply the additive genetic variances for each trait. The off-diagonal elements in **G** represent the additive genetic covariance for each pair of traits, or the degree to which trait values are co-inherited. Table 9.4a shows each element in **G** and **P** matrices for the case of two traits.

To see how the response to selection for one trait is a special case for a single trait because there are no

Table 9.4 Examples of response to selection for two phenotypes with the possibility of phenotypic or additive genetic covariance. The elements of the phenotypic variance-covariance matrix (**P**), the additive genetic variance-covariance matrix (**G**), the vector of selection differentials (**s**), and the vector of predicted changes in mean phenotype ($\Delta \bar{z}$) are shown in A.

A.

G =	Trait A	Trait B
Trait A	h^2	genetic cov(A, B)
Trait B	genetic cov(A, B)	h^2

P =	Trait A	Trait B
Trait A	variance(A)	phenotypic cov(A, B)
Trait B	phenotypic cov(A, B)	variance(B)

s = [selection differential trait A, selection differential trait B]
$\Delta \bar{z}$ = [change in mean of trait A, change in mean of trait B]

B.

G = 0.5	0		**P** = 1.0	0
0	0.5		0	1.5

s = 0.5, 0.5
$\Delta \bar{z}$ = 0.25, 0.1667

C.

G = 0.5	0		**P** = 1.0	0.6
0	0.5		0.6	1.5

s = 0.5, 0
$\Delta \bar{z}$ = 0.3289, −0.1316

D.

G = 0.5	0.6		**P** = 1.0	0
0.5	0.5		0	1.5

s = 0.5, 0
$\Delta \bar{z}$ = 0.25, 0.20

phenotypic or genetic correlations, the breeder's equation can be re-written as

$$R = (\mathbf{G}/\mathbf{P})\mathbf{s} \qquad (9.23)$$

where $\mathbf{G}$ is the additive genetic variance, $\mathbf{P}$ is the total phenotypic variance, and $\mathbf{s}$ is the selection differential. In the breeder's equation for one trait, $\mathbf{P}$ is implicitly set to one so that the response to selection is then expressed per unit of phenotypic variance. For example, if $\mathbf{G} = 0.5$, $\mathbf{P} = 1$, and $\mathbf{s} = 0.5$ phenotypic units for a single trait, then the population would have an increase of $R = 0.25$ phenotypic units in the next generation. A larger value of $\mathbf{P}$ would yield a smaller response to selection (e.g. $\mathbf{P} = 2$ yields $R = 0.125$), whereas a smaller value of $\mathbf{P}$ would produce a greater response to selection (e.g. $\mathbf{P} = 0.5$ yields $R = 0.5$). The phenotypic variance plays a role in response to selection since the selection differential is relative to the trait variance. In order to compare traits or populations with different values of $\mathbf{P}$ and $\mathbf{G}$, the additive genetic variance is scaled to the total phenotypic variance. This change of scale assures that the additive genetic variance is then equal to the heritability, a quantity that is independent of the magnitude of the total phenotypic variance. To see this scaling, imagine that $\mathbf{P}_1 = 1.0$ for one trait and $\mathbf{P}_2 = 5$ for another trait but that both traits have $\mathbf{G} = 0.3$. The additive genetic variance relative to the total phenotypic variance is $\mathbf{G}/\mathbf{P}_1 = 0.3/1 = 0.3$ for trait one and $\mathbf{G}/\mathbf{P}_2 = 0.3/5 = 0.06$ for trait two. Dividing $\mathbf{G}$ by $\mathbf{P}$ expresses the additive genetic variance per unit of phenotypic variance to give the heritability. In this example, the heritabilities are $h^2 = 0.3$ for trait one and $h^2 = 0.06$ for trait two, so, clearly, trait one would show a greater response to selection for a given value of the selection differential.

With the potential for different phenotypic variances for each trait, covariance between phenotypes as well as genetic correlation, the breeder's equation needs to be extended. For two or more traits, the change in mean phenotype for each trait is predicted by

$$\Delta \bar{\mathbf{z}} = \mathbf{G} \mathbf{P}^{-1} \mathbf{s} \qquad (9.24)$$

where $\mathbf{P}^{-1}$ indicates a matrix inverse (Lande 1979; Lande and Arnold 1983). For more than one trait, the change in the mean trait value is represented by the vector $\mathbf{z}$ rather than the scalar R since there are as many means as there are traits. The selection differential is still symbolized as $\mathbf{s}$ even though it is now a vector.

G matrix: The genetic additive variance/ covariance matrix that quantifies the additive genetic variance of each trait (the diagonal elements) as well as the genetic covariance between all pairs of traits (the off-diagonal elements).

P matrix: The phenotypic variance/ covariance matrix that quantifies the variance of each trait (the diagonal elements) as well as the covariance between each trait (the off-diagonal elements).

s: The selection differential, written as a vector of values when there are two or more traits.

z̄: The phenotypic mean, written as a vector of values when there are two or more traits.

Some examples will help illustrate how Eq. 9.21 serves to combine the direct effect of selection on each trait along with the indirect effects of genetic and phenotypic correlations between traits to predict the total change in trait mean. In Table 9.4b, there is no genetic correlation and also no phenotypic correlation, making the traits completely independent. Both traits have heritabilities of 0.5 and selection differentials of 0.5. The response to selection is exactly we would predict from the single-trait version of the breeder's equation for each trait separately. Note that trait B has a lower response to selection because it has more phenotypic variance, making the selection differential of 0.5 effectively weaker.

In the example of Table 9.4c, there is a fairly strong positive phenotypic correlation between the two traits and natural selection applied only to trait A but still no genetic correlation between traits. The mean of trait A is predicted to increase since it is experiencing natural selection and has a nonzero heritability. Trait B also shows response to selection, in this case, a reduction in mean value. This change in mean is due to the correlation between the two traits alone and not due to any direct natural selection in trait B since its selection differential is zero.

In the final example in Table 9.4d, there is a strong positive genetic correlation between the two traits and natural selection is acting to increase the average of trait A. There is now no phenotypic correlation between the traits. The means of both traits are predicted to increase in this case. The mean of trait A will increase because of selection acting directly on it. At the same time, the mean of trait

B will also increase. This occurs not because selection is acting on trait B, after all it has a selection differential of zero, but rather because the two traits are genetically correlated. The change in genotype and allele frequency caused by response to selection on trait A has also changed genotype and allele frequencies that influence the mean of trait B. Direct selection for an increase in the mean of A indirectly causes an increase in the mean of B due to a genetic correlation between the traits. To distinguish direct and indirect effects of natural selection, the change in a trait mean due to a direct effect is called **selec-**

Interact box 9.2 Response to natural selection on two correlated traits

Solving the equation $\Delta \bar{z} = GP^{-1}s$ requires the use of matrix algebra. For those who have access to the program Matlab, the text web page has a short program that can be used to define **G**, **P**, and **s** and then solve for $\Delta \bar{z}$.

As an exercise, change the sign of both the phenotypic and genetic covariances shown in the examples of Table 9.4. How does predicted response to selection change?

---

tion for while **selection of** describes a change in a trait mean caused by an indirect response to selection on a genetically correlated trait.

When traits are genetically correlated, natural selection, and any response, is potentially not as simple as it would be with a single trait in that is completely independent of all other traits (see Morrissey et al. 2010). This is particularly true of quantitative traits related to Darwinian fitness of individuals. First, natural selection experienced by one trait could lead to a response to selection in another trait that does not directly experience natural selection. This means that natural selection on correlated traits is capable of indirectly changing traits that have little or no relationship with fitness. Second, when two traits respond to selection over time, it may lead to the evolution of a negative genetic correlation that prevents further response to selection. If two traits are both related to fitness and each experience natural selection, then we expect alleles at any loci that independently cause variation in either trait to

become fixed by long-term response to selection. However, any loci that have opposite effects on the two traits will not experience fixation or loss caused by natural selection. For example, if increased frequencies of AA genotypes increase fitness of trait one but simultaneously lead to decreases fitness for trait two, natural selection should result in neither the A nor an allele fixing. Alleles at such loci with contrasting effects cannot be fixed by selection because changing the allele frequency to increase fitness for one trait simultaneously causes a decrease in fitness for the other trait. Genetic variation at loci with such contrasting effects on traits causes a negative genetic correlation and prevents further change in trait means by natural selection.

Many examples of correlated responses to natural selection have been observed in agricultural organisms since artificial selection is routinely practiced to alter quantitative traits to increase yield and improve growth and harvest phenotypes. In one experiment, pigs were subjected to one generation of artificial selection for increased litter size in sows (Estany et al. 2002). During the course of this artificial selection experiment, the progeny of the selected females were measured for a number of morphological and behavioral phenotypes between the ages of 75 and 165 days old. A direct response to artificial selection was observed, increasing the litter size of the selected sows by an average of 0.46 piglets per litter compared to unselected controls. At the same time, the progeny of the selected sows showed a number of phenotypic differences from the control progeny of unselected sows. The pigs from the selected sows deposited fat more rapidly, grew at different rates, and gained less weight per kilogram of feed. The pigs in the selected and control populations also showed behavioral differences, with the progeny of the selected sows visiting feeding stations less frequently but spending more time eating and eating more per feeding bout. Since only the number of piglets per litter was under artificial selection, all of the other changes observed in quantitative traits were the result of correlated responses to selection caused by genetic correlations.

Long-term response to selection

The breeder's equation predicts response to selection over single-generation intervals. Extrapolating the breeder's equation to longer time periods implicitly assumes that h^2 and **s** remain constant through time. However, when natural or artificial selection continues for many generations, the assumptions of the breeder's equation may no longer hold. In particular, additive genetic variation may be consumed by response to selection over time as allele frequencies at the multiple loci that cause quantitative trait variation change over time. How rapidly additive genetic variation is exhausted by selection depends critically on the number of loci that underlie a quantitative trait as well as the percentage of trait variation that is caused by each locus (Figure 9.12).

Genotypic variation in quantitative traits can sometimes be caused by the alleles segregating at a relatively small number of loci. When quantitative trait variation is caused by a small number of loci, additive genetic variance for a trait is depleted over time by response to selection. The decline in heritability occurs because response to selection causes allele frequencies at the loci that cause genotypic variation to move toward fixation and loss. When only a few loci explain genetic variation in a quantitative trait, changes in the trait mean from one generation to the next come about due to substantial changes in the allele frequencies of those few loci. For example, when there is artificial selection for an increased trait mean as in Figure 9.10, alleles that confer higher values of the trait at each locus are increased in frequency each generation. The greater a response to selection that has occurred, due to either continued selection over time or stronger selection, the more likely that allele frequencies at each locus will have been altered toward fixation and loss (Figure 9.12A). In addition, genotypic variation in all traits will decrease over time in finite populations due to genetic drift, a process that is accelerated by selection since selection itself leads to reduced effective population sizes because not all individuals in the population contribute alleles to the next generation of progeny.

An alternative model is that the additive genetic variation in a quantitative trait is caused by a very large number of individual loci, and all of these many loci have equal very small effects on a quantitative trait. Under these assumptions, response to selection may continue for a long time before changes occur in amount of additive genetic variance (Figure 9.12B). Under this **infinitesimal model**, as the number of loci grows very large, then the amount of trait variation explained by any one locus approaches zero. Under the many loci assumption, when a quantitative trait responds to selection, the trait mean changes but there is almost no change in the allele frequencies of the individual loci that cause genotypic variation because each locus explains such a small fraction of the additive genetic variation. Under the infinitesimal model then, response to selection can occur for many generations without causing substantial changes in the heritability required for response to selection. Note that even under the infinitesimal model response to selection acts to increase gametic disequilibrium for the loci that cause genotypic variation, potentially slowing response to selection over time.

> **Infinitesimal model**: A model of the genetic basis of quantitative traits that assumes a very large number of independent loci contribute equally to trait genotypic variation so that the impact of each locus on trait genotypic variation approaches zero.

The amount of additive genetic variation for a quantitative trait over many generations is not simply a function of the amount of standing genetic variation before natural selection starts. Rather, the amount of additive genetic variation over time is the net outcome of natural selection and genetic drift working to reduce genetic variation balanced by mutation introducing novel genetic variation. Depending on the number of loci influencing a trait and the mutation rate, depletion of additive genotypic variance in quantitative traits caused by long-term selection can be counteracted by the addition of new alleles through mutation. Figure 9.12C shows an example of the impact of recurrent mutation on a quantitative trait that is the product of 100 loci. Even though 90 of the loci start out fixed for one allele, mutation at these loci over time produces enough additive genetic variation to maintain trait variation and a linear response to selection over many generations. Ultimately, the steady-state level of additive genetic variation for a quantitative trait is a product of genetic drift, mutation, and selection all acting on the loci that cause variation in the trait. Because of this, long-term response to selection will depend on numerous parameters in a population such as the effective population size, the mutation rate, and the selection differential along with the number and distribution of effects of the loci that underlie the quantitative trait.

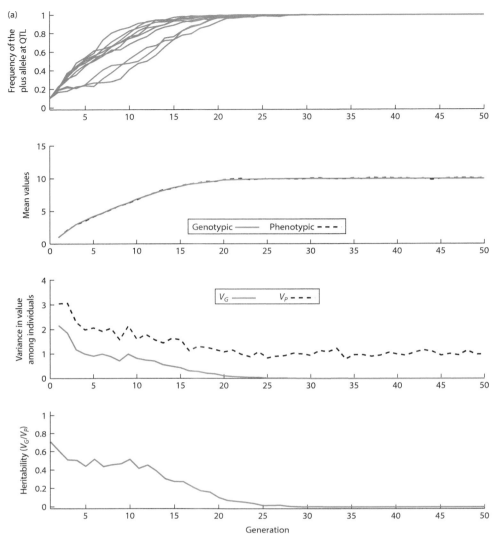

Figure 9.12 Simulations of directional selection on a quantitative trait with strictly additive genetic variance under three scenarios for the genetic architecture of the trait. From top to bottom, the graphs show the plus allele frequencies for 10 loci (in panels B and C 10 loci are shown of the total 100 loci that influence the trait), the genotypic and phenotypic mean values for the trait, genotypic and phenotypic trait variance (V_P and V_G), and the narrow-sense heritability. In panel A, there are 10 loci each with equal and large effects on the trait (10% of V_G). In panel B, 100 loci have equal small effects (each 1% of V_G) on the trait. In panel C, there are two loci with large effects (20% of V_G each) 98 loci with small effects as well as recurrent mutation between plus and minus alleles ($\mu = 0.001$ mutations gamete^{-1} generation^{-1}). In panel C, the initial allele frequencies for the two loci of large effect (dashed allele frequency lines) and eight loci of small effect are initially 0.1, while the remaining 90 loci are fixed for the minus allele. A selection plateau is reached in A when all of the loci causing genotypic variation in the trait reach fixation (V_G and h^2 drop to zero). Simulations in B and C never reach selection plateaus nor exhaust all trait variation because at least some loci that influence trait variation remain segregating. Genetic variation is maintained in B because selection in finite populations is not able to fix alleles at all loci with small effects. Even though the two loci with major effects fix rapidly in C, recurrent mutation at the many loci with small effects maintains some additive genetic variation for continued response to selection. Selection was accomplished by forming the next generation with the 50 individuals with the largest phenotypic values. In all simulations, the maximum genotypic value was 10.0, $V_E = 0.1$, and the truncation point for natural selection each generation was the 50th percentile of phenotypic value.

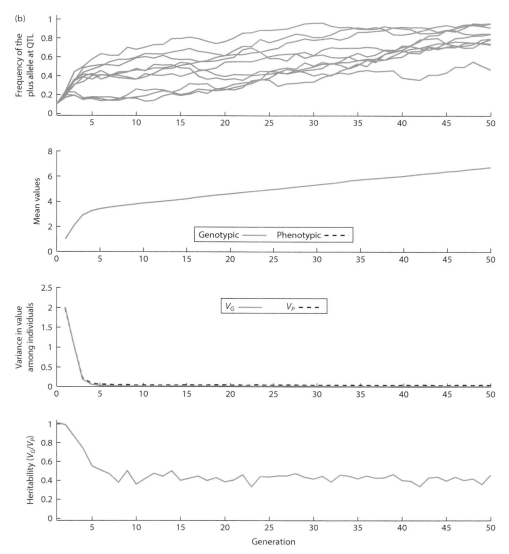

Figure 9.12 (*continued*)

Interact box 9.3 Response to selection and the number of loci that cause quantitative trait variation

Use the **text simulation web site** to simulate response to selection for a quantitative trait that has genotypic variation caused by either few loci or by many loci. The key parameter to vary is the number of loci that underlie the quantitative trait, nQTL. Select the radio button for **All loci have equal effects** and compare several runs of nQTL = 10 with several runs of nQTL = 200. Set N_e = 1000 to minimize drift and set the natural selection phenotypic value truncation point = 0.5 (only individuals in the upper 50% of phenotypic value reproduce), but leave the other simulation parameters at their default values. A run with nQTL = 200 may take a few minutes, so be patient.

How does the heritability change over time with 10 or 200 QTL? How does an increase in environmental variation (such as V_E = 1.0) influence the pattern?

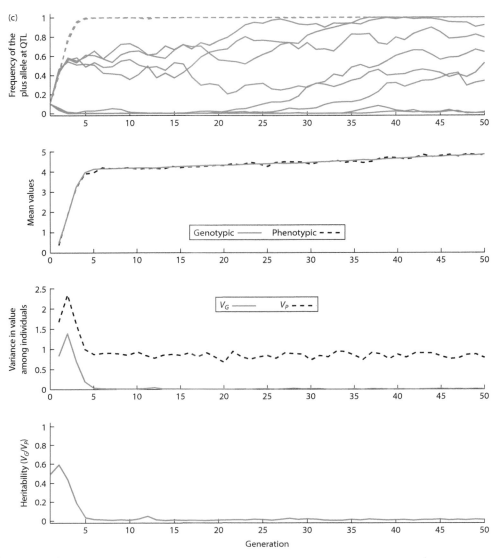

Figure 9.12   (*continued*)

The Illinois Long-Term Selection experiment provides one of the longest records of continuous response to selection for the phenotypes of oil content and percentage of protein in corn kernels (reviewed by Moose et al. 2004). This long-term selection experiment was initiated in 1896 with the goal of determining whether artificial selection could be used to develop strains of corn with kernel phenotypes that were improved from the perspective of animal feed and crop processing. Divergent selection for both higher and lower oil and protein content has been practiced for over 100 generations by using the highest and lowest scoring 20% of ears each generation to form the next generation (Figure 9.13). From an initial value of 4.7% oil

content, response to selection steadily changed oil content to about 22% in the high line and to 0% in the low line. Response to selection was similarly linear for protein content, starting at 10.9% and reaching 32.1% in the high line and 4.2% in the low line.

The Illinois Long-Term Selection experiment showed that response to selection is relatively steady and linear over a long period of time. (Slight fluctuations in the mean phenotype over time may be explained by some variation in the selection differential each generation as well as by environmental variation.) The observed change in phenotypic means in the Illinois Long-Term Selection experiment looks similar to the predicted response to selection on traits

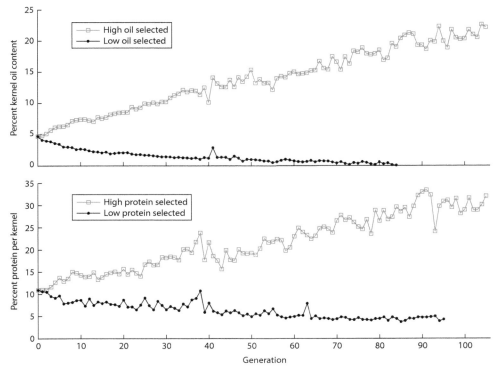

Figure 9.13 Phenotypic means for oil and protein content for high and low selected lines of the Illinois Long-Term Selection experiment initiated in 1896. Response to selection has been nearly linear over time, consistent with additive genetic variation that is caused by a relatively large number of loci each with a small effect on the phenotype. The low oil selected line was discontinued in generation 89 since oil content was too low to be measured reliably and plants had poor viability. Source: Data kindly provided by J.W. Dudley and S.P. Moose.

with many loci in the simulation results shown in Figure 9.12B and C. The results are, therefore, consistent with genetic variation in the oil- and protein-content phenotypes being caused by a relatively large number of loci and the possibility that mutation may have contributed some genetic variation over time.

Another example is long-term selection carried out for 70 generations to increase the percentage of body muscle, measured as protein content, in mice at 42 days old (Bünger et al. 1998). Figure 9.14A shows the mean amount of protein per individual over the course of the experiment. Selection increased protein content rapidly at first, but then response to selection slowed as the experiment continued. This is a classic example of a **selection plateau** where continued natural selection shows a diminishing response over time such that the trait mean asymptotes toward a constant value even though selection is still being applied to increase the trait mean. The additive genetic variance (V_A) and the heritability (h^2) shown in Figure 9.14B explain why response to selection decreased over

time. The additive genetic variance, and, therefore, the heritability, decreased through time so that the response to selection was not constant over the 70 generations of the experiment. The observed reduction over time in the response to selection and the selection plateau are similar to the predicted response to selection on traits with few loci of large effects shown in Figure 9.12A. The response to long-term selection for protein content in mice is, therefore, consistent with genetic variation caused by a relatively small number of loci having relatively large effects on the phenotype.

Two additional phenomena can cause limits to selection response. The first process is the accumulation of gametic disequilibrium that alters the amount of additive genetic variance. The additive genetic variance (V_A) can be decomposed into two parts:

$$V_A = V_a + D \qquad (9.25)$$

where V_a is the additive genic variance that is not impacted by gametic disequilibrium, and D is the gametic disequilibrium coefficient. When there is

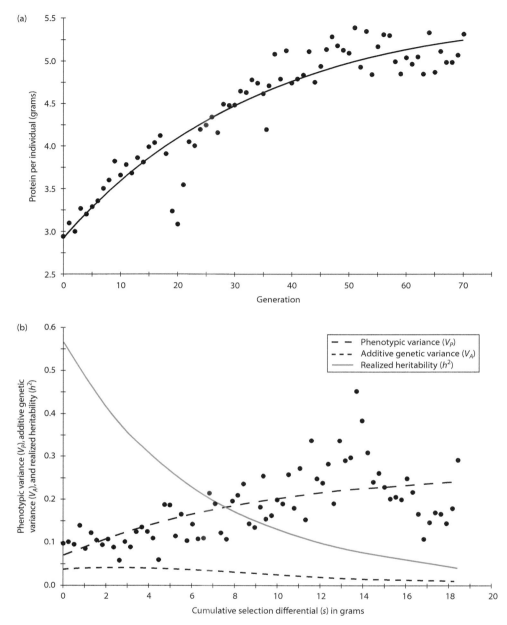

Figure 9.14 Long-term selection for muscle mass in mice (measured as protein content per individual) over 70 generations. Panel A shows the phenotypic mean over time, with a pronounced asymptote that indicates a diminishing response to selection over time or selection plateau. The total phenotypic variance, the additive genetic variance, and realized heritability are shown in panel B. Even though the selection differential is constant over time, the heritability declines steadily, as expected for a quantitative trait where genetic variation is caused by relatively few loci. The dip in protein content during generations 18–20 was an artifact due to an environmental effect. While the phenotypic variance increases over the experiment (panel B), this is caused largely by the increase in the phenotypic mean (the coefficient of variation of V_P stays nearly constant). Source: From Bünger, L., Renne, U., Dietl, G., and Kuhla, S. (1998) Long-term selection for protein amount over 70 generations in mice. *Genetical Research* 72: 93–109. Reprinted with the permission of Cambridge University Press.

gametic equilibrium ($D = 0$), then the additive genetic variance is all caused by variance in alleles, or genic variance. However, when there is some level of gametic disequilibrium, then the additive genetic variance can be reduced (negative D) or increased

(positive D). Directional and stabilizing natural selection tend to cause negative gametic disequilibrium because individuals with similar phenotypic values tend to have a correlated set of alleles at each of the loci that contribute to the trait (also see the example in Chapter 2). In contrast, disruptive selection tends to cause positive gametic disequilibrium. If selection is strong relative to recombination, then gametic disequilibrium will alter the additive genetic variance and thereby the heritability. A reduction in the response to selection or a selection plateau can occur when negative gametic disequilibrium caused by selection has accumulated such that V_A declines even though the loci the underlie the trait have not all reached fixation and loss. For more details on this complex topic, consult Bulmer (1985) and Walsh and Lynch (2018).

The other process that can limit response to selection is called **antagonistic pleiotropy**. It occurs when response to selection changes the mean of one trait (selection for) as well as the mean of a correlated trait (selection of). While selection for increased fitness drives change in the mean of one trait over time, the correlated change in the mean of the other trait may actually decrease fitness. After some response to selection has occurred, the fitness trade-offs between the two traits may reach a point where further change in the mean one trait that increases fitness is offset by the negative fitness consequences of change in the mean of a correlated trait. When such fitness trade-offs for correlated traits exist, they will ultimately limit response to selection even when additive genotypic variation exists for both traits. Examples of such trade-offs for correlated traits have been hypothesized or observed for a range of traits. One possible trade off maintained by antagonistic pleiotropy are alleles that tend to increase reproduction at early ages but decrease lifespan (Williams 1957). In support of the hypothesis that survival and reproduction have a fitness trade off, Silbermann and Tatar (2000) have shown that a heat-induced protein expressed by the *hsp70* locus influences both egg hatching rate and survival rate in *Drosophila melanogaster*. Higher levels of *hsp70* expression lead to longer life spans but at the same time lead to lower rates of egg hatching. If reproduction and survival in *D. melanogaster* are associated by antagonistic pleiotropy via *hsp70*, then response to any selection acting on these traits will not be able to exhaust all the additive genetic variation if the highest fitness is a balance of intermediate levels of both traits.

The genetic variance/covariance or **G** matrix has also become a focus of research to better understand how the genetic basis of potentially nonindependent quantitative traits changes over time with selection, genetic drift, and mutation (reviewed by Steppan et al. 2002; Jones et al. 2003). The **G** matrix represents the genetic inter-relationships among multiple traits so estimating it in a range of species and for a range of traits is a prerequisite to understand how phenotypes might respond to long-term natural selection. Understanding how the **G** matrix changes over time is also fundamental to understanding long-term response to selection. The predicted trait means in Eq. 9.21 rely on the constancy of the **G** matrix over time. If **G** changes rapidly or unpredictably over time, then predicted changes in trait means are not applicable over long periods of time. Rapid changes in **G** would also reduce the ability to infer past patterns of response to selection given a current estimate of **G**.

Neutral evolution of quantitative traits

Considering a quantitative trait as selectively neutral is useful to predict the action of basic processes that will reduce as well as contribute to additive genetic variation. If a quantitative trait is neutral, each locus that contributes to variation in the trait should be neutral as well. Therefore, we can employ expressions already developed to predict the consequences of genetic drift. Additive genotypic variation is expected to decrease with genetic drift over time because the alleles at the loci that form the basis of V_A will progress toward fixation and loss. The expected rate of decrease in V_A by genetic drift is

$$V_A^t = V_A^o \left(1 - \frac{1}{2N_e}\right)^t \qquad (9.26)$$

where V_A^o is the initial additive genotypic variation, N_e is the effective population size, and t is the number of generations that have elapsed. As explained in Chapter 3, the $\left(1 - \frac{1}{2N_e}\right)$ term in this equation predicts the decline in heterozygosity through time by genetic drift. Additive genetic variation is greater when heterozygosity is greater because alleles are at intermediate frequencies.

Next, we can predict the balance between the additive genotypic variation lost by drift and new additive variation gained due to new mutations at the loci that influence a quantitative trait. Let's start by reworking the equation for decline in V_A over an arbitrary time interval to instead predict the change in additive genotypic variation each generation due to drift:

$$\Delta V_A = -\frac{V_A}{2N_e} \qquad (9.27)$$

This equation is obtained from Eq. 9.26 from setting $t = 1$, multiplying the right side out, and then subtracting V_A^o from both sides. Then, assume that the amount of new additive genotypic variance caused by mutation each generation is V_A^M. The sum of additive variation lost by drift and additive variation gained by mutation is

$$\Delta V_A = -\frac{V_A}{2N_e} + V_A^M \qquad (9.28)$$

This defines the net change in additive genotypic variation due to the action of both drift and mutation. If we assume that a population is at drift–mutation equilibrium, then $\Delta V_A = 0$, allowing Eq. 9.24 to be rearranged to give

$$V_A = 2N_e V_A^M \qquad (9.29)$$

(see Lynch and Hill 1986; Bürger and Lande 1994).

The expected additive genotypic variance for a neutral quantitative trait at drift–mutation equilibrium was the basis of a much debated recommendation that $N_e = 500$ would be sufficient to maintain additive genotypic variation in populations of endangered species (Franklin 1980; Lande 1995). This recommendation came from rearranging Eq. 9.26 to solve for the effective population size:

$$N_e = \frac{V_A}{2V_A^M} \qquad (9.30)$$

Assuming that total phenotypic variation is caused only by V_A and V_E and using the definition of the narrow-sense heritability, V_A can be expressed as $V_A = \frac{h^2}{1 - h^2} V_E$. If $h^2 = 0.5$, then V_A in the numerator on the right side of Eq. 9.27 can be replaced with V_E. The final step was to utilize an estimate of the input of additive variation due to mutation available at the time that suggested $V_A^M \approx 0.001 V_E$. Putting this all together results in

$$N_e \approx \frac{V_E}{2(0.001 V_E)} \approx 500 \qquad (9.31)$$

as an estimate of the effective population size expected to maintain a heritability of about one-half.

Interact box 9.4 Effective population size and genotypic variation in a neutral quantitative trait

In the Quantitative Trait Loci (QTL) simulation module of the text web simulation web site, explore how the effective population size influences the level of genotypic variation for a selectively neutral quantitative trait. The key parameter to adjust is the number of individuals in the population. Try $N_e = 50$ and then $N_e = 500$ with nQTL = 10 and the other parameters at their default settings.

How much genotypic variance (V_G) do you find with these two effective population sizes? How do the levels of V_G over time with $N = 50$ or $N = 500$? How does the mutation rate and the environmental variance (V_E) in the trait influence the long-term heritability for each effective population size?

This recommendation for an effective population size target of 500 sparked a debate that centered around the numerous assumptions about the nature and causes of quantitative trait variation in natural populations. Lande (1995) pointed out that $V_A^M \approx 0.001 V_E$ was probably too high since the mutation rate needed to be discounted for the number of mutations that were highly deleterious and a more realistic assumption was $V_A^M \approx 0.0001 V_E$ by counting only nearly neutral mutations. This change in the mutational input of quantitative genetic variance each generation substantially increases the required effective population size to approximately 5000. Franklin and Frankham (1998) countered that assuming a heritability of 0.5 was larger than necessary because heritabilities for fitness-related traits were often around 0.1. They reasoned that even if $V_A^M \approx 0.0001 V_E$, an effective population size of about 550 was sufficient to maintain a heritability of 0.1. In contrast, inferred $\frac{V_A}{V_A^M}$ ratios were observed in a range of organisms to be between 30 and 300 which implies that N_e is between 15 and 150 (Houle et al. 1996; Lynch and Lande 1998). Since this effective population size is definitely too low for some of the species studied by Houle et al. (1996), such as

Drosophila, these findings then called into question the very assumptions of neutral drift–mutation equilibrium. An alternative explanation is that stabilizing natural selection is operating and causing reduced variation in quantitative traits and could explain the $\frac{V_A}{V_A^M}$ ratios rather than such low effective population sizes.

The perspective of neutral evolution for quantitative traits has also been employed to make inferences about the nature of the total genotypic variation (V_G). Genetic drift that occurs during genetic bottlenecks and founder events as well as consanguineous mating impacts the components of genotypic variation for neutral quantitative traits. In a seeming contradiction, genetic drift can produce a transient *increase* in additive genotypic variance for neutral quantitative traits that exhibit dominance (V_D) or epistasis (V_I) genotypic variance (Robertson 1952; Goodnight 1987, 1988; Willis and Orr 1993; Barton and Turelli 2004). For quantitative traits that exhibit only V_A and V_D, higher heterozygote frequencies produce more V_D and less V_A when the dominance coefficient (d) increases. Genetic drift causes an increase in homozygosity as allele frequencies change toward fixation and loss on average under random sampling. An increase in homozygosity is also a decrease in heterozygosity and thereby a decrease in V_D. For a single generation bottleneck of $N_e = 2$ and assuming no epistasis, Willis and Orr (1993) showed that the expected value of V_A over many replicate populations increases if $d > 0.29$. They also showed that as N_e during a bottleneck increases; the threshold dominance coefficient for V_A to increase approaches $d > 0.20$. The increase in V_A is greater for larger d and for lower initial frequencies of the recessive allele. These increases in V_A translate into increases in the heritability that also depend on the magnitude of V_E for the trait. Mating among relatives also acts to increase homozygosity while not altering allele frequencies, and so can also cause a reduction in V_D that leads to a relative increase in V_A.

These predictions about changes in heritability after genetic drift or consanguineous mating lead to a test of whether quantitative traits have genotypic variation that is exclusively additive or is a combination of additive and nonadditive genotypic variation. Van Buskirk and Willi (2006) reviewed the results of numerous studies that estimated heritability and V_A from both small or inbred populations as well as large randomly mated populations. They found that phenotypes closely associated with fitness (e.g. viability, fecundity, and body size) often did show an increase in heritability after consanguineous mating or population bottlenecks. In contrast, phenotypes not associated with fitness (e.g. morphological traits, bristle number, and oil content) showed only declines in heritability after consanguineous mating or population bottlenecks. This result is consistent with the explanation that V_G in fitness-related phenotypes was caused by both allele and genotype variation, whereas V_G in nonfitness phenotypes was caused by alleles alone in the organisms studied.

9.3 Quantitative trait loci (QTL)

- QTL mapping with single marker loci.
- QTL mapping with multiple marker loci.
- Limitations of QTL mapping studies.
- Genome-wide association studies (GWAS).
- Biological significance of identifying QTL.

Thus, so far in this chapter, quantitative traits have been described based on the mean and variance of values in a population. Quantities like the components of the total genotypic variance and the heritability illuminate population-level average qualities of phenotypes. However, the numerous loci that each contribute to such population variation have not been identified nor described. This final section of the chapter introduces the concepts and some methods needed to identify the individual loci that ultimately contribute to variation in quantitative traits.

Using the basic framework of phenotypic values and the population mean already established and joining it with genetic marker data for individuals, it is possible to identify individual regions in a genome that contribute to quantitative trait variation. The genomic regions that contribute to variation in quantitative traits are called **quantitative trait loci** or **QTLs**. In the simplest idealized case, individual QTLs are single genes that contribute to the value of a quantitative trait and have alternate alleles with different effects on phenotypic value. The trait mean and variance would then be the sum of the mean and variance contributed by each gene that affects the trait. In reality, QTLs are not necessarily individual genes but can be larger chromosomal regions held together by linkage that may contain several or many genes. It is these linkage blocks, each of which contains a genetic marker, that can be associated with an effect on the variance of a quantitative trait.

> **Candidate loci:** Loci that have known or inferred function and therefore are hypothesized to be causal contributors to genetic variation in a quantitative trait.
> **Genetic architecture:** The number of loci that underlie a quantitative trait and the magnitude of their contributions to quantitative trait genetic variation.
> **Quantitative trait locus or QTL:** A genome region associated by linkage, a gene, or even a single nucleotide possessing multiple alleles that affect the average value of a quantitative trait in a defined population associated through linkage with a genetic marker.

A major goal in identifying and describing the individual QTLs that cause quantitative genetic variation is to understand the **genetic architecture** of continuous phenotypes. By one definition, genetic architecture is all of the genetic and environmental factors that contribute to a quantitative trait, as well as their magnitude and their interactions (National Institute of General Medical Sciences 1998). More narrowly, genetic architecture often refers to the number of QTLs and the size of their effects on a quantitative trait. Identification of the number and phenotypic effects of QTLs has applications in many areas of biology. Identification of QTLs helps test the role, if any, of **candidate loci** (identified through independent molecular biology research or sequence analyses, for example) in explaining a portion of the genetic variance in quantitative traits. The reverse is also true, since loci identified by QTL mapping (which is described below) as causing some of the genetic variation in a quantitative trait are often further studied to better understand their function. In clinical settings, QTL mapping helps identify genes and alleles that cause disease conditions as well as genetic markers associated with disease QTLs that can be used to screen for disease risk. Identifying QTLs can improve the efficiency of animal and plant breeding, such as in the development of genetic markers associated with QTLs that can be used to screen individuals early in life for traits that may only be manifest later in life. One example would be screening tree seedlings for mature wood characteristics and planting those individuals with genetic marker genotypes associated with desirable phenotypes that appear only after many years of growth.

The number of QTLs and the size of their effects on a quantitative trait also have profound implications for evolutionary change such as the response to natural selection and the amount of variation generated by mutation. QTL mapping has the potential to deconstruct phenomena in quantitative genetics that have traditionally only been analyzed via variance components. In principle at least, QTL mapping can distinguish the specific causes of genetic correlations (pleiotropy or linkage), identify the loci and alleles that demonstrate dominance and epistasis, and show how specific loci involved in genotype-by-environment interactions respond to their environments.

QTL mapping with single marker loci

The process of identifying quantitative trait loci is called **QTL mapping** because it is based on the technique of linkage mapping that establishes the linear order of loci on chromosomes based on recombination frequencies. QTL mapping takes basic linkage mapping one step further by determining whether any of the mapped loci are associated with variation in a quantitative trait. QTL mapping requires variation for a quantitative trait within or between populations and numerous polymorphic genetic marker loci that are spread across the genome of the organism. The mapping is carried out for the marker loci, since, after all, the QTLs are unknown. QTLs are identified by differences in the mean phenotype of groups of individuals with different marker locus genotypes. Any association between phenotypic means and marker genotypes is caused by gametic disequilibrium between QTLs and marker loci when the two types of loci are close enough on a chromosome to be linked. Thus, QTL mapping uses known marker loci to detect the phenotypic signature of unknown QTLs that are in the same linkage block.

Genetic marker loci are a critical ingredient in QTL mapping. First and foremost, marker loci should be both independent of the phenotype(s) being mapped (i.e. not a QTL themselves) and also selectively neutral so that genotype frequencies are determined only by mating and recombination and not by viability or fecundity selection. Marker loci must be polymorphic because they are only informative when individuals with different phenotypic values possess distinct marker locus genotypes. Codominant marker loci where all possible genotypes are detectable present the easiest case to consider, although mating designs exist that can utilize dominant marker loci.

Before the present age of genomics, only loci with phenotypic effects or protein polymorphisms were used as markers, and QTL mapping was generally limited by the small number and low polymorphism of such marker loci. In the present day, QTL mapping is carried out using single nucleotide polymorphisms (SNPs) obtained with high-throughput sequencing, although other genetic markers such as anonymous length polymorphism (AFLP) and microsatellite or short tandem repeat (STR) loci can also be employed, and several types of marker loci can be combined.

Experimental mating designs are a major method of QTL mapping. One of the most basic breeding schemes for QTL mapping in organisms that can be raised and mated in captivity is called the **F2 design** or **recombinant inbred line design** (Figure 9.15). It starts with a population, perhaps, in the wild, that has considerable phenotypic variation as well as genetic variation at numerous molecular marker loci. From this original population, individuals are sampled to start inbred lines or subpopulations. The inbred lines are maintained by some type of consanguineous mating (e.g. selfing and brother–sister mating) for numerous generations. The inbred lines eventually contain individuals with a high probability of homozygosity for different alleles at the loci that cause variation in the phenotype (Q_1 and Q_2) as well as for the different alleles present at the genetic marker loci (M_1 and M_2). An individual is sampled from each of two inbred lines that exhibit different values for the phenotype(s) of interest as well as different homozygous genotypes at the marker locus. These individuals form the P1 generation in Figure 9.15.

When the P1 individuals are mated, they produce progeny that are all heterozygotes for both the QTL and the marker locus. Note that recombination events do not alter the gametes produced because each P1 individual is a double homozygote. The progeny of the P1 generation form the F1 generation. Figure 9.15 shows the four gametes that are produced by the F1 individuals and transmitted after random mating to their progeny in the F2 generation. Notice now that recombination between the QTL and the marker locus does influence gamete frequencies and thereby the genotype frequencies in the F2 generation. The QTL mapping analysis is based on two types of data for all F2 individuals: (i) the marker locus genotype is known and (ii) the phenotypic value is known.

The 10 unique two-locus genotypes in the F2 population are shown in Table 9.5, grouped into three

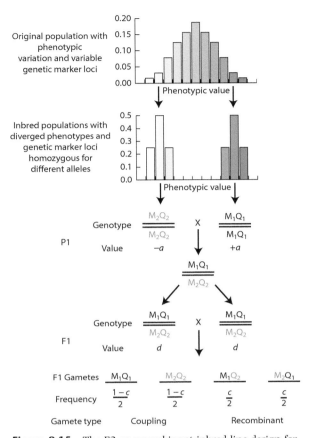

Figure 9.15 The F2 or recombinant inbred line design for QTL mapping assuming a on quantitative trait locus (Q) and one genetic marker locus (M). The top phenotypic distribution represents the variance in value in the population that is subject to QTL mapping. Individuals at the edges of this phenotypic distribution are then sampled to start lines to be inbred for 5 (self-fertilization) to 10 (sibling mating) generations in order to achieve high homozygosity. Individuals are then sampled from inbred lines with diverged phenotypes to form the P1 generation. The progeny of the P1 individuals are all double heterozygotes and form the F1 generation. Gamete frequencies produced by F1 individuals depend on recombination rates between loci. Random mating among F1 individuals produces the F2 individuals which are genotyped for the marker loci and measured for their phenotypic values.

classes based on the marker-locus genotype. Let's work through the steps needed to obtain the expected genotypic value of those F2 individuals with a M_1/M_1 marker locus genotype. The mean genotypic value is obtained by determining the frequency and the genotypic value of each QTL genotype that is associated with an M_1/M_1 marker genotype. There are three possible ways to make an M_1/M_1 marker-locus genotype in an F2 individual: (i) combine two M_1Q_1 coupling gametes, (ii) combine an M_1Q_1 coupling gamete with an

Table 9.5 Derivation of the expected phenotypic value for the three marker-locus genotypes when QTL mapping with a single marker locus associated with a single QTL. The three genetic marker genotypes may be associated with any of the three possible QTL genotypes because of recombination during the formation of F2 gametes. The difference between the M1/M1 and M2/M2 marker class means (expressions in the Marker-class mean value column) is equal to $2\hat{a}$. The phenotypic value of each marker locus genotype is a function of both the additive and dominance effects of the QTL (a and d) as well as the recombination rate (c). Unless there is no dominance and no recombination, estimates of QTL effects from single-marker-locus mapping are always minimum estimates. The gametes and expected gamete frequencies are given in Figure 9.15.

Gametes	F2 genotype	Genotype	Genotypic value	Frequency-weighted genotypic value	Marker genotype	Marker class contribution to F2 population mean value	Marker genotype frequency in F2 population	Marker class mean value
c/c	$\dfrac{M_1Q_1}{M_1Q_1}$	$\left(\dfrac{1-r}{2}\right)^2$	$+a$	$a\left(\dfrac{1-r}{2}\right)^2$	$\dfrac{M_1}{M_1}$	$\overline{G}^{pop}_{M_1M_1}=\dfrac{a(1-2r)}{4}+\dfrac{2dr(1-r)}{4}$	¼	$\overline{G}_{M_1M_1}=a(1-2r)+2dr(1-r)$
c/r	$\dfrac{M_1Q_1}{M_1Q_2}$	$(2)\dfrac{r}{2}\left(\dfrac{1-r}{2}\right)$	d	$2d\dfrac{r}{2}\left(\dfrac{1-r}{2}\right)$				
r/r	$\dfrac{M_1Q_2}{M_1Q_2}$	$\left(\dfrac{r}{2}\right)^2$	$-a$	$-a\left(\dfrac{r}{2}\right)^2$				
c/c	$\dfrac{M_1Q_1}{M_2Q_2}$	$(2)\left(\dfrac{1-r}{2}\right)^2$	d	$2d\left(\dfrac{1-r}{2}\right)^2$	$\dfrac{M_1}{M_2}$	$\overline{G}^{pop}_{M_1M_2}=\dfrac{d\left[(1-r)^2+r^2\right]}{2}$	½	$\overline{G}_{M_1M_2}=d[(1-r)^2+r^2]$
r/r	$\dfrac{M_1Q_2}{M_2Q_1}$	$(2)\left(\dfrac{r}{2}\right)^2$	d	$2d\left(\dfrac{r}{2}\right)^2$				
c/r	$\dfrac{M_1Q_1}{M_2Q_1}$	$(2)\dfrac{r}{2}\left(\dfrac{1-r}{2}\right)$	$+a$	$2a\dfrac{r}{2}\left(\dfrac{1-r}{2}\right)$				
r/c	$\dfrac{M_1Q_2}{M_2Q_2}$	$(2)\dfrac{r}{2}\left(\dfrac{1-r}{2}\right)$	$-a$	$-2a\dfrac{r}{2}\left(\dfrac{1-r}{2}\right)$				
c/c	$\dfrac{M_2Q_2}{M_2Q_2}$	$\left(\dfrac{1-r}{2}\right)^2$	$-a$	$-a\left(\dfrac{1-r}{2}\right)^2$	$\dfrac{M_2}{M_2}$	$\overline{G}^{pop}_{M_2M_2}=\dfrac{-a(1-2r)}{4}+\dfrac{2dr(1-r)}{4}$	¼	$\overline{G}_{M_2M_2}=-a(1-2r)+2dr(1-r)$
c/r	$\dfrac{M_2Q_2}{M_2Q_1}$	$(2)\dfrac{r}{2}\left(\dfrac{1-r}{2}\right)$	d	$2d\dfrac{r}{2}\left(\dfrac{1-r}{2}\right)$				
r/r	$\dfrac{M_2Q_1}{M_2Q_1}$	$\left(\dfrac{r}{2}\right)^2$	$+a$	$a\left(\dfrac{r}{2}\right)^2$				

M_1Q_2 recombinant gamete, and (iii) combine two M_1Q_2 recombinant gametes. Each of these possible genotypes will have a frequency in the F2 population that is a product of the respective gamete frequencies. Expected F2 genotype frequencies are, therefore, (i) $(\frac{1-c}{2})^2$ for the combination of two coupling gametes, (ii) $(\frac{c}{2})(\frac{1-c}{2})$ for the combination of a coupling and a recombinant gamete, and (iii) $(\frac{c}{2})^2$ for the combination of two recombinant gametes.

Each F2 genotype frequency must also be weighted to account for a genotype's relative abundance. Constructing a 4×4 Punnett square using the four possible F1 gametes that are shown in Figure 9.15 reveals that in the F2 population all double homozygote genotypes (e.g. $\frac{M_1Q_1}{M_1Q_1}$) have a frequency of 1/16, all Q locus heterozygotes (e.g. $\frac{M_1Q_1}{M_1Q_2}$) have a frequency of 2/16, and all double heterozygotes (e.g. $\frac{M_1Q_2}{M_2Q_1}$) also have a frequency of 2/16. Given an M_1/M_1 marker-locus genotype, the Q locus homozygote genotypes $\frac{M_1Q_1}{M_1Q_1}$ and $\frac{M_1Q_2}{M_1Q_2}$ occur half as often as the Q locus heterozygote $\frac{M_1Q_1}{M_1Q_2}$. Therefore, the Q locus heterozygote frequency is weighted by a factor of 2, as shown in Table 9.5.

Each of the individuals carrying an M_1/M_1 marker locus genotype will have a genotypic value that depends on the genotype at the QTL locus. It is customary to use an arbitrary scale of measurement for the genotypic values (see Chapter 10). On this scale, the genotypic value of the Q_1Q_1 genotype is assigned $+a$, the genotypic value of the Q_2Q_2 genotype is assigned $-a$, and the genotypic value of the Q_1Q_2 genotype is assigned d (refer to Figure 10.1). The point exactly mid-way between the genotypic values of the homozygotes, called the midpoint, is defined to be zero. When the genotypic value of the heterozygote is at the midpoint, $d = 0$. The degree of dominance can be expressed as the ratio of d/a. Using this scale to measure genotypic values, the $\frac{M_1Q_1}{M_1Q_1}$ genotype has a value of $+a$, the $\frac{M_1Q_1}{M_1Q_2}$ genotype a value of d, and the $\frac{M_1Q_2}{M_1Q_2}$ genotype a value of $-a$. The expected frequency of each of the three genotypic values within the M_1/M_1 marker-locus

genotype is then the product of the genotype frequency and the corresponding genotypic value.

The portion of the mean genotypic value for the *entire* F2 population due to those individuals with an M_1/M_1 marker genotype, let us call it $\overline{G}^{pop}_{M_1M_1}$, is the sum of the genotype-frequency-weighted genotypic values for each QTL genotype associated with an M_1/M_1 marker genotype

$$\overline{G}^{pop}_{M_1M_1} = a\left(\frac{1-c}{2}\right)^2 + 2d\left(\frac{c}{2}\right)\left(\frac{1-c}{2}\right) - a\left(\frac{c}{2}\right)^2$$
(9.32)

Each term can be multiplied out and the first term separated to give

$$\overline{G}^{pop}_{M_1M_1} = \frac{a(1-2c)}{4} + \frac{ac^2}{4} + \frac{2dr(1-c)}{4} - \frac{ac^2}{4}$$
(9.33)

which after addition simplifies to

$$\overline{G}^{pop}_{M_1M_1} = \frac{a(1-2c)}{4} + \frac{2dc(1-c)}{4}$$
(9.34)

as the expected genotypic mean of all individuals with an M_1/M_1 marker locus genotype.

Since ¼ of all individuals in the F2 population are expected to have an M_1/M_1 marker genotype by Hardy–Weinberg, we can multiply Eq. 9.34 by a factor of 4 to obtain the actual genotypic mean of that subset of individuals in the F2 population with M_1/M_1 genotypes

$$\overline{G}_{M_1M_1} = a(1-2c) + 2dc(1-c)$$
(9.35)

The same logic can be used to obtain expressions for $\overline{G}_{M_1M_2}$ and $\overline{G}_{M_2M_2}$ for the M_1/M_2 and M_2/M_2 marker genotypes, respectively. Note that environmental variation in phenotype does not need to be accounted for explicitly because it is not expected to change the *average* phenotypic value. However, variation in phenotype about the average will be caused by both V_G and V_E.

Examine the expression for the mean genotypic values for each marker genotype in Table 9.5. If there is free recombination between the QTL and the marker locus, $c = 0.5$ and the marker locus segregates independently of the QTL that influences phenotype. With free recombination between the M and Q loci, the marker-class genotypic means should not differ because the marker genotype is not associated with the QTL genotype. You can see that this is true by solving each marker-class mean for $c = 0.5$ and finding that the expected mean for all marker genotypes is $0.5d$. Therefore, when M is not associated by linkage with a QTL locus, there should be no difference between the marker-class means $\overline{G}_{M_1M_1}$ and $\overline{G}_{M_2M_2}$. If, on the other hand, the average phenotypic values of those individuals with different marker genotypes are different, then this is evidence that there is a QTL linked to the marker. In other words, a difference between $\overline{G}_{M_1M_1}$ and $\overline{G}_{M_2M_2}$ is evidence that the marker locus is located inside a linkage block that also contains a QTL.

Given a difference between $\overline{G}_{M_1M_1}$ and $\overline{G}_{M_2M_2}$, it is the possible to estimate the values of a and d for the linked QTL. We will need to make a distinction between the true values of a and d and the estimated values of $\hat{a}$ and $\hat{d}$ since recombination between the marker and QTL will cause $a \neq \hat{a}$ and $d \neq \hat{d}$. The difference between the two most extreme phenotypic values in the F2 population defines the genotypic scale from $+a$ to $-a$, as in Figure 9.15. Therefore, the difference between the M_1/M_1 and M_2/M_2 marker-class means is equal to 2. In an equation, this can be stated as

$$2\hat{a} = \overline{G}_{M_1M_1} - \overline{G}_{M_2M_2} \qquad (9.36)$$

Substituting the definitions of $\overline{G}_{M_1M_1}$ and $\overline{G}_{M_2M_2}$ from Table 9.5 gives

$$2\hat{a} = [a(1 - 2c) + 2dc(1 - c)] \\ - [-a(1 - 2c) + 2dc(1 - c)] \qquad (9.34)$$

which then simplifies to

$$\hat{a} = a(1 - 2c) \qquad (9.37)$$

The expected degree of dominance of the QTL is also obtained using the definition of d from the genotypic scale of measurement as the difference between the heterozygote genotypic value and the midpoint between $+a$ and $-a$. For QTL mapping data, the value of the midpoint is half the difference between the M_1M_1 and M_2M_2 marker-class mean values and the heterozygote genotypic value is the M_1M_2 marker-class mean value. The estimated degree of dominance is therefore

$$\hat{d} = \overline{G}_{M_1M_2} - \left(\frac{\overline{G}_{M_1M_1} + \overline{G}_{M_2M_2}}{2} \right) \qquad (9.38)$$

Using the definitions of $\overline{G}_{M_1M_2}$, $\overline{G}_{M_1M_1}$, and $\overline{G}_{M_2M_2}$ from Table 9.5 yields

$$\hat{d} = d\left[(1 - 2c)^2 + c^2\right] \\ - \frac{a(1 - 2c) + 2dc(1 - c) - a(1 - 2c) + 2dc(1 - c)}{2} \qquad (9.39)$$

which after simplifying the rightmost term gives

$$\hat{d} = d\left[(1 - 2c)^2 + c^2\right] - 2dc(1 - c) \qquad (9.40)$$

This equation is then simplified by factoring out d, multiplying through, and adding terms to obtain the expression

$$\hat{d} = d(1 - 2c)^2 \qquad (9.41)$$

A numerical example will help to illustrate the marker-class phenotypic mean values and how they result in estimates of $\hat{a}$ and $\hat{d}$. Different breeds of dog exhibit a very wide range of body sizes. An allele of the insulin-like growth factor 1 gene (IGF1) has been shown to be frequent in small dog breeds (<9 kg) but to have a frequency of near zero in large dog breeds (>30 kg; Sutter et al. 2007). Therefore, the IGF1 locus is likely to be a **major gene** (a QTL that explains a large amount of quantitative trait variation) for body size in dogs. Let's assume there are two IGF1 alleles segregating within a single randomly mating population of dogs and that body size ranges between a minimum of 9 kg and a maximum of 30 kg (see Figure 10.1).

Imagine that an F2 QTL mapping design was carried out along the lines of Figure 9.15 using large (30 kg) and small (9 kg) dogs as the P1 individuals. Suppose that the marker-class means were = 20 kg, = 18 kg, and = 12 kg. Also, make the unrealistic assumption for the moment that there is no recombination between the QTL and the marker locus ($c = 0$). The observed value of $2\hat{a}$ is the difference between $\overline{G}_{M_1M_1}$ and $\overline{G}_{M_2M_2}$, which is $2\hat{a} = 8$ kg or = 4 kg. To determine whether the QTL shows dominance, we first need to compute the midpoint value, which is $12 + (20-12)/2 = 16$ kg. The M_1M_2 marker-class mean is 18 kg and is 2 kg above the midpoint. The degree of dominance is $\hat{d}/\hat{a}$, which in this case is $2/4 = 0.5$. Therefore, in this example, the QTL allele that increases body mass is 50% dominant to the allele that decreases body mass.

The difference in body mass between the and marker-class means is 20 kg − 12 kg = 8 kg. The total difference in body weight in these dogs is 30 kg − 9 kg = 21 kg. The QTL linked to this genetic marker, therefore, accounts for 8 kg of the total 21 kg difference in body weight, or 8 kg/21 kg = 38% of the total body mass difference between large and small dog breeds. The percentage of the total difference in the phenotypic value of the two P1 individuals is called the **effect size** of the QTL. In this hypothetical example, the QTL would be considered a major gene because its effect size is large. The QTL mapping results also tell us that there is partial dominance for the two QTL alleles that have been examined in this study based on the mean phenotypic value of individuals heterozygous for the marker locus.

It is important to notice that the expressions for $\hat{a}$ and $\hat{d}$ are both functions of the recombination rate between the QTL and the marker locus. This is because the perceived true effect of the QTL is confounded with the recombination rate in a single-marker QTL mapping analysis. The true additive effect of a QTL (or a) based on and the recombination rate can be found by rearranging Eq. 9.35 to give

$$a = \frac{\hat{a}}{(1 - 2c)} \qquad (9.42)$$

Similarly, the true dominance effect of a QTL (or d) based on the recombination rate can be found by rearranging Eq. 9.39 to give

$$d = \frac{\hat{d}}{(1 - 2c)^2} \qquad (9.43)$$

So unless the recombination rate between the marker locus and QTL is zero, which it almost never is, a and d are actually larger than the estimates of $\hat{a}$ and $\hat{d}$ from a single-marker QTL mapping experiment.

To see that $\hat{a}$ and $\hat{d}$ are minimum estimates, let us reconsider the example above but assume a different recombination rate. The estimated phenotypic effect of the QTL ($\hat{a}$) is a constant. The observed difference in body mass between the $\overline{G}_{M_1M_1}$ and $\overline{G}_{M_2M_2}$ marker-class means is 8 kg, yielding an estimate of $2\hat{a} = 8$ or = 4 by Eq. 9.37. If the recombination rate is $c = 0$, then by Eq. 9.40, $a = 4$ kg/(1−2[0]) = 4 kg. However, if the recombination rate between the marker locus and the QTL is $c = 0.25$ instead, then by Eq. 9.35, $a = 4$ kg/(1−2[0.25]) = 8 kg. For dominance, if $c = 0$, then by Eq. 9.43, $d = 2$ kg/(1−2[0])² = 2 kg. However, if $c = 0.25$, then by

Eq. 9.43, $d = 2$ kg/(1−2[0.25])² = 8 kg. Therefore, the inferred true phenotypic effect of the QTL (a) and its degree of dominance (d) are both a function of the recombination rate between the marker locus and the QTL. The degree of dominance (d/a) inferred with $c = 0$ is 2/4 = 0.5. In contrast, the degree of dominance inferred with $c = 0.25$ is 8/8 = 1.0.

In this example, a recombination rate of 0.25 rather than zero *doubles* both the perceived effect of the QTL and the perceived degree of dominance. With a higher recombination rate, the QTL has a larger effect on the phenotype but not all of this effect is reflected in the marker-class means since recombination breaks down the association between QTL genotypes and marker locus genotypes. As the randomizing effect of recombination between a QTL and a marker locus increases, the smaller the difference between $\overline{G}_{M_1M_1}$ and $\overline{G}_{M_2M_2}$ becomes for a QTL. With free recombination between the QTL and the marker locus, $\overline{G}_{M_1M_1}$ and $\overline{G}_{M_2M_2}$ are expected to be equal because there is no association between the QTL and the marker locus genotypes.

Problem box 9.2
Compute the effect and dominance coefficient of a QTL

Sax (1923) was the first to carry out a QTL mapping analysis. He showed evidence for a QTL explaining continuous variation in the trait of seed weight of the common bean (*Phaseolus vulgaris*) based on differences in the phenotypic means of plants that differed in seed color. The P1 individuals had seed weights of 48 and 21 centigrams (cg) and were homozygous for different alleles (*P* and *p*) for a codominant gene that affects seed color. Using seed color patterns to determine genetic marker classes, the mean seed weights in the F2 individuals were

$$\overline{G}_{M_1M_1} = 30.7 \text{ cg} \quad \overline{G}_{M_1M_2} = 28.3 \text{ cg}$$
$$\overline{G}_{M_2M_2} = 26.4 \text{ cg}$$

Using the data collected by Sax, estimate $\hat{a}$ and $\hat{d}$ for the QTL for seed weight assuming that $c = 0.2$ and $c = 0$. What is the effect of the QTL in terms of the percentage of phenotypic difference between the P1 individuals? Did Sax identify a major gene for seed weight?

QTL mapping with multiple marker loci

QTL mapping that utilizes numerous genetic marker loci is now routine. It is then possible to utilize all pairs of genetic markers for QTL mapping. The advantage of such **flanking-marker QTL analysis** or **interval mapping** is that the resulting estimates of $\hat{a}$ and $\hat{d}$ are not confounded with the recombination rate. Interval mapping is carried out with an F2 mating design like that shown in Figure 9.15 to produce F1 individuals heterozygous at the two marker loci as well as the QTL. Figure 9.16 shows the arrangement of two marker loci that flank a QTL in an F1 individual along with the eight types of gametes (two coupling, four single recombinants, and two double recombinants) that can be produced by an F1 individual. These eight F1 gametes can be combined to make nine two-locus marker genotypes and 28 possible F2 genotypes. The recombination rate between the marker loci, c, can be estimated from the two-locus marker genotype frequencies in the F2 progeny. Assuming no interference (or that recombination rates c_A and c_B are independent), then $c = c_A + c_B - 2c_A c_B$.

Using the expected gamete frequencies shown in Figure 9.16, the mean phenotypic value of each marker genotype can be derived. Table 9.6 shows the derivation of the expected phenotypic value for the marker genotypes $A_1A_1B_1B_1$ and $A_1A_2B_1B_2$. As for QTL mapping based on a single genetic marker, the portion of the mean genotypic value for the *entire* F2 population is referred to as $\overline{G}^{pop}_{A_1A_1B_1B_1}$ and $\overline{G}^{pop}_{A_1A_2B_1B_2}$. These population mean genotypic values are the sum of the genotype-frequency-weighted genotypic values for each QTL genotype associated with a two-locus marker genotype. Since $\dfrac{(1-c)^2}{4}$ of all individuals in the F2 population are expected to have an $A_1A_1B_1B_1$ marker genotype, we can multiply $\overline{G}^{pop}_{A_1A_1B_1B_1}$ by a factor of $\dfrac{4}{(1-c)^2}$ to obtain the actual genotypic mean of that subset of individuals in the F2 population with $A_1A_1B_1B_1$ marker genotypes or $\overline{G}_{A_1A_1B_1B_1}$. The expected frequency of the $A_1A_2B_1B_2$ genotype is used to obtain the actual genotypic mean of F2 individuals with $A_1A_2B_1B_2$ marker genotypes or $\overline{G}_{A_1A_2B_1B_2}$. The expected genotypic means of the remaining seven possible marker-class means as well as a regression

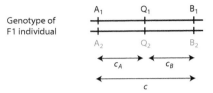

Figure 9.16 Interval mapping utilizes two maker loci (A and B) that sit on either side of a QTL. An individual with the genotype shown can produce two types of gametes if there is no recombination, four type of gametes if there is a single recombination event, and two types of gametes from a double recombination event. The expected gamete frequencies are a function of the recombination rates.

method to estimate $\hat{a}$ and $\hat{d}$ are given in Haley and Knott (1992).

Like QTL mapping based on a single genetic marker, the phenotypic value of the $A_1A_1B_1B_1$ marker genotype is a function of both the additive and dominance effects of the QTL. In contrast, the expected phenotypic value of the $A_1A_2B_1B_2$ marker genotype is a function only of d. Therefore, the $A_1A_2B_1B_2$ marker-class mean value provides an estimate of the dominance coefficient independent of a. Once the value of d is estimated, then other marker-class means can be used to estimate the value of a. As with single-marker QTL analysis, a statistically significant difference between marker-class mean values indicates that a QTL is present between a pair of genetic marker loci.

A hypothetical example of data produced by interval QTL mapping in an F2 design is shown in Figure 9.17. This example illustrates the difference in the value of the homozygous marker-class means (e.g. the difference between $\overline{G}_{A_1A_1B_1B_1}$ and $\overline{G}_{A_2A_2B_2B_2}$) for each of 17 genetic marker loci. The difference in marker-class means is given on the y axis while the

Table 9.6 Derivation of the expected phenotypic value for two genetic marker genotypes when QTL mapping with pairs of marker loci that flank a QTL. There are a total of nine genetic marker genotypes possible with two genetic marker loci. Like QTL mapping based on a single genetic marker, the phenotypic value of the $A_1A_1B_1B_1$ marker genotype is a function of both the additive and dominance effects of the QTL. In contrast, the expected phenotypic value of the $A_1A_2B_1B_2$ marker genotype is a function only of d. Therefore, the $A_1A_2B_1B_2$ marker-class mean value provides an estimate of the dominance coefficient independent of a. The gametes and expected gamete frequencies are given in Figure 9.16.

Marker genotype	Marker genotype frequency	F2 genotype	F2 genotype Frequency	F2 genotypic value	Frequency-weighted F2 genotypic value
$A_1A_1B_1B_1$	$\dfrac{(1-c)^2}{4}$	$\dfrac{A_1Q_1B_1}{A_1Q_1B_1}$	$\left(\dfrac{(1-c_A)(1-c_B)}{2}\right)^2$	$+a$	$\dfrac{a(1-c_A)^2(1-c_B)^2}{4}$
		$\dfrac{A_1Q_1B_1}{A_1Q_2B_1}$	$2\left(\dfrac{(1-c_A)(1-c_B)}{2}\right)\left(\dfrac{c_Ac_B}{2}\right)$	d	$\dfrac{2dc_Ac_B(1-c_A)(1-c_B)}{4}$
		$\dfrac{A_1Q_2B_1}{A_1Q_2B_1}$	$\left(\dfrac{c_Ac_B}{2}\right)^2$	$-a$	$\dfrac{-ac_A^2c_B^2}{4}$

$$\overline{G}^{pop}_{A_1A_1B_1B_1}=\frac{a(1-c_A)^2(1-c_B)^2}{4}+\frac{2dc_Ac_B(1-c_A)(1-c_B)}{4}+\frac{-ac_A^2c_B^2}{4}$$

$$\overline{G}_{A_1A_1B_1B_1}=\frac{\dfrac{a(1-c_A)^2(1-c_B)^2}{4}+\dfrac{2dc_Ac_B(1-c_A)(1-c_B)}{4}+\dfrac{-ac_A^2c_B^2}{4}}{(1-c)^2}=a\frac{(1-c_A)^2(1-c_B)^2-c_A^2c_B^2}{(1-c)^2}+d\frac{2c_Ac_B(1-c_A)(1-c_B)}{(1-c)^2}$$

Marker genotype	Marker genotype frequency	F2 genotype	F2 genotype Frequency	F2 genotypic value	Frequency-weighted F2 genotypic value
$A_1A_2B_1B_2$	$\dfrac{(1-c)^2+c^2}{4}$	$\dfrac{A_1Q_1B_1}{A_2Q_2B_2}$	$\left(\dfrac{(1-c_A)(1-c_B)}{2}\right)^2$	d	$\dfrac{d(1-c_A)^2(1-c_B)^2}{4}$
		$\dfrac{A_1Q_1B_2}{A_2Q_2B_1}$	$\left(\dfrac{(1-c_A)c_B}{2}\right)^2$	d	$\dfrac{d(1-c_A)^2c_B^2}{4}$
		$\dfrac{A_2Q_1B_1}{A_1Q_2B_2}$	$\left(\dfrac{c_A(1-c_B)}{2}\right)^2$	d	$\dfrac{dc_A^2(1-c_B)^2}{4}$
		$\dfrac{A_1Q_2B_1}{A_2Q_1B_2}$	$\left(\dfrac{c_Ac_B}{2}\right)^2$	d	$\dfrac{dc_A^2c_B^2}{4}$
		$\dfrac{A_1Q_1B_1}{A_2Q_1B_2}$	$2\left(\dfrac{(1-c_A)(1-c_B)}{2}\right)\left(\dfrac{c_Ac_B}{2}\right)$	$+a$	$\dfrac{2ac_Ac_B(1-c_A)(1-c_B)}{4}$
		$\dfrac{A_2Q_2B_2}{A_1Q_2B_1}$	$2\left(\dfrac{(1-c_A)(1-c_B)}{2}\right)\left(\dfrac{c_Ac_B}{2}\right)$	$-a$	$\dfrac{-2ac_Ac_B(1-c_A)(1-c_B)}{4}$

$$\overline{G}^{pop}_{A_1A_2B_1B_2}=\frac{d(1-c_A)^2(1-c_B)^2}{4}+\frac{d(1-c_A)^2c_B^2}{4}+\frac{dc_A^2(1-c_B)^2}{4}+\frac{dc_A^2c_B^2}{4}+\frac{2ac_Ac_B(1-c_A)(1-c_B)}{4}+\frac{-2ac_Ac_B(1-c_A)(1-c_B)}{4}$$

$$\overline{G}_{A_1A_2B_1B_2}=\frac{\dfrac{d(1-c_A)^2(1-c_B)^2}{4}+\dfrac{d(1-c_A)^2c_B^2}{4}+\dfrac{dc_A^2(1-c_B)^2}{4}+\dfrac{dc_A^2c_B^2}{4}+\dfrac{2ac_Ac_B(1-c_A)(1-c_B)}{4}+\dfrac{-2ac_Ac_B(1-c_A)(1-c_B)}{4}}{(1-c)^2+c^2}$$

$$=d\frac{c_Ac_B+c_A(1-c_B)^2+(1-c_A)c_B^2+(1-c_A)(1-c_B)}{c^2+(1-c)^2}$$

Problem box 9.3
Derive the expected marker-class means for a backcross mating design

A commonly employed QTL mapping design is the backcross of an F1 individual to one of the P1 individuals shown in Figure 9.15. One drawback of a backcross mating design for QTL mapping is that the resulting estimates of $\hat{a}$ depend on the value of d. To see that this is the case, consider the backcross $\dfrac{A_1 Q_1 B_1}{A_2 Q_2 B_2} \times \dfrac{A_1 Q_1 B_1}{A_1 Q_1 B_1}$ and the marker-class means in the population of progeny. Derive the expected marker-class means for $A_1A_1B_1B_1$ and $A_1A_2B_1B_2$ marker genotypes (call these $\overline{G}^{BC}_{A_1 A_1 B_1 B_1}$ and $\overline{G}^{BC}_{A_1 A_2 B_1 B_2}$, respectively) and then compute the expected value of $\hat{a} = \overline{G}^{BC}_{A_1 A_1 B_1 B_1} - \overline{G}^{BC}_{A_1 A_2 B_1 B_2}$.

To work through this problem, first notice that the P1 individual $\left(\dfrac{A_1 Q_1 B_1}{A_1 Q_1 B_1}\right)$ will produce only one type of gamete, whereas an F1 individual $\left(\dfrac{A_1 Q_1 B_1}{A_2 Q_2 B_2}\right)$ will produce eight types of gametes with the frequencies given in Figure 9.16. This problem is made easier by the commonly invoked assumption that the marker loci are close enough on the chromosome such that double recombination events are so rare they can be ignored. Therefore, the F1 parent gametes $A_2Q_1B_2$ and $A_1Q_2B_1$ can be left out of the marker-class means.

Start by constructing a table with two rows for the two F1 gametes produced by no recombination and four rows for the F1 gametes produced by a single recombination event. Then, following the model of Tables 9.5 and 9.6, fill in columns for the six categories of backcross progeny. The column headings are F1 parent gamete, expected F1 gamete frequency, backcross progeny genotype, backcross progeny genotypic value, frequency-weighted genotypic value, backcross progeny marker genotype expected frequency, and marker-class genotype mean. When adding the two terms that make up the $A_1A_1B_1B_2$ or $A_1A_2B_1B_1$ frequency-weighted genotypic values, any terms that contain $c_A c_B$ can be crossed out because of the assumption that double recombination events are so rare they can be ignored. The marker-class means are obtained by dividing the frequency-weighted genotypic value by the marker genotype expected frequency.

position of each marker locus on a chromosome is given on the x axis. The marker-class mean differences are near zero for marker loci not in gametic disequilibrium with a QTL. Marker loci near QTLs show some difference in marker-class means, but the marker loci and the QTLs experienced frequent recombination and so are not strongly associated. As marker loci closer to the QTLs are considered, the amount of gametic disequilibrium between a marker locus and a QTL increases, producing a greater difference in marker-class means. In this hypothetical example, the genetic markers at 33 and 85 map units lie closest to QTLs, as indicated by peak values for marker-class mean differences. The marker-class means at these marker loci differ by over one phenotypic standard deviation, indicating a statistically meaningful difference given

sampling error and multiple statistical tests. (The threshold for statistical significance of marker-class mean differences is often judged by a log of odds or LOD score and differs depending on the details of each QTL study; see Van Ooijen (1999).) The QTL near 33 map units increases the mean value while the QTL near 85 map units decreases the mean value. These two hypothetical QTLs have opposite effects on the trait, a situation sometimes called dispersion. Dispersion can lead to downwardly biased estimates of QTL effects because each marker locus is associated with two QTLs with opposite phenotypic effects. The perceived marker-class mean difference is therefore the net phenotypic effect caused by association with two QTLs.

A large number of QTL mapping breeding designs and estimation methods exist. Some methods are

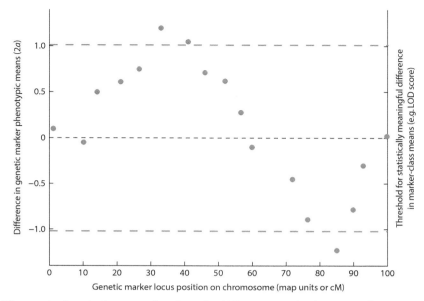

Figure 9.17 The difference in phenotypic mean values for each of 17 genetic marker loci versus the position of each marker locus on a chromosome. In this hypothetical example, there are two genetic markers that lie closest to QTL indicated by two peak values for marker-class mean differences. The phenotypes of the homozygote classes for the marker loci at 33, 41, and 85 map units differ by over one phenotypic standard deviation. The QTL near 33 cM increases the mean value, while the QTL near 85 cM decreases the mean value. Marker-class mean value differences greater than ± one phenotypic standard deviation (dashed lines) are considered statistically meaningful in this example. Marker-class mean differences smaller than one standard deviation could be different due to chance alone. The close proximity of these two QTL with opposite effects on the trait would lead to reduced estimates of QTL effects at all marker loci. This occurs because each marker locus is partly linked to a QTL of both positive and negative effect, so the perceived marker-class mean difference is the net effect of the two QTL. Genetic marker locus positions are established by observed recombination rates. One map unit or centimorgan (cM) distance along a chromosome is equal to a 1% recombination rate.

related to interval mapping but utilize more than just pairs of marker loci (**composite interval mapping**) or utilize all of the linked markers on individual chromosomes (**multipoint mapping**). Another approach is to test for associations between marker genotypes and phenotypic means in only those individuals that show extreme phenotypic differences in an F2 population such as the individuals at the upper and lower tails of the phenotypic distribution. In such **bulked segregant analyses**, the marker loci near QTL are expected to be in gametic disequilibrium because of linkage (e.g. Itoh et al. 2019). Therefore, the individuals in the lower tail of the phenotypic distribution would have one marker genotype and the individuals in the upper tail of the phenotypic distribution would have another marker genotype if a marker locus were very tightly linked to a QTL. In contrast, a marker locus independent of any QTL would show all genotypes at equal frequencies in the individuals that represented the upper and lower tails of the phenotypic distribution.

Limitations of QTL mapping studies

QTL mapping results are highly context sensitive, just like heritability estimates. The number and effects of QTL identified in one population may not be representative of QTL effects in another population of the same species. Mapping only identifies QTL that are segregating in the population at the time of mapping. Further, QTL mapping by the F2 design can only detect and estimate the phenotypic effects of two alleles at a QTL. The two QTL alleles detected are those fixed when inbreeding lines to form P1 in Figure 9.15. In reality, there may be more than two alleles segregating in a population for any QTL and detecting these requires screening replicate P1 crosses. Alleles at QTL that are fixed or lost in the original population or become fixed or lost by genetic drift during the formation of inbred lines cannot be identified. Likewise, the estimates of QTL effect sizes are always relative to the other QTL loci segregating in a population. For example, QTL X could explain 10% of the phenotypic difference between marker-class means in one population, whereas QTL Y has even bigger effect but happens to be fixed for one allele. If QTL Y is segregating instead, then QTL X will have a smaller perceived effect on the phenotype in a mapping study. This occurs because when two alleles are segregating at QTL Y, there will be a greater difference in phenotype between P1 individuals than when QTL Y is fixed.

Several types of statistical power limitations impact QTL mapping results. Actual QTLs of small effect cannot be identified easily because it is difficult to show that a small difference between marker-class means is statistically meaningful given inherent environmental variation, experimental measurement error, and corrections required when carrying out a very large number of statistical tests. The effect sizes of QTLs that are identified as statistically meaningful can be substantially inflated since the QTLs with small effects are not identified. This so-called **Beavis effect** is pronounced if the number of progeny in a mapping study is about 100 and modest with about 500 progeny (Beavis 1994; Xu 2003; Slate 2013; King and Long 2017). The number of QTLs identified in mapping studies is likely to underestimate the true number of QTLs that cause variation in a trait (Otto and Jones 2000). Under conditions similar to many actual QTL mapping studies, no more than about a dozen QTLs will be identified as statistically meaningful (Hyne and Kearsey 1995). In addition, two or more QTLs that are adjacent in the genome may appear as a single QTL due to disequilibrium. If the effects of two linked QTLs are in the same direction, the perceived single QTL will have an inflated effect. On the other hand, if the two linked QTLs have contrasting (or antagonistic) effects on the trait, then a single perceived QTL will be detected that has a downwardly biased effect size that is therefore less likely to be detected as statistically meaningful. The number and spacing of genetic markers also influence the perception of QTL numbers and effect sizes, since widely spaced markers (relative to the recombination rate) may miss QTLs or aggregate the effects of multiple QTLs. High-throughput sequencing methods have expanded the ability to map QTL with increased numbers of loci, improving precision of estimates and allowing QTL mapping in non-model species (Slate 2005; Jamann et al. 2015).

The granularity of QTL mapping studies is a function of the amount of recombination that occurs during the crossing scheme used to generate individuals for analysis. More opportunity for recombination generates smaller regions of the genome in linkage disequilibrium and finer QTL mapping resolution. The term **quantitative trait region** or **QTR** is sometimes used to describe an association between a marker locus and a marker-class mean difference since multiple linked QTLs may exist within the genome interval mapped. Further, fine-scale mapping with genetic markers spaced at smaller intervals along the chromosome in the specific chromosomal regions around QTR can be used to identify true single QTLs (e.g. Kroymann and Mitchell-Olds 2005). At the finest scale of resolution, it is possible to identify the nucleotides at a single nucleotide sites that cause trait variance. These are termed **quantitative trait nucleotides** or **QTN**. Variation in phenotypes among individuals may also be caused by levels of mRNA transcripts or protein, and the genetic variants can be identified as **expression quantitative trait loci** or **eQTLs** (Nica and Dermitzakis 2013).

Genome-wide association studies

GWAS use a suite of genetic marker loci, usually SNPs, that are distributed across the entire genomes, to test for disequilibrium between genetic markers and phenotypes (McCarthy et al. 2008; Korte and Farlow 2013). GWAS commonly employ a case–control design, where a group of individuals are classified according to their phenotype, such as presence and absence of a disease, and are also genotyped for a genome-wide set of genetic markers. The genetic markers are then tested for disequilibrium – strong allele frequency differences – between the case and control populations. Those marker loci with strong disequilibrium may be linked to a segment of the genome that is a causal contributor to the phenotypic difference. GWAS studies have been facilitated by relatively rapid techniques to screen genome-wide sets of genetic markers, such as genotyping-by-sequencing (Scheben et al. 2017) or SNP microarrays that detect allelic variants based on hybridization (LaFramboise 2009). The power of GWAS to detect associations between traits and marker loci is determined by the sample size of individuals and the allele frequencies (Visscher et al. 2017).

GWAS serve to identify QTL, but are a distinct approach that does not necessarily employ mating design to sample alleles from natural populations and establish linkage relationships between genetic marker loci and QTL. Therefore, GWAS have the potential to overcome some limitations of QTL mapping – limited sampling of allelic diversity, resolution limited by recombination, and the ability to carry out selective breeding for several generations. Further, GWAS and QTL mapping can be used together, with the former used to identify regions of the genome or QTR and the latter used with a denser set of genetic markers in the QTR regions to identify candidate QTL or QTN. An example of the complementary use of GWAS followed by QTL mapping was a study of the genetic basis of resistance to *Fusarium* ear rot disease in maize where 15 QTL were identified that explained 3–15% of the trait variation (Chen et al. 2013).

Employment of GWAS has led to the discovery of many genome regions that are associated with variation among individuals in quantitative phenotypes and has greatly expanded the number of candidate loci available for detailed study. Two generalizations have emerging from empirical GWAS (Visscher et al. 2017; Josephs et al. 2017; Sella and Barton 2019). One is that, for many traits studied, the heritable component of phenotypic variance is explained by many loci with additive effects that are small and often below the limits of detection given sample sizes. Another is that pleiotropy is common, and some loci contribute to variance among individuals in multiple phenotypes. These conclusions are conditional because GWAS studies have limits to their inferential power and have been carried out more commonly in humans and plant taxa. It is also true that connecting GWAS patterns to models of population genetics is ongoing, so understanding GWAS patterns in light processes such as neutral population genetic differentiation or natural selection is not yet fully realized.

Biological significance of identifying QTL

Identification of QTL in model and domesticated species has expanded greatly, facilitated by the large numbers of molecular markers generated with high-throughput genotyping techniques, as well as by genome and genetic linkage mapping projects. Table 9.7 shows some examples of the number of QTLs identified for various phenotypes in a range of species. In Table 9.7, the results range from one or a few QTLs with large effects on phenotypic variation (dog body size, human stature, etc.) to a large number of QTLs with individual small effects on

phenotypic variation (kernel oil content). This mirrors the overall trend in QTL mapping results, where the number of QTLs and their effect sizes are strongly dependent on the species, population, and phenotype studied (e.g. Santure et al. 2015). Whereas part of this diversity in QTL mapping results is likely caused by methodological and statistical power differences among studies, some of it likely reflects actual variability in the number and effects of QTLs that cause phenotypic variation.

QTL mapping studies not only identify effect sizes but can also quantify the degree of dominance and epistasis for QTLs. QTLs have been observed to have dominance that ranges from zero (additive gene action), through all degrees of partial dominance to complete dominance, to cases of overdominance. One classic example is the wide range of the d/a ratio (a is the estimated additive effect of an allele, and d is the estimated heterozygote value), spanning -2.0 to $+2.0$, observed for 74 QTLs detected for 11 phenotypes in tomato (de Vincente and Tanksley 1993). Based on these results, dominance frequently contributes to the genotypic variation of quantitative traits. Evidence for interaction variance caused by epistasis has been equivocal. Interactions between or among loci did not often explain much of the observed genotypic variation in early QTL mapping studies. However, it is generally more difficult to detect epistasis for QTLs due to statistical and sample size limitations (see Carlborg and Haley 2004). Effort has been directed toward testing for epistasis in QTL studies, and numerous empirical studies have identified statistically meaningful interactions between two or more QTLs as often as additive effects of QTLs (reviewed by Malmberg and Mauricio 2005).

The genetic architecture of quantitative traits – the effect size and gene action (additivity, dominance, or epistasis) of QTL underlying a trait – plays a crucial role in how quantitative traits will respond to natural selection. As an illustration, imagine that dog body mass is under natural selection for larger size such that $s = 0.3$ with $h^2 = 1.0$. The breeder's equation $R = h^2s$ predicts response to selection without reference to the genetic architecture of the quantitative trait. In this example, $R = (1.0)(0.3) = 0.3$ or a predicted increase of 0.3 standard deviations per generation. Now, imagine that the additive genetic variation in body mass is caused by a number of QTLs. The strength of selection on the trait is divided across the independent loci that cause the additive genetic variation in the trait. The pressure of natural selection on one QTL is then only as large as its role

Table 9.7 Examples of QTL identified by mapping with genetic marker loci.

Organism	Phenotype	Number marker loci	Number of QTL	References
Arabidopsis thaliana	# buds at flowering	65	28	Kearsey et al. (2003)
	Rosette size at 21 days		4	
	Biomass	105	6	Lisec et al. (2007)
	Flowering time	237	8	Salomé et al. (2011)
Dogs	Body size	116	1	Sutter et al. (2007)
	Body size, morphological traits	60 968	≤3	Boyko et al. (2010)
Drosophila	Sex comb bristles	~10 000	3	Cloud-Richardson et al. (2016)
	Prezygotic reproductive isolation	32	6	Moehring et al. (2006)
Humans	Taste sensitivity to Phenylthiocarbamide	50	1	Kim et al. (2003)
	Stature	>253	3	Perola et al. (2007)
Louisiana irises	Flowering time	>414	17	Martin et al. (2007)
Stickleback fish	Bony plates	160	4	Colosimo et al. (2004)
	Caudal peduncle length	14 998	22	Yang et al. (2016)
White spruce (*Picea glauca*)	Bud flush	836	33	Pelgas et al. (2011)
	Bud set		52	
	Height growth		52	
Eucalyptus cladocalyx	Wood density	130	3	Valenzuela et al. (2019)
Zea mays	Kernel oil concentration	488	>50	Laurie et al. (2004)
	100-kernel weight	29 927	10	Su et al. (2017)

in causing additive genetic variance in the phenotype. Said another way, the selection experienced by one QTL is proportional to its effect size. The breeder's equation can be modified for a single QTL by adjusting the selection differential for the proportion of additive genetic variation explained by a QTL according to

$$R = h^2[s(\text{QTL effect size})] \qquad (9.44)$$

To see how this modified breeder's equation works, let's return to the example of the QTL that explains 38% of the total body mass difference between large and small dog breeds. That one QTL experiences 38% of the selection pressure that is exerted on the entire phenotype because it causes 38% of the additive genetic variation. Response to selection will be accomplished by relatively large changes in the allele frequencies at that one QTL with a 38% effect. In fact, the selection differential on that one QTL is $s = (0.3)(0.38) = 0.114$. This also leads to a predicted response to selection of 0.144 standard deviations for that one QTL alone.

The predictions of the nearly neutral theory (see Chapter 8) show the consequences of this division of the selection differential among QTL in proportion to their effect size. In finite populations, the balance between genetic drift and natural selection can be predicted with the quantity $4N_es$ where s is the selection differential. When $4N_es >> 1$, then selection is the primary determinant of allele frequency; when $4N_es << 1$, then genetic drift is the main process influencing allele frequency; and when $4N_es$ is on the order of 1, then both selection and genetic drift determine allele frequency. For a QTL of 38% effect experiencing a selection differential of $s = 0.114$, the response to selection will depend on N_e. In the context of an effective population size greater than about 25, the frequencies of the alleles at a QTL with a 38% effect would be expected to be dictated exclusively by natural selection. Only if the effective population size were less than 10 would we expect genetic drift to exclusively dictate the fate of allele frequencies at the QTL.

Under more realistic circumstances, many QTLs will have effect sizes much smaller than 38% and traits will most commonly have heritabilities less than 1.0. To take another example, imagine a QTL with a 2% effect for a trait with $h^2 = 0.3$ that is experiencing an identical selection differential of $s = 0.3$. The effective response to selection for this QTL of small effect is $R = (0.3)((0.3)(0.02)) = 0.0018$ or a predicted increase of 0.0018 standard deviations per generation. This QTL of small effect experiences a very weak pressure from natural selection precisely because it is a weak cause of additive genetic variation in a trait that has a modest heritability. According to the $4N_es$ rule, this QTL would have to experience selection in the context of an effective population size greater than about 2000 for natural selection alone to dictate allele frequencies over time.

Interact box 9.5 Effect sizes and response to selection at QTLs

Use the **QTL** simulation module of the text web simulation website to simulate response to selection for a quantitative trait that has QTL with variable effect sizes. The key parameters to vary are the **Number of Quantitative Trait Loci (QTL)** and **QTL Architecture**, which offer three options for the distribution of QTL effect sizes.

Using a QTL Architecture setting of All QTL have equal effects, compare response to selection on mean phenotypic value for 10 or 200 QTL. How do allele frequencies change? How does the heritability change? Also, compare response to selection for the default N_e of 20 and a larger N_e of 500.

Use the modified breeder's equation given in Eq. 9.42 to compute the net response to selection for the QTL with large and small effects in this simulation. When the selection truncation point is 0.50 in the simulation, the selection differential is $s \approx 0.399$ (see the text web page for more details). The heritability is also $h^2 \geq 0.5$. Using these parameter values, what would you predict about the patterns of allele frequency change for QTL of large and small effect sizes?

Try out different values for the Natural selection phenotypic value truncation point or N_e and predict the consequences using Eq. 9.44. Then, run the simulation to test your predictions.

These observations about how the effective population size influences response to selection on individual QTLs shed light on an implicit assumption of the infinitesimal model of the genetic basis of quantitative traits. It is only in the context of large effective population sizes that the allele frequencies of QTLs with small effect sizes will respond to natural selection. In a finite population, as the number of QTLs grows large and the effect size approaches zero, the net response to selection will shrink and each locus will be subject to genetic drift rather than natural selection. Therefore, the infinitesimal model must also assume that the effective population size is inversely related to the QTL effect size. Drift–selection balance for QTLs in finite populations also serves to explain the simulation results in

Figure 9.12 that relate to long-term response to selection. In Figure 9.12B, there are many QTLs of small effect that clearly remain segregating for a longer period of time than the QTLs of large effect in Figure 9.12A. The reason why the QTL of small effect remain segregating longer is that they experience a relatively weak net selection pressure. The allele frequencies at the loci in Figure 9.12B are clearly spreading out toward fixation and loss as we would expect of many replicate neutrally evolving loci (see Chapter 3), although perhaps the probability of fixation is greater than it would be under pure genetic drift.

Observations suggest that QTLs experience a combination of both genetic drift and natural selection. The QTLs of small effect identified from the Illinois Long-Term Selection study clearly show the effects of both genetic drift and natural selection. Among the QTLs identified for kernel oil concentration, about 20% show effects that are opposite to the direction of natural selection, such as QTLs that reduce oil concentration segregating in the lines selected for high oil concentration (Laurie et al. 2004). Since selection is acting against these QTLs with opposite effects, they must have escaped loss due to strong genetic drift in the context of an effective population size of about 10. Results similar to kernel oil concentration have been observed for tomato QTLs.

Individual QTL of major effect have not been observed for many human quantitative trait phenotypes such as height, IQ, and schizophrenia, despite these traits having high heritabilities. This lack of individual QTL to explain the genetic component of trait variation leads to the dilemma of **missing heritability** (Maher 2008). Numerous explanations that are not mutually exclusive have been offered to explain the missing heritability (Gibson 2012). One explanation based on the causes of trait variation is that epigenetic mechanisms cause a substantial portion of resemblance between parents and offspring (e.g. Trerotola et al. 2015). However, modeling suggests that epigenetic marks must persist for numerous generations and make substantial contributions to identity by descent to explain missing heritability (Slatkin 2008). If epialleles did persist for many generations, then it becomes more likely that they would be in linkage disequilibrium with other loci such as SNPs and be detected in QTL mapping studies. Further, there is also little evidence that epigenetic mechanisms play any substantial role in adaptation via response to natural selection acting on additive genetic variation (Charlesworth et al. 2017).

A second possible explanation is that the genetic variation for these traits does follow an additive model, but the effect sizes of individual QTL are very small because the trait variation is caused by hundreds or thousands of loci. This would mean that detecting individual QTL would require larger sample sizes of individuals and loci. In support of this explanation, a simulation study showed that more complete marker coverage of the genome improved QTL detection, and most of the missing heritability was due to the inability to detect QTL variants with a moderate effect on phenotypic variation but that little of the heritability was due to very rare variants (Caballero et al. 2015). Empirical studies also lend support to this explanation for missing heritability. For example, 697 QTN found in 423 loci were identified in a meta-analysis of GWAS for adult height in humans covering 79 studies and including 253 288 individuals of European ancestry. These 697 QTN explained about 16% of the phenotypic variation suggesting that genetic human height is caused by a large number of loci of very small effect.

Chapter 9 review

- Variation in quantitative trait values among individuals within a population (V_P) has both genetic and environmental causes. The genetic causes are due to genotypic variance (V_G) that can be partitioned into the distinct components of additive (V_A), dominance (V_D), and epistasis (V_I) variance.

- Phenotypic variation can be caused by genotype-by-environment interaction ($V_{G \times E}$) where genotypes express heterogeneous phenotypic values in response to different environmental conditions.

- The additive component (V_A) of the genotypic variance (V_G) is caused by the sum of the phenotypic effects of alleles when they are assembled into genotypes. Phenotypic effects of alleles cause the resemblance of parents and offspring as well as the resemblance among relatives.

- Dominance (V_D) and epistasis (V_I) components of V_G are caused by the effects of genotypes. V_D and V_I do not contribute to phenotypic resemblance between parents and offspring because particulate inheritance breaks up genotypes (additive by additive epistasis is an exception).

- The proportion of the total genotypic variance (V_G) due to the additive effects of alleles is measured by the narrow-sense heritability $h^2 = V_A/V_P$. Parent–offspring regression is one method to estimate h^2.

- Response to selection over one generation depends on the force of natural selection and the heritability and is predicted by the breeder's equation $R = h^2 s$.

- Because traits show both genetic and phenotypic correlations, response to selection on one trait may change the mean of other correlated traits or be constrained by correlations with other traits.

- Long-term response to natural selection depends on the number of loci that underlie a quantitative trait. Linear response to selection over many generations is expected when many loci with small effects underlie a trait, consistent with the infinitesimal model. In contrast, selection plateaus are consistent with fewer loci of larger effect since selection causes fixation and loss at these loci. Depending on the rate, mutation may replace variation lost due to fixation and loss caused by response to selection.
- The neutral evolution of genotypic variance depends on the balance of genetic variation lost by genetic drift and mutation that replaces variation.
- The individual loci that cause variation in quantitative traits, or QTLs, can be identified by taking advantage of gametic disequilibrium between QTLs and genetic marker loci.
- In an F2 mating design, comparing the phenotypic means of F2 individuals bearing different marker genotypes identifies marker loci near QTLs.
- QTL mapping can only identify alleles that are segregating in the individuals used to found the mapped populations. QTL mapping tends to underestimate the true number of QTLs and overestimate the true effects of QTLs.
- QTL mapping quantifies the genetic architecture of quantitative traits by estimating the number of loci that cause quantitative trait variation, the distribution of QTL phenotypic effects, and the physical organization of QTLs on chromosomes.

Further reading

A review of quantitative trait mapping methods, empirical results, and challenges can be found in:

Mackay, T.F.C., Stone, E.A., and Ayroles, J.F. (2009). The genetics of quantitative traits: challenges and prospects. *Nature Reviews Genetics* 10: 565–577.

For a review of the progress over the last 100 years on fundamental questions in quantitative genetics and the evolution of quantitative traits, see:

Roff, D.A. (2007). A centennial celebration for quantitative genetics. *Evolution* 61: 1017–1032.

Much more details on statistical estimators as well as experimental designs used to estimate heritability and map QTLs can be found in:

Walsh, B. and Lynch, M. (2018). *Evolution and Selection of Quantitative Traits*. Oxford, United Kingdom: Oxford University Press.

The response to selection predicted by the breeder's equation is actually an approximation that neglects four other types of parent–offspring phenotype relationship, as explained in:

Heywood, J.S. (2005). An exact form of the breeder's equation for the evolution of a quantitative trait under natural selection. *Evolution* 59: 2287–2298.

An overview of heritability and how it remains a central concept in population genetics is provided by:

Visscher, P.M., Hill, W.G., and Wray, N.R. (2008). Heritability in the genomics era – concepts and misconceptions. *Nature Reviews Genetics* 9: 255–266.

A critical appraisal of the motivations for and the results from QTN estimation studies is provided by

Rockman, M.V. (2012). The QTN program and the alleles that matter for evolution: all that's gold does not glitter. *Evolution* 66: 1–17.

End-of-chapter exercises

1   What are the possible causes of phenotypic variation among individuals? What is the subset of factors that influence the rate of response to natural selection on quantitative traits?

2   With reference to the different causes of phenotypic variation, explain how epigenetic variation can both contribute to phenotypic differences among individuals yet contribute very little to trait heritability.

3   A species of Darwin's finch, *Geospiza fortis*, was studied for a number of years (Boag and Grant 1981). A drought altered the size of seeds available as food for the birds and increased mortality. Before the drought, birds had an average beak depth of 9.42 mm. Beak depth was 9.96 mm for surviving birds a year later. If is also known that a regression of father-offspring beak depth values had a slope of 0.47 (Boag 1983). What is the selection differential? What is the heritability? What response to selection is predicted for the beak depth phenotype? What are the assumptions of the estimate for response to selection?

4   For the situation in question 1, the population size of the birds plummeted from an initial size of about 1400 individuals to about 200 after the drought. How might a change in population size like this influence the response to selection? How would the genetic architecture for beak depth – the number and effect sizes of QTL – influence response to selection in a population?

5   Search the literature for a recent research paper that utilizes one or more of the population genetic predictions covered in this chapter. The topic can be any organism, application, or process but the paper must include a hypothesis test involving a topic such as components of variance in continuous traits, heritability, response to selection on quantitative traits, epistasis, epigenetic variation, a selection plateau, or QTL. Summarize the main hypothesis, goal, or rationale of the paper. Then, explain how the paper utilized a population genetic prediction

from this chapter and then summarize the results and the conclusions based on the prediction.

6   Construct a simulation model of parent–offspring to estimate heritability, or a simulation model of natural selection on a quantitative trait. Links to instructions to build spreadsheet models can be found on the text web site. These instructions can also be implemented in a programming language such as Python or R.

Problem box answers

Problem box 9.1 answer

For three diallelic loci each with codominance, the expected genotype frequencies in the population can be obtained by expanding

$$\left(\frac{1}{4}AA + \frac{1}{2}Aa + \frac{1}{4}aa\right)\left(\frac{1}{4}BB + \frac{1}{2}Bb + \frac{1}{4}bb\right)\left(\frac{1}{4}CC + \frac{1}{2}Cc + \frac{1}{4}cc\right)$$

or by constructing an 8 × 8 Punnett square (the gametes are ABC, AbC, ABc, Abc, aBC, abC, aBc, and abc). The phenotypes are in an expected ratio of 1 : 6 : 15 : 20 : 15 : 6 : 1 for phenotypes of 1, $2\frac{2}{6}$, $3\frac{4}{6}$, 5, $6\frac{2}{6}$, $7\frac{4}{6}$, and 9 as shown in Figure 9.18.

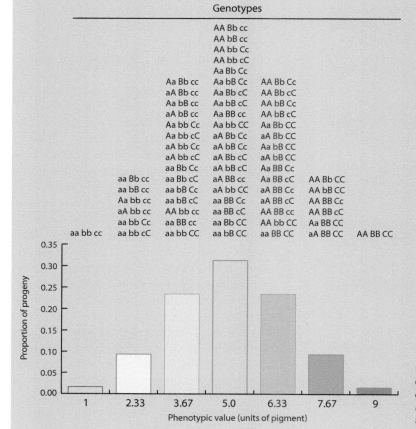

Figure 9.18 The genotypes and distribution of phenotypic values for a trait caused by three codominant diallelic loci in a randomly mating population where all alleles at each locus have a frequency of ½.

Table 9.8 Derivation of the expected phenotypic values for marker genotypes used to estimate $\hat{a}$ in a P1 × F1 backcross mating design when QTL mapping with pairs of marker loci that flank a QTL.

F1 parent gamete	F1 gamete frequency	BC progeny genotype	BC progeny genotypic value	Frequency-weighted genotypic value	BC progeny marker genotype	Marker genotype frequency	Marker-class mean $\left(\overline{G}^{BC}_{A_xA_xB_xB_x}\right)$
$A_1Q_1B_1$	$\dfrac{(1-c)}{2}$	$\dfrac{A_1Q_1B_1}{A_1Q_1B_1}$	$+a$	$a\dfrac{(1-c)}{2}$	$A_1A_1B_1B_1$	$\dfrac{(1-c)}{2}$	a
$A_1Q_1B_2$	$\dfrac{(1-c_A)c_B}{2}$	$\dfrac{A_1Q_1B_1}{A_1Q_1B_2}$	$+a$	$\dfrac{ac_B + ac_Ac_B + dc_A - dc_Ac_B}{2} \approx \dfrac{ac_B + dc_A}{2}$	$A_1A_1B_1B_2$	$\dfrac{c}{2}$	$\dfrac{ac_B + dc_A}{c}$
$A_1Q_2B_2$	$\dfrac{c_A(1-c_B)}{2}$	$\dfrac{A_1Q_1B_1}{A_1Q_2B_2}$	d				
$A_2Q_2B_1$	$\dfrac{(1-c_A)c_B}{2}$	$\dfrac{A_1Q_1B_1}{A_2Q_2B_1}$	d	$\dfrac{ac_A + ac_Ac_B + dc_B - dc_Ac_B}{2} \approx \dfrac{ac_A + dc_B}{2}$	$A_1A_2B_1B_1$	$\dfrac{c}{2}$	$\dfrac{ac_A + dc_B}{c}$
$A_2Q_1B_1$	$\dfrac{c_A(1-c_B)}{2}$	$\dfrac{A_1Q_1B_1}{A_2Q_1B_1}$	$+a$				
$A_2Q_2B_2$	$\dfrac{(1-c)}{2}$	$\dfrac{A_1Q_1B_1}{A_2Q_2B_2}$	d	$d\dfrac{(1-c)}{2}$	$A_1A_2B_1B_2$	$\dfrac{(1-c)}{2}$	d

Problem box 9.2 answer

The difference between the homozygote marker-class means is 30.7–26.4 = 4.3 cg so that $\hat{a}$= 2.15 cg assuming that $c = 0$, $a = 2.15$ cg. Assuming instead that $c = 0.2$, then $a = 2.15/(1-2[0.2]) = 3.58$ cg. The midpoint value is $26.4 + (30.7-26.4)/2 = 28.55$ cg. The M_1M_2 marker-class mean is less than the midpoint so $d = 28.3-28.55 = -0.25$. With $c = 0.2$, $d = -0.25/(1-2[0.2])^2 = -0.69$. The coefficient of dominance is $-0.25/2.15 = -0.12$ if $c = 0$ and $-0.69/3.58 = -0.19$ if $c = 0.2$. The difference in seed weight between the Q_1Q_1 and Q_2Q_2 genotypes accounts for 4.3 cg/27 cg or 16% of the total seed weight difference between the two parental lines. The phenotypic effect is large enough to be considered a major gene.

Problem box 9.3 answer

The backcross design results in $\hat{a} = a - d$ and, therefore, is a biased estimate of the additive effect of a QTL unless there is no dominance ($d = 0$). See Table 9.8.

CHAPTER 10

The Mendelian basis of quantitative trait variation

10.1 The connection between particulate inheritance and quantitative trait variation

- Establishing a scale for genotypic values
- Defining genotypic values as $+a$, d, and $-a$
- Phenotypic values as population averages
- Why we can neglect environmental variation (V_E)

This chapter will develop the concepts needed to understand the detailed connections between quantitative trait variation and particulate inheritance. Although the components of quantitative trait variation were described in Chapter 9 as population-level phenomena, the variance is ultimately caused by different alleles and genotypes possessed by individuals. The goal of this chapter is to show how additive and dominance components of variation in quantitative traits (V_A and V_D) are caused by allele and genotype frequencies in a population as well as by the nature of gene action when alleles are combined into genotypes. To accomplish this goal, we will work with a hypothetical quantitative trait that is the product of a single locus with two alleles throughout the chapter. While the use of a single diallelic locus as an example does not approximate the multilocus basis of quantitative traits and the multiallelic state of many loci, it greatly simplifies the resulting mathematical expressions while still illustrating key biological concepts. Bear in mind that the epistatic component of genetic variance (V_I) arises due to interaction between two or more loci and, therefore, cannot be represented in a single-locus model. Therefore, the use of a single locus is an implicit assumption that V_I is zero. This chapter will start by constructing expressions that predict the phenotypic mean value of some types of populations. The population of interest will initially be all individuals and later be only those individuals with genotypes that contain a certain allele. The population mean value will also be divided into components due to the additive action of alleles and the dominance effect of genotypes. Ultimately, these mean values will be used to build expressions for the variance in phenotypic values expected in a population, specifically the additive and dominance components of genetic variation (V_A and V_D). The chapter will conclude with a section on the expected phenotypic resemblance among populations of relatives based on the probabilities that related individuals share alleles or genotypes in common.

Scale of genotypic values

The hypothetical single locus used throughout this section will have two alleles, A_1 and A_2. By convention, the A_1 allele contributes to larger phenotypic values and the A_2 allele to smaller phenotypic values. A conceptual scale of measurement is used to represent the genotypic values of each of the three genotypes (Figure 10.1A). On this scale, the genotypic value of the A_1A_1 genotype is $+a$, whereas the genotypic value of the A_2A_2 genotype is $-a$. The genotypic value of the A_1A_2 genotype is d, and it is always measured relative to the midpoint as the A_1A_2 genotypic value minus the midpoint value. The midpoint on this scale of genotypic measurement, or the point exactly mid-way between the genotypic values of the homozygotes, is defined to be

(a)

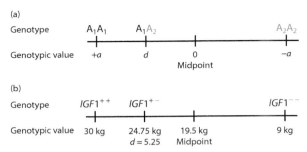

Figure 10.1 The genotypic scale of measurement for quantitative traits. The variables $+a$ and $-a$ define genotypic values of the A_1A_1 and A_2A_2 homozygotes, respectively. The genotypic value of the heterozygote is defined as d and is measured relative the midpoint ($d = A_1A_2$ genotypic value minus the midpoint). The degree of dominance is expressed as the ratio d/a. These genotypic measurements are illustrated for the $IGF1$ gene in dogs, which is a major gene contributing to body size differences. Since the degree of dominance is unknown for $IGF1$ genotypes, the genotypic value of the heterozygote is hypothetical.

zero. Notice that when the genotypic value of the heterozygote is at the midpoint, $d = 0$. This corresponds to the heterozygote having a genotypic value that is exactly the average of the two homozygotes as expected under additive gene action. In contrast, when the genotypic value of the heterozygote is not at the midpoint, then there is some type of dominance. When $d = +a$ or $d = -a$, there is complete dominance. If the genotypic value of the heterozygote is outside the range of the homozygotes, then there is overdominance ($d > +a$) or underdominance ($d < -a$). The degree of dominance can be expressed as the ratio of d/a.

Throughout this section, body size of domestic dogs will be used as an example to illustrate concepts and computations as in Chapter 9. Comparing large (>30 kg) and small (<9 kg) dog breeds, Sutter et al. (2007) showed that an allele of the insulin-like growth factor 1 gene ($IGF1$) was common to all small breeds and had a frequency of near zero in large breeds. Thus, the $IGF1$ locus is likely to be a **major gene** (a single locus that explains a large amount of quantitative trait variation) for body size in dogs. For the sake of illustration, assume that there are two $IGF1$ alleles segregating within a single randomly mating population of dogs and that body size ranges between a minimum of 9 kg and a maximum of 30 kg. Genotypic values for $IGF1$ under these assumptions are shown on the genotypic scale of measurement in Figure 10.1B.

Problem box 10.1
Compute values on the genotypic scale of measurement for *IGF*1 in dogs

Using the genotypic values of 30, 24.75, and 9 kg for the three $IGF1$ genotypes, compute a, the midpoint, d, and the degree of dominance.

In Chapter 9, the term **phenotypic value** was introduced to refer to the phenotype of an individual in the same units that it is measured in. Alternatively, phenotypic value refers to the *average phenotype* of a population of individuals. In the sense of the population average value, P refers to the average phenotypic value, G refers to the average phenotype for many individuals with a given genotype called the **genotypic value**, and E refers to the average deviation in phenotype among many individuals caused by all environmental factors combined. This leads to an equation for the causes of mean phenotypic value:

$$P = G + E \qquad (10.1)$$

In words, this equation states that the mean phenotype in a population is the sum of the mean genotypic value plus the average change in phenotype due to the environment.

Environmental deviation: The change in the mean phenotype of a population caused by all nongenetic influences. Often assumed to be zero because environment is equally likely to cause an increase or decrease in the individual phenotypic value, yielding a net change of zero.
Genotypic value: The average or expected phenotype of a population of individuals all possessing an identical genotype.
Phenotypic value: The average or expected phenotype of a population of individuals; the observed phenotype of a single individual.

It is common to assume that the mean environmental deviation is zero so that it can be ignored when deriving genotypic values. An example can be seen in Figure 9.3, where the genotypic value of an aabb genotype is 1 unit of pigment, and the genotypic value of an AaBb genotype is 5 units of pigment. The effect of the environment on individuals of a given genotype is an equal probability of deviating from the genotypic value by plus or minus 1 unit of pigment, which averages to an environmental deviation of 0 units of pigment.

With the definition of phenotypic value as a population average phenotype established, we can now develop expressions for the mean genotypic value based on genotypic values and population allele frequencies. Such mean values are prerequisites for determining how the phenotypic value of a population will change over time. As we will see eventually, mean phenotypic values are also needed to determine components of phenotypic variance since a variance is always a function of the mean.

10.2 Mean genotypic value in a population

- Deriving the population mean phenotypic value
- The population mean phenotypic value under random mating
- The population mean phenotypic value with nonrandom mating

The mean phenotypic value of a population (symbolized M) is dictated by the genotypic values and the genotype frequencies. Table 10.1 shows the frequencies, phenotypic values, and frequency-weighted phenotypic values for the three genotypes under the assumption of random mating. The average phenotype of the entire population is just the sum of the phenotypic values of all individuals. If the average environmental deviation for all genotypes is zero, then the mean phenotype of the population is the sum of the three frequency-weighted genotypic values:

$$M = p^2 a + 2pqd + \left(-q^2 a\right) \quad (10.2)$$

This can be simplified by factoring an a out of the first and third terms:

$$M = a\left(p^2 - q^2\right) + 2pqd \quad (10.3)$$

Then, notice that $(p^2 - q^2) = (p + q)(p - q)$ and also that $p + q = 1$ for a diallelic locus. Making this substitution leads to

$$M = a(p - q) + 2pqd \quad (10.4)$$

This population mean genotypic value is identical to the population mean phenotypic value if the mean environmental deviation is zero. It is important to note that M is measured as a deviation from the midpoint on the scale of genotypic values. Therefore, M must be added to the midpoint value to obtain the absolute population genotypic mean value.

This expression for the mean genotypic value in a population leads to two important biological conclusions. First, it shows that the mean of a quantitative trait depends on allele frequencies in the population because these dictate genotype frequencies. Second, the division of the expression for the mean phenotype into two terms is informative. The $a(p - q)$ term shows that changes in allele frequencies shift the mean up or down by some fraction of a depending on which of the two homozygotes is more frequent. At the extremes of allele frequency, when $p = 1$, then the mean phenotype is equal to $+a$, and if $q = 1$, then the mean phenotype is $-a$. The $2pqd$ term shows that the frequency of the heterozygote alone determines the impact of dominance on the mean phenotype. When the phenotypic value of the heterozygote is exactly at the midpoint between the homozygotes ($d = 0$), the heterozygote genotype has no impact on the population mean. Completely additive gene action exists when $d = 0$.

The role of allele frequency on the population mean can be seen in an example. Imagine a population of dogs where mating is random and the $IGF1^+$ and $IGF1^-$ alleles are both segregating. For $IGF1$ in dogs, $a = 10.5$ kg, $d = 5.25$ kg, and the midpoint is

Table 10.1 The population mean phenotype (M) obtained from genotype frequencies under random mating, genotypic values, and frequency-weighted genotypic values for a diallelic locus. These expectations assume that the environmental deviation is zero for each genotype.

Genotype	Frequency	Genotypic value	Frequency-weighted genotypic value
A_1A_1	p^2	a	$p^2 a$
A_1A_2	$2pq$	d	$2pqd$
A_2A_2	q^2	$-a$	$-q^2 a$
			$M = a(p - q) + 2pqd$

19.5 kg (see Problem box 10.1). If the two *IGF1* alleles are at equal frequencies, then $p = q = 0.5$. The mean phenotypic value would be

$$M = 10.5(0.5 - 0.5) + 2(0.5)(0.5)(5.25) = 2.625 \text{ kg}$$
$$(10.5)$$

as a deviation from the midpoint. In absolute terms, the mean phenotype in the population would be $2.625 + 19.5 = 22.125$ kg. Since both homozygotes are equally frequent, their effect on the average cancels out. That leaves 50% of the population as heterozygotes, shifting the mean above the midpoint by $\frac{1}{2}d$. If the two *IGF1* alleles are at unequal frequencies, say $p = 0.9$ and $q = 0.1$, the mean phenotypic value would be

$$M = 10.5(0.9 - 0.1) + 2(0.9)(0.1)(5.25) = 9.345 \text{ kg}$$
$$(10.6)$$

as a deviation from the midpoint and 28.845 kg in absolute terms. At these allele frequencies, the A_1A_1 homozygote composes 81% of the population, while the A_2A_2 homozygote is only 1% of the population, so the balance is strongly in favor of $+a$. The remaining 18% of the population is made up of heterozygotes, which also shift the average toward $+a$ since d is positive.

The role of gene action on the population mean phenotype can also be seen in this example. Suppose that the *IGF1* alleles are $p = 0.9$ and $q = 0.1$ but that $d = 0$. The mean phenotypic value is then

$$M = 10.5(0.9 - 0.1) + 2(0.9)(0.1)(0) = 8.4 \text{ kg}$$
$$(10.7)$$

as a deviation from the midpoint and 27.9 kg in absolute terms. This population mean phenotype is lower than the result in Eq. 10.6 because the genotypic value of the heterozygotes is zero when $d = 0$. Therefore, the 18% of the population made up of heterozygotes does not contribute to any deviation from the midpoint in addition to that caused by the frequencies of the two homozygotes.

So far, random mating has been assumed when predicting the mean phenotype. However, consanguineous mating is also common in populations and will change the mean phenotype in predictable ways. Using the results of Eq. 2.20, the expected genotype frequencies can include the impact of nonrandom mating. The mean phenotype with the possibility of nonrandom mating is given by

$$M = \left(p^2 + fpq\right)a + \left(2pq - f2pq\right)d + \left(q^2 + fpq\right)(-a)$$
$$(10.8)$$

where f is the inbreeding coefficient. Changes in autozygosity ($f \neq 0$) alter all genotype frequencies but will only impact the population mean phenotype when there is some degree of dominance ($d \neq 0$). The impact of the inbreeding coefficient is identical on each of the homozygotes, increasing or decreasing them by the same amount fpq with no impact on the mean. In contrast, changes in the autozygosity will impact the mean phenotype if there is dominance because the frequency of heterozygotes will change by the amount $-f2pq$.

Extending the *IGF1* example further, imagine that there is consanguineous mating so that $f = 0.2$ when allele frequencies are $p = 0.9$ and $q = 0.1$ and $d = 5.25$. This would lead to $f2pq = 0.036$ or a 3.6% deficit of heterozygotes and a 1.8% excess of both homozygotes compared to random mating. The mean phenotype is then

$$M = (0.81 + 0.018)(10.5) + (0.18 - 0.036)(5.25)$$
$$+ (0.01 + 0.018)(-10.5)$$
$$(10.9)$$

which equals 9.156 kg as a deviation from the midpoint or 28.656 kg. This final example shows that changes in genotype frequencies caused by nonrandom mating, even when allele frequencies remain constant, can impact the population mean phenotype.

10.3 Average effect of an allele

- Deriving the average phenotypic effect of an allele in a population
- The average effect as a deviation from the population mean
- The average effect as substitution of one allele in a genotype for another

To describe the inheritance of quantitative phenotypes across generations, it is necessary to think in terms of alleles. Although genotypes determine genotypic values, alleles and not genotypes are inherited by individuals. Thus, a new measure of value is needed that can be used to link the genotypic values of one generation with the genotypic values of their progeny in the next generation. The concept of **average effect** is used to assign a value to an *allele* and predict how it impacts the mean genotypic value of the population. As we will see, the average effect depends on the genotypic values a and d as well as the allele and genotype frequencies in the population.

Average effect of an allele: The mean phenotypic deviation from the population mean of that group of individuals which received a particular allele from one parent and the other allele from a parent drawn at random from the population.

One way to visualize the basis of the average effect is to think of a slot machine used in gambling (also called a poker or fruit machine). Mechanical slot machines normally have three wheels that are each labeled with symbols having different frequencies on the wheels. When the wheels are spun, they will each stop and display a random symbol. An imaginary average-effect slot machine has just two wheels that represent the two alleles in a diploid genotype. One of the wheels is broken and does not spin, so a single symbol is always displayed. For the average effect of the A_1 allele, for example, A_1 would always appear on the broken wheel. The other wheel has symbols for all of the alleles in the population in proportion to their frequency in the population. Spinning this slot machine many times with one allele on the broken wheel and recording the value of the genotype for each spin would give expected frequencies of all of the genotypes in the population that contain an A_1 allele under random mating. The average of all of the genotypic values from many spins would give the average value of all the genotypes that contain a given allele. This average value of genotypes containing one allele may very well be different than the average value of the population. This difference in averages is the average effect of the allele that is present on the broken wheel.

The derivation of the mean value of all genotypes that contain either an $A_1 (M_{A_1})$ or an $A_2 (M_{A_2})$ allele is shown in Table 10.2. This logic could be used for a locus with any number of alleles. If mating is random, we expect an A_1 allele to be paired in a genotype with another A_1 allele p percent of the time and with an A_2 allele q percent of the time. The average value of all genotypes that contain an A_1 allele is the frequency-weighted sum of the values of each genotype that has at least one A_1 allele. In symbols, this mean value is

$$M_{A_1} = pa + qd \tag{10.10}$$

for the A_1 allele. This quantity can be thought of as the mean value of a large number of individuals that all inherit the same allele from the same parent but that inherit the other allele in their genotype from a parent drawn at random from the population. Notice that Hardy–Weinberg expected genotype frequencies are a key assumption in Table 10.2, since expected genotype frequencies would be different if any of the Hardy–Weinberg assumptions did not hold.

The average effect is a *deviation* that measures the difference between the value of all genotypes that contain a given allele and the population mean. Therefore, the population mean must be subtracted from the mean values of genotypes produced by a given allele that are shown in Table 10.2. The average effect is then

$$\alpha_x = M_{Ax} - M \tag{10.11}$$

using the α to represent the average effect and x to indicate any allele. The average effect is determined by a regression line (see Figure 10.4, below) that minimizes the deviations (rather than the squared deviations) of individual values from the line. In many cases, when genotypes are in Hardy–Weinberg frequencies, the simple relationship of

Table 10.2 The mean value of all genotypes that contain either an A_1 (M_{A_1}) or an A_2 (M_{A_2}) allele. The average effect of an allele (α_x) is the difference between the mean value of the genotypes that contain a given allele and the population mean ($\alpha_x = M_{Ax} - M$).

	Genotypes and values			
Allele	A_1A_1	A_1A_2	A_2A_2	Mean value of all genotypes that contain a given allele
	$+a$	d	$-a$	
A_1	p	q	0	$M_{A_1} = pa + qd$
A_2	0	p	q	$M_{A_2} = pd - qa$

Eq. 10.11 suffices to describe the average effect. In more complicated situations such as when Hardy–Weinberg is not met ($f \neq 0$), the line that minimizes the deviations is determined by a least-squares fit.

Substituting the expression for the population mean from Eq. 10.4 and the mean value of all genotypes containing the A_1 allele from Table 10.2, the average effect of the A_1 allele is

$$\alpha_1 = pa + qd - (a(p-q) + 2pqd) \quad (10.12)$$

which simplifies to

$$\alpha_1 = q(a + d(q-p)) \quad (10.13)$$

as shown in Math box 10.1. The average effect of the A_2 allele is

$$\alpha_2 = -p(a + d(q-p)) \quad (10.14)$$

based on the difference between the mean values of genotypes produced by A_2 and the population mean.

Working through several examples based on the *IGF*1 locus in dogs will help illustrate the average effect and what it measures. Table 10.3 gives four examples of the average effect for the A_1 allele for the four combinations of two allele frequencies and two levels of dominance. In all of the examples, bear in mind that the A_1-containing genotypes are A_1A_1

Table 10.3 Examples of the average effect for the *IFG*1 locus in dogs. All cases assume that $a = 10.5$ kg as shown in the genotypic scale in **Figure 10.1**. For each set of allele frequencies and dominance, the table shows the population mean (M), the mean value of all genotypes that contain an A_1 allele (M_{A1}), the average effect of an allelic replacement (α), and the average effect of an A_1 allele (α_1). Values are all in kilograms and relative to the midpoint value of 19.5 kg.

A. $d = 0.0$, $p = 0.5$, $q = 0.5$
$M = 10.5(0.5 - 0.5) + 2(0.5)(0.5)(0.0) = 0.0$
A_1 $M_{A1} = pa + qd = (0.5)(10.5) + (0.5)(0.0) = 5.25$
$\alpha_1 = M_{A1} - M = 5.25 - 0.0 = 5.25$
$\alpha_1 = q[a + d(q-p)] = 0.5[10.5 + 0.0(0.5 - 0.5)]$
$= 5.25$
$\alpha = a + d(q-p) = 10.5 + 0.0(0.5 - 0.5) = 10.5$
$\alpha_1 = q\alpha = (0.5)(10.5) = 5.25$

B. $d = 0.0$, $p = 0.9$, $q = 0.1$
$M = 10.5(0.9 - 0.1) + 2(0.9)(0.1)(0.0) = 8.4$
A_1 $M_{A1} = pa + qd = (0.9)(10.5) + (0.1)(0.0) = 9.45$
$\alpha_1 = M_{A1} - M = 9.45 - 8.4 = 1.05$
$\alpha_1 = q[a + d(q-p)] = 0.1[10.5 + 0.0(0.1 - 0.9)]$
$= 1.05$
$\alpha = a + d(q-p) = 10.5 + 0.0(0.1 - 0.9) = 10.5$
$\alpha_1 = q\alpha = (0.1)(10.5) = 1.05$

C. $d = 5.25$, $p = 0.5$, $q = 0.5$
$M = 10.5(0.5 - 0.5) + 2(0.5)(0.5)(5.25) = 2.625$
A_1 $M_{A1} = pa + qd = (0.5)(10.5) + (0.5)(5.25)$
$= 7.875$
$\alpha_1 = M_{A1} - M = 7.875 - 2.625 = 5.25$
$\alpha_1 = q[a + d(q-p)] = 0.5[10.5 + 5.25(0.5 - 0.5)]$
$= 5.25$
$\alpha = a + d(q-p) = 10.5 + 5.25(0.5 - 0.5) = 10.5$
$\alpha_1 = q\alpha = (0.5)(10.5) = 5.25$

D. $d = 5.25$, $p = 0.9$, $q = 0.1$
$M = 10.5(0.9 - 0.1) + 2(0.9)(0.1)(5.25) = 9.345$
A_1 $M_{A1} = pa + qd = (0.9)(10.5) + (0.1)(5.25)$
$= 9.975$
$\alpha_1 = M_{A1} - M = 9.975 - 9.345 = 0.630$
$\alpha_1 = q[a + d(q-p)] = 0.1[10.5 + 5.25(0.1 - 0.9)]$
$= 0.630$
$\alpha = a + d(q-p) = 10.5 + 5.25(0.1 - 0.9) = 6.3$
$\alpha_1 = q\alpha = (0.1)(6.3) = 0.63$

Math box 10.1 The average effect of the A_1 allele

Start with the difference between the mean value of all genotypes produced by A_1 allele in Table 10.2 and the population mean in Eq. 10.4:

$$\alpha_1 = pa + qd - (a(p-q) + 2pqd) \quad (10.15)$$

Then, expand $a(p - q)$ to give

$$\alpha_1 = pa + qd - pa + qa - 2pqd \quad (10.16)$$

Cancel terms to obtain

$$\alpha_1 = qd + qa - 2pqd \quad (10.17)$$

Then, factor out q

$$\alpha_1 = q(d + a - 2pd) \quad (10.18)$$

Inside the parentheses, two terms contain d in common that can be factored to give

$$\alpha_1 = q(a + d(1 - 2p)) \quad (10.19)$$

The final step is to notice that $p + q = 1$ so that $(p + q)$ can be substituted:

$$\alpha_1 = q(a + d(p + q - 2p)) \quad (10.20)$$

After adding p to $-2p$, the simplified equation for the average effect of the A_1 allele is

$$\alpha_1 = q(a + d(q-p)) \quad (10.21)$$

and A_1A_2 and that all values are relative to a midpoint value of 19.5 kg.

In example (a) in Table 10.3, the A_1 and A_2 allele frequencies are equal. Since there is no dominance, the heterozygotes have a mean value equal to the midpoint. The two homozygotes are equally frequent, and their average values are equal but have opposite signs. This results in a population mean value of zero. Given an A_1 allele, when sampling a second allele from this population to make genotypes, it is equally likely to obtain either allele. Thus, 50% of the A_1 containing genotypes are A_1A_1 with a value of $+a = 10.5$, and 50% are A_1A_2 with a value of $d = 0$. In total, the mean value of the A_1-containing genotypes is $0.5(a)$ or 5.25 kg. This is the same as the average effect since the mean value of all three genotypes is zero. At these allele frequencies, the population mean is exactly at the midpoint and the A_1-containing genotypes serve to increase the average value of the population by 5.25 kg.

The situation is different in example (b) in Table 10.3 since the frequency of A_1 is high and the frequency of A_2 is low. Given an A_1 allele, when a second allele is drawn at random from this population to make genotypes, it is much more likely to be another A_1 allele rather than an A_2 allele. Thus, the mean of the A_1-containing genotypes is nearly the same as a (9.45 kg), the genotypic value of A_1A_1. However, the average effect is small (1.05 kg) since the mean value of all three genotypes is also large (8.4 kg). At these allele frequencies, the population mean is near its upper limit due to the low frequency of the A_2 allele and the resulting low frequency of A_2-containing genotypes that would reduce the average value of the population.

Comparing cases (a) and (c) in Table 10.3 is informative since allele frequencies are identical but the degree of dominance is different. With dominance in example (c), 50% of the A_1-containing genotypes are A_1A_1 with a value of $+a = 10.5$ and 50% are A_1A_2 with a value of $d = 5.25$. The mean value of the A_1-containing genotypes is now larger at 0.5 $(a) + 0.5(d)$ or 7.875 kg. However, because of the dominance, the mean value of the total population is also greater at 2.625 kg. Thus, the difference between the mean value of the A_1-containing genotypes and the mean value of the entire population remains at 5.25 kg, exactly as it was in example (a). Comparing examples (b) and (d) in Table 10.3

illustrates that the average effect changes with a shift in the dominance value. Both the population mean and the mean of the A_1-containing genotypes change with a change in dominance at those allele frequencies.

A critical lesson to take from these examples is the contextual nature of the average effect. Average effects depend on allele and genotype frequencies and are, therefore, specific to the population and time point when they are measured. This is sometimes a difficult point to grasp when considering the genetic basis of phenotypic variation. Even though the genotypic values and the degree of dominance may remain constant among populations, the average effect of an allele shifts depending on the allele and genotype frequencies. This is a part of the major distinction between identifying genes and alleles that impact phenotype viewed from the perspective of Mendelian genetics (e.g. an allele causes a certain phenotype) and understanding how such alleles shape phenotypic distributions within and between populations (e.g. an allele has a large average effect).

Another way to understand the average effect when there are two alleles at a locus is to suppose that it is possible to randomly sample genotypes from a population and then be able to substitute *one* of the alleles in a genotype for another. When there are two alleles, it would be possible to replace one A_2 allele initially present in the genotype with one A_1 allele or vice versa. The average effect can be thought of in terms of the change in the mean value of the population when one allele replaces another in all the sampled genotypes. To measure the average effect in this manner requires determining the change in value that comes about by allelic replacement as well as the frequency with which the change in value occurs. Summing these frequency-weighted value changes will give the mean change in value, that is, the average effect.

Imagine that we elect to replace an A_2 allele with A_1 allele in genotypes sampled at random from the population *that contains at least one* A_2 allele (see Figure 10.2). Let p be the frequency of the A_1 allele and q be the frequency of the A_2 allele. Changing an A_1A_2 genotype to an A_1A_1 genotype will change the value from d to $+a$, so the difference in value is $a - d$. Since the frequency of the A_1 allele that remains intact in the genotype is p, the frequency of changing an A_1A_2 genotype to an A_1A_1 genotype is p. The frequency-weighted change in value when making this allelic substitution is, therefore, $p(a - d)$. Similarly, when changing an A_2A_2 genotype to an A_1A_2 via replacement of one A_2 allele, the value changes from $-a$ to d, so the difference in value is $d - (-a)$ or $d + a$. The frequency of the A_2 allele that remains intact in the second genotype is now q, so the frequency-weighted change in value when making this allelic substitution is, therefore, $q(d + a)$. The total average change in value when replacing an A_2 allele with A_1 allele is the sum of these two separate changes in value or $p(a - d) + q(d + a)$. (Note that the same result, except multiplied by -1, can be obtained by electing to replace an A_1 allele with A_2 allele. The negative sign comes about because A_2 is the allele that decreases value.)

Using algebraic manipulation similar to that in Math Box 10.1, the expression for the average change in value due to an allelic replacement simplifies to

$$\alpha = a + d(q - p) \qquad (10.22)$$

with α not bearing a subscript denoting the average change in value caused by an allele replacement. Notice that $a + d(q - p)$ also appears in Eqs. 10.13 and 10.14. On a strictly mathematical basis, we could substitute α in these two equations to restate the expressions for the average effects of the A_1 and A_2 alleles

$$\alpha_1 = q\alpha \qquad (10.23)$$

and

$$\alpha_2 = -p\alpha \qquad (10.24)$$

However, this reformulation makes intuitive sense in terms of our imagined ability to replace one allele with another in genotypes used to derive Eq. 10.22. When replacing an A_2 allele with A_1 allele, the frequency of the A_2 allele in the population is q, and this is the frequency of allelic replacements under random mating (see Figure 10.2). Likewise, p is the frequency in the population of the A_1 allele being replaced with A_2 alleles. The expression for α_2 is negative because A_2 is the allele that decreases value. Table 10.3 also gives the computations of the average effect using Eqs. 10.23 and 10.24 with identical results.

Problem box 10.2
Compute average effects for *IGF*1 in dogs

Using Table 10.3 as a guide, compute the average effect in three ways: (1) as the difference between and the population mean M, (2) by the formula $\alpha_2 = -p(a + d (q - p))$ from Eq. 10.14, and (3) by $\alpha_2 = -p\alpha$ from Eq. 10.24. Be sure to compare your results with the average effects for the A_1 allele given in Table 10.3 and explain why the average effects are the same or different with reference to dominance and the allele frequencies.

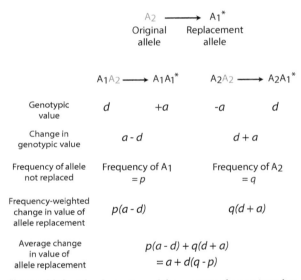

Figure 10.2 The derivation of the average change in value caused by replacing one A_2 allele with an A_1 allele in those genotypes sampled at random from the population that contain at least one A_2 allele. The mean change in value caused by an allelic replacement forms the basis of the average effect once it is multiplied by the frequency of the allele that is replaced.

10.4 Breeding value and dominance deviation

- Deriving breeding values in a population
- Breeding values under random and nonrandom mating
- Deriving dominance deviations in a population
- Dominance deviations under random and nonrandom mating

The next step in understanding the Mendelian basis of quantitative trait variation is to move back to the level of the genotype. In Chapter 9, both additive (V_A) and dominance (V_D) genetic variations were identified as separate components of the total genetic variation in quantitative traits. In terms of population mean values, we can divide the genotypic mean value into its components due to additive effects of alleles and the dominance effects of genotypes

$$G = A + D \qquad (10.25)$$

In words, this equation says that the mean genotypic value (G) is the sum of the mean breeding value (A) and the mean dominance deviation (D). As pointed out above, this assumes that there is no interaction genetic variance due to epistasis because we are working with a one-locus example. If there were epistasis, then there would also be a mean value for an interaction deviation due to the mean value of interactions among loci that make up genotypes.

This subsection will address how the mean phenotype of a population of progeny depends on mating among the population of parents. It is important to first understand the motivation to predict what is called the **breeding value** of a genotype. When natural selection occurs, it is essentially the differential mating success of certain phenotypes. Humans achieve the same thing in domestic plants and animals through artificial selection by allowing only those individuals with preferred phenotypes to breed. To the extent that phenotype is a function of genotype, natural and artificial selections allow some genotypes to breed more often than others. A full understanding of how and why a given mean phenotype occurs in a population of progeny, therefore, requires an understanding of the consequences of mating in the parental generation. Here, we will start with the genotypic value of an individual, and then track the frequencies and genotypic values of its progeny to predict the mean genotypic value of its progeny.

> **Breeding value of an individual:** Twice the mean value of the progeny that would be produced by a single genotype under random mating expressed as a difference from the population mean.

Parents pass on alleles and not genotypes to their progeny. (From the progeny point of view, individuals inherit one allele from each of two parents rather than a diploid genotype.) The impact of a single parent on the mean value of the progeny population could be measured in a manner akin to the average effect. Now, instead of a special slot machine, we could just take a single individual and have it mate with many individuals drawn at random from the parental population. Each progeny from these matings would inherit one allele from the focal parent and another allele from an individual drawn at random. The resulting population of progeny would have a mean value that could be measured. The **breeding value** of a genotype is two times the difference between the progeny mean value and the parental population mean value (M). Expressed as an equation,

$$\textit{Breeding value} = 2\left(M_{\text{progeny A}x\text{A}x} - M\right) \qquad (10.26)$$

where $M_{\text{progeny A}x\text{A}x}$ is the mean value of progeny produced when an individual of genotype A_xA_x is mated to many individuals drawn at random from the parental population. The difference between the means is multiplied by 2 because a parent's genotype possesses two alleles but its progeny inherits only one allele at a time.

The components that lead to an expression for M_{progeny} for the A_1A_1 genotype are shown in Table 10.4. When the A_1A_1 genotype mates, it will encounter and mate with individuals of a given genotype in proportion to the frequency of that genotype in the population. An A_1A_1 individual is expected to mate with A_1A_1, A_1A_2, and A_2A_2 individuals with frequencies of p^2, $2pq$, and q^2, respectively, since these are the Hardy–Weinberg expected genotype frequencies in the population. Each of the matings between an A_1A_1 genotype and another genotype will produce progeny with one or two genotypes. The phenotypic values of each of the progeny genotypes are also known. To obtain an expression for the mean phenotypic value of the progeny that result from the A_1A_1 genotype mating at random in the population ($M_{\text{progeny A1A1}}$), add up all of the progeny phenotypic values after weighting each one by its relative frequency among all of the progeny and by the frequency of mating pairs to obtain

Table 10.4 The mean phenotypic value of progeny that result when an individual of the genotype A_1A_1 mates randomly. All genotypes in the population have Hardy–Weinberg expected frequencies. Therefore, each of the mating pairs has an expected frequency of p^2, $2pq$, or q^2. The mean value of all progeny produced by the A_1A_1 genotype is the frequency-weighted sum of the progeny phenotypic values. $M_{\text{progeny A1A1}}$ forms the basis of the breeding value since the breeding value for A_1A_1 is $M_{\text{progeny A1A1}} - M$.

Focal genotype	A_1A_1			
Mate genotypes	A_1A_1		A_1A_2	A_2A_2
Mating frequency	p^2		$2pq$	q^2
Progeny genotype and relative frequency from each mating	A_1A_1	½ A_1A_1	½ A_1A_2	A_1A_2
Progeny values	$+a$	$+a$	d	d
Progeny mean value	$M_{\text{progeny A1A1}} = p^2a + 2pq(\frac{1}{2}a + \frac{1}{2}d) + q^2d = ap + dq$			

$$M_{\text{progeny A1A1}} = p^2a + 2pq(\tfrac{1}{2}a + \tfrac{1}{2}d) + q^2d \tag{10.27}$$

This expression can be simplified by first expanding the middle term

$$M_{\text{progeny A1A1}} = p^2a + pqa + pqd + q^2d \tag{10.28}$$

and then factoring to give

$$M_{\text{progeny A1A1}} = ap(p + q) + dq(p + q) \tag{10.29}$$

and then noticing that $p + q = 1$ so that

$$M_{\text{progeny A1A1}} = ap + dq. \tag{10.30}$$

The same logic can be applied to obtain the progeny mean values for the A_1A_2 and A_2A_2 genotypes under random mating.

Prepared with the expression for $M_{\text{progeny A1A1}}$, we can now obtain an equation for the breeding value of the A_1A_1 genotype. Using the definition of the breeding value given in Eq. 10.26,

$$Breeding\ value\ A_1A_1 = 2\big(M_{\text{progeny A1A1}} - M\big) \tag{10.31}$$

and then substituting in the expression for mean progeny value in Eq. 10.30 and the expression for the population mean in Eq. 10.4 gives

$$Breeding\ value\ A_1A_1 = 2(ap + dq - (a(p - q) + 2pqd)) \tag{10.32}$$

Rearrangement of this equation yields

$$Breeding\ value\ A_1A_1 = 2q(a + d(q - p)) \tag{10.33}$$

(Readers carrying out the algebra can refer to Math Box 10.1 for the steps that show $d(1 - 2p) = d(q - p)$). Notice that the $q(a + d(q - p))$ term is equal to the definition of the average effect of the A_1 allele given in Eq. 10.13. Therefore, the breeding value of the A_1A_1 genotype is simply equal to two times the average effect of the A_1 allele.

Based on the definition of the breeding value as the difference between the progeny mean and the population mean, the breeding value of any genotype is simply the sum of the average effects of each of the alleles that make up the genotype. We can, therefore, use the definitions of the average effect in the last section to define breeding values for the three genotypes at a diallelic locus

Breeding value $A_1A_1 = \alpha_1 + \alpha_1 = q\alpha + q\alpha = 2q\alpha$

$$(10.34)$$

Breeding value $A_1A_2 = \alpha_1 + \alpha_2 = q\alpha + -p\alpha = (q-p)\alpha$

$$(10.35)$$

Breeding value $A_2A_2 = \alpha_2 + \alpha_2 = -p\alpha + -p\alpha = -2p\alpha$

$$(10.36)$$

Because breeding values are made up of the average effects of the two alleles in a genotype, breeding values also depend on the population context in which they are measured. Thus, breeding values are not universal features of genotypes but rather depend on the population allele and genotypes frequencies.

The breeding value expressions tell us the mean value of the progeny that would be produced by each genotype under random mating. Look at the example breeding values for the three genotypes at the *IGF*1 locus in dogs (Table 10.5). Under the allele frequencies and zero dominance in case A, a population of progeny from an individual with an A_1A_1 genotype would have a mean of 10.5 kg relative to the midpoint. (In the absolute, the progeny would have a mean value of 30 kg, which is identical to the A_1A_1 genotypic value.) In case A with $p = q = 0.5$, an A_1 allele would be paired with an A_1 allele in half of the progeny and with an A_2 allele in half of the progeny to make equal frequencies of A_1A_1 and A_1A_2 progeny. The mean value of these progeny, and hence the average effect of an A_1 allele, would be

$0.5(10.5) + 0.5(0.0) = 5.25$ kg. The breeding value of an A_1A_1 genotype is twice the average effect of the A_1 allele because it has two A_1 alleles. Therefore, we double the average effect of an A_1 allele to determine that the breeding value of the A_1A_1 genotype is $2(5.25) = 10.5$ kg. Figure 10.3 also shows two examples of the breeding value for the *IGF*1 locus in dogs in relation to the genotypic scale of measurement and the population mean.

As another example focuses on cases (a) and (c) of Table 10.5 and consider why the heterozygotes both have breeding values of zero. When heterozygotes breed, they contribute equal proportions of A_1 and A_2 alleles to their progeny. With a dominance value of zero and equal allele frequencies, it makes intuitive sense in case (a) that the heterozygotes have a breeding value of zero. Each of the A_1 and A_2 alleles that a heterozygote contributes to its progeny is paired with A_1 and A_2 alleles in equal frequency from the population. That would form an equal number of A_1A_1 and A_2A_2 genotypes with a mean value of zero, whereas all heterozygous progeny would also have a mean value of zero since the heterozygote genotypic value is zero. In example (c), the progeny genotypes and frequencies are identical to example (a) since the allele frequencies are the same. However, there is now dominance such that the heterozygote has a genotypic value of 5.25 kg. Still, the breeding value of a heterozygote is zero. An equal frequency of A_1A_1 and A_2A_2 genotypes in the progeny gives a mean value of zero. The 50% of progeny that are heterozygotes have a value of 5.25 because of

Table 10.5 Examples of breeding values for the three *IFG*1 locus genotypes in dogs. Values are all in kilograms and relative to the midpoint value of 19.5 kg.

	Breeding values		
	A_1A_1	A_1A_2	A_2A_2
A. $d = 0.0$, $p = 0.5$, $q = 0.5$, $M = 0.0$, $\alpha = 10.5$	$2(0.5)(10.5) = 10.5$	$(0.5 - 0.5)(10.5) = 0.0$	$-2(0.5)(10.5) = -10.5$
B. $d = 0.0$, $p = 0.9$, $q = 0.1$, $M = 8.4$, $\alpha = 10.5$	$2(0.1)(10.5) = 2.1$	$(0.1 - 0.9)(10.5) = -8.4$	$-2(0.9)(10.5) = -18.9$
C. $d = 5.25$, $p = 0.5$, $q = 0.5$, $M = 2.625$, $\alpha = 10.5$	$2(0.5)(10.5) = 10.5$	$(0.5 - 0.5)(10.5) = 0.0$	$-2(0.5)(10.5) = -10.5$
D. $d = 5.25$, $p = 0.9$, $q = 0.1$, $M = 9.345$, $\alpha = 6.3$	$2(0.1)(6.3) = 1.26$	$(0.1 - 0.9)(6.3) = -5.04$	$-2(0.9)(6.3) = -11.34$

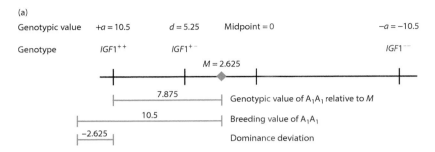

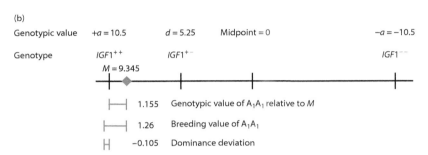

Figure 10.3 Illustration of dominance deviation for the *IGF1* gene in dogs. The dominance deviation is the difference between the genotypic value (measured relative to the population mean, *M*) and the breeding value. The dominance deviation is a consequence of the heterozygote genotypic value not falling at the midpoint. Panels A and B correspond to cases C and D, respectively, in Tables 10.3, 10.5, and 10.7.

dominance. Nonetheless, $0.5(5.25) = 2.625$ exactly equals the population mean of 2.625, so the heterozygous progeny also do not alter the population mean.

Another way to understand the breeding values is to determine the total breeding value in a population where all three genotypes are mating at random. Let the population of parents have Hardy–Weinberg expected genotype frequencies of p^2, $2pq$, and q^2. To find the average breeding value of all three genotypes in the parental population, multiply the breeding value of each genotype by its corresponding genotype frequency. This gives

$$\text{Mean breeding value of all genotypes} = p^2 2q\alpha + 2pq(q-p)\alpha - q^2 2p\alpha \tag{10.37}$$

This equation can be simplified by factoring $2pq$ from each term and expanding $(q - p)\alpha$ in the middle term to give

$$\text{Mean breeding value of all genotypes} = 2pq(p\alpha + q\alpha - p\alpha - q\alpha) = 0 \tag{10.38}$$

The conclusion is that the mean breeding value of all three genotypes mating at random is zero. This result makes intuitive sense because when a large parental population composed of genotypes in Hardy–Weinberg expected frequencies mates at random, the mean value of the progeny population should be exactly the same since the progeny genotype frequencies are exactly the same as in the parental population. It is only when genotypes mate more or less often than expected by random mating that the progeny population mean value differs from the parental population mean value. Note that with natural or artificial selection, mating is by definition nonrandom, and some genotypes mate more frequently than others. It is the over- or underrepresentation of parental genotypes in mating that causes the mean value of progeny to differ from the mean value of their parents.

Dominance deviation

With the breeding value established, we can now focus on the dominance deviation that makes up the second portion of the total mean genotypic value in $G = A + D$ (Eq. 10.25). While the breeding value measures the mean value of alleles passed to progeny by a given genotype, these same progeny also possess genotypes. Due to dominance, genotypes may not completely reflect the combinations of the alleles that make them up. For example, with complete dominance of the A_1 allele, an A_1A_2 genotype masks the fact that it has one A_2 allele since its phenotype is indistinguishable from that of an A_1A_1 genotype. When there is no dominance, the average value of progeny is a perfect representation of the parental genotypic value. However, dominance changes the average value of progeny and can make the breeding value different than the parental genotypic values. The difference between the genotypic value and the breeding value caused by dominance is called the **dominance deviation**.

> **Dominance deviation:** The difference between the genotypic value and the breeding value where the genotypic value is measured relative to the mean value of the population.

Expressions for the dominance deviations are obtained by taking the difference between the genotypic value and the breeding value for each genotype. We have already obtained all of the expressions needed for the dominance deviation. We do, however, need to obtain one new equation. Since breeding values are expressed relative to the population mean, we have to start by also expressing genotypic values relative to the population mean rather than relative to the midpoint. For the A_1A_1 genotype, the difference between the genotypic value and the population mean in Eq. 10.4 is

$$a - (a(p - q) + 2pqd) \qquad (10.39)$$

which simplifies to

$$2q(a - dp) \qquad (10.40)$$

as the genotypic value of A_1A_1 relative to the population mean.

Using the A_1A_1 genotype breeding value of $2q\alpha$ where $\alpha = a + d(q - p)$, the difference between the

genotypic value relative to the population mean and the breeding value is

$$A_1A_1 \; Dominance \; deviation = 2q(a - dp) - 2q(a + d(q - p)) \qquad (10.41)$$

This equation can be expanded to

$$A_1A_1 \; Dominance \; deviation = \\ 2qa - 2qdp - 2qa - 2q^2d + 2qdp \qquad (10.42)$$

and then canceling terms gives

$$A_1A_1 \; Dominance \; deviation = -2q^2d \qquad (10.43)$$

Table 10.6 gives the expressions for all three genotypic values relative to the population mean as well as the dominance deviations derived using this same reasoning.

The dominance deviations for the four *IGF1* examples are shown in Table 10.7. An examination of Table 10.7 shows that the genotypic values of the heterozygotes (measured from the population mean) and breeding values in the table are always zero when there is no dominance. In both cases (a) and (b), the A_1A_2 genotypic value is at the midpoint ($d = 0$), giving dominance deviations of zero regardless of the allele frequencies. Another way to describe this is to say that when there is no dominance, a genotype has a breeding value that is identical to its genotypic value (measured relative to the population mean) since genotypic values are determined by the addition of average effects alone. Without dominance, genotypic values in progeny as measured by the breeding value can be predicted perfectly from the combination of average effects.

In cases (c) and (d), the A_1A_2 genotypic values are not at the midpoint ($d \neq 0$), resulting in dominance deviations that are not zero. Let us examine the dominance deviation for the A_1A_1 genotype in case (c) of Table 10.7 as illustrated in Figure 10.3A. The genotypic value of the A_1A_1 genotype when measured relative to the population mean is $10.5 - 2.625 = 7.875$. The breeding value of the A_1A_1 genotype is 10.5 *based on the average effects of its two alleles*. With this breeding value, the mean of the progeny from an A_1A_1 genotype would be outside the largest phenotypic value, which is physically impossible. The breeding value has in essence assumed that the population mean is zero, as it would be at $p = q = 0.5$ without dominance. But

Table 10.6 Expressions for genotypic values relative to the population mean, breeding values, and dominance deviations. Genotypic values can be expressed relative to the population mean by subtracting the population mean [$M = a(p - q) + 2pqd$] from a genotypic value measured relative to the midpoint. The dominance deviation is the difference between the genotypic value expressed relative to the population mean (M) and the breeding value.

	Genotypes		
	A_1A_1	A_1A_2	A_2A_2
Genotypic value relative to midpoint	$+a$	d	$-a$
Genotypic value relative to population mean	$2q(a - dp)$	$a(p + q) + d(1 - 2pq)$	$-2p(a - dp)$
	$2q(\alpha - qd)$	$(q - p)\alpha + 2pqd$	$-2p(\alpha + pd)$
Breeding value	$2q[a + d(q - p)]$	$(q - p)[a + d(q - p)]$	$-2p[a + d(q - p)]$
	$2q\alpha$	$(q - p)\alpha$	$-2p\alpha$
Dominance deviation	$-2q^2d$	$2pqd$	$-2p^2d$

Table 10.7 Genotypic values, breeding values, and dominance deviations for the three *IFG*1 locus genotypes in dogs. Genotypic values, breeding value, and dominance deviation values are all given relative to the population mean, *M*. All values are in kilograms.

	Genotype		
	A_1A_1	A_1A_2	A_2A_2
A. $d = 0.0$, $p = 0.5$, $q = 0.5$, $M = 0.0$, $\alpha = 10.5$			
Genotype frequency	0.25	0.5	0.25
Genotypic value	10.5	0.0	−10.5
Breeding value	10.5	0.0	−10.5
Dominance deviation	0.0	0.0	0.0
B. $d = 0.0$, $p = 0.9$, $q = 0.1$, $M = 8.4$, $\alpha = 10.5$			
Genotype frequency	0.81	0.18	0.01
Genotypic value	2.1	−8.4	−10.5
Breeding value	2.1	−8.4	−10.5
Dominance deviation	0.0	0.0	0.0
C. $d = 5.25$, $p = 0.5$, $q = 0.5$, $M = 2.625$, $\alpha = 10.5$			
Genotype frequency	0.25	0.5	0.25
Genotypic value	7.875	2.625	−13.125
Breeding value	10.5	0.0	−10.5
Dominance deviation	−2.625	−2.625	−2.625
D. $d = 5.25$, $p = 0.9$, $q = 0.1$, $M = 9.345$, $\alpha = 6.3$			
Genotype frequency	0.81	0.18	0.01
Genotypic value	1.155	−4.095	−19.845
Breeding value	1.26	−5.04	−11.34
Dominance deviation	−0.105	0.945	−8.505

there is dominance, so the population mean is greater than it would be without dominance. The A_1A_1 breeding value needs to be adjusted downward so that it does not exceed the largest phenotypic value. This adjustment is the dominance deviation. In this particular case, the population mean is 2.625 rather than zero because of dominance shown by the A_1A_2 genotype (the A_1A_2 genotype has a value of 5.25 and makes up 50% of the population at $p = q = 0.5$). Therefore, the dominance deviation is -2.625. Note that, in general, the magnitude of dominance deviations changes with population allele frequencies just as the magnitude of genotypic values measured from the population mean and breeding values do.

The dominance deviation for the A_1A_1 genotype in case (d) of Table 10.7 is also diagrammed in Figure 10.3. At the allele frequencies in case (d), the dominance deviation is smaller because the frequency of A_1A_2 individuals in the population is smaller, causing less dominance contribution to the population mean. That ultimately results in a smaller difference between the genotypic value of A_1A_1 measured relative to M and the breeding value of A_1A_1.

We can also determine the mean dominance deviation in a population where all three genotypes are mating at random. As for the average breeding value, the average dominance deviation of all three genotypes is obtained by multiplying the dominance deviation of each genotype by its corresponding genotype frequency. This gives

$$
\begin{aligned}
&\textit{Mean dominance deviation of all genotypes} \\
&= p^2\left(-2q^2d\right) + 2pq(2pqd) - q^2\left(-2p^2d\right)
\end{aligned}
\tag{10.44}
$$

which simplifies to zero since the first and third terms cancel with the middle term. The mean dominance deviation of all three genotypes mating at random is zero. Just as for the mean breeding value, a mean dominance deviation makes intuitive sense because when a large parental population in Hardy–Weinberg expected genotype frequencies mates at random, the mean value of the progeny population should be exactly the same since the progeny genotype frequencies are exactly the same as in the parental population. Thus, the impacts of dominance on the progeny

means of each genotype counteract each other to give an overall mean of zero under random mating.

One last illustration will help show the connection between the genotypic values and the genotype frequencies on one hand and the breeding values and dominance deviations on the other. Figure 10.4 shows the least-squares regression line between the genotypic values and the number of A_1 alleles in a genotype for a population of 100 individuals (since genotype frequencies equal Hardy–Weinberg expectations, the population contains 25 A_1A_1, 50 A_1A_2, and 25 A_2A_2 individuals). The slope of the regression line gives the average phenotypic effect of a change in the number of A_1 alleles in the genotype. So, for example, changing all A_2A_2 genotypes to A_1A_2 genotypes (replacing one A_2 allele with an A_1 allele) would change the

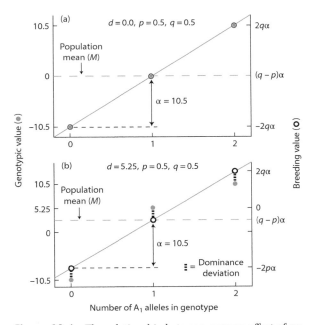

Figure 10.4 The relationship between average effect of an allele replacement, genotypic values, breeding values, and dominance deviations. The solid line represents the least squares regression of genotypic value on number of A_1 alleles in a genotype in a population of 100 individuals with Hardy–Weinberg genotype frequencies. The slope of the line is equal to the effect of an allelic replacement or α. There is no dominance in A and partial dominance in B. The breeding value is equal to the genotypic value adjusted by the effects of dominance on the progeny population mean as measured by the dominance deviation. All values are deviations from the population mean.

phenotypic value of that group of individuals from -10.5 to 0. Likewise, changing all A_1A_2 genotypes to A_1A_1 genotypes would change the phenotypic value of that group of individuals from 0 to 10.5. The average phenotypic effect of increasing the number of A_1 alleles is, therefore, 10.5 (or the slope of the regression line), so $\alpha = 10.5$ as shown in Figure 10.4.

Given the average effect of an allele replacement, we can then predict the expected phenotypic value of the progeny of any one genotype. Let us use the A_2A_2 genotype as an example. When A_2A_2 individuals mate, the mean value of the progeny population will *decrease* because progeny of A_2A_2 individuals receive the A_2 allele that confers the lower genotypic value. The amount of the decrease in value is $-2q\alpha$, which in Figure 10.4A equals -2 (0.5) (10.5) $=$ -10.5. Figure 10.4B shows the impact of dominance. Even with dominance, when A_2A_2 individuals mate, the mean value of the progeny population still decreases because their progeny receive the A_2 allele. The decrease in phenotypic value is $-2q\alpha = -10.5$. But dominance has caused the population mean to be 2.625 rather than zero. Another consequence of dominance is that when A_2A_2 individuals mate, some of the A_2 alleles will pair with A_1 alleles to make heterozygotes with the phenotype of d. Therefore, the average effects of two A_2 alleles need to be adjusted by the dominance deviation. The dominance deviation for A_2A_2 is $-2p^2d$, which in panel (a) equals zero and in panel (b) equals $-2(0.5)^2(5.25) = -2.625$.

10.5 Components of total genotypic variance

- Deriving the additive (V_A) and dominance (V_D) components of genotypic variation
- V_A and V_D are related to allele frequencies and genotypic values

We are at last in a position to obtain expressions for the components of variance in phenotype due to additive genetic variation and dominance genetic variation. Recall from earlier in the chapter that additive genetic variation (V_A) contributes to the resemblance between parents and offspring. Using the terminology of this section, we can say that V_A is due to the *variance* in breeding values. Similarly, earlier in the chapter, dominance genetic variation

(V_D) was described as being due to the effect, if any, of combining different alleles into a heterozygote genotype. We can now recognize V_D as the *variance* in dominance deviations. These variances describe the spread or range of values in a population caused by breeding value or by dominance.

The previous subsections devoted considerable effort to developing expressions for average values. In particular, we obtained expressions for the average breeding value and average dominance deviation of each genotype. Obtaining these averages was important because an average is a critical part of a variance (see the Appendix for the definition of a variance). As was shown above, in a randomly mating population, both the mean breeding value and the mean dominance deviation are zero. These are extremely useful results because they greatly simplify the expressions for the variance in breeding value and the variance in dominance deviation.

The additive variance, or V_A, is the variance of breeding values in a randomly mating population. Since the mean breeding value taken over all genotypes is zero, the variance in breeding value is simply the square of the mean breeding value for each genotype multiplied by the frequency of each genotype

$$V_A = p^2(2q\alpha)^2 + 2pq((q-p)\alpha)^2 + q^2(-2p\alpha)^2 \tag{10.45}$$

By expanding each term, factoring out $2pq\alpha^2$, and then canceling it, the equation simplifies to

$$V_A = 2pq\alpha^2 \tag{10.46}$$

or after substituting the definition of α from Eq. 10.22, to

$$V_A = 2pq(a + d(q-p))^2 \tag{10.47}$$

Similarly, the dominance variance, or V_D, is the variance in dominance deviation values in a randomly mating population. As before, the mean dominance deviation taken over all genotypes is zero, so the variance in breeding value is the square of the mean dominance deviation for each genotype multiplied by the frequency of each genotype:

$$V_D = p^2 \left(-2q^2 d \right)^2 + 2pq(2pqd)^2 + q^2 \left(-2p^2 d \right)^2$$
$$(10.48)$$

Expanding each term gives $4p^2q^2d^2(p^2 + 2pq + q^2)$, which then gives the equation for the dominance deviation variance as

$$V_D = (2pqd)^2 \qquad (10.49)$$

The separate expressions for V_A and V_D give us the means to estimate the total genotypic variance or V_G as

$$V_G = V_A + V_D \qquad (10.50)$$

which after substitution of the expressions for V_A and V_D becomes

$$V_G = 2pq(a + d(q-p))^2 + (2pqd)^2 \qquad (10.51)$$

Note that V_G is commonly referred to as the total genetic variance, even though it is the variance in genotypic values.

The components of the genetic variance and their relationship to the allele frequencies and the genotypic values can be seen in Figure 10.5. In all cases, V_G is the greatest at intermediate allele frequencies. When there is no dominance as in Figure 10.5A, V_G is made up exclusively of additive genetic variation. Without any dominance, V_A is the greatest at intermediate allele frequencies where the total frequency of the two homozygotes equals the frequency of the heterozygotes under random mating. When there is dominance, as in panels (b) and (c), V_G is made up of both additive and dominance genetic variation. Dominance variance is the greatest when the frequency of all heterozygotes is the greatest, or $p = q = 0.5$ for a diallelic locus under random mating. Also, notice that, with dominance, V_G is low when the frequency of the recessive allele (A_2) is low because the population has very few individuals

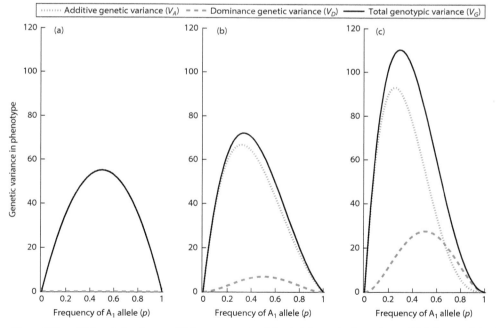

Figure 10.5 The additive (V_A) and dominance (V_D) components of the total genotypic variance (V_G). In panel A, there is no dominance ($d = 0.0$). Panel B has partial dominance ($d = 5.25$), and panel C has complete dominance ($d = 10.5$). In B and C, the dominance component of the total genotypic variance is greatest when allele frequencies are equal because the frequency of heterozygotes is greatest. In all cases, the value of $+a = 10.5$.

with the $-a$ genotypic value. This means that V_G is made up mostly of the variance in $+a$ and d genotypic values that are very similar or identical because of dominance. Since all of the illustrations in Figure 10.5 use a genotypic value of $a = 10.5$, each panel represents one case of the components of genotypic variance for *IGF1* depending on the degree of dominance.

Across the three graphs in Figure 10.5, the maximum V_G progressively increases and the allele frequency where maximum V_G occurs also shifts to lower values of p. This increase in maximum V_G corresponds to an increase in the maxima of both V_A and V_D. Complete dominance causes more total genotypic variation because the genotypic values in the population are only the extremes of $+a$ and $-a$ without a genotypic value that is intermediate. The variance in genotypic values is at a maximum when there are equal frequencies of the $+a$ and $-a$ genotypic values. At $p \approx 0.29$ and $q \approx 0.71$, the combined frequencies of the A_1A_1 and A_1A_2 genotypes are equal to the frequency of the A_2A_2 genotypes, thus maximizing V_G.

Interact box 10.2 Components of total genotypic variance, V_G

The additive and dominance components of the total genotypic variance can be interactively graphed using an Excel spreadsheet model. Values of a and d can be set in the model. Set $a = 10.5$ as in the *IGF1* example. Then, set d to 0.0, 5.25, and 10.5, and view the graphs in each instance. The graphs should be identical to those in Figure 10.5.

What would the graph of the components of the total genotypic variance look like with strong overdominance? Predict what the graph might look like and sketch it on paper. Then, set d to a value that constitutes overdominance and view the graph. Was your prediction correct? What is the impact of strong overdominance on heritability? If a population with strong overdominance and allele frequencies near $p = q = 0.5$ experienced genetic drift or consanguineous mating, how would V_A and the heritability change?

Math box 10.2 Deriving the total genotypic variance, V_G

While the method of adding V_A and V_D to obtain the total genotypic variance seems reasonable, it does have an important assumption. The total genetic variance, V_G, is the sum of $V_A + V_D$ plus twice the covariance between V_A and V_D

$$V_G = V_A + V_D + 2\text{cov}_{AD} \tag{10.52}$$

This covariance can be estimated by summing the genotype frequency-weighted product of the breeding value and the dominance deviation for each of the genotypes

$$\text{cov}_{AD} = p^2(2q\alpha)(-2q^2d) + 2pq((q-p)\alpha) \\ (2pqd) + q^2(-2p\alpha)(-2p^2d) \tag{10.53}$$

After multiplying out each term to get

$$\text{cov}_{AD} = -4p^2q^3\alpha d + 4p^2q^2((q-p)\alpha d \\ + 4p^2q^3\alpha d \tag{10.54}$$

and then factoring out $4p^2q^2\alpha d$ to get

$$\text{cov}_{AD} = -4p^2q^2\alpha d(-q+q-p+p) \tag{10.55}$$

it is then apparent that the covariance between the breeding values and the dominance deviations is zero.

The same overall result of the total genetic variance being a function of $V_A + V_D + 2\text{cov}_{AD}$ can be shown by starting with the expression for the total genotypic variance based on the frequency-weighted variances for each genotypic value

$$V_G = p^2(a-M)^2 + 2pq(d-M)^2 \\ + q^2(-a-M)^2 \tag{10.56}$$

and then using definitional substitutions and algebra to obtain an equation with terms that represent the variance of the breeding values plus the variance of the dominance deviations and the covariance between the breeding values and dominance deviations.

10.6 Genotypic resemblance between relatives

- Additive (V_A) and dominance (V_D) components of genotypic variation and resemblance of relatives
- Resemblance of relatives depends on the probability that alleles or genotypes are shared
- The covariance between mid-parent and offspring genotypic values

Obtaining expressions for the components of the total genotypic variation is ultimately useful because the variance components provide the basis to predict the resemblance of genotypic values between populations of related individuals. This relationship can also be reversed, and the phenotypic resemblance between relatives can be used to estimate the components of the total genotypic variance for specific phenotypes. This section shows how the variance within or covariance between relatives is related to the components of the genotypic variance. In practice, V_A is usually estimated from the resemblance between relatives to then determine the heritability of a phenotype. See the Appendix for background on the covariance, if necessary.

The expected covariance in genotypic values between relatives can be determined using the autozygosity of the relatives compared (Cotterman 1974, 1983; Crow and Kimura 1970). The expected covariance between related individuals is

$$\mathrm{cov}(x, y) = rV_A + uV_D \qquad (10.57)$$

where x and y represent the phenotypic values in a population of individuals, and r and u are fractions determined by probabilities of identity by descent for the individuals in groups x and y and for their parents. If the parents of individuals in group x are A and B and the parents of individuals in group y are C and D, then

$$r = 2f_{xy} \qquad (10.58)$$

if the parental population has not experienced mating among relatives ($f = 0$) and

$$u = f_{AC}f_{BD} + f_{AD}f_{BC} \qquad (10.59)$$

where f is the probability of autozygosity based on the probability of identity by descent as explained in Chapter 2.

These coefficients have a clear biological interpretation based on the pedigrees of individuals x and y (Figure 10.6). The r coefficient is twice the probability that individuals x and y inherit an allele that is

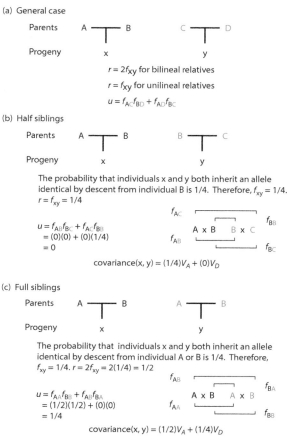

Figure 10.6 The expected covariance in genotypic values for relatives based on probabilities that individuals share alleles (r) and genotypes (u) that are identical by descent. The pedigree for the general case is shown in panel A. Half siblings (half brothers and sisters, panel B) share parent B in common and are thus unilineal relatives. Full siblings (full brothers and sisters, panel C) share both parents in common and are thus bilineal relatives. In general, bilineal relatives include dominance components in their expected covariances.

identical by descent. The u coefficient is the probability that individuals x and y inherit the same genotype. The probability of inheriting the same *genotype* depends on the probability that *both* alleles in the genotypes of x and y are identical by descent. Two alleles could be identical by descent through the parents in the left positions of each pedigree (f_{AB}) and at the same time through the parents in the right positions of each pedigree (f_{BD}), giving $f_{AC}f_{BD}$ as the probability that the genotype is identical by descent. Alternatively, two alleles could be identical by descent through the parents in the left and right positions of each pedigree (f_{AC}) and through the two parents in the right and left positions of each pedigree (f_{BD}), giving $f_{AD}f_{BC}$ as the probability that the genotype is identical by descent. Since both outcomes can occur in the populations of x and y individuals, the total probability of genotypes being identical by descent is $f_{AC}f_{BD} + f_{AD}f_{BC}$. In all cases, if one parent of x and one parent of y are unrelated, there is zero probability that the alleles they transmit to their progeny can be identical by descent.

Half and full siblings make instructive examples of how to determine the expected covariance between relatives. Figure 10.6 shows the pedigrees for these two cases. For both half and full siblings, the probability that individuals x and y inherit an allele identical by descent is ¼. Given that x and y have a parent in common, there is a probability of ½ that a given allele is transmitted from a common parent to individual x and an independent probability of ½ that a copy of the same allele is transmitted to individual y giving $f_{xy} = (1/2)(1/2) = ¼$. For half siblings, parents A, B, and C are unrelated so there is zero probability that they transmitted alleles identical by descent to their offspring. There is, however, a probability of ½ that both half siblings inherited the same allele from parent B, but this is only one allele and not a genotype. For full siblings, while A and B are unrelated, A and B are the parents of both x and y. It is, therefore, possible that x and y inherited the same allele from parent A and the same allele from parent B to produce genotypes that are identical by descent.

Half siblings resemble each other since 50% of individuals share one allele that is identical by descent. Full siblings have an even greater degree of resemblance since 50% of individuals share an *allele* identical by descent and at the same time 25% of individuals share a *genotype* identical by descent. The variance in genotypic value caused by alleles and genotypes corresponds to V_A and V_D. Therefore, the genotypic values of half siblings have a covariance ½V_A, while the genotypic values of full siblings have a greater covariance of ½V_A + ¼V_D. Table 10.8 gives additional examples of the expected covariance between various relatives based on the

same logic used to obtain the covariances for half and full siblings.

Unilineal relatives such as half siblings can share only one allele in their genotypes that is identical by descent. This is the case since one of the parents of each individual is related and can provide an avenue for inheritance of an allele that is identical by descent. The other allele in the genotype cannot be identical by descent because it is inherited from two different parents who are unrelated. In contrast, bilineal relatives, such as full siblings, share both parents in common or have parents who are related. Bilinear relatives, therefore, have dominance components in their expected covariances because they have a chance of inheriting both alleles and genotypes that are identical by descent. Sharing alleles in common leads to phenotypic resemblance due to the additive phenotypic effects of alleles while sharing genotypes in common leads to phenotypic resemblance due to the phenotypic effects of genotypes (dominance and epistasis).

With the knowledge of the genetic basis of covariance in genotypic values among relatives, we can look back to Figure 9.8 to better understand why Abney et al. (2001) decided to estimate heritabilities in a Hutterite population. The 806 individuals in that study descended from only 64 ancestors. That means that all individuals in the study had a nonzero probability of sharing two alleles that were identical by descent. The consequence is that the u coefficient in Eq. 10.57 was nonzero for all pairs of individuals in the study. This lead to improved precision for estimates of dominance variance because comparisons between all pairs of individuals could be used to estimate the covariance between phenotype and the u coefficient. In contrast, the u coefficient is zero in randomly mating populations except for between pairs of full siblings, making estimates of dominance variance imprecise because the number of full siblings in one family is quite small.

Table 10.8 Expected covariance in genotypic values between groups of relatives.

Relatives		Covariance in genotypic values
Offspring (x)	One parent (y)	½V_A
Offspring (x)	Mid-parent (y)	½V_A
Half siblings		¼V_A
Full siblings		½V_A + ¼V_D
Nephew/Niece (x)	Uncle/Aunt (y)	¼V_A
First cousins		1/8V_A
Monozygotic twins		$V_A + V_D$

Since the expected covariance between the mid-parent value and progeny values forms the basis of the parent–offspring regression, it is worth working through an additional method to obtain the covariance between mid-parent and offspring values. Deriving this covariance relies on a mathematical property of the covariance, or what some might call a math trick. The covariance can be expressed in a different form as

$$\mathrm{cov}(x, y) = \frac{1}{n}\sum_{i=1}^{n}(x_i - \bar{x}) - (y_i - \bar{y}) = \frac{1}{n}\sum_{i=1}^{n}(x_iy_i) - (\overline{xy})$$

(10.60)

In terms of the mid-parent and offspring values, the covariance is then

$$\mathrm{cov}(O, \bar{P}) = (\overline{OP}) - M^2$$ (10.61)

where O is the value of the offspring from each parental mating, $\bar{P}$ is the mid-parent value of a mating between two parental genotypes, and M is the population mean. In the case of parents and offspring, the average taken by multiplication by $1/n$ in Eq. 10.60 is instead accomplished by multiplying the product of $\bar{P}$ and O by the expected frequency of each parental mating. Table 10.9 gives the expected frequencies for each union of parental genotypes under random mating, along with the expected mid-parent and progeny values. Multiplying the appropriate three quantities from each row in Table 10.9 and then summing across rows

$$= p^4a^2 + 4p^3q((a^2 + 2ad + d^2)) + 2p^2q^2(0)$$
$$+ 4p^2q^2(d^2) + 4pq^3((a^2 - 2ad + d^2)) + q^4a^2 - M^2$$

(10.62)

With some algebraic manipulation, canceling of terms and substitution of the definition of the

population mean, the covariance between the mid-parent and offspring values is

$$= pq(a + d(q - p))^2$$ (10.63)

Since $\alpha = a + d(q - p)$ and $V_A = 2pq\alpha^2$, or after substituting in the definition of α,

$$= V_A$$ (10.64)

Chapter 10 review

- Genotypic values can be expressed on an arbitrary scale with $+a$ and $-a$ representing the values of the homozygotes with respect to a midpoint between $+a$ and $-a$ of zero. The value of the heterozygote is represented by d.
- The mean phenotypic value of a population with Hardy–Weinberg expected genotype frequencies is $M = a(p - q) + 2pqd$ where p is the frequency of the allele that increases value and q is the frequency of the allele the decreases value.
- The average effect of an allele is the difference between the mean value of that subset of genotypes that contain a given allele and the mean value of the entire population. Because alleles are inherited and genotypes are not, the average effect describes the mean value of those progeny that inherit a certain allele from one parent and the other allele sampled at random from the population.
- The genotypic mean is the sum of the breeding value and the dominance deviation.
- The breeding value of an individual is the mean value of the progeny produced by a given genotype assuming random mating. The breeding value is the sum of the average effects of the alleles in an individual's genotype.

Table 10.9 Frequencies and mean values for parents and progeny used to derive the covariance between the average value of parents (mid-parent value) and the average value of the progeny from each parental mating.

Parental mating	Parental mating frequency	Mid-parent value ($\bar{P}_i$)	Progeny genotype frequencies			Progeny value (O_i)
			A_1A_1	A_1A_2	A_2A_2	
$A_1A_1 \times A_1A_1$	p^4	a	1	—	—	a
$A_1A_1 \times A_1A_2$	$4p^3q$	$\frac{1}{2}(a + d)$	$\frac{1}{2}$	$\frac{1}{2}$	—	$\frac{1}{2}(a + d)$
$A_1A_1 \times A_2A_2$	$2p^2q^2$	$a + (-a) = 0$	—	1	—	$a + (-a) = 0$
$A_1A_2 \times A_1A_2$	$4p^2q^2$	d	$\frac{1}{4}$	$\frac{1}{2}$	$\frac{1}{4}$	$\frac{1}{2}d$
$A_1A_2 \times A_2A_2$	$4pq^3$	$\frac{1}{2}(-a + d)$	—	$\frac{1}{2}$	$\frac{1}{2}$	$\frac{1}{2}(-a + d)$
$A_2A_2 \times A_2A_2$	q^4	$-a$	—	—	1	$-a$

- The dominance deviation is the difference between the genotypic value of a given genotype and the breeding value. When there is dominance ($d \neq 0$), those progeny that are heterozygotes will have a value that is not the sum of the two alleles in their genotype.
- Since the mean breeding value of all genotypes is zero in a population under random mating, the mean value of a population should not change from one generation to the next under the many assumptions of Hardy–Weinberg.
- The variance in genotypic values around the population mean under random mating, commonly called the total genetic variance V_G, is the sum of the squared breeding values and the squared dominance deviations or $V_G = V_A + V_D$.
- The resemblance in genotypic value between relatives caused by V_A and V_D can be related to the probability of identity by descent. The expected covariance in genotypic values is a function of the probability that individuals share an allele that is identical by descent plus the probability that individuals share a genotype identical by descent. This covariance in genotypic values forms the basis of the parent–offspring method to estimate heritability.

Further reading

For more details on the Mendelian basis of interaction variance (V_I) as well as numerous perspectives on the role of epistasis in the evolutionary change of phenotypes, see chapters in:

Wolf, J.B., Brodie, E.D. III, and Wade, M.J. (eds.) (2000). *Epistasis and the Evolutionary Process*. Oxford: Oxford University Press.

End-of-chapter exercises

1   Search the literature for a recent research paper that utilizes one or more of the population genetic concepts covered in this chapter. The topic can be any organism, application, or process, but the paper must employ the concepts of genetic value or dominance deviation, the average effect of an allele, the breeding value, or the resemblance between related individuals. Summarize the main hypothesis, goal, or rationale of the paper and explain how the paper utilized a population concept from this chapter.

Problem box answers

Problem box 10.1 answer

The entire range of genotypic values is $2a = 30 - 9 = 21$. Therefore, $a = 10.5$ kg. The midpoint is then either $30 - 10.5 = 19.5$ or $9 + 10.5 = 19.5$. The genotypic value of the heterozygote relative to the midpoint is $d = 24.75 - 19.5 = 5.25$. The degree of dominance is $5.25/10.5 = 0.50$ or 50%.

Problem box 10.2 answer

Case (a):

$$A_2 = pd - qa = (0.5)(0.0) - (0.5)(10.5) = -5.25 \text{ kg}$$

$$= -5.25 - 0.0 = -5.25 \text{ kg}$$

$$\alpha_2 = -p(a + d(q-p)) = -0.5(10.5 + 0.0(0.5 - 0.5))$$

$$= -5.25 \text{ kg}$$

$$\alpha = a + d(q-p) = 10.5 + 0.0(0.5 - 0.5) = 10.5 \text{ kg}$$

$$\alpha_1 = -p\alpha = -(0.5)(10.5) = -5.25 \text{ kg}$$

Case (b):

$$A_2 = pd - qa = (0.9)(0.0) - (0.1)(10.5) = -1.05 \text{ kg}$$
$$= -1.05 - 8.4 = -9.45 \text{ kg}$$
$$\alpha_2 = -p(a + d(q-p)) = -0.9(10.5 + 0.0(0.1 - 0.9))$$
$$= -9.45 \text{ kg}$$
$$\alpha = a + d(q-p) = 10.5 + 0.0(0.1 - 0.9) = 10.5 \text{ kg}$$
$$\alpha_1 = -p\alpha = -(0.9)(10.5) = -9.45 \text{ kg}$$

Case (c):

$$A_2 = pd - qa = (0.5)(5.25) - (0.5)(10.5) = -2.625 \text{ kg}$$
$$= -2.625 - 2.625 = -5.25 \text{ kg}$$
$$\alpha_2 = -p(a + d(q-p)) = -0.5(10.5 + 5.25(0.5 - 0.5))$$
$$= -5.25 \text{ kg}$$
$$\alpha = a + d(q-p) = 10.5 + 5.25(0.5 - 0.5) = 10.5 \text{ kg}$$
$$\alpha_1 = -p\alpha = -(0.5)(10.5) = -5.25 \text{ kg}$$

Case (d):

$$A_2 = pd - qa = (0.9)(5.25) - (0.1)(10.5) = 3.675 \text{ kg}$$
$$= 3.675 - 9.345 = -5.67 \text{ kg}$$
$$\alpha_2 = -p(a + d(q-p)) = -0.9(10.5 + 5.25(0.1 - 0.9))$$
$$= -5.67 \text{ kg}$$
$$\alpha = a + d(q-p) = 10.5 + 5.25(0.1 - 0.9) = 6.3 \text{ kg}$$
$$\alpha_1 = -p\alpha = -(0.9)(6.3) = -5.67 \text{ kg}$$

Appendix

Statistical concepts arise in this book in several places. Chapter 1 points out the distinction between idealized parameters which are exact values and parameter estimates obtained through sampling from populations that have uncertainty. Both the variance and the covariance are important concepts that appear in a number of chapters, especially in Chapters 9 and 10 that cover quantitative genetics. This appendix is meant to provide a basic introduction to statistical concepts relevant to these topics for readers without much prior background.

Imagine drawing a random sample of objects, say a handful of jellybeans from a candy dish and weighing each one. The weights will not be identical, but will have some values that occur most often and some range of values. Plotting these values on a graph such as that in Figure A.1 would show how often each one occurs between the lowest and highest values, or their frequency distribution (often truncated to just "distribution" in conversation). We often use the average or the mode (the most frequently occurring value) to describe the central tendency or middle of a frequency distribution.

Let us examine a hypothetical case that will show the distinction between a parameter and a parameter estimate as well as illustrate a common means to quantify uncertainty caused by sampling variance. Imagine we would like to estimate the frequency of the A allele (for a locus with two alleles) in a population of mice. These mice inhabit barns in an area where there are many isolated farms, each with a suitable barn. Therefore, the mice are found in many discrete populations that make up a larger total population. An example of this type of population is diagrammed in Figure A.2.

We would like to estimate the frequency of the A allele in the entire population, which we will call $\hat{p}$. The entire population has an exact allele frequency, the parameter p, which we could only know if we determine the genotype of every mouse in the population. Since it would be very difficult to sample *every* mouse, we take samples from a number of distinct, independent populations in order to estimate the allele frequency within each population. The average of allele frequencies in sampled populations will be our estimate of the parameter p. Call the estimate of allele frequency for each barn $\hat{p}_i$, where the subscript i is just an index of which barn the value came from. For simplicity, we assume in this illustration that each value of $\hat{p}_i$ within a barn is known without error.

A common quantitative tool to measure and express the range of values within a sample is called the **variance** (symbolized by σ^2 or a lower case "sigma" squared). In plain language, the variance is a standardized measure of the range of observed values relative to the average. It is simple to obtain the average allele frequency among all of the sampled populations by adding up all of the $\hat{p}_i$ observations and dividing by the total number of observations. In notation, the mean or average allele frequency among the populations sampled would be

$$\bar{p} = \frac{\sum_{i=1}^{n} \hat{p}_i}{n} \qquad (A.1)$$

where the bar over $\bar{p}$ (pronounced "p-bar") indicates an average, $\sum_{i=1}^{n}$ means summing each of the $\hat{p}_i$ observations starting with the first (there are a total of n, as indicated above the summation symbol or "sigma"), and n is the sample size. The variance is an average of the square of how much each

Population Genetics, Second Edition. Matthew B. Hamilton.
© 2021 John Wiley & Sons, Inc. Published 2021 by John Wiley & Sons, Inc.
Companion website: www.wiley.com/go/hamilton/populationgenetics

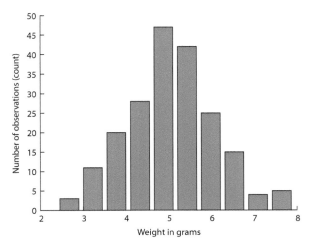

Figure A.1 The frequency distribution of hypothetical weights for 200 jellybeans. The mean is 5.06 and the variance is 1.06.

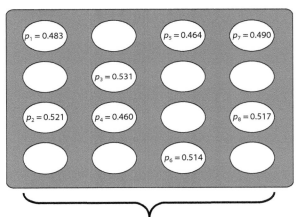

The total population allele frequency = p

Figure A.2 An abstract representation of 16 mouse populations. The total population (the entire rectangle) is composed of a series of smaller, discrete populations (the individual circles), sometimes called subpopulations. Each subpopulation has its own frequency of the A allele indicated by $\hat{p}_i$. The entire population also has an exact value for the frequency of the A allele, which can be estimated by sampling some subpopulations and then taking the average and variance of the $\hat{p}_i$ values. Here, eight subpopulations are sampled to estimate the allele frequency in the total population as $\bar{p} = 0.4976$, SD = 0.0272, and the SE of the mean = 0.0096.

observation differs from the mean. The variance is taken by summing the squared differences between each estimate and the average and then dividing by one less than the sample size,

$$\operatorname{var}(\hat{p}) = \sigma^2(\hat{p}) = \frac{\sum_{i=1}^{n} (\hat{p}_i - \bar{p})^2}{n-1}. \quad (A.2)$$

The square in the denominator comes about because $(\hat{p}_i - \bar{p})$ for observations less than the mean will be

negative for observations less than the mean but positive for observations greater than the mean. Squaring will make all differences positive so that positive and negative differences will not cancel each other out when summed. You should also note that distributions with larger means will have larger variances even if the spread of observations is identical in the two distributions. This makes comparing variance values difficult without reference to the mean.

> **Variance (σ^2):** the sum of the squared deviations from the mean divided by one less than the sample size.
> **Standard Deviation (σ):** the square root of the variance; the average deviation from the mean for a single observation. Quantifies the range of values around the mean seen in a sample.
> **Standard Error of an Average (SE):** the product of the standard deviation and the square root of the sample size divided by the sample size; how far the true population average (a parameter) may be from the sample average (a parameter estimate) by chance.

The variance estimator shown in Eq. A.2 is sometimes called the sampling variance, and it is an unbiased estimator of the variance in a very large population (unbiased means that the expected value of the variance is equal to the true value of the variance). An alternative form of the variance exists where the sums of squares is divided by n rather than by $n-1$. Dividing by n estimates what is sometimes called the parametric variance, or a variance for a finite population of size n where all n individuals have been used to compute the variance. The parametric variance is employed in idealized situations where every individual in a population can be measured or sampled. In practice, this distinction makes little difference for large n.

If we are willing to assume that our average is drawn from a frequency distribution called the "normal" distribution (which Interact Box A.1 suggests is not unreasonable), then we can use the variance in Eq. A.2 as a measure of uncertainty in our estimate of the average allele frequency. The standard deviation is symbolized by σ (a lower case "sigma") or "SD" and is simply the square root of the variance (taking the square root returns the variance back to the original units of measurement). The SD

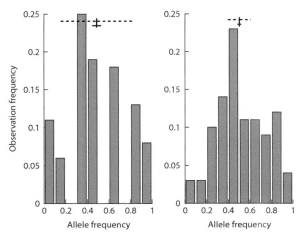

Figure A.3 Two frequency distributions of 100 data points each with nearly identical means (0.498 on the left and 0.506 on the right) but different degrees of variance among the observations (the variance is 0.0293 on the left and 0.0025 on the right). The lines at the top indicate the position of the mean (vertical line) and two standard deviations on either side of the mean (horizontal dashed line) and two standard errors on either side of the mean (horizontal solid line). Like allele frequencies, each distribution is on the interval of 0 to 1.

measures the average deviation from the mean for a single observation. Normal distributions are very useful because the standard deviation corresponds to the probability that an observation is some distance from the average. For an ideal normal distribution, about 68% of the observations fall within ± one SD of the mean, about 95% of the observations fall within ± two SD of the mean, and about 99% of the observations fall within ± three SD of the mean.

The SD of a single observation can also be used to quantify uncertainty in the average of all observations. The standard error, or SE, for the sum of all observations (the denominator in Eq. A.1) is $\sqrt{\text{sample size}}$ multiplied by the SD. Since the SD measures the average spread of an observation from the mean, then the SE of the sum adds these individual deviations up. The square root of the sample size is the multiplier because the SE grows more slowly than the sample size itself as observations are added to the sum. The SE of the sum can be related to the average by taking the average of the SE of the sum, or

$$\text{SE of average} = \frac{(\sqrt{n})SD}{n} = \frac{SD}{\sqrt{n}}. \qquad (A.3)$$

(the $\sqrt{n}$ term in the numerator of the middle equation cancels because $n = \sqrt{n}^2$ in the denominator). Notice that as the sample size increases, the SE of

the mean decreases. Like the SD, the SE of the average defines a probability range around the mean due to chance events in the sample. The probability intervals defined by the SE serve to establish what are called confidence intervals. By convention, 95% confidence intervals are frequently used to quantify the chances that the parameter estimate ($\overline{p}$) plus or minus 2 SE covers the true parameter value in the population (p).

To return to our mouse population example diagrammed in Figure A.2, $\overline{p} = 0.4976$, the variance is 0.00074 and the SD is 0.0272. The SE of the sum is ($\sqrt{8}$) (0.0272) = 0.0769, and the SE of the average is [($\sqrt{8}$) (0.0272)]/8 = 0.0096. Therefore, the 95% confidence interval for the mean is 0.4976 − (2•0.0096) to 0.4976 + (2•0.0096) or 0.4784–0.5168. Thus, we would expect 95 times out of 100 that the range of allele frequency between 0.4784 and 0.5168 would include the actual allele frequency of the population, p. You will be reassured to note that the values in Figure A.2 are random numbers generated by a computer using a distribution with a mean of 0.50, equivalent to the true parameter value. The value of $\overline{p}$ is very close to this parameter value, and the parameter value also falls inside the 95% confidence interval for $\overline{p}$ defined by plus or minus 2SE of the average.

Note that, in this example, the allele frequencies for each population ($\hat{p}_i$) can take on any value between zero and one and are, therefore, **continuous variables**. The estimate of allele frequency within each individual population is based on counting up alleles that can take only one of the two forms, a and A, called a **binomial variable**.

Problem A.1
Estimating the variance

Refer to Figure A.2 and replace the values in the figure for the allele frequency within each population with 0.6333, 0.5074, 0.4880, 0.3960, 0.5368, 0.3330, 0.5893, and 0.7029. What is the estimate of allele frequency in the entire population and what is the 95% confidence interval? How does the variance in this case compare with the variance for the original values given in Figure A.2? How does the range of allele frequencies that include the true population parameter compare to the original values in Figure A.2?

Interact box A.1 The central limit theorem

The central limit theorem predicts that the distribution of means will approach a normal distribution no matter what the shape of the distribution the original data sampled to compute the mean. The central limit theorem is the justification for using the standard deviation as an estimate of the confidence in a mean such as the average allele frequency (see main text). The central limit theorem also demonstrates that certainty in parameter estimates is directly proportional to the size of samples used to make such estimates. Instead of accepting the central limit theorem as a given, let us use simulation to explore it as a prediction that can be tested.

Step 1: On the book's web page, under the heading Appendix, click on the link for **Central limit theorem simulation**.

Step 2: A web page titled **Sampling Distributions** will open. After a moment to load, a **Begin** button will appear at the top left of the window. Press the button, and a new simulation window will open. Once the Java code for the window loads, you will see four sets of axes with some buttons and pop-out menus down the right side of the window. The top set of axes will contain a normal distribution with the mean, median, and standard deviation ("sd") given in three colors to the left of the distribution, and their positions are also indicated in those same colors below the x-axis. Please read the **Instructions** text and refer to the simulation window so that you understand the controls.

Step 3: This simulation samples individual data points from the distribution at the top. Pressing the **Animated** button on the right will show the sampling process with five data points at a time on the axes labeled **Sample Data**. Press the button and watch as five bars drop to indicate the five data points. Once the sample is complete, a single data point, the mean of five individual data points, will appear on the axes labeled **Distribution of Means**. Press the **Animated** button a few times to get a sense of how the two middle windows display the data. Be sure you can distinguish between the sample size (the **N** = menu right if the axes) and the number of samples (tabulated as **Reps** = left of the axes).

Step 4: Set N = 25 (the lower three graphs will clear), and press the **Animated** button five times (notice how **Reps** = increases by one each time). Now, press the **5** button once. This is like sampling five more samples of 25 without the animation. Notice how reps is now 10 since the repetitions add up. Click on the **Fit normal** checkbox to compare the histogram in the distribution of means to an ideal normal distribution. Ideal normal distributions have skew and kurtosis (measures of asymmetry about the mean) values of zero, like the normal distribution at the top.

Step 5: How does the sample size and the number of samples influence the distribution of means? To find out, simulate a range of values of one variable while holding the other variable constant (try N = 2 and N = 20 with 20, 50, 100, and 1000 reps). Look at the histograms, and write down the skew and kurtosis values for each combination of the sample size and number of samples. Do larger samples improve the approach to normality of the distribution of means?

Step 6: Is the distribution of means still a normal distribution even if the parent population (the top-most distribution) is not a normal distribution? Compare the skew and kurtosis values from samples taken from three different parent populations (changed with the pop-out menu to the right of the top-most axes) for a range of identical sample sizes and numbers of samples (try N = 5 with 20, 50, 100, and 1000 reps).

Step 7: Feel free to explore the topic further by clicking on the **Exercises** link below the **Begin** button.

If at any point, you would like to obtain a new simulation window with default starting values, just close the simulation window and then open a new one with the **Begin** button.

A.1   Covariance and Correlation

In population genetics and in all of biology, there are many situations where our goal is to understand the relationship between two variables. To extend the jellybean example used earlier, imagine that each jellybean was weighed and its length was also measured. Each individual jellybean then has two values for the two variables that describe it. One question we might ask is if the length and mass of jellybeans is related or if jellybeans of any mass can exhibit any length. These questions about jellybean mass and length can be answered by the determining the **covariance** and **correlation** between two variables.

The spread or scatter of two variables, call them x and y, viewed simultaneously is their **joint distribution**. Figure A.4 illustrates three different joint distributions of x and y values. The degree to which the values of two variables tend to vary in the same direction is measured by the covariance. The covariance is

$$\text{cov}(x, y) = \frac{1}{n}\sum_{i=1}^{n}(x_i - \overline{x})(y_i - \overline{y}) \qquad (A.4)$$

where the deviations of each observation from the mean for each variable are multiplied, these products are summed over all observations, and then the sum is divided by the number of observations to achieve an average. Notice that the deviations from the mean are not squared as they are for the variance. This means that the covariance measures the direction of the deviations from the mean as well as their magnitude.

The joint distributions in Figure A.4 illustrate a range of covariance values even though the variance of the x variable is constant. In Figure A.4Aa, the x and y values have zero covariance and the two variables are, therefore, independent. When a covariance is zero, then the scatter of each variable can be described by its variance alone without reference to the other variable. In both Figures A.4B and C, the values of x and y are not independent but rather tend to vary together. A positive covariance between x and y is shown in Figure A.4B, telling us that higher values of one variable tend to be associated with higher values of the other variable. Figure A.4C shows negative covariance between x and y telling us that higher values of one variable tend to be associated with lower values of the other variable and vice versa.

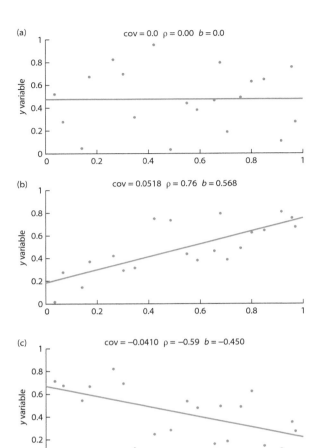

Figure A.4   Examples of joint distributions between two variables x and y. The covariance between x and y is zero in A, positive in B, and negative in C. In all three panels, the variance of x remains constant at 0.0912. Trends between the variables can be expressed as correlation coefficients (ρ) or as the slopes of the least-squares regression of y on x (b).

Using the jellybean analogy, a zero covariance says that length and weight are independent of each other, a positive covariance says that heavier jellybeans tend to be longer, and a negative covariance says that heavier jellybeans tend to be shorter.

The covariance forms the basis of the **correlation coefficient**, a summary measure of the strength and direction of the linear relationship between two variables. The Pearson product–moment correlation, symbolized by ρ (pronounced "roe"), is given by

$$\hat{\rho}_{x,y} = \frac{\text{cov}(x, y)}{\sqrt{\text{var}(x)\text{var}(y)}} \qquad (A.5)$$

where the $\sqrt{\text{var}(x)}$ is really the same thing as the standard deviation of x. The correlation is a dimensionless quantity that takes on values between -1 and $+1$. A perfect positive linear relationship between x and y gives a correlation of $+1$, a perfect negative linear relationship between x and y gives a correlation of $+1$, and a correlation of zero indicates independence of x and y. The correlation coefficients are also given for the joint distributions in Figure A.4. A number of other non-parametric correlation measures exist that are appropriate for data that are not normally distributed such as Spearman's rank-order correlation.

It is important to recognize that the correlation coefficient measures only associations between variables but does not provide any information on how that association came into being. Unfortunately, correlations between variables are commonly misinterpreted as indicating a cause-and-effect relationship between variables. A hypothetical example might be a positive correlation between sales of lemonade and sales of baseballs. Consumption of lemonade does not cause people to also buy a baseball. Rather both events are tied to the weather in a cause and effect relationship; in warm weather, people drink more lemonade and also play baseball. A classic demonstration of the weakness of the correlation as summary measure was made by Anscombe (1973), who concocted four data sets with the identical means, standard deviations, and correlation coefficients. These four data sets illustrate how non-normality, non-linearity, and outliers in data can produce high correlations but very poor summaries of the relationship between two variables.

The covariance also plays a fundamental role in regression analysis, which is used to estimate the resemblance between parents and offspring, as described in Chapter 9. Assume that y is a response or dependent variable and x is an independent variable. The slope (a) and intercept (b) of a regression line for the variables x and y are represented by the equation

$$y = a + bx \tag{A.6}$$

This equation can be rewritten using the average values of x and y

$$\bar{y} = a + b\bar{x}. \tag{A.7}$$

The difference between any individual observation y and the average of y is then

$$y - \bar{y} = a + bx - a - b\bar{x} \tag{A.8}$$

which simplifies to

$$y - \bar{y} = b(x - \bar{x}). \tag{A.9}$$

Multiplying both sides of Eq. A.9 by $(x - \bar{x})$ gives

$$(y - \bar{y})(x - \bar{x}) = b(x - \bar{x})^2. \tag{A.10}$$

Notice that the left-hand side looks a lot like the covariance in Eq. A.4 and the right-hand side looks a lot like the variance in Eq. A.2. If we sum the quantities on both sides of Eq. A.10 and then divide the sums by $1/n$ to make them averages, then Eq. A.10 becomes

$$\frac{1}{n}\sum_{i=1}^{n}(x_i - \bar{x})(y_i - \bar{y}) = b\frac{1}{n}\sum_{i=1}^{n}(x_i - \bar{x})^2 \tag{A.11}$$

which is the same as

$$\text{cov}(x, y) = b\,\text{var}(x). \tag{A.12}$$

Solving for the slope of the regression line gives

$$b = \frac{\text{cov}(x, y)}{\text{var}(x)}. \tag{A.13}$$

Therefore, we see that the slope of a regression line is the covariance between x and y divided by the variance in x.

The regression line slopes for the joint distributions in Figure A.4 can be computed from the covariances of x and y along with the variance in x. In all three panels of Figure A.4, $\text{var}(x) = 0.0912$. In the top panel, $\text{cov}(x,y) = 0$ so that the slope of the regression line is also zero. In the middle panel, $\text{cov}(x,y) = 0.0518$ so that $b = 0.0518/0.0912 = 0.568$. In the bottom panel, $\text{cov}(x,y) = -0.0410$ so that $b = -0.0410/0.0912 = -0.450$.

Further reading

One approachable beginning statistics text is

Freedman, D., Pisani, R., and Purves, R. (1997). *Statistics*, 3e. New York: Norton.

Two classic biologically oriented statistics texts are

Sokal, R.R. and Rohlf, F.J. (1995). *Biometry: The Principles and Practice of Statistics in Biological Research*, 3e. New York: W. H. Freeman & Company.
Zar, J.H. (1999). *Biostatistical Analysis*, 4e. Upper Saddle River, NJ: Prentice Hall.

For a humorous take on how basic graphs and statistics can lead to miscommunication and a book that will improve your own presentation of statistical information, see

Huff, D. (1954). *How to Lie with Statistics*. New York: W. W. Norton.

Problem box answers

Problem box A.1 answer

Using the new values produces a mean of 0.5233, a variance of 0.0147, and a standard deviation of 0.1212. The SE of the sum is $(\sqrt{8})(0.1212) = 0.3428$, and the SE of the average is $[(\sqrt{8})(0.0272)]/8 = 0.0429$. Therefore, the 95% confidence interval for the mean is $0.5233 - (2)(0.0429)$ to $0.5233 + (2)(0.0429)$ or 0.4375–0.6091. Thus, we would expect 95 times out of 100 that the range of allele frequency between 0.4375 and 0.6091 would include the actual allele frequency of the population, p. The confidence interval about $\hat{p}$ is much larger in this case since the variance is larger – all caused by the observations being more spread out around the mean. In this case, the allele frequency parameter is estimated with more uncertainty since the underlying observations used for the estimate have a much greater range of values.

Bibliography

Abbott, R.J., Gomes, M.F.M., Richard, H. et al. (1989). Population genetic structure and outcrossing rate of *Arabidopsis thaliana* Heynh. *Heredity* 62: 411–418. https://doi.org/10.1038/hdy.1989.56.

Abney, M., McPeek, M.S., and Ober, C. (2001). Broad and narrow heritabilities of quantitative traits in a founder population. *American Journal of Human Genetics* 68: 1302–1307.

Ackerman, M.S., Johri, P., Spitze, K. et al. (2017). Estimating seven coefficients of pairwise relatedness using population-genomic data. *Genetics* 206: 105–118.

Adams, R.I., Brown, K.M., and Hamilton, M.B. (2004). The impact of microsatellite electromorph size homoplasy on multilocus population structure estimates in a tropical tree (*Corythophora alta*) and an anadromous fish (*Morone saxatilis*). *Molecular Ecology* 13: 2579–2588.

Adeyemo, A. and Rotimi, C. (2010). Genetic variants associated with complex human diseases show wide variation across multiple populations. *Public Health Genomics* 13: 72–79.

Adriaensen, F., Chardon, J.P., De Blust, G. et al. (2003). The application of 'least-cost' modelling as a functional landscape model. *Landscape and Urban Planning* 64: 233–247.

Agrawal, A.F. and Whitlock, M.C. (2012). Mutation load: the fitness of individuals in populations where deleterious alleles are abundant. *Annual Review of Ecology, Evolution, and Systematics* 43: 115–135.

Akey, J.M., Zhang, G., Zhang, K. et al. (2002). Interrogating a high-density SNP map for signatures of natural selection. *Genome Research* 12: 1805–1814.

Alcala, N. and Rosenberg, N.A. (2017). Mathematical constraints on F_{ST}: Biallelic markers in arbitrarily many populations. *Genetics* 206: 1581–1600.

Allen, J.A. and Clarke, B.C. (1984). Frequency dependent selection: homage to E. B. Poulton. *Biological Journal of the Linnean Society* 23: 15–18.

Allison, A. (1956). The sickle-cell and haemoglobin C genes in some African populations. *Annals of Human Genetics* 21: 67–89.

Altshuler, D.M., Gibbs, R.A., Peltonen, L. et al. (2010). Integrating common and rare genetic variation in diverse human populations. *Nature* 467: 52–58.

Alvarez-Castro, J.M. and Alvarez, G. (2005). Models of general frequency-dependent selection and mating-interaction effects and the analysis of selection patterns in *Drosophila* inversion polymorphisms. *Genetics* 170: 1167–1179.

Andersen, E.C., Gerke, J.P., Shapiro, J.A. et al. (2012). Chromosome-scale selective sweeps shape *Caenorhabditis elegans* genomic diversity. *Nature Genetics* 44: 285–290.

Andolfatto, P. (2001). Contrasting patterns of X-linked and autosomal nucleotide variation in *Drosophila melanogaster* and *Drosophila simulans*. *Molecular Biology and Evolution* 18: 279–290.

Angers, B., Castonguay, E., and Massicotte, R. (2010). Environmentally induced phenotypes and DNA methylation: how to deal with unpredictable conditions until the next generation and after. *Molecular Ecology* 19: 1283–1295.

Anscombe, F.J. (1973). Graphs in statistical analysis. *American Statistician* 27: 17–21.

Arbogast, B.S., Edwards, S.V., Wakeley, J. et al. (2002). Estimating divergence times from molecular data on phylogenetic and population genetic timescales. *Annual Review of Ecology and Systematics* 33: 707–740.

Armbruster, P. and Conn, J. (2006). Geographic variation of larval growth in north American *Aedes albopictus* (Diptera: Culicidae). *Annals of the Entomological Society of America* 99: 1234–1243.

Armbruster, P. and Reed, D.H. (2005). Inbreeding depression in benign and stressful environments. *Heredity* 95: 235–242.

Armbruster, P., Hutchinson, R.A., and Linvell, T. (2000). Equivalent inbreeding depression under laboratory and field conditions in a tree-hole-breeding mosquito. *Proceedings of the Royal Society of London, Series B: Biological Sciences* 267: 1939–1945.

Ashley-Koch, A., Yang, Q., and Olney, R. (2000). Sickle hemoglobin (Hb S) allele and sickle cell disease: a

HuGE review. *American Journal of Epidemiology* 151: 839–845.

Avise, J.C., Arnold, J., Ball, R.M. et al. (1987). Intraspecific Phylogeography: the mitochondrial DNA bridge between population genetics and systematics. *Annual Review of Ecology and Systematics* 18: 489–522.

Balkenhol, N., Cushman, S.A., Storfer, A.T., and Waits, L.P. (eds.) (2015). *Landscape Genetics*. Chichester, UK: Wiley.

Banta, J.A. and Richards, C.L. (2018). Quantitative epigenetics and evolution. *Heredity* 121: 210–224.

Barrett, S.C.H. and Charlesworth, D. (1991). Effects of a change in the level of inbreeding on the genetic load. *Nature* 352: 522–524.

Barton, N.H. and Turelli, M. (2004). Effects of genetic drift on variance components under a general model of epistasis. *Evolution* 58: 2111.

Baudry, E., Derome, N., Huet, M., and Veuille, M. (2006). Contrasted polymorphism patterns in a large sample of populations from the evolutionary genetics model Drosophila simulans. *Genetics* 173: 759–767.

Bayes, T. (1763). LII. An essay towards solving a problem in the doctrine of chances. By the late Rev. Mr. Bayes, F. R. S. Communicated by Mr. Price, in a letter to John Canton, A. M. F. R. S. *Philosophical Transactions of the Royal Society of London* 53: 370–418.

Bazin, E., Glémin, S., and Galtier, N. (2006). Population size does not influence mitochondrial genetic diversity in animals. *Science* 312 (5773): 570–572. https://doi.org/10.1126/science.1122033.

Beaumont, M.A. (2005). Adaptation and speciation: what can F_{ST} tell us? *Trends in Ecology & Evolution* 20: 435–440.

Beaumont, M.A. and Rannala, B. (2004). The bayesian revolution in genetics. *Nature Reviews Genetics* 5: 251–261.

Beavis, W. (1994). The power and deceit of QTL experiments: lessons from comparative QTL studies. In: *Proceedings of the Forty-Ninth Annual Corn and Sorghum Industry Research Conference*, pp. 250–266.

Beck, N.R., Double, M.C., and Cockburn, A. (2003). Microsatellite evolution at two hypervariable loci revealed by extensive avian pedigrees. *Molecular Biology and Evolution* 20: 54–61.

Becquet, C., Patterson, N., Stone, A.C. et al. (2007). Genetic structure of chimpanzee populations. *PLoS Genetics* 3: e66.

Begun, D.J. and Aquadro, C. (1992). Levels of naturally occurring DNA polymorphism correlate with recombination rates in *D. melanogaster*. *Nature* 356: 519–520.

Bergelson, J., Stahl, E., Dudek, S., and Kreitman, M. (1998). Genetic variation within and among populations of *Arabidopsis thaliana*. *Genetics* 148: 1311–1323.

Bernstein, F. (1925). Zusammenfassende Betrachtungen über die erblichen Blutstrukturen des Menschen. *Zeitschrift für Induktive Abstammungs- und Vererbungslehre* 37: 237–270.

Besenbacher, S., Hvilsom, C., Marques-Bonet, T. et al. (2019). Direct estimation of mutations in great apes reconciles phylogenetic dating. *Nature Ecology & Evolution* 3: 286–292.

Bierne, N. (2010). The distinctive footprints of local hitchhiking in a varied environment and global hitchhiking in a subdivided population. *Evolution* 64: 3254–3272.

Bigler, J., Whitton, J., Lampe, J.W. et al. (2001). CYP2C9 and UGT1A6 genotypes modulate the protective effect of aspirin on colon adenoma risk. *Cancer Research* 61: 3566–3569.

Bininda-Emonds, O.R.P., Cardillo, M., Jones, K.E. et al. (2007). The delayed rise of present-day mammals. *Nature* 446 (7135): 507–512. https://doi.org/10.1038/nature05634.

Birchler, J.A., Dawe, R.K., and Doebley, J.F. (2003). Marcus Rhoades, preferential segregation and meiotic drive. *Genetics* 164: 835–841.

Birky, C.W. and Walsh, J. (1988). Effects of linkage on rates of molecular evolution. *Proceedings of the National Academy of Sciences of the United States of America* 85: 6414–6418.

Bitbol, A.-F. and Schwab, D.J. (2014). Quantifying the role of population subdivision in evolution on rugged fitness landscapes. *PLoS Computational Biology* 10: e1003778.

Bittles, A.H. and Black, M.L. (2010). Consanguinity, human evolution, and complex diseases. *Proceedings of the National Academy of Sciences* 107: 1779–1786.

Boag, P.T. (1983). The heritability of external morphology in Darwin's ground finches (*Geospiza*) on Isla Daphne Major, Galapagos. *Evolution* 37: 877–894.

Boag, P.T. and Grant, P.R. (1981). Intense natural selection in a population of Darwin's finches (Geospizinae) in the Galápagos. *Science* 214: 82–85.

Bodmer, W. (1965). Differential fertility in population genetics. *Genetics* 51: 411–424.

Bonin, A., Bellemain, E., Bronken Eidesen, P. et al. (2004). How to track and assess genotyping errors in population genetics studies. *Molecular Ecology* 13: 3261–3273.

Booker, T.R., Yeaman, S., and Whitlock, M.C. (2020) Variation in recombination rate affects detection of FST outliers under neutrality. *bioRxiv preprint*.

Bosshard, L., Dupanloup, I., Tenaillon, O. *et al.* (2017). Accumulation of deleterious mutations during bacterial range expansions. *Genetics*, genetics.300144.2017.

Boyko, A.R., Quignon, P., Li, L. et al. (2010). A simple genetic architecture underlies morphological variation in dogs. *PLoS Biology* 8: e1000451.

Brandt, D.Y.C., César, J., Goudet, J., and Meyer, D. (2018). The effect of balancing selection on population differentiation: a study with HLA genes. *G3 Genes|Genomes|Genetics* 8: 2805–2815.

Braverman, J.M., Hudson, R.R., Kaplan, N.L. et al. (1995). The hitchhiking effect on the site frequency spectrum of DNA polymorphisms. *Genetics* 140 (2): 783–796.

Brisson, D. (2018). Negative frequency-dependent selection is frequently confounding. *Frontiers in Ecology and Evolution* 6: 1–9.

Brodie, E.D. III (2000). Why evolutionary genetics does not always add up. In: *Epistasis and the Evolutionary Process* (eds. J. Wolf, E.D. Brodie III and M.J. Wade), 3–19. Oxford: Oxford University Press.

Bromham, L., Phillips, M.J., and Penny, D. (1999). Growing up with dinosaurs: molecular dates and mammalian radiation. *Trends in Ecology and Evolution* 14: 113–118.

Bromham, L., Penny, D., Rambaut, A., and Hendy, M. D. (2000). The power of relative rates tests depends on the data. *Journal of Molecular Evolution* 50: 1–42.

Brown, A.H.D.D.H. (1979). Enzyme polymorphism in plant populations. *Theoretical Population Biology* 1 (15): 1–42.

Brown, K.M., Baltazar, G.A., and Hamilton, M.B. (2005). Reconciling nuclear microsatellite and mitochondrial marker estimates of population structure: breeding population structure of Chesapeake Bay striped bass (*Morone saxatilis*). *Heredity* 94: 606–615.

Budowle, B., Shea, B., Niezgoda, S., and Chakraborty, R. (2001). CODIS STR loci data from 41 sample populations. *Journal of Forensic Science* 46: 453–489.

Bulmer, M. (1985). *The Mathematical Theory of Quantitative Genetics*. Oxford: Clarendon Press.

Bulmer, M. (1998). Galton's law of ancestral heredity. *Heredity* 81: 579–585.

Bünger, L., Renne, U., Dietl, G., and Kuhla, S. (1998). Long-term selection for protein amount over 70 generations in mice. *Genetical Research* 72: 93–109.

Buonagurio, D.A., Nakada, S., Parvin, J.D. et al. (1986). Evolution of human influenza a viruses over 50 years: uniform rate of change in NS gene. *Science* 232: 980–982.

Burger, R. and Lande, R. (1994). On the distribution of the mean and variance of a quantitative trait under mutation-selection-drift balance. *Genetics* 138: 901–912.

Buri, P. (1956). Gene frequency in small populations of mutant *Drosophila*. *Evolution* 10: 367–402.

Butler, J.M. (2006). Genetics and genomics of core short tandem repeat loci used in human identity testing. *Journal of Forensic Sciences* 51: 253–265.

Byers, D.L. and Waller, D.M. (1999). Do plant populations purge their genetic load? Effects of population size and mating history on inbreeding depression. *Annual Review of Ecology and Systematics* 30: 479–513.

Caballero, A., Tenesa, A., and Keightley, P.D. (2015). The nature of genetic variation for complex traits revealed by GWAS and regional heritability mapping analyses. *Genetics* 201: 1601–1613.

Carlborg, Ö. and Haley, C.S. (2004). Epistasis: Too often neglected in complex trait studies? *Nature Reviews Genetics* 5 (8): 618–625. https://doi.org/10.1038/nrg1407.

Carr, D.E., Dudash, M.R.M., and Bayliss, M.W. (2003). Recent approaches to the genetic basis of inbreeding depression in plants. *Philosophical Transactions of the Royal Society of London B Biological Sciences* 358 (1434): 1071–1084. https://doi.org/10.1098/rstb.2003.1295.

Carruthers, T., Sanderson, M.J., and Scotland, R.W. (2019). The implications of lineage-specific rates for divergence time estimation. *Systematic Biology* 69: 660–670.

Casares, P. (2007). A corrected Haldane's map function to calculate genetic distances from recombination data. *Genetica* 129: 333–338.

Cavalli-Sforza, L.L. and Bodmer, W.F. (1971). *The Genetics of Human Populations*. San Francisco: W. H. Freeman.

Cavalli-Sforza, L.L., Menozzi, P., and Piazza, A. (1994). *The History and Geography of Human Genes*. Princeton, N.J: Princeton University Press.

Ceballos, F.C., Joshi, P.K., Clark, D.W. et al. (2018). Runs of homozygosity: windows into population history and trait architecture. *Nature Reviews. Genetics* 19: 220–234.

Chakraborty, R., Meagher, T.R., and Smouse, P.E. (1988). Parentage analysis with genetic markers in natural populations. I. the expected proportion of offspring with unambiguous paternity. *Genetics* 118: 527–536.

Chan, A.T., Tranah, G.J., Giovannucci, E.L. et al. (2005). Genetic variants in the UGT1A6 enzyme, aspirin use, and the risk of colorectal adenoma. *JNCI Journal of the National Cancer Institute* 97: 457–460.

Chao, L. and Carr, D. (1993). The molecular clock and the relationship between population size and the generation time. *Evolution* 47: 688–690.

Charlesworth, D. (2006). Balancing selection and its effects on sequences in nearby genome regions. *PLoS Genetics* 2: e64.

Charlesworth, B. (2009). Fundamental concepts in genetics: effective population size and patterns of molecular evolution and variation. *Nature Reviews. Genetics* 10: 195–205.

Charlesworth, B. (2012). The effects of deleterious mutations on evolution at linked sites. *Genetics* 190: 5–22.

Charlesworth, B. (2013). Why we are not dead one hundred times over. *Evolution* 67: 3354–3361.

Charlesworth, B. and Charlesworth, D. (1997). Rapid fixation of deleterious allele can be caused by Muller's ratchet. *Genetical Research* 70: 63–73.

Charlesworth, B. and Charlesworth, D. (1998). Some evolutionary consequences of deleterious mutation. *Genetica* 102: 3–19.

Charlesworth, B. and Charlesworth, D. (1999). The genetic basis of inbreeding depression. *Genetical Research* 74: 329–340.

Charlesworth, B. and Charlesworth, D. (2017). Population genetics from 1966 to 2016. *Heredity* 118: 2–9.

Charlesworth, B., Morgan, M.T., and Charlesworth, D. (1993). The effect of deleterious mutations on neutral molecular variation. *Genetics* 134: 1289–1303.

Charlesworth, D., Charlesworth, B., and Morgan, M.T. (1995). The pattern of neutral molecular variation under the background selection model. *Genetics* 141: 1619–1632.

Charlesworth, B., Charlesworth, D., and Barton, N.H. (2003). The effects of genetic and geographic structure on neutral variation. *Annual Review of Ecology and Systematics* 34: 99–125.

Charlesworth, D., Barton, N.H., and Charlesworth, B. (2017). The sources of adaptive variation. *Proceedings of the Royal Society B: Biological Sciences* 284: 20162864.

Chen, J., Shrestha, R., Ding, J. et al. (2016). Genome-wide association study and QTL mapping reveal genomic loci associated with *Fusarium* ear rot resistance in tropical maize germplasm. *G3 Genes|Genomes|Genetics* 6: 3803–3815. https://doi.org/10.1534/g3.116.034561.

Cheptou, P.-O. and Donohue, K. (2011). Environment-dependent inbreeding depression: its ecological and evolutionary significance. *New Phytologist* 189: 395–407.

Cheverud, J.M. (2000). Detecting epistasis among quantitative trait loci. In: *Epistasis and the Evolutionary Process* (eds. J.B. Wolf, E.D. Brodie and M.J. Wade), 344. Oxford University Press.

Cheverud, J.M. and Routman, E.J. (1995). Epistasis and its contribution to genetic variance components. *Genetics* 139: 1455–1461.

Christensen, K., Arnbjerg, J., and Andresen, E. (1985). Polymorphism of serum albumin in dog breeds and its relation to weight and leg length. *Hereditas* 102: 219–223. https://doi.org/10.1111/j.1601-5223.1985.tb00618.x.

Clark, A.G. and Feldman, M. (1986). A numerical simulation of the one-locus multiple allele fertility model. *Genetics* 113: 161–176.

Clarke, B. (1973a). The effect of mutation on population size. *Nature* 242: 196–197.

Clarke, B. (1973b). Mutation and population size. *Heredity* 31: 367–379.

Clausen, J., Keck, D.D., and Hiesey, W. (1948). *Experimental Studies on the Nature of Species. III. Environmental Responses of Climatic Races of* Achillea. Washington, D.C.: Carnegie Institution of Washington.

Cloud-Richardson, K.M., Smith, B.R., and Macdonald, S.J. (2016). Genetic dissection of intraspecific variation in a male-specific sexual trait in *Drosophila melanogaster*. *Heredity* 117: 417–426.

Cockerham, C. (1954). An extension of the concept of partitioning hereditary variance for analysis of covariance among relatives when epistasis is present. *Genetics* 39: 859–882.

Cockerham, C. (1956). Effects of linkage on the covariances between relatives. *Genetics* 41: 138–141.

Cohen, J. (2001). AIDS origins: disputed AIDS theory dies its final death. *Science* 292: 615a–615a.

Colosimo, P.F., Peichel, C.L., Nereng, K. et al. (2004). The genetic architecture of parallel armor plate reduction in threespine sticklebacks. *PLoS Biology* 2: 635–641.

Comeron, J.M., Williford, A., and Kliman, R.M. (2008). The Hill–Robertson effect: evolutionary consequences of weak selection and linkage in finite populations. *Heredity* 100: 19–31.

Corander, J., Marttinen, P., Sirén, J., and Tang, J. (2008). Enhanced Bayesian modelling in BAPS software for learning genetic structures of populations. *BMC Bioinformatics* 9: 539.

Cordell, H. (2002). Epistasis: what it means, what it doesn't mean, and statistical methods to detect it in humans. *Human Molecular Genetics* 11: 2463–2468.

Cotterman, C. (1974). A calculus for statistico-genetics. In: *Genetics and Social Structure: Mathematical Structuralism in Population Genetics and Social Theory* (ed. P. Ballonoff), 157–272. Stroudsburg, PA: Dowden, Hutchinson and Ross.

Cotterman, C. (1983). Relationship and probability in Mendelian populations. *American Journal of Medical Genetics* 16: 393–440.

Cox, J.T. and Durrett, R. (2002). The stepping stone model: new formulas expose old myths. *The Annals of Applied Probability* 12: 1348–1377.

Coyne, J.A. and Orr, H. (2004). *Speciation*. Sunderland, MA: Sinauer Associates.

Coyne, J.A., Barton, N.H., and Turelli, M. (1997). Perspective: a critique of Sewall Wright's shifting balance theory of evolution. *Evolution* 51: 643–671.

Coyne, J.A., Barton, N.H., and Turelli, M. (2000). Is Wright's shifting balance process important in evolution? *Evolution* 54: 306–317.

Crandall, K.A., Posada, D., and Vasco, D. (1999). Effective population sizes: missing measures and missing concepts. *Animal Conservation* 2: 317–319.

Crawford, T. (1984). The estimation of neighborhood parameters for plant populations. *Heredity* 52: 273–283.

Crnokrak, P. and Barrett, S.C.H. (2002). Perspective: purging the genetic load: a review of the experimental evidence. *Evolution; International Journal of Organic Evolution* 56: 2347–2358.

Crow, J. (1958). Some possibilities for measuring selection intensities in man. *Human Biology* 30: 1–13.

Crow, J. (1972). Darwinian and non-Darwinian evolution. In: *Proceedings of the Sixth Berkeley Symposium on Mathematical Statistics and Probability*, pp. 1–22. vol. V, University of California Press, Berkeley, CA.

Crow, J.F. (1988). Eighty years ago: the beginnings of population genetics. *Genetics* 119: 473–476.

Crow, J. (1993a). Mutation, mean fitness, and genetic load. *Oxford Surveys in Evolutionary Biology* 9: 3–42.

Crow, J.F. (1993b). Felix Bernstein and the first human marker locus. *Genetics* 133: 4–7.

Crow, J.F. (1997). The high spontaneous mutation rate: is it a health risk? *Proceedings of the National Academy of Sciences of the United States of America* 94: 380–386.

Crow, J. (2002). Perspective: here's to Fisher, additive genetic variance, and the fundamental theorem of natural selection. *Evolution* 56: 1313–1316.

Crow, J.F. and Aoki, K. (1984). Group selection for a polygenic behavioral trait: estimating the degree of population subdivision. *Proceedings of the National Academy of Sciences of the United States of America* 81: 6073–6077.

Crow, E.W. and Crow, J.F. (2002). 100 years ago: Walter Sutton and the chromosome theory of heredity. *Genetics* 160: 1–4.

Crow, J.F. and Denniston, C. (1988). Inbreeding and variance effective population numbers. *Evolution* 42: 482–495.

Crow, J.F. and Kimura, M. (1970). *An Introduction to Population Genetics Theory*. New York: Harper and Row.

Crow, J.F. and Kimura, M. (1979). Efficiency of truncation selection. *Proceedings of the National Academy of Sciences of the United States of America* 76: 396–399.

Crow, J.F., Engels, W.R., and Denniston, C. (1990). Phase three of Wright's shifting-balance theory. *Evolution* 44: 233.

Culter, D. (2000a). Understanding the overdispersed molecular clock. *Genetics* 154: 1403–1417.

Culter, D. (2000b). Estimating divergence times in the presence of an overdispersed molecular clock. *Molecular Biology and Evolution* 17: 1647–1660.

Cummings, C.L., Alexander, H.M., Snow, A.A. et al. (2002). Fecundity selection in a sunflower crop-wild study: can ecological data predict crop allele changes? *Ecological Applications* 12: 1661–1671.

Cutter, A.D. and Payseur, B.A. (2013). Genomic signatures of selection at linked sites: unifying the disparity among species. *Nature Reviews Genetics* 14: 262–274.

Darwin, C. (1859). *On the Origin of Species by Means of Natural Selection, or, the Preservation of Favoured Races in the Struggle for Life*. London: John Murray.

Darwin, C. (1876). *The Effects of Cross and Self Fertilisation in the Vegetable Kingdom / by Charles Darwin*. London: J. Murray.

De Meester, L., Gómez, A., Okamura, B., and Schwenk, K. (2002). The monopolization hypothesis and the dispersal–gene flow paradox in aquatic organisms. *Acta Oecologica* 23: 121–135.

Denny, M.W. and Gaines, S. (2000). *Chance in Biology: Using Probability to Explore Nature*. Princeton, NJ: Princeton University Press.

Devlin, B. and Ellstrand, N.C. (1990). The development and application of a refined method for estimating gene flow from angiosperm paternity analysis. *Evolution* 44: 248–259.

DeWitt, T.J. and Scheiner, S.M. (2004). *Phenotypic Plasticity: Functional and Conceptual Approaches*. Oxford: Oxford University Press.

Dickerson, R. (1971). The structure of cytochrome c and the rates of molecular evolution. *Journal of Molecular Evolution* 1: 26–45.

Dobzhansky, T. (1955). A review of some fundamental concepts and problems of population genetics. *Cold Spring Harbor Symposia on Quantitative Biology* 20: 1–15.

Donnelly, P. and Kurtz, T.G. (1999). Particle representations for measure-valued population models. *The Annals of Probability* 27: 166–205.

dos Reis, M., Donoghue, P.C.J., and Yang, Z. (2016). Bayesian molecular clock dating of species divergences in the genomics era. *Nature Reviews Genetics* 17: 71–80.

Doukhan, L. and Delwart, E. (2001). Population genetic analysis of the protease locus of human immunodeficiency virus type 1 quasispecies undergoing drug selection, using a denaturing gradient-heteroduplex tracking assay. *Journal of Virology* 75: 6729–6736.

Dow, B.D. and Ashley, M.V. (1996). Microsatellite analysis of seed dispersal and parentage of saplings in bur oak, *Quercus macrocarpa*. *Molecular Ecology* 5: 615–627.

Dowling, D.K., Friberg, U., and Lindell, J. (2008). Evolutionary implications of non-neutral mitochondrial genetic variation. *Trends in Ecology & Evolution* 23: 546–554.

Drake, J.W. (1991). A constant rate of spontaneous mutation in DNA-based microbes. *Proceedings of the National Academy of Sciences* 88: 7160–7164.

Drake, J.W., Charlesworth, B., Charlesworth, D., and Crow, J.F. (1998). Rates of spontaneous mutation. *Genetics* 148: 1667–1686.

Dudash, M.R.M. (1990). Relative fitness of selfed and outcrossed progeny in a self-compatible, protandrous species, *Sabatia angularis* L. (Gentianaceae): a comparison in three environments. *Evolution* 44: 1129–1139.

Edmonds, C.A., Lillie, A.S., and Cavalli-Sforza, L.L. (2004). Mutations arising in the wave front of an expanding population. *Proceedings of the National Academy of Sciences* 101: 975–979.

Edwards, A. (1994). The fundamental theorem of natural selection. *Biological Reviews* 61: 335–337.

Edwards, A.W.F. (2002). The Fundamental Theorem of Natural Selection. *Theoretical Population Biology* 61 (3): 335–337. https://doi.org/10.1006/tpbi.2002.1570.

Edwards, A.W.F. (2008). G. H. Hardy (1908) and Hardy-Weinberg equilibrium. *Genetics* 179: 1143–1150.

Edwards, A.W.F. (2012). Reginald Crundall Punnett: first Arthur Balfour professor of genetics, Cambridge, 1912. *Genetics* 192: 3–13.

Eldon, B. and Wakeley, J. (2006). Coalescent processes when the distribution of offspring number among individuals is highly skewed. *Genetics* 172 (4): 2621–2633. https://doi.org/10.1534/genetics.105.052175.

Ellegren, H. (2000). Microsatellite mutations in the germline: implications for evolutionary inference. *Trends in Genetics:* 16: 551–558.

Emara, M.G., Kim, H., Zhu, J. et al. (2002). Genetic diversity at the major histocompatibility complex (B) and microsatellite loci in three commercial broiler pure lines. *Poultry Science* 81: 1609–1617.

Estany, J., Villalba, D., Tor, M. et al. (2002). Correlated response to selection for litter size in pigs: I growth, fat deposition, and feeding behavior traits. *Journal of Animal Science* 80: 2556–2565.

Estes, S., Phillips, P.C., Denver, D.R. et al. (2004). Mutation accumulation in populations of varying size: the distribution of mutational effects for fitness correlates in Caenorhabditis elegans. *Genetics* 166: 1269–1279.

Evanno, G., Regnaut, S., and Goudet, J. (2005). Detecting the number of clusters of individuals using the software structure: a simulation study. *Molecular Ecology* 14: 2611–2620.

Ewens, W. (1982). On the concept of the effective population size. *Theoretical Population Biology* 21 (3): 373–378.

Ewens, W.J. (2004). *Mathematical Population Genetics I. Theoretical Introduction*. New York, NY: Springer.

Excoffier, L., Smouse, P.E., and Quattro, J.M. (1992). Analysis of molecular variance inferred from metric distances among DNA haplotypes: application to human mitochondrial DNA restriction data. *Genetics* 131: 479–491.

Excoffier, L., Hofer, T., and Foll, M. (2009a). Detecting loci under selection in a hierarchically structured population. *Heredity* 103: 285–298.

Excoffier, L., Foll, M., and Petit, R.J. (2009b). Genetic consequences of range expansions. *Annual Review of Ecology, Evolution, and Systematics* 40: 481–501. https://doi.org/10.1146/annurev.ecolsys.39.110707.173414.

Eyre-Walker, A. and Keightley, P.D. (2007). The distribution of fitness effects of new mutations. *Nature Reviews Genetics* 8: 610–618.

Fabian, D.K., Kapun, M., Nolte, V. et al. (2012). Genome-wide patterns of latitudinal differentiation among populations of *Drosophila melanogaster* from North America. *Molecular Ecology* 21: 4748–4769.

Fabiani, A., Galimberti, F., Sanvito, S., and Hoelzel, A. (2004). Extreme polygyny among southern elephant seals on Sea Lion Island, Falkland Islands. *Behavioral Ecology* 15: 961–969.

Fairhurst, R.M. and Casella, J. (2004). Homozygous hemoglobin C disease. *New England Journal of Medicine* 350: 26.

Falconer, D.S. and Mackay, T. (1996). *Introduction to Quantitative Genetics*. Harlow: Longman.

Falush, D., Stephens, M., and Pritchard, J.K. (2003). Inference of population structure using multilocus genotype data: linked loci and correlated allele frequencies. *Genetics* 164: 1567–1587.

Faria, N.R., Rambaut, A., Suchard, M.A. et al. (2014). The early spread and epidemic ignition of HIV-1 in human populations. *Science* 346: 56–61.

Farlow, A., Long, H., Arnoux, S. et al. (2015). The spontaneous mutation rate in the fission yeast *Schizosaccharomyces pombe*. *Genetics* 201: 737–744.

Fay, J.C. and Wu, C.I. (2000). Hitchhiking under positive Darwinian selection. *Genetics* 155: 1405–1413.

Feldman, M.W. (1966). On the offspring number distribution in a genetic population. *Journal of Applied Probability* 3: 129–141.

Feldman, M.W., Christiansen, F.B., and Liberman, U. (1983). On some models of fertility selection. *Genetics* 105: 1003–1010.

Felsenstein, J. (1971). Inbreeding and variance effective numbers in populations with overlapping generations. *Genetics* 68: 581–597.

Felsenstein, J. (1974). The evolutionary advantage of recombination. *Genetics* 78: 737–756.

Felsenstein, J. (1981). Evolutionary trees from DNA sequences: a maximum likelihood approach. *Journal of Molecular Evolution* 17: 368–376.

Felsenstein, J. (1985). Phylogenies and the comparative method. *The American Naturalist* 125: 1–15.

Feng, Y., Zhang, Y., Ying, C. et al. (2015). Nanopore-based fourth-generation DNA sequencing technology. *Genomics, Proteomics & Bioinformatics* 13: 4–16.

Feng, C., Pettersson, M., Lamichhaney, S. et al. (2017). Moderate nucleotide diversity in the Atlantic herring is associated with a low mutation rate. *eLife* 6: e23907.

Fenster, C.B., Galloway, L.F., and Chao, L. (1997). Epistasis and its consequences for the evolution of natural populations. *Trends in Ecology and Evolution* 12: 282–286.

Fijarczyk, A. and Babik, W. (2015). Detecting balancing selection in genomes: limits and prospects. *Molecular Ecology* 24: 3529–3545.

Fisher, R. (1918). The correlation between relatives on the supposition of Mendelian inheritance. *Transactions of the Royal Society of Edinburgh* 52: 399–433.

Fisher, R.A. (1999). *The Genetical Theory of Natural Selection: A Complete Variorum Edition* (ed. H. Bennett). Oxford: Oxford University Press.

Fitch, W. (1976). Molecular evolutionary clocks. In: *Molecular Evolution* (ed. F.J. Ayala), 160–178. Sunderland, MA: Sinauer Associates.

Flanagan, S.P. and Jones, A.G. (2018). The future of parentage analysis: from microsatellites to SNPs and beyond. *Molecular Ecology* 28: 544–567.

Flint-Garcia, S.A., Thornsberry, J.M., and Buckler, E.S. (2003). Structure of linkage disequilibrium in plants. *Annual Review of Plant Biology* 54: 357–374.

Foll, M. and Gaggiotti, O. (2008). A genome-scan method to identify selected loci appropriate for both dominant and codominant markers: a Bayesian perspective. *Genetics* 180: 977–993.

Ford, E.B. (1964). *Ecological Genetics*. London: Chapman and Hall.

Ford, E.B. (1975). *Ecological Genetics*. London: Chapman and Hall.

Ford, M.J. (2002). Applications of selective neutrality tests to molecular ecology. *Molecular Ecology* 11: 1245–1262.

Fragata, I., Blanckaert, A., Dias Louro, M.A. et al. (2019). Evolution in the light of fitness landscape theory. *Trends in Ecology & Evolution* 34: 69–82.

François, O. and Durand, E. (2010). Spatially explicit Bayesian clustering models in population genetics. *Molecular Ecology Resources* 10 (5): 773–784. https://doi.org/10.1111/j.1755-0998.2010.02868.x.

François, O. and Waits, L.P. (2015). Clustering and assignment methods in landscape genetics. In: *Landscape Genetics*, 114–128. Chichester, UK: Wiley.

Frankham, R. (1995). Effective population size/adult population size ratios in wildlife: a review. *Genetical Research* 66: 95–107.

Frankham, R. (1998). Inbreeding and extinction: island populations. *Conservation Biology* 12: 665–675.

Franklin, I.R. (1980). Evolutionary change in small populations. In: *Conservation Biology – an Evolutionary-Ecological Perspective* (eds. M. Soule and B. Wilcox), 135–149. Sunderland, Massachusetts: Sinauer Associates.

Franklin, I.R. and Frankham, I.R. (1998). How large must populations be to retain evolutionary potential? *Animal Conservation* 1: 69–70.

Frankham, R. and Franklin, I.R. (1998). Response to Lynch and Lande. *Animal Conservation* 1: 73.

Frichot, E., Schoville, S., Bouchard, G., and François, O. (2012). Correcting principal component maps for effects of spatial autocorrelation in population genetic data. *Frontiers in Genetics* 3: –254.

Fry, J.D. and Heinsohn, S. (2002). Environmental dependence of mutational parameters for viability in *Drosophila melanogaster*. *Genetics* 161: 1155–1167.

Fu, Y.X. (1995). Statistical properties of segregating sites. *Theoretical Population Biology* 48: 172–197.

Fu, Y.X. (1996). New statistical tests of neutrality for DNA samples from a population. *Genetics* 145: 557–570.

Fu, Y.X. (1997). Statistical tests of neutrality of mutations against population growth, hitchhiking and background selection. *Genetics* 147: 915–925.

Fu, Y.X. and Huai, H. (2003). Estimating mutation rate: how to count mutations? *Genetics* 164: 797–805.

Fu, Y.X. and Li, W.-H. (1993). Statistical tests of neutrality of mutations. *Genetics* 133: 693–709.

Furstenau, T.N. and Cartwright, R.A. (2016). The effect of the dispersal kernel on isolation-by-distance in a continuous population. *PeerJ* 4: e1848.

Galtier, N., Nabholz, B., Glémin, S., and Hurst, G.D.D. (2009). Mitochondrial DNA as a marker of molecular diversity: a reappraisal. *Molecular Ecology* 18: 4541–4550.

Gao, L. (2005). Microsatellite variation within and among populations of *Oryza officinalis* (Poaceae), an endangered wild rice from China. *Molecular Ecology* 14: 4287–4297.

Gao, F., Yue, L., White, A.T. et al. (1992). Human infection by genetically diverse SIVSM-related HIV-2 in West Africa. *Nature* 358: 495–499.

Garrigan, D. and Hedrick, P.W. (2003). Perspective: detecting adaptive molecular evolution, lesions from the MHC. *Evolution* 57: 1707–1722.

Gaut, B.S. (1998). Molecular clocks and nucleotide substitution rates in higher plants. *Evolutionary Biology* 30: 93–120.

Gavrilets, S. (2004). *Fitness Landscapes and the Origin of Species*. Princeton, NJ: Princeton University Press Geertjes GJ, Postema J, Kamping A, van Delden W.

Geertjes, G.J., Postema, J., Kamping, A. et al. (2004). Allozymes and RAPDs detect little genetic population substructuring in the Caribbean stoplight parrotfish Sparisoma viride. *Marine Ecology Progress Series* 279: 225–235. https://doi.org/10.3354/meps279225.

Getty, T. (1999). What do experimental studies tell us about group selection in nature? *American Naturalist* 154: 596–598.

Gibson, G. (2012). Rare and common variants: twenty arguments. *Nature Reviews. Genetics* 13: 135–145.

Gibson, G. (2018). Population genetics and GWAS: a primer. *PLoS Biology* 16: e2005485.

Gillespie, J. (1989). Lineage effects and the index of dispersion of molecular evolution. *Molecular Biology and Evolution* 6: 636–647.

Gillespie, J. (1991). *The Causes of Molecular Evolution*. New York: Oxford University Press.

Gillespie, J. (1994). Substitution processes in molecular evolution. II. Exchangeable models from population genetics. *Evolution* 48: 1101–1113.

Gillespie, J. (1995). On Ohta's hypothesis: most amino acid substitutions are deleterious. *Journal of Molecular Evolution* 40: 64–69.

Gillespie, J. (2000). The neutral theory in an infinite population. *Gene* 261: 11–18.

Gillespie, J.H. (2000). Genetic drift in an infinite population. The pseudohitchhiking model. *Genetics* 155: 909–919.

Gillespie, J.H. and Langley, C. (1979). Are evolutionary rates really variable? *Journal of Molecular Evolution* 13: 27–34.

Gillespie, J.H.J. (2001). Is the population size of a species relevant to its evolution? *Evolution* 55: 2161–2169.

Gillham, N. (2001). Sir Francis Galton and the birth of eugenics. *Annual Review of Genetics* 35: 83–101.

Gillooly, J.F., Allen, A.P., West, G.B., and Brown, J. (2005). The rate of DNA evolution: effects of body size and temperature on the molecular clock. *Proceedings of the National Academy of Sciences of the United States of America* 102: 140–145.

Glazko, G.V. and Nei, M. (2003). Estimation of divergence times for major lineages of primate species. *Molecular Biology and Evolution* 20: 424–434.

Glémin, S., Bazin, E., and Charlesworth, D. (2006). Impact of mating systems on patterns of sequence polymorphism in flowering plants. *Proceedings of the Royal Society B: Biological Sciences* 273: 3011–3019.

Goldringer, I. and Bataillon, T. (2004). On the distribution of temporal variations in allele frequency: consequences for the estimation of effective population size and the detection of loci undergoing selection. *Genetics* 168: 563–568.

Goldstein, D.B. and Pollock, D. (1997). Launching micro-satellites: a review of the mutation processes and methods of phylogenetic inference. *Journal of Heredity* 88: 335–342.

Goldstein, D.B., Ruiz Linares, A., Cavalli-Sforza, L.L., and Feldman, M.W. (1995). An evaluation of genetic distances for use with microsatellite loci. *Genetics* 139: 463–471.

Goodman, S.J. (1997). RST Calc: a collection of computer programs for calculating estimates of genetic differentiation from microsatellite data and determining their significance. *Molecular Ecology* 6: 881–885.

Goodnight, C. (1987). On the effect of founder events on epistatic genetic variance. *Evolution* 41: 80–91.

Goodnight, C. (1988). Epistasis and the effect of founder events on the additive genetic variance. *Evolution* 42: 441–454.

Goodnight, C. (2000). Modeling gene interaction in structured populations. In: *Epistasis and the Evolutionary Process* (eds. J.B. Wolf, D.B. Edmund III and M.J. Wade), 129–145. Oxford: Oxford University Press.

Goodnight, C.J. and Stevens, L. (1997). Experimental studies of group selection: what do they tell us about group selection in nature? *American Naturalist* 150: 59–79.

Goodnight, C.J. and Wade, M. (2000). The ongoing synthesis: a reply to Coyne, Barton, and Turelli. *Evolution* 54: 317–324.

Gou, L., Bloom, J.S., and Kruglyak, L. (2019). The genetic basis of mutation rate variation in yeast. *Genetics* 211: 731–740.

Goudsmit, J., de Ronde, A., Ho, D.D., and Perelson, A. (1996). Human immunodeficiency virus fitness in vivo: calculations based on a single zidovudine resistance mutation at codon 215 of reverse transcriptase. *Journal of Virology* 70: 5662–5664.

Gralka, M., Stiewe, F., Farrell, F. et al. (2016). Allele surfing promotes microbial adaptation from standing variation (B Blasius, Ed). *Ecology Letters* 19: 889–898.

Graustein, A., Gaspar, J.M., Walters, J.R., and Palopoli, M.F. (2002). Levels of DNA polymorphism vary with mat- ing system in the nematode genus *Caenorhabditis*. *Genetics* 161: 99–107.

Griffiths, R.C. (1991). The two-locus ancestral graph. In: *Selected Proceedings of the Symposium on Applied Probability* (eds. I.V. Basawa and R.L. Taylor), 100–117. Hayward, CA: Institute of Mathematical Statistics.

Griffiths, R.C. and Marjoram, P. (1997). An ancestral recombination graph. In: *Progress in Population Genetics and Human Evolution*, 257–270. New York NY USA: Springer.

Gu, X. and Li, W.-H. (1992). Higher rates of amino acid substitution in rodents than in humans. *Molecular Phylogenetics and Evolution* 1: 211–214.

Guillot, G., Mortier, F., and Estoup, A. (2005). Geneland: a computer package for landscape genetics. *Molecular Ecology Notes* 5: 712–715.

Haasl, R.J. and Payseur, B.A. (2016). Fifteen years of genomewide scans for selection: trends, lessons and unaddressed genetic sources of complication. *Molecular Ecology* 25: 5–23.

Hadeler, K.P. and Liberman, U. (1975). Selection models with fertility differences. *Journal of Mathematical Biology* 2: 19–32.

Hague, M.T.J. and Routman, E.J. (2016). Does population size affect genetic diversity? A test with sympatric lizard species. *Heredity* 116: 92–98.

Haldane, J.B.S. (1924). A mathematical theory of natural and artificial selection. Part ii the influence of partial self-fertilisation, inbreeding, assortative mating, and selective fertilisation on the composition of mendelian populations, and on natural selection. *Biological Reviews* 1: 158–163.

Haldane, J. (1937). The effect of variation on fitness. *American Naturalist* 71: 337–349.

Haldane, J. (1940). The conflict between selection and mutation of harmful recessive genes. *Annals of Eugenics* 10: 417–421.

Haldane, J. (1957). The cost of natural selection. *Journal of Genetics* 55: 511–524.

Haley, C.S. and Knott, S.A. (1992). A simple regression method for mapping quantitative trait loci in line crosses using flanking markers. *Heredity* 69: 315–324.

Hallatschek, O., Hersen, P., Ramanathan, S., and Nelson, D.R. (2007). Genetic drift at expanding frontiers promotes gene segregation. *Proceedings of the National Academy of Sciences* 104: 19926–19930.

Halligan, D.L. and Keightley, P.D. (2009). Spontaneous mutation accumulation studies in evolutionary genetics. *Annual Review of Ecology, Evolution, and Systematics* 40: 151–172.

Hamilton, M.B. and Miller, J.R. (2002). Comparing relative rates of pollen and seed gene flow in the island model using nuclear and organelle measures of population structure. *Genetics* 162: 1897–1909.

Hamilton, M.B., Braverman, J.M., and Soria-Hernanz, D.F. (2003). Patterns and relative rates of nucleotide and insertion/deletion evolution at six chloroplast intergenic regions in new world species of the Lecythidaceae. *Molecular Biology and Evolution* 20: 1710–1721.

Hamilton, M.B., Tartakovsky, M., and Battocletti, A. (2018). SpEED-Ne: software to simulate and estimate effective population size (ne) from linkage disequilibrium observed in single samples. *Molecular Ecology Resources* 18: 714–728.

Hanski, I., Simberloff, D., and Simberloff, D. (1997). The metapopulation approach, its history, conceptual domain and application to conservation. In: *Metapopulation Biology: Ecology, Genetics, and Evolution* (eds. I. Hanski and M.E. Gilpin), 5–26. San Diego, CA: Academic Press.

Hardy, G.H. (1908). Mendelian proportions in a mixed population. *Science* 28: 49–50.

Harrison, R.G. (1989). Animal mitochondrial DNA as a genetic marker in population and evolutionary biology. *Trends in Ecology & Evolution* 4: 6–11.

Hasegawa, M., Kishino, H., and Yano, T. (1985). Dating of the human-ape splitting by a molecular clock of mitchondrial DNA. *Journal of Molecular Evolution* 22: 160–174.

Hastings, A. (1981). Disequilibrium, selection, and recombination: limits in two-locus, two-allele models. *Genetics* 98: 659–668.

Hastings, A. (1985). Stable equilibria at two loci in populations with large selfing rates. *Genetics* 109: 215–228.

Hastings, A. (1986). Limits to the relationship among recombination, disequilibrium and epistasis in two-locus models. *Genetics* 113: 177–185.

Hastings, A. (1987). Monotonic change of the mean phenotype in two-locus models. *Genetics* 117: 583–585.

Heather, J.M. and Chain, B. (2016). The sequence of sequencers: the history of sequencing DNA. *Genomics* 107: 1–8.

Heckel, G., Burri, R., Fink, S. et al. (2005). Genetic structure and colonization processes in European populations of the common vole, Microtus arvalis. *Evolution* 59: 2231–2242.

Hedges, S.B., Marin, J., Suleski, M. et al. (2015). Tree of life reveals clock-like speciation and diversification. *Molecular Biology and Evolution* 32: 835–845.

Hedrick, P.W. (1985). Coat variants in cats. Gametic disequilibrium between unlinked loci. *The Journal of Heredity* 76: 127–131.

Hedrick, P.W.P. (1987). Gametic disequilibrium measures: proceed with caution. *Genetics* 117: 331–341.

Hedrick, P. (2004). Estimation of relative fitnesses from relative risk data and the predicted future of haemoglobin alleles S and C. *Journal of Evolutionary Biology* 17: 221–224.

Hedrick, P.W. (2005). A standardized genetic differentiation measure. *Evolution* 59: 1633–1638.

Hedrick, P.W. (2006). Genetic polymorphism in heterogeneous environments: the age of genomics.

Annual Review of Ecology Evolution and Systematics 37: 67–93.

Hedrick, P., Jain, S., and Holden, L. (1978). Multilocus systems in evolution. In: *Evolutionary Biology*, vol. 11 (eds. M.K. Hecht, W.C. Steere and B. Wallace), 104–184. New York: Plenum Press.

Hein, J., Schierup, M.H., and Wiuf, C. (2005). *Gene Genealogies, Variation and Evolution*. New York: Oxford University Press.

Heller, R. and Siegismund, H.R. (2009). Relationship between three measures of genetic differentiation G_{ST}, D_{EST} and G'_{ST}: how wrong have we been? *Molecular Ecology* 18: 2080–2083. discussion 2088-91.

Hellmann, I., Ebersberger, I., Ptak, S.E. et al. (2003). A neutral explanation for the correlation of diversity with recombination rates in humans. *The American Journal of Human Genetics* 72: 1527–1535.

Hellmann, I., Prüfer, K., Ji, H. et al. (2005). Why do human diversity levels vary at a megabase scale? *Genome Research* 15: 1222–1231.

Hermisson, J. and Pennings, P.S. (2005). Soft sweeps. *Genetics* 169: 2335–2352.

Hermisson, J. and Pennings, P.S. (2017). Soft sweeps and beyond: understanding the patterns and probabilities of selection footprints under rapid adaptation. *Methods in Ecology and Evolution* 8: 700–716.

Heywood, J. (1986). The effect of plant size variation on genetic drift in populations of annuals. *American Naturalist* 127: 851–861.

Heywood, J.S. (2005). An exact form of the breeder's equation for the evolution of a quantitative trait under natural selection. *Evolution* 59: 2287–2298.

Hill, W.G. and Robertson, A. (1966). The effect of linkage on limits to artificial selection. *Genetical Research* 8: 269–294.

Hill, W.G. and Robertson, A. (1968). Linkage disequilibrium in finite populations. *Theoretical and Applied Genetics* 38: 226–231.

Hill, W.G., Goddard, M.E., and Visscher, P.M. (2008). Data and theory point to mainly additive genetic variance for complex traits (TFC Mackay, Ed). *PLoS Genetics* 4: e1000008.

Ho, S.Y.W. and Duchêne, S. (2014). Molecular-clock methods for estimating evolutionary rates and timescales. *Molecular Ecology* 23: 5947–5965.

Hodel, R.G.J., Segovia-Salcedo, M.C., Landis, J.B. et al. (2016). The report of my death was an exaggeration: a review for researchers using microsatellites in the 21st century. *Applications in Plant Sciences* 4: 1600025.

Hoffman, J.I. and Amos, W. (2004). Microsatellite genotyping errors: detection approaches, common sources and consequences for paternal exclusion. *Molecular Ecology* 14: 599–612.

Holsinger, K.E. and Feldman, M.W. (1985). Selection in complex genetic systems VI. Equilibrium properties of two locus selection models with partial selfing. *Theoretical Population Biology* 28: 117–132.

Holsinger, K.E. and Weir, B.S. (2009). Genetics in geographically structured populations: defining, estimating and interpreting F_{ST}. *Nature Reviews Genetics* 10: 639–650.

Hori, M. (1993). Frequency-dependent natural selection in the handedness of scale-eating cichlid fish. *Science* 260: 216–219.

Hotelling, H. (1933). Analysis of a complex of statistical variables into principal components. *Journal of Educational Psychology* 24: 417–441.

Houle, D. (1992). Comparing evolvability and variability of quantitative traits. *Genetics* 130: 195–204.

Houle, D., Morikawa, B., and Lynch, M. (1996). Comparing mutational variabilities. *Genetics* 143: 1467–1483.

Hu, X.-S. and Ennos, R. (1999). Impacts of seed and pollen flow on population genetic structure for plant genomes with three contrasting modes of inheritance. *Genetics* 152: 441–450.

Huang, W. and Mackay, T.F.C. (2016). The genetic architecture of quantitative traits cannot be inferred from variance component analysis. *PLoS Genetics* 12: 1–15.

Hubby, J.L. and Lewontin, R.C. (1966). A molecular approach to the study of genic heterozygosity in natural populations. I. the number of alleles at different loci in *Drosophila pseudoobscura*. *Genetics* 54: 577–594.

Hubisz, M. and Siepel, A. (2020). Inference of ancestral recombination graphs using ARGweaver. In: *Statistical Population Genomics. Methods in Molecular Biology*, vol. 2090 (ed. J. Dutheil), 231–266. New York, NY: Humana.

Hubner, R.A., Muir, K.R., Liu, J.-F. et al. (2006). Genetic variants of UGT1A6 influence risk of colorectal adenoma recurrence. *Clinical Cancer Research* 12: 6585–6589.

Hudson, R.R. (1983). Properties of a neutral allele model with intragenic recombination. *Theoretical Population Biology* 23: 183–201.

Hudson, R.R. (1985). The sampling distribution of linkage disequilibrium under an infinite allele model without selection. *Genetics* 109: 611–631.

Hudson, R.R. (1990). Gene genealogies and the coalescent process. *Oxford Surveys in Evolutionary Biology* 7: 1–44.

Hudson, R. (1994). How can low levels of DNA sequence variation in regions of the *Drosophila* genome with low recombination be explained? *Proceedings of the National Academy of Sciences of the United States of America* 91: 6815–6818.

Hudson, R.R. and Kaplan, N.L. (1988). The coalescent process in models with selection and recombination. *Genetics* 120: 831–840.

Hudson, R.R., Kreitman, M., and Agude, M. (1987). A test of neutral molecular evolution based on nucleotide data. *Genetics* 116: 153–159.

Hudson, R.R., Slatkin, M., and Maddison, W.P. (1992). Estimation of levels of gene flow from DNA sequence data. *Genetics* 132: 583–589.

Hunter, D. (2005). Gene-environment interactions in human diseases. *Nature Reviews Genetics* 6: 54–70.

Husband, B.C. and Schemske, D.W. (1996). Evolution of the magnitude and timing of inbreeding depression in plants. *Evolution* 50: 54.

Huxley, J. (1942). *Evolution: The Modern Synthesis*. London: Allen and Unwin.

Hyne, V. and Kearsey, M.J. (1995). QTL analyses-further uses of marker regression. *Theoretical and Applied Genetics* 91: 471–476.

Imhof, M. and Schlotterer, C. (2001). Fitness effects of advantageous mutations in evolving Escherichia coli populations. *Proceedings of the National Academy of Sciences of the United States of America* 98: 1113–1117.

Ingvarsson, P.K. (2004). Population subdivision and the Hudson–Kreitman–Aguade test: testing for deviations from the neutral model in organelle genomes. *Genetical Research* 83: 31–39.

Innan, H. (2006). Modified Hudson–Kreitman–Aguadé test and two-dimensional evaluation of neutrality tests. *Genetics* 173: 1725–1733.

Innan, H. and Stephan, W. (2003). Distinguishing the hitchhiking and background selection models. *Genetics* 165: 2307–2312.

Innan, H., Zhang, K., Marjoram, P. et al. (2005). Statistical tests of the coalescent model based on the haplotype frequency distribution and the number of segregating sites. *Genetics* 169: 1763–1777.

Itoh, N., Segawa, T., Tamiru, M. et al. (2019). Next-generation sequencing-based bulked segregant analysis for QTL mapping in the heterozygous species *Brassica rapa*. *Theoretical and Applied Genetics* 132: 2913–2925.

Jacquard, A. (1975). Inbreeding: One word, several meanings. *Theoretical Population Biology* 7: 338–363.

Jakobsson, M., Edge, M.D., and Rosenberg, N.A. (2013). The relationship between F_{ST} and the frequency of the most frequent allele. *Genetics* 193: 515–528.

Jamann, T.M., Balint-Kurti, P.J., and Holland, J.B. (2015). QTL mapping using high-throughput sequencing. In: *Methods in Molecular Biology* (ed. N.J. Clifton) 1284, 257–285. https://doi.org/10.1007/978-1-4939-2444-8_13.

Janecek, L.L., Honeycutt, R.L., Adkins, R.M., and Davis, S.K. (1996). Mitochondrial gene sequences and the molecular systematics of the artiodactyl subfamily Bovinae. *Molecular Phylogenetics and Evolution* 6: 107–119.

Janes, J.K., Miller, J.M., Dupuis, J.R. et al. (2017). The $K = 2$ conundrum. *Molecular Ecology* 26: 3594–3602.

Jay, F., François, O., Durand, E.Y., and Blum, M.G.B. (2015). POPS : a software for prediction of population genetic structure using latent regression models. *Journal of Statistical Software* 68.

Jensen, J.D., Payseur, B.A., Stephan, W. et al. (2019). The importance of the neutral theory in 1968 and 50 years on: a response to Kern and Hahn 2018. *Evolution* 73: 111–114.

Jombart, T., Devillard, S., Dufour, A.-B., and Pontier, D. (2008). Revealing cryptic spatial patterns in genetic variability by a new multivariate method. *Heredity* 101: 92–103.

Jombart, T., Pontier, D., and Dufour, A.-B. (2009). Genetic markers in the playground of multivariate analysis. *Heredity* 102: 330–341.

Jonas, A., Taus, T., Kosiol, C. et al. (2016). Estimating the effective population size from temporal allele frequency changes in experimental evolution. *Genetics* 204: 723–735.

Jones, A.G. and Ardren, W.R. (2003). Methods of parentage analysis in natural populations. *Molecular Ecology* 12: 2511–2523.

Jones, M.W. and Hutchings, J.A. (2002). Individual variation in Atlantic salmon fertilization success: implications for effective population size. *Ecological Applications* 12: 184–193.

Jones, K.N. and Reithel, J.S. (2001). Pollinator-mediated selec-tion on a flower color polymorphism in experimental populations of Antirrhinum (Scrophulariaceae). *American Journal of Botany* 88: 445–447.

Jones, A.G., Arnold, S.J., and Burger, R. (2003). Stability of the G-matrix in a population experiencing pleiotropic mutation, stabilizing selection, and genetic drift. *Evolution* 57: 1747–1760.

Jones, A.G., Small, C.M., Paczolt, K.A., and Ratterman, N.L. (2010). A practical guide to methods of parentage analysis. *Molecular Ecology Resources* 10: 6–30.

Jónsson, H., Sulem, P., Kehr, B. et al. (2017). Parental influence on human germline de novo mutations in 1,548 trios from Iceland. *Nature* 549: 519–522.

Jorde, L. (1997). Inbreeding in human populations. In: *Encyclopedia of Human Biology*, vol. 5 (ed. R. Dulbecco), 1–13. San Diego, CA: Academic Press.

Jorde, P.E. and Ryman, N. (2007). Unbiased estimator for genetic drift and effective population size. *Genetics* 177: 927–935.

Josephs, E.B., Stinchcombe, J.R., and Wright, S.I. (2017). What can genome-wide association studies tell us about the evolutionary forces maintaining genetic variation for quantitative traits? *New Phytologist* 214: 21–33.

Joshi, A., Bo, M.H., and Mueller, L.D. (1999). Poisson distribution of male mating success in laboratory populations of Drosophila melanogaster. *Genetical Research* 73: 239–249.

Jost, L. (2008). G_{ST} and its relatives do not measure differentiation. *Molecular Ecology* 17: 4015–4026.

Jost, L. (2009). *D* vs. G_{ST} : response to Heller and Siegismund (2009) and Ryman and Leimar (2009). *Molecular Ecology* 18: 2088–2091.

Jost, L., Archer, F., Flanagan, S. et al. (2018). Differentiation measures for conservation genetics. *Evolutionary Applications* 11: 1139–1148.

Judson, H.F.H. (2001). Talking about the genome. *Nature* 409: 769.

Jukes, T.H. and Cantor, C.R. (1969). Evolution of protein molecules. In: *Mammalian Protein Metabolism* (ed. H.N. Munro), 21–132. New York: Academic Press.

Kaeuffer, R., Réale, D., Coltman, D.W., and Pontier, D. (2007). Detecting population structure using STRUCTURE software: effect of background linkage disequilibrium. *Heredity* 99: 374–380.

Kamath, P.L., Haroldson, M.A., Luikart, G. et al. (2015). Multiple estimates of effective population size for monitoring a long-lived vertebrate: an application to Yellowstone grizzly bears. *Molecular Ecology* 24: 5507–5521.

Kaplan, N.L., Darden, T., and Hudson, R.R. (1988). The coalescent process in models with selection. *Genetics* 120: 819–829.

Kaplan, N.L., Hudson, R.R., and Langley, C.H. (1989). The "hitchhiking effect" revisited. *Genetics* 123: 887–899.

Karageorgi, M., Groen, S.C., Sumbul, F. et al. (2019). Genome editing retraces the evolution of toxin resistance in the monarch butterfly. *Nature* 574: 409–412.

Katju, V. and Bergthorsson, U. (2019). Old trade, new tricks: insights into the spontaneous mutation process from the partnering of classical mutation accumulation experiments with high-throughput genomic approaches (K Makova, Ed). *Genome Biology and Evolution* 11: 136–165.

Kay, K.M., Whittall, J.B., and Hodges, S.A. (2006). A survey of nuclear ribosomal internal transcribed spacer substitution rates across angiosperms: an approximate molecular clock with life history effects. *BMC Evolutionary Biology* 6: 36.

Kearsey, M.J., Poni, H.S., and Syed, N.H. (2003). Genetics of quantitative traits in *Arabidopsis thaliana*. *Heredity* 91: 456–464.

Keele, B.F., Van Heuverswyn, F., Li, Y. et al. (2006). Chimpanzee reservoirs of pandemic and nonpandemic HIV-1. *Science (New York, N.Y.)* 313: 523–526.

Keightley, P.D. and Eyre-Walker, A. (1999). Terumi Mukai and the riddle of deleterious mutation rates. *Genetics* 153: 515–523.

Keightley, P.D., Ness, R.W., Halligan, D.L., and Haddrill, P.R. (2014). Estimation of the spontaneous mutation rate per nucleotide site in a Drosophila melanogaster full-sib family. *Genetics* 196: 313–320.

Keinan, A. and Reich, D. (2010). Human population differentiation is strongly correlated with local recombination rate (DJ Begun, Ed). *PLoS Genetics* 6: e1000886.

Kerns, J.A., Cargill, E.J., Clark, L.A. et al. (2007). Linkage and segregation analysis of black and brindle coat color in domestic dogs. *Genetics* 176: 1679–1689.

Kerr, W.E. and Wright, S. (1954). Experimental studies of the distribution of gene frequencies in very small populations of *Drosophila melanogaster*. III. Aristapedia and spineless. *Evolution* 8: 293–302.

Kim, U.-K., Jorgenson, E., Coon, H. et al. (2003). Positional cloning of the human quantitative trait locus underlying taste sensitivity to phenylthiocarbamide. *Science* 299: 1221–1224.

Kim, S., Plagnol, V., Hu, T.T. et al. (2007). Recombination and linkage disequilibrium in *Arabidopsis thaliana*. *Nature Genetics* 39: 1151–1155.

Kimura, M. (1953). "Stepping stone" model of population. *Annual Report of the National Institute of Genetics, Japan* 3: 62–63.

Kimura, M. (1955). Solution of a process of random genetic drift with a continuous model. *Proceedings of the National Academy of Sciences of the United States of America* 41: 144–150.

Kimura, M. (1960). Optimum mutation rate an degree of dominance as determined by the principle of genetic load. *Journal of Genetics* 57: 21–34.

Kimura, M. (1962). On the probability of fixation of mutant genes in a population. *Genetics* 47: 713–719.

Kimura, M. (1967). On the evolutionary adjustment of spontaneous mutation rates. *Genetical Research* 9: 23–34.

Kimura, M. (1968). Evolutionary rate at the molecular level. *Nature* 217: 624–626.

Kimura, M. (1970). The length of time required for a selectively neutral mutant to reach fixation through random frequency drift in a finite population. *Genetical Research Cambridge* 15: 131–133.

Kimura, M. (1980). A simple model for estimating evolutionary rates of base substitutions through comparative studies of nucleotide sequences. *Journal of Molecular Evolution* 16: 111–120.

Kimura, M. (1983). *The Neutral Theory of Molecular Evolution*. Cambridge: Cambridge University Press.

Kimura, M. (1989). The neutral theory of molecular evolution and the world view of neutralists. *Genome* 31: 24–31. https://doi.org/10.1139/g89-009.

Kimura, M. and Crow, J.F. (1964). The number of alleles that can be maintained in a finite population. *Genetics* 49: 725–738.

Kimura, M. and Maruyama, T. (1966). The mutational load with epistatic gene interactions in fitness. *Genetics* 54: 1337–1351.

Kimura, M. and Ohta, T. (1969a). The average number of generations until fixation of a mutant gene in a finite population. *Genetics* 61: 763–771.

Kimura, M. and Ohta, T. (1969b). The average number of generations until extinction of an individual mutant gene in a finite population. *Genetics* 63: 701–709.

Kimura, M. and Ohta, T. (1978). Stepwise mutation model and distribution of allelic frequencies in a finite population. *Proceedings of the National Academy of Sciences of the United States of America* 75: 2868–2872.

Kimura, M. and Weiss, G.H. (1964). The stepping stone model of population structure and the decrease of genetic correlation with distance. *Genetics* 49: 561–576.

Kimura, M., Maruyama, T., and Crow, J.F. (1963). The mutation load in small populations. *Genetics* 48: 1303–1312.

King, J.L. and Jukes, T.H. (1969). Non-Darwinian evolution. *Science* 164: 788–798.

King, E.G. and Long, A.D. (2017). The Beavis effect in next-generation mapping panels in *Drosophila melanogaster*. *G3: Genes|Genomes|Genetics* 7: 1643–1652.

Kingman, J.F.C. (1982a). On the genealogy of large populations. In: *Essays in Statistical Science Papers in Honor of P. A. P. Moran*, 27–43. London: Applied Probability Trust.

Kingman, J.F.C. (1982b). The coalescent. *Stochastic Processes and their Applications* 13: 235–248.

Klopfstein, S., Currat, M., and Excoffier, L. (2006). The fate of mutations surfing on the wave of a range expansion. *Molecular Biology and Evolution* 23: 482–490.

Kondrashov, A.S. (1995). Contamination of the genome by very slightly deleterious mutations: why have we not died 100 times over? *Journal of Theoretical Biology* 175: 583–594.

Kondrashov, A.S. and Crow, J.F. (1988). King's formula for the mutation load with epistasis. *Genetics* 120: 853–856.

Korte, A. and Ashley, F. (2013). The advantages and limitations of trait analysis with GWAS: A review. *Plant Methods* 9 (1): 29.

Kraaijeveld-Smit, F.J.L., Beebee, T.J.C., Griffiths, R.A. et al. (2005). Low gene flow but high genetic diversity in the threatened Mallorcan mid-wife toad Alytes muletensis. *Molecular Ecology* 14: 3307–3315. https://doi.org/10.1111/j.1365-294X.2005.02614.x.

Krebs, J.E., Goldstein, E.S., and Kilpatrick, S.T. (2017). *Lewin's Genes XII*. Jones and Bartlett.

Kreitman, M. (1983). Nucleotide polymorphism at the alcohol dehydrogenase locus of *Drosophila melanogaster*. *Nature* 304: 412–417.

Krimbas, C.B. and Tsakas, S. (1971). The genetics of dacus oleae. V. Changes of esterase polymorphism in a natural population following insecticide control-selection or drift? *Evolution* 25: 454–460.

Kronforst, M.R. and Flemming, T.H. (2001). Lack of genetic differentiation among widely spaced subpopulations of a butterfly with home range behavior. *Heredity* 86: 243–250.

Kroymann, J. and Mitchell-Olds, T. (2005). Epistasis and balanced polymorphism influencing complex trait variation. *Nature* 435: 95–98.

Kucukyildirim, S., Long, H., Sung, W. et al. (2016). The rate and spectrum of spontaneous mutations in *Mycobacterium smegmatis*, a bacterium naturally devoid of the postreplicative mismatch repair pathway. *G3: Genes|Genomes|Genetics* 6: 2157–2163.

Kumar, S., Stecher, G., Suleski, M., and Hedges, S.B. (2017). TimeTree: a resource for timelines, Timetrees, and divergence times. *Molecular Biology and Evolution* 34: 1812–1819.

LaBella, A.L., Opulente, D.A., Steenwyk, J.L. et al. (2019). Variation and selection on codon usage bias across an entire subphylum. *PLoS Genetics* 15: e1008304.

LaFramboise, T. (2009). Single nucleotide polymorphism arrays: a decade of biological, computational and technological advances. *Nucleic Acids Research* 37: 4181–4193.

Lande, R. (1979). Quantitative genetic analysis of multi-variate evolution, applied to brain-body size allometry. *Evolution* 33: 402–416.

Lande, R. (1995). Mutation and conservation. *Conservation Biology* 9: 782–791.

Lande, R. and Arnold, S. (1983). The measurement of selection on correlated characters. *Evolution* 37: 1210–1226.

Lande, R. and Schemske, D.W. (1985). The evolution of self-fertilization and inbreeding depression in plants. I. Genetic models. *Evolution* 39: 24.

Lang, G.I., Rice, D.P., Hickman, M.J. et al. (2013). Pervasive genetic hitchhiking and clonal interference in forty evolving yeast populations. *Nature* 500: 571–574.

Langley, C.H. and Fitch, W. (1974). An examination of the constancy of the rate of molecular evolution. *Journal of Molecular Evolution* 3: 161–177.

Laroche, J., Li, P., and Bousquet, J. (1995). Mitochondrial DNA and monocot-dicot divergence time. *Molecular Biology and Evolution* 12: 1151–1156.

Latter, B.D. (1973). The island model of population differentiation: a general solution. *Genetics* 73: 147–157.

Laurie, C.C., Chasalow, S.D., LeDeaux, J.R. et al. (2004). The genetic architecture of response to long-term artificial selection for oil concentration in the maize kernel. *Genetics* 168: 2141–2155.

Lawson, D.J., van Dorp, L., and Falush, D. (2018). A tutorial on how not to over-interpret STRUCTURE and ADMIXTURE bar plots. *Nature Communications* 9: 3258.

Gigord, L.D.B., Macnair, M.R., and Smithson, A. (2001). Negative frequency-dependent selection maintains a dramatic flower color polymorphism in the rewardless orchid *Dactylorhiza sambucina* (L.) Soò. *Proceedings of the National Academy of Sciences of the United States of America* 98: 6253–6255.

LeBlois, R., Estoup, A., and Rousset, F. (2009). IBDSim: a computer program to simulate genotypic data under isolation by distance. *Molecular Ecology Resources* 9: 107–109.

Lee, H., Popodi, E., Tang, H., and Foster, P.L. (2012). Rate and molecular spectrum of spontaneous mutations in the bacterium *Escherichia coli* as determined by whole-genome sequencing. *Proceedings of the National Academy of Sciences* 109: E2774–E2783.

Leffler, E.M., Bullaughey, K., Matute, D.R. et al. (2012). Revisiting an old riddle: what determines genetic diversity levels within species? *PLoS Biology* 10: e1001388.

Lehe, R., Hallatschek, O., and Peliti, L. (2012). The rate of beneficial mutations surfing on the wave of a range expansion. *PLoS Computational Biology* 8: e1002447.

Leitner, T. and Albert, J. (1999). The molecular clock of HIV-1 unveiled through analysis of a known transmission history. *Proceedings of the National Academy of Sciences of the United States of America* 96: 10752–10757.

Lemey, P., Pybus, O.G., Wang, B. et al. (2003). Tracing the origin and history of the HIV-2 epidemic. *Proceedings of the National Academy of Sciences* 100: 6588–6592.

Levin, D.A. (1977). The organization of genetic variability in *Phlox drummondii*. *Evolution* 31: 477.

Levin, D.A. (1978). Genetic variation in annual phlox: self-compatible versus self-incompatible species. *Evolution* 32: 245–263.

Lewontin, R.C. (1964). The interaction of selection and linkage. I. General considerations; heterotic models. *Genetics* 49: 49–67.

Lewontin, R. (1974). *The Genetic Basis of Evolutionary Change*. New York: Columbia University Press.

Lewontin, R.C. (1985a). Population genetics. *Annual Review of Genetics* 19: 81–102.

Lewontin, R.C. (1985b). Population genetics. In: *Evolution: Essays in Honour of John Maynard Smith* (eds. P.J. Greenwood, P.H. Harvey and M. Slatkin), 3–18. Cambridge: Cambridge University Press.

Lewontin, R.C. (1988). On measures of gametic disequilibrium. *Genetics* 120: 849–852.

Lewontin, R.C. and Hubby, J.L. (1966). A molecular approach to the study of genic heterozygosity in natural populations. II. Amount of variation and degree of heterozygosity in natural populations of Drosophila pseudoobscura. *Genetics* 54: 595–609.

Lewontin, R.C. and Krakauer, J. (1973). Distribution of gene frequency as a test of the theory of the selective neutrality of polymorphisms. *Genetics* 74: 175–195.

Lewontin, R.C. and White, M.J.D. (1960). Interaction between inversion polymorphisms of two chromosome pairs in the grasshopper, *Moraba scurra*. *Evolution* 14: 116–129.

Li, H. (2011). A new test for detecting recent positive selection that is free from the confounding impacts of demography. *Molecular Biology and Evolution* 28: 365–375.

Li, H. and Durbin, R. (2011). Inference of human population history from individual whole-genome sequences. *Nature* 475: 493–496.

Li, W.-H., Tanimura, M., and Sharp, P.M. (1987). An evaluation of the molecular clock hypothesis using mammalian DNA sequences. *Journal of Molecular Evolution* 25: 330–342.

Li, W.-H., Ellsworth, D.L., Krushkal, J. et al. (1996). Rates of nucleotide sub-stitution in primates and rodents and the generation-time effect hypothesis. *Molecular Phylogenetics and Evolution* 5: 182–187.

Libiger, O., Nievergelt, C.M., and Schork, N.J. (2009). Comparison of genetic distance measures using human SNP genotype data. *Human Biology* 81: 389–406.

Linck, E. and Battey, C.J. (2019). Minor allele frequency thresholds strongly affect population structure inference with genomic data sets. *Molecular Ecology Resources* 19: 639–647.

Lisec, J., Meyer, R.C., Steinfath, M. et al. (2007). Identification of metabolic and biomass QTL in *Arabidopsis thaliana* in a parallel analysis of RIL and IL populations. *The Plant Journal* 53: 960–972.

Liu, Y., Nickle, D.C., Shriner, D. et al. (2004). Molecular clock-like evolution of human immunodeficiency virus type 1. *Virology* 329: 101–108.

Lohmueller, K.E., Mauney, M.M., Reich, D., and Braverman, J.M. (2006). Variants associated with common disease are not unusually differentiated in frequency across populations. *The American Journal of Human Genetics* 78: 130–136.

Lohmueller, K.E., Albrechtsen, A., Li, Y. et al. (2011). Natural selection affects multiple aspects of genetic variation at putatively neutral sites across the human genome. *PLoS Genetics* 7: e1002326.

Long, H., Winter, D.J., Chang, A.Y.-C. et al. (2016). Low base-substitution mutation rate in the germline genome of the ciliate *Tetrahymena thermophil*. *Genome Biology and Evolution*, evw223: 10.1093/gbe/evw223.

Luria, S.E. and Delbrück, M. (1943). Mutations of Bacteria from virus sensitivity to virus resistance. *Genetics* 28: 491–511.

Lynch, C.B. (1977). Inbreeding effects upon animals derived from a wild population of Mus musculus. *Evolution* 31: 526.

Lynch, M. (2010). Evolution of the mutation rate. *Trends in Genetics* 26: 345–352. https://doi.org/10.1016/j.tig.2010.05.003.

Lynch, M. (2011). The lower bound to the evolution of mutation rates. *Genome Biology and Evolution* 3: 1107–1118.

Lynch, M. (2016). Mutation and human exceptionalism: our future genetic load. *Genetics* 202: 869–875.

Lynch, M. and Gabriel, W. (1990). Mutation load and the survival of small populations. *Evolution* 44: 70–72.

Lynch, M. and Hill, W.G. (1986). Phenotypic evolution by neutral mutation. *Evolution* 40: 915–935.

Lynch, M. and Lande, R. (1998). The critical effective size for a genetically secure population. *Animal Conservation* 1: 70–72.

Lynch, M., Ackerman, M.S., Gout, J.-F. et al. (2016). Genetic drift, selection and the evolution of the mutation rate. *Nature Reviews. Genetics* 17: 704–714.

Mackay, T. (2001). The genetic architecture of quantitative traits. *Annual Review of Genetics* 35: 303–339.

Mackay, T.F.C., Stone, E.A., and Ayroles, J.F. (2009). The genetics of quantitative traits: challenges and prospects. *Nature Reviews Genetics* 10: 565–577.

Maddison, W. (1997). Gene trees in species trees. *Systematic Biology* 46: 523–536.

Maher, B. (2008). Personal genomes: the case of the missing heritability. *Nature* 456: 18–21.

Maier, E., Tollrian, R., Rinkevich, B., and Nürnberger, B. (2005). Isolation by distance in the Scleractinian coral *Seriatopora hystrix* from the Red Sea. *Marine Biology* 147 (5): 1109–1120. https://doi.org/10.1007/s00227-005-0013-6.

Majerus, M. (1998). *Melanism: Evolution in Action*. New York: Oxford University Press.

Malécot, G. (1969). *The Mathematics of Heredity*. San Francisco: W. H. Freeman.

Malmberg, R.L. and Mauricio, R. (2005). QTL-based evidence for the role of epistasis in evolution. *Genetical Research Cambridge* 86: 89–95.

Manchenko, G.P. (2002). *Handbook of Detection of Enzymes on Electrophoretic Gels*. Boca Raton: CRC Press.

Manchenko, G. (2003). *Handbook of Detection of Enzymes on Electrophoretic Gels*. Boca Raton, FL: CRC Press 1993, 4087–4091.

Manel, S., Schwartz, M.K., Luikart, G., and Taberlet, P. (2003). Landscape genetics: combining landscape ecology and population genetics. *Trends in Ecology & Evolution* 18: 189–197.

Marigorta, U.M., Lao, O., Casals, F. et al. (2011). Recent human evolution has shaped geographical differences in susceptibility to disease. *BMC Genomics* 12: 55.

Marriage, T.N., Hudman, S., Mort, M.E. et al. (2009). Direct estimation of the mutation rate at dinucleotide microsatellite loci in *Arabidopsis thaliana* (Brassicaceae). *Heredity* 103: 310–317.

Marrotte, R.R. and Bowman, J. (2017). The relationship between least-cost and resistance distance. *PLoS One* 12: e0174212.

Martin, A.P. and Palumbi, S.R. (1993). Body size, metabolic rate, generation time, and the molecular clock. *Proceedings of the National Academy of Sciences* 90: 4087–4091.

Martin, N.H., Bouck, A.C., and Arnold, M.L. (2007). The genetic architecture of reproductive isolation in Louisiana irises: flowering phenology. *Genetics* 175: 1803–1812.

Maruyama, T. and Kimura, M. (1980). Genetic variability and effective population size when local extinction and recolonization of subpopulations are frequent. *Proceedings of the National Academy of Sciences of the United States of America* 77: 6710–6714.

Mather, N., Traves, S.M., and Ho, S.Y.W. (2020). A practical introduction to sequentially Markovian coalescent methods for estimating demographic history from genomic data. *Ecology and Evolution* 10: 579–589.

Maynard Smith, J. (1978). *The Evolution of Sex*. Cambridge: Cambridge University Press.

Maynard Smith, J. and Haigh, J. (1974). The hitchhiking effect of a favourable gene. *Genetical Research* 23: 23–35.

McCarthy, M.I., Abecasis, G.R., Cardon, L.R. et al. (2008). Genome-wide association studies for complex traits: Consensus, uncertainty and challenges. *Nature Reviews Genetics* 9 (5): 356–369. https://doi.org/10.1038/nrg2344.

McCauley, D.E., Raveill, J., and Antonovics, J. (1995). Local founding events as determinants of genetic structure in a plant metapopulation. *Heredity* 75: 630–636.

McDonald, J.H. and Kreitman, M. (1991). Adaptive protein evolution at the *Adh* locus in *Drosophila*. *Nature* 351: 652–654.

McDonald, D.B. and Potts, W.K. (1997). DNA microsatellites as genetic markers at several scales. In: *Avian*

Molecular Evolution and Systematics (ed. D. Mindell), 29–49. New York: Academic Press.

McRae, B.H. (2006). Isolation by resistance. *Evolution* 60: 1551–1561.

McRae, B.H. and Beier, P. (2007). Circuit theory predicts gene flow in plant and animal populations. *Proceedings of the National Academy of Sciences of the United States of America* 104: 19885–19890.

McRae, B.H., Dickson, B.G., Keitt, T.H., and Shah, V.B. (2008). Using circuit theory to model connectivity in ecology, evolution, and conservation. *Ecology* 89: 2712–2724.

McVean, G.A.T. (2002). A genealogical interpretation of linkage disequilibrium. *Genetics* 162: 987–991.

McVean, G. (2009). A genealogical interpretation of principal components analysis. *PLoS Genetics* 5.

McVean, G.A.T. and Cardin, N.J. (2005). Approximating the coalescent with recombination. *Philosophical Transactions of the Royal Society of London. Series B, Biological Sciences* 360: 1387–1393.

Meagher, T.T.R. (1986). Analysis of paternity within a natural population of *Chamaelirium luteum*. 1. Identification of most-likely male parents. *The American Naturalist* 128: 199–215.

Meagher, S., Penn, D.J., and Potts, W.K. (2000). Male-male competition magnifies inbreeding depression in wild house mice. *Proceedings of the National Academy of Sciences* 97: 3324–3329.

Meirmans, P.G. (2012). The trouble with isolation by distance. *Molecular Ecology* 21: 2839–2846.

Meirmans, P.G., Liu, S., and van Tienderen, P.H. (2018). The analysis of Polyploid genetic data (F Allendorf, Ed). *The Journal of Heredity* 109: 283–296.

Meneely, P.M. (2016). Pick your Poisson: an educational primer for Luria and Delbrück's classic paper. *Genetics* 202: 371–375.

Menozzi, P., Piazza, A., and Cavalli-Sforza, L. (1978). Synthetic maps of human gene frequencies in Europeans. *Science* 201: 786–792.

Miller, J.R. (2010). Survival of mutations arising during invasions. *Evolutionary Applications* 3: 109–121.

Miller, H.C., Moore, J.A., Nelson, N.J., and Daugherty, C.H. (2009). Influence of major histocompatibility complex genotype on mating success in a free-ranging reptile population. *Proceedings of the Royal Society B: Biological Sciences* 276: 1695–1704.

Mlodinow, L. (2008). *The Drunkard's Walk: How Randomness Rules our Lives*. New York, NY: Pantheon Books.

Modiano, D., Luoni, G., Sirima, B.S. et al. (2001). Haemoglobin C protects against clinical *Plasmodium falciparum* malaria. *Nature* 414: 305–308.

Moehring, A.J., Llopart, A., Elwyn, S. et al. (2006). The genetic basis of prezygotic reproductive isolation between Drosophila santomea and D. yakuba due to mating preference. *Genetics* 173: 215–223.

Möhle, M. and Sagitov, S. (2001). A classification of coalescent processes for haploid exchangeable population models. *Annals of Probability* 29 (4): 1547–1562. https://doi.org/10.1214/aop/1015345761.

Moll, R.H., Lindsey, M.F., and Robinson, H.F. (1964). Estimates of genetic variances and level of dominance in maize. *Genetics* 49: 411–423.

Moorjani, P., Amorim, C.E.G., Arndt, P.F., and Przeworski, M. (2016). Variation in the molecular clock of primates. *Proceedings of the National Academy of Sciences* 113: 10607–10612.

Moose, S.P., Dudley, J.W., and Rocheford, T.R. (2004). Maize selection passes the century mark: a unique resource for 21st century genomics. *Trends in Plant Science* 9: 358–364.

Moran, P.A.P. (1958). Random processes in genetics. *Mathematical Proceedings of the Cambridge Philosophical Society* 54: 60.

Moran, P. and Watterson, G. (1959). The genetic effects of family structure in natural populations. *Australian Journal of Biological Sciences* 12: 1.

Moreno-Estrada, A., Gignoux, C.R., Fernandez-Lopez, J.C. et al. (2014). The genetics of Mexico recapitulates native American substructure and affects biomedical traits. *Science* 344: 1280–1285.

Morrissey, M.B., Kruuk, L.E.B., and Wilson, A.J. (2010). The danger of applying the breeder's equation in observational studies of natural populations. *Journal of Evolutionary Biology* 23: 2277–2288.

Morton, N. (1971). Population genetics and disease control. *Social Biology* 8: 243–251.

Mousseau, T.A. and Roff, D.A. (1987). Natural selection and the heritability of fitness components. *Heredity* 59: 181–197.

Mousset, S., Derome, N., and Veuille, M. (2004). A test of neutrality and constant population size based on the mismatch distribution. *Molecular Biology and Evolution* 21: 724–731.

Mueller, A.K., Chakarov, N., Krüger, O., and Hoffman, J.I. (2016). Long-term effective population size dynamics of an intensively monitored vertebrate population. *Heredity* 117: 290–299.

Mukai, T. (1964). The genetic structure of natural populations of *Drosophila melanogaster*. I. Spontaneous mutation rate of polygenes controlling viability. *Genetics* 50: 1–19.

Mukai, T., Chigusa, S.I., Mettler, L.E., and Crow, J.F. (1972). Mutation rate and dominance of genes affecting viability in *Drosophila melanogaster*. *Genetics* 72: 335–355.

Muller, H.J. (1932). Some genetic aspects of sex. *The American Naturalist* 66: 118–138.

Muller, H.J. (1950). Our load of mutations. *American Journal of Human Genetics* 2: 111–176.

Muller, H.J. (1964). The relation of recombination to mutational advance. *Mutation Research* 106: 2–9.

Muse, S.V. and Weir, B. (1992). Testing for the equality of evolutionary rates. *Genetics* 132: 269–276.

Myers, J.R. (2004). An alternative possibility for seed coat color determination in Mendel's experiment. *Genetics* 166: 1137.

Myers, S., Spencer, C.C.A., Auton, A. et al. (2006). The distribution and causes of meiotic recombination in the human genome. *Biochemical Society Transactions* 34: 526–530.

Myles, S., Davison, D., Barrett, J. et al. (2008). Worldwide population differentiation at disease-associated SNPs. *BMC Medical Genomics* 1: 22.

Nabholz, B., Mauffrey, J.-F., Bazin, E. et al. (2008). Determination of mitochondrial genetic diversity in mammals. *Genetics* 178: 351–361.

Narain, P. (1970). A note on the diffusion approximation for the variance of the number of generations until fixation of a neutral mutant gene. *Genetical Research Cambridge* 15: 251–255.

Narasimhan, V.M., Rahbari, R., Scally, A. et al. (2017). Estimating the human mutation rate from autozygous segments reveals population differences in human mutational processes. *Nature Communications* 8: 303.

Nathan, R., Klein, E., Robledo-Arnuncio, J.J., and Revilla, E. (2012). Dispersal kernels: review. In: *Dispersal Ecology and Evolution* (eds. J. Clobert, M. Baguette, T.G. Benton and J.M. Bullock), 186–210. Oxford University Press.

National Institute of General Medical Sciences (1998). *The Genetic Architecture of Complex Traits Workshop*. https://www.nigms.nih.gov/News/reports/archivedreports1999-1996/Pages/genetic_arch.aspx.

Nei, M. (1972). Genetic distance between populations. *The American Naturalist* 106: 283–292.

Nei, M. (1973). Analysis of gene diversity in subdivided populations. *Proceedings of the National Academy of Sciences of the United States of America* 70: 3321–3323.

Nei, M. (1978a). Estimation of average heterozygosity and genetic distance from a small number of individuals. *Genetics* 89: 583–590.

Nei, M. (1978b). The theory of genetic distance of the evolution of human races. *Japanese Journal of Human Genetics* 23: 341–369.

Nei, M. (2005). Selectionism and neutralism in molecular evolution. *Molecular Biology and Evolution* 22 (12): 2318–2342. https://doi.org/10.1093/molbev/msi242.

Nei, M. (2013). *Mutation-Driven Evolution*. Oxford: Oxford University Press.

Nei, M. and Chakravarti, A. (1977). Drift variances of F_{ST} and G_{ST} statistics obtained from a finite number of isolated populations. *Theoretical Population Biology* 11: 307–325.

Nei, M. and Kumar, S. (2000). *Molecular Evolution and Phylogenetics*. New York: Oxford University Press.

Nei, M. and Li, W.-H. (1979). Mathematical model for studying genetic variation in terms of restriction endonucleases. *Proceedings of the National Academy of Sciences of the United States of America* 76: 5269–5273.

Nei, M. and Maruyama, T. (1975). Letters to the editors: Lewontin-Krakauer test for neutral genes. *Genetics* 80: 395.

Nei, M. and Roychoudhury, A.K. (1974). Sampling variances of heterozygosity and genetic distance. *Genetics* 76: 379–390.

Nei, M. and Tajima, F. (1981). Genetic drift and estimation of effective population size. *Genetics* 98: 625–640.

Nei, M., Chakravarti, A., and Tateno, Y. (1977). Mean and variance of F_{ST} in a finite number of incompletely isolated populations. *Theoretical Population Biology* 11: 291–306.

Ness, R.W., Morgan, A.D., Vasanthakrishnan, R.B. et al. (2015). Extensive *de novo* mutation rate variation between individuals and across the genome of *Chlamydomonas reinhardtii*. *Genome Research* 25: 1739–1749.

Neuhauser, C. (1999). The ancestral graph and gene genea- logy under frequency-dependent selection. *Theoretical Population Biology* 56: 203–214.

Neuhauser, C. and Krone, S. (1997). The genealogy of samples in models with selection. *Genetics* 145: 519–534.

Newby, J. (1980). *Mathematics for the Biological Sciences*. Oxford: Oxford University Press.

Nica, A.C. and Dermitzakis, E.T. (2013). Expression quantitative trait loci: present and future. *Philosophical Transactions of the Royal Society of London. Series B, Biological Sciences* 368: 20120362.

Nordborg, M. (1997). Structured coalescent processes on different time scales. *Genetics* 146: 1501–1514.

Novitski, C.E. (2004a). Revision of Fisher's analysis of Mendel's garden pea experiments. *Genetics* 166: 1139–1140.

Novitski, E. (2004b). On Fisher's criticism of Mendel's results with the garden pea. *Genetics* 166: 1133–1136.

Nunez-Farfan, J. and Schlichting, C.D. (2001). Evolution in changing environments: the "synthetic" work of Clausen, Keck, and Hiesey. *Quarterly Review of Biology* 76: 433–457.

Nunney, L. (1993). The influence of mating system and overlapping generations on effective population size. *Evolution* 47: 1329–1341.

Ohta, T. (1972). Evolutionary rate of cistrons and DNA divergence. *Journal of Molecular Evolution* 1: 263–286.

Ohta, T. (1992). The nearly neutral theory of molecular evolution. *Annual Review of Ecology and Systematics* 23: 263–286.

Ohta, T. (1993). An examination of the generation-time effect on molecular evolution. *Proceedings of the National Academy of Sciences of the United States of America* 90: 10676–10680.

Ohta, T. (1995). Synonymous and nonsynonymous substitutions in mammalian genes and the nearly neutral theory. *Journal of Molecular Evolution* 40: 56–63.

Ohta, T. and Gillespie, J.H. (1996). Development of neutral nearly neutral theories. *Theoretical Population Biology* 49 (2): 128–142. https://doi.org/10.1006/tpbi.1996.0007.

Ohta, T. and Kimura, M. (1969a). Linkage disequilibrium at steady state determined by random genetic drift and recurrent mutation. *Genetics* 63: 229–238.

Ohta, T. and Kimura, M. (1969b). Linkage disequilibrium due to random genetic drift. *Genetical Research* 13: 47–55.

Ohta, T. and Kimura, M. (1971). On the constancy of the evolutionary rate of cistrons. *Journal of Molecular Evolution* 1: 18–25.

Ollivier, L. and James, J.W. (2004). Predicting annual effective size of livestock populations. *Genetical Research* 84: 41–46.

Oppold, A.-M. and Pfenninger, M. (2017). Direct estimation of the spontaneous mutation rate by short-term mutation accumulation lines in Chironomus riparius. *Evolution Letters* 1: 86–92.

Orel, V. (1996). *Gregor Mendel: The First Geneticist*. Oxford: Oxford University Press.

Orr, H. (1998). The population genetics of adaptation: the distribution of factors fixed during adaptive evolution. *Evolution* 52: 935–949.

Orr, H. (2003). The distribution of fitness effects among beneficial mutations. *Genetics* 163: 1519–1526.

Orsini, L., Vanoverbeke, J., Swillen, I. et al. (2013). Drivers of population genetic differentiation in the wild: isolation by dispersal limitation, isolation by adaptation and isolation by colonization. *Molecular Ecology* 22: 5983–5999.

Ossowski, S., Schneeberger, K., Lucas-Lledó, J.I. et al. (2010). The rate and molecular spectrum of spontaneous mutations in *Arabidopsis thaliana*. *Science (New York, N.Y.)* 327: 92–94.

Otto, S.P. and Day, T. (2007). *A Biologist's Guide to Mathematical Modeling in Ecology and Evolution*. Princeton, NJ: Princeton University Press.

Otto, S.P. and Jones, C. (2000). Detecting the undetected: estimating the total number of loci underlying a quantitative trait. *Genetics* 156: 2093–2107.

Ouma, J.O., Marquez, J.G., and Krafsur, E.S. (2005). Macro-geographic structure of the tsetse fly, *Glossina pallidipes* (Diptera: Glossinadae). *Bulletin of Entomological Research* 95: 437–447.

Paetkau, D., Calvert, W., Stirling, I., and Strobeck, C. (1995). Microsatellite analysis of population structure in Canadian polar bears. *Molecular Ecology* 4: 347–354.

Paetkau, D., Slade, R., Burden, M., and Estoup, A. (2004). Genetic assignment methods for the direct, real-time estimation of migration rate: a simulation-based exploration of accuracy and power. *Molecular Ecology* 13: 55–65.

Page, R. and Holmes, E. (1998). *Molecular Evolution: A Phylogenetic Approach*. Malden, MA: Blackwell Science.

Palstra, F.P. and Ruzzante, D.E. (2008). Genetic estimates of contemporary effective population size: what can they tell us about the importance of genetic stochasticity for wild population persistence? *Molecular Ecology* 17: 3428–3447.

Pannell, J.R. and Charlesworth, B. (2000). Effects of metapopulation processes on measures of genetic diversity. *Philosophical Transactions of the Royal Society of London. Series B: Biological Sciences* 355: 1851–1864.

Park, S.Y., Love, T.M.T., Perelson, A.S. et al. (2016). Molecular clock of HIV-1 envelope genes under early immune selection. *Retrovirology* 13: 38.

Patterson, N., Price, A.L., and Reich, D. (2006). Population structure and eigenanalysis. *PLoS Genetics* 2: e190.

Pearson, K. (1901). LIII. On lines and planes of closest fit to systems of points in space. *The London, Edinburgh, and Dublin Philosophical Magazine and Journal of Science* 2: 559–572.

Peck, S.L., Ellner, S.P., and Gould, F. (1998). A spatially explicit stochastic model demonstrates the feasibility of Wright's shifting balance theory. *Evolution* 52: 1834–1839.

Peck, S.L., Ellner, S.P., and Gould, F. (2000). Varying migration and deme size and the feasibility of the shifting balance. *Evolution* 54: 324–327.

Peischl, S. and Excoffier, L. (2015). Expansion load: recessive mutations and the role of standing genetic variation. *Molecular Ecology* 24: 2084–2094.

Peischl, S., Dupanloup, I., Kirkpatrick, M., and Excoffier, L. (2013). On the accumulation of deleterious mutations during range expansions. *Molecular Ecology* 22: 5972–5982.

Peischl, S., Dupanloup, I., Bosshard, L., and Excoffier, L. (2016). Genetic surfing in human populations: from genes to genomes. *Current Opinion in Genetics & Development* 41: 53–61. https://doi.org/10.1016/j.gde.2016.08.003.

Pelgas, B., Bousquet, J., Meirmans, P.G. et al. (2011). QTL mapping in white spruce: gene maps and genomic regions underlying adaptive traits across pedigrees, years and environments. *BMC Genomics* 12: 145.

Penn, D. and Potts, W. (1998). MHC-disassortative mating preferences reversed by cross-fostering. *Proceedings of the Royal Society of London, Series B: Biological Sciences* 265: 1299–1306.

Penrose, L. (1949). The meaning of fitness in human populations. *Annals of Eugenics* 14: 301–304.

Perola, M., Sammalisto, S., Hiekkalinna, T. et al. (2007). Combined genome scans for body stature in 6,602 European twins: evidence for common Caucasian loci. *PLoS Genetics* 3: e97.

Phillips, P. (1998). The language of gene interactions. *Genetics* 149: 1167–1171.

Phillips, P.C., Otto, S.P., and Whitlock, M. (2000). Beyond the average: the evolutionary importance of gene interactions and variability of epistatic effects. In: *Epistasis and the Evolutionary Process* (eds. J.B. Wolf, E.D. Brodie and M.J. Wade), 20–38. New York: Oxford University Press.

Pigliucci, M. (2001). *Phenotypic Plasticity*. Baltimore, MD: Johns Hopkins University Press.

Pitman, J. (1999). Coalescents with multiple collisions. *The Annals of Probability* 27: 1870–1902.

Pitra, C., Hansen, A.J., Lieckfeldt, D., and Arctander, P. (2002). An exceptional case of historical outbreeding in African sable antelope populations. *Molecular Ecology* 11: 1197–1208.

Plotkin, J.B. and Kudla, G. (2011). Synonymous but not the same: the causes and consequences of codon bias. *Nature Reviews Genetics* 12: 32–42.

Pollak, E. (1978). With selection for fecundity the mean fitness does not necessarily increase. *Genetics* 90: 383–389.

Pollak, E. (1983). A new method for estimating the effective population size from allele frequency changes. *Genetics* 104: 531–548.

Posada, D. and Crandall, K. (1998). Modeltest: testing the model of DNA substitution. *Bioinformatics* 14: 817–818.

Posada, D. and Crandall, K. (2001). Selecting the best-fit model of nucleotide substitution. *Systematic Biology* 50: 560–580.

Pritchard, J.K. and Przeworski, M. (2001). Linkage disequilibrium in humans: models and data. *The American Journal of Human Genetics* 69: 1–14.

Pritchard, J.K., Stephens, M., and Donnelly, P. (2000). Inference of population structure using multilocus genotype data. *Genetics* 155: 945–959.

Provine, W. (1971). *The Origins of Theoretical Population Genetics*. Chicago, IL: University of Chicago Press.

Provine, W. (1986). *Sewall Wright and Evolutionary Biology*. Chicago, IL: University of Chicago Press.

Przeworski, M., Charlesworth, B., and Wall, J.D. (1999). Genealogies and weak purifying selection. *Molecular Biology and Evolution* 16: 246–252.

Ptak, S.E. and Przeworski, M. (2002). Evidence for popula- tion growth in humans is confounded by fine-scale population structure. *Trends in Genetics* 18: 559–563.

Qiao, Z., Powell, J., and Evans, D. (2018). MHC-dependent mate selection within 872 spousal pairs of European ancestry from the health and retirement study. *Genes* 9: 53.

Rand, D.M. (1994). Thermal habit, metabolic rate and the evolution of mitochondrial DNA. *Trends in Ecology and Evolution* 9: 125–131.

Rasmann, S., De Vos, M., Casteel, C.L. et al. (2012). Herbivory in the previous generation primes plants for enhanced insect resistance. *Plant Physiology* 158: 854–863.

Rasmussen, M.D., Hubisz, M.J., Gronau, I., and Siepel, A. (2014). Genome-wide inference of ancestral recombination graphs. *PLoS Genetics* 10 (5) https://doi.org/10.1371/journal.pgen.1004342.

Renwick, A., Bonnen, P.E., Trikka, D. et al. (2003). Sampling properties of estimators of nucleotide diversity at discovered SNP sites. *International Journal of Applied Mathematics and Computer Science* 13: 385–394.

Richards, E.J. (2006). Inherited epigenetic variation – revisiting soft inheritance. *Nature Reviews Genetics* 7: 395–401.

Richards, E.J. (2008). Population epigenetics. *Current Opinion in Genetics & Development* 18 (2): 221–226. https://doi.org/10.1016/j.gde.2008.01.014.

Robertson, A. (1952). The effect of inbreeding on variation due to recessive genes. *Evolution* 37: 1195–1208.

Robertson, A. (1975a). Letters to the editors: remarks on the Lewontin-Krakauer test. *Genetics* 80: 396.

Robertson, A. (1975b). Gene frequency distributions as a test of selective neutrality. *Genetics* 81: 775–785.

Robinson, L.M., Boland, J.R., and Braverman, J.M. (2016). Revisiting a classic study of the molecular clock. *Journal of Molecular Evolution* 82: 110–116.

Rockman, M.V. (2012). The QTN program and the alleles that matter for evolution: all that's gold does not glitter. *Evolution* 66: 1–17.

Rodríguez, F., Oliver, J.L., Marín, A., and Medina, J. (1990). The general stochastic model of nucleotide substitution. *Journal of Theoretical Biology* 142: 485–501.

Roff, D.A. (2001). *Life History Evolution*. Sunderland, MA: Sinauer Associates.

Roff, D.A. (2007). A centennial celebration for quantitative genetics. *Evolution* 61: 1017–1032.

Roff, D.A. and Mousseau, T. (1987). Quantitative genetics and fitness: lessons from Drosophila. *Heredity* 58: 103–118.

Rogers, A.R. (2014). How population growth affects linkage disequilibrium. *Genetics* 197: 1329–1341.

Rogers, A.R. and Harpending, H. (1992). Population growth makes waves in the distribution of pairwise genetic differences. *Molecular Biology and Evolution* 9: 552–569.

Rogers, A.R. and Huff, C. (2009). Linkage disequilibrium between loci with unknown phase. *Genetics* 182: 839–844.

Romero, C., Pedryc, A., Munoz, V. et al. (2003). Genetic diversity of different apricot geographical groups determined by SSR markers. *Genome* 46: 244–252.

Rosenberg, N.A., Pritchard, J.K., Weber, J.L. et al. (2002). Genetic structure of human populations. *Science* 298: 2381–2385.

Roughgarden, J. (1996). *Theory of Population Genetics and Evolutionary Ecology: An Introduction*. Upper Saddle River, NJ: Prentice Hall.

Rousset, F. (1997). Genetic differentiation and estimation of gene flow from F-statistics under isolation by distance. *Genetics* 145: 1219–1228.

Rousset, F. (2008a). Dispersal estimation: demystifying Moran's I. *Heredity* 100: 231–232.

Rousset, F. (2008b). genepop'007: a complete re-implementation of the genepop software for Windows and Linux. *Molecular Ecology Resources* 8: 103–106.

Rousset, F. (2013). Exegeses on maximum genetic differentiation. *Genetics* 194: 557–559.

Roychoudhury, A.K. and Nei, M. (1988). *Human Polymorphic Genes: World Distribution*. New York: Oxford University Press.

Rozen, D.E., de Visser, J.A.G.M., and Gerrish, P.J. (2002). Fitness effects of fixed beneficial mutations in microbial populations. *Current Biology* 12: 1040–1045.

Ruel, J.J. and Ayres, M.P. (1999). Jensen's inequality predicts effects of environmental variation. *Trends in Ecology & Evolution* 14: 361–366.

Ruse, M. (1996). Are pictures really necessary? The case of Sewall Wright's "adaptive landscapes". In: *Picturing Knowledge: Historical and Philosophical Problems Concerning the Use of Art in Science* (ed. B.S. Baigrie), 303–337. Toronto, Canada: University of Toronto Press.

Russell, L.B. and Russell, W. (1996). Spontaneous mutations recovered as mosaics in the mouse specific-locus test. *Proceedings of the National Academy of Sciences of the United States of America* 93: 13072–13077.

Ruzzante, D.E., McCracken, G.R., Parmelee, S. et al. (2016). Effective number of breeders, effective population size and their relationship with census size in an iteroparous species, salvelinus fontinalis. *Proceedings of the Royal Society B: Biological Sciences* 283.

Sagitov, S. (1999). The general coalescent with asynchronous mergers of ancestral lines. *Journal of Applied Probability* 36: 1116–1125.

Ryman, N. and Leimar, O. (2009). G_{ST} is still a useful measure of genetic differentiation – a comment on Jost's D. *Molecular Ecology* 18: 2084–2087.

Sagitov, S. and Jagers, P. (2005). The coalescent effective size of age-structured populations. *The Annals of Applied Probability* 15: 1778–1797.

Salemi, M., Strimmer, K., Hall, W.M. et al. (2001). Dating the common ancestor of SIVcpz and HIV-1 group M and the origin of HIV-1 subtypes using a new method to uncover clock-like molecular evolution. *FASEB Journal* 15: 276–278.

Salomé, P.A., Bomblies, K., Laitinen, R.A.E. et al. (2011). Genetic architecture of flowering-time variation in *Arabidopsis thaliana*. *Genetics* 188: 421–433.

Sancristobal, M. and Chevalet, C. (1997). Error tolerant parent identification from a finite set of individuals. *Genetical Research* 70: 53–62.

Sanjuán, R., Moya, A., Elena, S.S.F. et al. (2004). The distribution of fitness effects caused by single-nucleotide substitutions in an RNA virus. *Proceedings of the National Academy of Sciences of the United States of America* 101: 8396–8401.

Santure, A.W., Poissant, J., De Cauwer, I. et al. (2015). Replicated analysis of the genetic architecture of quantitative traits in two wild great tit populations. *Molecular Ecology* 24: 6148–6162.

Sarich, V.M. and Wilson, A. (1967). Immunological time scale for hominid evolution. *Science* 158: 1200–1203.

Sarkar, S., Ma, W.T., and GVH, S. (1992). On fluctuation analysis: a new, simple and efficient method for computing the expected number of mutants. *Genetica* 85: 173–179.

Sax, K. (1923). The association of size difference with seed-coat pattern and pigmentation in *Phaseolus vulgaris*. *Genetics* 8: 552–560.

Scally, A. and Durbin, R. (2012). Revising the human mutation rate: implications for understanding human evolution. *Nature Reviews Genetics* 13: 745–753.

Schaal, B. (1980). Measurement of gene flow in *Lupinus texensis*. *Nature* 284: 450–451.

Scheben, A., Batley, J., and Edwards, D. (2017). Genotyping-by-sequencing approaches to characterize crop genomes: choosing the right tool for the right application. *Plant Biotechnology Journal* 15: 149–161.

Schemske, D.W. and Bierzychudek, P. (2001). Perspective: evolution of flower color in the desert annual *Linanthus parryae*: Wright revisited. *Evolution* 55: 1269–1282.

Schierup, M.H., Vekemans, X., and Christiansen, F. (1998). Allelic genealogies in sporophytic self-incompatibility systems in plants. *Genetics* 150: 1187–1198.

Schlager, G. and Dickie, M.M. (1971). Natural mutation rates in the house mouse. Estimates for five specific loci and dominant mutations. *Mutation Research* 11: 89–96.

Schneider, S. and Excoffier, L. (1999). Estimation of past demographic parameters from the distribution of pairwise differences when the mutation rates vary among sites: application to human mitochondrial DNA. *Genetics* 152: 1079–1089.

National Research Council, Commission on DNA Forensic Science (1996). *The Evaluation of Forensic DNA Evidence*. Washington, D.C.: National Academies Press.

Sella, G. and Barton, N.H. (2019). Thinking about the evolution of complex traits in the era of genome-wide association studies. *Annual Review of Genomics and Human Genetics* 20: 461–493.

Selvin, S. (1980). Probability of nonpaternity determined by multiple allele codominant systems. *American Journal of Human Genetics* 32: 276–278.

Serbezov, D., Jorde, P.E., Bernatchez, L. et al. (2012). Short-term genetic changes: evaluating effective population size estimates in a comprehensively described brown trout (*salmo trutta*) population. *Genetics* 191: 579–592.

Seyfert, A.L., Cristescu, M.E.A., Frisse, L. et al. (2008). The rate and spectrum of microsatellite mutation in *Caenorhabditis elegans* and *Daphnia pulex*. *Genetics* 178: 2113–2121.

Sharma, S., Dutta, T., Maldonado, J.E. et al. (2013). Forest corridors maintain historical gene flow in a tiger metapopulation in the highlands of Central India. *Proceedings of the Royal Society B: Biological Sciences* 280: 20131506–20131506.

Shringarpure, S. and Xing, E.P. (2009). mStruct: inference of population structure in light of both genetic admixing and allele mutations. *Genetics* 182: 575–593.

Sick, K. (1965). Haemoglobin polymorphism of cod in the north sea and the North Atlantic Ocean. *Hereditas* 54: 49–69.

Silbermann, R. and Tatar, M. (2000). Reproductive costs of transgenic hsp70 overexpression in *Drosophila*. *Evolution* 54: 2038–2045.

Sjödin, P., Kaj, I., Krone, S. et al. (2005). On the meaning and existence of an effective population size. *Genetics* 169: 1061–1070.

Skipper, R. (2004). The heuristic role of Sewall Wright's 1932 adaptive landscape diagram. *Philosophy of Science* 71: 1176–1188.

Slate, J. (2005). Quantitative trait locus mapping in natural populations: progress, caveats and future directions. *Molecular Ecology* 14: 363–379.

Slate, J. (2013). From Beavis to beak color: a simulation study to examine how much QTL mapping can reveal about the genetic architecture of quantitative traits. *Evolution* 67: 1251–1262.

Slatkin, M. (1977). Gene flow and genetic drift in a species subject to frequent local extinctions. *Theoretical Population Biology* 12: 253–262.

Slatkin, M. (1987a). Gene flow and the geographic structure of natural populations. *Science* 236: 787–792.

Slatkin, M. (1987b). The average number of sites separating DNA sequences drawn from a subdivided population. *Theoretical Population Biology* 32: 42–49.

Slatkin, M. (1991). Inbreeding coefficients and coalescence times. *Genetical Research* 58: 167–175.

Slatkin, M. (1995). A measure of population subdivision based on microsatellite allele frequencies. *Genetics* 139: 457–462.

Slatkin, M. (2008). Linkage disequilibrium – understanding the evolutionary past and mapping the medical future. *Nature Reviews. Genetics* 9: 477–485.

Slatkin, M. and Barton, N.H. (1989). A comparison of three indirect methods for estimating average levels of gene flow. *Evolution* 43: 1349–1368.

Slatkin, M. and Hudson, R. (1991). Pairwise comparisons of mitochondrial DNA sequences in stable and exponentially growing populations. *Genetics* 129: 555–562.

Slatkin, M. and Voelm, L. (1991). F(ST) in a hierarchical island model. *Genetics* 127 (3): 627–629.

Smeds, L., Qvarnström, A., and Ellegren, H. (2016). Direct estimate of the rate of germline mutation in a bird. *Genome Research* 26 (9): 1211–1218. https://doi.org/10.1101/gr.204669.116.

Snow, A.A. and Palma, P. (1997). Commercialization of transgenic plants: potential ecological risks. *BioScience* 47: 86–96.

Solberg, O.D., Mack, S.J., Lancaster, A.K. et al. (2008). Balancing selection and heterogeneity across the classical human leukocyte antigen loci: a meta-analytic review of 497 population studies. *Human Immunology* 69: 443–464.

Spear, S.F., Balkenhol, N., Fortin, M.-J. et al. (2010). Use of resistance surfaces for landscape genetic studies: considerations for parameterization and analysis. *Molecular Ecology* 19: 3576–3591.

Speed, D. and Balding, D.J. (2015). Relatedness in the post-genomic era: is it still useful? *Nature Reviews Genetics* 16: 33–44.

Spence, J.P., Steinrücken, M., Terhorst, J., and Song, Y. S. (2018). Inference of population history using coalescent HMMs: review and outlook. *Current Opinion in Genetics & Development* 53: 70–76.

Springer, N.M. and Schmitz, R.J. (2017). Exploiting induced and natural epigenetic variation for crop improvement. *Nature Reviews Genetics* 18: 563–575.

Stapley, J., Feulner, P.G.D., Johnston, S.E. et al. (2017). Variation in recombination frequency and distribution across eukaryotes: patterns and processes.

Philosophical Transactions of the Royal Society, B: Biological Sciences 372: 20160455.

Stearns, F.W. and Fenster, C.B. (2016). The effect of induced mutations on quantitative traits in *Arabidopsis thaliana*: natural versus artificial conditions. *Ecology and Evolution* 6: 8366–8374.

Steinberg, E.K., Lindner, K.R., Gallea, J. et al. (2002). Rates and patterns of microsatellite mutations in pink Salmon. *Molecular Biology and Evolution* 19: 1198–1202.

Stephan, W. (2010). Genetic hitchhiking versus background selection: the controversy and its implications. *Philosophical Transactions of the Royal Society of London. Series B, Biological Sciences* 365: 1245–1253.

Stephan, W. (2019). Selective sweeps. *Genetics* 211: 5–13.

Steppan, S.J., Phillips, P.C., and Houle, D. (2002). Comparative quantitative genetics: evolution of the G matrix. *Trends in Ecology and Evolution* 17: 320–327.

Stigler, S.M. (1986). *The History of Statistics: the Measurement of Uncertainty Before 1900*. Cambridge, Mass: Belknap Press of Harvard University Press.

Strobeck, C. (1987). Average number of nucleotide differences in a sample from a single subpopulation: a test for population subdivision. *Genetics* 117: 149–153.

Su, C., Wang, W., Gong, S. et al. (2017). High density linkage map construction and mapping of yield trait QTLs in maize (*Zea mays*) using the genotyping-by-sequencing (GBS) technology. *Frontiers in Plant Science* 8.

Sutter, N.B., Bustamante, C.D., Chase, K. et al. (2007). A single IGF1 allele is a major determinant of small size in dogs. *Science* 316: 112–115.

Sved, J.A., Cameron, E.C., and Gilchrist, A.S. (2013). Estimating effective population size from linkage disequilibrium between unlinked loci: theory and application to fruit Fly outbreak populations. *PLoS One* 8: e69078.

Tajima, F. (1989a). Statistical method for testing the neutral mutation hypothesis by DNA polymorphism. *Genetics* 123: 585–595.

Tajima, F. (1989b). The effect of change in population size in DNA polymorphism. *Genetics* 123: 597–601.

Tajima, F. (1993a). Simple methods for testing molecular clock hypothesis. *Genetics* 135: 599–607.

Tajima, F. (1993b). Measurement of DNA polymorphism. In: *Mechanisms of Molecular Evolution* (eds. N. Takahata and A.G. Clark), 37–59. Sunderland, MA: Sinauer Associates.

Tajima, F. (1996). The amount of DNA polymorphism maintained in a finite populations when the neutral mutation rate varies among sites. *Genetics* 143: 1457–1465.

Takahata, N. (1983). Gene identity and genetic differentiation of populations in the finite island model. *Genetics* 104: 497–512.

Takahata, N. and Nei, M. (1984). F and G statistics in the finite island model. *Genetics* 107: 501–504.

Takezaki, N. and Nei, M. (1996). Genetic distances and reconstruction of phylogenetic trees from microsatellite DNA. *Genetics* 144: 389–399.

Tatsumoto, S., Go, Y., Fukuta, K. et al. (2017). Direct estimation of de novo mutation rates in a chimpanzee parent-offspring trio by ultra-deep whole genome sequencing. *Scientific Reports* 7: 13561.

Tellier, A. and Lemaire, C. (2014). Coalescence 2.0: a multiple branching of recent theoretical developments and their applications. *Molecular Ecology* 23: 2637–2652.

Templeton, A.R. and Read, B. (1994). Inbreeding: one word, several meanings, much confusion. In: *Conservation Genetics* (eds. V. Loeschcke, S.K. Jain and J. Tomiuk), 91–105. Basel: Birkhäuser Basel.

Thompson, E.A. (1988). Two-locus and three-locus gene identity by descent in pedigrees. *IMA Journal of Mathematics Applied in Medicine and Biology* 5: 261–279.

Thompson, M.J. and Jiggins, C.D. (2014). Supergenes and their role in evolution. *Heredity* 113: 1–8.

Tishkoff, S.A. and Verrelli, B. (2003). Patterns of human genetic diversity: implications for human evolution-ary history and disease. *Annual Review of Genomics and Human Genetics* 4: 293–340.

Travis, J.M.J., Munkemuller, T., Burton, O.J. et al. (2007). Deleterious mutations can surf to high densities on the wave front of an expanding population. *Molecular Biology and Evolution* 24: 2334–2343.

Trerotola, M., Relli, V., Simeone, P., and Alberti, S. (2015). Epigenetic inheritance and the missing heritability. *Human Genomics* 9: 17.

Turelli, M., Schemske, D.W., and Bierzychudek, P. (2001). Stable two-allele polymorphisms maintained by fluctuating fitnesses and seed banks: protecting the blues in *Linanthus parryae*. *Evolution; International Journal of Organic Evolution* 55: 1283–1298.

Turner, J.R. (1981). "Fundamental theorem" for two loci. *Genetics* 99: 365–369.

Turner, J. (1992). Stochastic processes in populations: the horse behind the cart? In: *Genes in Ecology: the 33rd Symposium of the British Ecological Society* (eds Crawford TJ, Hewitt GM), p. Blackwell Scientific Publications, Oxford.

Uchimura, A., Higuchi, M., Minakuchi, Y. et al. (2015). Germline mutation rates and the long-term phenotypic effects of mutation accumulation in wild-type laboratory mice and mutator mice. *Genome Research* 25: 1125–1134.

Uyenoyama, M. (1997). Genealogical structure among alleles regulating self-incompatibility in natural populations of flowering plants. *Genetics* 147: 1389–1400.

Valenzuela, C., Ballesta, P., Maldonado, C. et al. (2019). Bayesian mapping reveals large-effect pleiotropic QTLs for Wood density and slenderness index in 17-year-old trees of *Eucalyptus cladocalyx*. *Forests* 10: 241.

Van Buskirk, J. and Willi, Y. (2006). The change in quantitative genetic variation with inbreeding. *Evolution* 60: 2428–2434.

Van Ooijen, J.W. (1999). LOD significance thresholds for QTL analysis in experimental populations of diploid species. *Heredity* 83: 613–624.

Van Valen, L. (1973). A new evolutionary law. *Evolutionary Theory* 1: 1–30.

Varvio, S.-L.L., Chakraborty, R., and Nei, M. (1986). Genetic variation in subdivided populations and conservation genetics. *Heredity* 57 (Pt 2): 189–198.

Vekemans, X. and Slatkin, M. (1994). Gene and allelic genealogies at a gametophytic self-incompatibility locus. *Genetics* 137: 1157–1165.

Verity, R. and Nichols, R.A. (2014). What is genetic differentiation, and how should we measure it- G_{ST}, D, neither or both? *Molecular Ecology* 23: 4216–4225.

Verity, R. and Nichols, R.A. (2016). Estimating the number of subpopulations (K) in structured populations. *Genetics* 203: 1827–1839.

Viard, F., Justy, F., and Jarne, P. (1997). Population dynamics inferred from temporal variation at microsatellite loci in the selfing snail Bulinus truncatus. *Genetics* 146: 973–982.

de Vicente, M.C. and Tanksley, S. (1993). QTL analysis of transgressive segregation in an interspecific tomato cross. *Genetics* 134: 585–596.

Vieira, M.L.C., Santini, L., Diniz, A.L., and de Freitas Munhoz, C. Microsatellite markers: what they mean and why they are so useful. *Genetics and Molecular Biology* 39: 312–328. https://doi.org/10.1590/1678-4685-GMB-2016-0027.

Vignieri, S.N. (2005). Streams over mountains: influence of riparian connectivity on gene flow in the Pacific jumping mouse (*Zapus trinotatus*). *Molecular Ecology* 14: 1925–1937.

Visscher, P.M., Hill, W.G., and Wray, N.R. (2008). Heritability in the genomics era – concepts and misconceptions. *Nature Reviews Genetics* 9 (4): 255–266. https://doi.org/10.1038/nrg2322.

de Visser, J.A.G.M. and Elena, S.F. (2007). The evolution of sex: empirical insights into the roles of epistasis and drift. *Nature Reviews Genetics* 8: 139–149.

de Visser, J.A.G.M. and Krug, J. (2014). Empirical fitness landscapes and the predictability of evolution. *Nature Reviews Genetics* 15: 480–490.

Vitti, J.J., Grossman, S.R., and Sabeti, P.C. (2013). Detecting natural selection in genomic data. *Annual Review of Genetics* 47: 97–120.

Wade, M. (1992). Epistasis. In: *Keywords in Evolutionary Biology* (eds. E.F. Keller and E.A. Lloyd), 87–91. Cambridge, MA: Harvard University Press.

Wade, M.J. (2013). Phase III of Wright's shifting balance process and the variance among demes in migration rate. *Evolution* 67: 1591–1597.

Wade, M.J. and Goodnight, C. (1991). Wright's shifting balance theory: an experimental study. *Science* 253: 1015–1018.

Wade, M.J. and Goodnight, C.J. (1998). Perspective: the theories of Fisher and Wright in the context of metapopulations: when nature does many small experiments. *Evolution* 52: 1537–1553.

Wade, M.J. and McCauley, D.E. (1988). Extinction and recolonization: their effects on the genetic differentiation of local populations. *Evolution* 42: 995–1005.

Wade, M.J., Goodnight, C.J., and Stevens, L. (1999). Design and interpretation of experimental studies of inter-demic selection: a reply to Getty. *American Naturalist* 154: 599–603.

Wakeley, J. (1998). Segregating sites in Wright's island model. *Theoretical Population Biology* 53: 166–174. https://doi.org/10.1006/tpbi.1997.1355.

Wakeley, J. (1999). Nonequilibrium migration in human history. *Genetics* 153: 1863–1871.

Wakeley, J. (2005). The limits of theoretical population genetics. *Genetics* 169: 1–7.

Wakeley, J. (2009). *Coalescent Theory: An Introduction*. Denver, CO.: Roberts and Company Publishers.

Wakeley, J. and Aliacar, N. (2001). Gene genealogies in a metapopulation. *Genetics* 159: 893–905.

Wakeley, J. and Sargsyan, O. (2009). Extensions of the coalescent effective population size. *Genetics* 181: 341–345.

Wallace, B. (1991). *Fifty Years of Genetic Load. An Odyssey*. Ithaca, NY: Cornell University Press.

Walsh, B. and Lynch, M. (2018). *Evolution and Selection of Quantitative Traits*. Oxford University Press.

Wang, J. (2017). The computer program structure for assigning individuals to populations: easy to use but easier to misuse. *Molecular Ecology Resources* 17: 981–990.

Wang, I.J. and Bradburd, G.S. (2014). Isolation by environment. *Molecular Ecology* 23: 5649–5662.

Wang, M. and Schreiber, A. (2001). The impact of habitat fragmentation and social structure on the population genetics of roe deer (*Capreolus capreolus* L.) in Central Europe. *Heredity* 86: 703–715.

Wang, J., Santiago, E., and Caballero, A. (2016). Prediction and estimation of effective population size. *Heredity* 117: 193–206.

Waples, R.S. (1989). A generalized approach for estimating effective population size from temporal changes in allele frequency. *Genetics* 121: 379–391.

Waples, R.S. (2005). Genetic estimates of contemporary effective population size: to what time periods do the estimates apply? *Molecular Ecology* 14: 3335–3352.

Waples, R.S. (2006). A bias correction for estimates of effective population size based on linkage disequilibrium at unlinked gene loci. *Conservation Genetics* 7 (2): 167–184. https://doi.org/10.1007/s10592-005-9100-y.

Waples, R.S. and Yokota, M. (2007). Temporal estimates of effective population size in species with overlapping generations. *Genetics* 175: 219–233.

Waples, R.S., Antao, T., and Luikart, G. (2014). Effects of overlapping generations on linkage disequilibrium estimates of effective population size. *Genetics* 197: 769–780.

Watterson, G. (1975). On the number of segregating sites in genetic models without recombination. *Theoretical Population Biology* 7: 256–276.

Weinberg, W. (1908). Über Vererbungsgesetze beim Menschen (S f. vaterl. Naturk. Wortem. 64: an English translation can be found in Boyer, Ed). *Zeitschrift für Induktive Abstammungs- und Vererbungslehre* 1: 440–460.

Weinreich, D.M., Watson, R.A., and Chao, L. (2005). Perspective: sign epistasis and genetic constraint on evolutionary trajectories. *Evolution* 59: 1165–1174.

Weinreich, D.M., Delaney, N.F., DePristo, M.A., and Hartl, D.L. (2006). Darwinian evolution can follow only very few mutational paths to fitter proteins. *Science* 312: 111–114.

Weir, B.S. (1996). *Genetic Data Analysis II: Methods for Discrete Population Genetic Data*. Sunderland, MA: Sinauer Associates.

Weir, B.S. and Cockerham, C.C. (1984). Estimating *F*-statistics for the analysis of population structure. *Evolution* 38: 1358.

Weir, B.S. and Goudet, J. (2017). A unified characterization of population structure and relatedness. *Genetics* 206: 2085–2103.

Weir, B.S., Cockerham, C.C., and Reynolds, J. (1980). The effects of linkage and linkage disequilibrium on the covariances of non-inbred relatives. *Heredity* 45: 351–359.

Westneat, D.F. and Stewart, I.R.K. (2003). Extra-pair paternity in birds: causes, correlates, and conflict. *Annual Review of Ecology, Evolution, and Systematics* 34: 365–396.

Whitlock, M.C. (2011). G'_{ST} and *D* do not replace F_{ST}. *Molecular Ecology* 20: 1083–1091.

Whitlock, M.C. and Barton, N.H. (1997). The effective size of a subdivided population. *Genetics* 146: 427–441.

Whitlock, M.C. and McCauley, D.E. (1990). Some population genetic consequences of colony formation and extinction: genetic correlations within founding groups. *Evolution* 44: 1717–1724.

Whitlock, M.C. and McCauley, D.E. (1999). Indirect measures of gene flow and migration: FST not equal to 1/(4Nm + 1). *Heredity* 82 (Pt 2): 117–125.

Whitlock, M.C., Phillips, P.C., Moore, F., and Tonsor, S.J. (1995). Multiple fitness peaks and epistasis. *Annual Review of Ecology and Systematics* 26: 601–629.

Wilkins, J.F. (2004). A separation-of-timescales approach to the coalescent in a continuous population. *Genetics* 168: 2227–2244.

Williams, G.C. (1957). Pleiotropy, natural selection, and the evolution of senescence. *Evolution* 11: 398–411.

Williams, G.C. (1966). *Adaptation and Natural Selection-a Critique of Some Current Evolutionary Thought*. Princeton, NJ: Princeton University Press.

Williams, G.C. (1992). *Natural Selection: Domains, Levels, and Challenges*. New York: Oxford University Press.

Willis, J.H. and Orr, H. (1993). Increased heritable variation following population bottlenecks: the role of dominance. *Evolution* 47: 949–957.

Wilton, P.R., Carmi, S., and Hobolth, A. (2015). The SMC' is a highly accurate approximation to the ancestral recombination graph. *Genetics* 200: 343–355.

Wiuf, C. and Hein, J. (1999). Recombination as a point process along sequences. *Theoretical Population Biology* 55: 248–259.

Wolf, J.B., Brodie, E., and Wade, M. (2000). *Epistasis and the Evolutionary Process*. Oxford: Oxford University Press.

Workman, P.L. (1964). The maintenance of heterozygosity by partial negative assortative mating. *Genetics* 50: 1369–1382.

Worobey, M., Santiago, M.L., Keele, B.F. et al. (2004). Contaminated polio vaccine theory refuted. *Nature* 428: 820–820.

Wright, S. (1921). Systems of mating. *Genetics* 6: 97–159.

Wright, S. (1922). Coefficients of inbreeding and relationship. *The American Naturalist* 56: 330–338.

Wright, S. (1931). Evolution in Mendelian populations. *Genetics* 16: 97–159.

Wright, S. (1932). The roles of mutation, inbreeding, cross-breeding and selection in evolution. *Proceedings of the Sixth International Congress of Genetics* 1: 356–366.

Wright, S. (1943a). An analysis of local variability of flower color in *Linanthus Parryae*. *Genetics* 28: 139–156.

Wright, S. (1943b). Isolation by distance. *Genetics* 28: 114–138.

Wright, S. (1946). Isolation by distance under diverse systems of mating. *Genetics* 31: 39–59.

Wright, S. (1951). The genetical structure of populations. *Annals of Eugenics* 15: 323–354.

Wright, S. (1978). *Evolution and the Genetics of Populations, Vol. 4: Variability Within and Among Natural Populations.* Chicago, IL: University of Chicago Press.

Wright, S. (1988). Surfaces of selective value revisited. *American Naturalist* 131: 115–123.

Wright, S. and Kerr, W.E. (1954). Experimental studies of the distribution of gene frequencies in very small populations of *Drosophila melanogaster*: II. Bar. *Evolution* 8: 225–240.

Wright, S.I., Lauga, B., and Charlesworth, D. (2003). Subdivision and haplotype structure in natural populations of *Arabidopsis lyrata*. *Molecular Ecology* 12: 1247–1263.

Xu, S. (2003). Theoretical basis of the Beavis effect. *Genetics* 165: 2259–2268.

Yang, J., Guo, B., Shikano, T. et al. (2016). Quantitative trait locus analysis of body shape divergence in nine-spined sticklebacks based on high-density SNP-panel. *Scientific Reports* 6: 26632.

Yang, S., Wang, L., Huang, J. et al. (2015). Parent-progeny sequencing indicates higher mutation rates in heterozygotes. *Nature* 523 (7561): 463–467. https://doi.org/10.1038/nature14649.

Yi, X. and Dean, A.M. (2019). Adaptive landscapes in the age of synthetic biology. *Molecular Biology and Evolution* 36: 890–907.

Zeng, K., Fu, Y.X., Shi, S., and Wu, C.I. (2006). Statistical tests for detecting positive selection by utilizing high-frequency variants. *Genetics* 174: 1431–1439.

Zharkikh, A. (1994). Estimation of evolutionary distances between nucleotide sequences. *Journal of Molecular Evolution* 39: 315–329.

Zuckerkandl, E. (1987). On the molecular clock. *Journal of Molecular Evolution* 26: 34–46.

Zuckerkandl, E. and Pauling, L. (1962). Molecular disease, evolution, and genetic heterogeneity. In: *Horizons in Biochemistry* (eds. M. Kasha and B. Pullman), 189–225. New York: Academic Press.

Zuckerkandl, E. and Pauling, L. (1965). Evolutionary divergence and convergence in proteins. In: *Evolving Genes and Proteins* (eds. V. Bryson and H.J. Vogel), 97–165. New York: Academic Press.

Index

Page numbers in *italic* denote figures. Page numbers in **bold** denote tables.

Population Genetics, Second Edition. Matthew B. Hamilton.
© 2021 John Wiley & Sons, Inc. Published 2021 by John Wiley & Sons, Inc.
Companion website: www.wiley.com/go/hamilton/populationgenetics